FOOD PACKAGING

FOOD PACKAGING

Nanotechnology in the Agri-Food Industry, Volume 7

Edited by

ALEXANDRU MIHAI GRUMEZESCU

Department of Science and Engineering of Oxide Materials and Nanomaterials, Faculty of Applied Chemistry and Materials Science, University Politehnica of Bucharest, Bucharest, Romania

ELSEVIER

AMSTERDAM • BOSTON • HEIDELBERG • LONDON • NEW YORK • OXFORD
PARIS • SAN DIEGO • SAN FRANCISCO • SINGAPORE • SYDNEY • TOKYO
Academic Press is an imprint of Elsevier

Academic Press is an imprint of Elsevier
125 London Wall, London EC2Y 5AS, United Kingdom
525 B Street, Suite 1800, San Diego, CA 92101-4495, United States
50 Hampshire Street, 5th Floor, Cambridge, MA 02139, United States
The Boulevard, Langford Lane, Kidlington, Oxford OX5 1GB, United Kingdom

Library of Congress Cataloging-in-Publication Data
A catalog record for this book is available from the Library of Congress

British Library Cataloguing-in-Publication Data
A catalogue record for this book is available from the British Library

ISBN: 978-0-12-804302-8

For information on all Academic Press publications
visit our website at https://www.elsevier.com/

Publisher: Nikki Levy
Acquisition Editor: Patricia Osborn
Editorial Project Manager: Jaclyn Truesdell
Production Project Manager: Caroline Johnson
Designer: Mark Rogers

Typeset by Thomson Digital

CONTENTS

LIST OF CONTRIBUTORS

Ahmad Allahbakhsh
Department of Polymer Engineering, Faculty of Chemical Engineering, Tarbiat Modares University, Tehran; Young Researchers and Elite Club, Shiraz Branch, Islamic Azad University, Shiraz, Iran

Cecilia I. Alvarez Igarzabal
Universidad Nacional de Córdoba, Medina Allende and Haya de la Torre, Ciudad Universitaria, Instituto Multidisciplinario de Biología Vegetal (IMBIV), CONICET, Facultad de Ciencias Químicas, Córdoba, Argentina

Ana Vera Alves Machado
University of Minho, Institute of Polymers and Composites (IPC) and Institute of Nanostructures, Nanomodelling, and Nanofabrication, Guimarães, Portugal

Marina Patricia Arrieta
Institute of Polymer Science and Technology, Madrid, Spain

Gabriela B. Baldino
Braskem, Polymer Technology Center, Triunfo, Brazil

Silvia E. Barbosa
Planta Piloto de Ingeniería Química, PLAPIQUI (UNS-CONICET), Bahía Blanca, Argentina

Nuria Burgos
University of Alicante, Department of Analytical Chemistry, Nutrition & Food Sciences, Alicante, Spain

Luciana A. Castillo
Planta Piloto de Ingeniería Química, PLAPIQUI (UNS-CONICET), Bahía Blanca, Argentina

Corinne Chappey
Normandie University; University of Rouen, Polymères, Biopolymères, Surfaces; UMR 6270, Mont Saint Aignan, France

Cintia B. Contreras
Universidad Nacional de Córdoba, Medina Allende and Haya de la Torre, Ciudad Universitaria, Instituto Multidisciplinario de Biología Vegetal (IMBIV), CONICET, Facultad de Ciencias Químicas, Córdoba, Argentina

Viviane Dalmoro
Institute of Chemistry, Federal University of Rio Grande do Sul, Porto Alegre, Brazil

Ana M. Díez-Pascual
Alcalá University, Analytical Chemistry, Physical Chemistry and Chemical Engineering Department, Faculty of Biology, Environmental Sciences and Chemistry, Madrid, Spain

Tungalag Dong
Inner Mongolia Agricultural University, College of Food Science and Engineering, Hohhot, Inner Mongolia, P.R. China

Karolth Espinosa
Planta Piloto de Ingeniería Química, PLAPIQUI (UNS-CONICET), Bahía Blanca, Argentina

Kateryna Fatyeyeva
Normandie University; University of Rouen, Polymères, Biopolymères, Surfaces; UMR 6270, Mont Saint Aignan, France

Emmanuel Flores-Huicochea
Instituto Politécnico Nacional, Centro de Desarrollo de Productos Bióticos, Departamento de Desarrollo Tecnológico, San Isidro. Yautepec, Mexico

Esther García-Serna
University of Alicante, Department of Analytical Chemistry, Nutrition & Food Sciences, Alicante, Spain

Isabel Gonçalves de Moura
University of Minho, Institute of Polymers and Composites (IPC) and Institute of Nanostructures, Nanomodelling, and Nanofabrication, Guimarães, Portugal

Agustín González
Universidad Nacional de Córdoba, Juan Filloy s/n, Ciudad Universitaria, Instituto de Ciencia y Tecnología de los Alimentos Córdoba (ICYTAC), CONICET, Facultad de Ciencias Químicas, Córdoba, Argentina

Tatjana Haramina
University of Zagreb, Faculty of Mechanical Engineering and Naval Architecture, FAMENA, Department of Materials, Laboratory for Polymers and Composites, Zagreb, Croatia

Tina Harifi
Amirkabir University of Technology, Department of Textile Engineering, Functional Fibrous Structures & Environmental Enhancement (FFSEE), Tehran, Iran

Alfonso Jiménez
University of Alicante, Department of Analytical Chemistry, Nutrition & Food Sciences, Alicante, Spain

José María Kenny
Institute of Polymer Science and Technology, Madrid, Spain

Alain Largeteau
University of Bordeaux, CNRS, Pessac, France

Daniel López
Institute of Polymer Science and Technology, Madrid, Spain

Juan López
Polytechnical University of Valencia, Instituto de Tecnología de Materiales, Alcoy, Spain

Ana Sofia Lemos Machado Abreu
University of Minho, Institute of Polymers and Composites (IPC) and Institute of Nanostructures, Nanomodelling, and Nanofabrication, Guimarães, Portugal

Giulio Malucelli
Politecnico di Torino, Department of Applied Science and Technology, Local INSTM Unit, Alessandria, Italy

Stéphane Marais
Normandie University; University of Rouen, Polymères, Biopolymères, Surfaces; UMR 6270, Mont Saint Aignan, France

Ana Cristina Mellinas
University of Alicante, Department of Analytical Chemistry, Nutrition & Food Sciences, Alicante, Spain

Pawan Kumar Mishra
Mendel University, Department of Wood Science, Brno, Czech Republic

Katarzyna Mitura
Koszalin University of Technology, Department of Biomedical Engineering, Koszalin, Poland

Majid Montazer
Department of Textile Engineering, Functional Fibrous Structures & Environmental Enhancement (FFSEE), Amirkabir Nanotechnology Research Institute (ANTRI), Amirkabir University of Technology, Tehran, Iran

Aleksandra R. Nesic
University of Belgrade, Vinca Institute for Nuclear Sciences, Belgrade, Serbia

Luciane Oliveira
Braskem, Polymer Technology Center, Triunfo, Brazil

Shweta Pandey
Jamia Hamdard, Department of Pharmaceutics, New Delhi, India

Laura Peponi
Institute of Polymer Science and Technology, Madrid, Spain

Marcia Pires
Braskem, Polymer Technology Center, Triunfo, Brazil

Mythili Prakasam
University of Bordeaux, CNRS, Pessac, France

J. Rodolfo Rendón-Villalobos
Instituto Politécnico Nacional, Centro de Desarrollo de Productos Bióticos, Departamento de Desarrollo Tecnológico, San Isidro. Yautepec, Mexico

Iva Rezić
University of Zagreb, Department of Applied Chemistry, Faculty of Textile Technology, Zagreb, Croatia

Tonči Rezić
University of Zagreb, Faculty of Biotechnology and Food Technology, Department of Biochemical Engineering, Laboratory for Biochemical Engineering, Industrial Microbiology and Malting and Brewing Technology, Zagreb, Croatia

Francisco Rodríguez-González
Instituto Politécnico Nacional, Centro de Desarrollo de Productos Bióticos, Departamento de Desarrollo Tecnológico, San Isidro. Yautepec, Mexico

Sanja I. Seslija
University of Belgrade, Institute of Chemistry, Technology and Metallurgy, Belgrade, Serbia

Harshita Sharma
Jamia Hamdard, Department of Pharmaceutics, New Delhi, India

Adriane Simanke
Braskem, Polymer Technology Center, Triunfo, Brazil

Javier Solorza Feria
Instituto Politécnico Nacional, Centro de Desarrollo de Productos Bióticos, Departamento de Desarrollo Tecnológico, San Isidro. Yautepec, Mexico

Iman Soltani
North Carolina State University, Fiber and Polymer Science Program, Raleigh, NC, United States

Richard J. Spontak
North Carolina State University, Department of Chemical & Biomolecular Engineering; North Carolina State University, Department of Materials Science & Engineering, Raleigh, NC, United States

Miriam C. Strumia
Universidad Nacional de Córdoba, Medina Allende and Haya de la Torre, Ciudad Universitaria, Instituto Multidisciplinario de Biología Vegetal (IMBIV), CONICET, Facultad de Ciencias Químicas, Córdoba, Argentina

Sushama Talegaonkar
Jamia Hamdard, Department of Pharmaceutics, New Delhi, India

Arsénio Vasconcelos de Sá
University of Minho, Institute of Polymers and Composites (IPC) and Institute of Nanostructures, Nanomodelling, and Nanofabrication, Guimarães, Portugal

Rupert Wimmer
Mendel University, Department of Wood Science, Brno, Czech Republic; University of Natural Resources and Life Sciences, Institute for Natural Materials Technology, Tulln, Austria

Konrad Wyrębski
PLASTMOROZ Sp. z o.o. Sp., Białogard, Poland

Xueyan Yun
Inner Mongolia Agricultural University, College of Food Science and Engineering, Hohhot, Inner Mongolia, P.R. China

Paweł K. Zarzycki
Koszalin University of Technology, Faculty of Civil Engineering, Environmental and Geodetic Sciences, Department of Environmental Technologies and Bioanalytics, Koszalin, Poland

Joao Henrique Zimnoch dos Santos
Institute of Chemistry, Federal University of Rio Grande do Sul, Porto Alegre, Brazil

SERIES FOREWORD

The emergence of nanotechnology has reached impressive heights in recent years and the development of special nanodevices and nanomaterials has found intriguing applications in agriculture and food sector. Most of the investigated nanotechnological approaches initially aimed to solve evolving problems in the agri-food industry in order to impact on the economic potential. Soon after the implementation of new technologies and approaches that were using nanostructured materials, the worldwide concern was rapidly extended to numerous applications that could be developed by using the science of nanosized materials. Smart materials, biosensors, packaging materials, nutraceuticals, and nanodevices have been designed to address numerous agri-food related issues with direct impact in health, economy, ecology, and industry. As the engineering of nanostructures has constantly progressed and extended its applications, there is virtually unlimited potential in this sector. However, the widely differing opinions on the applicability and usefulness of nanotechnology between both specialists and the general public has hampered progress. The main concern manifested by people is related to the potential risk for health and the environmental impact of the recently developed nanoengineered materials and devices. Therefore, current approaches are strictly considering these concerns when designing nanotechnological solutions for agriculture and food sectors.

This multivolume series was developed by the constant need to discover current inquiries and approaches on the field of agri-food science and also to learn about the most recent progress, approaches, and applications that have emerged through nanotechnology.

As agriculture is the backbone of most developing countries, nanotechnology has the potential to revolutionize the agriculture and food sector by promoting productivity through genetic improvement of plant and animal foods. It can also ensure the delivery of drugs, genes, and pesticides to specific sites at cellular levels in targeted plants and animals, by limiting side effects. Nanotechnology can be used to evaluate gene expression under different stress condition for both plant and animal foods through the development of nanoarray-based gene-technologies. Additionally, this technology can detect fertilizers, pesticides with high precision by smart nanosenors for an adequate management of the natural resources. Moreover, numerous industrial-related applications with direct impact on economy have emerged. For example, nano- and micro-structured arrays can detect the early presence of pathogens, contaminants, and food spoilage factors. Other applications for this technology are smart integration systems for food processing and packaging, as well as nanoemulsion-based decontaminants for food equipment and storage compartments, and nanoparticles that facilitate the bioavailability and delivery of nutrients directly to cells.

The potential benefits of nanotechnology for agriculture, food, fisheries, and aquaculture were identified and supported by many countries, which invested a significant amount of money in the development of applications. Also, numerous campaigns are currently trying to increase awareness on the developing process and recent technologies in order to influence the acceptance of customers. Although nanoagri-food industrialized concept could help to find a sustainable solution for the current global food crisis, the offered advantages should balance the concerns regarding soil, water, environment, and health related issues that such approach could bring.

The series entitled *Nanotechnology in the Agri-Food Industry* brings comprehensive and recent knowledge regarding the impact of the science of nanometer-sized materials on the field of agriculture and food industry, but also discuss the current inquiries regarding risks of these applications in all relevant fields such as environment and health, aiming to increase awareness to a wider amount of readers.

Dr Alina Maria Holban
Department of Microbiology and Immunology, Faculty of Biology, University of Bucharest, Romania

Dr Monica Cartelle Gestal
Harvill Lab, University Of Georgia, Athens, GA, United States

SERIES PREFACE

About the Series (Volumes I–X)

In a permanently changing society, health and well being remain the key drivers for the food industry. Despite the technological progress made in the agri-food industry, a true food crisis emerges in several areas of the globe. This can be explained by insufficient food but mostly by inadequate food for a very distinct range of consumers. In this context, innovative technologies represent the core throughout the whole food chain from raw materials/ingredient sourcing, food processing, quality control of finished products, and packaging. Nanotechnology, coupled with novel interdisciplinary approaches and processing methods, has enabled some important advances recently flourishing in many of these areas. The science of nanosized materials can improve and even resolve the huge challenges faced by the food and bioprocessing industries for developing and implementing systems that can produce qualitative and quantitative foods that are safe, sustainable, environment friendly, and efficient. This emerging tool finds its applications in various fields and represents an endless approach for the development of innovative strategies in food development, processing, and packaging.

This multivolume set aims to bring together the most recent and innovative applications of nanotechnology in the agri-food industry, but also to present the future perspectives in the design of new or alternative foods.

The series contains 200 chapters organized in 10 volumes, prepared by outstanding research groups that made significant impacts on the field of nanotechnology and food-related research sectors. This comprehensive set represents an updated and highly structured material for undergraduate and postgraduate students in food science, biotechnological, engineering fields, but also a valuable resource of recent scientific progress, along with most known applications of nanomaterials on the food industry to be used by researchers, engineers, and academia. Moreover, novel opportunities and ideas for developing or improving technologies in the agri-food industry by innovative companies, biotechnological industries, and other economical structures are highlighted and their potential is widely dissected. This series may be also valuable for the wide audience interested in recent nanotechnological progress in the agri-food field worldwide.

These 10 volumes cover almost all aspects related to the applications of *Nanotechnology in the Agri-Food Industry* and are named as:

Volume I Novel Approaches
Volume II Encapsulations
Volume III Emulsions
Volume IV Nutraceuticals
Volume V Nutrient Delivery
Volume VI Food Preservation
Volume VII Food Packaging
Volume VIII NanoBioSensors
Volume IX Water Purification
Volume X New Pesticides and Soil Sensors

Each volume contains 20 chapters, which were carefully composed and illustrated to highlight the most innovative and intensively investigated applications of nanotechnology on particular wide interest domains of the agri-food industry field.

Alexandru Mihai Grumezescu
University Politehnica of Bucharest,
Department of Science and Engineering
of Oxide Materials and Nanomaterials,
Faculty of Applied Chemistry and
Materials Science, Bucharest, Romania
http://grumezescu.com/

VOLUME PREFACE

Food Packaging focuses on the development of novel nanobiomaterials, the enhancement of barrier performance of nondegradable and biodegradable plastics as well as their fabrication and application in industrial packaging. This book brings together fundamental information and the most recent advances in the synthesis, design, and impact of alternative food packaging. A special attention is offered to smart materials and nanodevices able to detect quality parameters in packaged food, such as freshness, degradation, and contamination. Also, ecological approaches aiming to obtain bioplastics packages from waste materials are highlighted and discussed as a novel approach in modern food packaging. Nonetheless, this volume presents the advances made in biodegradable and bioactive packaging utilized for preserving flavor, nutritious ingredients, and therapeutic food compounds.

Volume VII contains 20 chapters, prepared by outstanding international researchers from the USA, Argentina, Brazil, Mexico, Portugal, Spain, France, Italy, Austria, Poland, Czech Republic, Croatia, Serbia, India, Iran, and China.

In Chapter 1, *Nanotechnological Strategies Yielding High-Barrier Plastic Food Packaging*, Iman Soltani et al. describe the coating and nanocomposite approaches developed to enhance barrier performance in food packaging, as well as to introduce developing trends and future needs.

Giulio Malucelli, in Chapter 2, entitled *High Barrier Composite Materials Based on Renewable Sources for Food Packaging Applications*, reviews the current state of the art and the future perspectives concerning the design of polymer nanocomposite materials based on renewable sources, which show high barrier properties toward different diffusing species and hence are expected to be suitable for packaging applications. Also, some case studies refer to the exploitation of composite materials based on renewable sources specifically for food packaging applications and some recent surface engineering approaches, such as layer-by-layer treatments, are described in detail, because they have a great potential in modern packagings and industry.

Chapter 3, prepared by Sushama Talegaonkar et al., entitled *Bionanocomposites: Smart Biodegradable Packaging Material for Food Preservation*, deals with methods of preparation of polymeric nanocomposites, types, and properties of nanoreinforcements, and details a few important biodegradable polymers that are commonly used for preparing nanocomposites for food packaging purposes. This chapter also gives an account of commercial achievements of food packaging materials that are currently used in the market.

Chapter 4, *Encapsulation of Sensors for Intelligent Packaging*, prepared by Viviane Dalmoro et al., gives an up-to-date overview about the concept of intelligent packaging that can sense environmental changes and notify consumers that the product is in a harmful condition to be consumed. Different technologies have been employed to produce intelligent packaging, among them the encapsulation of sensors/indicators. Also among them the authors highlighted indicators sensitive to pH alteration caused by degradation of food products, such as acid or alkaline gases, which are the most promising ones. Examples of pH indicators, free or encapsulated in silica-based materials, are discussed and exemplified together with the legal aspects dealing with the regulation of the use of such devices in food.

Xueyan Yun et al., in Chapter 5, *Fabrication of High Barrier Plastics and its Application in Food Packaging*, focus on the enhancement of barrier performance of nondegradable and biodegradable plastics as well as their fabrication and application in food packaging. The fabrication techniques include the orientation, coating, nanocompositing, blending, and layer-by-layer (LBL) nanoassemblies based on synthetic and biobased polymers. The major emphasis is on the enhancement of barrier properties of oxygen, carbon dioxide, and water vapor, which is the key for the food packaging plastics. The fabrication of new high barrier plastics with direct impact in the development and potential application of new biodegradable materials in the food packaging industry is also discussed.

Chapter 6, *Biodegradable Food Packaging Nanocomposites Based on ZnO-Reinforced Polyhydroxyalkanoates*, prepared by Ana M. Díez-Pascual et al., gives an overview about fabrication and characterization of polyhydroxyalkanoates-based bionanocomposites reinforced with ZnO nanoparticles with a view to use them as food packaging materials.

Isabel Gonçalves de Moura et al., in Chapter 7, *Bioplastics From Agro-Wastes for Food Packaging Applications*, focus on the preparation of new functional biopolymers for packaging based on extracted cellulose, which exhibits broadly tunable thermomechanical properties and biodegradation. Therefore, this contribution shows the potential of agro-wastes to produce new cellulose-based bioplastics for food packaging applications.

Agustín González et al., in Chapter 8, *Study of the Structure/Property Relationship of Nanomaterials for Development of Novel Food Packaging*, provide an overview of the latest innovations in food packaging based on polymer nanomaterials, focusing on two main objectives: improvement of mechanical and barrier properties of the materials and search of functional properties designed and oriented to prepare active/intelligent food packaging. This chapter also reviews the use of different types of nanosystems to develop new nanosized materials, which enhances the final properties of synthetic and bio-polymers for food packaging.

Katarzyna Mitura et al., in Chapter 9, *Bioactive Food Packaging With Nanodiamond Particles Manufactured by Detonation and Plasma-Chemical Methods*, summarize the recent advances on research and applications of diamonds and nanodiamonds particles in the food industry, medicine, and cosmetology. Moreover, potential toxicological risk of nanodiamonds modified food packaging layers is highlighted and discussed within this chapter.

Ana Sofia Abreu et al., in Chapter 10, *Biodegradable Polymer-nanocomposites for Packaging Applications*, discuss and present examples of biodegradable polymers obtained from different sources and the development of new materials and nanocomposites based on these materials for food packaging applications.

Mythili Prakasam et al., in Chapter 11, *Flexible Packaging for Nonthermal Decontamination by High Hydrostatic Pressure*, give an up-to-date overview about nonthermal processing, high hydrostatic pressure processing, currently used flexible polymers for food packaging, and the influence of high hydrostatic pressure processing on the properties of these polymers.

Marina Patricia Arrieta et al., in Chapter 12, *An Overview of Nanoparticles Role in the Improvement of Barrier Properties of Bioplastics for Food Packaging Applications*, explain the role of nanoparticles on the polymers permeation and diffusion processes focusing on food-safety and environmental concerns.

Chapter 13, *Biopolymer/Clay Nanocomposites as the High Barrier Packaging Material: Recent Advances*, prepared by Kateryna Fatyeyeva et al., reviews different types of edible and biodegradable biobased nanocomposites with special emphasis on the advantages of nanotechnology application for improving the barrier properties of the polymer matrices. These properties include rigidity, stiffness or flexibility, durability, temperature, moisture stability, and the barrier properties against water, oxygen, and other gases.

In Chapter 14, *Barrier Properties Improvement Using Additives*, J. Rodolfo Rendón-Villalobos et al. discuss how nanotechnology can provide solutions for modifying the permeation behavior of foils, increasing barrier properties, improving mechanical and heat-resistance properties, developing active antimicrobial and antifungal surfaces, and sensing, as well as signaling microbiological and biochemical changes and creating the nanobiodegradable packaging.

Chapter 15, prepared by Iva Rezić et al., *Metal Nanoparticles and Carbon Nanotubes—Perfect Antimicrobial Nanofillers in Polymer-Based Food Packaging Materials*, deals with some of the most important questions related to metallic nanoparticles in food packaging (health and environmental considerations, performance and costs), with special emphasis on their applicable and toxicological properties.

Majid Montazer et al., in Chapter 16, *New Approaches and Future Aspects of Antibacterial Food Packaging: From Nanoparticles Coating to Nanofibers and Nanocomposites With Foresight to Address the Regulatory Uncertainty*, discuss and highlight the impact of different antibacterial nanostructures used in food packaging, from the earliest application of single layer nanoclay/polymer to the current use of nanofibers, polymer nanocomposites, and bionanocomposites for creating potent antibacterial systems. In this contribution, nanopackaging materials with antibacterial functions have the potential to influence the food packaging sector from boosting food protection, safety, and quality, to extend food shelf life and to solve environmental problems.

Nuria Burgos et al., in Chapter 17, *Nanoencapsulation of Flavor and Aromas in Food Packaging*, present an up-to-date state-of-the-art regarding nanoencapsulated systems as an efficient support of flavors or aromas incorporated with (bio)polymers for food packaging.

Chapter 18, *Polypropylene/Talc Nanocomposites Films as Low-Cost Barrier Materials for Food Packaging*, prepared by Karolth Espinosa et al., give a detailed analysis of polypropylene/talc nanocomposite films and present the feasibility of using them in food packaging.

Aleksandra R. Nesic et al., in Chapter 19, *The Influence of Nanofillers on Physical-Chemical Properties of Polysaccharide-Based Film Intended for Food Packaging*, provide a review of implementation potentials and development of the nanocomposite materials based on polysaccharides (thermoplastic starch, chitosan, cellulose, alginate, pectin) in combination with various nanofillers emphasizing mechanical, thermochemical, and barrier properties.

Chapter 20, prepared by Ahmad Allahbakhsh et al., *High Barrier Graphene/Polymer Nanocomposite Films*, gives an overview about barrier graphene/polymer nanocomposites and their permeability performances and also discusses the future perspectives of graphene/polymer nanocomposites in the food packaging industry and the potential challenges that should be addressed when aiming to utilize such packaging.

Alexandru Mihai Grumezescu
University Politehnica of Bucharest,
Department of Science and Engineering
of Oxide Materials and Nanomaterials,
Faculty of Applied Chemistry and
Materials Science, Bucharest, Romania
http://grumezescu.com/

1

NANOTECHNOLOGICAL STRATEGIES YIELDING HIGH-BARRIER PLASTIC FOOD PACKAGING

Iman Soltani*, Richard J. Spontak,†**
**North Carolina State University, Fiber and Polymer Science Program, Raleigh, NC, United States; **North Carolina State University, Department of Chemical & Biomolecular Engineering, Raleigh, NC, United States; †North Carolina State University, Department of Materials Science & Engineering, Raleigh, NC, United States*

1 Introduction

In light of their lower weight and price (due to source materials and production) and their inert nature and robust mechanical properties, polymeric materials continue to replace traditional materials such as metals and ceramics in food packaging applications. A recent study by Franklin Associates (2014) indicates that the food industry earns about 38% of the market share for flexible plastic packaging used in the United States last year. Within this category, the top three subcategories requiring flexible plastic packaging include meat products (16%), baked goods (13%), and snack foods (11%). Moreover, the top three US market share categories for rigid beverage containers include carbonated soft drinks (31%), water (23%), and beer (14%). A complementary breakdown on the basis of mass is provided in Fig. 1.1. In addition to the considerations listed earlier, use of plastics in food packaging can significantly reduce material and energy expenditures, as well as greenhouse pollution, relative to alternative packaging materials. For example, plastics employed in the production of beverage containers account for annual energy savings of about 1.2×10^{11} MJ and decreased CO_2 emissions by 9.7×10^6 metric tons (Franklin Associates, 2014; Lilienfeld, 2014). Although they exhibit distinct advantages compared to their inorganic counterparts in food packaging, polymeric substitutes generally remain

Food Packaging. http://dx.doi.org/10.1016/B978-0-12-804302-8.00001-7

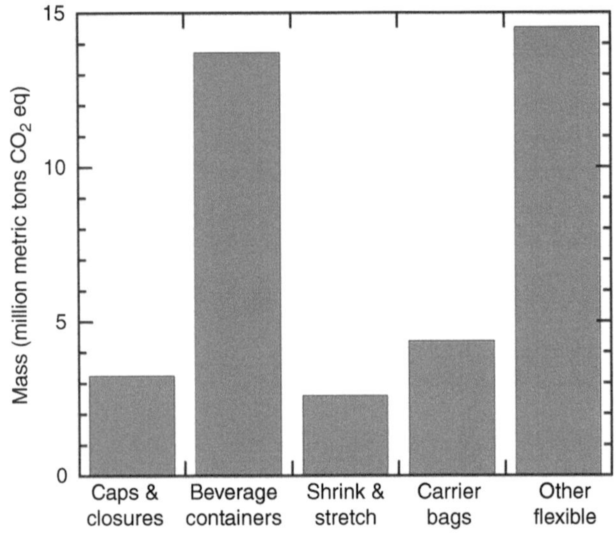

Figure 1.1. The 2014 mass quantities (on a CO_2 equivalent basis) of commercially relevant polymeric materials employed in various food packaging functions in the United States. Data reported by Franklin Associates.

inferior in terms of their barrier performance. Thus, identification of economically viable strategies by which to improve the barrier properties of polymers to extend the transportation and shelf lives of food products has become an important fundamental and technological objective. One of the primary approaches commercially developed for this purpose depends on the postprocessing application of impermeable metal or ceramic coatings, which must be sufficiently thin to remain flexible and avoid cracking. This tactic nonetheless is in need of further improvement to satisfy ongoing requirements in food packaging applications (Inagaki et al., 1999; da Silva Sobrinho et al., 2000; Czeremuszkin et al., 2001; Hanika et al., 2003; Lewis, 2006; Jiang et al., 2010; Carcia et al., 2009; Zhang, 2010; Priolo et al., 2010a).

The development of thin inorganic coatings designed to improve the barrier properties of polymeric films can be traced back to the 1970s when vacuum metallization was utilized to evaporate a thin opaque layer of aluminum on polyester (Chatham, 1996). In fact, coatings based on physical evaporation methods applied under vacuum were considered as the most effective means by which to enhance the barrier performance of polymeric films for several subsequent decades (Cueff et al., 1995; Imai et al., 1995; Deng et al., 2000; Erlat et al., 1999, 2000; Leterrier, 2003; Low and Xu, 2005). A practical drawback to single-layer thin coatings was the existence of discrete defects, which expectedly had an adverse effect on their otherwise impermeable characteristics. Solutions to this nontrivial shortcoming prompted the development of multilayer coatings, such as those afforded by layer-by-layer (LBL) procedures (Keller et al., 1995; Lvov et al., 1996; Affinito et al., 1996; Iler, 1966; Affinito et al., 1997; Kotov, 1999; Hammond, 2004; He et al., 2004; Jang and Grunlan, 2005; Zhang et al., 2007; Jang et al., 2008; Cini et al., 2010; Orozco et al., 2010; Yang et al., 2010; Priolo et al., 2010a, 2010b, 2011; Shchepelina et al., 2011), and almost defect-free atomic layer deposition (ALD) coatings (Elam et al., 2003; Wilson et al., 2005; Groner et al., 2006; Langereis et al., 2006; Peng et al., 2007; Wilson et al., 2008; Carcia et al., 2009; Carcia et al., 2010; George, 2010; Spagnola et al., 2010). Inorganic modification of bulk polymers promoted the introduction of polymer nanocomposites (PNCs) as attractive alternatives to coating processes. Such hybrid materials were effectively free from the typical problems that plagued inorganic coatings (eg, cracking and delamination), and the incorporation of nanoscale objects into a polymer matrix concurrently augmented other physical, often mechanical, properties. Lastly, the processing required to generate PNCs at large, industrial scales was facile and less expensive relative to high-vacuum coating processes, and it provided a largely unexplored route by which to tune barrier properties via tortuosity considerations at nanoscale dimensions (Alexandre and Dubois, 2000; Zeng and Lee, 2001; Ray et al., 2002, 2006; Li et al., 2003; Shah and Paul, 2004; Lertwimolnun and Vergnes, 2005; Zhong et al., 2005; Ray, 2006; Bousmina, 2006; Tokihisa et al., 2006; Utracki et al., 2006; Jacquelot et al., 2006; Shah et al., 2006; Hamzehlou and Katbab, 2007; Moncada et al., 2007). In the following sections, we address fundamental aspects of various polymer coating and nanocomposite strategies designed for improved barrier performance, and discuss their efficacy in the food packaging industry.

2 Existing Demands and Challenges

Food and medical packaging materials must be designed with high barrier properties against reactive gases such as oxygen to increase the transportation and shelf life of the packaged contents. According to the results of Hanika et al. (2003) presented in Fig. 1.2a, the oxygen transmission rate (OTR) required for most food products should lie between 0.1 and 30 cm^3/m^2-day-bar (assuming STP conditions), depending on the oxygen sensitivity of the food type. Baby food and beer are much more susceptible to oxidation than, for instance, ketchup and nuts. Two principal

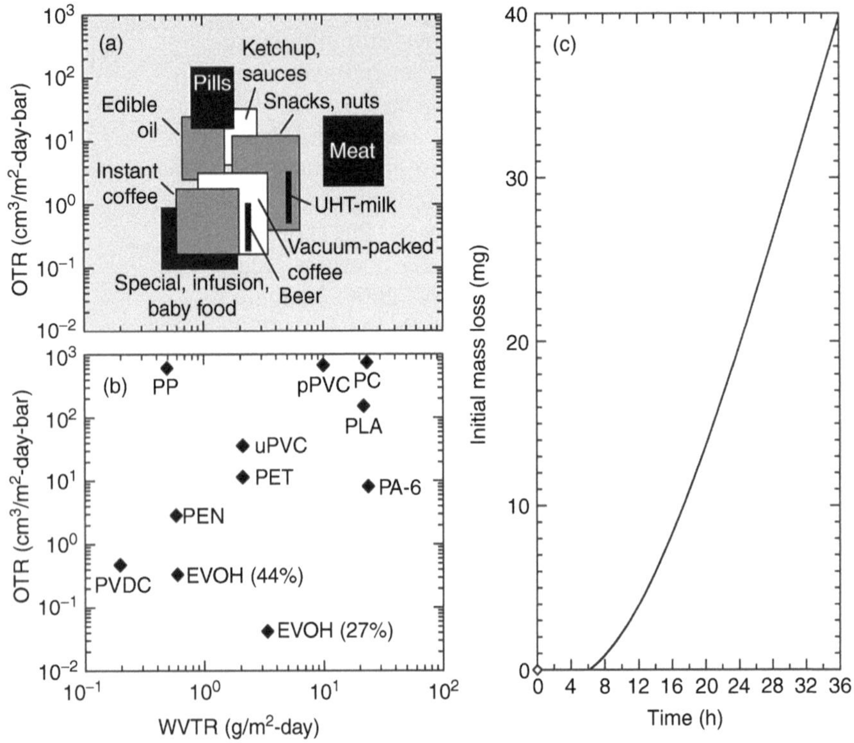

Figure 1.2. (a) Oxygen barrier requirements for various food packaging at 23°C and (b) typical OTR values of commercially relevant polymer films measuring 100 μm thick at 23°C as functions of the WVTR. Included here are ethylene vinyl alcohol (EVOH) with two different degrees of substitution, poly(vinylidene chloride) (PVDC), poly(ethylene naphthalate) (PEN), poly(ethylene terephthalate) (PET), nylon-6 (PA-6), unplasticized poly(vinyl chloride) (uPVC), plasticized poly(vinyl chloride) (pPVC), poly(lactic acid) (PLA), polycarbonate (PC), and polypropylene (PP). (c) The initial loss of CO_2 from a PET bottle containing a carbonated soft drink measured at 24°C and a gas pressure of 3 bar. (a, b) Adapted from Hanika et al. (2003) with permission from Wiley; (c) Adapted with permission from the Composite Agency.

factors govern the OTR: polymer thickness and polymer chemistry, the latter of which dictates the diffusivity (D) and solubility (S) [and, hence, permeability ($P = D \times S$)] of penetrant molecules through a dense polymer matrix. Polymer chemistry must also be considered on the basis of safety because some polymers may contain residual monomer (eg, bisphenol A) that can leach into and contaminate food products upon contact (Schecter et al., 2010). Reported OTR values for polymer films measuring 100 μm thick (Hanika et al., 2003; Groner et al., 2006) are also displayed as a function of water vapor transmission rate (WVTR) in Fig. 1.2b and reveal that most commercially relevant polymers, with notable exception of liquid crystalline polymers (not shown) and some ethylene vinyl alcohol (EVOH) copolymers, exhibit an OTR greater than 0.1 cm^3/m^2-day-bar, which makes most of these polymers generally suitable for food packaging applications.

In the specific case of carbonated beverages packaged in melt-blown bottles and maintained at elevated pressures of ≈ 3 bar (Composite Agency), however, retention of carbon dioxide requires high barrier properties against an acid gas that is known to act as a plasticizer for many thermoplastic polymers (Fried et al., 1989). Although polyethylene terephthalate (PET) is widely employed as a substitute for traditional glass bottles, its inferior barrier properties promote a significantly shorter shelf life (Ray et al., 2006). In fact, the experimental results provided in Fig. 1.2c indicate that pressurized carbon dioxide permeates through conventional PET soft drink bottles after only a few seconds, resulting in a mass loss of ≈20 mg after only 24 h. After a single year, the mass of CO_2 remaining decreases by 75%. This barrier performance is responsible for a PET-bottled carbonated soft drink typically having a shelf life of 3–4 months (Boyer and McKinney, 2013). Recent considerations that have exacerbated this shelf-life problem include the use of thinner walls to reduce bottle weight and the infusion of recycled PET to reduce solid landfill (University of South Carolina Research Foundation, 2003; Hamzehlou and Katbab, 2007; Aionicesei et al., 2008; Ozcam, 2011; Markočič et al., 2011). Therefore, for this and related reasons, the need for effective barrier properties in the food packaging industry has prompted numerous polymer coating and nanocomposite technologies (Jang et al., 2008). Due to complications arising from, for instance, specific interactions between penetrant molecules and the matrix polymer or the presence of nano/microscale interfaces, however, improving barrier properties has proven to be scientifically challenging. Advances in barrier properties achieved on the basis of new or existing material strategies have been less than the progress devoted to developing and commercializing the approaches themselves (Chatham, 1996; Sorrentino et al., 2006; Mittal, 2010a).

3 Coating Methods for Improved Barrier Properties

One of the primary classifications of methods intended to improve the barrier properties of polymer substrates involves the deposition of a thin high-barrier inorganic coating. Such coatings can be further categorized on the basis of the number of layers as either single-layered or multilayered. In the following sections we discuss both coating categories.

3.1 Single-Layer Coatings

Single-layer barrier coatings for plastics are conveniently divided into metal and oxide/nitride coatings on the basis of their optical transparency. Opaque metal coatings have been commercialized since the early 1970s and are responsible for the aluminized PET and polypropylene (PP) packaging commonly encountered in snack food and beverage packaging. These vacuum-metalized plastics, in particular, are produced at an industrial scale through evaporation and deposition of a 10–100 nm thick Al film on a moving polymeric substrate maintained in a roll-to-roll configuration inside a large vacuum chamber (Graff et al., 2005). The OTR of a PET film measuring 13 μm thick and coated with a thin (20 nm) Al coating decreases by about 2 orders of magnitude from about 100 to 1 cm^3/m^2-day-bar (Jamieson and Windle, 1983). Satisfactory performance of such laminates inherently depends on good adhesion between the coating and polymer substrate through the formation of chemical bonds during deposition (Silvain et al., 1992; Milde et al., 1996). Two obvious shortcomings of metalized coatings are that they are optically opaque and recycling can be problematic (Graff et al., 2005). Conversely, transparent oxide and nitride coatings derived from SiO_x, AlO_x, TiO_x, and Si_3N_4 are frequently used in polymer packaging to facilitate product viewing and microwaving. Polymers modified with such coatings are easier to recycle, thereby reducing environmental concerns. The primary advantage of transparent oxide/nitride coatings relative to metal coatings is that they provide higher barrier properties wherein gas permeation is restricted to existing defects. Thus, the properties of these barrier coatings strongly depend on the density of defects that develop during fabrication due to factors such as deposition conditions and coating thickness (Erlat et al., 2000; Leterrier, 2003). The series of atomic force microscopy (AFM) images presented in Fig. 1.3 (Erlat et al., 1999) demonstrates how the coating uniformity changes with thickness due to the transverse propagation of defects. Since metal oxides and nitrides are dielectric and possess high melting points, high-throughput vacuum deposition can become technologically challenging, especially when performed on polymer substrates that must be maintained at temperatures below 100°C to avoid softening, distortion, or undesirable expansion (Gonzalez et al., 1994; Graff et al., 2005).

Generally speaking, transparent oxide and nitride coatings are deposited on polymer substrates through the use of three commercial processes: evaporation (thermal or electron beam), sputtering, and chemical vapor deposition (CVD) (Cueff et al., 1995; da Silva Sobrinho et al., 1998, 2000; Erlat et al., 2000; Hanika

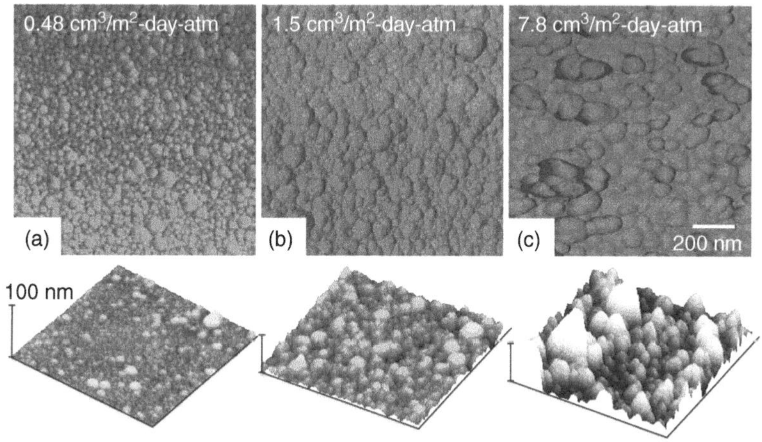

Figure 1.3. A series of 2- and 3-dimensional AFM height images (top and bottom rows, respectively) acquired from SiO_x coatings varying in OTR (labeled) on PET substrates at 30°C. Adapted from Erlat et al. (1999), with permission from the American Chemical Society.

et al., 2003; Low and Xu, 2005; Carcia et al., 2009). Evaporation is a physical vapor deposition (PVD) coating process that is divided into thermal and electron beam-assisted methods. It is performed on the scale of molecular flow, in which case the distance between the source undergoing evaporation and the substrate is less than the mean free path of the source species in the gas state. As a result of thermal heating, molecules become sufficiently energized to leave the source and, after traveling through the evacuated deposition chamber, condense on the substrate surface (Deng et al., 2000). While strictly thermal evaporation has been conducted with many different coatings, the evaporative deposition of high-melting metal oxides often requires assistance from a focused electron beam (Imai et al., 1995; Chatham, 1996; Czeremuszkin et al., 2001). In sputtering, the source material is bombarded by a plasma discharge, and source molecules are deposited as a vapor (Hanika et al., 2003). Due to greater parameter tunability available in sputtering relative to evaporation, the properties of sputtered coatings can be more precisely manipulated (Cueff et al., 1995). Moreover, sputtering requires a lower vacuum for deposition and provides a uniform film on a larger area with a wider choice of source materials than evaporation, and sputtered coatings tend to possess better adhesion to polymer substrates. CVD operates on the principle that vapor-phase metal-containing molecules migrate from the source to the substrate in a manner that differs from PVD. Source gas molecules undergo convective diffusion close to the substrate surface, where they absorb and react to release gaseous products that desorb from the surface. Unlike evaporation and sputtering, CVD does not rely on line of sight and thus yields a conformal coating. In the case of polymer substrates, plasma-enhanced CVD (PECVD) involves the use of microwave- or radio frequency–activated plasma to activate the source material at low temperatures (Inagaki et al., 1999; Erlat et al., 1999; da Silva Sobrinho et al., 2000). Plasmas excited by microwave radiation yield greater ionization and thus higher deposition rates than radio frequency plasmas, whereas radio frequency plasmas are excited at lower frequencies and permit more precise control over the ion energy and flux. This feature is particularly beneficial for polymer substrates that possess a relatively low glass transition temperature and cannot be subjected to very much heat (da Silva Sobrinho et al., 1998).

The last and most recently developed single-layer coating process included here is ALD, which was introduced by Suntola and Autson (1977) under the name of atomic layer epitaxy to describe the deposition of ZnS on glass at 320°C. In 2003, ALD was applied for the first time to deposit Al_2O_3 on a polymer substrate as nucleating base layer (Elam et al., 2003). In the same fashion as CVD, ALD is deposited through a chemical reaction, not physical bombardment, and provides conformal coverage, which makes it particularly valuable for coating complex three-dimensional shapes (Peng et al., 2007; Spagnola, 2010). During ALD performed under vacuum, two species are alternatively introduced to react with a surface of interest and each deposit a single atomic layer. As schematically depicted in Fig. 1.4a for the deposition of Al_2O_3 onto a hydrophilic polymer substrate, trimethylaluminum (TMA) is used to react with surface hydroxyl groups to form a hydrophobic layer (dose mode). Excess (unreacted) TMA is removed by purging with an inert gas (purge mode) and water vapor is introduced into the chamber to react with the newly formed layer and make the surface hydrophilic again. In this fashion, the surface is prepared for subsequent reaction with TMA. This highly controlled deposition cycle can be repeated until a coating of desired thickness is achieved with atomic precision, as evidenced by the thickness measurements displayed in Fig. 1.4b (Peng et al., 2007). Separation of the reactive species by an inert gas purge prevents the possibility of gas-phase reactions, which increases coating uniformity relative to CVD (George, 2010; Spagnola et al., 2010). Thus far, ALD has been extensively investigated as a means by which to deposit Al_2O_3 coatings at relatively low temperatures (typically ≈120°C) on a wide variety of thermoplastic polymers including PET, PP, polyethylene, polyimide (PI), polystyrene (PS), poly(vinyl alcohol) (PVA), and poly(methyl methacrylate) (Wilson et al., 2005;

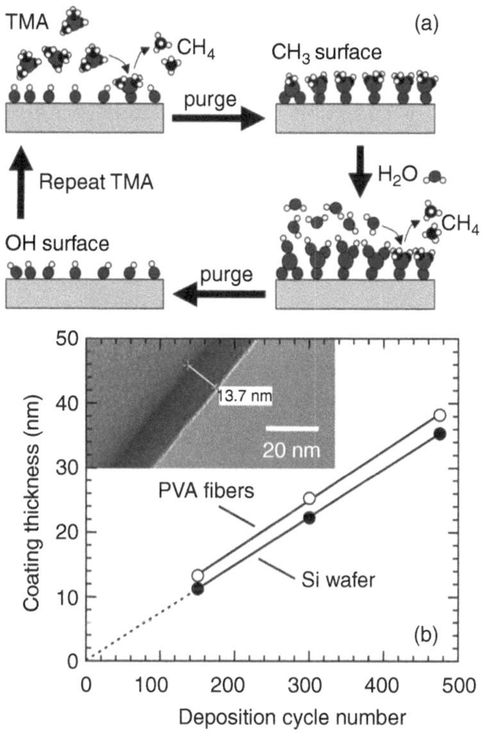

Figure 1.4. (a) The sequence of events corresponding to ALD conducted over a single cycle using TMA as the reactive precursor to generate an Al_2O_3 coating. (b) Thickness as a function of deposition cycle number for Al_2O_3 coatings on PVA fibers and silicon wafers (labeled). The lines are linear regressions to the data. The inset is a TEM image of the coating (about 14 nm thick) on a PVA fiber. Adapted from Peng et al. (2007), with permission from the American Chemical Society.

Groner et al., 2006; Peng et al., 2007; Wilson et al., 2008; Carcia et al., 2009, 2010).

3.2 Multilayer Coatings

Despite the improved barrier properties attained with polymer substrates coated with single metal and metal oxide layers, a drawback to single layer coatings deposited by methods such as PVD and CVD that is responsible for inconsistent results is the formation of defects. Such defects, usually associated with inefficient particle packing, can become increasingly problematic, especially as the coating thickness is increased and associated stresses develop. This issue, coupled with commercial costs related to high-throughput processing, is largely responsible for motivating complementary studies of alternative deposition strategies, such as multilayers composed of organic and inorganic species, for enhancing the barrier properties of polymeric substrates (Hanika et al., 2003; Graff et al., 2005). The polymer/metal oxide/polymer design, for instance, refers to a hybrid coating composed of alternating polymer and ceramic layers and deposited through methods such as polymer multilayer (PML), sputtering and evaporation. In the PML process, monomers are first flash-evaporated under vacuum and permitted to condense on the target substrate. This liquid layer is then solidified (cured) through exposure to UV radiation or a focused electron beam. A metal oxide layer is subsequently deposited on the newly formed polymeric coating via sputtering or evaporation, and this cycle is repeated to generate a multilayer architecture. An important benefit of this approach is that the first polymeric coating increases the smoothness of the target substrate, thereby reducing the risk of pinhole defects as additional layers are deposited. Moreover, the topmost polymeric coating completely covers the final metal oxide layer, protecting it against stresses that arise during handling. Although the number and size of the defects that develop as this organic/inorganic multilayer is fabricated are less numerous and problematic than those encountered in single-layer processes, this approach cannot altogether eliminate the possibility of defect formation. In addition, the polymeric coating process is somewhat complicated from a commercial standpoint, requiring both the vacuum evaporation and curing of monomers in highly controllable fashion (Jang et al., 2008; Affinito et al., 1996, 1997).

LBL deposition, a relatively new technology intended for fabricating multilayer thin films, is performed through alternating immersion of a polar substrate in an oppositely charged aqueous solution or suspension. Relative to the multilayer framework discussed earlier, the LBL strategy results in a larger number of thinner layers. The nanostructure and properties of LBL assemblies depend sensitively on the molecular weight of the components employed (Ozcam, 2011), their chemistry (Yang et al., 2010), ionic strength and pH (Priolo et al., 2010a). A typical LBL assembly consists of a multilayer with alternating layers of a charged organic binder and a natural (ie, not organically modified) clay nanoparticles, and the thickness can range from a few nanometers to several hundreds of nanometers (Jang et al., 2008). This process is relatively straightforward and effectively relies on constructing the multilayer from the ground up, resulting in a one-dimensional sandwich arrangement. One strategy by which to control the deposition of nanoscale organic layers on a solid substrate relies on the Langmuir-Blodgett (LB) technique for the formation of a monolayer on the surface of water, followed by attachment of the monolayer to the substrate. This methodology, which reproducibly yielded stearate coatings measuring less than 3 nm thick (Blodgett and Langmuir, 1937), has become ubiquitous in the study of thin organic layers (Kuhn and Möbius, 1971). According to Decher (2012), drawbacks associated with the LB technique in multilayer fabrication are twofold: (1) frequent rearrangement of deposited layers due to insufficiently firm spatial constraints, and (2) limited choice of molecular species that are suitable for this technique. These disadvantages can be overcome in the design of LBL assemblies through the use of oppositely charged species, as demonstrated by Iler (1966), who generated stable LBL multilayers by alternating the deposition of positively charged silica and negatively charged alumina colloidal particles. The use of electrostatic interactions to produce charged multilayers of organic compounds has been largely pioneered by Decher (Decher and Hong, 1991a, 1991b; Decher et al., 1992; Decher, 1997, 2012), who started with the alternating deposition of anionic and cationic amphiphiles and later introduced simple polyelectrolytes. Other charged species such as functional polymers, biological nanoparticles, and inorganic nanoparticles have been gradually incorporated into the LBL methodology. Although LBL coatings were initially applied to planar substrates, further developments have established that multilayer deposition on small nonplanar surfaces is also possible. For example, multilayers of polycations and anionic exfoliated zirconium phosphate have been successfully deposited on high-surface area silica (Keller et al., 1995).

While electrostatic interaction between oppositely charged species is widely accepted as the main driving force in the preparation of LBL assemblies, other mechanisms that can be used for this purpose include hydrogen bridging (Hammond, 2004; Orozco et al., 2010; Shchepelina et al., 2011), covalent bonds (Cini et al., 2010), hydrophobic interactions (Kotov, 1999; He et al., 2004), coordination interactions (Zhang et al., 2007), and complementary postchemical reactions (Decher, 2012). In the case of multilayer coatings generated via electrostatic interactions, LBL deposition is further divided into three primary methods. According to the first approach, linear polyions constitute the alternating species that stack on top of each other, thereby forming a polyelectrolyte multilayer composed of oppositely charged polymers (Decher et al., 1992). The second type of electrostatically driven LBL assembly is based on the use of charged particles, such as globular proteins, successively deposited on linear polyions (He et al., 2004; Orozco et al., 2010). In the last category, charged particles are replaced with charged platelets exemplified by exfoliated natural clays such as montmorillonite (MMT) deposited from suspension (Lvov et al., 1996; He et al., 2004; Jang et al., 2008; Yang et al., 2010). This latter approach, schematically depicted in Fig. 1.5a, has been found to yield coated polymer substrates exhibiting remarkable barrier properties. In fact, use of natural nanoclays such as MMT in conjunction with polyelectrolytes (eg, polyethylene imine and polyacrylamide) has yielded coated PET substrates possessing extremely high barrier properties after deposition of relatively few layers. Independent studies (Yang et al., 2010; Priolo et al., 2010a, 2011) report OTR values lower than the detection limit of oxygen permeation sensors capable of measuring 0.005 $cm^3/(m^2\text{-day-bar})$.

Methods developed to deposit alternating layers in LBL assemblies have evolved over the years, but dipping, first introduced for producing multilayers from oppositely charged polyelectrolytes in 1992, still remains the most widely used deposition method.

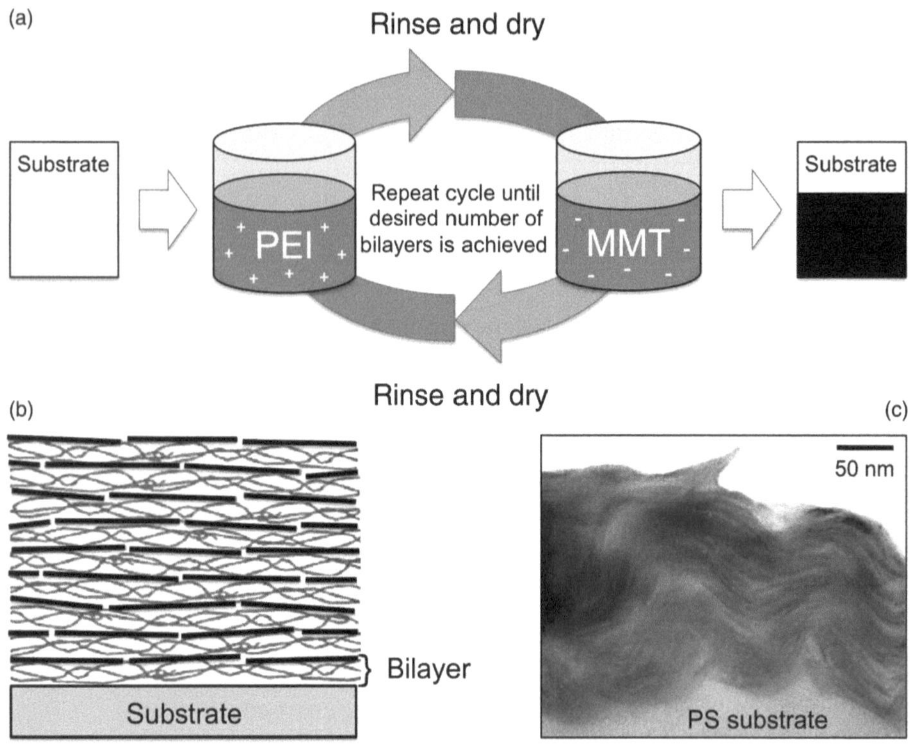

Figure 1.5. (a) The sequence of events corresponding to LBL deposition conducted with alternating immersion of the substrate in an aqueous poly(ethylene imine) (PEI) polycation solution and an aqueous MMT suspension. (b) Schematic illustration of the bilayers formed by LBL deposition onto a polymeric substrate. (c) Cross-sectional TEM image confirming the existence of numerous layers of clay platelets on the surface of a PS substrate. Adapted from Priolo et al. (2010a), with permission from the American Chemical Society.

Several complementary deposition strategies that have likewise been explored rely on spraying (Schlenoff et al., 2000) and spin-coating (Chiarelli et al., 2001, 2002; Jiang et al., 2004; Von Klitzing et al., 2012). Multilayer preparation by dipping requires alternating immersion of the polymer substrate in aqueous oppositely charged solutions or suspensions. It immediately follows that the surface to be coated must possess an excess charge that is either naturally occurring or introduced (by plasma or corona treatment) to bind with the organic layers. Previous studies (Jang et al., 2008; Yang et al., 2010; Priolo et al., 2010a, 2011) have established that this interaction is significantly improved through the use of adhesive polyelectrolyte interlayers composed of branched polyethyleneimine (PEI). In a typical LBL process conducted by dipping (Fig. 1.5a), immersion of the substrate in a PEI solution of predetermined concentration for a predetermined period of time is followed by rinsing with deionized water (DIW) and drying with filtered air. The cycle continues with immersion of the PEI-coated substrate in an aqueous clay suspension, which is also followed by rinsing in DIW and drying with filtered air. These steps are repeated until a desired number of bilayers, depicted in Fig. 1.5b, is deposited. To achieve more uniform multilayers and consistent properties, several studies have demonstrated that the dipping process can be fully automated (Félix et al., 2009; Yang et al., 2010; Priolo et al., 2010a, 2011).

The impressive orientation and exfoliation of nanoclay particles observed in transmission electron microscopy (TEM) micrographs of LBL assemblies (cf Fig. 1.5c) are unique features of these multilayer coatings. Due to the strong electrostatic attraction of positively charged PEI, the negatively charged nanoclay platelets absorb with their largest dimension parallel to the PEI-coated surface, resulting in a highly laminar, oriented, and packed arrangement of exfoliated nanoclay platelet layers that promote unique barrier and optical properties that are not achievable by any other method of combining polymers and nanoclay (Priolo et al., 2010b). The rate at which the LBL assembly is constructed layer-by-layer is classified as either linear or superlinear. On one hand, *linear growth* usually occurs in solutions with low ionic strength at low temperature, such as the growth of bilayers composed of PEI and MMT presented in Fig. 1.6a (Yang et al., 2010; Cini et al., 2010; Priolo et al., 2011; Von Klitzing et al., 2012). *Superlinear growth*, on the other hand, occurs in solutions with high ionic strength and at high temperature, and is characterized by an exponential increase due to free diffusion of the polyelectrolyte through the entire multilayer. In quadlayers composed of PEI, poly(acrylic acid), PEI, and MMT, the rate by which the thickness

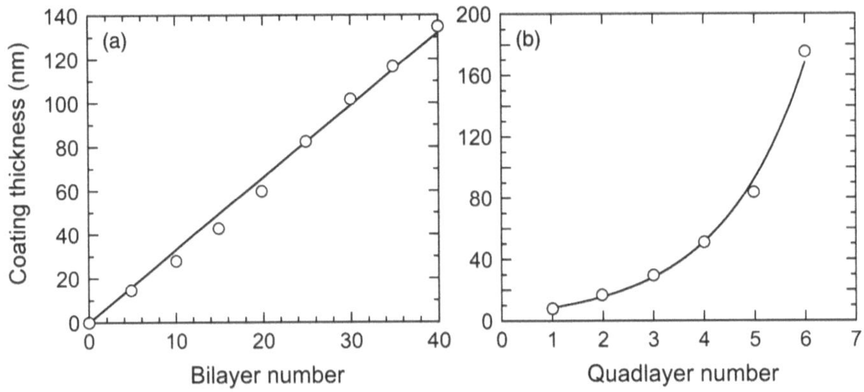

Figure 1.6. LBL coating thickness presented as a function of the number of (a) bilayers and (b) quadlayers for depositions employing either PEI alone or alternating immersion in PEI and poly(acrylic acid). Lines correspond to linear (a) and exponential (b) regressions of the data from Priolo et al. (2010a, 2010b), respectively.

increases appears closer to exponential in Fig. 1.6b, attaining a thickness of 174 nm after only six deposition steps, due to interdiffusion of the polyelectrolyte layers during deposition (Priolo et al., 2010b). In marked contrast, 40 bilayers in Fig. 1.6a produces a multilayer thickness of <140 nm.

Compared to other coating technologies developed to enhance the barrier properties of polymer substrates, LBL is generally more facile, as it is conducted under ambient conditions in the presence of aqueous solutions/suspensions (Jang et al., 2008). Thus, LBL deposition is considered to be a low-cost and environmentally friendly process method that is capable of depositing a wide spectrum of different species, ranging from small organic molecules and inorganic nanoparticles to polymers and colloids, on polymer substrates to fabricate precise nano/microscale assemblies possessing desired properties. From nanoscale particles and platelets to macroscale substrates, nearly every substrate with a solvent-accessible surface can be subjected to LBL deposition (Decher, 2012). Conversely, in traditional compounding, the negligibly small entropy of mixing associated with long polymer chains and generally endothermic interactions thwart homogenous mixing even at low additive concentrations. Due to the strong electrostatic attraction between oppositely charged species in the LBL approach, cyclic deposition of layers, coupled with the formation of ionic complexes (an exothermic process), help to make nanoscale mixtures possible, and the composition of the multilayer assembly can be controlled through tunable variations in solution/suspension factors such as pH and ionic strength (Keller et al., 1995; Hammond, 2004).

4 Polymer Nanocomposites for Improved Barrier Properties

Despite their ability to improve the barrier properties of thermoplastic substrates, the various coating technologies mentioned earlier inherently possess nontrivial shortcomings for many commercial packaging applications, especially those encountered in the food industry. First and foremost, surface coatings are vulnerable to the development of defects promoted by stresses introduced during handling over the course of their service life (Chatham, 1996; da Silva Sobrinho et al., 1998, 2000; Erlat et al., 1999, 2001; Czeremuszkin et al., 2001; Hanika et al., 2003; Leterrier, 2003; Graff et al., 2005; Lewis, 2006b). In addition, the possibility always exists that a deposited surface coating delaminates and completely peels from the substrate due to insufficient adhesion. Another important technological consideration that warrants mention at this juncture is that coated plastic substrates cannot be processed by conventional processing routes such as extrusion and injection molding. One strategy to overcome all of these drawbacks involves the production of PNCs in which the bulk, rather than surface, properties are modified through the physical or chemical incorporation of impermeable nanoparticles (Ray et al., 2002; University of South Carolina Research Foundation, 2003; Ray and Okamoto, 2003; Osman et al., 2004; Jacquelot et al., 2006; Hamzehlou and Katbab, 2007). While nanoparticles conveniently serve as obstacles to molecular diffusion, they can serve other important purposes. For instance, as the global need for polymer recycling continues to escalate due to growing environmental concerns, the intrinsic barrier performance and overall processability of recycled thermoplastics such as PET become increasingly compromised as a consequence of repeated chain scission, which tends to reduce molecular weight and the melt strength required for processes such as blow molding and flat-die extrusion. Inclusion of highly dispersed clay nanoparticles can increase the melt viscosity of recycled polymers and, thus, improve their processability (Krishnamoorti and Yurekli, 2001; Lim and Park, 2001; Kim et al., 2003; Li et al., 2003; Zhong et al., 2005; Shen et al., 2005; Kráčalík et al., 2007a; Bhattacharya et al., 2008; Song and Jin, 2010; Beall and Powell, 2011). Therefore, PNCs are rightfully considered a technologically attractive approach by which to enhance the barrier properties, as well as other physical properties, of polymeric materials in a wide variety of packaging applications.

By decreasing the size of typical micrometer-sized particles down to the nanometer range, the surface-to-volume ratio can be dramatically increased by orders of magnitude, thereby promoting a corresponding increase in interfacial contact with the polymer matrix (Mittal, 2010b). As the degree of contact between the particles and polymer is increased, important material properties (eg, barrier, flame resistance, and mechanical) require a systematically lower particle loading level, which constitutes a critical consideration in terms of weight reduction and the energy required to transport packaged goods (Sawyer et al., 2007). Among the different types of PNCs that have been successfully developed, polymer-clay nanocomposites (PCNs) containing layered silicates, initially introduced in 1985 (Okada and Usuki, 2006) to improve the mechanical properties of polymeric materials, remain one of the most popular (Beall and Powell, 2011). For such organic/inorganic hybrid materials to be considered nanocomposites, the silicate layers must be dispersed throughout the polymer matrix in such fashion that at least one dimension of the layers is less than 100 nm (Kawasumi, 2004; Mittal, 2010b). Since the individual platelets comprising a layered silicate typically measure a few nanometers thick and, depending on the type of clay, greater than 100 nm across, the result upon exfoliation (discussed later) is a PNC with nanoscale sheets possessing a relatively high aspect ratio. For this reason, PCNs can be implemented to create a highly tortuous pathway for molecular diffusion and, by doing so, generate barrier properties that are vastly superior to many other PNCs. Moreover, layered silicates are considered nontoxic by the US Food and Drug Administration (FDA) and hold a "generally recognized as safe" (GRAS) classification. Another material property that merits consideration especially in food packaging is optical transparency, because consumers frequently desire to inspect the contents prior to purchase or consumption. Since the platelets of layered silicates are very thin and the refractive indices of clay nanoparticles and typical thermoplastics are not very dissimilar, the resulting PCNs generally retain their transparency (Mittal, 2010a). Taken together, these advantages unequivocally make PCNs an ideal class of candidates for food packaging.

Hybrid materials composed of a polymer matrix and a layered silicate nanofiller are categorized into three groups on the basis of their degree of mixing, as depicted in Fig. 1.7a (Alexandre and Dubois, 2000; Hussain et al., 2006). If the attractive force between adjacent clay platelets is sufficiently strong or the degree of shear mixing is low, long polymer chains are unable to diffuse into the interlayer spaces (galleries) between the platelets. In this case, the level of mixing is low, and layered silicate stacks that remain intact form a distinctly separate phase in the polymeric matrix. Such hybrid composites, classified as conventional microcomposites due to their large size scale and relatively low interfacial

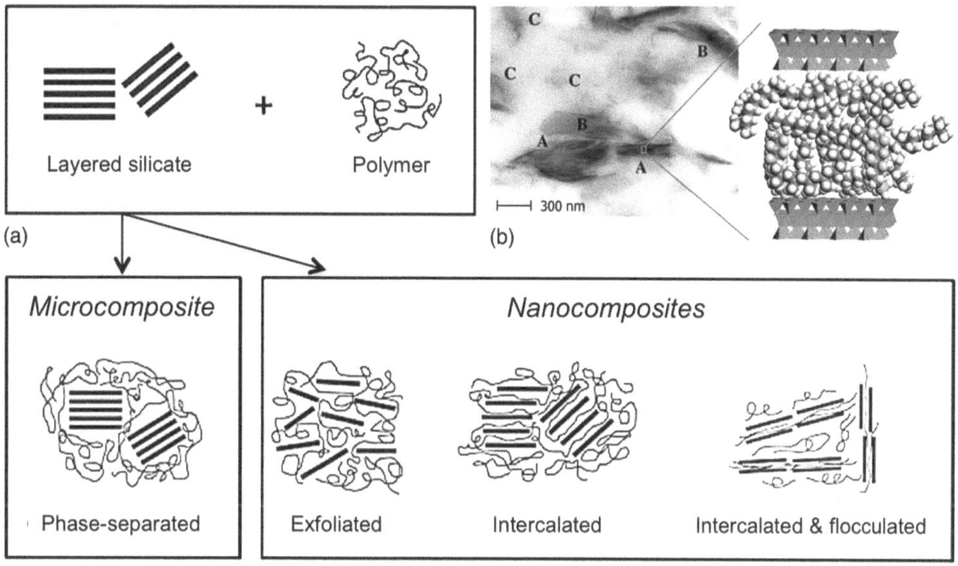

Figure 1.7. (a) Schematic illustration showing the results of incorporating layered silicates into polymeric matrices to yield either phase-separated microcomposites or different types of nanocomposites (labeled), depending on the level of dispersion achieved. (b) TEM image of a nanocomposite composed of dispersed clay platelets in a PP matrix. Here, A, B, and C identify intercalated stacks, disordered intercalated tactoids and exfoliated platelets, respectively. A schematic representation of an intercalated clay stack is included wherein parallel platelets are separated by PP chains. (A) Adapted from Galimberti (2012). Reproduced from Manias (2001), with permission from the Materials Research Society.

area, lack the significant property improvement afforded by nanocomposites. When the polymer chains diffuse between adjacent clay platelets in layered silicates and increase their basal spacing without disrupting their parallel alignment, the resulting PCNs are considered to be intercalated, in which crystallographic regularity is preserved over distances of a few nanometers. Finally, if the polymer chains overcome the electrostatic interactions of the clay platelets and completely disrupt platelet stacking, individual platelets are forced to disperse randomly throughout the polymer matrix to yield exfoliated PCNs wherein the average distance between platelets is composition-dependent. Due to the level of dispersion achieved and the interfacial area generated, exfoliated PCNs hold the most promise for remarkable property improvement and are therefore of greatest technological interest (Alexandre and Dubois, 2000; Krishnamoorti and Yurekli, 2001; Ray and Okamoto, 2003; Galimberti, 2012). Realistically, exfoliated and intercalated morphologies frequently coexist in PCNs, and classification reflects the more prevalent level of dispersion. For example, as observed in Fig. 1.7b (Manias, 2001), a PCN composed of PP contains intercalated clay platelets separated by 2–3 nm and exfoliated platelets containing up to ~3 clay platelets separated by more than 10 nm in the PP matrix. In this case, polymer intercalation induces limited platelet disordering that falls far short of complete exfoliation, resulting in the existence of an intermediate state (Fujiyama, 2010).

Layered silicates used to produce PNCs are commonly aluminosilicates that include such minerals as MMT, mica, and vermiculite. Of these, MMT is one of the most popular aluminosilicates due to its high aspect ratio and, hence, surface area, high cation exchange capacity, and surface reactivity. This latter attribute is particularly important for the fabrication of PCNs from most commercial polymers, which are intrinsically hydrophobic. To achieve satisfactory levels of intercalation or exfoliation in such polymer matrices, the hydrophilic MMT platelet surfaces are organically modified to increase their hydrophobicity and polymer compatibility. This conversion requires replacement of the surface cations of water-swollen MMT platelets with organic cations (such as alkyl ammonium) that serve to decrease the surface energy and increase the interlayer distance (Giannelis et al., 1999; Ray and Okamoto, 2003; Beall and Powell, 2011). Although quaternary ammonium ions are traditionally employed to modify MMT and promote exfoliation for improved barrier performance, the resulting PCNs are not approved as GRAS by the FDA (Beall and Powell, 2011). This disadvantage, coupled with (1) the low thermal stability of such organically modified clays (Frankowski et al., 2007) and (2) the high processing temperatures required for many engineering polymers such as PET, hinders the widespread utilization of so-called organoclays in food packaging and other clean applications (Kráčalík et al., 2007b; Ozcam, 2011).

4.1 Methods to Prepare PCNs

On the basis of the starting materials and processes to be used, the methods developed to produce PCNs can be broadly divided into three approaches: solution mixing, in situ polymerization, and melt compounding. In the first, a solvent is introduced for two concurrent purposes: to swell/disperse the clay platelets and dissolve the polymer chains. In an idealized scenario, the polymer chains located between swollen platelets remain in place upon solvent removal, thereby yielding an intercalated or exfoliated PCN insofar as the platelets do not undergo sedimentation. This problem can be largely prevented by ultrasonication, coupled with the increase in solution viscosity that accompanies solvent drying (Ray and Okamoto, 2003; Hussain et al., 2006; Fujiyama, 2010; Beall and Powell, 2011). For in situ polymerization, the layered silicates of interest are directly incorporated into liquid or dissolved monomers. Polymerization of the monomers produces a continuous polymeric matrix containing intercalated or exfoliated clay platelets. In this method (which was initially used to prepare PCNs consisting of nylon and MMT), the problem associated with coercing large macromolecules to diffuse between clay platelets to induce intercalation simply does not exist (Messersmith and Giannelis, 1993; Zeng and Lee, 2001; Ray and Okamoto, 2003; Hussain et al., 2006; Yilmazer and Ozden, 2006; Fujiyama, 2010). Both of these liquid-based approaches can result in uniformly dispersed PCNs wherein the silicates are either intercalated or exfoliated (preferred) because the processes start with highly swollen platelets. In melt compounding, however, a premade, high-molecular-weight polymer is typically mixed with an MMT-based organoclay that possesses a higher interlayer spacing than natural MMT. Mixing proceeds in the presence of high shear during extrusion at elevated temperatures in the melt state of the polymer. As mentioned earlier, the temperatures must be sufficiently low to preclude degradation of the organic surfactants attached to the clay platelets (Frankowski et al., 2007). Significant advantages of this method are (1) its compatibility with high-throughput production processes such as injection and blow molding, (2) its environmental friendliness because organic solvents are avoided, and (3) its applicability for a wide range of commercially relevant commodity and engineering polymers. Although caution must be exercised to minimize damage to the organoclay, this method of intercalating and exfoliating clay platelets in polymers has become an expedient method of fabricating PCNs with excellent barrier properties for packaging (Zeng and Lee, 2001; Lim and Park, 2001; Kim et al., 2003; Ammala et al., 2008; Rousseaux et al., 2011).

4.2 Methods to Characterize PCNs

One of the most important features of PCNs relates to the level of dispersion of layered silicates in the polymer matrix, because it plays an important role in determining the resultant mechanical and barrier properties. Thus, the morphological characterization of PCNs by microscopy techniques is critical. Due to this intimate structure-property design relationship, the development of nanocomposites parallels the development of high-resolution microscopy techniques, such as AFM, TEM, and scanning electron microscopy. Due to its inherently high spatial resolution and ability to reveal the internal nanostructure of (cryo)sectioned or spin/dip-coated specimens that are electron-transparent, TEM remains the primary microscopy technique utilized to examine the level of nanoparticle dispersion and, in particular, the degree of exfoliation of layered silicates embedded within a polymer matrix. To complement the real-space images acquired from TEM, X-ray diffractometry (XRD) provides a reciprocal-space measurement of intercalation. While XRD is capable of quantitating the interplanar spacing of layered silicates in ordered (intercalated or immiscible) systems and inferentially identifying exfoliation (due to the absence of a scattering peak from the layered silicate), TEM can qualitatively assess any type of PCN morphology and affords a more definitive exfoliation classification. In conventional phase-contrast TEM images of PCNs, clay platelets appear as dark, often linear features residing in a lighter polymer matrix due to the higher electron density of (and greater scattering from) the platelets, as evidenced by the TEM image presented in Fig. 1.7b (Morgan and Gilman, 2003; Osman et al., 2004; Zhong et al., 2005). Typically, polymers that possess polar groups are more prone to interact with hydrophilic clay platelets and promote exfoliation. An excellent example of a polymer that permits a high level of platelet dispersion is nylon-6 (Shah and Paul, 2004; Utracki et al., 2006), which comprised the matrix in the first reported PCN. In marked contrast, nonpolar polymers such as polystyrene (PS) and PP undergo weaker molecular interactions with natural clay and are much less likely to induce exfoliation (Li et al., 2003). In fact, even when organically modified clay or compatibilizers such as maleic anhydride are incorporated to promote dispersion during melt mixing, nonpolar polymer matrices frequently generate intercalated, rather than exfoliated, PCNs.

5 Molecular Transport in Polymers and Nanocomposites

Permeation of a penetrant species through a homogeneous medium proceeds via a three-step process that commences with adsorption of the species onto one side of the medium and is followed by diffusion of the species through the medium before desorption of the species at the opposite side of the medium. The driving force for this process is unbalanced chemical potentials for the penetrant species on the opposing sides of the permeable medium. In the limit of ideal gas mixtures, the chemical potentials can be expressed in terms of partial pressures of the penetrant species so that permeation is attributable to a pressure difference and occurs in the direction from high to low pressure. The permeability or permeation coefficient (P) serves as the principal metric for systematically evaluating the ability of a gas (or liquid) to undergo molecular transport through a membrane and is given by

$$P = \frac{(\text{amount of gas under a stated condition})(\text{membrane thickness})}{(\text{membrane area})(\text{time})(\text{pressure difference})} \quad (1.1)$$

Although CGS and SI units can be written for the permeability, a named unit that remains in common use for this purpose is the Barrer, which is defined as (Comyn, 1985)

$$1\,\text{Barrer} = \frac{10^{-10}\,(\text{cm}^3\,\text{STP})\,\text{cm}}{\text{cm}^2\,\text{s}\,\text{cmHg}} \quad (1.2)$$

where STP corresponds to the *s*tandard *t*emperature and *p*ressure conditions of 25°C and 1.013 bar, respectively (Hanika et al., 2003).

5.1 Single-Gas Permeability in Dense Polymers

Although single-component gas permeabilities can be measured through the direct use of compound specific sensors (typically used in oxygen and water transport studies), reported values are routinely determined from the pressure change that accompanies the molecular transport of a gas through a polymer membrane in an apparatus generically referred to as a constant-volume variable-pressure permeation cell. The cell is divided into upstream and downstream sides relative to the membrane to be analyzed. The downstream (receiving) volume is known and the associated gas pressure is measured as a function of time (t). The upstream side of cell is charged with a gas maintained at constant pressure (p). Due to permeation of the gas through the membrane, the downstream pressure expectedly increases. Although the downstream pressure initially increases nonlinearly with time, a linear $p(t)$ response eventually develops when the permeation reaches steady-state conditions. In this experimental arrangement, the permeability can be calculated from

$$P = \frac{VL}{ART\,\Delta p}\frac{\mathrm{d}p}{\mathrm{d}t} \quad (1.3)$$

where V is the downstream volume, L is the membrane thickness, A represents the exposed membrane area, R is the universal gas constant, T denotes absolute temperature, Δp corresponds to the difference between the constant upstream and initial downstream pressures, and $\mathrm{d}p/\mathrm{d}t$ is the linear change in pressure with respect to time at steady state (Patel et al., 2003; Hu et al., 2006; Perez-Blanco et al., 2010).

5.2 Barrier Properties of Polymer Nanocomposites

When impermeable particles are present in a permeable polymeric matrix, diffusing species are forced to change their trajectory so that they can proceed around such particles. If these impermeable particles are numerous, a tortuous path that affects the barrier properties of the organic/inorganic hybrid material develops inside the membrane. Attempts to model the permeability through such heterogeneous materials have yielded different results on the basis of the assumptions employed. A classic example is the mathematical relationship derived by Maxwell (1873) to describe the dielectric properties of a composite consisting of spherical objects suspended in a continuous dielectric medium. Surprisingly, the functional form of this relationship can be used to obtain accurate predictions for the permeability of a single gas through a dense polymer membrane containing a dispersion of impermeable spheres, insofar as the spheres measure on macroscale, not nanoscale, dimensions (Merkel et al., 2002). Expressed in terms of permeabilities, this relationship takes the form of

$$P_c = P_m\left(\frac{1-\phi_f}{1+\phi_f/2}\right) \tag{1.4}$$

where P_c and P_m identify the composite and matrix permeabilities, respectively, and ϕ_f is the volume fraction of spherical objects hereafter referred to as the "filler" (Patel et al., 2003; Hu et al., 2006). To account for the high aspect ratio encountered with clay platelets, a molecular tortuosity factor (τ), defined as the path length of molecular diffusion to the membrane thickness, has been introduced by Nielsen (1967). In the context of this formalism, P_c can be conveniently written as (Klute, 1959; Nielsen, 1967; Fredrickson and Bicerano, 1999)

$$P_c = P_m\left(\frac{1-\phi_f}{\tau}\right) \tag{1.5}$$

Comparing Eqs. (1.4) and (1.5) immediately reveals that $\tau = 1 + \phi_f / 2$ in the limit as the filler aspect ratio approaches unity. A shortcoming of Eq. (1.5), however, is that the change in permeant concentration as it passes around a filler particle is neglected, thereby making the predicted permeability higher than its experimentally measured value. A maximum in τ is achieved if (1) the impermeable filler is envisaged as either rectangular or circular platelets that are uniformly and completely dispersed and oriented parallel to the membrane surface and (2) minimal interactions

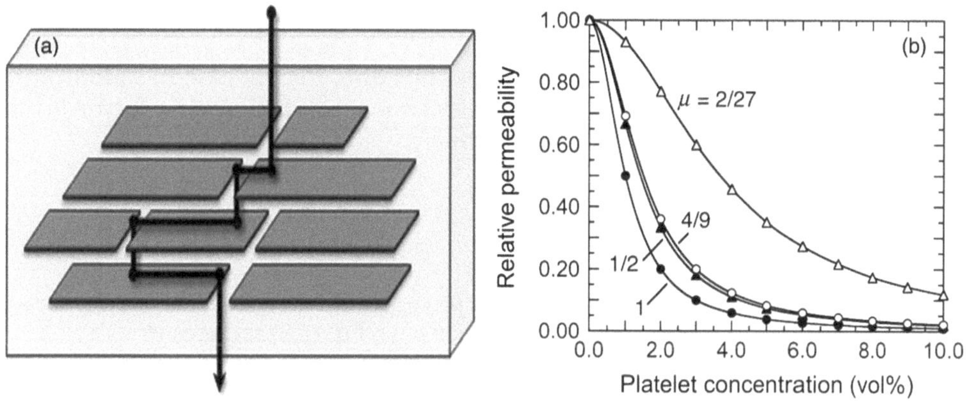

Figure 1.8. (a) Schematic diagram illustrating the diffusion pathway for a penetrant species through platelet-shaped nanoscale fillers embedded in a polymer matrix. (b) Predictions of the relative permeability of PCNs as a function of plate concentration according to the Cussler model Eq. (1.8) in which $\alpha = 100$ and μ is assigned discrete values (labeled) according to the spatial arrangements and platelet descriptions discussed in the text. Adapted from Galimberti (2012).

exist between the platelets and polymer matrix (Nielsen, 1967). Also, minimal interaction between clay and polymer is inherently assumed. In light of these assumptions, τ can be approximated by

$$\tau = 1 + \left(\frac{L}{2W}\right)\phi_{\mathrm{f}} \tag{1.6}$$

where L and W represent the characteristic length and thickness of the platelets, respectively, that are schematically depicted in Fig. 1.8a. Substituting Eq. (1.6) into Eq. (1.5) yields

$$P_{\mathrm{c}} = P_{\mathrm{m}}\left(\frac{1-\phi_{\mathrm{f}}}{1+\alpha\phi_{\mathrm{f}}/2}\right) \tag{1.7}$$

where α (= L/W) is defined as the aspect ratio. In the case of a spherical particle, $L = W$ and $\alpha = 1$ so that Eq. (1.7) simplifies to Eq. (1.4). Moreover, in the limit as $\phi_{\mathrm{f}} \to 0$ for $\alpha \approx 1$, $P_{\mathrm{c}} \to P_{\mathrm{m}}$ as intuitively expected. Within this same limit, however, the product $\alpha\phi_{\mathrm{f}}$ in the denominator of Eq. (1.7) becomes nonnegligible if $\alpha >> 1$ (DeRocher et al., 2005). It therefore follows that, at very low concentrations, impermeable platelets are effective at reducing gas permeation when they are highly anisotropic and the aspect ratio is very large (Nielsen, 1967).

Within the framework of the Nielsen model, the uniformly distributed platelets are presumed to possess equal size and aspect ratio, as well as spatial arrangement, which are obvious oversimplifications of real PCNs. To account for variations in size, shape and positional order, Cussler et al. (1988) have considered several different scenarios and, for the case of randomly spaced thin platelets, introduced a second geometric factor (μ) in their expression for P_c, namely,

$$P_c = P_m\left(\frac{1-\phi_f}{1-\phi_f+\mu\alpha^2\phi_f^2}\right) \tag{1.8}$$

In the limit of low concentrations ($\phi_f \rightarrow 0$) of highly anisotropic filler particles (with $\alpha >> 1$), Eq. (1.8) reduces to

$$P_c = P_m\left(\frac{1}{1+\mu\alpha^2\phi_f^2}\right) \tag{1.9}$$

Predictions from this equation are in favorable agreement with experimental data in the semidilute regime where $\phi_f << 1$ for the case of $\alpha\phi_f > 1$. While the geometric factor μ accounts for a combination of several different factors (although the interplay of these factors is not presently understood), experimental results provide sensible estimates. In the case of randomly distributed clay platelets whose shape lies between ribbons and hexagons (or disks), μ is anticipated to vary from 2/27 to 1/2 (Moggridge et al., 2003; DeRocher et al., 2005; Jang, 2008). If these are periodically distributed, corresponding values of μ are 1 (ribbons) and 4/9 (hexagons). According to the predictions obtained from Eq. (1.9) and displayed in Fig. 1.8b, the relative permeability (P_c/P_m) decreases, and the barrier performance improves, with increasing clay content and the value of μ at constant α.

One reason for deviation between experimental results and the Nielsen model can be traced back to its assumption that all platelets within the polymer matrix are perfectly aligned horizontal to the membrane surface, which is not the case. A modified Nielsen model has also been proposed (Bharadwaj, 2001; Gatos and Karger-Kocsis, 2007) in which an order parameter, $S = (3\cos^2\theta - 1)/2$, is incorporated to account for variations in platelet orientation (cf Fig. 1.9a):

$$P_c = P_m\left(\frac{1-\phi_f}{1+(2/3)(\alpha/2)(S+1/2)\phi_f}\right) \tag{1.10}$$

In the event that all the clay platelets are horizontally oriented parallel to the PCN surface, $\theta = 0°$ and $S = 1$, which allows Eq. (1.10) to simplify to Eq. (1.7). Conversely, if the platelets are oriented normal to the membrane surface (and the diffusive pathway through

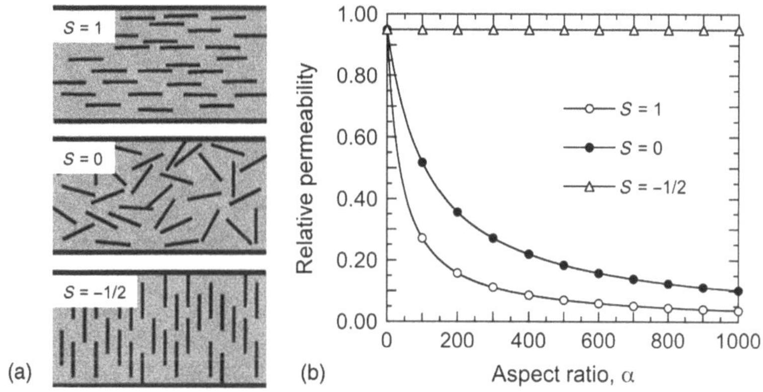

Figure 1.9. (a) Graphical representations of three different orientations (labeled with the corresponding order parameter, S, that clay platelets can adopt in a PCN. (b) Predictions for the relative permeability of PCNs containing 5 vol.% clay as a function of the platelet aspect ratio for the three values of S (labeled) that are schematically depicted in (a). Adapted from Bharadwaj (2001), with permission from the American Chemical Society.

the polymer matrix is unobstructed), then $\theta = 90°$, $S = -1/2$ and the parenthetical factor in Eq. (1.10) reduces to $1 - \phi_f$. In this scenario as $\phi_f \rightarrow 0$, $P_c \rightarrow P_m$. The predicted relative permeabilities included in Fig. 1.9b for PCNs with 5 vol.% clay filler demonstrate how platelet orientation influences barrier performance for the three physically relevant values of S: 1 (horizontal orientation), 0 (random orientation), and $-1/2$ (vertical orientation). As anticipated from Eq. (1.10), predictions corresponding to the random orientation converge with those from the horizontal orientation at large α.

6 Case Studies of Enhanced Barrier Performance

In this section, we examine the results of several studies that have either elucidated beneficial design rules or shown commercial promise for developing polymeric materials with improved barrier performance. While such materials might have originally focused on applications other than food packaging, the technological innovations responsible for the materials can, in most cases, be directly translated to the food packaging industry.

6.1 Single-Layer PVD/CVD Coatings

As mentioned earlier, several contemporary production strategies are available for depositing single inorganic coatings on polymer surfaces to enhance barrier performance. One of these is reactive sputtering, and an excellent example of this approach is the deposition of AlO_xN_y (aluminum oxynitride) on PET at medium frequency (40 kHz) in an AC magnetron with a commercial Al source and N_2 as the primary reactive gas. Due to the low density of defects in the resultant barrier coating (which is an intrinsic property of AlO_xN_y), OTR values as low as ≈1 cm³/(m²-day-atm) at 30°C and 0% relative humidity (RH) have been reported (Erlat et al., 2001). Relative to polar polymers such as PET, however, nonpolar polymers such as PP tend to adhere poorly to inorganic coatings and, as a consequence, achieve less effective barrier improvement. The study conducted by Deng et al. (2000) directly compares aluminum oxide coatings measuring 10 nm thick and evaporated on PET and PP substrates. Their results unequivocally establish that the permeability of oxygen decreases by more than 20× through surface-coated PET but displays no significant change through surface-coated PP. This difference is attributed to the presence of the PET carbonyl groups, which promote nucleation and growth of the inorganic coating on the PET surface. Without the existence of such specific interactions (as in the case of PP), a more defective and hence permeable coating develops.

Similarly impressive results have been documented for the PECVD of SiO_2 and SiN on a biaxially oriented PET substrate from hexamethyldisiloxane (HMDSO) in the presence of O_2 and SiH_4 in the presence of NH_3, respectively. Over a range of coatings varying from 8 to 200 nm in thickness, a minimum OTR of 0.5 cm³/(m²-day), which corresponds to a barrier improvement of 1000×, can be achieved when the coating thickness exceeds a critical value of 15 nm for SiO_2 and 8 nm for SiN (da Silva Sobrinho et al., 1998). By substituting a tetramethoxysilane/O_2 mixture as the gaseous precursor for radio-frequency- (RF-) PECVD on PET, Inagaki et al. (1999) have measured an OTR of 0.1 cm³/(m²-day-atm) under their best experimental conditions. An independent study employing two different organosilane monomers, HMDSO, and trimethylsilane, for the PECVD of SiO_x coatings on PET reveal no significant differences in coating characteristics or barrier efficacy (Erlat et al., 1999). Because permeation depends on molecular diffusion and solubility (both of which are thermally activated processes), the permeation of noninteracting gases that are important in the food packaging industry (eg, O_2) through dense polymer membranes is expected to obey an Arrhenius expression of the form

$$P = P_0 \exp\left(\frac{-\Delta G^{\ddagger}}{RT}\right) \tag{1.11}$$

where P_0 is the permeation at infinite temperature (a preexponential factor) and $\Delta G^{\ddagger}$ denotes the thermal activation energy.

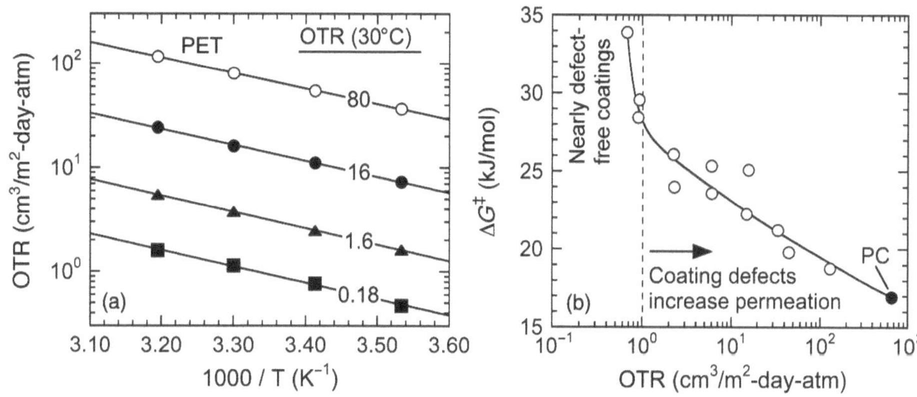

Figure 1.10. (a) Arrhenius-type temperature dependence of the OTR of four SiO_x coatings (labeled) deposited by PECVD on PET substrates. The regressed lines correspond to Eq. (1.11), and the resulting slopes yield the thermal activation energy, which is provided as a function of OTR for SiO_x coatings deposited on PC substrates in (b). The solid line serves as a guide for the eye, whereas the dashed line approximates the position of the abrupt change in slope. (a) Reproduced from Erlat et al. (1999), with permission from the American Chemical Society; (b) adapted from Erlat et al. (2000), with permission from the Materials Research Society.

This behavior is evident from the experimental data included in Fig. 1.10a. Interestingly, Arrhenius behavior is likewise apparent for SiO_x-coated PET over the same temperature range (Erlat et al., 1999, 2000; da Silva Sobrinho et al., 2000). Since $\Delta G^‡$ provides a measure of the energy barrier that must be overcome for permeation to occur, it embodies several factors, ranging from free-volume considerations in dense polymers to defects in inorganic coatings, which together influence permeation. Changes in the magnitude of $\Delta G^‡$ under isothermal conditions therefore provide indirect evidence of variations in the permeation mechanism. In Fig. 1.10b, the reduction in OTR — to below 0.8 $cm^3/(m^2$-day-atm)—induced by varying the coating thickness of SiO_x deposited by RF-PECVD on PC is accompanied by a substantial increase in $\Delta G^‡$, indicating that the pores in the coating that permit molecular diffusion are approaching the cross-sectional size of the diffusing O_2 molecules. In contrast, systematic changes in $\Delta G^‡$ are not observed in SiO_x-, SiO_2-, or SiN-coated PET, which suggests that molecular transport through the PET matrix, and not the coating, is rate-limiting most likely due to sufficiently large pores that exist as defects in the coatings.

Other important considerations regarding the deposition of single-layer coatings that have not yet been addressed include the shape and surface roughness of the polymer substrate. For uniformly thick coatings on nonplanar substrates (eg, beverage bottles), CVD is more effective than PVD. The line-of-sight deposition process by which PVD operates becomes increasingly problematic as the surface becomes more curved and/or rougher. Studies endeavoring to elucidate the molecular mechanisms of the two deposition processes have demonstrated that a gaseous species deposited by CVD is able to undergo several collisions with the substrate before reacting (thereby achieving better conformal coverage), whereas the species deposited by PVD immediately condenses on the surface after initial contact (da Silva Sobrinho et al., 2000; Hanika et al., 2003; Low and Xu, 2005). Comparison of the barrier improvement factor (BIF = 1 / relative permeability) measured for Al-based PVD coatings and SiO_x-based CVD coatings in Table 1.1 confirms that the latter materials provide better barrier performance against O_2 (by more than an order of magnitude in terms of their BIF values) due to their inherently conformal coverage. Another reason why PVD is believed to produce inferior surface coatings is the inadequate formation of an organic/inorganic interphase layer, which develops during CVD. Over this interphase layer measuring up to a few tens of nanometers thick, the composition gradually transitions from inorganic coating to organic polymer. In PECVD processes, this layer forms during the first few seconds of exposure to the plasma as volatile organic compounds are released from the polymer surface and interact with the plasma and its associated gases. The presence of this interphase greatly improves adhesion between the deposited coating and polymer substrate, thereby yielding superior barrier properties relative to PVD coatings (da Silva Sobrinho et al., 1998; Czeremuszkin et al., 2001).

Since aluminum metal and oxides of silicon and aluminum are known to have excellent barrier properties against water and oxygen, they are good candidates for barrier coatings. When such superbarrier materials are coated as single layers on polymeric substrates, however, the resulting material exhibits only two to three orders of magnitude improvement in barrier properties compared to the neat polymers (Chatham, 1996; Lewis, 2006; Cini et al., 2010). The reason for such meager performance enhancement is the existence of nanoscale defects and pores that develop during deposition (cf Fig. 1.11). By analyzing the defects and corresponding barrier properties of coated polymers, da Silva Sobrinho et al. (1998) have concluded that the mechanism of permeation changes from classic solution/diffusion in polymers to porous media transport in thin barrier coatings. Silicon-derived coatings deposited by PECVD possess more mechanical flexibility compared to those prepared by PVD, and such mechanical robustness is at least partly responsible for fewer defects and longer retention of enhanced barrier properties, even though such coatings suffer from low crack resistance and are prone to crack

Table 1.1 Comparison Between the Barrier Properties of Different Polymer-Based Materials[a]

Deposition[b]	Polymer	Coating	P_m (cm^3/m^2-day-atm)	P_c (cm^3/m^2-day-atm)	BIF	T(°C)	Ref.
Sputtering	PET	80 nm AlO_xN_y	1.8×10^{-3}	5.0×10^{-5}	36	23	Erlat et al. (2001)
Evaporation	PET [d)]	17 nm Al_2O_3	2.1×10^{-3}	5.6×10^{-5}	38	30	Roberts et al. (2002)
PECVD	PET	100 nm SiO_x	1.7×10^{-5}	3.8×10^{-6}	440	30	Inagaki et al. (1999)
PECVD	PET	20-200 nm SiO_x	2.0×10^{-3}	4.5×10^{-6}	444	30	Erlat et al. (1999)
ALD	PET	10 nm Al_2O_3	—	$<2.5\times10^{-7}$	—	23	Carcia et al. (2009)
Al_2O_3	PET	1.25 μm	6.0×10^{-2}	$<3.1\times10^{-5}$	>1935	—	Affinito et al. (1996)
LBL (30)	PET	571 nm	1.5×10^{-3}	$<9\times10^{-7}$	>1700	23	Jang et al. (2008)
LBL (30)	PET	118 nm	1.5×10^{-3}	$<9\times10^{-7}$	>1700	23	Yang et al. (2010)
LBL (70)	PET	231 nm	1.5×10^{-3}	$<9\times10^{-7}$	>1700	23	Priolo et al. (2010a)
LBL (24)	PET	120 nm	1.5×10^{-3}	$<9\times10^{-7}$	>1700	23	Priolo et al. (2011)
PNC	Epoxy	5 vol.%	1.5×10^{-3}	4.2×10^{-4}	3.6	23	Osman et al. (2004)
PNC	PLA	4 wt.%	2.0×10^{-2}	1.8×10^{-2}	1.13	20	Ray et al. (2002)
PNC	HDPE	5 wt.%	4.1×10^{-1}	2.46×10^{-1}	1.68	20	Jacquelot et al. (2006)

[a] Except for the CO_2 data of Jacquelot et al. (2006), all other data are provided for O_2.
[b] In the case of LBL deposition, the number in parentheses identifies the number of bilayers.

failure (Czeremuszkin et al., 2001). Small, nanoscale defects in thin barrier coatings tend to form as a consequence of two types of stress encountered during nucleation and growth in PVD and CVD processes. The first is the intrinsic stress that develops due to the nonequilibrium nature of the deposition processes and the disordered state of the coating, while the second includes thermal stresses that arise because of differences in thermal contraction between the coating and substrate upon cooling. The extent to which these defects compromise barrier improvement is dependent on coating thickness, as illustrated in Figs. 1.12a, 1.12b for barrier coatings differing in thickness. To avoid cracking induced by differences in intrinsic stresses between the coating and

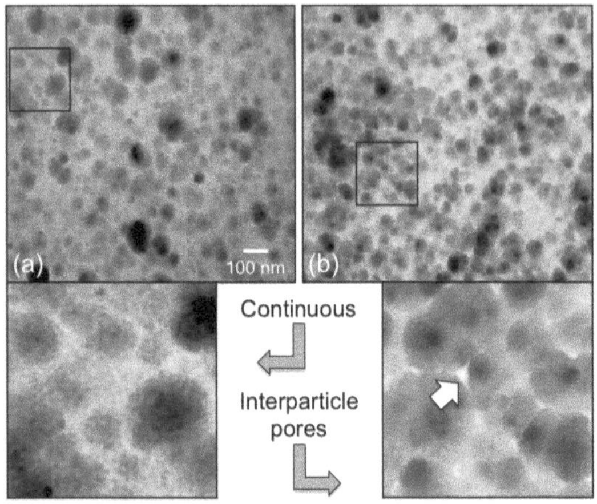

Figure 1.11. A pair of TEM images collected from SiOx coatings deposited by PECVD on PET substrates and varying in OTR (in cm^3/m^2-day-atm): (a) 0.52 and (b) 16. The enlargements of the boxed regions reveal the contiguity of the coatings, with interparticle pore defects clearly visible. Adapted from Erlat et al. (1999), with permission from the American Chemical Society.

polymer, coating thicknesses should not exceed ca. 100 nm. For a variety of coatings, an optimum thickness that provides maximum barrier performance has been reported. In the specific case of SiO_x coatings, those greater than 40 nm thick have been found to display evidence of cracks that diminish their barrier improvement (Hanika et al., 2003; Jiang et al., 2010). Large defects are frequently

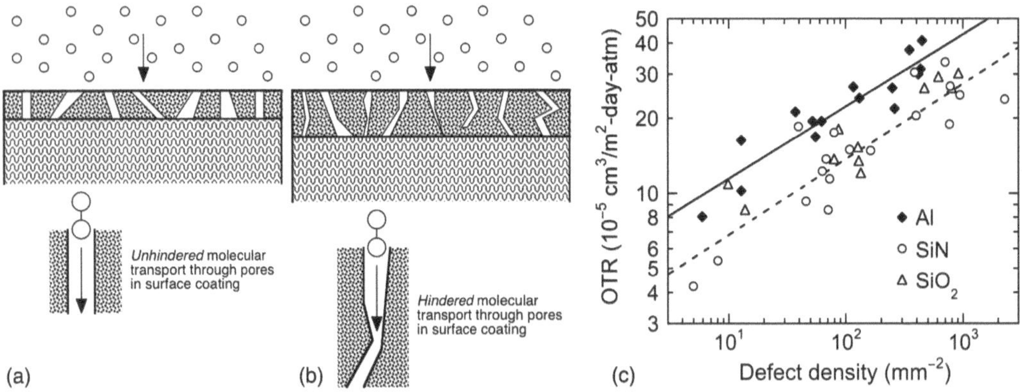

Figure 1.12. Schematic diagrams of molecular transport through (a) thin and (b) thick single-layer coatings illustrating the effect of defect size and shape. In (c), OTR values are provided as a function of defect density for three different single-layer coatings (labeled) measured above their critical thicknesses by da Silva Sobrinho et al. (2000). The solid lines are power-law regressions to the data and are provided as guides for the eye. (a, b) Adapted from Erlat et al. (2000), with permission from the Materials Research Society; (c) from da Silva Sobrinho et al. (2000).

introduced into barrier coatings by poor handling or substrate/equipment cleanliness (Erlat et al., 2001). With the importance of large defects notwithstanding, the density of nanoscale defects incurred during coating fabrication is dependent on the intrinsic properties of the coating and the deposition conditions. According to the results presented on dual-logarithmic coordinates in Fig. 1.12c, the barrier performance, expressed in terms of the OTR, scales directly with the defect density (da Silva Sobrinho et al., 2000).

6.2 Single-Layer ALD Coatings

Recent advances in ALD as a commercially competitive process by which to enhance the barrier properties of polymer substrates have been very promising. Groner et al. (2006) report that Al_2O_3 coatings measuring 5 nm thick and deposited by ALD on polymide (PI) with a glass transition temperature (T_g) of 300°C and poly(ethylene naphthalate) with T_g = 125°C yield OTR values below the minimum detectable limit of the MOCON OX-TRAN 2/21 instrument (with a detection limit of $<5 \times 10^{-3}$ cm^3/m^2-day). For comparison with barrier coatings achieved by PECVD, we hasten to point out that a SiN coating measuring 50 nm thick and an Al_2O_3 coating measuring 10 nm thick have also resulted in nondetectable OTR values on the same instrument, which indicates that the barrier performance of some coatings must be characterized by other means (Carcia et al., 2009). In ALD, control over the temperature of the substrate is critically important to ensure suitable precursor chemical deposition. If, on one hand, the temperature is lower than an optimum level, the growth rate either increases due to surface condensation of gas precursors or decreases due to incomplete removal of unreacted precursor during the purge step. If the temperature is excessive, on the other hand, the growth rate can be reduced because of reevaporation of deposited species or increased as a consequence of thermal decomposition of gaseous precursors prior to deposition (Carcia et al., 2010; Spagnola et al., 2010). If thermal control is adequate and the ALD growth rate is precisely regulated, this relatively new deposition process is poised to elucidate the multitude of factors governing the evolution of barrier properties provided by thin inorganic coatings on polymer substrates.

6.3 Multilayer LBL Coatings

In similar fashion as ALD, the LBL deposition process is now sufficiently developed to achieve a uniform coating assembly composed of alternating nanoscale layers of polymer and highly

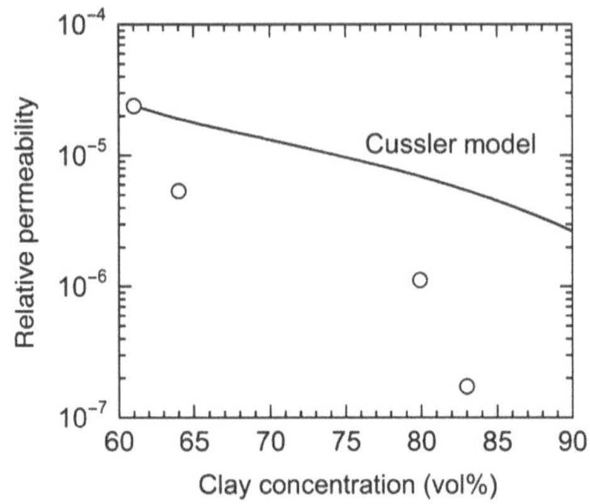

Figure 1.13. Experimental relative permeability values of LBL multilayer coatings as a function of clay concentration, as well as corresponding predictions from the Cussler model. Adapted from Priolo et al. (2011), with permission from the American Chemical Society.

exfoliated and oriented impermeable clay platelets. This type of transparent sandwich structure is capable of promoting very high barrier performance. The idea underpinning LBL coatings is essentially to construct, through repetitive dipping action, a nanoscale brick wall that requires permeating gas molecules to pass through long, tortuous polymer layers that are separated by oriented layers of impermeable clay platelets. The polymer layers must therefore be suitably thick to increase the residence time of penetrant molecules and, in doing so, provide excellent barrier performance with just a few layers. In this case, exfoliated clay platelets with a high aspect ratio must be positioned with their large dimension parallel to the substrate surface to increase the diffusion path length, and concurrently decrease the permeability, of permeating species (Yang et al., 2010; Priolo et al., 2011). While this strategy relies on the same increase in tortuosity sought in the design of PCNs, existing models fail to predict the very low relative permeability of LBL assemblies, as evidenced by the results provided in Fig. 1.13 (DeRocher et al., 2005; Priolo et al., 2011). Using this conceptual approach, Jang et al. (2008) have generated a high-barrier LBL assembly composed of polyacrylamide (0.25 wt.%) and Na^+MMT (0.20 wt.%) and measuring 571 nm thick with 30 bilayers on a PET substrate. They report OTR values below 0.005 $cm^3/(m^2$-day-atm) at 23°C and 0% RH. Interestingly, with only 20 bilayers, they measure an OTR value of 0.217 $cm^3/(m^2$-day-atm), which is comparable to the barrier performance of SiO_x-coated PET (Erlat et al., 1999; Jang et al., 2008). It should be recognized here that, although the multilayer thickness usually increases linearly with an increasing number of layers, the barrier performance increases superlinearly (eg, exponentially), indicating that LBL assemblies with a large number of layers can serve as a viable alternative to metal oxide coatings in food and electronic packaging applications. In this spirit, a record-setting LBL assembly with 70 bilayers fabricated from PEI (0.1 wt.%) and MMT (0.2 wt.%) on PET yields an oxygen permeability of $<0.2 \times 10^{-8}$ $cm^3/(m^2$-day-atm), which is the lowest ever reported for a polymer composite (Roberts et al., 2002; Priolo et al., 2010a). An important consideration in the production of LBL assemblies is the immersion time within the polymer solution and clay suspension. Even with 10 s in a 0.1 wt.% PEI solution and 5 min in a 1 wt.% MMT suspension, an assembly with 30 bilayers has been observed to exhibit an undetectable OTR. Short immersion times and complete automation are crucial for the commercialization of LBL processing in the food and beverage packaging industries (Yang et al., 2010; Priolo et al., 2010a).

6.4 Polymer-Clay Nanocomposites

Early studies of intercalated PCNs prepared by twin-screw extrusion from biodegradable poly(lactic acid) (PLA) and 4 wt.% organically modified MMT exhibit a BIF of only 1.13 relative to neat PLA (Ray et al., 2002). By using MMT modified with benzyldibutyl(2-hydroxyethyl) ammonium chloride, benzylbis(2-hydroxyethyl) butyl ammonium chloride, and benzyltriethanol ammonium chloride, however, Osman et al. (2004) have demonstrated that the barrier properties of PCNs containing epoxy are significantly improved compared to pure epoxy due to greater chemical interactions and the likelihood of exfoliation. Higher BIF levels have been reported for PCNs composed of a thermoplastic: 1.45 through the melt extrusion of nylon-6 with 1.6 wt.% MMT (Hasegawa et al., 2003), 1.48 through the in situ polymerization of styrene in the presence of 5 wt.% organically modified MMT (Nazarenko et al., 2007), and 1.71 through the melt extrusion of PP with about 6 wt.% organically modified MMT and maleic anhydride (Golebiewski and Galeski, 2007). While these BIF values fall substantially short of those achieved through the use of single or multilayer coatings, an important advantage of PCNs is that their physical (ie, mechanical and flammability-resistant) properties are concurrently and substantially improved upon incorporation of the clay nanoparticles.

The theoretical models introduced earlier for PCNs provide valuable guidance with regard to the factors influencing barrier performance. Deviations from model predictions are often attributable to two unrealistic model assumptions: complete exfoliation

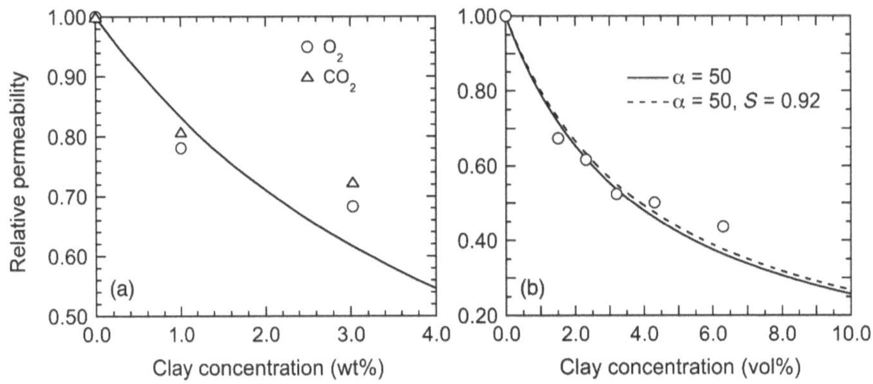

Figure 1.14. (a) Relative permeabilities of O_2 and CO_2 (labeled) in LDPE PCNs as a function of MMT concentration. The solid line corresponds to predictions from the Nielsen model in which $\alpha = 120$. (b) Relative permeability of O_2 in PCNs prepared from hydrogenated acrylonitrile butadiene rubber with predictions from the original Nielsen model (solid line with $\alpha = 50$) and the Bharadwaj modification (dashed line with $\alpha = 50$ and $S = 0.92$) illustrating the marginal improvement achieved by adding the order parameter. (a) Adapted from Shah et al. (2006), with permission from Elsevier; (b) adapted from Gatos and Karger-Kocsis (2007), with permission from Elsevier.

and invariant orientation (parallel to the substrate surface). As Shah et al. (2006) have demonstrated in Fig. 1.14a, predictions from the Nielsen tortuosity model Eq. (1.7), which implicitly invokes the first assumption, agree favorably with experimental observations for O_2, N_2, and CO_2 in the case of melt-processed LDPE containing 1 wt.% organically modified MMT possessing an aspect ratio of 120. At higher MMT loading levels, however, the predictions underestimate the relative permeability. Deviation from the second assumption can be taken into account through the use of an order parameter, introduced by Bharadwaj (2001) in Eq. (1.10). In Fig. 1.14b, setting $S = 0.92$ yields predictions that fit the experimental data reasonably well, implying that most, but not all, of the platelets are oriented parallel to the substrate surface. Since complete parallel orientation of the platelets corresponds to $S = 1.0$, this relatively minor reduction in S explains why the original Nielsen model not only helps to explain experimental results (especially at low loading levels where the first assumption is likely to be valid) but also remains a widely used predictive tool (Gatos and Karger-Kocsis, 2007).

7 Conclusions and Outlook

Improving the barrier properties of commodity and engineering thermoplastics commonly used in food packaging applications to contend with growing economic demands for increased shelf life and decreased weight (and energy costs associated with transportation) has been achieved through the development of various inorganic surface coatings and PNC designs. Single-layer metal and metal oxide coatings are routinely deposited by PVD and CVD methods, with the latter providing more conformal coatings on rough surfaces or nonplanar (three-dimensional) objects. In addition, PVD coatings are generally less effective as barrier materials than CVD coatings due to a relatively higher defect density caused by lower mechanical flexibility and the absence of an interphase layer. Silicon-based CVD coatings are, however, prone to cracking as a consequence of the stresses that develop in the system. Attempts to increase the coating thickness to reduce transverse defects and improve barrier performance fail beyond an optimum thickness because of stress mismatch between the coating and substrate. ALD constitutes a very attractive emerging technology designed to overcome the drawbacks associated with single-layer deposition processes. Recent studies have demonstrated that thin, nearly defect-free ALD coatings measuring only 5 nm thick exhibit OTR values below the detectable limit of conventional permeation measurements. Although ALD is classified here as a single-layer process because the result is a chemically homogeneous coating, ALD coatings actually consist of numerous layers that are deposited step-by-step at atomic dimensions by controlled reaction and purge sequences. Since the resulting coating is nearly defect-free, it comes closest to serving as an idealized impenetrable barrier intended to prevent (not hinder) diffusion. In similar fashion but at larger length scales, LBL deposition is a true multilayer process that likewise holds immense commercial promise for the barrier enhancement of polymer substrates by introducing extensive tortuosity into the diffusion pathway of permeating gases. Since the nanoscale brick wall composed of oppositely charged clay "bricks" and polymer "mortar" is fabricated by repeated alternating immersion into a clay suspension and a polymer solution, the clay layers comprising these transparent coatings are highly oriented. This deposition process can be performed at ambient temperature and pressure, and yields coatings that exhibit OTR levels below 0.005 $cm^3/(m^2\text{-day-atm})$.

An appealing alternative to deposition methods involves PCNs, in which a natural or organically modified clay is introduced into a polymer matrix. A very important commercial consideration is that this approach is capable of generating large volumes of products with various shapes and sizes through the use of conventional polymer processing technologies, such as extrusion or blow/injection molding, which are routinely employed in today's food packaging industry. Since exfoliation is required to achieve barrier improvement in PCNs and since most polymers employed in the preparation of PCNs are nonpolar, natural clays have been chemically modified with organic surfactants or physically modified with compatibilizing agents to improve the dispersability of the clay platelets within the polymer matrix. At high levels of exfoliation with about 5 wt.% clay, the barrier improvement factor is ≈3×, which pales in comparison to CVD, ALD, and LBL deposition methods with improvement factors in excess of 2–3 orders of magnitude. Due to their relatively facile processability, mechanical robustness and lengthy stability, however, PCNs remain a commercially important class of polymer-based barrier materials, and several theoretical formalisms have been developed to predict the barrier efficacy of PCNs. The original Nielsen model, as well as its later versions proposed to account for variations in platelet orientation, and the Cussler model account for the high aspect ratio of clay platelets and other geometric factors that influence the diffusion pathway of permeating molecules. All the methods discussed in this chapter are intended to improve the barrier performance of food packaging through the enlightened use of hybrid materials in which polymeric substrates/matrices are coupled with impermeable inorganic coatings or additives. Alternate material designs capable of providing comparable or superior barrier performance in economical and sustainable fashion are of tremendous fundamental and technological interest.

Acknowledgments

We express our sincere gratitude to the United Resource Recovery Corporation (Spartanburg, SC, USA) for supporting this work.

References

Affinito, J.D., Eufinger, S., Gross, M.E., Graff, G.L., Martin, P.M., 1997. PML/oxide/PML barrier layer performance differences arising from use of UV or electron beam polymerization of the PML layers. Thin Solid Films 308, 19–25.

Affinito, J.D., Gross, M.E., Coronado, C.A., Graff, G.L., Greenwell, I.N., Martin, P.M., 1996. A new method for fabricating transparent barrier layers. Thin Solid Films 290, 63–67.

Aionicesei, E., Škerget, M., Knez, Ž., 2008. Measurement of CO_2 solubility and diffusivity in poly(L-lactide) and poly(D,L-lactide-*co*-glycolide) by magnetic suspension balance. J. Super. Fluids 47, 296–301.

Alexandre, M., Dubois, P., 2000. Polymer-layered silicate nanocomposites: preparation, properties and uses of a new class of materials. Mater. Sci. Eng. R 28, 1–63.

Ammala, A., Bell, C., Dean, K., 2008. Poly(ethylene terephthalate) clay nanocomposites: improved dispersion based on an aqueous ionomer. Compos. Sci. Technol. 68, 1328–1337.

Beall, G.W., Powell, C.E., 2011. Fundamentals of Polymer-Clay Nanocomposites, first ed. Cambridge University Press, Cambridge, UK.
Bharadwaj, R.K., 2001. Modeling the barrier properties of polymer-layered silicate nanocomposites. Macromolecules 34, 9189–9192.
Bhattacharya, S.N., Kamal, M.R., Gupta, R.K., 2008. Polymeric Nanocomposites: Theory and Practice. Hanser, Munich, Germany.
Blodgett, K.B., Langmuir, I., 1937. Built-up films of barium stearate and their optical properties. Phys. Rev. 51, 964–982.
Bousmina, M., 2006. Study of intercalation and exfoliation processes in polymer nanocomposites. Macromolecules 39, 4259–4263.
Boyer, R., McKinney, J., 2013. Food storage guidelines for consumers, Publication 348-960. Virginia Cooperative Extension, Virginia Polytechnic Institute and State University, Blacksburg, VA, pp.1–12.
Carcia, P.F., McLean, R.S., Groner, M.D., Dameron, A.A., George, S.M., 2009. Gas diffusion ultrabarriers on polymer substrates using Al_2O_3 atomic layer deposition and SiN plasma-enhanced chemical vapor deposition. J. Appl. Phys. 106, 023533.
Carcia, P.F., McLean, R.S., Reilly, M.H., 2010. Permeation measurements and modeling of highly defective Al_2O_3 thin films grown by atomic layer deposition on polymers. Appl. Phys. Lett. 97, 221901–221903.
Chatham, H., 1996. Oxygen diffusion barrier properties of transparent oxide coatings on polymeric substrates. Surf. Coat. Technol. 78, 1–9.
Chiarelli, P.A., Johal, M.S., Casson, J.L., Roberts, J.B., Robinson, J.M., Wang, H., 2001. Controlled fabrication of polyelectrolyte multilayer thin films using spin-assembly. Adv. Mater. 13, 1167–1171.
Chiarelli, P.A., Johal, M.S., Holmes, D.J., Roberts, J.B., Robinson, J.M., Wang, H., 2002. Polyelectrolyte spin-assembly. Langmuir 18, 168–173.
Cini, N., Tulun, T., Decher, G., Ball, V., 2010. Step-by-step assembly of self-patterning polyelectrolyte films violating (almost) all rules of layer-by-layer deposition. J. Am. Chem. Soc. 132, 8264–8265.
Comyn, J., 1985. Introduction to polymer permeability and the mathematics of diffusion. In: Comyn, J. (Ed.), Polymer Permeability. Elsevier, London, pp. 1–10.
Cueff, R., Baud, G., Benmalek, M., Besse, J.P., Butruille, J.R., Dunlop, H.M., Jacquet, M., 1995. Characterization and adhesion study of thin alumina coatings sputtered on PET. Thin Solid Films 270, 230–236.
Cussler, E.L., Hughes, S.E., Ward, III, W.J., Aris, R., 1988. Barrier membranes. J. Memb. Sci. 38, 161–174.
Czeremuszkin, G., Latrèche, M., Wertheimer, M.R., da Silva Sobrinho, A.S., Latrèche, M., 2001. Ultrathin silicon-compound barrier coatings for polymeric packaging materials: an industrial perspective. Plasm. Polym. 6, 107–120.
Da Silva Sobrinho, A.S., Czeremuszkin, G., Latreche, M., Wertheimer, M.R., 2000. Defect-permeation correlation for ultrathin transparent barrier coatings on polymers. J. Vac. Sci. Technol. A 18, 149–157.
Da Silva Sobrinho, A.S., Latreche, M., Czeremuszkin, G., Klemberg-Sapieha, J.E., Wertheimer, M.R., 1998. Transparent barrier coatings on polyethylene terephthalate by single-and dual-frequency plasma-enhanced chemical vapor deposition. J. Vac. Sci. Technol. A 16, 3190–3198.
Decher, G., 1997. Fuzzy nanoassemblies: toward layered polymeric multicomposites. Science 277, 1232–1237.
Decher, G., 2012. Layer-by-layer assembly. In: Decher, G., Schlenoff, J.B. (Eds.), Multilayer Thin Films. Wiley-VCH, Weinhem, pp. 1–22.
Decher, G., Hong, J.D., 1991a. Buildup of ultrathin multilayer films by a self-assembly process: I. consecutive adsorption of anionic and cationic bipolar

amphiphiles on charged surfaces. Makromol. Chem. Macromol. Symp. 46, 321–327.

Decher, G., Hong, J.D., 1991b. Buildup of ultrathin multilayer films by a self-assembly process: II. consecutive adsorption of anionic and cationic bipolar amphiphiles and polyelectrolytes on charged surfaces. Berich. Bunseng. Phys. Chem. 95, 1430–1434.

Decher, G., Hong, J.D., Schmitt, J., 1992. Buildup of ultrathin multilayer films by a self-assembly process: III. consecutively alternating adsorption of anionic and cationic polyelectrolytes on charged surfaces. Thin Solid Films 210, 831–835.

Deng, C., Assender, H.E., Dinelli, F., Kolosov, O.V., Briggs, G.A.D., Miyamoto, T., Tsukahara, Y., 2000. Nucleation and growth of gas barrier aluminium oxide on surfaces of poly(ethylene terephthalate) and polypropylene: effects of the polymer surface properties. J. Polym. Sci. B 38, 3151–3162.

DeRocher, J.P., Gettelfinger, B.T., Wang, J., Nuxoll, E.E., Cussler, E.L., 2005. Barrier membranes with different sizes of aligned flakes. J. Memb. Sci. 254, 21–30.

Elam, J.W., Wilson, C.A., Schuisky, Sechrist, Z.A., George, S.M., 2003. Improved nucleation of TiN atomic layer deposition films on SiLK low-k polymer dielectric using an Al_2O_3 atomic layer deposition adhesion layer. J. Vac. Sci. Technol. B 21, 1099–1107.

Erlat, A.G., Henry, B.M., Ingram, J.J., Mountain, D.B., McGuigan, A., Howson, R.P., Grovenor, C.R.M., Briggs, G.A.D., Tsukahara, Y., 2001. Characterisation of aluminium oxynitride gas barrier films. Thin Solid Films 388, 78–86.

Erlat, A.G., Spontak, R.J., Clarke, R.P., Robinson, T.C., Haaland, P.D., Tropsha, Y., Harvey, N.G., Vogler, E.A., 1999. SiO_x gas barrier coatings on polymer substrates: morphology and gas transport considerations. J. Phys. Chem. B 103, 6047–6055.

Erlat, A.G., Wang, B.-C., Spontak, R.J., Tropsha, Y., Mar, K.D., Montgomery, D.B., Vogler, E.A., 2000. Morphology and gas barrier properties of thin SiO_x coatings on polycarbonate: correlations with plasma-enhanced chemical vapor deposition conditions. J. Mater. Res. 15, 704–717.

Félix, O., Zheng, Z., Cousin, F., Decher, G., 2009. Are sprayed LbL-films stratified? A first assessment of the nanostructure of spray-assembled multilayers by neutron reflectometry. Compt. Rend. Chim. 12, 225–234.

Franklin Associates, 2014. Impact of plastics packaging on life cycle energy consumption and greenhouse gas emissions in the United States and Canada. Available from: http://plastics.americanchemistry.com/Education-Resources/Publications/Impact-of-Plastics-Packaging.pdf

Frankowski, D.J., Capracotta, M.D., Martin, J.D., Khan, S.A., Spontak, R.J., 2007. Stability of organically modified montmorillonites and their polystyrene nanocomposites after prolonged thermal treatment. Chem. Mater. 19, 2757–2767.

Fredrickson, G.H., Bicerano, J., 1999. Barrier properties of oriented disk composites. J. Chem. Phys. 110, 2181–2188.

Fried, J.R., Liu, H.C., Zhang, C., 1989. Effect of sorbed carbon dioxide on the dynamic mechanical properties of glassy polymers. J. Polym. Sci. C 27, 385–392.

Fujiyama, M., 2010. Morphology development in polyolefin nanocomposites. In: Mittal, V. (Ed.), Optimization of Polymer Nanocomposite Properties. Wiley-VCH, Weinheim, pp. 67–92.

Galimberti, M., 2012. Rubber clay nanocomposites. In: Boczkowska, A. (Ed.), Advanced Elastomers —Technology, Properties and Applications. InTech, Rijeka, Croatia, Available from: http://www.intechopen.com/books/advanced-elastomers-technology-properties-and-applications/rubber-clay-nanocomposites

Gatos, K.G., Karger-Kocsis, J., 2007. Effect of the aspect ratio of silicate platelets on the mechanical and barrier properties of hydrogenated acrylonitrile butadiene rubber (HNBR)/layered silicate nanocomposites. Eur. Polym. J. 43, 1097–1104.

George, S.M., 2010. Atomic layer deposition: an overview. Chem. Rev. 110, 111–131.

Giannelis, E.P., Krishnamoorti, R., Manias, E., 1999. Polymer-silicate nanocomposites: model systems for confined polymers and polymer brushes. In: Granick, S. (Ed.), Polymers in Confined Environments. Springer, Berlin, pp. 107–147.

Golebiewski, J., Galeski, A., 2007. Thermal stability of nanoclay polypropylene composites by simultaneous DSC and TGA. Compos. Sci. Technol. 67, 3442–3447.

Gonzalez, P., Garcia, E., Pou, J., Fernandez, D., Serra, J., Leon, B., Pérez-Amor, M., Gonzfilez, P., Garcia, E., Pou, J., Fernfindez, D., Serra, J., 1994. Silicon oxide thin films grown by Xe2 excimer lamp chemical vapour deposition: the role of the substrate temperature and the window-substrate distance. Thin Solid Films 241, 348–351.

Graff, G.L., Burrows, P.E., Williford, R.E., Praino, R.F., 2005. Barrier layer technology for flexible displays. In: Crawford, G. (Ed.), Flexible Flat Panel Displays. Wiley, Hoboken, pp. 57–77.

Groner, M.D., George, S.M., McLean, R.S., Carcia, P.F., 2006. Gas diffusion barriers on polymers using Al_2O_3 atomic layer deposition. Appl. Phys. Lett. 88, 51903–51907.

Hammond, P.T., 2004. Form and function in multilayer assembly: new applications at the nanoscale. Adv. Mater. 16, 1271–1293.

Hamzehlou, S., Katbab, A.A., 2007. Bottle-to-bottle recycling of PET via nanostructure formation by melt intercalation in twin screw compounder: improved thermal, barrier, and microbiological properties. J. Appl. Polym. Sci. 106, 1375–1382.

Hanika, M., Langowski, H., Moosheimer, U., Peukert, W., 2003. Inorganic layers on polymeric films: influence of defects and morphology on barrier properties. Chem. Eng. Technol. 26, 605–614.

Hasegawa, N., Okamoto, H., Kato, M., Usuki, A., Sato, N., 2003. Nylon 6/Na-montmorillonite nanocomposites prepared by compounding nylon 6 with Na-montmorillonite slurry. Polymer 44, 2933–2937.

He, P., Hu, N., Rusling, J.F., 2004. Driving forces for layer-by-layer self-assembly of films of SiO_2 nanoparticles and heme proteins. Langmuir 20, 722–729.

Hu, X., Tang, J., Blasig, A., Shen, Y., Radosz, M., Merkel, T.C., Bondar, V.I., Nagai, K., Freeman, B.D., Pinnau, I., 2006. CO_2 permeability, diffusivity and solubility in polyethylene glycol-grafted polyionic membranes and their CO_2 selectivity relative to methane and nitrogen. J. Memb. Sci. 281, 130–138.

Hussain, F., Hojjati, M., Okamoto, M., Gorga, R.E., 2006. Polymer-matrix nanocomposites, processing, manufacturing, and application: an overview. J. Comp. Mater. 40, 1511–1575.

Iler, R.K., 1966. Multilayers of colloidal particles. J. Colloid Interf. Sci. 21, 569–594.

Imai, N., Sekiguchi, M., Kano, M., Krug, T., Steiniger, G., Meier, A., 1995. Evaporation method of forming transparent barrier film. U.S. Patent #5,378,506, issued January 3, 1995.

Inagaki, N., Tasaka, S., Hiramatsu, H., 1999. Preparation of oxygen gas barrier poly(ethylene terephthalate) films by deposition of silicon oxide films plasma-polymerized from a mixture of tetramethoxysilane and oxygen. J. Appl. Polym. Sci. 71, 2091–2100.

Jacquelot, E., Espuche, E., Gérard, J., Duchet, J., Mazabraud, P., 2006. Morphology and gas barrier properties of polyethylene-based nanocomposites. J. Polym. Sci. B 44, 431–440.

Jamieson, E.H.H., Windle, A.H., 1983. Structure and oxygen-barrier properties of metallized polymer film. J. Mater. Sci. 18, 64–80.
Jang, W.-S., 2008. Layer-by-layer assembly of clay-filled polymer nanocomposite thin films. PhD Dissertation, Texas A&M University.
Jang, W.-S., Grunlan, J.C., 2005. Robotic dipping system for layer-by-layer assembly of multifunctional thin films. Rev. Sci. Instrum. 76, 103904.
Jang, W.-S., Rawson, I., Grunlan, J.C., 2008. Layer-by-layer assembly of thin film oxygen barrier. Thin Solid Films 516, 4819–4825.
Jiang, C., Markutsya, S., Tsukruk, V.V., 2004. Compliant, robust, and truly nanoscale free standing multilayer films fabricated using spin assisted layer by layer assembly. Adv. Mater. 16, 157–161.
Jiang, J., Benter, M., Taboryski, R., Bechgaard, K., 2010. Oxygen barrier coating deposited by novel plasma-enhanced chemical vapor deposition. J. Appl. Polym. Sci. 115, 2767–2772.
Kawasumi, M., 2004. The discovery of polymer clay hybrids. J. Polym. Sci. A 42, 819–824.
Keller, S.W., Johnson, S.A., Brigham, E.S., Yonemoto, E.H., Mallouk, T.E., 1995. Photoinduced charge separation in multilayer thin films grown by sequential adsorption of polyelectrolytes. J. Am. Chem. Soc. 117, 12879–12880.
Kim, T.H., Lim, S.T., Lee, C.H., Choi, H.J., Jhon, M.S., 2003. Preparation and rheological characterization of intercalated polystyrene/organophilic montmorillonite nanocomposite. J. Appl. Polym. Sci. 87, 2106–2112.
Klute, C.H., 1959. Diffusion of small molecules in semicrystalline polymers: water in polyethylene. J. Appl. Polym. Sci. 1, 340–350.
Kotov, N.A., 1999. Layer-by-layer self-assembly: the contribution of hydrophobic interactions. Nanostruct. Mater. 12, 789–796.
Kráčalík, M., Studenovský, M., Mikešová, J., Kovárˇová, J., Sikora, A., Thomann, R., Friedrich, C., 2007a. Recycled PET organoclay nanocomposites with enhanced processing properties and thermal stability. J. Appl. Polym. Sci. 106, 2092–2100.
Kráčalík, M., Studenovský, M., Mikešová, J., Sikora, A., Thomann, R., Friedrich, C., Fortelný, I., Simoník, J., 2007b. Recycled PET nanocomposites improved by silanization of organoclays. J. Appl. Polym. Sci. 106, 926–937.
Krishnamoorti, R., Yurekli, K., 2001. Rheology of polymer layered silicate nanocomposites. Curr. Opin. Colloid Interf. Sci. 6, 464–470.
Kuhn, H., Möbius, D., 1971. Systems of monomolecular layers assembling and physicochemical behavior. Angew. Chem. Int. Ed. 10, 620–637.
Langereis, E., Creatore, M., Heil, S.B.S., Van de Sanden, M.C.M., Kessels, W.M.M., 2006. Plasma-assisted atomic layer deposition of Al_2O_3 moisture permeation barriers on polymers. Appl. Phys. Lett. 89, 081915.
Lertwimolnun, W., Vergnes, B., 2005. Influence of compatibilizer and processing conditions on the dispersion of nanoclay in a polypropylene matrix. Polymer 46, 3462–3471.
Leterrier, Y., 2003. Durability of nanosized oxygen-barrier coatings on polymers. Prog. Mater. Sci. 48, 1–55.
Lewis, J., 2006. Material challenge for flexible organic devices. Mater. Today 9, 38–45.
Li, J., Zhou, C., Wang, G., Zhao, D., 2003. Study on rheological behavior of polypropylene/clay nanocomposites. J. Appl. Polym. Sci. 89, 3609–3617.
Lilienfeld, R., 2014. Packaging myths and mantras. Use-Less-Stuff Rep. 12, 1–3.
Lim, Y.T., Park, O.O., 2001. Phase morphology and rheological behavior of polymer/layered silicate nanocomposites. Rheolog. Acta 40, 220–229.
Low, H., Xu, Y., 2005. Moisture barrier of Al_xO_y coating on poly(ethylene terephthalate), poly(ethylene naphthalate) and poly(carbonate) substrates. Appl. Surf. Sci. 250, 135–145.

Lvov, Y., Ariga, K., Ichinose, I., Kunitake, T., 1996. Molecular film assembly via layer-by-layer adsorption of oppositely charged macromolecules (linear polymer, protein and clay) and concanavalin A and glycogen. Thin Solid Films 284, 797–801.

Manias, E., 2001. A direct-blending approach for polypropylene/clay nanocomposites enhances properties. Mater. Res. Soc. Bull. 26, 862–863.

Markočič, E., Škerget, M., Knez, Ž., 2011. Solubility and diffusivity of CO_2 in poly(L-lactide)–hydroxyapatite and poly(D,L-lactide-*co*-glycolide)–hydroxyapatite composite biomaterials. J. Super. Fluids 55, 1046–1051.

Maxwell, J.C., 1873. A Treatise on Electricity and Magnetism. Oxford University Press, London.

Merkel, T.C., Freeman, B.D., Spontak, R.J., He, Z., Pinnau, I., Meakin, P., Hill, A.J., 2002. Ultrapermeable, reverse-selective nanocomposite membranes. Science 296, 519–522.

Messersmith, P.B., Giannelis, E.P., 1993. Polymer-layered silicate nanocomposites: in situ intercalative polymerization of ε-caprolactone in layered silicates. Chem. Mater. 5, 1064–1066.

Milde, F., Goedicke, K., Fahland, M., 1996. Adhesion behavior of PVD coatings on ECR plasma and ion beam treated polymer films. Thin Solid Films 279, 169–173.

Mittal, V., 2010a. Barrier resistance generation in polymer nanocomposites. In: Mittal, V. (Ed.), Optimization of Polymer Nanocomposite Properties. Wiley-VCH, Weinheim, pp. 173–192.

Mittal, V., 2010b. Polymer nanocomposites: synthesis, microstructure, and properties. In: Mittal, V. (Ed.), Optimization of Polymer Nanocomposite Properties. Wiley-VCH, Weinheim, pp. 1–20.

Moggridge, G.D., Lape, N.K., Yang, C., Cussler, E.L., 2003. Barrier films using flakes and reactive additives. Prog. Organ. Coat. 46, 231–240.

Moncada, E., Quijada, R., Retuert, J., 2007. Comparative effect of metallocene and ziegler-natta polypropylene on the exfoliation of montmorillonite and hectorite clays to obtain nanocomposites. J. Appl. Polym. Sci. 103, 698–706.

Morgan, A.B., Gilman, J.W., 2003. Characterization of polymer layered silicate (clay) nanocomposites by transmission electron microscopy and X-ray diffraction: a comparative study. J. Appl. Polym. Sci. 87, 1329–1338.

Nazarenko, S., Meneghetti, P., Julmon, P., Olson, B.G., Qutubuddin, S., 2007. Gas barrier of polystyrene montmorillonite clay nanocomposites: effect of mineral layer aggregation. J. Polym. Sci. B 45, 1733–1753.

Nielsen, L.E., 1967. Models for the permeability of filled polymer systems. J. Macromol. Sci. – Chem. 1, 929–942.

Okada, A., Usuki, A., 2006. Twenty years of polymer-clay nanocomposites. Macromol. Mater. Eng. 291, 1449–1476.

Orozco, V.H., Kozlovskaya, V., Kharlampieva, E., López, B.L., Tsukruk, V.V., 2010. Biodegradable self-reporting nanocomposite films of poly(lactic acid) nanoparticles engineered by layer-by-layer assembly. Polymer 51, 4127–4139.

Osman, M.A., Mittal, V., Morbidelli, M., Suter, U.W., 2004. Epoxy-layered silicate nanocomposites and their gas permeation properties. Macromolecules 37, 7250–7257.

Ozcam, A.E., 2011. Surface modification and characterization of polymeric materials for targeted functionality. PhD Dissertation, North Carolina State University.

Patel, N.P., Miller, A.C., Spontak, R.J., 2003. Highly CO_2 permeable and selective polymer nanocomposite membranes. Adv. Mater. 15, 729–733.

Peng, Q., Sun, X.-Y., Spagnola, J.C., Hyde, G.K., Spontak, R.J., Parsons, G.N., 2007. Atomic layer deposition on electrospun polymer fibers as a direct route to Al_2O_3 microtubes with precise wall thickness control. Nano Lett. 7, 719–722.

Perez-Blanco, M., Hammons, J.R., Danner, R.P., 2010. Measurement of the solubility and diffusivity of blowing agents in polystyrene. J. Appl. Polym. Sci. 116, 2359–2365.
Priolo, M.A., Gamboa, D., Grunlan, J.C., 2010a. Transparent clay–polymer nano brick wall assemblies with tailorable oxygen barrier. ACS Appl. Mater. Interf. 2, 312–320.
Priolo, M.A., Gamboa, D., Holder, K.M., Grunlan, J.C., 2010b. Super gas barrier of transparent polymer–clay multilayer ultrathin films. Nano Lett. 10, 4970–4974.
Priolo, M.A., Holder, K.M., Gamboa, D., Grunlan, J.C., 2011. Influence of clay concentration on the gas barrier of clay–polymer nanobrick wall thin film assemblies. Langmuir 27, 12106–12114.
Ray, S.S., 2006. Rheology of polymer/layered silicate nanocomposites. J. Indust. Eng. Chem. 12, 811–842.
Ray, S.S., Okamoto, M., 2003. Polymer/layered silicate nanocomposites: a review from preparation to processing. Prog. Polym. Sci. 28, 1539–1641.
Ray, S., Quek, S.Y., Easteal, A., Chen, X.D., 2006. The potential use of polymer-clay nanocomposites in food packaging. Int. J. Food Eng. 2, 5.
Ray, S.S., Yamada, K., Okamoto, M., Ueda, K., 2002. Polylactide-layered silicate nanocomposite: a novel biodegradable material. Nano Lett. 2, 1093–1096.
Roberts, A.P., Henry, B.M., Sutton, A.P., Grovenor, C.R.M., Briggs, G.A.D., Miyamoto, T., Kano, M., Tsukahara, Y., Yanaka, M., 2002. Gas permeation in silicon-oxide/polymer (SiO_x/PET) barrier films: role of the oxide lattice, nano-defects and macro-defects. J. Memb. Sci. 208, 75–88.
Rousseaux, D.D.J., Sallem-Idrissi, N., Baudouin, A.-C., Devaux, J., Godard, P., Marchand-Brynaert, J., Sclavons, M., 2011. Water-assisted extrusion of polypropylene/clay nanocomposites: a comprehensive study. Polymer 52, 443–451.
Sawyer, L.C., Grubb, D.T., Meyers, G.F., 2007. Polymer Microscopy, third ed. Springer, New York.
Schecter, A., Malik, N., Haffner, D., Smith, S., Harris, T.R., Paepke, O., Birnbaum, L., 2010. Bisphenol A (BPA) in U.S. food. Environ. Sci. Technol. 44, 9425–9430.
Schlenoff, J.B., Dubas, S.T., Farhat, T., 2000. Sprayed polyelectrolyte multilayers. Langmuir 16, 9968–9969.
Shah, R.K., Krishnaswamy, R.K., Takahashi, S., Paul, D.R., 2006. Blown films of nanocomposites prepared from low density polyethylene and a sodium ionomer of poly (ethylene-*co*-methacrylic acid). Polymer 47, 6187–6201.
Shah, R.K., Paul, D.R., 2004. Nylon 6 nanocomposites prepared by a melt mixing masterbatch process. Polymer 45, 2991–3000.
Shchepelina, O., Drachuk, I., Gupta, M.K., Lin, J., Tsukruk, V.V., 2011. Silk on silk layer by layer microcapsules. Adv. Mater. 23, 4655–4660.
Shen, L., Lin, Y., Du, Q., Zhong, W., Yang, Y., 2005. Preparation and rheology of polyamide-6/attapulgite nanocomposites and studies on their percolated structure. Polymer 46, 5758–5766.
Silvain, J.F., Veyrat, A., Ehrhardt, J.J., 1992. Effect of the temperature on the adhesion and the morphology of thin metal films evaporated on polyethylene terephthalate. Thin Solid Films 221, 114–119.
Song, M., Jin, J., 2010. Rheological behavior of polymer nanocomposites. In: Mittal, V. (Ed.), Optimization of Polymer Nanocomposite Properties. Wiley-VCH, Weinheim, pp. 93–122.
Sorrentino, A., Tortora, M., Vittoria, V., 2006. Diffusion behavior in polymer-clay nanocomposites. J. Polym. Sci. B 44, 265–274.
Spagnola, J.C., 2010. Atomic layer deposition on fiber forming polymers and nonwoven fiber structures. PhD Dissertation, North Carolina State University.

Spagnola, J.C., Gong, B., Arvidson, S.A., Jur, J.S., Khan, S.A., Parsons, G.N., 2010. Surface and sub-surface reactions during low temperature aluminium oxide atomic layer deposition on fiber-forming polymers. J. Mater. Chem. 20, 4213–4222.
Suntola, T. Autson, J., 1977. Method for producing compound thin films. US Patent1977 4,058,430, issued November 25, 1977.
Tokihisa, M., Yakemoto, K., Sakai, T., Utracki, L.A., Sepehr, M., Li, J., Simard, Y., 2006. Extensional flow mixer for polymer nanocomposites. Polym. Eng. Sci. 46, 1040–1050.
University of South Carolina Research Foundation, 2003. Polymer Nanocomposite Technology Brief. Available from: http://ip.research.sc.edu/PDF/Polymer Nanocomposite Technology Brief v1.pdf.
Utracki, L.A., Sepehr, M., Li, J., 2006. Melt compounding of polymeric nanocomposites. Int. Polym. Process. 21, 3–16.
Von Klitzing, R., Kohler, R., Chenigny, C., 2012. Neutron reflectometry at polyelectrolyte multilayers. In: Decher, G., Schlenoff, J.B. (Eds.), Multilayer Thin Films. Wiley-VCH, Weinheim, pp. 219–268.
Wilson, C.A., Grubbs, R.K., George, S.M., 2005. Nucleation and growth during Al2O3 atomic layer deposition on polymers. Chem. Mater. 17, 5625–5634.
Wilson, C.A., McCormick, J.A., Cavanagh, Goldstein, D.N., Weimer, A.W., George, S.M., 2008. Tungsten atomic layer deposition on polymers. Thin Solid Films 516, 6175–6185.
Yang, Y.-H., Malek, F.A., Grunlan, J.C., 2010. Influence of deposition time on layer-by-layer growth of clay-based thin films. Indust. Eng. Chem. Res. 49, 8501–8509.
Yilmazer, U., Ozden, G., 2006. Polystyrene–organoclay nanocomposites prepared by melt intercalation, in situ, and masterbatch methods. Polym. Compos. 27, 249–255.
Zeng, C., Lee, L.J., 2001. Poly (methyl methacrylate) and polystyrene/clay nanocomposites prepared by in-situ polymerization. Macromolecules 34, 4098–4103.
Zhang, H., 2010. Scale-up of extrusion foaming process for manufacture of polystyrene foams using carbon dioxide, MS Thesis, University of Toronto.
Zhang, X., Chen, H., Zhang, H., 2007. Layer-by-layer assembly: from conventional to unconventional methods. Chem. Commun. 14, 1395–1405.
Zhong, Y., Zhu, Z., Wang, S.-Q., 2005. Synthesis and rheological properties of polystyrene/layered silicate nanocomposite. Polymer 46, 3006–3013.

2

HIGH BARRIER COMPOSITE MATERIALS BASED ON RENEWABLE SOURCES FOR FOOD PACKAGING APPLICATIONS

Giulio Malucelli
Politecnico di Torino, Department of Applied Science and Technology, Local INSTM Unit, Alessandria, Italy

1 Introduction

Nowadays, the demand for packaging materials that provide even greater protection to their contents, especially in the food, beverage, and pharmaceutical industries, is rapidly increasing. At present, the widespread of the use of plastic materials in the packaging sector has stimulated some discussions about their effective capability to limit and/or block the diffusion of vapors and gases that are responsible for the shortening of the shelf life of the packaged products: indeed, packaging materials are mainly aimed at containing a product, preserving and protecting the product's integrity and quality, providing its presentation and identification, as well as the required information, to consumers.

Furthermore, packaging materials and packaging technology play a key role in the market and their economical importance is growing more and more: according to recent Smithers Pira Statistics (Smithers Pira, 2013), the global packaging market stood at \$799 billion in 2012, increasing by 1% over 2011. Sales were found to increase by 3% in real terms for 2013 to about \$820 billion; in addition, an annual growth of 4% per year to 2018 in the world packaging market, with sales to reach more than \$1 trillion by 2018, is forecasted. Sales of packaging are concentrated in Asia, which accounted for 36% of the total in value terms in 2012; in the same year, North America and Western Europe (which resulted the fourth largest consumer of packaging with a global share of

Food Packaging. http://dx.doi.org/10.1016/B978-0-12-804302-8.00002-9

6%, closely followed by South and Central America with 5%) totally shared 23% and 22%, respectively. In the same period, the Middle East gained 3% of the global demand for packaging, while Africa and Australasia each had a 2% share.

Specifically concerning the food packaging, the preservation of food safety and quality has always represented a very important issue: indeed, on the one hand, food needs to be preserved from contamination and spoilage by microorganisms and, on the other hand, from wastage and spoilage when exposed to moisture, atmospheric gases, and light for a prolonged period of time.

In other words, four basic functions of food packaging (Yam et al., 2005) should be fulfilled:

- *Containment.* This is the most basic function of a package and it is important for its easy transportation or handling.
- *Protection.* This function is as important as the previous one: indeed, food packaging keeps food products in a reduced volume, preventing it to leak or break up and protecting it against possible contaminations and changes.
- *Convenience.* Food should be freely enjoyed by consumers, at their convenience. To this aim, food packages should be designed on the basis of individual lifestyles, taking care, as an example, of portability and availability of multiple/single portions.
- *Communication.* Food packaging should provide the consumers with important information about the contained food product and its nutritional content, together with guidelines about preparation and preservation.

From a market point of view, food packaging accounted in 2009 for 30.0% of the global packaging market and a significant growth is expected in the next years, with a projected value of about $193 billion.

Over the past decades, thanks to the huge research work carried out both in academia and industry, packaging materials showed a significant evolution and at present they have reached outstanding performances mainly due to the enhancements of their optical, mechanical, thermal, and barrier features; in addition, recent nanotechnological breakthroughs and emerging technologies have offered the perspective of developing a new generation of intelligent packaging systems, specifically designed for food, owing to the possibility of sensing, detecting, or recording changes in the product, the package or its environment (Vanderroost et al., 2014).

Despite all these achievements, the development of improved packaging materials has faced several limitations and weaknesses. First of all, the most commonly used packaging materials are usually petrochemical-based plastics (eg, polyolefins, polyesters, and polyamides), which exhibit acceptable barrier features but they are totally nonbiodegradable. Thus, serious ecological problems can be foreseen, as thousands of tons of these plastic packagings are discarded daily and most of them are usually put in landfills (Avella et al., 2005). As a consequence, in consideration of these packaging waste disposal problems, materials having a high barrier capacity and a potential biodegradability when compared to the products derived from petroleum, have been sought and their possible use for the design of sustainable packaging assessed. For this purpose, biopolymers have been one of the favorable alternatives to be exploited and developed into ecofriendly food packaging systems, due to their biodegradability, abundance, and renewability (Tang et al., 2012; Liu et al., 2005; Muratore et al., 2005): indeed, they show the capability to be degraded or broken down through the action of naturally existing organisms, leading to complete mineralization in the environment (Jayaramudu et al., 2013; Majeed et al., 2013) and leaving behind by-products such as water and carbon dioxide, which can be considered safe toward the environment.

Fig. 2.1 reports a schematic classification of biopolymers.

Unfortunately, the use of biopolymers as food packaging materials has clearly shown some drawbacks: compared to the conventional nonbiodegradable materials derived from petroleum, biopolymers exhibit poorer mechanical (eg, lower tensile strength), thermal (eg, inferior continuous use temperature), and barrier (eg, higher water and oxygen permeability) properties. In order to overcome these limitations, many research efforts have been made: in particular, these latter include the exploitation of the nanocomposite concept, which has been established as a promising route for improving the mechanical and barrier features of biopolymers (Azeredo et al., 2011; Othman, 2014; Abdollahi et al., 2012; Arfat et al., 2014; Di Maio et al., 2014; Rhim et al., 2013a; Kanmani and Rhim, 2014; Reddy and Rhim, 2014; Sanuja et al., 2014).

Indeed, the use of a multiphase material comprising two or more constituents (ie, at least a biopolymer matrix representing the continuous phase, and a nanofiller as the discontinuous phase) allows obtaining a new valuable product, in which the applied (mechanical, thermal, environmental) stresses are transferred from the matrix to the reinforcement through the boundary between them, thus achieving enhanced thermomechanical and barrier properties of the matrix itself. The aforementioned nanocomposite concept belongs to the definition of nanomaterials, in accordance with ISO TS 27687:2008 (ISO, 2008), ISO/TS 80004-1 (ISO, 2010) and ISO TS 80004-4:2011 standards, schematically reported in Fig. 2.2.

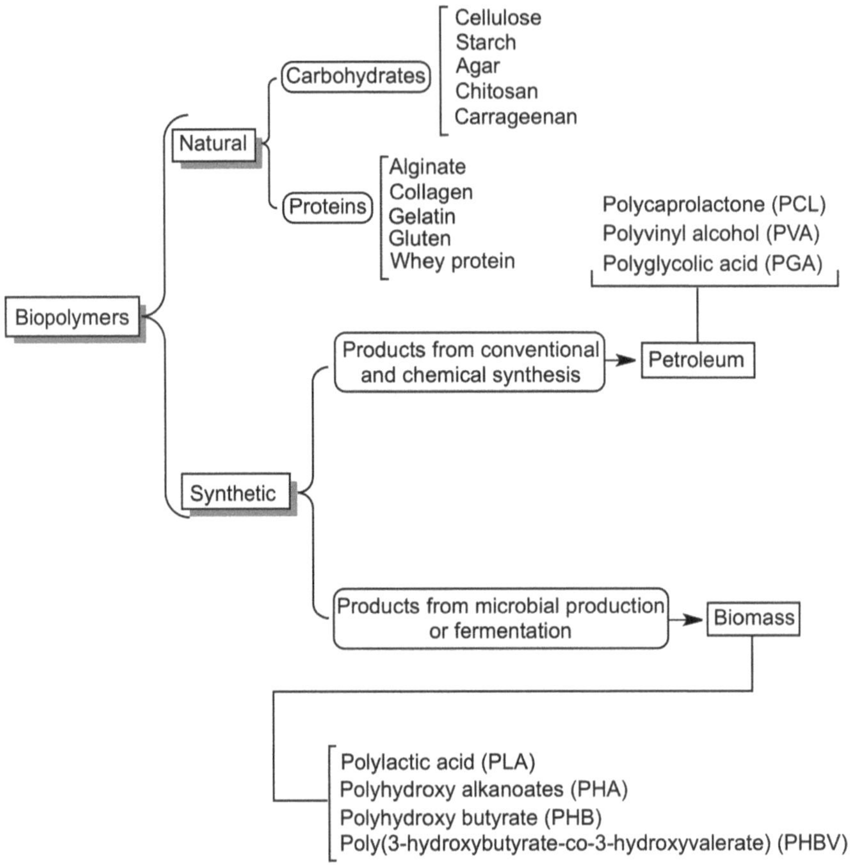

Figure 2.1. Classification of biopolymers.

In many cases, the dispersion of nanofillers such as, for instance, silicate, clay, and titanium dioxide into different biopolymers not only promotes an increase of their overall thermomechanical and barrier features, but may give rise to multifunctional materials also acting as antimicrobial agents (Rhim et al., 2013b), biosensors, and oxygen scavengers in food packaging applications. Thus, the obtained bionanocomposites behave as active food packaging whereby the food packaging (or its inner surface) is able to somehow interact with food by releasing beneficial compounds like antimicrobial and/or antioxidant agents, or by removing some unfavorable elements (like oxygen or water vapor).

Therefore, the development of bionanocomposite materials for food packaging applications is important not only to limit the environmental problems but also to provide the food packaging materials with multifunctionality, hence approaching the concept

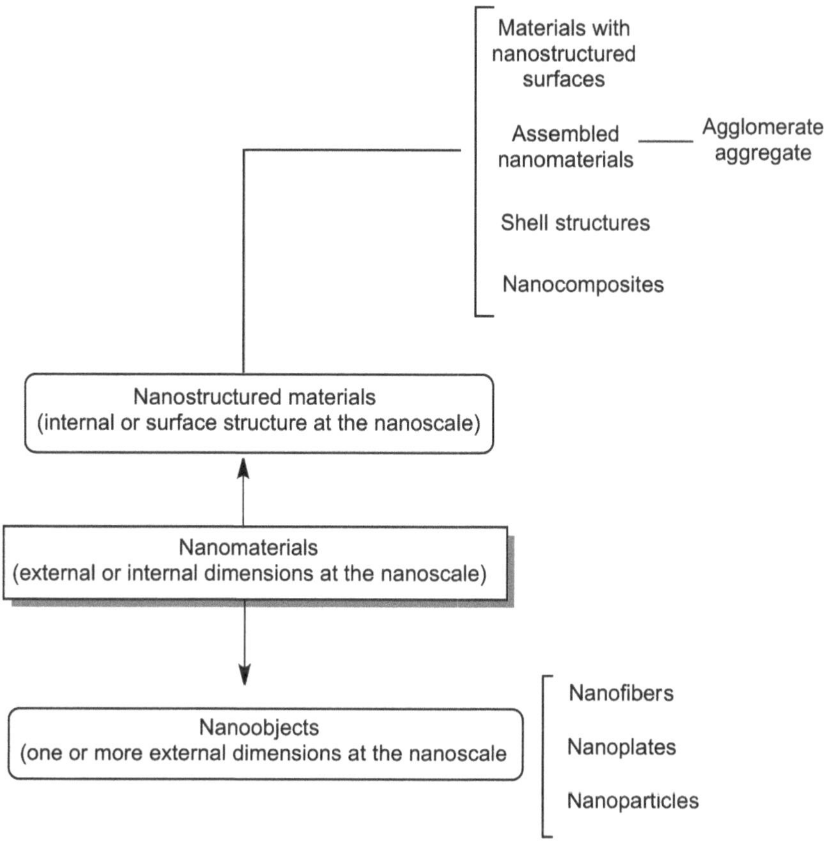

Figure 2.2. Overall classification of nanomaterials.

of "smart packaging system": this may represent a turning point and a clear step forward as compared to the technological level that marked the first plastic packaging materials designed in the past two decades.

On the basis of the aforementioned issues, this chapter thoroughly classifies the nanocomposites derived from renewable sources, which are currently used in food packaging applications, as a valid alternative to the conventional nonbiodegradable materials derived from petroleum. The focus will be specifically devoted to the investigation of the structure/morphology peculiarities that allow achieving the required high barrier performances. In addition, some case studies referring to the exploitation of nanocomposite materials based on renewable sources specifically for food packaging applications are presented in detail. Finally, some recent surface engineering approaches exploited for designing nanostructured "green" assemblies that are able to provide high barrier properties are exhaustively reported: in particular, the potentialities of the layer-by-layer (LBL) approach are thoroughly discussed.

2 Industrial Polymeric Matrices From Renewable Sources for the Development of Food Packaging Systems

This section is aimed to summarize the state of the art concerning the polymer matrices from renewable sources, which have been the research focus at a lab-scale in the packaging field and are currently explored and exploited for food packaging applications at an industrial level.

Polymers from renewable resources can be roughly classified into three groups:

- natural polymers, like starch, protein and cellulose
- synthetic polymers from natural monomers, like polylactic acid
- polymers from microbial fermentation, like polyhydroxybutyrate

The biodegradable polymers here addressed include polylactide (polylactic acid, PLA), poly(hydroxyalkanoate)s (PHAs) such as poly(ß-hydroxybutyrate) (PHB), and poly(3-hydroxybutyrate-co-3-hydroxyvalerate) (PHBV), and natural renewable polymers such as starch and proteins. Their structures are schematized in Fig. 2.3.

2.1 Polylactic Acid (Polylactide, PLA)

PLA belongs to the family of synthetic aliphatic polyesters and is considered as biodegradable and compostable. It is a thermoplastic polymer showing high strength and modulus that can be derived from annually renewable resources. In the past four decades, PLA has been identified as a valuable material for biomedical applications including drug delivery, medical implants, resorbable sutures, and scaffolds for tissue engineering; nowadays it is one of the most used matrices for food packaging applications.

PLA shows several advantages in relation to the petroleum-based polymers usually employed for packaging, namely:

- good transparency, which is slightly higher than that of poly(ethylene terephthalate) and polystyrene.
- Significant degradation in biological environment such as soil or compost: this phenomenon takes place in two main stages,

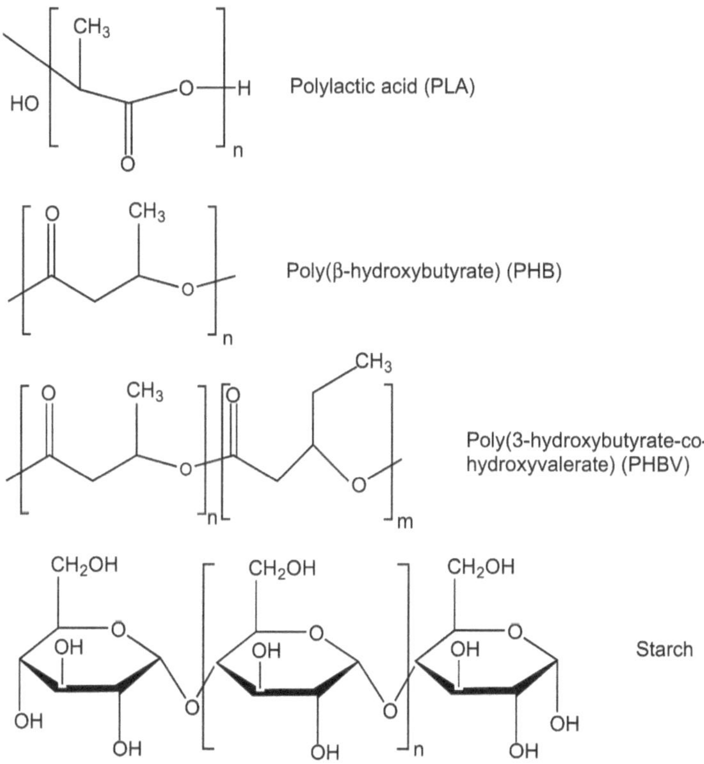

Figure 2.3. Chemical structures of some biodegradable polymers.

which are defined as the hydrolytic and enzymatic degradation. The former starts with the diffusion of water into the polymer, leading to random nonenzymatic chain scissions of the ester groups. The latter occurs when a molecular weight of about 10,000–20,000 g mol^{-1} is achieved: in these conditions microbes can be active in the metabolization process, hence converting the hydrolyzed polymer into carbon dioxide, water and humus. Unlike the first hydrolytic step, which may occur in the bulk polymer through the diffusion of water into the polymer, the second step caused by microbes occurs at the surface only.

- Biocompatibility: PLA has been demonstrated to be biocompatible and to degrade into non-toxic components and it has been approved by the Food and Drug Administration (FDA) even for implantation in the human body.
- Processability: the main conversion approaches of PLA are based on melt processing. Commercial grades of this polymer can typically be processed using a conventional twin-screw

extruder; however, its thermal degradation, which can involve hydrolysis, depolymerization, oxidative random-chain scission, inter- and intramolecular transesterification reactions, is the main drawback for processing in molten state: therefore, the melt processing has to be carefully carried out.

Its monomer, that is, lactic acid, can be obtained by the fermentation of carbohydrate materials like glucose, which is derived by hydrolysis from starch. The structure of lactic acid contains one asymmetric carbon, and can therefore exist as two stereoisomers, L and D. L-lactic acid is naturally present in several organisms, while D-lactic acid is very rare in nature.

Two methods are currently used to obtain polylactic acid, that is, via polycondensation of lactic acid or via ring opening of a dimer of lactic acid (lactide).

The direct synthesis of PLA by polycondensation shows the typical drawbacks of step growth polymerization: more specifically it yields to low molecular weight PLA that is not suitable for most thermoplastic applications. Conversely, high molecular weight PLA is obtained in relatively low yields; in addition, this polymer is very sensitive to the presence of impurities like ethanol or acetic acid arising from the fermentation process. However, it is possible to synthesize high molecular weight PLA (around 300,000 g mol^{-1}) by using highly pure lactic acid and carefully removing the water formed during the polycondensation. Another possibility refers to the use of chain extenders that can couple the formed oligomeric structures, thus giving rise to high molecular weight products.

The exploitation of ring opening polymerization of lactic acid for synthesizing PLA was started by Carothers in 1932 and was further developed by DuPont and Ethicon (Peterson et al., 1999).

Compared to the other biodegradable polymers, PLA is preferable because of its availability and low cost.

Cargill-Dow (USA) is the major company that offers a series of PLA grades (NatureWorks) deriving from renewable agricultural resources like corn or sugar beets. Other companies such as Mitsui Chemicals (Japan) and Shimadzu (Japan) also manufacture PLA with smaller production capacities.

Polylactic acid shows good mechanical and thermal properties, similar to poly(ethylene terephtalate) or polystyrene, depending on the selected properties. Table 2.1 lists some thermal and physicomechanical data for polylactic acid.

Referring to the content of the two stereoisomers (D and L) in PLA grades, the properties of the polymer are strictly related to the ratio between the two mesoforms: indeed, as compared to D-PLA, L-PLA shows a higher crystallinity, which can lead to increased melting temperatures and lower toughness. Furthermore, PLA can

Table 2.1 Some Physicomechanical and Thermal Properties of PLA

Density (g cm^{-3})	1.25
Elastic modulus (tensile, GPa)	3.5
Tensile strength (MPa)	50
Elongation @break (%)	6
Flexural modulus (GPa)	4
Flexural strength (MPa)	80
Shear modulus (GPa)	2.4
Glass transition temperature (°C)	60
Specific heat capacity (J kg^{-1} K^{-1})	1800
Thermal conductivity (W m^{-1} K^{-1})	0.13
Heat deflection temperature @ 455 kPa (°C)	65

Adapted from Harper (2000). Modern Plastic Handbook. McGraw-Hill, New York.

be plasticized using different products like polyethylene glycol, tricthyl citrate, and partial fatty acid esters. Although the barrier properties (in particular toward water vapor and oxygen) are quite moderate, as compared to those of polystyrene, the most significant limitations of PLA use refer to its high density and polarity, poor heat resistance and low toughness. Neat PLA is currently being used in packaging as films, thermoformed and blow molded containers, food service ware and short shelf-life bottles, in competition with polystyrene and/or poly(ethylene terephtalate).

2.2 Starch-Based Materials

Among the different industrial biobased polymers, starch-based products are very appealing: indeed, unlike most of the other biobased polymers, for which expensive and time-consuming synthesis steps are usually needed, they can be easily produced directly by plasticizing the native starch. In addition, the use of starch as a polymeric material represents a good opportunity for the starch industry to extend the growth of nonfood applications sector. Starch is usually extracted from different agricultural products, such as corn, potato, wheat, cassava, tapioca and rice. Its structure is that of a polysaccharide containing two different

Table 2.2 Amylose Content of Different Starches

Starch Source	Amylose (wt.%)
Waxy rice	0
Waxy sorghum	0
Cassava	17
Sweet potato	18
Potato	20
Arrowroot	21
Sago	26
Wheat	26
Corn	28
High amylose corn	70

macromolecules, amylose and amylopectin, based on glucose units.

Amylose is a linear homopolymer with a molecular weight usually between 100,000 and 500,000 g mol^{-1}. Its content in starches varies in a quite wide range, as presented in Table 2.2.

Amylopectin is a highly branched polymer with very high molecular weight higher than millions (the number of glucose units ranges between 2000 and 200,000); its inner chains are made of 20–24 glucose subunits. The amylopectin content in starches is about 70% by weight, though it varies depending on the source (eg, it results higher in medium-grain rice to 100% in glutinous rice, waxy potato starch, and waxy corn, and lower in long-grain rice, amylomaize, and russet potatoes).

The amylose/amylopectin ratio depends on the starch botanical origin ranging from rich-amylose starch to rich-amylopectin starch (the so-called waxy corn). After being extracted, native starch shows a granular structure consisting of alternating amorphous and crystalline domains and it cannot be used directly as a polymeric material. For this purpose, native starch is subjected to a preliminary treatment, which promotes its destructurization either by gelatinization or by applying thermomechanical stresses using a continuous processing method (usually extrusion).

In both cases a plasticizer is employed in order to break up the hydrogen bonds between the macromolecules, hence decreasing the melting and glass transition temperatures below the decomposition temperature. To this aim, different kinds of plasticizers can be used: among all the suitable products, water and glycerol are the most currently exploited. The so-plasticized derivative (usually called *thermoplastic starch*, TPS, or *plasticized starch*) is homogeneous and shows a substantial amorphous structure, although some crystalline domains can be still observed: similarly to oil-based thermoplastics, TPS can be used for different industrial applications, particularly referring to the packaging industry. Indeed, for this application, starch-based polymers show a great potential ascribed to their biodegradability, their combustibility, but also the natural abundance and the renewability of the native material. In addition, TPS exhibits a relatively low cost, which makes it an attractive alternative to such polymers derived from petrochemicals as polyolefins.

Since the beginning of its industrial utilization in the plastics industry, starch-based polymers have gained a significant importance in the biobased polymer market. In Europe, the production capacity of starch plastics increased from 30,000 tons in 2003 to 130,000 tons in 2007, with an average annual growth of about 50% (Shen et al., 2009).

The most common applications of starch-based polymers refer to packaging industry for the manufacturing of soluble films, loose fills and films for bags and sacks. The most important starch materials manufacturers (Shen et al., 2009) are Novamont (60,000 tons $year^{-1}$, Italy), Rodenburg (40,000 tons $year^{-1}$, The Netherlands), Biotec (20,000 tons $year^{-1}$, Germany), Limagrain (10,000 tons $year^{-1}$, France), and Cereplast (10,000 tons $year^{-1}$, USA).

2.3 Poly(Hydroxyalkanoate)s (PHAs)

Poly(hydroxyalkanoates) can be accumulated as a carbon or energy source in several microorganisms (it is well-known that more than 300 microorganisms can synthesize and then accumulate these polymers) under the condition of limiting nutritional elements (Lee et al., 1999). Polyhydroxy-butyrate (PHB) and (polyhydroxybutyrate-co-3-hydroxyvalerate) (PHBV) are the most common PHA biopolymers: the former is accumulated in the form of granules (the size of which is about 0.5 μm) as a carbon and energy source in the cytoplasm (under suitable conditions, up to 90% of PHB can be accumulated with respect to the bacteria dry mass). The latter, PHBV, is a copolymer of hydroxyl-butyrate (HB) and hydroxyl-valerate (HV), successfully synthesized, for the first time, by using specific additives in the growth medium in the early 1970s.

The isolation and recovery of the PHAs requires breaking the cell walls by means of mechanical shear or enzymatic digestion, followed by polymer extraction that can be performed by centrifugation.

The stereo-chemical regularity of PHB shows some drawbacks, as it favors a progressive crystallization with aging, hence making the polymer very brittle: this drawback has been overcome by incorporating comonomers by grafting or by using suitable formulations.

As far as copolymerization is considered, the synthesis of PHBV is not a cost-effective process, because of the high costs of the copolymer; furthermore, its toxicity toward the bacteria promotes lower production yields and its presence affects PHB crystallization kinetics, which results in longer processing cycle times. However, PHB could be toughened by performing annealing processes in an oven, thus widening its application potentialities. Unlike PHB, which melts at 175–180°C, the melting point of the copolymer is shifted toward lower temperatures (ie, 137°C) in the presence of just 25% hydroxyvalerate units: as a consequence, the overall processability is substantially improved, and the mechanical stability as well.

PHAs are produced by Metabolix under the trademark Biopol; bacterial PHBV is manufactured by PHB Industrial SA (Brazil) and Tianan (China). Quite recently, Procter and Gamble has developed a wide range of PHB cohydroxyalkanoates under the trademark Nodax.

Packaging materials made from PHAs show excellent film forming capabilities and coating properties. The properties of the obtained films can be adjusted by changing the HB/HV ratio: more specifically, a high PHB content leads to the obtainment of a strong and stiff product, whereas the presence of HV units enhances flexibility and toughness.

Tables 2.3 and 2.4 collect some thermal and mechanical properties of PHB and PHVB copolymers, respectively.

PHAs are less hydrophilic than the polysaccharide-based counterparts, thus showing better moisture barrier properties; conversely, at the same time, they have poor gas barrier properties; furthermore, they are biodegradable in soil and have excellent processability. Despite the high cost of production, several processes for producing PHA from cheap carbon sources have been developed (Lee et al., 1999).

2.4 Proteins

Wheat gluten is one of the most cost-effective proteins for packaging applications: it is a low-cost by-product of the wheat

Table 2.3 Some Thermal and Mechanical Properties of Polyhydroxy-butyrate (PHB)

T_g (°C)	1
T_m (°C)	172
Crystallinity (%)	62
Young modulus (GPa)	2.05
Tensile strength (MPa)	285
Tensile strain @ break (%)	2.5
Notched Izod impact strength (J m^{-1})	22
Hardness shore D	69

Adapted from Chaijamrus and Udpuay (2008) and Pachekoski et al. (2009).

Table 2.4 Typical Properties of Polyhydroxybutyrate-co-3-hydroxyvalerate (PHBV)

Density (g cm^{-3})	1.25
T_g (°C)	−10÷30
T_m (°C)	120÷160
MFI @190°C/2.16 kg (g 10 min^{-1})	15÷30
Young modulus (GPa)	2.3
Tensile strength (MPa)	39
Tensile strain @ break (%)	4

Adapted from Gerard et al. (2014).

starch industry, with very interesting viscoelastic properties and low water solubility. Gluten can be readily prepared by gently washing dough under a stream of running water. This process is able to remove the bulk of the soluble and particulate matter to leave a proteinaceous mass, which maintains its cohesiveness on stretching.

Gluten comprises about 75% protein on a dry weight basis; most of the remainder is represented by starch and lipids.

In addition, the vast majority of the proteins are of a single type called prolamins, which are the major storage proteins present in the starchy endosperm cells of the grain, where they are synthesized and deposited via the secretory system. Wheat prolamins are known as gliadins and glutenins and together form gluten (Shewry et al., 1986).

Besides wheat gluten, soy, pea, and whey proteins among all the protein sources have shown a great potential in the design of bio-based packaging materials, because of their excellent film forming features. In particular, whey protein components (α-lactalbumin and β-lactoglobulin) have been widely studied for their potential use as agents able to form edible and biodegradable films based on a waste stream from the cheese industry (Kokoszka et al., 2010).

Usually, WPI products are obtained by ion exchange, followed by ultrafiltration or microfiltration and a subsequent ultrafiltration: this interest has been motivated by their high mechanical properties and excellent oxygen barrier properties (Sothornvit et al., 2007). The film-forming properties of WPI have been exploited for obtaining transparent, flexible, colorless, and odorless films. As already mentioned for starch, useable gluten films, as well as whey proteins can be only manufactured in the presence of plasticizers: this also leads to two main detrimental effects, that is, a significant decrease of the mechanical properties and a higher sensitivity to water.

3 Biocomposite Materials Based on Renewable Sources for Food Packaging Applications: A Recent Literature Survey

This section reviews the current state of the art, as derived from the recent open scientific literature, about nanocomposite materials (based on matrices from renewable sources), which seem very promising in food packaging applications and are worthy to further investigate for assessing the feasibility of their scale-up at a pilot, semiindustrial or industrial level, on the basis of the "proof of concept" already demonstrated at a lab-scale level.

Generally speaking, as compared to their unfilled counterparts, nanocomposites show remarkable enhancements in properties with very low nanofiller content (usually <5 wt.%): these improvements can comprise high tensile modulus, increased strength and heat resistance and superior barrier properties toward different permeants. Nanocomposites also offer other benefits such low density, transparency, acceptable surface properties, and recyclability.

As mentioned earlier, the matching of renewable matrices with suitable environmentally friendly nanofillers allows developing and managing a new food packaging concept, based on the combination of components that exhibit, apart from the required enhanced barrier and mechanical features, also other peculiarities, like improved oxidation stability, decreased solvent uptake, self-extinguishment (flame retardancy) and, eventually, tunable biodegradability due to high aspect ratio and high surface area of nanofillers.

Nanofillers can be grouped into three-dimensional spherical and polyhedral particles (like colloidal silica and polyhedral oligomeric silsesquioxanes), two-dimensional nanofibers (carbon nanotubes), or one-dimensional platelets (nanoclays).

Such layered inorganic solids as clay minerals, graphite, and metal phosphates represent the most common (and sometimes the cheapest) class of materials used as nanofillers. In particular, clay minerals like montmorillonite, hectorite, saponite, and laponite have been proved to be very effective for providing the (bio) polymeric matrices, which they are added to, with different peculiar features, thanks to their unique structure and properties (Alexandre and Dubois, 2000; Sinha Ray and Okamoto, 2003; Sinha Ray and Bousmina, 2005).

Thanks to the natural origin of clay, biobased polymers can be reinforced with these clay minerals in order to enhance their mechanical and barrier properties while maintaining their biodegradability. Similarly, cellulose nanofibers or whiskers are currently of high interest for biobased polymers (Eichhorn et al., 2010), as they provide the opportunity to make use of renewable resources. Therefore the potential replacement of conventional packaging materials such as plastics obtained from oil or petrochemicals with biopolymer-based nanocomposites having enhanced properties is currently involving several research groups coming from both academia and industry.

3.1 Biocomposite Materials Based on PLA Matrices

Quite recently, Armentano et al. (2013) thoroughly reviewed the current step forward in the development of PLA bionanocomposites for packaging applications. More specifically, it was found that the most followed strategy for creating a tortuous pathway in PLA is the incorporation of nanofillers and, in particular, of layered silicates and cellulose nanowhiskers, exploiting their environmental friendliness and natural abundance. From an overall point of view, PLA/organoclay nanocomposites prepared by traditional methods (ie, extrusion blow molding) showed a maximum of 60%

reduction of the oxygen permeability coefficient for a clay content ranging in between 5 and 10 wt.%; conversely, higher nanoclay contents led to the formation of aggregates that hindered further barrier improvements.

Molinaro et al. (2013) prepared nanocomposite films by blending semicrystalline PLA with different types of organoclays (namely, Cloisite 20A and Cloisite 30B) using a two-step procedure: first, PLA was blended with a nanoclay (nominal clay content: 5 wt.%) using a corotating intermeshing twin-screw extruder. Then, the obtained PLA pellets containing the organoclays were casted into films by using a single screw extruder cast film machine. Both water vapor and oxygen transmission rates (WVTR and OTR, respectively) were measured on the nanocomposite films and compared to the unfilled PLA: it was found that the incorporation of Cloisite 30B organoclay improved the OTR by 50%, while the presence of Cloisite 20A enhanced the WVTR by 53%.

Recently, Huang et al. (2014) homogeneously dispersed and fully exfoliated graphene oxide nanosheets in PLA. A strong reduction of gas permeability coefficients of PLA films was achieved: indeed, both O_2 and CO_2 permeability coefficients were lowered by about 45% and 68%, respectively, at a very low graphene oxide loading (1.37 vol.%). These findings were attributed to the outstanding barrier properties of the nanofiller, as well as to the strong interfacial adhesion between graphene oxide nanosheets and the biopolymer matrix, rather than to the changes in crystallinity degree and crystalline structure of PLA.

Park et al. (2012) assessed the use of chitosan or/and clay (Cloisite 30B) coatings as a valuable alternative oxygen barrier material to PLA films. To this aim, PLA films were individually coated with each chitosan or chitosan/clay solution using an automatic film applicator and then dried at room conditions for 24 h. It was found that the WVTR values of PLA films coated with chitosan/clay nanocomposite decreased by 14% as compared with uncoated PLA films and PLA films coated with chitosan (PLA film thickness: 20 μm; chitosan and chitosan/clay coating thickness: 2 μm). This finding was ascribed to the presence of ordered dispersed nanoparticle layers with large aspect ratios in the chitosan matrix. The best barrier effect exerted by the coating was found when the OTR values of PLA films treated with the chitosan/clay nanocomposite were measured: indeed, OTR decreased by 99.5% as compared with the neat PLA films. This strong improvement was attributed to the combination of two different phenomena: a decreased area available for diffusion, due to the replacement of preamble polymer by the impermeable nanoclay lamellae, and an increased distance, the permeant has to travel across the film as it follows a tortuous path around the impermeable nanoclay platelets.

Bang and Kim (2012) included biodegradable poly(lactic acid) as an organic component into the inorganic silica networks obtained by using a sol–gel method, in order to prepare environmentally friendly hybrid coating materials with improved gas barrier properties. More specifically, a PLA film obtained by melt extrusion casting process was used as a substrate for coating with the prepared PLA/SiO_2 hybrids. The obtained films showed a very high transparency (the optical transmittance of films was above 92%), and outstanding oxygen and water vapor barrier properties: indeed, these latter were found to increase by 69.7% and 45.7%, respectively, over those of neat PLA films. In addition, the hybrid coating materials enhanced their barrier properties when subjected to an aging process, due to the formation of dense network structures.

Siracusa et al. (2012) investigated four commercial PLA films, natural- and surface-modified, aiming to make a correlation between the mechanical, thermal, and barrier properties as derived from three different surface modification, the biopolymer was previously subjected to (ie, SiO_x coating, anti-UV coating, varnish). More specifically, the carbon dioxide and oxygen transmission rates, measured at two different temperatures (namely, 7 and 23°C) were found to strongly decrease as a consequence of the presence of any surface treatment: indeed, the gas transmission rates were lowered by more than two orders of magnitude after the surface modification of the PLA films, irrespective of the type of treatment selected.

Recently, Girdthep et al. (2015) developed biodegradable polymer nanocomposites as materials suitable for packaging of moisture-sensitive products. More specifically, different amounts of silver-loaded kaolinite (1–5 phr), synthesized via chemical reduction, were blended with PLA in the presence of poly(butylene adipate-co-terephthalate) as a flexibility enhancer, in the presence of small amounts of tetrabutyl titanate as a compatibilizer. The obtained nanocomposites containing 4 phr of modified clay were found to decrease the elongation at break from 213% to 53.8%, enhancing, at the same time, the thermal stability of the polymer matrix; furthermore, a significant 42% reduction of water vapor permeability (WVP) was assessed and ascribed to the kaolinite platelets embedded in the matrix, which behave like a barrier for water vapor through the creation of a longer and more tortuous path during permeation.

3.2 Biocomposite Materials Based on PHA Matrices

Fabra et al. (2013) fabricated multilayered composite assemblies consisting of polyhydroxybutyrate-co-valerate (valerate content: 12 wt.%) and zein nanofibers obtained through an electrospinning process. These latter derive from a natural prolamine, obtained from maize, that exhibits a more hydrophobic character than other proteins because of the high content of apolar amino acids, namely proline and glutamine, which provide this material with peculiar properties in terms of water resistance, viscosity, and thermal resistance. It was noted that the decrease of WVP was significantly higher in the composite assembly where the external layers were obtained by compression-molding rather than by casting. The presence of zein nanostructured interlayers significantly enhanced the oxygen barrier properties of the multilayer films prepared by using both the methods: in particular, the oxygen permeability of the prepared multilayer structures was significantly improved for the longest deposition time by up to 58% and 76% in films prepared by casting and compression-molding methods, respectively. Furthermore, the influence of these interlayers on the water vapor and limonene (chosen as a standard system to test aroma barrier) permeabilities was found to strictly depend on the zein content: the higher was the zein content, the better were the achieved barrier properties.

Pursuing this research work, the same group (Fabra et al., 2014a,b) very recently compared the influence of different types of interlayers, namely consisting of electrospun whey protein isolate (WPI), pullulan (eg, a polysaccharide polymer consisting of maltotriose units), zein, and zein blended with WPI and pullulan, on the barrier properties of polyhydroxybutyrate-co-valerate (with a 3 wt.% valerate content). It was found that the decrease of both oxygen and water permeability of the obtained multilayered assemblies was strictly correlated with parameters such as morphology, thickness, and intrinsic permeability of the selected interlayers: more specifically, significant improvements were achieved only when fibrillar morphologies (ie, pullulan and/or zein fibers) were obtained and used. In detail, unlike the electrospun WPI nanostructures, showing a beaded morphology, for which no significant decrease of the water vapor and oxygen permeability values was assessed, a reduction of oxygen and WVP values between 38 and 48% and 28 and 35% was observed in multilayer systems prepared by the addition of electrospun zein and/or pullulan nanofibers, respectively.

Finally, the multilayer PHA-based materials containing the aforementioned electrospun zein interlayer were also stored for

3 months in order to evaluate any possible modification of their physicochemical properties and microbiological safety (for this latter by monitoring the growth of an inoculum of *Listeria monocytogenes* over time). The storage period did not significantly modify the optical and barrier properties of multilayer films, which, conversely, became stiffer and less deformable.

Pardo-Ibanez et al. (2014) melt-compounded polyhydroxyalkanoates with a keratin additive derived from poultry feathers. Just 1 wt.% of keratin additive turned out to significantly lower the water, limonene, and oxygen permeabilities of the unfilled polymer to less than a half of its initial value.

3.3 Biocomposite Materials Based on Starch Matrices

Recently, Salaberria et al. (2015) compared the role of different nanochitin morphologies on the structural and functional properties of TPS-based films prepared through a casting-evaporation method. It was found that both the selected morphology (ie, nanocrystals, CHNC, or nanofibers, CHNF) and content of nanochitin strongly affected the barrier properties of the starch matrix: in particular, the oxygen permeability of the TPS matrix turned out to lower by about 30 and 25% for the nanocomposite films prepared with CHNC and CHNF, respectively. As far as the WVP is considered, water vapor transmission values decreased in the presence of up to 10 wt.% of chitin nanocrystals. Beyond this concentration, poor barrier properties were observed. This finding was ascribed to different reasons: (1) the presence, at the surface of the chitin nano-objects, of excessive residual amino groups, which show more affinity for water than hydroxyls, and (2) the occurrence of agglomeration phenomena, promoted by an excessive content of CHNC or CHNF, which may decrease the effective amount of chitin nanoobjects in the matrix, hence their barrier action.

Nafchi et al. (2013) prepared bionanocomposite films based on sago starch and bovine gelatin, filled with zinc oxide nanorods (nanofiller loading: 1–5 wt.% total solid). All the films were plasticized with 40% (w/w of total solid) of a combination of sorbitol/glycerol, using 3:1 ratio of the two products. The presence of the nanofiller was found to enhance the barrier properties of the biopolymer matrix: indeed, the incorporation of 5% of ZnO nanorods to starch and gelatin films markedly decreased the permeability to oxygen by 40% and 55%, respectively.

Bhat et al. (2013) enhanced the poor mechanical and barrier properties of starch-based films by incorporating different amounts of lignin (1–5 wt.%), isolated from oil palm black liquor waste. A significant decrease in WVP (of about −35%) was found in the composite materials containing 1 or 2 wt.% of the filler: this finding was attributed to the compatibility and hydrophobicity of starch/lignin combinations, resulting in the reduction in diffusion of water vapor across the prepared films.

Recently, Hietala et al. (2013) incorporated different amounts of cellulose nanofibers (up to 20 wt.%) in a TPS, by using continuous twin-screw extrusion, aiming to decrease the moisture sensitivity of the biopolymer matrix. Indeed, the moisture diffusion coefficient of starch (1.66×10^{-9} cm^2 s^{-1}) was found to significantly decrease with increasing the amount of cellulose nanofibers (1.27×10^{-9} cm^2 s^{-1}, for the biocomposites containing 20 wt.% of the nanofiller). This finding was ascribed to several factors, like: (1) the good compatibility of cellulose with starch; (2) the higher hygroscopicity of starch with respect to cellulose because of the higher crystallinity of this latter; (3) the significant reduction of the swelling of starch as a consequence of the creation of a fibrous cellulose network that lowers the moisture penetration.

Dash and Swain used nanosilicon carbide at different loadings (up to 10 wt.%) for preparing starch-based composites, exploiting a solution method (Dash and Swain, 2013): in the presence of the highest nanofiller concentration, the oxygen permeability of the biopolymer matrix was found to decrease by 3.5 times.

Pica et al. (2012) prepared composite films based on potato starch plasticized with glycerol and different amounts of exfoliated zirconium phosphate (1–6 wt.%). As compared with unfilled plasticized starch, a strong reduction in water uptake was observed in the composite films, due to the orientation and the high aspect ratio of the zirconium phosphate particles embedded in the polymer matrix; as a consequence, the water permeability of starch was expected to decrease at a large extent.

3.4 Biocomposite Materials Based on Other Matrices

Dash et al. (2012) prepared bionanocomposites based on albumin bovine with different amounts (1, 2, 5, 8, and 10 wt.%) of an organoclay (namely, Cloisite 93A), exploiting a solution technique using $CuSO_4$/glycine chelate complex as the catalyst. The oxygen barrier properties of the obtained nanocomposites were measured: a one-third reduction in the permeability was observed when the clay loading was 10 wt.%: this finding was ascribed mainly to intercalation phenomena of albumin bovine into clay galleries that limit the oxygen permeation.

Luduena et al. (2012) investigated the influence of different lignocellulosic fillers (namely, agricultural cotton, cellulose obtained from cotton and hydrolyzed cellulose) and content on the morphology and crystallization behavior of biodegradable ecocomposites based on polycaprolactone; furthermore, the barrier properties of the obtained products were evaluated in view of their use for packaging applications. Unfortunately, irrespective of the type of lignocellulosic filler used, the WVP values were found to increase with increasing the filler loading: this behavior was ascribed to the tendency of the fillers to agglomerate.

Follain et al. (2013) prepared poly(ε-caprolactone)-based nanocomposite films containing cellulose nanocrystals (3–12 wt.% content), isolated from *Luffa cylindrica* by using a film casting/evaporation method carried out in organic media. The influence of the cellulose nanocrystals loading on the barrier features of PCL was assessed through water sorption and water permeation processes: the nanoscale rod-like morphology of cellulose nanocrystals was found to make longer the diffusion path of diffusing molecules because of tortuosity effects, hence remarkably reducing water permeability.

Rodriguez et al. (2012) prepared nanocomposites based on cellulose acetate and containing 5 wt.% of different organoclays (ie, montmorillonite/alkylammonium and montmorillonite/chitosan, which have been synthesized using a cationic exchange process), exploiting a solvent casting method. As compared to the unfilled polymer, the incorporation of the nanofillers turned out to lower the oxygen transmission by 50%. Conversely, the presence of nanofillers was not able to reduce the water permeability because of the high water affinity of cellulose acetate the strongly limited the effect of nanofillers as barrier additives.

Swain et al. (2012) incorporated different organoclay (namely, Cloisite 30B) amounts, ranging from 1 to 8 wt.%, into a soy protein matrix, using a solution intercalation method. In the presence of the highest nanoclay loading, the oxygen permeability was found to decrease by six times with respect to the unfilled polymer.

Rhim et al. (2011) prepared agar-based nanocomposite films containing different types of nanoclays (namely, Cloisite Na^+, Cloisite 30B, and Cloisite 20A), exploiting a solvent casting method. Compared to those of the unloaded agar film (about 2.22×10^{-9} g m m^{-2} s^{-1} Pa^{-1}) the WVP values of the obtained nanocomposites were found to depend on the type of nanoclay used. More specifically, the nanocomposite films containing Cloisite Na^+ showed a decreased WVP (about 1.68×10^{-9} g m m^{-2} s^{-1} Pa^{-1}), unlike the films containing the organically modified nanoclays, for which a slight increase was observed. These findings were mainly ascribed to the surface hydrophilicity or hydrophobicity of the clay types used.

Pursuing this research, Rhim (2012) prepared binary blend films with different mixing ratios of agar and κ-carrageenan (an anionic sulfated polysaccharide extracted from red seaweeds), using a solution casting method. The effect of the presence of Cloisite Na^+ (5 wt.% clay content) on the water vapor barrier properties was assessed: in particular, it was found that the WVP of the intercalated agar/κ-carrageenan/clay nanocomposite film (about 1.34×10^{-9} g m m^{-2} s^{-1} Pa^{-1}) decreased significantly by about 26% as compared with the unfilled counterpart. This behavior was ascribed to the tortuous pathway for moisture vapor diffusion due to the impermeable clay layers dispersed in the polymer matrix.

Very recently, Gonzalez and Alvarez Igarzabal (2015) reinforced soy protein isolate films with starch nanocrystals (SNC) at different loadings (ie, 5, 20, and 40 wt.%), exploiting a simple casting method. The incorporation of the nanofiller was found to lower the water solubility and swelling of the protein film, as well as its WVP: these findings were ascribed to the decreased affinity of the films for water, due to the presence of starch nanocrystals.

Kumar et al. (2010) exploited melt extrusion for preparing bionanocomposite films based on soy protein isolate and different amounts of montmorillonite (Cloisite Na^+) ranging from 5 to 15 wt.%. The presence of the nanoclay was responsible for a significant decrease of the WVP values: indeed, a significant 22.1% decrease of WVP was achieved by embedding just 5 wt.% of nanofiller in the polymer matrix. This reduction was ascribed to the creation of a tortuous pathway for water vapor to diffuse out of the bionanocomposite, due to the homogeneous distribution and high aspect ratio of the employed nanoclay.

Abdollahi et al. (2013) exploited a solvent casting technique for preparing alginate-based nanocomposites containing two different nanofillers, namely montmorillonite and cellulose nanoparticles. Both the nanofillers, the concentration of which ranged in between 1 and 5 wt.%, turned out to significantly affect the water solubility and hence the WVP of the polymer matrix, which decreased by 20 and 18% in alginate/montmorillonite and in alginate/cellulose nanoparticles films, respectively.

Kochumalayil et al. (2013) prepared oriented bionanocomposite coatings based on xyloglucan (known also as galactoxyloglucan, a polysaccharide with the main chain identical to that of cellulose) and montmorillonite (Cloisite Na^+, at different loadings, namely, 4.3, 8.9, and 20 wt.%), exploiting continuous water-based processing. It was found that the presence of the highest nanoclay loading in the biopolymer matrix was very effective in achieving an oxygen permeability of just 1.44 cm^3 μm m^{-2} d^{-1} kPa^{-1} that makes the bionanocomposite an ideal candidate material for the replacement of aluminum gas barrier films in packaging.

4 Layer-by-Layer (LBL) Assemblies: A Recent Surface Engineering Approach for Enhancing the Barrier Properties of Bioplastic Substrates

Nowadays, the number of food packaging applications that need a suitable atmosphere inside the package (eg, moisture content, oxygen concentration) to be ensured for the entire shelf-life of the enclosed product, is growing more and more every year.

As it is well known that the typical use of complex multilayer films allows achieving high barrier properties toward the permeation of gases and vapors, but, specifically referring to plastics recyclability in the packaging sector (that represents more than 39% of the worldwide production of plastics) (PlasticsEurope, 2012), the large use of multilayered plastics resulted in high volumes of discarded packaging, which are difficult to recycle because of their complex structure.

In order to overcome these limitations, the first approach was to exploit nanotechnology for producing layered silicates nanocomposites, which often exhibit enhanced barrier properties due to the creation of a tortuous path that is able to slow down and sometimes strongly limit the diffusion of a gas molecule through the polymer, as schematized in Fig. 2.4.

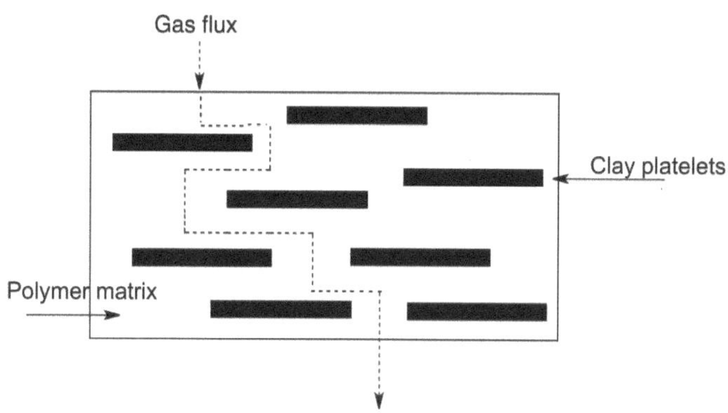

Figure 2.4. Scheme of tortuous paths in layered silicates polymer nanocomposites.

The morphology of the nanoparticles within the polymer matrix, as far as their dispersion and orientation are concerned, strongly affects the barrier features of the composite material: in other words, the higher the nanoplatelets are exfoliated and perpendicularly oriented to the direction of the gas flow, the lower is the final permeability achieved, because of the tortuous paths, through which diffusion occurs. However, it is very difficult to create such a morphology in the nanocomposite material, as the nanoclay lamellae are more likely to orient in a random way. As a consequence, this latter morphology results in a limited effect on the final barrier properties achieved.

The surface deposition of high aspect ratio nanoparticles represents an alternative and more efficient way for obtaining highly oriented lamellar nanocomposites: in particular, a new approach called layer-by-layer (LBL) can be successfully exploited for depositing very thin layers made of nanoparticles almost perfectly arranged parallel to the surface of the polymer substrate, hence maximizing the barrier effect. This technique, although first described by Iler (1966), was practically developed in the early 1990s by Decher and Hong (1991).

From a general point of view, the LBL technique simply consists in an alternate adsorption of electrically oppositely charged chemical species on a chosen substrate. Therefore, it is a step-by-step coating build-up mainly based on electrostatic forces, although different interactions (covalent bonds, hydrogen bonds, etc.) can be also employed. The LBL assembly through electrostatic interactions (schematically depicted in Fig. 2.5) simply requires the alternate immersion of the substrate into oppositely charged polyelectrolyte solutions (or dispersions). Thus, an assembly of positively and negatively charged layers piled up on the substrate surface is obtained, exploiting a total surface charge reversal after each immersion step. The suitable reagents involved in the process can be chosen from a wide range of products: cationic or anionic polyelectrolytes, metallic or oxidic colloids, layered silicates. Each of these categories can be further divided into subcategories so that, from a potential point of view, infinite combinations of multilayered assemblies can be easily designed and selected on the basis of the application or desired properties (Carosio et al., 2013; Malucelli et al., 2014).

Compared to other more traditional thin film deposition techniques, the LBL approach exhibits several advantages: (1) it allows an easy incorporation of functional materials, (2) the process takes place under ambient (eg, atmospheric pressure, room temperature), and (3) environmentally friendly conditions (solvent is mostly water, the concentrations of the polyelectrolytes solutions or nanoparticles dispersions are usually below 1 wt.%).

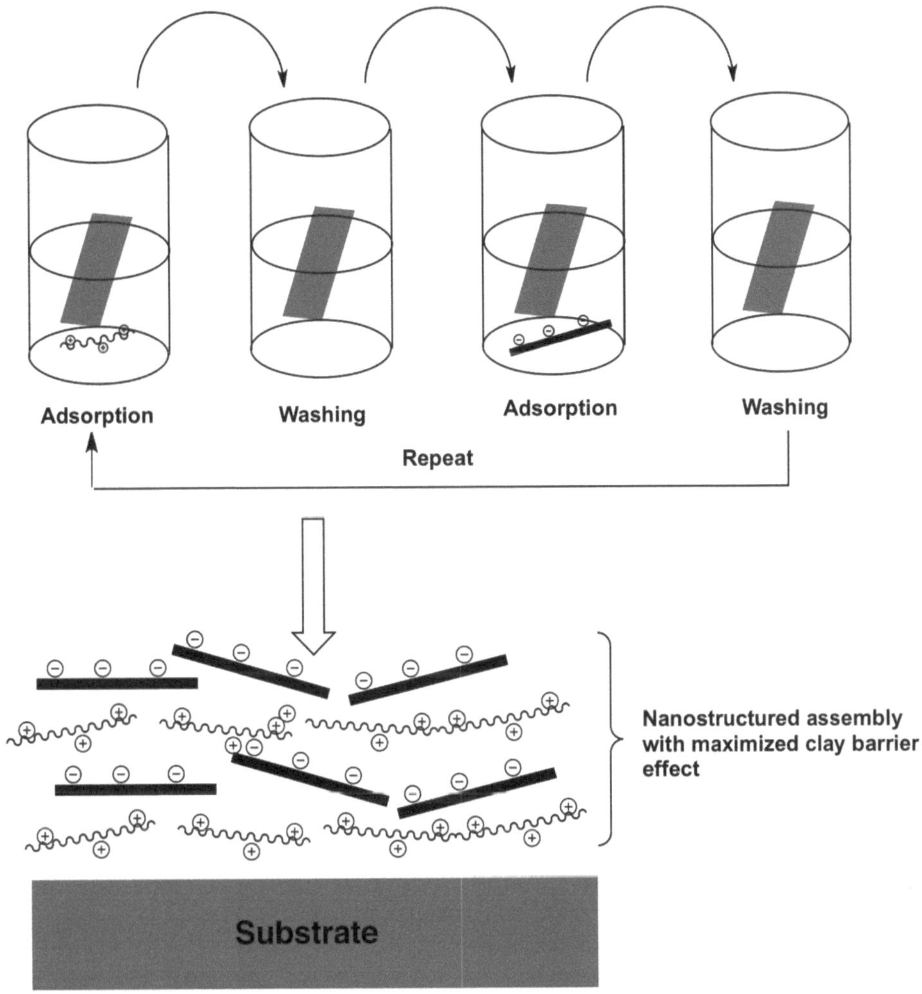

Figure 2.5. Schematic representation of the layer-by-layer (LBL) approach.

Apart from its use for the build-up of coatings showing extreme oxygen barrier properties, the LBL approach has been exploited also to impart flame retardancy, antireflection, electrical conductivity, and antibacterial properties to different substrates.

Specifically referring to the use of clay nanoplatelets and polyelectrolytes, it is thus possible to make assemblies, with overall thicknesses ranging from 100 to 1000 nm, where each clay nanoplatelet is positioned perpendicular to the gas flux, thus providing the highest barrier effect possible, without changing such substrate properties as transparency and flexibility. In other words, the LBL technique can be considered as a surface engineering strategy,

Table 2.5 Comparison of the OTR Values of LBL Assemblies With Other Barrier Technologies

System	Oxygen Transmission Rate [$cm^3\ m^{-2}\ day^{-1}\ atm^{-1}$]
EVOH	0.05
PET	85.0
Metallized	1.20
Nanocomposite	60.0
LBL-treated	<0.005*

* Film OTR is below the low-end detection limit [0.005 $cm^3\ m^{-2}\ day^{-1}\ atm^{-1}$)] of the used apparatus.
Adapted from Carosio et al. (2013).

which does not promote any change of the bulk features of the treated substrate. In order to highlight the advantages provided by this surface method, Table 2.5 compares the oxygen transmission rates (OTR) taken from the scientific literature for LBL treated plastics with those of conventional polymer nanocomposites, multilayer films, and commercially available products.

Hereafter, some recent advances concerning the application of the LBL method for the design of high barrier coatings, using renewable sources either as components in the LBL assembly or as substrates to protect, will be summarized.

Carneiro-da-Cunha et al. (2010) exploited the LBL approach using two polysaccharides with opposite charges, chitosan and sodium alginate, which were deposited onto aminolyzed/charged PET: the resulting multilayered assemblies significantly reduced the water permeability of the substrate. Indeed, as compared with the values reported in the open scientific literature for alginate and chitosan edible films (24.6×10^{-11} g m^{-1} s^{-1} Pa^{-1} (Hambleton et al., 2009) and 3.8×10^{-11} g m^{-1} s^{-1} Pa^{-1} (Souza et al., 2009), respectively), the prepared LBL assemblies were able to strongly reduce the water permeability, reaching 0.85×10^{-11}g m^{-1} s^{-1} Pa^{-1}. These findings were explained on the basis of strong interactions taking place in between adjacent alginate and chitosan layers (Jang et al., 2008), which enhance the tortuosity of the material, thus decreasing its permeability to water molecules.

The same approach was exploited by this group to generate a five layer nanostructured assembly of alternating k-carrageenan and chitosan molecules on aminolyzed/charged PET (Pinheiro et al., 2012). The presence of the LBL assembly promoted a significant decrease of both water and oxygen permeabilities, which were as low as 0.02×10^{-11}g m^{-1} s^{-1} Pa^{-1} and 0.043×10^{-11}g m^{-1} s^{-1} Pa^{-1}, respectively.

Laufer et al. (2012) paired positively charged chitosan, at two different pH levels (3 and 6), with anionic montmorillonite clay nanoplatelets, exploiting the LBL approach. More specifically, the 30-bilayer assembly made of chitosan (at pH 6) and nanoclay (the overall thickness of the assembly was about 100 nm) was found to strongly lower the oxygen permeability of a 0.5-mm-thick polylactic acid film by four orders of magnitude, below the low-end detection limit (0.005 cm^3 m^{-2} day^{-1} atm^{-1}) of the used apparatus. This finding was attributed to the as defined "nanobrick wall" structure, consisting of clay nanoplatelets as bricks surrounded by a polyelectrolyte mortar, which is responsible for creating an extremely tortuous path slowing down the diffusion of oxygen molecules.

Pursuing this research, the same group (Laufer et al., 2013) prepared multilayered thin assemblies made of chitosan, carrageenan, and sodium montmorillonite clay, exploiting the LBL technique, aiming to fabricate fully renewable polysaccharide-based thin films with high barrier properties toward oxygen. For this purpose, trilayers or quadlayers assemblies were designed. In detail, 10 trilayers of chitosan/nanoclay/carrageenan (<40 nm thick) were first deposited on PET films: the oxygen permeability of these latter (1.76×10^{-15} cm^3 cm cm^{-2} s^{-1} Pa^{-1}) was lowered by one order of magnitude under dry conditions testing. The use of an additional layer of chitosan to the trilayer sequence, that is, the building up of a "quadlayer" film (made of chitosan/carrageenan/chitosan/nanoclay), the thickness of which resulted below 60 nm, allowed reducing oxygen permeability of PET by even two orders of magnitude under the same conditions. These findings were ascribed to the particular nanobrick wall structure of the prepared coatings, as well as to a strong association among the oppositely charged polysaccharides.

Carosio et al. (2013) deposited two different types of LBL polymer-nanoclay architectures on commercially available PLA films: more specifically, either 20 bilayers (BL) with a 2-layer (polymer-clay) repeating unit or 6 quad-layers (QL) with a 4-layer (3 polymer layers and 1 of clay) repeating unit were used. Furthermore, the use of a spray apparatus, as a possible (more industrial) alternative to the conventional dipping process was investigated. As compared to untreated PLA (which showed a permeability to oxygen in dry conditions of 14.5 cm^3 mm m^{-2} day^{-1} atm^{-1}), the presence of 20 BL or 6 QL significantly decrease the permeability of the renewable substrate (1.7 and 0.09 cm^3 mm m^{-2} day^{-1} atm^{-1} permeability values were achieved, respectively). In particular, the QL nanoassemblies exhibited the best performances with a permeability three orders of magnitude lower than that of untreated PLA. This finding was attributed to the clay, which was perpendicularly oriented to the direction of the gas flow, on the biopolymer surface, as well as to an increased interlayer distance, with respect to the BL system, between each clay layer, due to the coating structure and to the interactions between the employed polymers that are able to create highly interdiffused assemblies with very low free volume available for oxygen diffusion. Furthermore, aiming to assess a possible industrial application, the potential of LBL spray deposition was investigated: to this aim, the same PLA substrates were spray-coated, using a commercially available apparatus, using the same solutions employed in the dipping process and building up the same assemblies (20 BL and 6 QL). The nanocoatings deposited by spraying turned out to be quite useful as barrier coatings, but, at the same time, less performing as compared to the dipping-deposited counterparts: indeed, 8.8 and 1.5 cm^3 mm m^{-2} day^{-1} atm^{-1} permeability values were measured in dry conditions, respectively for 20 BL and 6 QL, notwithstanding the shorter deposition time required (eg, 10 min instead of 40 min for creating the assembly of 6 QL).

5 Conclusions and Future Perspectives

This chapter has clearly demonstrated that significant and high impact advances in the use of nanotechnology for food packaging applications based on the use of renewable sources have been achieved in the last decade. Indeed, the overall progress derived from both the use of bionanocomposite materials and the exploitation of surface engineering strategies has significantly enhanced the quality and the required performances of the packaging materials, opening a new pathway toward a more sustainable and sometimes cost-effective approach.

Nevertheless, it has to be taken into account that our society is becoming increasingly complex: as a consequence, such users like food suppliers, food manufactures, logistic operators, retailers, and, obviously, consumers are continuously demanding innovative and creative food packaging to guarantee food safety, quality, and traceability.

The fulfilment of these requests needs the development of appropriate technologies that can be integrated in the food packaging world. In particular, the food packaging innovations to be commercially viable and successfully accepted by the target groups should meet the ever increasing regulatory necessities, showing, at the same time, a clear final beneficial outcome that compensates the possible extra expenses of using the new technologies.

In addition, the food packaging innovations that are expected to be designed and then adopted, should also aim at lowering the environmental pressure that surrounds the food packaging field: in particular, a broad range of sustainability issues, like process optimization, waste prevention, efficient use of resources, recycle, reuse, should be considered and thoroughly assessed.

To date, the research efforts on the development of food packaging biobased materials have been primarily focused on the production and the mechanical properties of the biobased polymers; conversely, little attention has been paid to investigate the interactions taking place between the polymer matrices and the food products. From the data reported in the open scientific literature, it is worth noting that only a few studies on food stability and quality when packaged in biobased materials have been published so far, thus indicating that the science of biobased packaging materials for foods is still very much at the beginning.

Another important issue that is worth highlighting refers to the main risk associated with the nanosized components, the bionanocomposites embed in, which can migrate into the food, that can potentially result in adverse health effects. Several studies have clearly demonstrated that some nanoparticles are very dangerous, as they are able to promote intracellular damages, vascular disease, and pulmonary inflammation: because health and safety features of many nanomaterials have not been fully understood yet, food safety should be the main concern when the feasibility and suitability of nanomaterials for food packaging applications are being investigated. This issue implies the need for a complete and detailed toxicological analysis aimed at elucidating the potential risks involved. It is worth mentioning that legislators have established food regulations that should protect the consumers from involuntarily exposure to risk: as an example, the European Regulation Act states that the overall migration of compounds from active and intelligent packaging to the food deserves investigation. In addition, the European Food Safety Authority has mandated that each component of any packaging system should not affect adversely on human health; similarly, any released or absorbed compound should not deceive the consumer. However, unequivocal methodologies able to detect and quantify the migrated nanomaterials from packaging are currently lacking: as a consequence, any regulation enforcement will be ineffective.

Cost effectiveness is for sure another important key point referring to nanotechnology applications for advanced packaging systems: indeed, providing packaging with smart features will likely increase the cost per package, especially during the early stage of product assessment and introduction on the market. The profit margin of food is relatively low compared to other consumer products and it has been demonstrated that the expected total packaging cost should not be above 10% of the product cost. This means that the use of nanotechnology for food packaging applications should be based on appropriate cost-benefit analyses to justify its implementation.

Consumer approval is another factor that food packaging companies dealing with nanotechnology should carefully consider. Indeed, consumers' "nanophobia" as regard to nanotechnologies can potentially lead to poor acceptance, irrespective of their role in enhancing the safety and quality of the food products.

In conclusion, despite that also the current research efforts on these new emerging nanotechnologies are still somewhat immature (also considering that some of the most promising lab-scale proofs of concept need to be upscaled and industrialized) and several problems still need a solution, it is expected that the next research challenges will be directed toward the development of a new generation of smart packaging systems, able to allow a better monitoring of the flow, safety, and quality of the food products, reducing, at the same time, the detrimental impact of packaging waste and food loss on the environment.

References

Abdollahi, M., Alboofetileh, M., Rezaei, M., Behrooz, R., 2013. Comparing physico-mechanical and thermal properties of alginate nanocomposite films reinforced with organic and/or inorganic nanofillers. Food Hydrocoll. 32, 416–424.

Abdollahi, M., Rezaei, M., Farzi, G., 2012. A novel active bio-nanocomposite film incorporating rosemary essential oil and nanoclay into chitosan. J. Food Eng. 111, 343–350.

Alexandre, M., Dubois, P., 2000. Polymer-layered silicate nanocomposites: preparation, properties and uses of a new class of materials. Mater. Sci. Eng. R 28, 1–63.

ISO/TS 27687: Nanotechnologies—Terminology and definitions for nano-objects—Nanoparticle, nanofibre and nanoplate, 2008.

ISO/TS 80004-1: Nanotechnologies—Vocabulary—Part 1 Core terms, 2010.

Arfat, Y.A., Benjakul, S., Prodpran, T., Sumpavapol, P., Songtipya, P., 2014. Properties and antimicrobial activity of fish protein isolate/fish skin gelatin film containing basil leaf essential oil and zinc oxide nanoparticles. Food Hydrocoll. 41, 265–273.

Armentano, I., Bitinis, N., Fortunati, E., Mattiolia, S., Rescignano, N., Verdejo, R., Lopez-Manchado, M.A., Kenny, J.M., 2013. Multifunctional nanostructured

PLA materials for packaging and tissue engineering. Prog. Polym. Sci. 38, 1720–1747.

Avella, M., Vlieger, J.J.D., Errico, M.E., Fischer, S., Vacca, P., Volpe, M.G., 2005. Biodegradable starch/clay nanocomposite films for food packaging applications. Food Chem. 93, 467–474.

Azeredo, H.M.C.D., Mattoso, L.H.C., McHugh, T.H., 2011. Nanocomposites in food packaging—A review. Advances in Diverse Industrial Applications of Nanocomposites. InTech, Rijeka.

Bang, G., Kim, S.W., 2012. Biodegradable poly (lactic acid)-based hybrid coating materials for food packaging films with gas barrier properties. J. Indust. Eng. Chem. 18, 1063–1068.

Bhat, R., Abdullah, N., Hj Din, R., Tay, G.-S., 2013. Producing novel sago starch based food packaging films by incorporating lignin isolated from oil palm black liquor waste. J. Food Eng. 119, 707–713.

Carneiro-da-Cunha, M., Cerqueira, M.A., Souza, B.W.S., Carvalho, S., Quintas, M.A.C., Teixeira, J.A., Vicente, A.A., 2010. Physical and thermal properties of a chitosan/alginate nanolayered PET film. Carbohydr. Polym. 82, 153–159.

Carosio, F., Fina, A., Alongi, J., Malucelli, G., Camino, G., 2013. Layer-by-layer coatings for enhanced shelf-life. Pack. Films 4, 8–10.

Chaijamrus, S., Udpuay, N., 2008. Production and characterization of polyhydroxybutyrate from molasses and corn steep liquor produced by *Bacillus megaterium* ATCC 6748. Agric. Eng. Int. X, 1–12.

Dash, S., Kisku, S.K., Swain, S.K., 2012. Effect of nanoclay on morphological, thermal, and barrier properties of albumin bovine. Polym. Compos. 33, 2201–2206.

Dash, S., Swain, S.K., 2013. Synthesis of thermal and chemical resistant oxygen barrier starch with reinforcement of nano silicon carbide. Carbohydr. Polym. 97, 758–763.

Decher, G., Hong, J.D., 1991. Buildup of ultrathin multilayer films by a self-assembly process. II. consecutive adsorption of anionic and cationic bipolar amphiphiles on charged surfaces. Makromol. Chem. Macromol. Symp. 46, 321–327.

Di Maio, L., Scarfato, P., Milana, M.R., Feliciani, R., Denaro, M., Padula, G., Incarnato, L., 2014. Bio-nanocomposite polylactic acid/organoclay films: functional properties and measurement of total and lactic acid specific migration. Pack. Technol. Sci. 27, 535–547.

Eichhorn, S.J., Dufresne, A., Aranguren, M., Marcovich, N.E., Capadona, J.R., Rowan, S.J., Weder, C., Thielemans, W., Roman, M., Renneckar, S., Gindl, W., Veigel, S., Keckes, J., Yano, H., Abe, K., Nogi, M., Nakagaito, A.N., Mangalam, A., Simonsen, J., Benight, A.S., Bismarck, A., Berglund, L.A., Peijs, T., 2010. Review: current international research into cellulose nanofibres and nanocomposites. J. Mater. Sci. 45, 1–33.

Fabra, M.J., Lopez-Rubio, A., Lagaron, J.M., 2013. High barrier polyhydroxyalcanoate food packaging film by means of nanostructured electrospun interlayers of zein. Food Hydrocoll. 32, 106–114.

Fabra, M.J., López-Rubio, A., Lagaron, J.M., 2014a. On the use of different hydrocolloids as electrospun adhesive interlayers to enhance the barrier properties of polyhydroxyalkanoates of interest in fully renewable food packaging concepts. Food Hydrocoll. 39, 77–84.

Fabra, M.J., Sanchez, G., Lopez-Rubio, A., Lagaron, J.M., 2014b. Microbiological and ageing performance of polyhydroxyalkanoate-based multilayer structures of interest in food packaging. Food Sci. Technol. 59, 760–767.

Follain, N.G., Belbekhouche, S., Bras, J., Siqueira, G., Marais, S.P., Dufresne, A., 2013. Water transport properties of bio-nanocomposites reinforced by *Luffa cylindrica* cellulose nanocrystals. J. Memb. Sci. 427, 218–229.

Gerard, T., Budtova, T., Podshivalov, A., Bronnikov, S., 2014. Polylactide/poly(hydroxybutyrate-co-hydroxyvalerate) blends: morphology and mechanical properties. eXPRESS Polym. Lett. 8, 609–617.

Girdthep, S., Worajittiphon, P., Molloy, R., Leejarkpaic, T., Punyodom, W., 2015. Formulation and characterization of compatibilized poly(lactic acid)-based blends and their nanocomposites with silver-loaded kaolinite. Polym. Int. 64, 203–211.

Gonzalez, A., Alvarez Igarzabal, C.I., 2015. Nanocrystal-reinforced soy protein films and their application as active packaging. Food Hydrocoll. 43, 777–784.

Hambleton, A., Frédéric Debeaufort, F., Bonnotte, A., Voilley, A., 2009. Influence of alginate emulsion-based films structure on its barrier properties and on the protection of microencapsulated aroma compound. Food Hydrocoll. 23, 2116–2124.

Harper, C., 2000. Modern Plastic Handbook. McGraw-Hill, New York.

Hietala, M., Mathew, A.P., Oksman, K., 2013. Bio-nanocomposites of thermoplastic starch and cellulose nanofibers manufactured using twin-screw extrusion. Eur. Polym. J. 49, 950–956.

Huang, H.D., Ren, P.G., Xu, J.Z., Xu, L., Zhong, G.J., Hsiao, B.S., Li, Z.M., 2014. Improved barrier properties of poly(lactic acid) with randomly dispersed graphene oxide nanosheets. J. Memb. Sci. 464, 110–118.

Iler, R.K., 1966. Multilayers of colloidal particles. J. Colloid Interface Sci. 21, 569–594.

Jang, W.-S., Rawson, I., Grunlan, J.C., 2008. Layer-by-layer assembly of thin film oxygen barrier. Thin Solid Films 516, 4819–4825.

Jayaramudu, J., Reddy, G.S.M., Varaprasad, K., Sadiku, E.R., Ray, S.S., Rajulu, A.V., 2013. Preparation and properties of biodegradable films from *Sterculia urens* short fiber/cellulose green composites. Carbohydr. Polym. 93, 622–627.

Kanmani, P., Rhim, J.-W., 2014. Properties and characterization of bio-nanocomposite films prepared with various biopolymers and ZnO nanoparticles. Carbohydr. Polym. 106, 190–199.

Kochumalayil, J.J., Bergenstråhle-Wohlert, M., Utsel, S., Wågberg, L., Zhou, Q., Berglund, L.A., 2013. Bioinspired and highly oriented clay nanocomposites with a xyloglucan biopolymer matrix: extending the range of mechanical and barrier properties. Biomacromolecules 14, 84–91.

Kokoszka, S., Debeaufort, F., Lenart, A., Voilley, A., 2010. Water vapour permeability, thermal and wetting properties of whey protein isolate based edible films. Int. Dairy J. 20, 53–60.

Kumar, P., Sandeep, K.P., Alavi, S., Truong, V.D., Gorga, R.E., 2010. Preparation and characterization of bio-nanocomposite films based on soy protein isolate and montmorillonite using melt extrusion. J. Food Eng. 100, 480–489.

Laufer, G., Kirkland, C., Cain, A.A., Grunlan, J.C., 2012. Clay–chitosan nanobrick walls: completely renewable gas barrier and flame-retardant nanocoatings. ACS Appl. Mater. Interfaces 4, 1643–1649.

Laufer, G., Kirkland, C., Cain, A.A., Grunlan, J.C., 2013. Oxygen barrier of multilayer thin films comprised of polysaccharides and clay. Carbohydr. Polym. 95, 299–302.

Lee, S.Y., Choi, J., Wong, H.H., 1999. Recent advances in polyhydroxyalkanoate production by bacterial fermentation: mini-review. Int. J. Biol. Macromol. 25, 31–36.

Liu, W., Misra, M., Askeland, P., Drzal, L.T., Mohanty, A.K., 2005. "Green" composites from soy based plastic and pineapple leaf fiber: fabrication and properties evaluation. Polymer 46, 2710–2721.

Luduena, L., Vázquez, A., Alvarez, V., 2012. Effect of lignocellulosic filler type and content on the behavior of polycaprolactone based eco-composites for packaging applications. Carbohydr. Polym. 87, 411–421.

Majeed, K., Jawaid, M., Hassan, A., Bakar, A.A., Khalil, H.P.S.A., Salema, A.A., Inuwa, I., 2013. Potential materials for food packaging from nanoclay/natural fibres filled hybrid composites. Mater. Design 46, 391–410.

Malucelli, G., Carosio, F., Alongi, J., Fina, A., Frache, A., Camino, G., 2014. Materials engineering for surface-confined flame retardancy. Mater. Sci. Eng. R-Reports. 84, 1–20.

Molinaro, S., Cruz Romero, M., Boaro, M., Sensidoni, A., Lagazio, C., Morris, M., Kerry, J., 2013. Effect of nanoclay-type and PLA optical purity on the characteristics of PLA-based nanocomposite films. J. Food Eng. 117, 113–123.

Muratore, G., Del Nobile, M.A., Buanocore, G.G., Lanza, C.M., Asmundo, C.N., 2005. The influence of using biodegradable packaging films on the quality decay kinetic of plum tomato. J. Food Eng. 67, 393–399.

Nafchi, A.M., Nassiri, R., Sheibani, S., Ariffin, F., Karim, A.A., 2013. Preparation and characterization of bio-nanocomposite films filled with nanorod-rich zinc oxide. Carbohydr. Polym. 96, 233–239.

Othman, S.H., 2014. Bio-nanocomposite materials for food packaging applications: types of biopolymer and nano-sized filler. Agric. Agric. Sci. Procedia 2, 296–303.

Pachekoski, W.M., Marcondes Agnelli, J.A., Belem, L.P., 2009. Thermal, mechanical and morphological properties of poly (hydroxybutyrate) and polypropylene blends after processing. Mater. Res.

Pardo-Ibanez, P., Lopez-Rubio, A., Martinez-Sanz, M., Cabedo, L., Lagaron, J.C., 2014. Keratin–polyhydroxyalkanoate melt-compounded composites with improved barrier properties of interest in food packaging applications. J. Appl. Polym. Sci. 131, 39947–39956.

Park, S.H., Lee, H.S., Choi, J.H., Jeong, C.M., Sung, M.H., Park, H.J., 2012. Improvements in barrier properties of poly(lactic acid) films coated with chitosan or chitosan/clay nanocomposite. J. Appl. Polym. Sci. 125, E675–E680.

Peterson, K., Nielsen, P.V., Bertelsen, G., Lawther, M., Olsen, M.B., Nilsson, N.H., Mortensen, G., 1999. Potential of biobased materials for food packaging. Trends Food Sci. Technol. 10, 52–68.

Pica, M., Donnadio, A., Casciola, M., 2012. Starch/zirconium phosphate composite films: hydration, thermal stability, and mechanical properties. Starch/Starke 64, 237–245.

Pinheiro, A.C., Bourbon, A.I., Medeiros, B.G.S., da Silva, L.H.M., da Silva, M.C.H., Carneiro-da-Cunha, M.G., 2012. Interactions between k-carrageenan and chitosan in nanolayered coatings e structural and transport properties. Carbohydr. Polym. 87, 1081–1090.

PlasticsEurope, 2012. Plastics: the facts 2012. An analysis of European plastics production, demand and waste data for 2011.

Reddy, J.P., Rhim, J.-W., 2014. Characterization of bio-nanocomposite films prepared with agar and paper-mulberry pulp nanocellulose. Carbohydr. Polym. 110, 480–488.

Rhim, J.-W., 2012. Physical-mechanical properties of agar/κ-carrageenan blend film and derived clay nanocomposite film. J. Food Sci. 77, N66–N73.

Rhim, J.-W., Lee, S.-B., Hong, S.-I., 2011. Preparation and characterization of agar/clay nanocomposite films: the effect of clay type. J. Food Sci. 76, N40-N48.

Rhim, J.-W., Park, H.-M., Ha, C.-S., 2013a. Bio-nanocomposites for food packaging applications. Prog. Polym. Sci. 38, 1629–1652.

Rhim, J.W., Wang, L.F., Hong, S.I., 2013b. Preparation and characterization of agar/silver nanoparticles composite films with antimicrobial activity. Food Hydrocoll. 33, 327–335.

Rodriguez, F.J., Galotto, M.J., Guarda, A., Bruna, J.E., 2012. Modification of cellulose acetate films using nanofillers based on organoclays. J. Food Eng. 110, 262–268.

Salaberria, A.M., Diaz, R.H., Labidi, J., Fernandes, S.C.M., 2015. Role of chitin nanocrystals and nanofibers on physical, mechanical and functional properties in thermoplastic starch films. Food Hydrocoll. 46, 93–102.

Sanuja, S., Agalya, A., Umapathy, M.J., 2014. Studies on magnesium oxide reinforced chitosan bio-nanocomposite incorporated with clove oil for active food packaging application. Int. J. Polym. Mater. Polym. Biomater. 63, 733–740.

Shen, L., Haufe, J., Patel, M.K., 2009. Product overview and market projection of emerging bio-based plastics, Group Science, Technology and Society (STS) Copernicus Institute for Sustainable Development and Innovation. Utrecht University, Utrecht

Shewry, P.R., Tatham, A.S., Forde, J., Kreis, M., Miflin, B.J., 1986. The classification and nomenclature of wheat gluten proteins: a reassessment. J. Cereal Sci. 4, 97–106.

Sinha Ray, S., Bousmina, M., 2005. Biodegradable polymers and their layered silicate nanocomposites: in greening the 21st century materials world. Prog. Mater. Sci. 50, 962–1079.

Sinha Ray, S., Okamoto, M., 2003. Polymer/layered silicate nanocomposites: a review from preparation to processing. Prog. Polym. Sci. 28, 1539–1641.

Siracusa, V., Blanco, I., Romani, S., Tylewicz, U., Rocculi, P., Dalla Rosa, M., 2012. Poly(lactic acid)-modified films for food packaging application: physical, mechanical, and barrier behavior. J. Appl. Polym. Sci. 125, E390–E401.

Smithers Pira, 2013. The future of global packaging. Smithers Pira International Market Report. Smithers Pira International Ltd, 2013.

Sothornvit, R., Olsen, C.W., McHugh, T.H., Krochta, J.M., 2007. Tensile properties of compression-molded whey protein sheets: determination of molding condition and glycerol-content effects and comparison with solution-cast films. J. Food Eng. 78, 855–860.

Souza, B.W.S., Cerqueira, M.A., Casariego, A., Lima, A.M.P., Teixeira, J.A., Vicente, A.A., 2009. Effect of moderate electric fields in the permeation properties of chitosan coatings. Food Hydrocoll. 23, 2110–2115.

Swain, S.K., Priyadarshini, P.P., Patra, S.K., 2012. Soy protein/clay bio-nanocomposites as ideal packaging materials. Polym.-Plast. Technol. Eng. 51, 1282–1287.

Tang, X.Z., Kumar, P., Alavi, S., Sandeep, K.P., 2012. Recent advances in biopolymers and biopolymer-based nanocomposites for food packaging materials. Crit. Rev. Food Sci. Nutr. 52, 426–442.

Vanderroost, M., Ragaert, P., Devlieghere, F., De Meulenaer, B., 2014. Intelligent food packaging: the next generation. Trends Food Sci. Technol. 39, 47–62.

Yam, K.L., Takhistov, P.T., Miltz, J., 2005. Intelligent packaging: concepts and applications. J. Food Sci. 70, R1–R10.

3

BIONANOCOMPOSITES: SMART BIODEGRADABLE PACKAGING MATERIAL FOR FOOD PRESERVATION

Sushama Talegaonkar*, Harshita Sharma*, Shweta Pandey*, Pawan Kumar Mishra, Rupert Wimmer**,†**

**Jamia Hamdard, Department of Pharmaceutics, New Delhi, India; **Mendel University, Department of Wood Science, Brno, Czech Republic; †University of Natural Resources and Life Sciences, Institute for Natural Materials Technology, Tulln, Austria*

1 Introduction

Food packaging is generally aimed to prevent the degradation of food by physical or chemical contaminants, microbial contamination, and loss of aroma while retaining the quality of product during its extended shelf life. To address aforementioned issues, a packaging must act as a gaseous barrier to hinder the exchange of moisture, carbon dioxide, oxygen, and aromatics along with acceptable optical, physical, and mechanical properties. A classical answer to the aforementioned questions was found in petroleum-based polymers. However, these petroleum-based polymers used for food packaging are not biodegradable, nonrenewable, and noncompostable, therefore presenting a significant serious environmental and disposal problem and waste generation issues worldwide. According to reports, packaging waste accounted for 71.6 million tons or 29.5% of the total municipal solid waste (MSW) in 2009 in the United States (http://www.epa.gov/osw/nonhaz/municipal/pubs/msw2009-fs.pdf) and 56.3 million tons or 25% of the MSW in 2006 in Europe (http://www.waste-management-world.com/index/display/article-display/304406/articles/waste-management-world/volume-8/issue-4/features/msw-management-in-europe.html) (Valdés et al., 2014).

Food Packaging. http://dx.doi.org/10.1016/B978-0-12-804302-8.00003-0

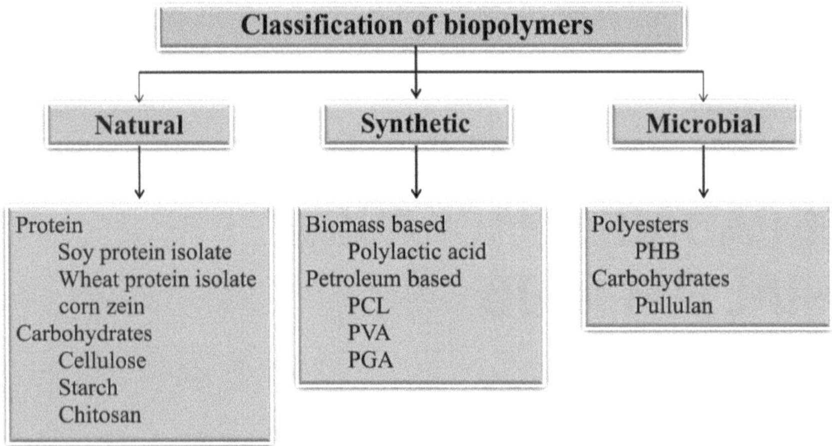

Figure 3.1. Classification of biopolymers. *PCL*, Polycaprolactone; *PVA*, polyvinyl alcohol, *PGA*, polyglycolic acid; *PHB*, polyhydroxybutyrate.

Growing concern over environmental issues and limited number of disposal methods drive the search for biodegradable polymers (biopolymers)-based alternatives to petroleum-based polymers. Biopolymers can be classified based on their origin as natural, synthetic, and microbial, as shown in Fig. 3.1 (Rhim et al., 2013).

Examples of biodegradable polymers include proteins such as soy protein isolate, wheat protein isolate, corn zein, wheat gluten, and gelatin, whereas carbohydrates include starch cellulose, chitosan, and agar. Polylactic acid, polycaprolactone, and polyvinyl alcohol are a few examples of synthetic biopolymers. Microbial polyester biopolymers include polyhydroxybutyrate (PHB) and carbohydrates include pullulan and curdlan (Rhim et al., 2013).

But polymers, if used alone, have poor tensile, mechanical, and barrier properties, which are responsible for their failure as strong and successful packaging materials. Polymers also fail to walk hand in hand with today's increasing demands and expectations of consumers. Thus, to improve their mechanical and barrier properties and to impart them with novel characteristics, polymers are reinforced with nanofillers that interact at various levels. Many terms have been used for advanced multifunctional packaging like smart, active, and intelligent packaging. The idea behind using these terminologies is to explain selective responsiveness of individual packaging when placed in different environments. This responsiveness enables the intelligent packaging to sense changes in certain properties of food products and report it to the customer (Robertson, 2006).

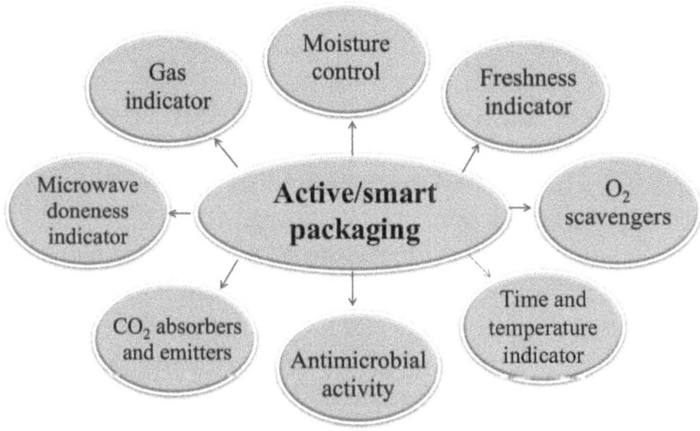

Figure 3.2. Ideal properties of smart packaging.

Active packaging materials comprising of bionanocomposites play multiple roles, including those of protecting the food item from external environment, as well as, increasing their shelf life by maintaining their quality for a longer period of time. These smart packaging materials prevent oxidation of the food product, thus delaying its deterioration. They may also act as moisture barriers, antimicrobial agents, freshness indicator, and so on (Fig. 3.2). There are several reports of active and intelligent packaging systems that are dealt with in detail with individual types of nanocomposites in the following sections (Brody et al., 2008).

2 Nanoreinforcements

2.1 Nanoclays

Nanoclay, also known as layered silicates, is one of the most widely used and studied nanoagents that is used to prepare nanocomposites for food packaging. Advantages of nanoclays include their wide availability, easy processability, good performance, and lower cost (Ray et al., 2006). The nanoclays that are used as nanofillers in nanocomposites are usually bidimensional platelets having lengths of several micrometers and a very minute thickness of around 1 nm (Azeredo et al., 2011).

Nanoclays are usually used to prepare nanocomposites of polymeric materials, such as starch. The interaction between nanoclays and polymers can produce two types of nanocomposites, the first type includes intercalated nanocomposites, which are formed due to the penetration of chains of polymer into the regions between the layers of the clay, forming an ordered multilayer structure with alternating layers of polymer and clay. The second type of nanocomposite is exfoliated nanocomposites, which result from extensive polymer penetration. In this type, the clay layers get delaminated and randomly dispersed in the polymer matrix. Reportedly, exfoliated nanocomposites are known to exhibit the best properties because of the optimal interactions between polymer and clay (Azeredo et al., 2011). Montmorillonite (MMT) is the clay that has been most widely used as a nanofiller for preparing nanocomposites for food packaging. Its chemical general formula is $Mx(Al_4\text{-}xMg_x)\ Si_8O_{20}(OH)_4$. MMT exhibits excellent reinforcing filler abilities because of its high surface area and large aspect ratio, which ranges between 50 and 1000 (Uyama et al., 2003).

Nielsen (1967) explained the reason behind the success of clay as a nanofiller for food packaging materials. According to him, clay, when used as a filler, improves the barrier properties of nanocomposites, probably by increasing the tortuosity of the diffusive path for permeants (such as, moisture, oxygen, and microbes), and presenting a longer path for them to travel to be able to diffuse through the film. This theory is further supported by other authors also (Adame and Beall, 2009; Mirzadeh and Kokabi, 2007). The mineral clays have also been explored as biocide carriers for inorganic biocides, such as, silver, zinc, copper, and magnesium (Patakfalvi and Dékány., 2004; Patil et al., 2005). Nanoclays have also been functionalized to impart antimicrobial properties (Thostenson et al., 2005). Chandrasekaran et al. (2011) produced aminopropyl functionalized magnesium phillosilicates (AMP clay) by sol–gel synthesis method. This nanocomposite was found to strongly inhibit the growth of microorganisms, such as, *Escherichia coli*, *Staphylococcus aureus*, and *Candida albicans* (Rhim et al., 2013).

2.2 Nanocellulose

Nanocellulose is also a suitable candidate to be used as a reinforcement material. It is useful in producing nanocomposites having high strength and low weight. Cellulose is also a low-cost material. Moreover, it is biodegradable, environment friendly, and easy to recycle by combustion (Azeredo et al., 2011; Helbert et al. 1996). Two types of reinforcements can be derived from cellulose, namely microfibrils and whiskers (Azizi Samir et al., 2005). Microfibrils are the nanofibers that are synthesized from cellulose in plants or animals. Microfibrils are actually the bundles of molecules, elongated and stabilized by hydrogen bonding. The diameters of microfibrils are in the nanosize range, around 2–20 nm, and lengths are in the range of micrometers. Each microfibril is made up of two parts: crystalline and amorphous. The crystalline parts of the microfibrils are the whiskers, which are also known as nanocrystals, nanorods, or rodlike cellulose microcrystals. Length of the whiskers range from 500 nm to 1–2 μm, and diameter from 8 to 20 nm or less, resulting in high aspect ratios. The whiskers can be isolated from microfibrils by several methods. The most commonly used method is acid hydrolysis. Acid hydrolysis method to separate crystalline part (ie, whiskers) from microfibrils basically consists of removing the amorphous parts of the fibrils by dissolving them in acid (usually sulfuric acid), leaving the crystalline regions intact (Azeredo, 2009).

Azeredo (2009) studied the effect of cellulose nanofibers on the physical properties of mango puree edible films and found that cellulose nanofibers, when used as filler, effectively increased the tensile strength and Young's modulus of the mango puree films. Helbert et al. (1996) explained this effect of cellulose nanofibers on the modulus. According to them, an increase in the modulus is not only a result of the geometry and stiffness of the nanofibers, but also of the formation of a fibrillar network within the matrix of the polymer. Cellulose nanofibers seem to link through hydrogen bonds.

2.3 Silicon Dioxide (SiO_2)

Nanoparticles of SiO_2, also known as silica, have been used as nanoreinforcements to prepare nanocomposites for food packaging to improve mechanical and/or barrier properties of polymer films (Azeredo, 2009). In a few studies, researchers observed that the presence of grafting polymers on the surface of SiO_2 improved the tailorability of the composites, which means that presence of different species of grafting monomers affects the tensile properties and interfacial interactions (Zhang and Rong, 2003; Wu et al., 2002). In the study conducted by Wu et al. (2002), it was observed that SiO_2 nanoparticles improved tensile properties of the material including strength and modulus as well as elongation, when added to polypropylene matrix. Xiong et al. (2008) also obtained similar results where tensile properties including elongation were improved for a starch matrix after the addition of SiO_2 nanoparticles. Water absorption by starch matrix was also found to be decreased. Vladimiriov et al. (2006) incorporated SiO_2 in an isotactic polypropylene (iPP) matrix using maleic anhydride grafted polypropylene (PP-g-MA) as a compatibilizer. SiO_2 was found to increase the storage modulus of the iPP matrix, thus making the material stiffer, and also improving the oxygen barrier of the matrix. Jia et al. (2007) prepared nanocomposites of polyvinyl alcohol and SiO_2 by radical copolymerization of vinyl SiO_2 nanoparticles and vinyl acetate. The nanocomposites possessed improved thermal and mechanical properties as compared to the pure polyvinyl alcohol. This was a result of strong interactions between SiO_2 and the polymer matrix via covalent bonding. Later, Tang et al. (2008) prepared biodegradable films composed of starch, polyvinyl alcohol, and SiO_2. As the concentration of SiO_2 was increased, the tensile properties and water resistance of the films were improved. An increase in the intermolecular hydrogen bonds, as well as formation of C–O–Si groups, between SiO_2 and starch, or SiO_2 and polyvinyl alcohol was also observed, which improved the compatibility and miscibility between the components of the film.

2.4 Carbon Nanotubes

Carbon nanotubes are nanoscale tubes that may consist of one atom–thick single wall nanotube, or several concentric tubes known as multiwalled nanotubes (Azeredo, 2009). Carbon nanotubes possess extraordinarily high aspect ratios and elastic moduli (Zhou et al., 2004). According to Lau and Hui (2002), carbon nanotubes reportedly have theoretical elastic modulus and tensile strength values as high as 1 TPa and 200 GPa, respectively. In an experiment conducted by Kim et al. (2008), carbon nanotubes were modified by introducing carboxylic acid groups on their surfaces with the purpose of enhancing intermolecular interactions between the poly(ethylene-2,6-naphtalene) matrix and carbon nanotubes. Carbon nanotubes were found to greatly improve the thermal stability, as well as, tensile strength and modulus of poly(ethylene-2,6-naphtalene), even at a concentration as low as 0.1 wt.%. Several other experiments have also been conducted to study the effect of carbon nanotubes on polymer films. Polyvinyl alcohol is one such polymer that has been found to have its tensile strength/modulus improved by the addition of carbon nanotubes (Bin et al., 2006; Chen et al., 2005). Other polymers that have been reinforced with carbon nanotubes to improve their tensile properties including polypropylene (Lopez Manchado et al., 2005) and polyamide (Zeng et al., 2006).

2.5 Other Agents

2.5.1 *Antimicrobial Agents*

Antimicrobial packaging is a type of active packaging that prevents or retards the growth of microbes in the food products. This is achieved by reinforcing the packaging films with antimicrobial agents. Antimicrobial food packaging materials are expected to extend the lag phase and reduce the growth rate of microorganisms, thus increasing the shelf life of the food products, while maintaining their quality and safety. Antimicrobial agents are mainly of two types, organic and inorganic. Organic antimicrobial agents are usually less stable, particularly at high temperature and pressure conditions, as compared to inorganic antimicrobial agents. Thus, organic ones are not suitable and present a major obstacle for the formulation of active or smart packaging films. Consequently, inorganic antimicrobial agents, such as, metals and metal oxides have gained attention recently. This is because of their ability to withstand harsh processing conditions, such as, high temperatures and pressures (Emamifar, 2011; Zhang et al., 2007).

Direct addition of antimicrobial agents might result in a loss of activity due to leaching into the food products and undesirable reactions with food components, such as, lipids or proteins. Therefore, the packaging films reinforced with antimicrobial agents can prove to be more efficient. This will help in achieving a controlled and sustained migration of the antimicrobial compound into the food, thus allowing for inhibition of microbes, as well as, residual activity over time during the transport, storage, and shelf life of the food product (Emamifar, 2011; Mauriello et al., 2005).

2.5.1.1 Silver Nanoparticles

Silver has been known for its antimicrobial properties for a long time and is used for the treatment of burns, wounds, and several bacterial infections. It has been used in the form of metallic silver, silver nitrate, silver sulfadiazine, and so on. Nanosizing of any metal changes its physical and chemical properties drastically, more often amplifying its therapeutic effects. Silver nanoparticles thus have great a potential as antimicrobial agents. Silver usually shows action against those bacteria that have also developed resistance against antibiotics (Emamifar, 2011; Rai et al., 2009). The probable mechanisms of antimicrobial activity of silver nanoparticles include generation of reactive oxygen species (ROS) that induce oxidative stress and cause the degradation of the membrane structure of the cell; and the release of ions from the surface of the nanoparticles that may cause death of the microbes due to their binding to cell membrane (Sharma et al. 2015; Emamifar et al., 2011). Silver ions that are released from the surface of nanoparticles are also reported to interact with thiol groups present in cell proteins, inducing bacterial inactivation, condensation of DNA molecules, and loss of their replication ability (Sharma et al. 2015; Emamifar et al., 2010).

2.5.1.2 Zinc Oxide Nanoparticles

Few of the ceramic substances, such as zinc oxide (ZnO), calcium oxide, and magnesium oxide are found to show antimicrobial properties and thus can be used to prepare smart packaging systems (Yamamoto, 2001). These ceramic compounds contain minerals that are essentially required by human bodies and they exhibit high potency (Sawai and Yoshikawa, 2004). ZnO nanoparticles are widely used in a number of applications such as paints, cosmetics, drug delivery systems, and fillings in medical materials. ZnO is known to exhibit antibacterial activity that increases as the particle size decreases. Antibacterial action of ZnO can be stimulated by visible light. ZnO nanoparticles have also been incorporated in various polymers, where they effectively absorb ultraviolet light without reemitting it as heat and therefore improves the stability of polymer composites (Martirosyan and Schneider, 2014). Also, ZnO is recommended as a generally recognized as safe (GRAS) material by the US Food and Drug Administration (FDA). Mechanisms of antimicrobial effect of ZnO nanoparticles are similar to that of silver nanoparticles, that is, induction of oxidative stress due to generation of reactive oxygen species, which may interact with proteins, DNA, and other components of bacterial cells leading to death; and binding of ZnO to the membrane of the microbial cells. Other mechanisms may include disorganization of cellular membrane due to the accumulation of ZnO nanoparticles (Sharma et al., 2015; Emamifar, 2011). Jin et al. (2009) studied several approaches including film, powder, PVP capped and coating, for the application of ZnO nanoparticles in food systems. They observed that ZnO nanoparticles exhibited antimicrobial effects against *Listeria monocytogenes* and *Salmonella enteritidis* in liquid egg white and in culture media.

2.5.1.3 Titanium Dioxide (TiO_2) Nanoparticles

Titanium dioxide (TiO_2), also called titania, is another nontoxic ceramic compound, which can be used as an antimicrobial agent in food packaging materials. TiO_2 is approved by the FDA for use in human food, drugs, cosmetics, and food contact materials. Cidal effects of TiO_2 have been reported against *E. coli*, *Salmonella choleraesuis*, *Vibrio parahaemolyticus*, *L. monocytogenes*, *Pseudomonas aeruginosa*, *S. aureus*, *Diaporthe actinidiae*, and *Penicillium expansum*. The development of TiO_2-coated or incorporated food packaging and food preparing equipment has also received attention (Emamifar, 2011).

Chawengkijwanich and Hayata (2008) developed a TiO_2 powder coated packaging film and tested its ability to inactivate *E. coli* both in vitro and in actual tests, using two different particle sizes and two types of illumination at different intensities. The results showed that both ultraviolet A (UVA) alone and TiO_2 coated OPP film combined with UVA reduced the number of *E. coli* cells in vitro, but that the reduction of *E. coli* cell numbers was greater by TiO_2-coated OPP film combined with UVA. The antimicrobial effect of TiO_2-coated film was dependent on the UVA light intensity (0, <0.05, and 1 mW cm^{-2}) and the kind of artificial light (black-light and daylight fluorescent bulbs), but it was independent of the particle size of TiO_2 coating on the surface of OPP film. The results of actual tests showed that the TiO_2-coated film could reduce the microbial contamination on the surface of solid food products and thus reduce the risks of microbial growth.

2.5.2 Moisture Control Agents

A certain amount of moisture is essential for most of the food items, but too much moisture can prove to be detrimental. Also, complete absence of moisture is vital for dry food products. Too much moisture can result in caking of powdered products, softening of crispy products, such as crackers, and moistening of hygroscopic products, such as, sweets and candy. Also, excessive moisture loss can lead to desiccation of the food product. Thus, moisture-control agents help in overcoming problems caused by presence of excessive moisture, as well as, excessive loss of the moisture from food products. These agents control and regulate moisture activity thereby preventing microbial growth, controlling rate of oxidation, and preventing caking and softening of food products (Brody et al., 2008). Polyurethane films have been found to show moisture-responsive properties in a number of studies (Huang et al., 2005; Yang et al., 2004). Sidorenko et al. (2007) used polyglycidyl methacrylate-based water-responsive hydrogels for constructing a responsive actuator. In another study, the composite films made up of polypyrrole and polyol-borate were found to show fast moisture-responsive properties (Ma et al., 2013). In a recent study, Zhang et al. (2015) used sustainable cellulose as the starting material to fabricate self-standing moisture-responsive films. The cellulose was modified by stearoyl moieties to produce cellulose stearoyl esters (CSEs), which possessed diverse degrees of substitution. The CSE films that had a low degree of substitution of 0.3 (CSE 0.3) showed moisture-responsive properties, while films of CSEs that had higher degrees of substitution of 1.3 or 3 (CSE 1.3 and CSE 3) did not show any moisture-responsive property. Also, CSE 0.3 films could be reversibly folded and unfolded in rhythmical bending motions as a result of the absorption and desorption of water molecules over the surface of the film. Moisture-responsive films having nonwetting surfaces were produced by coating the nanoparticles of CSE 3 onto the films of CSE 0.3. These films also possessed the characteristic of performing quick reversible bending movements and shape transition. Moreover, the bilayer films that consisted of a layer of CSE 0.3 at one side and a layer of CSE 3 at the other were found to exhibit dual characters of moisture, as well as, temperature responsiveness.

2.5.3 Oxygen Scavengers

The presence of oxygen in any food package can prove to be harmful as oxygen triggers the oxidative reactions and facilitates growth of certain microbes, all resulting in spoilage or deterioration in quality of food, foul odors, unacceptable color change, change in flavor, reduction in nutritional values, and so on. These problems, which are caused by the presence of oxygen, can be avoided by the use of oxygen scavengers. Oxygen scavengers are the agents that reduce the concentration of oxygen by reacting with it, there by inhibiting various oxidative reactions that could deteriorate the quality of food. The most commonly and widely used oxygen scavenger is ferrous oxide. Other oxygen scavengers are catechols, ascorbic acid, sulfites, photosensitive dyes, nylons, unsaturated hydrocarbons, enzymes such as glucose oxidase, and so on (Brody et al., 2008).

3 Synthesis of Nanocomposites

There are various approaches that can be used to synthesize nanocomposites of biopolymers. The preparation method poses a significant effect on the properties of the nanocomposites.

Following are the three major pathways through which nanocomposites may be formed:

1. *In situ polymerization*: In situ polymerization, also known as interlamellar method, involves the combination and swelling of fillers by absorption of a liquid monomer, which results in penetration of monomer units in between the layers of filler (eg, silicates). After penetration of monomer, polymerization is initiated by heat, radiation, or incorporation of an initiator, which ideally locks the exfoliated filler particles in the resulting polymer matrix.
2. *Solvent intercalation/exfoliation method*: The solvent intercalation method is used to form both intercalated and exfoliated nanocomposites. In this method, the filler particles are first swollen in a solvent and the polymer (intercalant) is also dissolved in the solvent separately. Next, the filler solution is

added to the polymer solution. The polymer chains then intercalate and displace the solvent within the interlayer of the fillers. The solvent is then allowed to evaporate.

3. *Melt intercalation/exfoliation method*: In melt intercalation method of nanocomposites formation, the filler and polymer are mixed together at a temperature above the melting point of the polymer. Then they may be kept at that particular temperature for a period of time, put under shear, or other necessary conditions to cause the intercalation and the exfoliation of the filler.

Out of these three commonly used methods for preparation of nanocomposites, the most suitable and appealing method is the melt intercalation/exfoliation method, owing to its versatility, its compatibility with current polymer processing equipment, such as, extrusion and injection molding, and its environmentally benign character due to the absence of solvents. Also, the melt intercalation method allows for the use of polymers that are otherwise not suitable for in situ polymerization or solution intercalation method (Brody et al., 2008; Rhim and Ng, 2007).

4 Cellulose Nanocomposites

Cellulose, the main component of the plant cell wall, is one of the most abundant biopolymers on earth. Cellulose inherently is a hydrophilic polymer of glucose and has been traditionally used in the paper industry. Nanocellulose adds a new dimension to the application of cellulose. Nanocellulose can be MFC/NFC/CNF (microfibrillated cellulose/nanofibrillated cellulose/cellulose nanofibers), CNC/CNW (cellulose nanocrystals/cellulose nanowhiskers), and BC (bacterial cellulose). CNF are prepared mechanically by high pressure homogenization or milling. They have both, amorphous and crystalline regions. CNFs are mainly produced by acid hydrolysis, during which amorphous regions are dissolved and only crystalline cellulose is left. BC is a biotechnological product and cellulose fibers are of bacterial origin (synthesized by bacteria). Due to the nature of their origin, they are of the highest purity and have been widely studied for their biomedical applications.

Nanocellulose has been produced from a wide range of lignocellulosic sources that include wood, lignocellulosic biomass of different origins, and even some by-products of food processing industry. One of the most specific characteristics of cellulose is that each of its glucose unit bears three hydroxyl groups, which endows nanocellulose a reactive surface covered with numerous active hydroxyl groups. The ability of these hydroxyl groups to form hydrogen bonds plays a major role in the formation of fibrillar and semicrystalline packing, which governs the important physical properties of this highly cohesive material. Nanocellulose-based composites in comparison to traditional composites show improved and tunable barrier properties while keeping their biodegradable nature intact. First, because of the smaller size of nanofibers, the surface area-to-volume ratio is relatively higher than that of microparticles; and second, the nanosize of the fibers improve the barrier properties of the film by forming dense network in combination with sufficiently strong particle-polymer bonds. The other ways to improve moisture resistance of inherently hydrophilic cellulose can be by pretreatment before homogenization, posttreatment, and chemical grafting (Dufresne, 2013). Gas permeability can be improved by forming nanocomposites with other fillers (eg, clay) or chemical modification of cellulose especially for utilization in moist conditions which has been a major limitation of cellulose.

The effect of applying edible coatings to mature-green tomatoes on the survival of *Salmonella montevideo* on the surface and in core tissue was investigated by Zhuang et al. (1996). A coating of hydroxypropylmethylcellulose (HPMC) significantly reduced the number of viable *S. montevideo* cells on the surface of tomatoes. However, only a 2 log unit reduction was achieved in the core tissue. The addition of citric acid, acetic acid, or sorbic acid at a concentration of 0.2–0.4% to HPMC did not significantly improve the bactericidal activity. Concentrations of 72–88% ethanol in HPMC were found to be most effective in inactivating *S. montevideo* on the surface of tomatoes. Similar inactivation of *S. montevideo* in the core tissue was achieved by dipping tomatoes in 55–88% ethanol solutions. Also, use of HPMC coating retarded the rate of loss of firmness and change in color of tomatoes stored at 20°C for up to 18 days. This study demonstrated the antimicrobial properties of cellulose and opened up new avenues for further investigation of cellulose in the field of food packaging.

Bruce et al. (2005) used primary cellular material at the laboratory scale to develop a method of producing purified cellulose cell wall fragments and cellulose microfibrils. They prepared novel composite materials with a range of matrix binders and determined their tensile properties. Swede root was processed and used to produce fragments of cell wall. The separation of microfibrils from the cell wall fragments was accomplished using a high-pressure food homogeniser. Four types of composite matrices were prepared from wet state cellulose. Out of the composites that were prepared from fragments of purified cell wall, the best strength was found to be possessed by an acrylic matrix (125 MPa) while the composite of purified cell wall fragments and polyvinyl alcohol showed the best stiffness (15.32 GPa). For the composites prepared from fibrilized material of the cell wall, polyvinyl alcohol gave the best results in terms of tensile strength (145 MPa) and tensile modulus (8.9 GPa). The composites made up of plant-derived hemicellulose matrix showed a tensile strength value of 100 MPa and a tensile modulus value of 6.0 GPa. As all the values obtained were in the acceptable range, the researchers successfully demonstrated the applicability of cellulose in preparation of composites.

Iwamoto et al. (2005) carried out the fibrillation of pulp fiber by two methods, a high-pressure homogenizer treatment and a grinder treatment. The grinder treatment was found to be a successful method for carrying out fibrillation of wood pulp fibers into nanofibers. The nanofibers demonstrated promising characteristics as reinforcement material for optically transparent composites. Due to the very small size, the nanofiber-reinforced composite retained the transparency of the matrix resin even at a high fiber concentration of 70 wt.%. Since the CNFs are an aggregate of semicrystalline-extended cellulose chains, their addition also results into significant improvement in the thermal expansion properties of plastics, as well as, maintaining the ease of bending.

Ifuku et al. (2007) acetylated the bacterial CNFs to enhance the properties of optically transparent composites of acrylic resin reinforced with the bacterial CNFs. Due to acetylation, the surface properties of the nanofibers were changed and the moisture content of the composite decreased by approximately 66% than that of unacetylated composite. Also, the acetylation reduced the thermal expansion coefficient of a BC sheet to one-third of the initial value.

In an interesting study, an all-cellulose composite, in which both the fibers and the matrix were cellulose, was prepared by changing the solubility of the matrix cellulose into the solvent from that of the fibers, by pretreatment of the matrix cellulose. The tensile strength of uniaxially reinforced all-cellulose composite was found to be 480 MPa at 25°C, and the dynamic storage modulus was found to be as high as 20 GPa at 300°C. These values are higher than those of conventional glass-fiber-reinforced composites. In addition, the linear thermal expansion coefficient was about 10^{-7} K^{-1}. This all-cellulose composite showed several considerable advantages, it was composed of sustainable resources; it was biodegradable; there were fewer interfaces between the fiber and the matrix; and it possessed excellent mechanical and thermal performance during use (Nishino et al., 2004).

Ho a et al. (2012) synthesized NFC with cationic functional groups. This trimethylammonium modified NFC (TMA-NFC) was used for the preparation of composites with various layered silicates. These belonged to the groups of MMT, kaolin, talc, vermiculite, and mica. The respective composites were prepared by high-shear homogenization followed by filtration and hot-pressing. Owing to the electrostatic attraction of cationic cellulose fibrils and anionic silicate layers, the interactions between TMA-NFC and the layered silicates were very pronounced. This mutual interaction between TMA-NFC and layered silicate, however, was influenced not only by layered silicate properties but also by the composite preparation method, as the researchers have described in the detailed study.

In another study performed by the same group (Ho et al. 2012), composites of TMA-NFC and layered silicates were prepared by high-shear homogenization followed by pressure filtration and vacuum hot-pressing, which gave rise to particularly homogeneous dispersion of the silicate particles. Thirteen different clays and micas were employed. Good interactions between TMA-NFC and layered silicates were obtained due to electrostatic attraction between cationic fibrils and anionic silicate layers, and even favored by high-shear homogenization process. Layered silicates exerted a pronounced influence on the water vapor barrier and mechanical properties. Importantly, it was found that the insertion of layered silicate particles did not improve automatically the barrier properties as indicated by the commonly used "fiber–brick composite" model. Mica R120 at a 50 wt.% loading in composites with TMA-NFC matrix showed 30 times improved water vapor permeability and 5 times higher E-modulus compared to commercially used base paper.

Liquid ammonia has been used to treat films of NFC, TMA-NFC, and their composites with vermiculite. Crystal structure, mechanical properties, water vapor permeation, and water vapor adsorption of the resulting materials were investigated. With the exception of TMA-NFC and vermiculite composites, pronounced effects on the addressed mechanical properties arose after exposure of the materials to ammonia. Furthermore, treatment of composite films with ammonia led to a distinct decrease in water vapor permeation. Surprisingly, TMA-NFC and vermiculite composites films showed the best water vapor barrier properties, highest tensile strength, and highest elastic modulus after treatment with liquid ammonia. This is considered as a consequence of electrostatic attraction between the positively charged ammonium groups in TMA-NFC and the anionic silicate layers of vermiculite (Ho et al., 2013).

Özgür Seydibeyoğlu and Oksman (2008) fibrillated cellulose from micro to nanoscale and evaluated how these microfibers and nanofibrils affected the mechanical and thermal properties of thermoplastic polyurethane. They prepared composite materials using compression molding, by stacking the cellulose fiber mats between polyurethane films. The results showed that both microfibers and nanofibrils reinforced the polyurethane and provided better heat stability. The addition of 16.5 wt.% of cellulose nanofibrils to polyurethane increased the strength by approximately 5 times and the stiffness by around 30 times. These results were very promising in terms of obtaining fibrils with a novel processing method and by improving the mechanical and thermal properties of polyurethane. This is also expected to expand the application areas of polyurethanes.

Edible cellulosic films made with HPMC have proven to be inadequate moisture barriers. To improve its water vapor barrier properties, Coma et al. (2001) incorporated different hydrophobic compounds into the HPMC matrix. They included few fatty acids and derivatives into the film-forming solution prior to film formation. Stearic acid was chosen because of its high capacity to considerably reduce the water vapor transmission rate. Antimicrobial action of edible HPMC film was obtained by the incorporation of nisin into the film-forming solution. Nisin is an antimicrobial peptide that is effective against Gram-positive bacteria. The inhibitory effect of this bacteriocin was tested for inhibition of *Listeria innocua* and *S. aureus*. The stearic acid was found to reduce the inhibitory effect of active HPMC film against both of the selected strains. The researchers explained this phenomenon by electrostatic interactions between the cationic nisin and the anionic stearic acid. Further studies showed that antimicrobial effect of the films varied with the nature of the hydrophobic compound incorporated; film without lipid exhibited the highest antimicrobial activity, methylstearate film, lower, and stearic acid film, the least. This further corroborated the idea of electrostatic interactions.

Recently, El-Wakil et al. (2015) developed bionanocomposites by casting/evaporation of wheat gluten, CNC, and TiO_2 nanoparticles. They studied the effect of addition of different percentages of CNC, and TiO_2 on tensile strength, Young's modulus, and water sensitivity. A significant improvement in the studied properties was observed when 7.5% CNC and 0.6% TiO_2 were added to wheat gluten. Wheat gluten/CNC (7.5%/0.6%) TiO_2 blend suspension was used to coat commercial packaging unbleached kraft paper sheets via one, two, and three coating layers. A significant enhancement of 56 and 53% in breaking length and burst index, respectively, was achieved for three layers coated paper. The antimicrobial activity of the coated papers, against *Saccharomyces cervisiae*, Gram-negative bacteria *E. coli*, and Gram-positive bacteria *S. aureus* was investigated and expressed in terms of percent reduction of surviving number of the tested organisms. More than 98.5% reduction of surviving number was observed against the test organisms as compared to TiO_2-free coated paper.

5 Starch Nanocomposites

Starch has been considered as the most promising and economic member of biodegradable polymers' category. This polymer owes its popularity and wide usage to its several advantages such as wide and easy availability, total compostability without formation of any toxic residues, and cost as low as less than 1 Euro kg^{-1}. Another major advantage of starch is its inherent complete biodegradability and it is even known to promote the biodegradability of a nonbiodegradable plastic when blended (Abreu et al., 2015; Sun et al., 2014; Goyal and Goyal, 2012; Avella et al., 2005).

Starch is a widely available semicrystalline form of carbohydrate, stored in granules as a reserve in most plants, such as, potato, rice, corn, and cereal grain to name a few. Starch is actually a mixture of two polymers of α-glucose, that is, linear amylose (poly-α-1,4-D-glucopyranoside) and highly branched amylopectin (poly-α-1,4-D-glucopyranoside and α-1,6-D-glucopyranoside). Amylose molecules consist of 200–20,000 glucose units forming a helix due to the bond angles present among the glucose units. Amylopectin is a highly branched polymer containing short side chains of 30 glucose units attached to every 20–30 glucose units along the chain. Amylopectin molecules may contain up to 2 million glucose units. The amylose remains linked by α-(1,4)-linkages whereas amylopectin has α-(1,4)-linked backbone and α-(1,6)-linked branches. Amylose forms the amorphous part of the starch, whereas the amylopectin predominantly imparts the crystallinity. The relative amounts of amylose and amylopectin varies with the source of the starch. For example, starch from corn granules contains approximately 30% amylase and 70% amylopectin. The different ratios of amylose to amylopectin impart very different properties to the materials (Tang et al., 2012; Ren et al., 2009; Avella et al., 2005).

Despite several advantages, starch alone does not make a good candidate for packaging material because of its poor mechanical properties, strong hydrophilic behavior (thus poor moisture barrier), and crystalline nature leading to poor process capability.

To improve mechanical properties of starch, plasticizers, such as, glycerol and other low molecular weight polyhydroxy compounds, urea and polyethers are added. Starch exhibits thermoplastic behavior when a plasticizer is added. Plasticizers efficiently provide stability to starch by reducing the intramolecular hydrogen bonding. Amount of plasticizer and amylase to amylopectin ratio can be varied to obtain thermoplastic starch (TPS) having different water absorbing tendency, water solubility, and viscosity. Though plasticizers improve the processibility and flexibility of TPS, it is still not suitable to be used in food packaging applications because of its poor mechanical properties and sensitivity to moisture (Tang et al., 2012). Thus, to overcome these problems, preparing nanocomposites of starch is a promising strategy and can be used in order to make it a suitable packaging material having properties of providing good moisture and oxygen barrier and being mechanically strong along with the inherent advantages that starch already possesses. In recent years, starch nanocomposites have been prepared and have demonstrated their success as a smart packaging material.

Recently, Abreu et al. (2015) prepared nanocomposites of quaternary ammonium salt modified MMT C30B with starch; nanocomposites of silver nanoparticles with starch; and nanocomposites of both, silver nanoparticles and C30B, with starch. They studied color and opacity, permeability toward water vapor and oxygen, contact angle, and dynamic mechanical analysis. Results were correlated to the observations with the incorporation of C30B and silver nanoparticles. Films demonstrated antimicrobial activity against *S. aureus, E. coli,* and *C. albicans* without any significant differences between silver nanoparticle concentrations. The presence of silver nanoparticles imparted bacteriostatic effect, as well as, also enhanced dispersion of C30B, and changed the surface polarity, too. Incorporation of C30B in starch film was found to improve the mechanical and gas barrier properties. The migration of components from the nanofilm of starch was found to be insignificant and within the legal limits. Thus, researchers demonstrated that the starch films incorporated with C30B and silver nanoparticles have the potential to be used as a suitable packaging nanostructured material.

Sun et al. (2014) carried out the characterization of edible corn starch films, which were impregnated with calcium carbonate nanoparticles. They found that the concentration of calcium carbonate nanoparticles significantly affected the tensile properties of the nanocomposite films. The addition of calcium carbonate nanoparticles to the corn starch films greatly increased tensile strength from 1.40 to 2.24 MPa, elongation from 79.21% to 118.98%, and Young's modulus from 1.82 to 2.41 MPa. Increase in these mechanical properties of corn starch films with the addition of calcium carbonate nanoparticles indicated that calcium carbonate nanoparticles could be used to improve strength and flexibility of the film to a great extent. The incorporation of calcium carbonate nanoparticles also increased the opacity of films, lowered the degree of water vapor permeability, and film solubility value as compared to those of the corn starch films. This suggested that a smooth and compact structure was formed between nanoparticles and corn starch, which could be confirmed by scanning electron microscopy. The addition of calcium carbonate nanoparticles to corn starch films also increased the melting temperature of films, as determined by differential scanning calorimetry. Results of this study indicated that the 0.06% concentration of calcium carbonate nanoparticles may serve as the best option for the desirable properties of edible films.

In another experiment, Sadegh-Hassani and Nafchi (2014) used the casting method to prepare bionanocomposite films of potato starch reinforced with halloysite nanoclay. The composition included potato starch with 40% (w/w) of a mixture of sorbitol/glycerol (weight ratio of 3:1 as plasticizer) with nanoclay (0–5% w/w). Results of the studies performed revealed that mechanical properties of films were improved by increasing the concentration of nanoclay. Tensile strength was found to be increased from 7.33 to 9.82 MPa, and elongation at break decreased from 68.0% to 44.0%. Solubility in water decreased from 35% to 23%, and heat seal strength increased from 375 to 580 N m^{-1}. Also incorporation of clay nanoparticles in the structure of biopolymer decreased permeability of the gaseous molecules.

Taghizadeh et al. (2013) investigated the effect of concentration of sodium montmorillonite clay (MMT-Na) on the physical properties and extent of enzymatic hydrolysis of polyvinyl alcohol: starch: carboxymethyl cellulose nanocomposites using the enzyme α-amylase. As revealed by the results of this work, the films with MMT-Na concentration of 5 wt.% exhibited a considerably reduced extent and rate of starch hydrolysis. The authors attributed these results to interactions between polyvinyl alcohol: starch: carboxymethyl cellulose and MMT-Na that further prevented enzymatic attack on the remaining starch phases within the blend. The rate of glucose production from each nanocomposite substrates was found to be decreased greatly as the content of MMT-Na increased from 0% to 5% (w/w). At this concentration of MMT-Na, the films showed the lowest rate of glucose production values. With the increase in the content of MMT-Na from 0% to 5%, the strain required to break decreased significantly (from 35.56% to 5.22%).

Fazli et al. supported the concept of utilization of the nanomaterials in polymer structure to improve their mechanical and physical properties. He proposed that nanocrystalline cellulose has the ability of improving mechanical and physical properties of polymeric films. They employed chemical methods to produce nanocrystalline cellulose and starch bionanocomposite. X-ray diffraction results of the starch biofilm showed that crystalline structure of nanocrystalline cellulose is maintained (Fazli et al. 2013).

Ren et al. (2009) prepared TPS and its nanocomposite (TPS/OMMT) with 15% carbamide, 15% ethanolamine, and different contents of organic activated montmorillonite (OMMT) by twin-screw extruder. Results of Fourier transform infrared spectroscopy and wide angle X-ray diffraction showed that the alkylamine in dodecyl benzyl dimethyl ammonium bromide could react with MMT through cation exchange reaction. Scanning electron microscope images revealed that the lower amounts of OMMT was dispersed well in the matrixes of TPS. The carbamide, ethanolamine, and OMMT could destroy the crystallization behavior of starch, but only MMT restrained this behavior, which is favorable for long-term storage. Evaluation of mechanical properties revealed that the tensile strength and moduli of TPS/MMT nanocomposites were better than those of TPS alone, while the elongation at break was decreased with the increasing concentration of OMMT. When the content of MMT was 4%, the tensile strength and modulus of TPS was improved from 4.2 and 42 MPa to 6.0 and 76 MPa, respectively.

Avella et al. (2005) prepared biodegradable starch/clay nanocomposite films for the purpose of being used as food packaging material. Using polymer melt processing techniques, the films were prepared by homogeneously dispersing the nanoparticles of MMT in various starch-based materials. The results of structural and mechanical characterization of starch and clay nanocomposite films revealed a good intercalation of the polymeric phase into clay interlayer galleries, along with an improvement in mechanical parameters, namely modulus and tensile strength. Migration tests were performed and the films were put in contact with vegetables and stimulants to verify the conformity of prepared samples with actual regulations and the European directives on biodegradable materials. Thus, these results demonstrated that biodegradable starch/clay nanocomposite films can be utilized in the food packaging sector owing to their low overall migration limit.

6 Chitosan Nanocomposites

Chitosan, one of the strongest natural polymers, is the second-most abundant natural polysaccharide. The favorable properties of chitosan which are accountable for its popularity include the ease of obtaining the polymer in different physical forms, its non-toxic nature, biocompatibility, biodegradability, and its ability to form films with antibacterial properties (Dias et al. 2014; Mesquita et al., 2010). Owing to these advantages and its huge abundance, it is extensively used for food packaging, biosensors, water treatment, and biomedical applications, such as, drug delivery, tissue engineering, wound healing, and for use in antimicrobial strategies (Mesquita et al., 2010).

Chitosan is a partially deacetylated derivative of chitin. It is a linear polysaccharide consisting of β-(1,4)-D-glucosamine residues (deacetylated unit) with a variable number of randomly located *N*-acetyl-glucosamine groups (acetylated unit) (Zuo et al., 2013). The formation of chitosan involves derivatization by elimination of the acetyl group of chitin, and the final sugar monomer obtained is the *N*-glucosamine. In its natural form, chitin is predominantly organized in extended chain conformation and assembled in the form of microfibrils. This structural organization is important for the mechanical function of cuticles and exoskeletons of insects and crustaceans. In addition, chitin structures provide support for tissues and organs such as eyes, muscles, throat, and so on. Chitosan is less common in nature, but is present as a cell wall component in filamentous fungi, in which chitosan biosynthesis takes place through deacetylation of chitin (Mushi et al., 2014).

Despite of its unique physical and inherent antibacterial properties, chitosan is not found to be suitable enough to satisfy conditions of an ideal packaging material. Chitosan has a hydrophilic nature and thus chitosan films exhibit significantly higher permeability to water. Its inherent water sensitivity and relatively low mechanical properties, as compared to petroleum-based polymers, restrict the use of chitosan films for food packaging applications. Thus, chitosan needs to be reinforced with nanofillers in order to improve its mechanical and moisture barrier properties, making it a suitable packaging material for food products (Zuo et al., 2013; Mesquita et al., 2010).

In a recent study, an antibacterial and conductive bionanocomposite film consisting of polypyrrole, ZnO nanoparticles, and chitosan was electrochemically synthesized on indium tin oxide glass substrate by electrooxidation of 0.1 M pyrrole in an aqueous solution containing appropriate amounts of ZnO nanoparticles uniformly dispersed in chitosan. This method enabled the room temperature electrosynthesis of bionanocomposite film consisting of ZnO nanoparticles incorporated within the growing polypyrrole/chitosan composite. The morphology of polypyrrole/ZnO/chitosan bionanocomposite was studied by scanning electron microscopy. Indium tin oxide—polypyrrole/chitosan and indium tin oxide—polypyrrole/ZnO/chitosan bioelectrodes were characterized using the Fourier transform infrared technique, X-ray diffraction, and thermogravimetric analysis. The electrical conductivity of nanocomposites was investigated by a four-probe method.

The prepared nanocomposites were also analyzed for antioxidant activity using the 2,2-diphenyl-1-picrylhydrazyl assay. The results demonstrated that by the addition of ZnO nanoparticles, the antioxidant activity of nanocomposites increased significantly. The electrical conductivity of films showed a sudden decrease for lower weight ratios of ZnO nanoparticles (5 wt.%), while it was increased gradually for higher ratios (10, 15, and 20 wt.%). The nanocomposites were analyzed for antibacterial activity against Gram-positive and Gram-negative bacteria. The results showed that the synthesized bionanocomposite was effective against all of the studied bacteria, and its effectiveness was higher for *P. aeruginosa*. The thermal stability and physical properties of bionanocomposite films were increased by an increase in the weight ratio of ZnO nanoparticles, thus promising novel food packaging applications for the electrically conductive polysaccharide-based nanocomposites, particularly owing to their antibacterial properties (Saeideh Ebrahimiasl et al., 2015).

Dias et al. (2014) developed nanocomposites of chitosan and characterized them using a full factorial design with different levels of MMT and encapsulated tocopherol. Encapsulated tocopherol influenced the intercalation of MMT in the chitosan matrix up to the concentration of 10%, resulting in formation of the films, which had reduced water vapor permeability, increased hydrophobicity, and lower equilibrium moisture content. These observations indicated toward the potential of chitosan: MMT nanocomposites as an active food packaging material. But, agglomeration of encapsulated tocopherol occurred at a content more than 10%. Transmission electron microscopy revealed that this agglomeration deteriorated the properties of the prepared films. As the tocopherol content was increased in the films, the films were found to become slightly more yellow, more irregular, and less transparent with a higher haze index.

According to the observation of Mushi et al. (2014), chitosan reinforcement by stiff particles or fibers is usually obtained at the expense of lowered ductility and toughness. Thus, they studied the reinforcement of chitosan film by a new type of native chitin nanofibers. Films were prepared by casting colloidal suspensions of chitin in dissolved chitosan. The nanocomposite films were actually chitin nanofiber networks in chitosan matrix. The polymer matrix nanocomposites were produced in volume fractions of 8, 22, and 56% chitin nanofibers. The results demonstrated favorable chitin–chitosan synergy for colloidal dispersion. Also, the prepared nanocomposites were found to possess lower moisture-absorbing tendency, which can probably be due to the favorable chitin–chitosan interface. The highest toughness (area under stress–strain curve) was observed at 8 vol.% chitin content.

In another experiment performed by Zuo et al. (2013), free-standing graphene oxide–chitosan (GO–chitosan) nanocomposite films were prepared. It was found that the GO–chitosan films were biologically compatible and mechanically reinforced. Through the formation of amide linkages between carboxylic acid groups of graphene oxide and amine groups of chitosan, graphene oxide could get evenly dispersed in the chitosan matrix. The tensile strength of GO–chitosan film was found to be improved by 2.5-fold and Young's modulus was increased by around 4.6-fold, as compared to pristine chitosan film. The glass transition temperature of GO–chitosan composite film changed from 118 to 158°C, as compared to the pristine chitosan. This indicated the enhanced thermal stability of the nanocomposite film.

Abdolmohammadi et al. (2012) investigated the effects of calcium carbonate nanoparticles on the mechanical and thermal properties and surface morphology of polycaprolactone/chitosan nanocomposites. The nanocomposites of polycaprolactone/chitosan/calcium carbonate were prepared using a melt blending technique. Results of tensile measurement studies revealed an increase in the tensile modulus after loading of the calcium carbonate nanoparticles. Tensile strength and elongation at break evaluation demonstrated a gradual improvement with the addition of up to 1 wt.% of calcium carbonate nanoparticles. But for the loading of more than 1 wt.% of calcium carbonate nanoparticles, performance of the aforementioned properties was found to be decreased. The thermal stability was observed to be most enhanced at loading of 1 wt.% of calcium carbonate nanoparticles. The fractured surface morphology of the polycaprolactone/chitosan blend was found to be more stretched and homogeneous in polycaprolactone/chitosan/calcium carbonate nanocomposite. Micrograph of transmission electron microscopy displayed good dispersion of calcium carbonate nanoparticles within the matrix of nanocomposite film.

In another experiment, silver nanoparticles of small sizes were synthesized using the wet chemical reduction method into the lamellar space layer of MMT/chitosan as an organomodified mineral solid support in the absence of any heat treatment. Silver nitrate, MMT, chitosan, and sodium borohydride were used as the silver precursor, the solid support, the natural polymeric stabilizer, and the chemical reduction agent, respectively. MMT was suspended in aqueous silver nitrate/chitosan solution. Silver nanoparticles were formed on the interlayer and external surface of MMT/chitosan. The antibacterial activity of silver nanoparticles in MMT/chitosan was investigated against Gram-positive bacteria, that is, *S. aureus* and methicillin-resistant *S. aureus* and Gram-negative bacteria, that is

E. coli, *E. coli* O157:H7, and *P. aeruginosa* by the disc diffusion method using Mueller Hinton agar at different sizes of silver nanoparticles. All of the synthesized silver/MMT/chitosan bionanocomposites were found to have high antibacterial activities (Shameli et al., 2011).

A new type of biodegradable nanocomposite was obtained by layer-by-layer (LBL) technique using highly deacetylated chitosan and eucalyptus wood (CNWs). Hydrogen bonds and electrostatic interactions between the negatively charged sulfate groups on the whisker surface and the ammonium groups of chitosan were the responsible factors for the growth of the multilayered films. Images of scanning electron microscopy demonstrated high density and homogeneous distribution of CNWs adsorbed on all the chitosan layers. Cross-section characterization of the assembled films indicated an average of approximately 7 nm of thickness/bilayer. These results indicated that the methodology used can be extended to different biopolymers for the design of new bio-based nanocomposites in a wide range of applications such as food packaging (Mesquita et al., 2010).

Apart from cellulose, starch, and chitosan, there are several other biopolymers that have been explored for the preparation of bionanocomposites for the applications in food packaging. A few of them are listed in Table 3.1 with examples.

7 Commercial Products in Active Packaging

Intelligent packaging applications are used in tampering detection (breaching of containment), quality indicators, time-temperature indicators, gas sensing devices, microbial growth, pathogen detection, traceability devices, radio frequency identification labels/tags/chips, product authenticity, and so on. A brief account of few such commercial examples is given in the next sections.

7.1 Moisture Control Packaging

Dri-loc absorbant pads from Cryovac, sealed air corporation is based on super-absorbent polymers located between two other plastic layers (http://www.cryovac.com/NA/EN/food-packaging-products/driloc-produce.aspx).

They offer three advantages:

1. Perforated with tiny one-way valves
2. Juices absorbed from the bottom
3. Prevents food from drying

Similar products, Fresh-R-Pax Absorbent pads, as shown in Fig. 3.3, and Fresh-R-Pax Absorbent trays are also available from Maxwell Chase Technologies, LLC.

Table 3.1 Other Bionanocomposites Used for Food Packaging

S. No.	Polymer	Reinforcement	Results	References
1	Wheat gluten	Cellulose nanocrystals; titanium dioxide nanoparticles	A significant improvement in the tensile strength, Young's modulus and water sensitivity was observed when 7.5% cellulose nanocrystals and 0.6% titanium dioxide were added to wheat gluten. An enhancement of 56 and 53% in breaking length and burst index, respectively, was achieved when wheat gluten/cellulose nanocrystals 7.5%/0.6% TiO_2 blend suspension was coated on commercial packaging unbleached kraft paper sheets. Also, the coated packaging paper sheets were found to have significant antimicrobial activity.	El-Wakil et al. (2015)
2.	Poly(3-hydroxybutyrate)	Zinc oxide nanoparticles	The Young's modulus, tensile and impact strength of the biopolymer were enhanced by up to 43, 32, and 26%, respectively. Nanocomposites exhibited reduced water uptake and superior gas and vapor barrier properties. They also showed antibacterial activity against both Gram-positive and Gram-negative bacteria, which was progressively improved on increasing zinc oxide concentration. The migration levels of composites in both nonpolar and polar stimulants also decreased with increasing nanoparticle content.	Díez-Pascual and Díez-Vicente (2014)
3.	Polylactic acid	Montmorillonite	The addition of organoclay (montmorillonite) improved mechanical and biodegradable properties of polylactic acid. Food contact test showed that the nanocomposites can be used as food packaging materials.	Pereira et al. (2014)
4.	Gelatin	Zinc oxide nanorods	Young's modulus and tensile strength of gelatin films were increased by 42 and 25%, respectively, with incorporation of 5% ZnO nanorods. The surface topography of the gelatin films indicated an increase in surface roughness with increasing concentration of ZnO nanorods. The conductivity of the films also significantly increased.	Rouhi et al. (2013)

Table 3.1 Other Bionanocomposites Used for Food Packaging (*cont.*)

S. No.	Polymer	Reinforcement	Results	References
5.	PHB/ polycaprolactone (PCL)	Anionic clay Mg-Al layered double hydroxide	Nanocomposite exhibited an enhancement of 300% in elongation at break and 66% tensile strength in the presence of 1.0 wt.% of the stearate Mg-Al layered double hydroxide. Clay improved compatibility between polymer matrix and the best ratio 80PHB/20PCL/1stearate Mg-Al LDH surface was well dispersed and stretched before it broke.	Liau et al. (2014)
6.	Polypropylene	Calcium carbonate nanoparticles	Results of mechanical analysis revealed that Young's modulus increased as a function of content of calcium carbonate nanoparticles. Presence of calcium carbonate nanoparticles drastically reduced the permeability of polypropylene to both oxygen and carbon dioxide.	Avella et al. (2006)

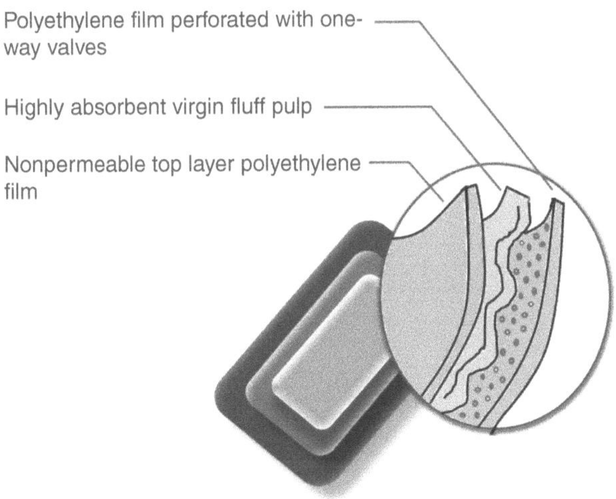

Figure 3.3. Dri-loc absorbant pads.

7.2 Oxygen Scavenging Packaging

In oxygen scavenging, existing technology is based on: iron powder oxidation, ascorbic acid oxidation, photosensitive dye oxidation, enzymatic oxidation (eg, glucose oxidase), unsaturated fatty acids, rice extract, or immobilized yeast on a solid substrate. They are available as labels, sachets, cards, or films, which can be incorporated into the packaging film. Majority of commerically available O_2 scavengers are based on the principle of iron oxidation:

$$Fe \rightarrow Fe^{2+} + 2e^-$$

$$\tfrac{1}{2}O_2 + H_2O + 2e^- \rightarrow 2OH^-$$

$$Fe^{2+} + 2OH^- \rightarrow Fe(OH)_2$$

$$Fe(OH)_2 + \tfrac{1}{4}O_2 + \tfrac{1}{2}H_2O \rightarrow Fe(OH)_3$$

AGELESS OMAC Oxygen Absorbing Film incorporates oxygen scavenger into a composite film (http://ageless.mgc-a.com/product/ageless-omac/), as shown in Fig. 3.4.

7.3 Carbon Dioxide Emitters and Scavengers

The function of carbon dioxide is to inhibit microbial growth and extend the shelf life of the product. Removal of O_2 or dissolution of CO_2 in the product creates a partial vacuum, which may

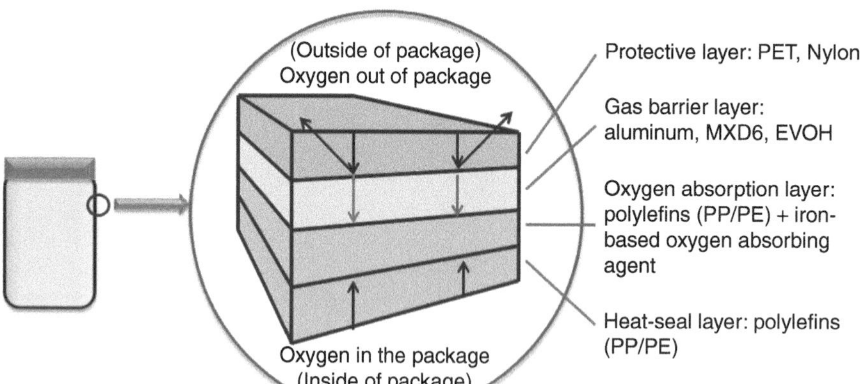

Figure 3.4. Oxygen scavenging film.

cause collapse of flexible packaging. Dual action CO_2 generators and O_2 scavengers work in such a way that package collapsing does not occur.

Verifrais package (SARL Codimer) extends the shelf life of fresh red meats; standard tray with a false, perforated bottom containing sachet of sodium bicarbonate/ascorbate. The juice dripping from the meat onto the sachet results in CO_2 emission, leads to dual functionalities of replacement of CO_2 absorbed by the meat, and prevention of package collapse.

The patented cellulose UltraZap XtendaPak pad from paper Pak industries, packed with super absorbent polymers, effectively wicks juice away from cut tomatoes while generating CO_2 gas to reduce the growth of microorganisms inside the package (http://paperpakindustries.com/products/ultrazap_xtendapak.php).

7.4 Oxygen Level Indicating Packaging

The AGELESS EYE, from Mitsubhi Gas Chemicals, is an example of a packaging system that indicates the presence of oxygen. It is an in-package monitor that indicates the presence of oxygen at a glance. The pink color of the indicator indicates oxygen-free packaging (less than 0.1%) whereas blue color indicates presence of oxygen (0.5% or more) (http://www.mgc.co.jp/eng/products/abc/ageless/eye.html).

8 Conclusions

Food packaging plays the role of a physicochemical barrier for the food products in order to maintain their integrity throughout their shelf life. Shortcomings of conventional packaging materials, such as, poor mechanical strength, low water resistance, inability to prevent degradation, and nonbiodegradability can be overcome by use of "bionanocomposites." A bionanocomposite is a nanodimensional multiphase material derived from the combination of two or more natural components. The use of bionanocomposites in the food packaging industry is a new strategy to attain markedly improved packaging properties, such as, increased modulus of strength, decreased gas and water vapor permeability, increased water resistance, improved thermal stability, and barrier properties against the migration of oxygen, carbon dioxide, aroma, and flavors. Furthermore, bionanocomposites offer an environmentally friendly approach. Biologically active compounds, such as, antimicrobials can also be added to impart desired functional properties. Cellulose, starch, and chitosan are some of the biopolymers that have been extensively studied and used to prepare bionanocomposites by reinforcing them with active agents, such as, nanoclays, SiO_2 nanoparticles, and silver nanoparticles, thus making them suitable to be used as smart packaging material by improving their mechanical properties and imparting "smart" properties, such as antimicrobial action. Thus, nanocomposites, when used as or are incorporated in packaging materials, provide smart properties to the packaging system, ultimately increasing the shelf life of the product and improving its integrity and quality.

References

Abdolmohammadi, S., Siyamak, S., Ibrahim, N.A., Yunus, W.M.Z.W., Zaki Ab Rahman, M., Azizi, S., Fatehi, A., 2012. Enhancement of mechanical and thermal properties of polycaprolactone/chitosan blend by calcium carbonate nanoparticles. Int. J. Mol. Sci. 13, 4508–4522.

Abreu, A.S., Oliveiraa, M., de Sáa, A., Rodriguesb, R.M., Cerqueirab, M.A., Vicenteb, A.A., Machado, A.V., 2015. Antimicrobial nanostructured starch based films for packaging. Carbohydr. Polym. 129, 127–134.

Adame, D., Beall, G.W., 2009. Direct measurement of the constrained polymer region in polyamide/clay nanocomposites and the implications for gas diffusion. Appl. Clay Sci. 42, 545–552.

Avella, M., De Vlieger, J.J., Errico, M.E., Fischer, S., Vacca, P., Maria Grazia Volpe, M.G., 2005. Biodegradable starch/clay nanocomposite films for food packaging applications. Food Chem. 93, 467–474.

Avella, M., Errico, M.E., Gentile, G., 2006. High performance iPP based nanocomposites for food packaging application. NSTI Nanotech. 2, 841–844.

Azeredo, H.M.C., 2009. Nanocomposites for food packaging applications. Food Res. Int. 42, 1240–1253.

Azeredo, H.M.C., Capparelli Mattoso, L.H., McHugh, T.H., 2011. Nanocomposites in food packaging: a review. In: Reddy, B. (Ed.), Advances in Diverse Industrial Applications of Nanocomposites, InTech. Available from: http://www.intechopen.com/books/advances-in-diverse-industrial-applications-ofnanocomposites/nanocomposites-in-food-packaging-a-review

Azizi Samir, M.A.S., Alloin, F., Dufresne, A., 2005. Review of recent research into cellulosic whiskers, their properties and their application in nanocomposite field. Biomacromolecules 6, 612–626.

Bin, Y., Mine, M., Koganemaru, A., Jiang, X., Matsuo, M., 2006. Morphology and mechanical and electrical properties of oriented PVA–VGCF and PVA–MWNT composites. Polymer 47, 1308–1317.

Brody, A.L., Bugusu, B., Han, J.H., Sand, C.K., Mchugh, T.H., 2008. Innovative food packaging solutions. J. Food Sci. 73, R107–R116.

Bruce, D.M., Hobson, R.N., Farrent, J.W., Hepworth, D.G., 2005. High-performance composites from low-cost plant primary cell walls. Compos. Part Appl. Sci. Manuf. 36, 1486–1493.

Chandrasekaran, G., Han, H.K., Kim, G.J., Shin, H.J., 2011. Antimicrobial activity of delaminated aminopropyl functionalized magnesium phyllosilicates. Appl. Clay Sci. 53, 729–736.

Chawengkijwanich, C., Hayata, Y., 2008. Development of TiO_2 powder-coated food packaging film and its ability to inactivate *Escherichia coli* in vitro and in actual tests. Int. J. Food Microbiol. 123, 288–292.

Chen, W., Tao, X., Xue, P., Cheng, X., 2005. Enhanced mechanical properties and morphological characterizations of poly(vinyl alcohol)–carbon nanotube composite films. Appl. Surf. Sci. 252, 1404–1409.

Coma, V., Sebti, I., Pardon, P., Deschamps, A., Pichavant, F.H., 2001. Antimicrobial edible packaging based on cellulosic ethers, fatty acids, and nisin incorporation to inhibit *Listeria innocua* and *Staphylococcus aureus*. J. Food Prot. 64, 470–475.

Dias, M.V., Azevedo, V.M., Borges, S.V., Soares, N.F.F., Fernandes, R.V.B., Marques, J.J., Medeiros, E.A.A., 2014. Development of chitosan/montmorillonite nanocomposites with encapsulated α-tocopherol. Food Chem. 165, 323–329.

Díez-Pascual, A.M., Díez-Vicente, A.L., 2014. Poly(3 hydroxybutyrate)/ZnO bionanocomposites with improved mechanical, barrier and antibacterial properties. Int. J. Mol. Sci. 15, 10950–10973.

Dufresne, A., 2013. Nanocellulose: a new ageless bionanomaterial. Mater. Today 16, 220–227.

El-Wakil, N.A., Hassan, E.A., Abou-Zeid, R.E., Dufresne, A., 2015. Development of wheat gluten/nanocellulose/titanium dioxide nanocomposites for active food packaging. Carbohydr. Polym. 124, 337–346.

Emamifar, A., 2011. Advances in Nanocomposite Technology. Applications of antimicrobial polymer nanocomposites in food packaging. In: Hashim A. (Ed.), InTech. Available from: http://www.intechopen.com/books/advances-in-nanocomposite-technology/applications-of-antimicrobialpolymer-nanocomposites-in-food-packaging

Emamifar, A., Kadivar, M., Shahedi, M., Solaimanianzad, S., 2010. Evaluation of nanocomposite packaging containing Ag and ZnO on the shelf life of fresh orange juice. Innov. Food Sci. Emerg. Technol. 11, 742–748.

Emamifar, A., Kadivar, M., Shahedi, M., Solaimanianzad, S., 2011. Effect of nanocomposite packaging containing Ag and ZnO on inactivation of *Lactobacillus plantarum* in orange juice. Food Cont. 22, 408–413.

Fazli, F.A., Babazadeh, A., Fazli, F.A., 2013. Crystalline structure of starch based nano composites. Int. J. Biol. Food. Vet. Agric. Eng. 7, 725–728.

Goyal, S., Goyal, G.K., 2012. Nanotechnology in food packaging: a critical review. Russ. J. Agric. and Socio-Econ. Sci. 10, 14–24.

Helbert, W., Cavaillé, C.Y., Dufresne, A., 1996. Thermoplastic nanocomposites filled with wheat straw cellulose whiskers. Process. Mech. Behav. 17, 604–611.

Ho a, T.T.T., Ko, Y.S., Zimmermann, T., Geiger, T., Caseri, W., 2012. Processing and characterization of nanofibrillated cellulose/layered silicate systems. J. Mater. Sci. 47, 4370–4382.

Ho, T.T.T., Zimmermann, T., Ohr, S., Caseri, W.R., 2012. Composites of cationic nanofibrillated cellulose and layered silicates: water vapor barrier and mechanical properties. ACS Appl. Mater. Interf. 4, 4832–4840.

Ho, T.T.T., Zimmermann, T., Caseri, W.R., Smith, P., 2013. Liquid ammonia treatment of (cationic) nanofibrillated cellulose/vermiculite composites. J. Polym. Sci. B 51, 638–648.

Huang, W.M., Yang, B., An, L., Li, C., Chan, Y.S., 2005. Water-driven programmable polyurethane shape memory polymer: demonstration and mechanism. Appl. Phys. Lett. 86, 114105.

Ifuku, S., Nogi, M., Abe, K., Handa, K., Nakatsubo, F., Yano, H., 2007. Surface modification of bacterial cellulose nanofibers for property enhancement of optically transparent composites: dependence on acetyl-group DS. Biomacromolecules 8, 1973–1978.

Iwamoto, S., Nakagaito, A.N., Yano, H., Nogi, M., 2005. Optically transparent composites reinforced with plant fiber-based nanofibers. Appl. Phys. A 81, 1109–1112.

Jia, X., Li, Y., Cheng, Q., Zhang, S., Zhang, B., 2007. Preparation and properties of poly(vinyl alcohol)/silica nanocomposites derived from copolymerization of vinyl silica nanoparticles and vinyl acetate. Eur. Polym. J. 43, 1123–1131.
Jin, T., Sun, D., Su, J.Y., Zhang, H., Sue, H.J., 2009. Antimicrobial efficacy of zinc oxide quantum dots against *Listeria monocytogenes*, *Salmonella enteritidis*, and *Escherichia coli* O157:H7. J. Food Sc. 74, 46–52.
Kim, J.Y., Han, II, S., Hong, S., 2008. Effect of modified carbon nanotube on the properties of aromatic polyester nanocomposites. Polymer 49, 3335–3345.
Lau, K.T., Hui, D., 2002. The revolutionary creation of new advanced materials: carbon nanotube composites. Compos. Part B 33, 263–277.
Liau, C.P., Ahmad, M.B., Shameli, K., Yunus, W.M.Z.W., Ibrahim, N.A., Zainuddin, N., Then, Y.Y., 2014. Preparation and characterization of polyhydroxybutyrate/polycaprolactone nanocomposites. Sci. World J. 2014, 9.
Lopez Manchado, M.A., Valentini, L., Biagotti, J., Kenny, J.M., 2005. Thermal and mechanical properties of single-walled carbon nanotubes–polypropylene composites prepared by melt processing. Carbon 43, 1499–1505.
Ma, M., Guo, L., Anderson, D.G., Langer, R., 2013. Bio-inspired polymer composite actuator and generator driven by water gradients. Science 339, 186–189.
Martirosyan, A., Schneider, Y.-J., 2014. Engineered nanomaterials in food: implications for food safety and consumer health. Int. J. Environ. Res. Public Health 11, 5720–5750.
Mauriello, G.D.L.E., La Storia, A., Villani, F., Ercolini, D., 2005. Antimicrobial activity of a nisin-activated plastic film for food packaging. Lett. Appl. Microbiol. 41, 464–469.
Mesquita, J.P., Donnici, C.L., Pereira, F.V., 2010. Biobased nanocomposites from layer-by-layer assembly of cellulose nanowhiskers with chitosan. Biomacromolecules 11, 473–480.
Mirzadeh, A., Kokabi, M., 2007. The effect of composition and draw-down ratio on morphology and oxygen permeability of polypropylene nanocomposite blown films. Eur. Polym. J. 43, 3757–3765.
Mushi, N.E., Utsel, S., Berglund, L.A., 2014. Nanostructured biocomposite films of high toughness based on native chitin nanofibers and chitosan. Front. Chem. 2, 99.
Nielsen, L.E., 1967. Models for the permeability of filled polymer systems. J. Macromol. Sci. A 1, 929–942.
Nishino, T., Matsuda, I., Hirao, K., 2004. All-cellulose composite. Macromolecules 37, 7683–7687.
Özgür Seydibeyoğlu, M., Oksman, K., 2008. Novel nanocomposites based on polyurethane and micro fibrillated cellulose. Compos. Sci. Technol. 68, 908–914.
Patakfalvi, R., Dékány, I., 2004. Synthesis and intercalation of silver nanoparticles in kaolinite/DMSO complexes. Appl. Clay Sci. 25, 149–159.
Patil, A.J., Muthusamy, E., Mann, S., 2005. Fabrication of functional protein organoclay lamellar nanocomposites by biomolecule-induced assembly of exfoliated aminopropyl functionalized magnesium phillosilicates. J. Mater. Chem. 15, 3838–3843.
Pereira, M., Abreu, A.S., Oliveira, M., Nobrega, J.M., Machado, A.V., 2014. Biodegradable nanocomposite for food packaging application. Food Chem. 93, 467–474.
Rai, M., Yadav, A., Gade, A., 2009. Silver nanoparticles as a new generation of antimicrobials. Biotechnol. Adv. 27, 76–83.
Ray, S., Easteal, A., Quek, S.Y., Chen, X.D., 2006. The potential use of polymer-clay nanocomposites in food packaging. Int. J. Food Eng. 2, 1556–3758.

Ren, P., Shen, T., Wang, F., Wang, X., Zhang, Z., 2009. Study on biodegradable starch/OMMT nanocomposites for packaging applications. J. Polym. Environ. 17, 203–207.
Rhim, J.W., Ng, P.K.W., 2007. Natural biopolymer-based nanocomposite films for packaging applications, critical reviews in food science and nutrition. Crit. Rev. Food Sci. Nutr. 47, 411–433.
Rhim, J.W., Park, H.M., Ha, C.S., 2013. Bio-nanocomposites for food packaging applications. Prog. Polym. Sci. 38, 1629–1652.
Robertson, G., 2006. Food Packaging Principles and Practices, 2nd ed. Taylor & Francis, Boca Raton, FL, 545.
Rouhi, J., Mahmud, S., Naderi, N., Ooi, C.R., Mahmood, M.R., 2013. Physical properties of fish gelatin-based bio-nanocomposite films incorporated with ZnO nanorods. Nano. Res. Lett. 8, 364.
Sadegh-Hassani, F., Nafchi, A.M., 2014. Preparation and characterization of bionanocomposite films based on potato starch/halloysite nanoclay. Int. J. Biol. Macromol. 67, 458–462.
Saeideh Ebrahimiasl, S., Zakaria, A., Kassim, A., Basri, S.N., 2015. Novel conductive polypyrrole/zinc oxide/chitosan bionanocomposite: synthesis, characterization, antioxidant, and antibacterial activities. Int. J. Nanomed. 10, 217–227.
Sawai, J., Yoshikawa, T., 2004. Quantitative evaluation of antifungal activity of metallic oxide powders (MgO, CaO and ZnO) by an indirect conductimetric assay. J. Appl. Microbiol. 96, 803–809.
Shameli, K., Ahmad, M.B., Zargar, M., Yunus, W.M.Z.W., Ibrahim, N.A., Shabanzadeh, P., Moghaddam, M.G., 2011. Synthesis and characterization of silver/montmorillonite/chitosan bionanocomposites by chemical reduction method and their antibacterial activity. Int. J. Nanomed. 6, 271–284.
Sharma, H., Mishra, P.K., Talegaonkar, S., Vaidya, B., 2015. Metal nanoparticles: a theranostic nanotool against cancer. Drug Discov. Today 20 (9), 1143–1151.
Sidorenko, A., Krupenkin, T., Taylor, A., Fratzl, P., Aizenberg, J., 2007. Reversible switching of hydrogel-actuated nanostructures into complex micropatterns. Science 315, 487–490.
Sun, Q., Xi, T., Li, Y., Xiong, Liu., 2014. Characterization of corn starch films reinforced with $CaCO_3$ nanoparticles. PLoS One 9, e106727.
Taghizadeh, M.T., Sabouri, N., Ghanbarzadeh, B., 2013. Polyvinyl alcohol:starch: carboxymethyl cellulose containing sodium montmorillonite clay blends; mechanical properties and biodegradation behaviour. SpringerPlus 2, 1–8.
Tang, S., Zou, P., Xiong, H., Tang, H., 2008. Effect of nano-SiO_2 on the performance of starch/polyvinyl alcohol blend films. Carbohydr. Polym. 72, 521–526.
Tang, X.Z., Kumar, P., Alavi, S., Sandeep, K.P., 2012. Recent advances in biopolymers and biopolymer-based nanocomposites for food packaging materials. Crit. Rev. Food Sci. Nutr. 52, 426–442.
Thostenson, E.T., Li.F C.Y., Chou, T.W., 2005. Nanocomposites in context. Compos. Sci. Technol. 65, 491–516.
Uyama, H., Kuwabara, M., Tsujimoto, T., Nakano, M., Usuki, A., Kobayashi, S., 2003. Green nanocomposite from renewable resources: plant oil-clay hybrid materials. Chem. Mater. 15, 2492–2494.
Valdés, A., Mellinas, A.C., Ramos, M., Garrigós, M.G., Jiménez, A., 2014. Natural additives and agricultural wastes in biopolymer formulations for food packaging. Front. Chem. 2 (6), 10.
Vladimiriov, V., Betchev, C., Vassiliou, A., Papageorgiou, G., Bikiaris, D., 2006. Dynamic mechanical and morphological studies of isotactic polypropylene/fumed silica nanocomposites with enhanced gas barrier properties. Compos. Sci. Technol. 66, 2935–2944.

Wu, C.L., Zhang, M.Q., Rong, M.Z., Friedrick, K., 2002. Tensile performance improvement of low nanoparticles filled-polypropylene composites. Compos. Sci. Technol. 62, 1327–1340.
Xiong, H.G., Tang, S.W., Tang, H.L., Zou, P., 2008. The structure and properties of a starch-based biodegradable film. Carbohydr. Polym. 71, 263–268.
Yamamoto, O., 2001. Influence of particle size on the antibacterial activity of zinc oxide. Int. J. Inorg. Mater. 3 (7), 643–646.
Yang, B., Huang, W.M., Li, C., Lee, C.M., Li, L., 2004. On the effects of moisture in a polyurethane shape memory polymer. Smart Mater. Struct. 13, 191–195.
Zeng, H., Gao, C., Wang, Y., Watts, P.C.P., Kong, H., Cui, X., Yan, D., 2006. In situ polymerization approach to multiwalled carbon nanotubes-reinforced nylon 1010 composites: mechanical properties and crystallization behavior. Polymer 47, 113–122.
Zhang, M.Q., Rong, M.Z., 2003. Performance improvement of polymers by the addition of grafted nano-inorganic particles. Chinese J. Polym Sci. 21, 587–602.
Zhang, L., Jiang, Y., Ding, Y., Povey, M., York, D., 2007. Investigation into the antibacterial behaviour of suspensions of ZnO nanoparticles (ZnO nanofluids). J. Nano. Res. 9, 479–489.
Zhang, K., Geissler, A., Standhardt, M., Mehlhase, S., Gallei, M., Chen, L., Thiele, C.M., 2015. Moisture-responsive films of cellulose stearoyl esters showing reversible shape transitions. Sci. Rep. 5, 11011.
Zhou, X., Shin, E., Wang, K.W., Bakis, C.E., 2004. Interfacial damping characteristics of carbon nanotube-based composites. Compos. Sci. Technol. 64, 2425–2437.
Zhuang, R., Beuchat, L.R., Chinnan, M.S., Shewfelt, R.L., Huang, Y.W., 1996. Inactivation of *Salmonella montevideo* on tomatoes by applying cellulose-based edible films. J. Food Protect. 59, 808–812.
Zuo, P.P., Feng, H.F., Xu, Z.Z., Zhang, L.F., Zhang, Y.L., Xia, W., Zhang, W.Q., 2013. Fabrication of biocompatible and mechanically reinforced graphene oxide-chitosan nanocomposite films. Chem. Cent. J. 7, 39.

4

ENCAPSULATION OF SENSORS FOR INTELLIGENT PACKAGING

Viviane Dalmoro*, Joao Henrique Zimnoch dos Santos*, Marcia Pires, Adriane Simanke**, Gabriela B. Baldino**, Luciane Oliveira****

**Institute of Chemistry, Federal University of Rio Grande do Sul, Porto Alegre, Brazil; **Braskem, Polymer Technology Center, Triunfo, Brazil*

1 Introduction

Lifestyle changes, market trends, and the growing level of demand by the consumers has boosted the packaging industry to focus on the quest for solutions that provide maximum food security, while keeping its nutritional value at competitive prices. Generally, packaging is considered to provide functions, such as, containment, protection, convenience, and communication. Advanced materials, such as intelligent packaging, are capable of maintaining the quality of the product, monitoring the history of the packaging, and communicating this information to the consumer. Another type of advanced packaging is the active packaging that can control, and even interact, with its content by changing the conditions inside the package to extend shelf life and maintain quality of food products. According to the regulations 1935/2004/EC and 450/2009/EC, active materials are defined as systems that extend the shelf life, maintain or even improve the condition of packaged food. In this way, active packaging has the critical function of protection, while intelligent packaging, communication.

1.1 Market

Packaging is one of the most important factors that maintains the quality of food during transportation, storage, and end-use. In this way, the most important function is guaranteeing safe delivery of food product until consumption. There are many trends and approaches in the food packaging industry that are used to achieve better food quality, shelf-life extension, and safety enhancement. Some of these trends are technologies are active- and

Food Packaging. http://dx.doi.org/10.1016/B978-0-12-804302-8.00004-2

intelligent packaging. Many developments have been based on these technologies, especially on the intelligent packaging that can notify consumers about harmful changes and provide logistic information to the whole chain (Han, 2005; Alavi et al., 2015).

In the past years, the discussions about active and intelligent packaging were made under scientific and academic points of view. Nowadays, companies' interests on this topic are growing, and consequently their investments, to develop new materials and packages using these technologies. A recent example is the creation of Active and Intelligent Packaging Association (AIPIA), which promotes an annual World Congress and Exhibition on Active and Intelligent Packaging to discuss the latest market solutions, the issues, and the impact packaging has. This is a consequence of predictions in packaging trends made by market surveys, which show the demand for intelligent packaging is expected to grow and could achieve $5.3 billion in 2017 (Realini and Marcos, 2014). The technologies will be based on printed electronics, quick response (QR) code, time/temperature indicators, labels, and others (Smart Packaging Case Study, 2011).

Studies presented by the AIPIA, published by Visiogain (a market survey company), showed significant growth for food traceability technologies. It was estimated to achieve US$11–15 billion in 2015. All these provisions still drive these developments, which need to be guided by industry investment capacity and regulation laws.

Specialists from the AIPIA assert that the European Union would be quicker to use the intelligent packaging in its market, based on the demand for reducing waste, traceability investments, and new regulations guided by campaigns such as "save food." However, other developing countries will not ignore this kind of technology in order to have better packaging standards; as we can find many research groups from Latin America working on intelligent sensors and intelligent packaging applications (Yoshida et al., 2014).

The intelligent technology available in the market involves communication through intelligent labels, QR codes, and radio frequency identification (RFID) systems, and it can be found in countries, such as, Asia and Europe. There are many opportunities for the use of intelligent technology, such as, food-spoilage detection, contamination detection, anticounterfeit products, and pharmaceutical traceability.

Some topics described by markets surveys highlight points involving intelligent packaging, such as:

1. Interrelated factors that drive demand for intelligent packaging, such as, concerns with safety and reducing waste;
2. Intelligent packaging that can enhance brand communication;
3. Companies that are looking for new ways to communicate with consumers;

4. Consumer suspicion and cost factors that may inhibit acceptance;
5. Performance of intelligent technology that needs to guarantee the low possibility of discarding food based solely on the indication of a color change if the food is still fresh.

Published studies and sensor companies present several opportunities for the application of intelligent packaging, such as, fruits using ripening sensors to help consumers to buy fruits as ripe as they like it; dairy food, meat, and juices using freshness indicators to communicate the quality of the food; in logistics control and drink packaging using time/temperature indicators; thermochromics inks for pizza, to control the time delivery; and others.

Specialized conferences and market evaluations from France identify intelligent technology as an opportunity to avoid food wastage by reconfiguring their marketing approach using an automatic system to reduce food prices when they are close to the expiration date (Smart Packaging Case Study, 2011).

Besides the opportunities on food waste reduction, development in intelligent packaging has also been an innovative drive for many industries. For instance, Keep-it Technologies recently received an award for the best sustainable solution for environment and health for its shelf-life indicator for fresh foods at the EAT Stockholm Food Forum.

Another approach for intelligent packaging lies in the sensory experience provided to consumers. Consumers are looking for experiences different from their routine, such as, self-cooling or self-heating packages. This trend can open many opportunities for the industry to work on new developments. Fig. 4.1 shows the evolution of the number of publications dealing with intelligent and active packaging.

According to Fig. 4.1, the number of studies devoted to active packaging is roughly five times higher than those involving intelligent ones. Perhaps this phenomenon may be due to the difficulty in developing an intelligent device, capable of interacting with the content within the packaging, stable enough to resist to packaging preparation, and capable of providing feasible and easy signal measurements that can be perceived by the consumers. In the following sections, the most common types of intelligent packaging are briefly discussed.

1.2 Classification of Intelligent Packaging

As already mentioned, intelligent packaging monitors some aspects of the food product or its environment and reports this information to the consumer or food manufacturer. An intelligent

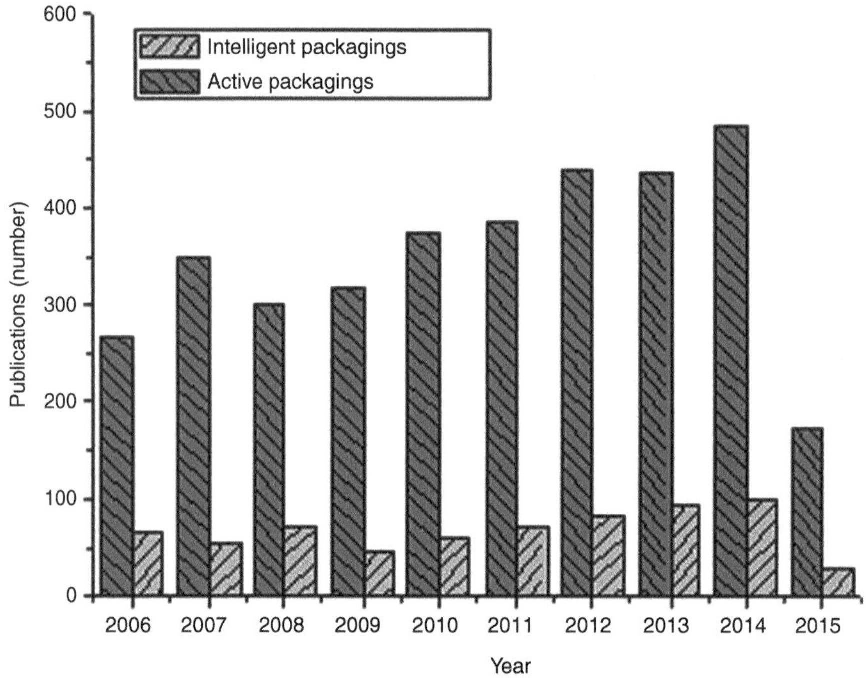

Figure 4.1. Number of publications in the period 2006–2015 involving active or intelligent packagings. Data from Scopus, www.scopus.com.

packaging system provides information concerning product quality, safety, or preservation (Yam et al., 2005). However, these systems must be activated easily and the rate of change in the device must be well correlated with the rate of food spoilage. Currently, the intelligent feature is obtained by three major technologies that are produced, sensors, indicators, and RFID (Realini and Marcos, 2014). These systems differ in their physical composition, amount, types of data transported, and the capture and distribution of the data. The technologies currently in the market are usually based on different principles: chemical, enzymatic, biological, and physical (Kerry et al., 2006). However, almost all of these technologies convey the alterations that have occurred in food by means of visual change, especially in color. The general purpose of this section is to provide an overview of the different types of intelligent packaging concepts for each major technology, including freshness indicators, integrity indicators, time–temperature indicators, RFIDs, and sensors.

1.2.1 Indicators

The indicators provide visual and qualitative information on packaged food. The information is generated through a color change, an increase in color intensity, or diffusion of a dye along a path. This information should be irreversible, because it is not possible to store the measurement data and time (Pavelková, 2012). They do not provide quantitative information, such as, temperature and concentration. Nowadays, most commercial systems employ this type of indicators for intelligent packaging (Vanderroost et al., 2014). The different classes of indicators currently used are discussed in the next sections.

1.2.1.1 Freshness Indicators

Freshness indicators determine direct information about the quality of fresh food. The alterations that take place in foods are due to microbial growth and/or oxidation of lipids which can used as indicators of food freshness (Kuswandi et al., 2013). The microbial growth generates certain metabolites, such as, glucose, organic acids, carbon dioxide, biogenic amines, and sulfuric compounds (Arvanitoyannis and Stratakos, 2012). The reaction, between the indicator and the metabolites from microbial growth or with chemicals compounds resulting from lipid oxidation, visually indicates the quality of the food. The most frequently used ones are based on detection of spoilage metabolites through visible color change. In general, these systems employ colorimetric indicators to detect the spoilage gases (sulfur, nitrogen, and carbon dioxide), providing a clue of the freshness of the food product (Biji et al., 2015). Freshness indicators are also used as fruit ripening indicators informing the consumer when the fruit is ripe and ready to eat.

1.2.1.2 Integrity Indicators

Package integrity is a key factor in preserving the quality and extending the shelf life of food stuffs in modified atmosphere packaging. One major problem, of package integrity in flexible plastic packages, is associated with leaking seals. Moreover, the gas composition inside a packaging easily changes due to the respiration of the food product in environmental conditions. It is worth mentioning that food products (fresh fruits and vegetables) continue to respire even when packed. The process consumes oxygen and produces water vapor and carbon dioxide. The reduction of the respiration rate is a key factor to maintain the taste, quality, and appearance of the edible product. This can be achieved by storage at low temperatures, reduction of the oxygen levels, or an increase in the carbon dioxide levels within the package.

The presence of oxygen is one major factor that determines food quality loss and reduced shelf life. The oxidative process in food products contributes to the deterioration in safety, flavor, odor, texture, and color of the products, as well as, growth of aerobic microorganisms. Lipidic oxidation is responsible for off-flavors and rancidity, which decrease nutritional and sensory qualities (Yam et al., 2005). Gas indicators can monitor changes in the composition of gases inside a package; providing means to monitor the quality and safety of food. The most common gas indicators for food packaging applications are the oxygen indicators that typically signal the presence or absence of oxygen. A number of indicators are designed to supply qualitative or semi-quantitative information about alterations in the gas concentrations through visual colorimetric changes (Vu and Won, 2013). Oxygen indicators monitor the atmosphere inside the package. The color of the indicator changes when oxygen is present and it can indicate improper sealing or adulteration. Gas indicators also are used to detect water vapor, ethanol, hydrogen sulfide, and carbon dioxide. Several such systems are based on organometallic complexes. A limitation, pertaining to finding a system that can be compatible with legislation for intelligent packaging, exists.

1.2.1.3 Time–Temperature Indicators

Temperature is an important environmental factor for food preservation. Temperature variations have a great influence on the microbial growth and on the kinetics of physical and chemical deteriorations in food products and may compromise the safety and the shelf life of perishable products. Time–temperature indicators (TTIs) provide visual information of the temperature history of a perishable food from the point of production to the end consumer. The systems available in the market are based on different principles, such as, the diffusion of a dye, enzymatic hydrolysis of a lipid substrate, pH change due to microbial growth, and solid-state polymerization that leads to the formation of a highly colored polymer (Kerry et al., 2006; Biji et al., 2015; Singh, 2000). In general, TTIs are small self-adhesive tags that inform when the product is subjected to abusive temperatures. TTIs are very useful for chilled or frozen products because they allow the monitoring of storage conditions and indicate the abuse of temperature for such foods. Therefore, TTIs can be used as shelf-life indicators and information providers about a cold chain break.

1.2.2 Radio Frequency Identification Tags

RFID is an automatic information technology that uses electromagnetic fields to acquire, store, and transmit diverse information without human intervention. RFIDs contain a tag, an antenna, a reader and a communication infrastructure. The tags have an integrated circuit connected to an antenna. The communication between the tag and the reader is performed through electromagnetic wave transmission and the information is stored on a chip. RFIDs differ from sensors and indicators because they does not provide qualitative or quantitative information about the quality of food (McFarlane and Sheffi, 2003). However, the stored information may be used for several applications, such as, identification, automation, asset tracking, security, traceability control, and supply-chain management (Hong et al., 2011). The combined use of RFID technology with sensors or indicators can provide information on the integrity of the packaging, the quality of the food, and conditions of storage and transport. This information provides more visibility and detects possible cold chain interruptions, resulting in an improvement in supply chain-management efficiency due to less waste production.

1.2.3 Sensors

A sensor is a device that detects, records, and transmits information into some sort of signal that is quantifiable. In general, two basic functional units form sensors: a receptor part and a transducer part. Some types of sensors used in intelligent packaging are physical sensors, chemical sensors, biosensors, and gas sensors (Kerry et al., 2006). However, sometimes this classification is very difficult because it is not possible to decide how the sensor operates unequivocally. Moreover, the term sensor in the literature sometimes refers to the detection of component/systems, that is, to the indicator itself. Sensor, in the common sense, is considered as a device that detects a given analyte, for instance. Thus, a device that is exposed to gases (and/or a given molecule) and suffers a color change that can be detected by naked eyes is named as a sensor.

In recent years, chemical sensors have received heightened visibility in the monitoring of food quality and integrity of the packaging because they can provide information about the chemical composition of the environment, which can be a liquid or gaseous phase (Yebo et al., 2012; Vanderroost et al., 2014; Azeredo, 2009). According to the IUPAC (Hulanicki et al., 1991), a chemical sensor is a device that may promote the transformation of the chemical information (ranging from the concentration of a given analyte to the total components analysis) in a detectable analytical signal. It consists of a receptor that recognizes a chemical analyte and a transducer that converts and generates the output signal. In recent years there has been an increase in studies regarding the use of nanomaterials in detection, especially for gas detection. Nanoparticles, such as, graphene, graphite, nanofibers, and nanotubes have received growing attention (Schedin et al., 2007; Eulises, 2013). These nanomaterials exhibit interesting properties, like high specific surface area, excellent electrical and mechanical properties, which provide an excellent detection sensitivity (Vanderroost et al., 2014). Due to these properties, nanomaterials have become suitable for use as chemical sensors.

A chemical sensor based on the recognition of biological material is a biosensor. Biosensors are devices able to detect a target metabolite produced during the food-deterioration process. These systems are constituted by bioreceptors that recognize the analyte and transducers that convert biochemical signals into an electronic response (Alocilja and Radke, 2003). They may be further differentiated according to the biological elements used in the receptor, which may be cells, proteins, enzymes, antibodies, antigens, hormones, DNA, RNA, or biogenic amines (de Corcuera and Cavalieri, 2003).

Gas sensors are used to detect the presence of gaseous analytes in the package. They can monitor volatile organic compounds (VOCs) and gas molecules (H_2, CO, NO_2, O_2, H_2S, NH_3, CO_2, CH_4, etc.), resulting from food spoilage and package leakage, and evaluate the product quality and the package integrity (Kerry et al., 2006).

An intelligent packaging with a sensor that monitors packaged food product and its environment throughout the entire supply chain is really helpful in ensuring the quality and safety of the packaged food for end consumers. Despite the growing demand of consumers for safety of food products, it is still necessary to overcome obstacles to apply sensors commercially on a large scale.

Several prototypes of commercial sensors for intelligent packaging have been developed but require further research for their integration in food packaging (Realini and Marcos, 2014; Vanderroost et al., 2014; Biji et al., 2015). Due to the potential hazardous effects of the chemical and biologic components in sensors on the food packaging, special attention and extensive research efforts are mandatory. Therefore, the integration of such devices is not an easy ordeal. Several aspects have to be taken into account. For instance, the components may not be put directly into contact with the food, and in addition, they may be prone to migrate into the food content. Furthermore, they can impinge on the texture, flavor, or taste of the edibles. From the durability point of view, they have to be stable enough to support the packaging conditions. In the next section, some aspects concerning packaging technology are briefly discussed.

2 Materials for Packaging and Packaging Properties

The use of intelligent technologies in packaging needs to maintain the packaging properties and to guarantee the packaging performance during the application. Based on these considerations many issues need to be taken into account in the development of new devices as intelligent packaging, such as, materials to be used, the final properties, process condition, to mention a few.

Several materials are used for food-packaging applications, such as, plastic, glass, and metal (Marsh and Bugusu, 2007). Plastics are the most commonly used materials for food packaging applications because they have a wide range of properties that can be tailored according to the product requirement. Low density, good processability, high volume production, and low cost are other factors that contribute in making plastics attractive for packaging applications (Robertson, 2013; Alavi et al., 2015).

The most common synthetic plastics used in packaging are polyethylene, polypropylene, polyamide, polyvinyl chloride, polyethylene terephthalate, and polystyrene. Besides them, biopolymers such as, starch, proteins, cellulose, and chitosan can also be used for this application (Weber et al., 2002). Although biopolymers are intended to be more sustainable, friendly, and biodegradable than synthetic plastics, they have not been able to replace them due to a number of technical and economic issues. Among the technical issues that limit their industrial application, it is possible to highlight the poor mechanical properties, low water resistance, and the fact that most of the biopolymers cannot be processed at high temperatures, because of their low stability. Furthermore, most of biopolymers cannot meet the barrier requirements to water vapor and oxygen (Mensitieri et al., 2011; Siracusa et al., 2008; Tang et al., 2012).

One strategy used to overcome these technical issues is through the development of bionanocomposites, which has been extensively studied and whose application in food packaging is expected to grow in the next years (Tang et al., 2012; Sorrentino et al., 2007; Lagaron and Lopez-Rubio, 2011; Bordes et al., 2009; Cabedo et al., 2006; Rhim and Ng, 2007). In the past decade, continuous studies regarding nanocomposites have proved that they can be used to develop new packaging with better properties than the conventional materials. Improvements in the physical properties of polymers, including mechanical strength, thermal stability, and gas-barrier properties are few of the advantages of their use in food packaging (Youssef, 2013; Azeredo, 2009; Arora and Pauda, 2010; de Abreu et al., 2007; Duncan, 2011; Ray et al., 2006). Although there are a number of benefits of using nanotechnology for food packaging applications, especially the large potential, further innovation into new packaging concepts is required as a number of challenges still remain to be addressed. Issues, such as, health and environmental safety, regulatory affairs, among others, can hinder its use in food-packaging applications (Siegrist et al., 2008).

The type of packaging depends on the nature of food product and the storage conditions suitable for it. A large variety of food-packaging types are available, from a monolayer film to a sophisticated combination of multilayered films, depending on the application. In the case of multilayered films, each layer has a specific function, such as, gas-barrier layer and sealing layer.

The choice of a specific polymer for food packaging application depends on the food product and the requirements for manufacturing, handling, and storage. The suitability of a specific polymer to be applied in food packaging is determined by polymer properties, which will depend on the polymer structure and polymer processing. Polymer structure depends on the used monomer, polymerization conditions, type of catalyst/initiator, number of used monomers (homopolymers, copolymers, or physical blends), and so on.

Improvements in polymer performance and its characteristics can also be achieved by the use of specific additives. The type and amount of additive to be used depends on the polymer, polymer process, and polymer application. There are two main reasons to use an additive in a polymer during processing and application, (1) to change some polymer properties that will help these aspects; and (2) to infer stability during these processes. Among the most common additives used in polymers are antioxidants, metal deactivators, ultraviolet (UV) absorbers, hindered amine-light stabilizers, lubricants, and nucleating-agents/clarifiers. Appropriate selection of additives helps in development of value-added plastics with improved durability and other advantages. In order to get an improved performance of the final product it is important to guarantee homogeneity and good dispersion of the additives in the polymer; otherwise, they can negatively affect the final properties of the polymer. Loss of optical properties due to the formation of gels, for example, is one of the common problems faced when different additives are included in polymers to make films. Additives can also migrate to polymer surface and influence other film properties, like sealing and blocking. Sealing performance is one of the most important characteristics of the polymer and it can be affected by the presence of additives at the surface of the polymer film.

Different processes can be used to transform plastics into end-use products. The most common processes used worldwide are extrusion, injection molding, blow molding, thermoforming, and compression molding, in which extrusion and injection molding are responsible for approximately 70% of the processes. Although in the majority of cases the finished article is produced in a single operation, an earlier additional process is generally required to transform the polymer into a more convenient form (eg, granules) to be processed and also to feed the required additives. In the extrusion process, the polymer is melted, homogenized, and forced through a capillary-shaped die. It emerges as a continuous strip, which is cooled in a long-water bath so that it may be chopped into short granules. The processing stages of heating, shaping, and cooling may be continuous or a repeated cycle of events, but in most cases the processes may be automated.

In the case of intelligent packaging, indicators can be used to communicate or indicate changes in food quality, for example, and can be directly added to the polymer. The indicators as other additives can be incorporated into the food packaging by different ways. One way is to put the additive during the granulation process when the polymer (in the form of powder, flakes, or spheres) is converted into pellets. Plastics have high-melting temperatures, demanding the use of high-process temperatures, which may be critical for some additives and indicators that could degrade at these high temperatures. Among the factors that must be taken into account in the selection of any chemical compounds to be incorporated in food packaging, it is possible to highlight the chemical nature of the additives, chemical nature of polymers, and process conditions. The choice of the additive and indicator is often limited by heat lability during extrusion and by incompatibility with the packaging material. High-extrusion temperatures can cause the decomposition of these chemical compounds, and by consequence, the expected performance. Another important factor is the compatibility of the additive/polymer depends on the polarity of both. Most of the additives used for packaging applications are polar, that are incompatible with nonpolar polymers, such as, polyethylene and polypropylene. This incompatibility may cause bad dispersion of the additive in the polymer-gel formation.

Due to the difficulty in dispersion of these chemical compounds in nonpolar polymers and at high temperatures for polyolefin processing, it is difficult to find studies regarding the use of olefin-based intelligent polymers (Tomoyoshi et al., 1986; Kleiböhmer et al., 1999). On the other hand, some studies regarding biobased-intelligent polymers can be found (Maciel et al., 2012), because such kinds of polymer suffer from thermal lability. The encapsulation of some indicators/sensors can provide enough thermal stability to resist the high temperature or the necessary resident time required for polyolefin processing in the production of intelligent packaging. Several approaches may be found in the literature for the development of solid sensor or for the encapsulation of an indicator within a solid matrix. Among them, the sol–gel technology provides a versatile tool in this direction, and is discussed in the next section.

3 Sol–Gel and Encapsulated Sensors

Chemical sensors, in different configurations doped with reagents of specific sensitivities to a wide variety of analytes, have been consistently reported. A widely employed strategy to produce chemical sensors is the encapsulation of an indicator within a polyelectrolyte (McShane et al., 2002) and materials produced by the sol–gel technology (Butler et al., 1998; Capeletti et al., 2010). Sol–gel technology provides a suitable approach to prepare, under mild conditions, stable, optically transparent host matrices for the design and development of chemical sensors (Tran-Thi et al., 2011; Melde et al., 2008).

The general procedure for sol–gel material synthesis includes: (1) sol (colloid suspension) preparation, (2) gelation (sol–gel transition), and (3) aging and drying of the gel (Schubert and Husing, 2012). Sol–gel polymerization of silica is governed by hydrolysis reactions (addition of water to reactive precursors, such as, alkoxysilanes) to form reactive Si–OH species, which undergo the condensation reaction via an alcohol-producing mechanism ("silanol– ether") or via a water-producing mechanism ("silanol–silanol") and which leads to the generation of Si–O–Si-bridges (Schubert and Husing, 2012; Morosanova, 2012). Therefore, a structure of crosslinked-siloxane bonds, that is, a silica network, is formed. Ether and alkyl chloride are some of the molecules employed in the production of silica through thermally induced non-hydrolytic sol–gel chemistry (Sanchez et al., 2011). The kinetics of hydrolysis and condensation reactions is slow; therefore common acids and bases (catalysts) have been employed to accelerate the sol–gel process.

The ratio of the rate constant of hydrolysis and condensation determines the physical properties and the degree of crosslinking of sol–gel materials (Schulz-Ekloff et al., 2002). More specifically, it determines the size and nature of the polycondensate structures in the sol, that is, the length of linear chains, the degree of branching, or the density (Schulz-Ekloff et al., 2002). The hydrolysis and condensation reactions are competitive and occur simultaneously. Under acidic conditions (pH <5) hydrolysis is favored and the condensation reaction is the rate-determining step. At pH >5 the faster reaction is condensation (Schubert and Husing, 2012). Therefore, the formation of sol–gel materials proceed by a distinct evolution depend on the catalyst type and the basic or acidic route, therefore the characteristics of produced materials differ. Since in the basic route the kinetics of condensation reactions is faster, spheres are generally produced, sometimes with nanometric size. As in the acidic route, first hydrolysis takes place and the growth of silica network occurs in the advanced stage. The resulting structure tends to be ramified.

Rheological characteristics of the colloidal dispersions can be tuned and sol–gel materials can be processed through many methodologies, such as, film-deposition methods, fiber extrusion, fiber pulling, electrospinning, electrochemical deposition, (soft)lithography-based techniques, aerosol or spray, and ink-jet printing as shown in Fig. 4.2 (Sanchez et al., 2011). It can be observed that sol–gel monoliths, thin films, mono- and polydisperse powders, fibers, and patterns are obtained by sol–gel technology.

The most noticeable feature of sol–gel process is its versatility that enables it to modulate the desired properties and morphology. This is achieved by tuning chemical and physical parameters of the sol–gel process, such as, employing suitable functionalized alkoxysilanes (containing amino, phenyl, methyl, vinyl, mercapto, and other groups); controlling the starting pH solution, the kind of solvent, and the water/precursor ratio; and the processing temperature and drying conditions.

Compared to organic polymers, sol–gel materials offer several advantages which include higher chemical, photochemical, and thermal stability; mechanical strength; optical transparency (down to 250 nm); compatibility with various sensitive molecules (Lin and Brown, 1997); more rigid environments, as well as, a higher migration stability (Schulz-Ekloff et al., 2002). Another advantage of sol–gel materials is their stability for processing, which enables protection of the encapsulated compounds. These features explain the intense utilization of sol–gel materials for sensor construction.

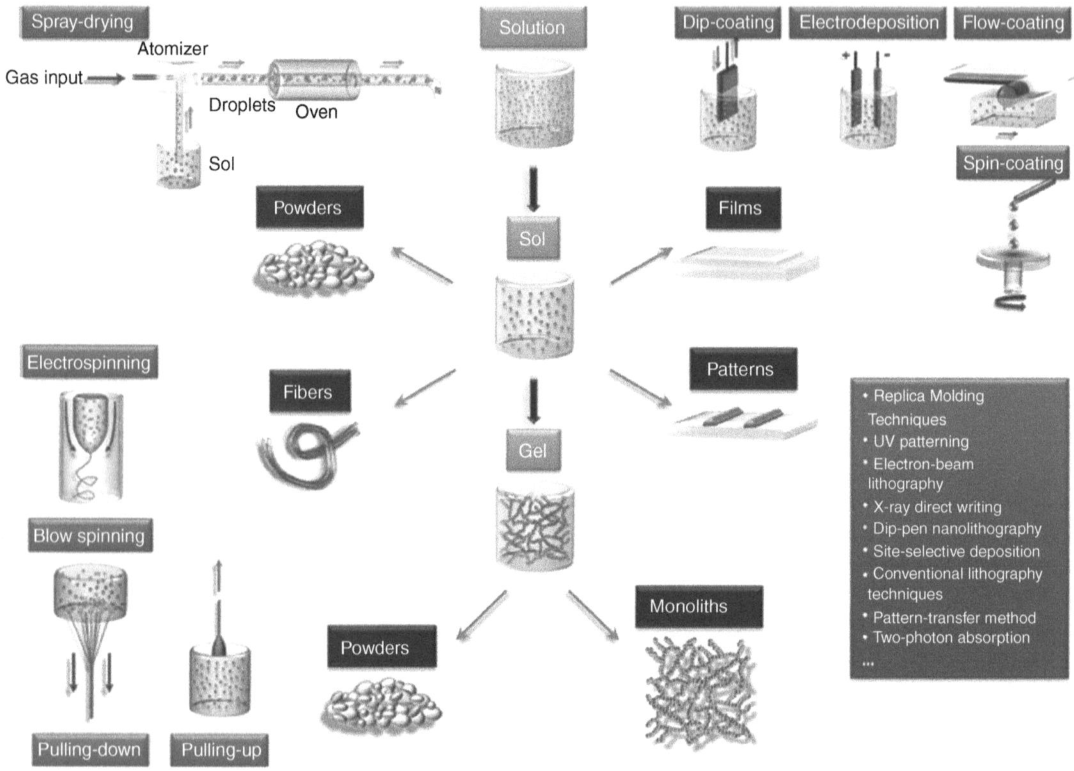

Figure 4.2. Processing routes to materials using sol–gel methods. Reproduced with permission from Sanchez et al. (2011).

Among chemical indicators, the colorimetric ones allow a quick and suitable device for response detection. As the preparation of porous silica can be conducted at room temperature, organic chromophores are incorporated without loss of their properties. The encapsulation of compounds susceptible to color change within the sol–gel matrix when exposed to contact with analyte may be a way to broaden and/or to increase the chemical and thermal properties of these compounds. For efficient performance of the sensor, the sol–gel materials should possess an appropriated porosity and permeability to allow the access of the analytes to the sensing centers. Nevertheless, in spite of the porosity being large enough to allow the analyte to permeate the sensor matrix, this porosity has to small enough to avoid leaching of the sensing molecules from the bulk of the sensor. Moreover, a good-colorimetric sensor must exhibit a low response time.

One of the first applications of sol–gel-derived materials has been the development of a pH sensor via the encapsulation of pH-sensitive indicator dyes within this porous-host matrix with the production of monolithic structures. Zaggout's group immobilized methyl orange (Zaggout et al., 2005) and thymol blue (Zaggout et al., 2006) by acid-catalyzed sol–gel reaction using tetraethylorthosilicate (TEOS) in the presence of HCl as the catalyst and obtained monolithic polysiloxane-solid materials. The entrapment of pH indicators affected the silica particle size and larger silica particles were produced. In this case, the interaction between the pH indicator and silica network was only physical. If this approach tends to preserve the identity of the encapsulated indicator, it may be prone to leaching because no chemical bonding occurs in the silica network.

Another strategy for the development of optical sensors consists entrapping of pH indicators in sol–gel films (Yari and Dinarvand, 2011; Miled et al., 2004). The sensor produced by doping of sol–gel films with bromopyrogallol red was shown to be stable for 10 months without a fall in the sensitivity, response time, and adsorption signal (UV–vis) (Yari and Dinarvand, 2011). In the same direction, mesostructured-silica sol doped with bromothymol blue was deposited by dip coating onto glass slides or optical fibers and tested using UV measurements and evanescent wave sensing, respectively (Miled et al., 2004). The results indicate that the device is sensitive to pH variation with greater sensitivity in the acidic range.

For production of wavelength-dependent pH optical sensors the technique layer-by-layer has been proposed (Raoufi et al., 2012). Layers of brilliant yellow dye, which is negatively charged due to sulfonate groups and the polycation poly(allylaminehydrochloride), were deposited at the end of a bare-silica core of an optical fiber. The sensitivity of this sensor is given in terms of wave-number shift obtained by the slope of a dose–response graph at the inflection point. The authors observed that the sensitivity of the sensors is improved by increasing the number of bilayers up to c. five to six bilayers. Layer numbers higher than six, reduced the sensitivity. The probe coated by six bilayers, displayed a 4.65 nm peak-wavelength shift per 0.2 pH units in the pH range from 6.80 to 9.00, showing the maximum sensitivity.

The preparation route taken during the fabrication of a sol–gel material governs the physical–chemical properties. Sol–gel attempts the use of a pH-indicating indicator encapsulation that can be produced by different sol–gel routes, such as, acidic (Schyrr et al., 2014; Yari and Dinarvand, 2011), basic (Maia et al., 2013; Capeletti et al., 2013), and two-steps (Wu et al., 2006), which affect the textures and morphological properties. The two-steps route consists of the addition of an acidic compound on the silica precursor/solvent mix as the first step and as the second step, the incorporation of an alkaline solution. Therefore, the hydrolysis governs the process in the first step and in the second one the condensation reaction is dominant.

Electrochemical analysis indicated that the interactions of alizarin red with the silica network generated through its encapsulation involves four sol–gel processes, namely hydrolytic acid-catalyzed, hydrolytic base-catalyzed, hydrolytic without catalyst, and nonhydrolytic sol–gel processes and were dependent on the employed sol–gel route (Capeletti et al., 2013). Using differential-pulse voltammetry technique, it was verified that the oxidation peak of encapsulated alizarin, via the acidic route and hydrolytic sol–gel route without catalyst, was dependent on the pH and a slope value of approximately 30 mV was observed. This behavior pointed out a process that implicates two electrons and one proton, suggesting a possible interaction of silica–indicator by the OH groups. For the hydrolytic base-catalyzed route, the electrochemical behavior was different from that of the sensor prepared by the acidic route. These results were associated with the change of silica–dye interactions and with the formation of a higher number of products during the chemical reactions and/or reactions on the electrode surface. Based on the decreasing cathodic-peak intensity for alizarin in the nonhydrolytic route compared to the acidic route, it was proposed that the consumption of alizarin red occurs in its quinone form during the process. In sum, the nature of the encapsulating route may affect the behavior of the resulting solid sensor.

The porosity of sol–gel material is influenced by the incorporation of surfactant during the process. The sol–gel reaction in the presence of surfactant sodium dodecyl sulfate (SDS) render the silica structures to more open, that is, a higher pore diameter leads to increased availability of entrapped indicator (methyl orange). Thus, the UV measurements of immobilized methyl orange/dodecylsulfate showed a stronger band than the immobilized methyl orange (Zaggout et al., 2005). The highly porous silica produced by employing surfactant is a better host for sensing molecules, since the analyte can easily diffuse toward the sensing centers (El Nahhal et al., 2012; Sanchez et al., 2003; Rottman et al., 1999). These features allow a faster kinetics of color change for the materials containing surfactant (El Nahhal et al., 2014).

Besides the influence in silica porosity, the surfactant used in the sol–gel process can favor the acidic, neutral, or basic form of the encapsulated indicator depending on its charge. For instance, the surfactant alkyl hydroxyethyl dimethyl ammonium (HY) protonated cresol red (CR) entrapped within silica, resulting in a strong electrostatic interaction between the protonated form of CR and the HY surfactant molecules (El Nahhal et al., 2012). Thus, only the neutral form of this indicator was observed in the UV spectra. On the other hand, the cetyl trimethyl ammonium bromide (CTAB) acts as a retarding agent for the protonation of CR by silanol groups present in the silica network and the UV band at 445 nm associated to the acid form presents a higher intensity in comparison to the system in absence of a surfactant.

Although the coentrapment of surfactants is efficient to adjust the sensor behavior, their leaching to the liquid media can affect the functionality of the active agents. Therefore, other strategies have been adopted to control the leaching and response of the pH indicator encapsulated, as discussed in the following paragraph.

Indicator leaching can occur at all times in a liquid environment. Thus, it is a problem for sensor applications with prolonged usage time or continuous monitoring. As previously mentioned, the leaching problem emerges because the physical dimensions of the pH indicators may be less than the typical pore diameter of the sol–gel material (Butler et al., 1998). For the modulation of indicator leaching on silica-based sol–gel materials, two effective strategies have been used: (1) covalent linkage between the indicator with large molecules before sol–gel entrapment (Plaschke et al., 1996); and (2) use of organosilanes to make hybrid gels (Kowada et al., 2005; Kassal et al., 2014). Nevertheless, the covalent linkage is usually complex and can reduce the sensitivity of the indicator, since the chemical identity of the indicator may be affected by this grafting reaction.

The introduction of organic functional groups into inorganic precursor results in an inorganic–organic hybrid material known as Ormosil (organically modified silicate) and affects the produced material's properties. Those modifications allow the control of certain physicochemical properties, such as, hydrophobicity, porosity, and ion-permeability, which considerably affect the sensing properties and mechanical stability of the resulting material (Kassal et al., 2014).

With the insertion of organic functionalized silica precursors the pore size and pore volume of sol–gel materials usually decrease and therefore minimizes indicator leaching (Wu et al., 2006). The environment around the indicator within the gel material has shown to be dependent on silica precursors. Increasing the proportion of methyltrimethoxysilane (MTMS) in MTMS/TEOS films increases the pK_a of indicators to higher values in comparison to indicators in an aqueous solution (Kowada et al., 2005). Such results suggest that MTMS contributes to the stabilization of the pH indicators in the gel films. Nevertheless with larger MTMS content, the pH indicators were strongly stabilized, compromising its pH response. For the sol–gel sensor prepared via acid catalysis of TEOS with one of three following organosilanes, namely MTMS, phenyltrimethoxysilane, or isobutyltrimethoxysilane in a 2:1 mole ratio, no appreciable leaching of the pH sensitive compound (bromocresol green or cresol red) was observed after 7 days (Makote and Collinson, 1999) of exposition, while the encapsulated dye leached completely out of the films prepared solely from TEOS in a relatively short period of time (12–48 h).

According to Butler et al. (1998) the key issues in determining the leaching of a dopant from sol–gel-derived materials include the pore diameter of the sol–gel material, as commented previously, and the solubility of the dopant in the leaching solution. The interaction between the sol–gel matrix and the pH indicator changes depending on the pH medium (Wu et al., 2006). In general, in alkaline solutions the indicators have a higher solubility than in acidic ones due to their deprotonation. These features may be partly responsible for their higher and faster release in alkaline solutions (Butler et al., 1998; Wu et al., 2006). Moreover, above pH 11 the leaching increases due to alkaline attacks resulting in silica dissolution (Butler et al., 1998; Wu et al., 2006). It has been reported that a reduced leaching of indicator molecules was obtained by increasing the aging time of the starting sol or by increasing the time and temperature of the drying step (Kraus et al., 1993).

A correlation of the surface free energy data obtained using contact-angle measurements with the pH sensing response of bromocresol green immobilized in a silica hybrid matrix was proposed by Kassal et al. (2014). They observed that the polar component of the film surface free energy (γ^p in Owens–Wendt and Wu models) gradually decreases from most polar organic modification (glycidoxypropyl group, 17.0 mJ m^{-2}) toward increasingly more lipophilic groups (methyl, 12.6 mJ m^{-2}; ethyl, 7.5 mJ m^{-2}, and phenyl 5.8 mJ mm^{-2}). A good correlation between the response time and an increase in lipophilicity of the organic group was verified, with response times being 7.5, 9.8, and 11.6 min for methyl, ethyl, and phenyl groups, respectively. This trend does not explain the results of glycidoxypropyl group, which, despite having the highest surface hydrophilicity, exhibited a relatively long response time (11.0 min). The glycidoxypropyl group (with two oxygen atoms) can become involved in hydrogen-bonding networks with other silanol groups and water molecules on the sol–gel surface. This effect has been attributed by the authors to be the essential reason for the prolonged-response times of this sensor.

The reactivity of the entrapped indicators depends mainly on their chemical nature and on the interactions between the indicators and the silica matrices (Wu et al., 2006). Inside the pores, their surface has both charges associated hydrogen bonds, which would clearly influence the passage of any ions (including small species such as the hydrogen ion) through the bulk of the silica. When the dye materials employed in pH sensors interact with the electrostatic network within the porous structure of silica, the activity of the dye toward the analyte is reduced, resulting in longer response times (Ismail et al., 2002; Kassal et al., 2014).

The response time in acid solution of methyl orange and phenyl red incorporated within TEOS/bis(trimethoxysilyl) hexane and TEOS/*n*-propyltriethoxysilane matrix was 2–4 min and 20–25 min, respectively. In an alkaline solution the response time response is c. 10 min (Wu et al., 2006). Methyl orange and methyl red exhibit similar chemical structures, however, only the entrapped methyl-orange system shows an acid–base response. Methyl orange has a SO_3 group, while methyl red has a COOH one. The latter may complex with silicon atoms and take part in the sol–gel process. The carboxylic acids can interact with ethoxysilane groups to form silyl esters, leading to the formation of Si–O–Si networks and carboxylic acid–ester by-products. Thus, the group responsible by the acid–base process of methyl red is involved in parallel reactions and that system does not respond to pH changes.

The entrapment of indicator in sol–gel matrix may shift the pH color change range (Wu et al., 2006). For example, methyl orange exhibits a pH indicating range of 3.1–4.4. Gels of methyl orange suffer from color changes from red (acidic color) to yellow (basic color) when pH solution values are higher than 7.0 for TEOS/bis(trimethoxysilyl) hexane system and higher than 9.0 for TEOS and TEOS/*n*-propyltriethoxysilane materials. The color changes from yellow (basic color) to red (acidic color) were verified in solutions with pH lower than 4.0–5.5. The same behavior was found for phenyl red. According to the researchers, the pH-color change values of the indicators entrapped in the gels present a hysteresis compared to the cases of free indicators in solutions (Wu et al., 2006).

External parameters, such as, humidity may also affect the response of the sensors. Humidity provokes the partial blocking of the sol–gel pores by water molecules and therefore the exposition of oxygen sensor to the humidity medium, reduces its response. In order to overcome such limitations, a more hydrophobic ethyltriethoxysilane-based matrix has been reported in the literature to be less sensitive to the interference effect of ambient humidity in comparison to the methyltriethoxysilane matrix (Wencel et al., 2005).

In sum, the reactivity and release of the indicators entrapped in the gels is regulated by a combination of factors including textural properties of the silica materials, indicator solubility, and interactions between the silica matrix and the indicators. After a careful study of the parameters of sol–gel process several factors can be improved that include, the inclusion of organofunctionalized silica precursor, pH of starting solution, and type of pH indicator. The consideration of these aspects can increase the performance of pH sensor, long-term stability, and shelf-life time can be achieved.

4 Application of Encapsulated Sensors in Food Science

For efficient development of a food spoilage indicator it is essential to know the characteristics of the quality-indicating metabolites (Nopwinyuwong et al., 2010). During storage of the food products in air packaging, the growth of aerobic and sometimes anaerobic microorganisms can occur. Organic acids, such as, lactic acid and acetic acid are the major compounds that play a role in glucose fermentation by lactic-acid bacteria. Ethanol is another major end product of the fermentative metabolism of lactic-acid bacteria. The loss of meat freshness results in the transformation of its protein into smaller compounds, such as free amino acids, by microorganisms with proteolytic activity or by hydrolysis (Rukchon et al., 2014). These amino acids can be degraded partially or totally into simple compounds, such as, carbon dioxide, water, ammonia, and hydrogen sulfide. Another indicator of microbiological spoilage of food proteins, mainly in seafood, is the high level of total volatile basic nitrogen, for example, ammonia, dimethylamine, and trimethylamine. Concerning foods, the sol–gel pH-sensor technology have been employed for detection of amine (Bueno et al., 2015; Oberg et al., 2006), formaldehyde (Meng et al., 2014), carbon dioxide species (McEvoy et al., 2004; Morris et al., 2004), and acetic acid (Kuswandi et al., 2013).

The detection of amine vapors is a particular interest in food quality control. For example, the degree of freshness of fish and meat can be estimated by control of biogenic amines produced by bacterial spoilage. Sensitive detection of amine vapors, based on the indicator bromocresol green adsorbed onto silica spheres matrix, was developed by Obberg (Oberg et al., 2006). The absorbance of diethylamine, triethylamine, and *tert*-butylamine was measured and the detection of amine compounds was performed, that verified a correlation between the response of sensor and the basicity of the amine on the silica surface. A sensor based on immobilization of methyl red onto cellulose membrane was proposed for real-time monitoring of chicken freshness (Kuswandi et al., 2014). The pH of fresh chicken meat is c. 5.50 and for spoilt chicken meat, pH > 6.0 is usually detected. The pH increment may occur during deterioration of meat products due to the increase in volatile amine concentration. The methyl red/cellulose membrane presents a pH in the range of 5.0–6.0, where the color changes (from red to yellow) as a result of a pH shift due to the increase in volatile-amine concentration, which in turn may indicate the spoilage. The sensor accurately responded to the increase in pH value in the package headspace, being detected after 8 h and after 7 days for room and chiller temperatures, respectively, indicating spoilage (Kuswandi et al., 2014).

The development of a plastic device to discriminate between different types of amines (triethylamine, isobutylamine, and isopentylamine) was recently reported (Bueno et al., 2015). The authors separately immobilized five pH indicators (namely alizarin, bromophenol blue, chlorophenol red, methyl red, and thymol blue) in solvent-cast cellulose acetate membrane with plasticizer Tween, without previous sol–gel encapsulation. To extract red, green, and blue (RGB) values before and after exposition of each individual amine a smartphone was employed and an inhouse software analyzed data by nonsupervised pattern recognition methods, multianalytical methods, hierarchical clustering analysis (HCA), and principal component analysis (PCA). Meat samples were contaminated with each amine in order to simulate the contamination with microorganisms that produce such volatiles. There are three clear clusters in the plots obtained from results that are presented in Fig. 4.3 indicating the possibility to discriminate meat samples contaminated with each of the amines. The explanation proposed by the authors, for the color changes (red to orange or green to violet, for instance) of the membranes according to the nature of the amines, is based on the difference of polarities. Generally the solubility decreases with the increase in the number of carbon atoms, which is a mechanism of color change controlled by the amines' solubility and acidity constants. It is worth mentioning that the sensor array did not change color in the presence of meat without contamination.

Concerning fish, Zaragozá et al. (2012, 2013) proposed an optoeletronic nose in the shelf-life assessment of fresh sea bream in cold storage. Chromogenic indicators with different chemical recognition properties were incorporated into inorganic materials (aluminium oxide and silica gel) forming an optoelectronic nose. The color change of optoeletronic array was measured for 11 days using RGB color and Cielab coordinates were correlated

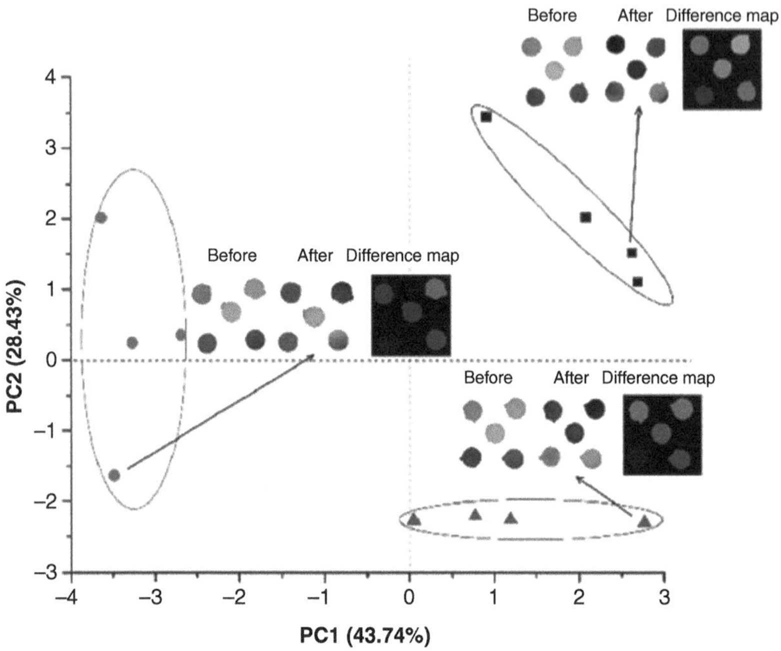

Figure 4.3. PCA-scores plot obtained from the RGB values extracted from the color membranes with five pH indicators, in contact with volatile phase, collected from a chamber containing meat contaminated with isobutylamine, triethylamine, and isopentylamine. Highlighted images show the real images of the color-change profile (red to orange or green to violet, for instance) of the sensor array for one of the experiments before and after the meat is contaminated with each amine and the difference map for each experiment respectively. Reproduced with permission from Bueno et al. (2015).

to physicochemical parameters and microbiological analysis was performed. They detected a clear loss of freshness from 4 to 7 days. Based on the model, which correlates the chromogenic data with concentrations of bacteria, the authors described the feasibility of this system to assist the development of optoelectronic noses for fish freshness monitoring. The use of a chromogenic array, based on several chromogenic indicators having a potential reactivity against volatile organic compound supported in silica and alumina, has been reported for the control of evolution of boiled-marinated turkey packaged in modified atmosphere packaging (MAP) (30% CO_2:70% N_2) (Salinas et al., 2014a). Moreover, the microbiological and sensorial evolution of the boiled-marinated turkey during the aging process was also evaluated. The colorimetric response of the array was studied by HCA. HCA considers a complete dimensionality of the data and provides a dendrogram as output. The resulting dendrogram showed three main clusters with the classification fresh (0–24 days), not-fresh (31–38 days), and not-edible (45 day); presenting quite a good relationship with sensory evaluation and microbiological analysis. A similar device was capable of monitoring meat spoilage of fresh pork sausages under modified MAP, offering characteristic-colorimetric fingerprints (Salinas et al., 2014b)

Sensors for detection of ammonia were prepared by deposition of phenyl-triethoxysilane, MTMS, dimethyl-dimethoxysilane, and aminopropyl-triethoxysilane sol–gel layer containing bromophenol blue on glass substrate. The sensitivity of the sensors calculated by reflectance measurements increased in the phenyl–methyl–dimethyl order (Markovics and Kovacs, 2013). Such results point out that the choice of silica precursor is a determinant for good performance of a sensor.

A very interesting study reported the employment of a tag in a polyamide laminated with linear low-density polyethylene packaging to detect the deterioration of eggs dessert (Nopwinyuwong et al., 2010) and chicken (Rukchon et al., 2014). The tag consists of two or more groups of pH-sensitive dyes added in hydropropyl methylcellulose with polyethyleneglicol as plastifier. The first group was a mixture of bromothymol blue and methyl red (M1 type) and the second one a mixture of bromothymol blue, bromocresol, and phenol red (P type) (Nopwinyuwong et al., 2010). The spoilage microorganisms of two analyzed foods generated carbon dioxide, which forms carbonic acid in the presence of moisture. Carbonic acid is diprotic and forms hydrogen ions (H^+) and bicarbonate ions (HCO_3^-). H^+ combines with water molecule to form a hydronium ion, (H_3O^+). This later reacts with the basic form (In^-) of the indicator label, resulting in an acid form (HIn), which in turn produces a color change of the indicator label. Fig. 4.4 depicts samples of skinless chicken breast with food spoilage indicator labels. For instance, indicator label may change from green (fresh) to yellow (spoilage) or violet (fresh) to green (spoilage) depending on the employed label. The sensibility and rate response was dependent on the groups of pH-sensitive dyes (Rukchon et al., 2014).

A sensor for formaldehyde was proposed comprising of an Ormosil coating, prepared from methyltriethoxysilane and polydimethylsiloxane through the sol–gel process as host matrix and immobilization of rose anthocyanin and hydroxylamine sulfate (Meng et al., 2014). The formaldehyde reacted with hydroxylamine sulfate producing acidic proton, which turned the pH of the Ormosil coating close to 1.5 and converted the hemiketal form (pale yellowish) of anthocyanin into flavylium form (deep rosy). The responses of the sensing coating to various gases, such as carbon monoxide, benzene, toluene, xylene, chloroform, hexane, ethanol, methanol, ammonia, and water were examined in parallel, and were found to have a good specificity to formaldehyde.

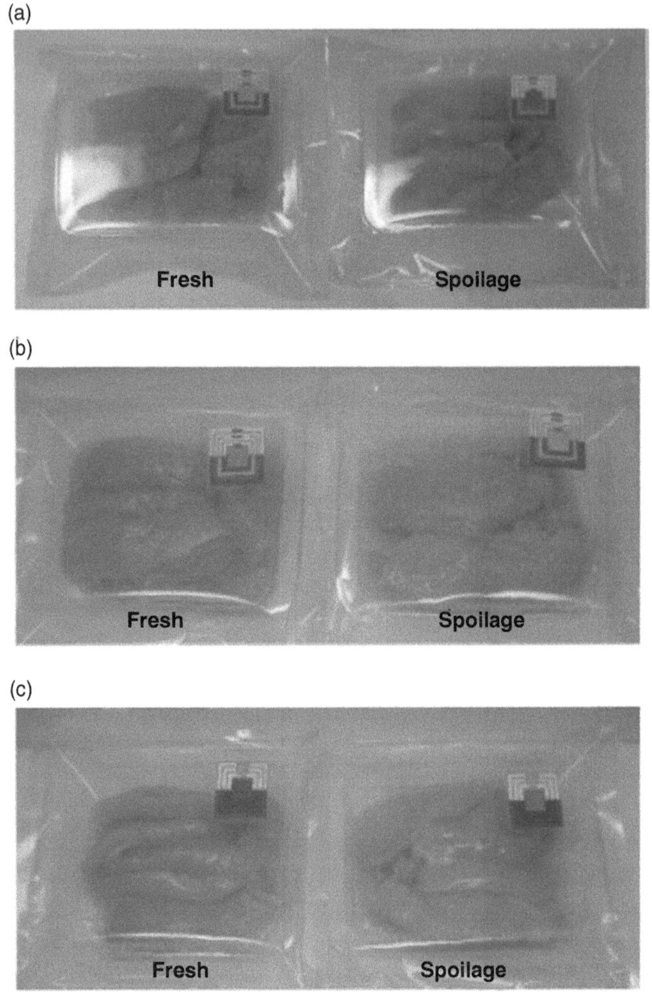

Figure 4.4. Packaged skinless-chicken breast with different food spoilage indicator labels. (a) green = fresh, yellow = spoilage; (b) green = fresh, orange–yellow = spoilage; and (c) violet = fresh, green = spoilage. Reproduced with permission from Rukchon et al. (2014).

For detection of guava deterioration, bromophenol blue was immobilized onto cellulose membrane (Kuswandi et al., 2013). The pH value of the guava varied initially from pH 3.94 at freshly ripe to pH 3.56 at an overripe stage after 7 days due to the production of acetic acid. Consequently, the bromophenol blue immobilized changed color from blue to green. The rate of color change of the pH indicator (intensity of blue in arbitrary unit) to various freshness states can be observed in the Fig. 4.5. A good correlation

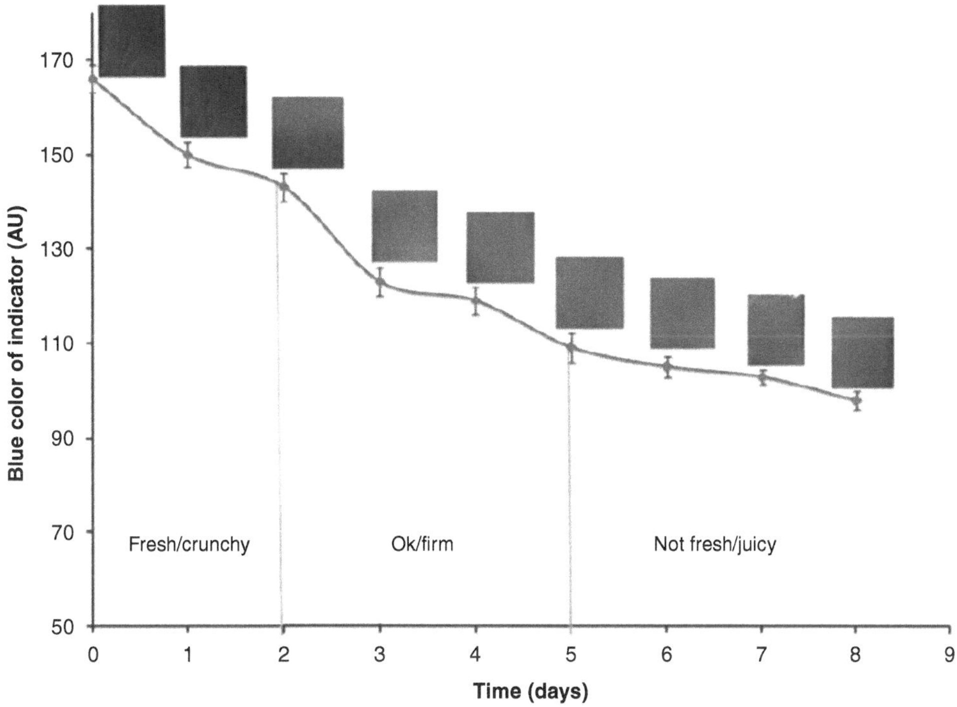

Figure 4.5. The rate of color change of the indicator response toward guava freshness state. Reproduced with permission from Kuswandi et al. (2013).

between the color indicator and pH values was found and the onset of overripeness was detected after 5 days.

A change in a gaseous bacterial metabolite concentration is indicative of bacterial growth in a perishable food product, as in the case of carbon dioxide production. A gas-permeable sensor, housed for the detection carbon dioxide, adapted to contain the pH indicator, and positioned in the interior of food packaging was patented (Morris et al., 2004). The sensor housing inhibits light degradation of the pH indicator and comprises one of silicone and a combination of agar and silicone products. In another patent, the carbon dioxide sensor comprises a pH indicator and a long-lived reference luminophore, which is doped in sol–gel particles and coimmobilized with the pH indicator in a porous sol–gel matrix, or immobilized in a separate oxygen impermeable layer, and the pH indicator in a sol–gel matrix being laid over the impermeable layer (McEvoy et al., 2004).

In spite of several successfully developed encapsulated systems, their commercialization meets some restrictions, especially concerning the chemical products involved in the sensing system, which may be present for the allowed products for a given type of food or which are yet to be submitted to a rigorous test set. In the following section, some aspects concerning the scope of the food regulatory agencies, regarding to the intelligent packaging are discussed.

5 Regulation for Nanotecnology and Active and Intelligent Packaging

As mentioned earlier, the basic functions of packaging have been classified into four categories: protection, communication, convenience, and containment (Robertson, 2013). According to these basic functions, the traditional-food packaging should be inert, ensure that the food-contact material does not pose a risk to health, and prevent the release of substances that change organoleptic properties of the food stuffs (taste, appearance, texture, or even smell). Guidelines and requirements of food contact materials legislations were built on the basis of this concept of packaging.

In Europe, the food contact-material legislation was revised in 2004 and the European Union's Regulation 1935/2004 introduced specific provisions concerning active and intelligent packaging. The former refers to a system that extends the shelf life of food, while the latter deals with systems that react when the food quality has deteriorated (packaging that changes color, for example). In addition to these general requirements, specific rules for both, active and intelligent materials and articles, are laid down in the Regulation (EC) No. 450/2009, adopted in 2009. One of the main aspects of these new regulations is that the safety assessment of a substance or of a combination of substances that constitute the components of active- and intelligent packaging should be carried out by the European Food Safety Authority (EFSA), after the submission of a valid application, in agreement with Articles 9 and 10 of Regulation (EC) No. 1935/2004. After the EFSA has performed a risk assessment, it will issue (as an opinion) the substances that should be included in the community list of authorized substances that may be used in active- and intelligent components.

According Regulation (EC) No. 450/2009, the intelligent packaging systems, by the monitoring of the condition of packaged food (or of its environment), are supposed to provide to the user with information on the conditions of the food and should not release their constituents into the food (Restuccia et al., 2010). These systems may be positioned on the outer surface of the package and may be separated from the food by a functional barrier, which is a barrier within food-contact materials or articles preventing the migration of substances from that barrier into the food. Moreover, specific requirements for active- and intelligent systems are contained in this regulation, such as, migration limits, rules concerning the declaration of compliance, and the specific-labeling rules.

In the United Sates, the safety of each individual substance that comprises the article is covered by some regulatory status, including an effective Food Contact Substance Notification (FCSN). Moreover, the substance of interest must be (1) safe; (2) not introduce any harmful products into the food; (3) not contaminate the food (physically); and (4) not affect taste or odor of the food. The safety of the used substance may be dictated by the dietary exposure resulting from the intended use.

If the proposed active or intelligent packaging does not involve any substance to be added to the food or has no technical effect (indirect additive), the US Food and Drug Administration (FDA) has dictated no special regulation. The substances are regulated as other food-contact substances. Conversely, if the added substance is a "direct additive," it should be analyzed according to the FDA.

No specific legislation for such packaging has been published in South America (Mercosur) until now. Indirect additives can be employed if included in the Positive List of additives for plastic materials (packaging and equipment manufacturing). Direct additives must follow the specific national regulation and have to be approved as food additives by the national legislation.

Many developed materials can be classified as nano- or nanostructured materials. The main concern regarding the use of these substances is to assess if the migration behavior from nanomaterials is different from those of regular substances and whether substances migrating are more reactive and have a differential toxicological profile. Considering this, different countries and regions are developing general strategies and requirements for the use of nanomaterials in food packaging materials. For instance, there are no US-specific legislations for nanomaterials and thereby nanomaterials fall under the competence of the existing authorities. Although there are no legislations in the United States, the main registration for chemical substances is regulated by the US Environmental Protection Agency (EPA) through the Toxic Substances Control Act [TSCA, 15 USC (C. 53) 2601–2692]. The TSCA has many parts, and governs the listing, risk assessment, and reporting of existing and new substances produced in volumes more than 10000 kg/year.

New substances are registered by the submission of a Premanufacturing Notice (PMN) to EPA at least 90 days before the start of manufacture or import. For an existing substance, EPA may issue a Significant New Use Notification (SNUR), requiring additional chemical toxicity and/or environmental information for a nanoscale version of the substance. For a new substance a PMN is required. EPA will assess whether the nanomaterial presents or may present an unreasonable risk to human health or to the environment, similar to other new chemicals.

The FDA regulates a wide range of products, including foods, cosmetics, drugs, devices, veterinary products, and tobacco products; some of which may utilize nanotechnology or contain nanomaterials. In Jun. 2014, the FDA issued three final guidance documents reflecting the FDA's current thinking on certain issues related to the use of nanotechnology in FDA-regulated products. These guidance documents have been issued as part of the FDA's ongoing implementation of recommendations from the FDA's 2007 Nanotechnology Task Force Report. The FDA issues draft guidance on the use of nanotechnology in food for animals (Food Drug Administration, 2015).

In Europe, the Commission adopted the Commission Recommendation (2011/696/EU) on the definition of nanomaterial on Oct. 18, 2011. The Recommendation responds to the increasing use of specific legislative provisions addressing nanomaterials and the need to ensure harmonized terminology across different pieces of legislation, as well as, a request to the European Parliament. The current EU definition for nanomaterial means a natural, incidental, or manufactured material containing particles, in an unbound state, as an aggregate, or as an agglomerate; where, for 50% or more of the particles in the number size distribution, one or more external dimensions, is in the size range 1–100 nm. This definition of nanomaterials will be integrated in the EU legislation, where appropriate.

Nanomaterials and their risks has been the subject of several opinions of the Scientific Committee on Emerging and Newly Identified Health Risks (SCENIHR). The overall conclusion so far is that, even though nanomaterials being not per se dangerous, scientific uncertainty about the safety of nanomaterials still exists for many aspects. Therefore the safety assessment of the substances must be done on a case-by-case basis and current risk assessment methods are applicable, even if work on particular aspects of risk assessment is still required. Some manufactured nanomaterials have already been in use for a long time (eg, carbon black, titanium dioxide) and show low toxicity.

In Europe, nanomaterials are also regulated by REACH (Registration, Evaluation, Authorization and Restriction of Chemicals) and CLP (Classification, Labeling, and Packaging) because they are covered by the definition of a chemical "substance" in both regulations. The general obligations in REACH include registering substances manufactured at 1 ton or more and providing information of the supply chain; that apply to nanomaterials as for any other substance.

Nanomaterials, as well as, other materials that fulfill the criteria for classification as hazardous under Regulation (EC) No. 1272/2008 on CLP of substances and mixtures must be notified to European Chemicals Agency (ECHA), classified and labeled. According to the EU Commission nowadays the REACH sets the best possible framework for the risk management of nanomaterials when they occur as substances or mixtures but more specific requirements for nanomaterials within the framework have proven necessary and therefore envisages modifications in some of the REACH Annexes (European Commission, 2015).

Until now, South America has no specific legislation for food packaging with nanomaterials. Although there is no time frame, in Brazil, the National Agency of Sanitary Control (ANVISA), establishes the general criteria and classification of materials for packaging and equipment in contact with food. ANVISA has identified the need of development of standards, requiring the applicant to report about the nature of nanotechnology products and processes, the need to establish standards or specific guides for the safety assessment, and the need to monitor the control of products (cosmetics, medicines, disinfectants, etc.) or nanotechnological processes in the ANVISA area of competence.

It is worth noting that the use of nanomaterials must be approved not only as a substance for use in food packaging materials but also in accordance with the specific regulations for use in the active- or intelligent packaging systems.

6 Final Remarks

Lifestyle changes, market trends, and the growing level of demand by the consumers has boosted the packaging industry to focus on the quest for solutions that provide maximum food security, while keeping its nutritional value at competitive prices. Generally, packaging is considered to provide functions, such as, containment, protection, convenience, and communication. Advanced materials, such as intelligent packaging, are capable of maintaining the quality of the product, monitoring the history of the packaging, and communicating this information to the consumer. Quality assurance of foods either within the processing chain or during their commercialization until the ingestion by consumer is of extreme importance. It involves an issue of public health, which is pushing a demanding society for more restrictive-regulatory instruments. Therefore, advanced packaging with enhanced functions has constantly been investigated and among them, intelligent packaging is becoming a reality in the present market.

Intelligent packaging systems provide information to the user on the conditions of the food or its environment, such as, temperature and pH. During storage in packaging, acids, amines, carbon dioxide, and aldehydes are produced depending on the mechanism of food-deterioration process. Thereby, monitoring of pH in food packaging provides information about the conservation of the state of food.

The development of pH sensors by pH-indicator encapsulation via sol–gel process is a strategy that improves its thermal stability, required in the production of polyolefin-intelligent packaging. This technology can be integrated in novel intelligent food-packaging system. Despite the versatility of the sol–gel process modulation of a fast detection with no pH-indicator leaching from the sol–gel matrix depends on the tuning of several parameters, such as, employment of organosilicon as precursors, suitable surfactants, and silica-route production. A more simple strategy is the direct incorporation of a pH indicator in the food packaging, in this case, usually by a biopolymer. A drawback of biopolymer packaging is the low-barrier properties to water vapor and oxygen, which in some circumstances falls short of the basic requirements of a good packaging.

Adoption of intelligent-packaging technology by the food industry will offer consumer, benefits and convenience, and undoubtedly result in a further improvement in the food security. Nevertheless, for industrial implementation of such technology, several parameters still demand investigations, such as, a balance in packaging quality, sensor functionality, and absence of toxicity in a necessary equilibrium to bring this technology to the consumer market; which is not a simple ordeal.

Acknowledgments

The authors acknowledge the financial support of CNPq (Project 403342/2013-9). V. Dalmoro thanks the CAPES (PNPD program) for the postdoctoral grant.

References

Alavi, S., Thomas, S., Sandeep, K.P., Kalarikkal, N., Varghese, J., Yaragalla, S., 2015. Polymers for Packaging Application, first ed. Apple Academic Press, Waretown, NJ.

Alocilja, E.C., Radke, S.M., 2003. Market analysis of biosensors for food safety. Biosens. Bioelectron. 18, 841–846.
Arora, A., Pauda, G.W., 2010. Nanocomposites in food packaging. J. Food Sci. 75, 43–49.
Arvanitoyannis, I.S., Stratakos, A.C., 2012. Application of modified atmosphere packaging and active/smart technologies to red meat and poultry: a review. Food Bioprocess. Tech. 5, 1423–1446.
Azeredo, H.M.C., 2009. Nanocomposites for food packaging applications. Food Res. Int. 42, 1240–1253.
Biji, K.B., Ravishankar, C.N., Mohan, C.O., Gopal, T.K.S., 2015. Smart packaging systems for food applications: a review. J. Food Sci. Technol. 52 (10), 6125–6135.
Bordes, P., Pollet, E., Avérous, L., 2009. Nano-biocomposites: biodegradable polyester/nanoclay systems. Prog. Polym. Sci. 34, 125–155.
Bueno, L., Meloni, G.N., Reddy, S.M., Paixão, T.R.L.C., 2015. Use of plastic-based analytical device, smartphone and chemometric tools to discriminate amines. RSC Adv. 5, 20148–20154.
Butler, T.M., MacCraith, B.D., McDonagh, C., 1998. Leaching in sol–gel-derived silica films for optical pH sensing. J. Non-Crystal. Sol. 224, 249–258.
Cabedo, L., Feijoo, J.L., Villanueva, M.P., Lagarón, J.M., Giménez, E., 2006. Optimization of biodegradable nanocomposites based on PLA/PCL blends for food packaging applications. Macromol. Symp. 233, 191–197.
Capeletti, L.B., Bertotto, F.L., Dos Santos, J.H.Z., Moncada, E., Cardoso, M.B., 2010. The effect of the sol–gel route on the characteristics of acid–base sensors. Sens. Actuat. B 151, 169–176.
Capeletti, L.B., Dos Santos, J.H.Z., Moncada, E., Da Rocha, Z.N., Pepe, I.M., 2013. Encapsulated alizarin red species: the role of the sol–gel route on the interaction with silica matrix. Powder Technol. 237, 117–124.
de Abreu, D.A.P., Losada, P.P., Angulo, I., Cruz, J.M., 2007. Development of new polyolefin films with nanoclays for application in food packaging. Eur. Polym. J. 43, 2229–2243.
De Corcuera, J.I.R., Cavalieri, R.P., 2003. Biosensors. In: Heldman, D. (Ed.), Encyclopedia of Agricultural, Food, and Biological Engineering. Marcel Dekker, New York, pp. 119–123.
Duncan, T.V., 2011. Applications of nanotechnology in food packaging and food safety: barrier materials, antimicrobials and sensors. J. Colloid Interface Sci. 363, 1–24.
El Nahhal, I.M., Zourab, S.M., Kodeh, F.S., Babonneau, F., Hegazy, W., 2012. Sol–gel encapsulation of cresol red in presence of surfactants. J. Sol-Gel Sci. Technol. 62, 117–125.
El Nahhal, I.M., Zourab, S.M., Kodeh, F.S., Abdelsalam, F.H., 2014. Sol–gel encapsulation of bromothymol blue pH indicator in presence of Gemini 12-2-12 surfactant. J. Sol-Gel Sci. Technol. 71, 16–23.
Eulises, U., 2013. Fullerenes and their applications in science and technology. Introduction to nanotechnology. Florida International University. EEE-5425, p. 1–5.
European Commission, 2015. Nanomaterials. Available from: http://ec.europa.eu/growth/sectors/chemicals/reach/nanomaterials/index_en.htm
Food Drug Administration, 2015. FDA issues three final guidances related to nanotechnology applications in regulated products, including cosmetics and food substances. Available from: http://www.fda.gov/ScienceResearch/SpecialTopics/Nanotechnology/ucm301093.htm
Han, J.H., 2005. Innovation in Food Packaging. Elsevier Academic Press, Winnipeg, Canada, pp. 3–5.

Hong, H., Dang, J., Tsai, Y., Liu, C., Lee, W., Chen, P., 2011. An RFID application in the food supply chain: a. case study of convenience stores in Taiwan. J. Food Eng. 106, 119–126.
Hulanicki, A., Glab, S., Ingman, F., 1991. Chemical sensors definitions and classification. Pure Appl. Chem. 63, 1247–1250.
Ismail, F., Malins, C., Goddard, N.J., 2002. Alkali treatment of dye-doped sol–gel glass films for rapid optical pH sensing. Analyst 127, 253–257.
Kassal, P., Šurina, R., Vrsaljko, D., Steinberg, I.M., 2014. Hybrid sol–gel thin films doped with a pH indicator: effect of organic modification on optical pH response and film surface hydrophilicity. J. Sol-Gel Sci. Technol. 69, 586–595.
Kerry, J.P., O'Grady, M.N., Hogan, S.A., 2006. Past, current and potential utilisation of active and intelligent packaging systems for meat and muscle-based products: a review. Meat Sci. 74, 113–130.
Kleiböhmer, G., Matzik, I., Heiduk, C.B., 1999. US Patent EP93,204,0A1. Method for checking the perishableness condition of foods
Kowada, Y., Ozeki, T., Minami, T., 2005. Preparation of silica-gel film with pH indicators by the sol–gel method. J. Sol-Gel Sci. Technol. 33, 175–185.
Kraus, S.C., Czolk, R., Reichert, J., Ache, H.J., 1993. Optimization of the sol-gel process for the development of optochemical sensors. Sens. Actuat. B 199, 199–202.
Kuswandi, B., Maryska, C., Abdullah, J.A., Heng, L.Y., 2013. Real time on-package freshness indicator for guavas packaging. Food Measure 7, 29–39.
Kuswandi, B., Oktaviana, R., Abdullah, J.A., Heng, L.Y., 2014. Novel on-package sticker sensor based on methyl red for real-time monitoring of broiler chicken cut freshness. Packag. Technol. Sci. 27, 69–81.
Lagaron, J.M., Lopez-Rubio, A., 2011. Nanotechnology for bioplastics: opportunities, challenges and strategies. Trends Food Sci. Technol. 22, 611–617.
Lin, J., Brown, C.W., 1997. Sol-gel glass as a matrix for chemical and biochemical sensing. Trends Anal. Chem. 16, 200–211.
Maciel, V.B.V., Yoshida, C.M.P., Franco, T.T., 2012. Development of a prototype of a colourimetric temperature indicator for monitoring food quality. J. Food Eng. 111, 21–27.
Maia, F., Tedim, J., Bastos, A.C., Ferreira, M.G.S., Zheludkevich, M.L., 2013. Nanocontainer-based corrosion sensing coating. Nanotechnology 24, 415502.
Makote, R., Collinson, M.M., 1999. Organically modified silicate films for stable pH sensors. Anal. Chim. Acta 394, 195–200.
Markovics, A., Kovacs, B., 2013. Fabrication of optical chemical ammonia sensors using anodized alumina supports and sol–gel method. Talanta 109, 101–106.
Marsh, K., Bugusu, B., 2007. Food packaging—roles, materials, and environmental issues. J. Food Sci. 72, 39–55.
McEvoy, A., MacCraith, B., McDonagh, C., von Bultzingslowen, C., von Bultzingslowen, O., 2004. US Patent 20,060,257,094. Optical CO_2 and combined O_2/CO_2 sensors.
McFarlane, D., Sheffi, Y., 2003. The impact of automatic identification on supply chain operations. Int. J. Logist. Manag. 14, 1–17.
McShane, M.J., Brown, J.Q., Guice, K.B., Lvov, Y.M., 2002. Polyelectrolyte microshells as carriers for fluorescent sensors: loading and sensing properties of a ruthenium-based oxygen indicator. J. Nanosci. Nanotechnol. 2, 411–416.
Melde, B.J., Johnson, B.J., Charles, P.T., 2008. Mesoporous silicate materials in sensing. Sensors 8, 5202–5228.
Meng, Q., Han, T., Wang, G., Zheng, N., Cao, C., Xie, S., 2014. Preparation of a natural dye doped Ormosil coating for the detection of formaldehyde in the optical gas sensor. Sens. Actuat. B 196, 238–244.

Mensitieri, G., Di Maio, E., Buonocore, G.G., Nedi, I., Oliviero, M., Sansone, L., Iannace, S., 2011. Processing and shelf life issues of selected food packaging materials and structures from renewable resources. Trends Food Sci. Technol. 22, 72–80.
Miled, O.B., Grosso, D., Sanchez, C., Livage, J., 2004. An optical fibre pH sensor based on dye doped mesostructured silica. J. Phys. Chem. Solids 65, 1751–1755.
Morosanova, E.I., 2012. Silica and silica–titania sol–gel materials: synthesis and analytical application. Talanta 102, 114–122.
Morris, R., McMorris, J.A., Acosta, G., Hill, J., Tank, A.R., Bishop, A., 2004. US Patent 20,040,115,319. Food-borne pathogen and spoilage detection device and method.
Nopwinyuwong, A., Trevanich, S., Suppakul, P., 2010. Development of a novel colorimetric indicator label for monitoring freshness of intermediate-moisture dessert spoilage. Talanta 81, 1126–1132.
Oberg, K.I., Hodyss, R., Beauchamp, J.L., 2006. Simple optical sensor for amine vapors based on dyed silica microspheres. Sens. Actuat. B 115, 79–85.
Pavelková, A., 2012. Intelligent packaging as device for monitoring of risk factors in food. J. Microbiol. Biotechnol. Food Sci. 2, 282–292.
Plaschke, M., Czolk, R., Reichert, J., Ache, H.J., 1996. Stability improvement of optochemical sol-gel film sensors by immobilisation of dye-labeled dextrans. Thin Solid Films 279, 233–235.
Raoufi, N., Surre, F., Sun, T., Rajarajan, M., Grattan, K.T.V., 2012. Wavelength dependent pH optical sensor using the layer-by-layer technique. Sens. Actuat. B 169, 374–381.
Ray, S.S., Easteal, A., Quek, S.Y., Chen, X.D., 2006. The potential use of polymer-clay nanocomposites in food packaging. Int. J. Food Eng. 2, 1–11.
Realini, C.E., Marcos, B., 2014. Active and intelligent packaging systems for a modern society. Meat Sci. 98, 404–419.
Restuccia, D., Spizzirri, U.G., Parisi, O.I., Cirillo, G., Curcio, M., Iemma, F., Puoci, F., Vinci, G., Picci, N., 2010. New EU regulation aspects and global market of active and intelligent packaging for food industry applications. Food Control 21, 1425–1435.
Rhim, J.W., Ng, P.K.W., 2007. Natural biopolymer-based nanocomposite films for packaging applications. Crit. Rev. Food Sci. Nutr. 47, 411–433.
Robertson, G.L., 2013. Food Packaging—Principles and Practice, third ed. CRC Press, Taylor & Francis Group, Boca Raton, FL.
Rottman, C., Grader, G., Hazan, Y.D., Melchior, S., Avinir, D., 1999. Surfactant-induced modification of dopants reactivity in sol-gel matrixes. J. Am. Chem. Soc. 121, 8533–8543.
Rukchon, C., Nopwinyuwong, A., Trevanich, S., Jinkarn, T., Suppakul, P., 2014. Development of a food spoilage indicator for monitoring freshness of skinless chicken breast. Talanta 130, 547–554.
Salinas, Y., Ros-Lis, J.V., Vivancos, J., Martínez-Máñez, R., Aucejo, S., Herranz, N., Lorente, I., Garcia, E., 2014a. A chromogenic sensor array for boiled marinated turkey freshness monitoring. Sens. Actuat. B 190, 326–333.
Salinas, Y., Ros-Lis, J.V., Vivancos, J., Martínez-Máñez, R., Marcos, M.D., Aucejo, S., Herranz, N., Lorente, I., Garcia, E., 2014b. A novel colorimetric sensor array for monitoring fresh pork sausages spoilage. Food Control 35, 166–176.
Sanchez, C., Lebeau, B., Chaput, F., Boilot, J.P., 2003. Optical properties of functional hybrid organic-inorganic nanocomposites. Adv. Mater. 15, 1969–1994.

Sanchez, C., Belleville, P., Popall, M., Nicole, L., 2011. Applications of advanced hybrid organic–inorganic nanomaterials: from laboratory to market. Chem. Soc. Rev. 40, 696–753.
Schedin, K., Geim, A.K., Morozov, S.V., Hill, E.W., Blake, P., 2007. Detection of individual gas molecules adsorbed on graphene. Nat. Mater. 6, 652–655.
Schubert, U., Husing, N., 2012. Synthesis of Inorganic Materials. Wiley, Weinheim.
Schulz-Ekloff, G., Wohrl, D., van Duffel, B., Schoonheydt, R.A., 2002. Chromophores in porous silicas and minerals: preparation and optical properties. Micropor. Mesopor. Mat. 51, 91–138.
Schyrr, B., Pasche, S., Scolan, E., Ischer, R., Ferrario, D., Porchet, J., Voirin, G., 2014. Development of a polymer optical fiber pH sensor for on-body monitoring application. Sens. Actuat. B 194, 238–248.
Siegrist, M., Stampfli, N., Kastenholz, H., Keller, C., 2008. Perceived risks and perceived benefits of different nanotechnology foods and nanotechnology food packaging. Appetite 51, 283–290.
Singh, R.P., 2000. Scientific principles of shelf life evaluation. In: Man, D., Jones, A. (Eds.), Shelf Life Evaluation of Food. second ed. Aspen Publishers, Gaithersburg, pp. 3–22.
Siracusa, V., Rocculi, P., Romani, S., Dalla Rosa, M., 2008. Biodegradable polymers for food packaging: a review. Trends Food Sci. Technol. 19, 634–643.
MarketLine, 2011. Smart packaging case study. Available from: http://pt.slideshare.net/devansh1221/innovation-in-packaging-27123616?related=1
Sorrentino, A., Gorrasi, G., Vittoria, V., 2007. Potential perspectives of bio-nanocomposites for food packaging applications. Trends Food Sci. Technol. 18, 84–95.
Tang, X.Z., Kumar, P., Alavi, S., Sandeep, K.P., 2012. Recent advances in biopolymers and biopolymer-based nanocomposites for food packaging materials. Crit. Rev. Food Sci. Nutr. 52, 426–442.
Tomoyoshi, O., Ichiro, T., Makoto, N., 1986. US Patent JP610,131,16A. Temperature indicating sheet.
Tran-Thi, T.H., Dagnelie, R., Crunaireza, S., Nicole, L., 2011. Optical chemical sensors based on hybrid organic–inorganic sol–gel nanoreactors. Chem. Soc. Rev. 40, 621–639.
Vanderroost, M., Ragaert, P., Devlieghere, F., Meulenaer, B.D., 2014. Intelligent food packaging: the next generation. Trends Food Sci. Technol. 39, 47–62.
Vu, C., Won, K., 2013. Novel water-resistant UV-activated oxygen indicator for intelligent food packaging. Food Chem. 140, 52–56.
Weber, C.J., Haugaard, V., Festersen, R., Bertelsen, G., 2002. Production and applications of biobased packaging materials for the food industry. Food Add. Contam. 19, 172–177.
Wencel, D., Higgins, C., Guckian, A., McDonagh, C., MacCraith, B.D., 2005. Novel hybrid sol-gel materials for smart sensor windows optical sensors laboratory. Proc. SPIE 5826, 696–705.
Wu, Z., Jiang, Y., Xiang, H., You, L., 2006. Understanding the mechanisms of reaction and release of acid–base indicators entrapped in hybrid gels. J. Non-Crystal. Sol. 352, 5498–5507.
Yam, K.L., Takhistov, P.T., Miltz, J., 2005. Intelligent packaging: concepts and applications. J. Food Sci. 70, R1–R10.
Yari, A., Dinarvand, M., 2011. Sol-gel film doped with bromopyrogallol red as a highly sensitive sensing element for a new pH optical sensor. J. Iran. Chem. Soc. 8, 1091–1097.
Yebo, N., Sree, S.P., Levrau, E., Detavernier, C., Hens, Z., Martens, J.A., 2012. Selective and reversible ammonia gas detection with nano porous film functionalized silicon photonic micro ring resonator. Opt. Express 20, 11855–11862.

Yoshida, C.M.P., Maciel, V.B.V., Mendonça, M.E.D., Franco, T.T., 2014. Chitosan biobased and intelligent films: monitoring pH variations. LWT Food Sci. Technol. 55, 83–89.
Youssef, A.M., 2013. Polymer nanocomposites as a new trend for packaging applications. Polym. Plast. Technol. 52, 635–660.
Zaggout, F.R., El-Ashgar, N.M., Zourab, S.M., El-Nahhal, I.M., Motaweh, H., 2005. Encapsulation of methyl orange pH-indicator into a sol-gel matrix. Mater. Lett. 59, 2928–2931.
Zaggout, F.R., El-Nahhal, I.M., Qaraman, A.E.A., Al Dahoudi, N., 2006. Behavior of thymol blue analytical pH-indicator entrapped into sol–gel matrix. Mater. Lett. 60, 3463–3467.
Zaragozá, P., Ribes, S., Fuentes, A., Vivancos, J., Fernández-Segovia, I., Ros-Lis, J.V., Barat, J.M., Martínez-Máñez, R., 2012. Fish freshness decay measurement with a colorimetric array. Procedia Eng. 47, 1362–1365.
Zaragozá, P., Fuentes, A., Fernández-Segovia, I., Vivancos, J., Rizo, A., Ros-Lis, J.V., Barat, J.M., Martínez-Máñez, R., 2013. Evaluation of sea bream (*Sparusaurata*) shelf life using an optoelectronic nose. Food Chem. 138, 1374–1380.

5

FABRICATION OF HIGH-BARRIER PLASTICS AND ITS APPLICATION IN FOOD PACKAGING

Xueyan Yun, Tungalag Dong
Inner Mongolia Agricultural University, College of Food Science and Engineering, Hohhot, Inner Mongolia, P.R. China

1 Introduction

Processing and packaging are the two important phases of operations in the food industry. The final step is the packaging stage. For a very long period of time, people ate only available food that could be found in their immediate surroundings. With the ending of nomadic lifestyle, people started to live in sheltered areas, and thus they needed containers to store food (Risch, 2009). Modern food packaging is believed to have begun in the 19th century. The development of food packaging has evolved as humans' lifestyles have changed. Food packaging has evolved from simply a container to hold food to something that can now play an active role in food quality (Marsh and Bugusu, 2007). A package can provide protection, tampering resistance, and some special physical, chemical, or biological needs. It also can bear a nutrition facts label and other information about food being offered for sale. The development of the food industry has greatly promoted the development of the food packaging industry. It is a complex process that enables food to circulate from the producer to the consumer. Factors influencing the quality deterioration of the packaged food include moisture, gas type, temperature, light, and pH in contact with the food. The gas permeability and moisture of the food packaging materials are the most important factors affecting the quality preservation of the contents (Arora and Padua, 2010; Lange and Wyser, 2003). If the packaging material cannot provide a suitable atmosphere around the food, microorganisms may contaminate the food and make it unsafe (Robertson, 2010). A good package should create an acceptable barrier between the food and external environment, particularly with respect to water vapor, oxygen, and microorganisms to

Food Packaging. http://dx.doi.org/10.1016/B978-0-12-804302-8.00005-4

protect food from the loss of nutrients, color, taste, and aroma, and preserve the general appearance that is expected by consumers. The shelf life greatly depends on the barrier ability of a package.

Polymers are currently the most widely used food packaging material. According to the world packaging organization, the revenues of the global packaging industry are more than $500 billion; the top four packaging materials are paper and cardboard (32%), plastic (28%), metal (24%), and glass (6%); the total accounted for about 90% of the food packaging industry materials. The usage of plastics is significant because it offers several desired features like softness, lightness, transparency, good barrier, and high durability at low cost. The barrier property of a polymer refers to its ability to allow a permeant through it. The permeant passes through the polymer in three steps: sorption to polymer, diffusion, and desorption from polymer (Fereydoon and Ebnesajjad, 2013). Barrier properties of a package depend on the characteristics of the penetrating molecules, temperature, and the permeant concentration inside and surrounding the package, and the characteristics of the polymer, such as the degree of crystallinity, orientation, and the thermal and mechanical histories of plastic films.

Over the past few years, plastic packaging materials have been optimized to match product-specific sensitivities. But in fact, monolayer films are generally quite permeable to a variety of gases and not able to meet all the properties required for food packaging. Increased demands have been placed on materials for both currently approved and novel emerging applications and interest continues to grow in improving the performance of existing polymers for food packaging and in developing new polymeric systems (Mikkonen and Tenkanen, 2012; Brody et al., 2008).

This section focuses on the enhancement of barrier performance of nondegradable and biodegradable plastics as well as their fabrication and application in food packaging. The fabrication techniques include the orientation, coating, nanocompositing, blending, and LBL nanoassembiles based on synthetic and biobased polymers. The major emphasis is on the enhancement of barrier properties against the diffusion of oxygen, carbon dioxide, flavor compounds, and water vapor, and that is the key for the plastics used in vacuum and MAP.

2 Factors That Influence the Shelf Life of Packaged Food

Knowledge of the main factors to influence shelf life of packaged food is the first step in developing food packaging that will minimize undesirable changes in quality and maximize the development and maintenance of desirable properties. Deteriorative reactions in foods are influenced by two factors: the nature of the food and its surroundings. Food packaging protection is a means of keeping the food always in a state that is ensured to last during complex storage circulation conditions. Packaging can be used to provide a barrier to damages that are brought about by physical or chemical agents, such as atmospheric O_2, moisture, light, and so on, which can lead to deterioration. It should be noted that protection can only relieve but not prevent biotic deterioration, such as ripening of fruit or respiration of vegetables. The factors that can be controlled or improved with suitable packaging technology and material are discussed in the following section.

2.1 Light

As it is well known, chemical reactions that change food nutrients occur for both processed and fresh foods all the time. Light is one of the most important factors, which has a great influence on food quality and shelf life. Light is essentially an electromagnetic vibration in the wavelength range between 4000 and 7000 Å; UV light refers to the part of the electromagnetic spectrum that ranges from 200 to 400 nm and is divided into three regions: short-wave ultraviolet (UV-C), from 200 to 280 nm; UV-medium wave (UV-B), from 280 to 320 nm; and UV-long wave (UV-A), from 320 to 400 nm (Kasahara et al., 2015). The catalytic effects of light are most pronounced in the lower wavelengths of the visible spectrum and in the ultraviolet region. When the packaged food reaches store shelves, packaging materials allow the light to pass through the packages and cause the photo-oxidation degradation of fat, proteins, vitamins, and pigments in food. This results in peculiar smell, the loss of good color, and nutrient loss (Hutchings, 1999).

2.1.1 The Mechanism of Photo-Oxidation

It has been reported that serious photooxidation occurred when milk was exposed to wavelengths longer than 575 nm with the existence of cholorophyllic compounds (Airado-Rodríguez et al., 2011). The main mechanism of light effects on food is photo-oxidation. The mechanism of photooxidation (Fig. 5.1) is described below:

There are three types of oxygen: atmospheric atmospheric triplet oxygen (the ground state $^3O_2$), dissociation state ($^2O_2$) and singlet oxygen ($^1O_2$). Some sensitizers in singlet state can absorb the energy of light in a very short time and then become excited (Sens1, such as riboflavin). Singlet sensitizers in excited state return to their ground state via emission of light, intersystem crossing, or internal conversion. Heat and fluorescence are produced

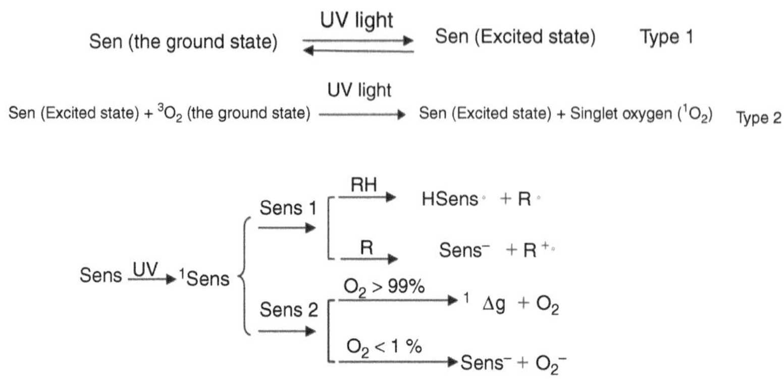

Figure 5.1. General mechanism of photosensitized oxidation.

by internal conversion and emission of light. Intersystem crossing results in excited triplet state of sensitizers. Excited triplet sensitizers could accept an electron or hydrogen from the substrate and produce free radicals (Type 1) (Choe and Min, 2006).

Excited triplet sensitizers can react with $^3O_2$, and electron transfer occurs and produces superoxide anion. Superoxide anion further produces hydrogen peroxide, one of the reactive oxygen species by spontaneous dismutation, and with the presence of transition metals, such as iron or copper, the reaction of hydrogen peroxide with superoxide results in singlet oxygen formation by Haber–Weiss reaction. The excitation energy of triplet sensitizers can be transferred onto nearby $^3O_2$ to form $^1O_2$ by triplet-triplet annihilation, then the sensitizers return to their ground singlet state (Sens 2, such as chlorophyll, Type 2). Photosensitized oxidation of food occurs in the presence of light, sensitizers, and atmospheric oxygen, in which $^1O_2$ is produced (Choe and Min, 2006).

2.1.2 Light and Lipid Oxidation

Oils and fats are important matters affecting the nutrition and flavor of food. The usual sensitizers in food are chlorophyll, pheophytin, riboflavin, and geoporphyrin. Since oils and fats are rich in oleic acid, linoleic acid, linolenic acid, arachidonic acid, and so forth, oxidation of fat may occur at the double bonds in the unsaturated fatty acids with the presence of sensitizer. The final decomposition products including aldehydes, alcohols, acids, hydrocarbons, and so forth, give rise to undesirable flavors. It is believed that the peroxidization of fatty acid is mainly responsible for the formation of undesirable volatile compounds.

2.1.3 Light and Protein

The function of light on proteins is divided into two cases: one is firsthand destruction to proteins, the other is damage caused by photo-oxidation because of the existence of sensitizers in food. The aromatic amino acids, tryptophan, tyrosine, and phenylalanine, and so on in proteins can absorb energy from the UV and lead to the break of peptide chains when exposed to the ultraviolet-medium wave less than 310 nm. Undesirable odors and pigment will produce during this process.

When exposed to ultraviolet radiation (wave $\geq$ 310 nm), the protein is damaged because of the photo-oxidative affect of light with the presence of sensitizer. It has been reported that photo-oxidation can cause significant decreases in color and could cause the modification of protein (Semagoto et al., 2014).

2.1.4 Light and Vitamins

Many dairy products are rich in vitamins and photosensitizer, which absorbs visible light and is known to be implicated in various photochemical processes that have a crucial role in the redox stability of food (Cardoso et al., 2012; Airado-Rodríguez et al., 2011). After absorption of photons transfer their photon energy to molecular oxygen, the excited delta singlet oxygen ($^1O_2$) is produced. $^1O_2$ has an effect on —C=C— of vitamin molecules. Photo-oxidation occurs with the presence of sensitizers and finally reduces the nutritive value of food.

Milk has riboflavin, a kind of sensitizer. In a study of Sattar et al. (1977), levels of vitamin fell as much as 80%, and even fell 100% when milk was exposed to sunlight for just 1 hour (Sattar et al., 1977). The higher the content of riboflavin in milk, the faster the vitamin loss (Jung et al., 1998).

2.1.5 Light and Food Colors

Color is one of the most important sensory attributes that determines food quality, and is highly affected by light. The color in processed food often comes from various synthetic dyestuff, whereas the color of natural food come from pigments and is related to healthy characteristics. Pigments in food are organic compounds, and have some chromophores, such as systems composed of several conjugated —C=C— double bonds, in which case a food may contain double bonds of impure atoms: —C=O—, —N=O— and —C=S—. Oxidation may occur at those double bonds and result in color change. Many pigments themselves are very good sensitizers, such as chlorophyll, when exposure to light causes formation of singlet oxygen and finally changes the color of food.

In addition, xanthophylls, heme, anthocyanin, proanthocyanidin, betacyanin, monascorubin, and some flavonoids in food are easy to oxidize and fade out under light.

Browning takes place readily when ham and sausage and any other pickled meats are exposed to visible light in storage and this adversely affects their quality. This is because the red color characteristic of dry-cured ham made with nitrate or nitrite is caused by nitrosylmyoglobin. Nitrosylmyoglobin is formed by the reaction of myoglobin with nitric oxide generated from nitrite. Visible light can make the pigment reverse into myoglobin and chromogen. The myoglobin and chromogen are further oxidated into the met-myoglobin and met-chromogen and finally make food brown. In summary, protecting food from light is very important.

2.2 Oxygen and Carbon Dioxide

Respiration is the taking in of oxygen (O_2) and the giving off of carbon dioxide (CO_2), so those two kinds of gas are extremely important for all lives, as well as for food. But the effect of O_2 on food in general is more complex. The concentration of gas in the environment surrounding food plays an important role on the growth of microorganisms, the respiration of fresh food, and the oxidation of food. Atmospheric O_2 generally has a detrimental effect on the nutritive quality of foods. Oxidation results in the formation of hydroperoxides, peroxides, and epoxides, loss of vitamins, and so on, therefore it is desirable to maintain many types of foods at a low O_2 atmosphere, or prevent the supply of O_2 into the package after being sealed. O_2 atmosphere of most packaged foods depends greatly on the nature of the package. Vacuum packing is a method of packaging that removes air from the package prior to sealing. This kind of packaging is often used for many processed foods, such as cheese and sausage. Fresh fruits and vegetables need O_2 to ripen and should also retain water, and therefore these packaging materials should allow O_2 to enter. Changes in the O_2 atmosphere of packaged foods is decided by both the respiration and permeability of packages. Flushing the inside of the package with a gas such as CO_2 or nitrogen (N_2) before sealing is the other basis of MAP. High concentrations of gases such as CO_2 are used to retard microbial growth and thus extend the shelf life of foods. MAP is increasing importance for the packaging of fresh fruits and vegetables, fresh foods, and bakery products. For example, fresh meat packed in MAP packages is usually high in CO_2, low in O_2, and then kept in low temperature. To keep the fixed atmosphere, the packaging materials should be a good barrier to CO_2 and O_2 and flexible. Thus the selection of materials for package has to balance barrier properties with many requirements, such as form, method of preservation, and subsequent reuse. Suitable packaging technology combined with suitable packaging material with a certain gas barrier property, effectively creates a suitable O_2 atmosphere for different food.

2.3 Moisture

Water is the origin of life, and physical and chemical deterioration of food are also related to the equilibrium of moisture content and of great importance for maintaining or extending shelf life. The effects of water activity in food preservation are mainly embodied in the following aspects.

The growth of microorganisms: The growth of microorganisms is affected by water activity, temperature, pH, and so on. Water activity is the most important factor, and has great influence on food quality and its expiration. Water activity is directly associated with the breeding of yeasts, molds, and bacteria in food. Relative humidity (RH) below 60% doesn't support mold growth, so dry foods are microbiologically stable.

Physical and chemical reactions: Besides effects on the growth of microorganisms, chemical and enzyme reactions are also affected by water activity. Water activity can lead to changes in the rates of chemical reactions. It accelerates chemical reactions such as food browning, lipid oxidation, protein denaturation, and finally causes loss of quality.

Physical properties: Food of a high water activity level is usually moist, juicy, tender, and rubbery. If reduced, the water level may cause changes in the texture, such as food becoming hard, dry, and insipid. Food of low water activity level is usually light and crisp and will become moist and insipid. For some food, the removal of moisture prevents the growth and reproduction of microorganisms that cause decay and minimizes many of the moisture-driven deterioration reactions (Joardder et al., 2014). For some other foods, the moisture content must be maintained during processing and production, and in the preservation of tea, grain, delicatessen items, and so on, it is important to avoid water permeation after packaging. However, for fresh food, it is important to avoid dehydration.

3 The Barrier Property of Polymer

As mentioned before, the barrier property of a polymer refers to its ability to transfer a permeant through it. The quantitative evaluation of the barrier properties of a plastic material is normally associated with the concept of permeability.

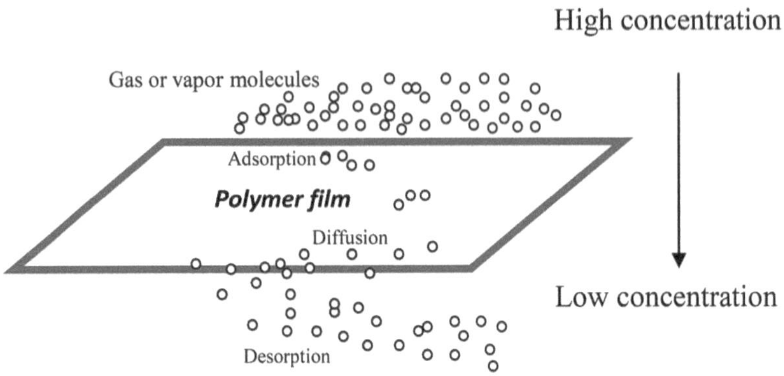

Figure 5.2. General mechanism of permeation through a polymer film.

3.1 General Mechanism of Permeation

Permeability is defined as the quantification of permeate transmission, gas or vapor, through a resisting material. The permeant passes through the polymer by net effect from the high concentration to the low concentration side with a permeant having different concentrations on its two sides, in three steps: sorption to polymer, diffusion, and desorption from polymer (Fereydoon and Ebnesajjad, 2013) as shown in Fig. 5.2.

3.2 Factors That Affect the Barrier Property of Polymer

The transmission rate (TR) at which a gas, vapor, or liquid passes through a polymer film has usually been used to evaluate the barrier properties. It is an important determinant of the packaging protection provided by barrier materials. TR is defined as: the amount of gas that permeates per unit of area and time in packaging materials. TRs of polymer films are controlled by properties of the polymer film, the permeating species, and the environmental conditions both inside and outside the package. It should be noted that since the different thickness and time can greatly affect the TR of polymers, the permeability coefficient is the rate of a quantity of permeant that passes through a polymer in a unit of time with a unit surface area having unit thickness with a unit pressure difference on both sides of the sample, that has been consistently calculated to compare the barrier abilities of different polymers.

3.3 Permeating Species and TR

As stated before, passage of a small molecule through a polymer film is thought to involve three steps: sorption, diffusion, and desorption. Permeability (*P*) is the product of diffusion coefficient (*D*) and solubility (*S*).

$$P = D \cdot S$$

Different kinds of permeating substances also have different molecular sizes, weight, and shape, and thus have different *D* and *S*. However, systematic change in *D* for changes in molecular sizes has an inverse relationship with the change of *S*. Thus there are very small changes in *P*. Compared with gases, more condensable vapors and liquids permeate faster.

The polarity of the permeating molecule also affects its ability to pass through a polymer. For example, polar permeating molecules are easily adsorbed on the surface of polymers with polar groups. Specific interactions between polar/hydroxyl functionalities on polymers and water leads to very high water adsorptions and permeabilities, but polarity leads to poor adsorptions of many gas molecules (eg, O_2, N_2). To the contrary, compared with permeating species with polarity, nonpolar molecules are relatively easy to get through hydrocarbon polymers. As shown in Table 5.1, a variety of polymer films exhibit selected permeabilities for O_2, N_2, and CO_2 (Yun et al., 2015).

3.4 Environmental Conditions and TR

The TR of some films is strongly affected somehow by environmental conditions, such as humidity, temperature, and pressure. Generally, it's hard for a gas molecule to travel through the crystallization region of polymers. Permeation occurs only in free volume of the amorphous region. The TR of most polymers increases with increasing humidity. That is because the water absorptivity of polymer chains results in slight increased volume among chains. Thus, the gas molecules get through the films easily. But in some special cases, the TR decreased with increasing humidity within a range, since the water vapor forms a barrier on the surface of films. With further increasing humidity, the TR begins increasing with increasing humidity (Dong et al., 2015a). The impact of temperature on TR is mainly on both the permeating molecule itself and the polymer. An increase in temperature corresponds to faster motion of the particles and the molecules can more easily get through. On the other hand, higher temperature leads to more intense motion of polymer chains, even causing glassy polymer to

Table 5.1 Selected Permeability of Some Polymer Films

Polymer	O_2 permeability ($\times 10^{-7}$mL m/m^2 d Pa)	CO_2 Permeability ($\times 10^{-7}$mL m/m^2 d Pa)	N_2 Permeability ($\times 10^{-7}$mL m/m^2 d Pa)	Water Vapor Permeability ($\times 10^{-7}$g m/m^2 d Pa)
LDPE	57.6	264	—	0.018
PCL	11.1	—	4.39	0.0027
PLLA	1.17	6.24	0.268	2345

Note: The data are original; oxygen permeability test condition: 25°C, 0% RH; water vapor permeability test condition: 25°C, 50% RH.

achieve a high elastic state, thus providing more free volume for the small molecule to pass through. Pressure mainly acts on the motion of gas molecules and has almost no effect on the polymer exception of ultra-high pressure. Pressure would force permeating molecules at high concentrations to get through the film at higher speeds. Furthermore, if the low concentration is treated with enough high pressure, the direction of penetration even can be changed.

4 High-Barrier Plastics

In recent years, there have been a number of instances of food corruption and decay because of lack of protection. These situations are associated with the barrier property of materials. Barrier property is a function that a packaging material must have, especially for food and medicine packages. To achieve a long shelf life, food and medicine packaging needs a high barrier material to prevent contamination by O_2, water vapor, and acid, and to prevent the loss of odor and organic vapor. In the case of vacuum and MAP, high barrier material can maintain the stability of the internal environment.

Monolayer film may be unable to meet all the barrier requirements of food packaging. Most food packages have multilayer constructions to achieve the desired functional requirements of the package in a cost-effective manner. Materials are selected according to the package function and cost consideration to design multilayer films to meet different food packaging requirements. Generally speaking, in any multilayer structure several layers are seen. An abuse-resistant outer layer is designed to protect or preserve contents from damage due to external forces. The polymers that are typically used as the abuse layer are polypropylene (PP), polyamide (PA), polyethylene terephthalate (PET), and poly(ethylene) (PE). A sealant layer of the multilayer structure is important for a multilayer structure. Copolymers with low melting points and low temperature heat sealants—such as PE, ethylene vinyl acetate (EVA), acid copolymers, and ionomers—are used as the seal layer. In the core layer, the polymers with high barrier are used, such as aliphatic PA6, ethylene vinyl alcohol (EVOH), poly(*m*-xylene adipamide) (MXD6), and poly(vinylidene chloride) (PVDC). Generally, transparent films with an oxygen transmission rate (OTR) of 25.4 μm film below 5 mL/m^2 d (oxygen permeability $< 14.55 \times 10^{-15}$ mL m/m^2 s Pa) are considered as the high barrier materials in food packaging (Zhou, 2001). Table 5.2 shows permeability of a number of plastic films that are high barriers. The data shows that the oxygen barrier properties of different plastic films even have orders of magnitude differences.

For the design of a multilayer structure, the selection of material for each layer is according to the preferred functional requirements of the package. For example, many meat products need to be protected against O_2. Materials such as ethylene-vinyl alcohol (EVOH) and poly(vinyl alcohol) (PVA) are well known as good oxygen barrier layers. For some food, such as milk and fried food, protection against light is also very important; an extremely thin aluminum layer has traditionally been used as one of the core layers of multilayer packaging film for this kind of food. Many food products need to be protected against the gain or loss of moisture,

Table 5.2 Some Plastic Films With Good Barriers

Material	O_2 Permeability ($\times 10^{-11}$ mL m/m^2 d Pa)	Water Vapor Permeability ($\times 10^{-9}$ g m/m^2 d Pa)
EHA	2.7	4.8
PA6	37.8	18.1
EVOH	4.4	2.5
PVA	0.000052	0.017
PA6/PE	195.0	4.6
EHA/PE	3.5	11.2
PPC	0.075	0.0443

Note: The data are original; oxygen permeability test condition: 23°C, 65% RH water vapor test condition: 23°C, 50% RH.

as well. Materials such as coated cellophane, PE, PP, poly(vinyl chloride) (PVC), and polyester films are excellent barriers to water vapor and are used to obstruct the transmission of water vapor through the film. These materials are often used on the outside layers of multilayer films.

4.1 Improvement of Barrier Properties of Films

With the development of society and the improvement of science and technology, the modification of polymers for food packaging is developing toward high performance functionality. The modification techniques include orientation, nanocompositing, coating, blending, and LBL nanoassembiles based on synthetic and biobased polymers.

4.1.1 Orientation

Stretching can result in a highly oriented film and lead to a reduction in the diffusion coefficient of gases through polymers (Tabatabaei et al., 2008, 2009; Lin et al., 2008; Chu and Kimura, 1996; Fereydoon et al., 2015). This reduces the gas and water-vapor permeability of films. The reduction in gas permeability could be attributed to an increase in crystallinity during stretching. Researchers have found that the barrier property of many polymers changed after orientating. For example, the gas permeability of PP is reduced by a factor of two after biaxial orientation without sacrificing clarity and toughness of the film (Lin et al., 2010). Orientation greatly reduces the gas permeability of PP (Somlai et al., 2005; Lin et al., 2008, 2010). The oxygen diffusivity, solubility, and consequently its permeability are reduced for PET films after uniaxial stretching, due to a reduction in the amount of excess-hole free volume in the polymer (Liu et al., 2002). Uniaxial orientation also significantly improved the O_2 and CO_2 barrier properties of poly(ether block amide) copolymers (Armstrong et al., 2012).

The oxygen permeability of the stretched MXD6 and MXD6/clay films increased with increasing draw ratio. But up to a critical value, while further stretching resulted in a reduction in oxygen permeation (Fereydoon et al., 2015). Similar result is also found in stretched and annealed poly(L-lactic acid) (PLLA) films in our study. Clay-loaded or ionically bonded gas barrier thin films easily crack with moderately stretching. But hydrogen-bonded PAA/poly(ethylene oxide) multilayer thin films remain crack-free and these nanocoatings still provide more than a 5-times reduction in OTR after 100% strain (Xiang et al., 2014a; Lutkenhaus et al., 2005). PVA hybrid films containing small amount of clay show that PVA hybrid films with 300% biaxial stretching exhibited the best improvements in O_2 barrier properties (Ham et al., 2013). The structural development of nylon films during uniaxial stretching is influenced by initial morphology, molecular orientation, ability to crystallize, type of crystalline structure and physical state of the material. Additionally, the final properties of stretched films are strongly dependent on processing conditions such as stretching rate, extent of deformation, and stretching temperature.

However, the relationship between orientation and gas permeability is not well understood. For example, permeability does not exhibit a simple relationship with optical birefringence, which reflects the total combined molecular orientation of the crystalline and amorphous phases (Taraiya et al., 1993; Masuko et al., 1970) whereas gas transport is seen as occurring only through the amorphous regions. Indeed, one report, with very limited data, demonstrated a correlation between oxygen permeability and the Herman's amorphous phase orientation function in the transport direction (Taraiya et al., 1993).

4.1.2 Nanocomposites

Polyolefins such as PP, PE, PET, poly(styrene) (PS), and PVC are most frequently used for food packaging; their major drawback is an inherent permeability to gasses and other small molecules. Nanocomposites are the latest materials aimed at solving the aforementioned problems, are created by dispersing an inert, nanoscale filler throughout a polymeric matrix. Filler materials can include clay (Khan et al., 2014; Reddy et al., 2013; Schauberger et al., 2013; Souza et al., 2012), silica (SiO_2) nanoparticles (Tsai et al., 2010), carbon nanotubes (Yu et al., 2014), cellulose-based nanofibers, or nanowhiskers (Abdollahi et al., 2013), and so on.

Nanocomposites are those that contain two or more phases, in which one phase is the continuous phase and the other is the disperse phase. The dispersal of nano-sized fillers into the polymer matrix affects the barrier properties of a homogeneous polymer film in two specific ways. The first way is by creating a tortuous path for gas diffusion. Because the filler materials are essentially impermeable inorganic crystals, gas molecules must diffuse around them rather than taking a straight-line path that lies perpendicular to the film surface. The result is a longer mean path for gas diffusion through the film in the presence of fillers. Essentially, the tortuous path allows the manufacturer to attain larger effective film thicknesses while using smaller amounts of polymer. The second way that nanoparticulate fillers influence the barrier properties is by causing changes to the polymer matrix itself at the interfacial regions. If the polymer-nanoparticle interactions are favorable, polymer strands are located in close proximity to each nanoparticle, thus attenuating hopping rates between free volume holes, or alternated density and/or size of holes. The result is that gas molecules find it difficult to travel through these interfacial zones.

It has been reported that the mechanical and thermal and other properties could be significantly improved for polyolefin-based nanocomposites compared with original polyolefins. Clay is the common inorganic nanofiller used in many studies. Given its large lateral dimensions (100–1000 nm), important specific surface area (800 m^2/g), and high swelling capacity, montmorillonite (MMT) clay is often considered a layered silicate of interest. Nanoclay dispersed in a polymer matrix is a key to improve the barrier property. During nanocomposites preparation, short-range order of nanoparticles can either be maintained by intercalation of polymer chains (Vaia et al., 1993, 1995), which leads to an expansion of the interlayer spacing, or disappear in the case of exfoliation (Giannelis, 1996). Exfoliated nanoclays are effective at improving gas barrier properties of polymeric materials. When dispersed into polymers, they create a maze structure that presents a tortuous path to moving gases, greatly slowing their permeation rate.

Polyurethane/nanoclay systems have also attracted attention for their barrier-like properties (Osman et al., 2003; Xia et al., 2005). Natural MMT is modified with various alkylammonium surfactants and incorporated into a polyurethane matrix. The gas permeation is highly sensitive to the preprocessing method. Membranes with sonicated particles in the polymer matrix had significantly lower permeation than membranes with stirred clay particles. This improvement in barrier properties is attributed to a better dispersion of the sonicated clays in the polymer matrix (Herrera-Alonso et al., 2009).

Clay has been added to different polymers including multilayered PET, epoxy composites, and PS/PE (Kim et al., 2005; Ryu et al., 2004). To improve the interaction with hydrophobic polymers such as high-density polyethylene (HDPE), the clay is organophilized, increasing polymer/clay affinity and the ability of forming exfoliated nanocomposites (Gorrasi et al., 2003; Yeh et al., 2006; Dadbin et al., 2008). Permeation study found that the additive modified organoclay showed a significant effect on the barrier properties of HDPE. A 2wt.% MMT additive in the polymer matrix, compared with the neat PE, the permeability of CO_2 decreased by 43.7% (Carrera et al., 2013).

The reacting surfactant group anchored on the silicate surface may improve the dispersion of the silicate platelets in the polymer matrix, eventually leading to improved gas barrier properties and mechanical properties of poly(urethane) (PU) (Dan et al., 2006;

Wang et al., 2002). The PS/clay hybrid systems have been prepared via in situ polymerization of styrene in the presence of unmodified MMT, both PS/VDAC-MMT, and PS/C18DMB nanocomposites displayed relative oxygen permeability noticeably smaller than that found for the conventional composite PS/Na-MMT system (Nazarenko, 2007). Barrier membranes of poly(*n*-butylacrylate) are also produced by the in situ polymerization of *n*-butylacrylate in the presence of surface modified natural MMT (Herrera-Alonso et al., 2010). A significant enhancement on barrier properties of grapheme oxide (GO)/PVA nanocomposite films is achieved; that is, oxygen and water vapor permeability of GO/PVA nanocomposite films decreased about 98% and 68% at a GO loading of 0.72 vol.%, respectively (Huang et al., 2012).

A good nanofiller dispersion system is achieved for PA6/MMT nanocomposite (Picard et al., 2007). Poly(*n*-butyl methacrylate)/clay and poly(*n*-butyl methacrylate-*co*-methyl methacrylate)/clay nanocomposites are synthesized via emulsion polymerization to act as barrier membranes (Herrera-Alonso et al., 2010). The improved water barrier of PA12 incorporating OMMT C30B are prepared by nanocompositing, the permeation results revealed limited water-barrier properties of the nanocomposite films and clearly showed a permeability transition at low silicate content (Alexandre et al., 2009). Incorporating clay in PA6 and MXD6 enhanced oxygen barrier by 40% and 46%, respectively (Fereydoon et al., 2015; Magniez et al., 2009).

Graphene, a relatively new material with a two-dimensional nanostructure, GO can be stably dispersed in an aqueous solution without any stabilizing agent. Adding highly exfoliated carbon layers can significantly alter mechanical and electrical properties of polymers at extremely small loading (Stankovich et al., 2006; Ramanathan et al., 2008). Unlike carbon nanotubes that can provide similar mechanical and electrical benefits, these impermeable two-dimensional sheets can reduce gas permeability of host membranes (Kim et al., 2010; Kim, 2008).

A novel thermal-sensitive PU/SiO_2 nano-hybrid membrane is prepared via in situ nano-hybrid method in wet phase inversion process. It found that, compared with the neat PU, the nano-hybrid membrane shows relatively lower water vapor permeability (WVP) at low temperatures and higher WVP at high temperatures (Hu et al., 2008). An improvement in water barrier of PET was achieved by incorporating monodisperse SiO_2 particles. Decreasing the SiO_2 particle size demonstrably improved the PET film hydrophobicity and lowered the water diffusion coefficient, that is, SiO_2 particles of 35 nm in size gave the greatest enhancement of water barrier properties (Ke et al., 2010). It was also found that the addition of smaller size SiO_2 particles effectively improved the transparent PP (Asuka and Liu, 2006). The multidispersion of near-monodisperse SiO_2 particles was effectively treated with either hydrophilic or hydrophobic reagents (Ke and Stroeve, 2005; Su et al., 2002).

A titanate coupling agent (TC9) and a 3-(Methacryloyloxy) propyltrimethoxysilane (KH570) are used for the surface modification of SiO_2 nanoparticles. It is found that both KH570 and TC9 coupling agents improve the nanoparticle dispersion compared with unmodified SiO_2 nanoparticles. These particles are well dispersed and formed nucleation centers in a polyester matrix, which obviously created good surface, thermal, and crystallization properties. The improvement in dispersion is found to be due to increased surface hydrophobicity of the treated SiO_2 nanoparticles (Zhang, 2014).

4.1.3 Coating

Organic and inorganic coatings with a nanolayer of various polymer substrate films has been reported to improve gas barrier properties (Kim, 2008; Keiji et al., 1999; Hwang et al., 2009; Nam and Kim, 2010). To get high-performance transparent gas barrier films, inorganic materials such as silicon oxide (Kwak et al., 2009; Iwamori et al., 2003a,b; Erlat et al., 1999; Gruniger et al., 2006; Silva Sobrinho et al., 1998; Inagaki et al., 1997), aluminum oxide (Hosseini et al., 2015; Groner et al., 2006; Henry et al., 2001), silicon nitride (Chen et al., 2006a), silicon oxynitride (Shim et al., 2008), and so forth, are used, which are usually prepared with various coating methods such as sputtering, plasma-enhanced chemical vapor deposition (PECVD), evaporation, atomic layer deposition, and sol-gel coating. Since the inorganic materials thus coated have a dense chemical structure, they have very high gas barrier properties.

Silicon oxide thin films are of increasing interest as diffusion barriers on packaging material. The deposition of these SiO_2-like [or silicon oxide (SiO_x)] films on polymeric webs by PECVD is examined intensely in the recent years. The deposition of silicon-oxide thin films as diffusion barriers on paper-based substrates is achieved by PECVD technique, it is shown that the existence of persistent pores in the substrate material is in direct relation to the success of a silicon oxide coating in terms of the reduction of the oxygen permeability (Grüniger and Rohr, 2003; Erlat et al., 1999). Barrier films consisting of SiO_x/silicon nitride (SiO_x/SiN_x) stacks are consecutively deposited onto the PA substrate and the results showed that the SiO_x and SiN_x films presented good barrier properties (Chen et al., 2006b; Dennler et al., 2006; Zhang et al., 2015a). A deposited oxide layer with an appropriate polymer can improve the barrier of base materials (Leterrier, 2003; Lange et al., 2015). The k-Al_2O_3/resin (Tseng and Chang, 2014) and TiO_2/PU (Kasanen et al., 2011) composite coating on PE are also developed to improving the gas barrier or the adhesion between the protective PU barrier layer and the PE substrate, respectively. Good gas barrier PP laminates usually contain thin aluminum foil. The oxygen permeation rate and water vapor permeation rate of aluminum-coated PP are reduced by 100-fold and 10-fold then uncoated PP, respectively (Prins and Hermans, 1959; Weiss et al., 1989).

Organic coating on the substrates also can be alternated polymeric films to enhance barrier properties. The PET is coated PVDC; compared to neat PET film, the OTR, and WVTR values are decreased to 6~7 cm^3/m^2 day and 3~4 g/m^2 day, respectively (Maikap et al., 2015). Spin coating has been shown to be a suitable method of preparing reproducible thin films of cellulose from a solution by removing the solvent with high-speed spinning (Kontturi et al., 2007). The cellulose nanocrystals hydrolized with sulfuric or hydrochloric acid are spin-coated on glass substrate; the resulting films showed that the hydrochloric acid cellulose crystals resulted in films with lower homogeneity, higher roughness, low permeability for N_2 and O_2, and also gas selectivity for some gas combinations (Herrera et al., 2014). The deposition of the pullulan/hybrid coatings decreased the OTR of the plastic substrate by up to two orders of magnitude under dry conditions (Farris et al., 2012).

4.1.4 Blending

Blending is a relatively lower cost way to develop a high barrier material. The barrier properties of a polymer blend depend not only on the intrinsic barrier property of each component but also on the final blend morphology. Droplet or fibril-type morphology gives only modest permeability decrease with low content of barrier polymer, whereas a dramatic increase in barrier properties is achieved by generating a laminar morphology within the polymer blend. The blend morphology is generally determined by many factors, such as rheological property of the components, interfacial tension, composition, stress level, mixing history, and procedure of blend preparation. In polymer blend system, the improvement in barrier properties is accomplished mainly through the formation of a laminar morphology of the dispersed phase, resulting in an increase of the tortuous path (Kamal et al., 1984; Subramanian, 1985; Subramanian and Mehra, 1987; Holsti-Meittinen et al., 1995; Lee and Kim, 1997). The laminar structure is generated in LLDPE/EVOH, PP/EVOH, and PP/PA blends with improved barrier properties by extrusion blown or biaxial orientation film process, and so on (Kim and Chun, 1999; Yeo et al., 2001; Jarus et al., 2002; Kim and Kim, 2003). The permeation test has revealed that the barrier of CO_2 and O_2 and permselectivity (P_{CO_2}/P_{O_2}) of PA6/EVOH films increase as the EVOH weight fraction is increased (Incarnato et al., 1999).

4.1.5 Layer-by-Layer Assembly

LBL assembly is currently gaining popularity as a relatively inexpensive thin film deposition technique, and to fabricate transparent, super gas barrier films (Priolo et al., 2015).

In a typical preparation, an appropriately prepared, often negatively charged substrate, such as metals, silicones, and glasses, is immersed into a mixture of positively charged material, such as poly(diallydimethylammonium chloride) (PDDA), poly(alylmine hydrochloride) (PAH), or polyethleneimine (PEI), for a given amount of time. The substrate is then removed from the mixture, rinsed with deionized water, optionally dried in a stream of filtered air (or nitrogen), and then immersed into a mixture of negatively charged material, such as poly(styrene sulfonate) (PSS), poly(vinyl sulfate), or poly(acrylic acid) (PAA), rinsed and dried; as a result, a double polyelectolyte layer (bilayer) is built up on the substrate. With such cyclic depositions, one can achieve multilayer films on the substrates with desired structures and thicknesses. Since the thickness of a single bilayer of polyelectrolyte is below 1 nm, the LBL technique allows for nanometer-scale control of the thickness of thin films (Decher, 1997; Decher and Hong, 1991). The thickness can also be controlled by altering the characteristics of a given deposition mixture, such as pH, buffer, ionic strength, temperature, or counter-ion, as well as molecular weight and environment conditions. Besides the electrostatic attraction, hydrogen bonding, hydrophobic force, hydrogen bonding, and covalent bonding can also be used as LBL self-assembly forces to form multilayered structures.

LBL can be used to deposit a variety of ingredients that include polymers, nanoparticles, organic dyes, dendrimers, and biological molecules. Many LBL systems show linear growth as a function of layers deposited, but those containing weak polyelectrolytes often exhibit exponential growth (Podsiadlo et al., 2008; Yoo et al., 2006). The LBL thin film assemblies of PEI and PAA are studied, in an effort to produce all-polymer thin films with a high oxygen barrier (Hagen et al., 2014a), the OTR of 8 bilayers 350 nm PEI/PAA film is lower 0.005 $cm^3/(m^2 day)$ (Yang et al., 2011a).

By altering the pH of PEI and PAA, large thickness variations can be realized. The LBL thin film created by PAH and PAA shows the ability to systematically control the thickness, density, composition, and wettability by simply changing the charge density of these polyelectrolytes (Yoo et al., 1998; Shiratori and Rubner, 2000; Choi and Rubner, 2005). The ability to tailor charge density with pH provides a more tailorable LBL process.

Despite having clear advantages over dipping, spraying has been unable to replace dipping as the most common method for layer-by-layer deposition because there is no consensus on the quality of spray-coated films. A major concern is that spray-coated assemblies are thinner than their dip-coated counterparts (Izquierdo et al., 2005; Kolasinska et al., 2008; Shukla et al., 2010). The common deposition time used for spraying is typically a few seconds, whereas dipping is 5~15 min for each layer. A 7-BL film prepared by PEI and PAA deposited under optimized spraying condition is proved to be thicker than a dip-coated film created with the same deposition time. Spray-assisted film thickness is found to be independent of spraying pressure, but strongly influenced by flow rate and spray time. A 7-bilayer PEI/PAA film prepared using the best spraying parameters (50 psi, 0.4 g/s, 5 s) resulted in a thickness of 538 nm and an OTR of 0.08 $cm^3/(m^2$ day atm). Preparing the same 7-BL film with dipping produced a thinner film with a 1 order of magnitude greater OTR.

The PEI/GO multilayer composite membranes absorbed on PET or silicon wafers substrates was fabricated by the LBL self-assembly technique (Stevens et al., 2014a; Yang et al., 2013), and the multilayer film reduces the hydrogen transmission rate of the uncoated PET film from 1357 to 24 cm^3/m^2 day 0.1MPa. Highly transparent and extremely flat LBL barrier films are easily assembled, using negatively charged GO monolayers and positively charged poly(alylmine hydrochloride), and can significantly reduce oxygen permeability (Lee et al., 2014). The amino-phenyl functionalized GO is alternatively complexed with the GO through an electrostatic LBL assembly process on the surface of polyimide (PI). The OTR is decreased from 220 to 78 cm^3/m^2 day atm, due to the barrier effect of the graphene films on the PI surface (Kim et al., 2013).

The individual clay flakes consist of approximately 1-nm thick plate-shaped nanoparticles, with an aspect ratio of 100:1000, and with a negative surface charge in deionized (DI) water (Ploehn and Liu, 2006; Priolo et al., 2010a). The key factor for the use of nanoclay in barrier layers is its alignment parallel to substrates. Clay-based assemblies often show a reduction in gas barrier (Jang et al., 2008; Priolo et al., 2010b; Holder et al., 2012), a self-aligned clay multilayer can be fabricated by the absorption of PVA onto MMT because of their absorption capabilities and strong hydrogen bonding. The LBL deposition of oppositely charged thin films containing self-assembled PVA and MMT, the permeability tests showed that their gas barrier characteristics, optical transparency are improved (Jin et al., 2012; Walther et al., 2010). The $AuCl_3$ was drop-casted onto a MMT(PVA)-PDDA multilayer film, which resulted in the formation of Au_2O_3. Au_2O_3 linked the MMT platelets together and prevented the permeation of water vapor through the gas barrier film (Song et al., 2015).

In an effort to impart light gas barrier to polymer substrates, the film growth and structure of assemblies made with cationic PEI and anionic MMT and PAA via LBL assembly, where one deposition sequence of PEI/PAA/PEI/MMT is referred to as a quadlayer (Priolo et al., 2010a,b, 2011; Stevens et al., 2014b). After depositing a 122 nm five layers coating on a 51 μm PS, the gas permeability of PS decreased three orders of magnitude compared with bare PS. The coated film even demonstrated better performance than thick-laminated EVOH film and metalized film (Tzeng et al., 2014). Further barrier improvement can be achieved simply by lowering the pH of the clay suspension in the MMT and PEI system (Hagen et al., 2014b). In this situation, the charge of the deposited PEI layer is increased in the clay suspension and causes more clay to be deposited. A quadlayer assembly consisting of three repeat units of branched PEI, PAA, PEI, and MMT is only 24 nm thick when deposited with 1 min exposure to each ingredient. The thickness of this film increases to 55 nm and meanwhile the OTR reduces to 0.05 $cm^3/(m^2$ day atm) by reducing the exposure time of polyelectrolytes to 5 s. The OTR is 2 orders of magnitude lower than the same film made using 1 min exposures (Xiang et al., 2014b).

4.2 Biodegradable Polymers

Traditional plastic materials exhibit a very long time of environmental decomposition, which represents a serious problem with the long life of plastic waste (Wendy et al., 1998; Chandra and Rustgi, 1998). The biodegradable polymers disposed in bioactive environments degrade by the enzymatic action of microorganisms such as bacteria, fungi, and algae. A wide range of biodegradable polymers are at present available. They are better suited for a number of applications such as shopping bags, food-service packaging materials, and agricultural mulch films (Nir et al., 1993; Kasirajan and Ngouajio, 2012). The barrier property of materials is especially important for atmosphere and vacuum packaging (Shi et al., 2013). The gas barrier of materials used presently is far better than most biodegradable materials. Therefore, improving the barrier performance of the biodegradable packaging materials is the key for the application of biodegradable materials in high barrier packaging (Dong et al., 2015b).

Poly(lactic acid) (PLA) is a kind of important degradable material, which is being increasingly used as biomaterials, thus it attracted much attention. PLA is brittle and exhibits low thermal stability, medium gas barrier, and low solvent resistance against water (Jongwhan et al., 2009). The intercalation of nanoplatelets such as layered silicate MMT can lead to an enhancement mechanical properties and gas barrier of PLA (Jongwhan et al., 2009; Ray et al., 2002; Sinha et al., 2003; Gain et al., 2005; Bhatia et al., 2012). Besides, the barrier properties of the PLA films are also enhanced when blended with other barrier polymers, such as poly(vinyl acetate-*co*-vinyl alcohol) (Razavi et al., 2012), EVOH (Sanchez-Garcia et al., 2011), aroma compound ethyl acetate (Colomines et al., 2010), and so forth. The stretched PLA film with different draw ratios was prepared by the melt-extrusion method in our work, the PLA film with draw ratio of $R = 3.5$ shows a good performance on barrier property (Dong et al., 2015a).

Transparent and bendable cellulose films regenerated from cellulose solutions in aqueous m aqueous alkali (NaOH or LiOH)/urea systems exhibit high oxygen barrier properties (Yang et al., 2011b). The oxygen permeability of PLA film surface-coated with TEMPO-oxidized cellulose nanofibers is decreased from 746 to 1 cm^3/m^2 day Pa (Fukuzumi et al., 2009). Thin film assemblies of PEI/nanofibrillated cellulose or PEI/carboxymethyl cellulose are adsorbed on PLA substrates using the LBL technique, the resulting films show high oxygen permeabilities of 0.34 and 0.71 cm^3 $\mu m/m^2$ day kPa at 23°C and 50% RH, respectively, which is in the same range as PVA and EVOH (Aulin et al., 2013). PCL and PPC are coated with cellophane via the intermolecular hydrogen bonds of CS with the both polymers, the oxygen barrier of polymer is increased three orders of magnitude, and the multilayer films still keep good water vapor barrier to some degree (Dong et al., 2014a). The LBL process is carried out by the alternate adsorption of CS and cellulose nanocrystals (Halasz et al., 2015), PEI and MMT (Carosio et al., 2014), sodium alginate and PEI (Gu et al., 2013) on PLA film to to improve the barrier properties.

Laminations of Gelatin or Wheat gluten as the middle layer in the PLA/Gelatin/PLA multilayer film also has a beneficial effect on the water vapor and oxygen barrier properties (Martucci and Ruseckaite, 2010; Cho et al., 2010). The confined crystallization of PEO in nanolayered films prepared by layer-multiplying coextrusion of thousands of alternating PS and PEO nanolayers, the crystallization habit imparted more than two orders of magnitude reduction in oxygen permeability of the PEO layers (Wang et al., 2009). The coated films are composed of natural materials based on biodegradable and biocompatible polymers, such as CS, clay, soy protein, and SiO_2 hybrid, which exhibit excellent oxygen-barrier properties compared with neat PLA films (Park et al., 2012; Rhim et al., 2007; Bang and Kim, 2012).

5 Plastics for Food Packaging Application

Early in 1994, researcher packaged fresh tilapia fillets in high barrier film bags under suitable atmosphere and showed that storing them at refrigeration temperatures could extend their shelf life to more than 25 days (Reddy et al., 1994). Nowdays, more than 30 types of plastics have been widely used as materials for packaging food, including PE, PVC, PDVC, PS, PA, PET, and EVOH (Lau and Wong, 2000). The effect of different barrier properties packaging materials (PE, PET, PVC) on the shelf life of shelled fresh bamboo shoots was studied. PET and PVC could inhibit the browning and retard the decomposition of total sugar, reducing sugar and protein, which maintained the shelled bamboo shoots' quality and prolonged their shelf life by at least 8 days. High-barrier PVC and PET packaging materials are important factors to extend the shelf life of bamboo shoots (Zhou et al., 2012). Four alternative films with different water and gas permeabilities are used for Swiss cheese packaging. The result indicated that the tight plastic film used as the industry standard could decrease the rRNA transcriptional activity of the cheese smear microflora within standard film and O_2 control using O_2-scavengers or O_2-tight film to prevent cheese smear defects. The development of cheese smear greatly depended on the permeability of the packaging materials along with a decrease in rRNA transcriptional activity of smear microorganisms (Amato et al., 2015).

Study has found that the packaging materials result in different shelf life of bananas. Four packaging materials including perforated low-density polyethylene (LDPE) bag, perforated HDPE bag, dried banana leaf, and teff straw are used for banana packaging and compared to a no-packaging-materials group. The results showed that unpackaged fruits remained marketable for only 15 days. Banana remained marketable for 36 days in the HDPE and LDPE bags (Hailu et al., 2014). The effects of nonvacuum and vacuum on biogenic amines and organic acids during storage for 180 days at 4°C are investigated in Kashar cheese compared with nonvacuum packaging, vacuum-packaged cheeses had more lactic, formic, acetic, and butyric acids than did cheeses packaged without vacuum (Andiç et al., 2011). Study has also demonstrated that vacuum-packaging with high-barrier material that guarantees the absence of oxygen inside the packaging contributed to the preservation of color by inhibiting the in-pack enzymatic browning. Combined with other preserving strategies this method can effectively extend the shelf of fresh-cut peaches to at least 21 days under 10°C refrigerated storage (Denoya et al., 2015).

A study on storage stability of durian fruit leather was carried out at room temperature using four types of packaging materials: laminated aluminium foil, HDPE, LDPE, and PP for 12 weeks. The result showed that the quality of products is closely related to the water vapor characteristics of the packaging materials. LAF had the lowest changes in moisture content, lowest decrease in color quality, and lowest increase in microbiol counts (Irwandi et al., 1998). Lamb chops and steaks are packaged in PVC film and vacuum packaged in high oxygen-barrier (HOB) film. The result showed that at 15 days of retail display, chops packaged in HOB films are still acceptable in overall appearance, but the PVC-film-wrapped chops are unacceptable after 3 days and would only have a realistic display life of 2 days (Henry et al., 2006). Vacuum-packaging combined with HOB film has greatly extended the shelf life of lamb chops. To prepare suitable barrier films, many modifications have been made based on commercial polymers. As is well known, milk is often packaged in translucent containers providing little protection against flavor degradation from light. The effectiveness of TiO_2 modifications of HDPE packaging in affecting light-induced oxidation of extended shelf-life milk is studied. Products are evaluated over a 35-day shelf life when stored under fluorescent light (2200 lux) at 4°C. The results showed that riboflavin concentration in milk decreased by greater than 40% without TiO_2 as light protected and 28.5% in the high TiO_2 packaged milk (Johnson, 2013). To assess the effect of light transmittance and oxygen permeability of various packaging materials on keeping the quality of low-fat pasteurized milk. The chemical and sensory changes in low-fat pasteurized milk stored under fluorescent light at 4°C in 500 mL bottles made of (1) multilayer pigmented (HDPE +20 g/kg TiO_2/HDPE + 40 g/kg carbon black/HDPE + 20 g/kg TiO_2), (2) monolayer pigmented (HDPE + 20 g/kg TiO_2), (3) clear PET, and (4) pigmented (PET + 20 g/kg TiO_2). Vitamin A losses recorded are 11% for multilayer and monolayer pigmented HDPE and pigmented PET bottles versus 16% for the paperboard cartons and 31% for the clear PET bottles after 7 days of storage. Respective losses for riboflavin are 28% for the multilayer pigmented HDPE bottle and the paperboard carton, 30% for the monolayer pigmented HDPE bottle, 33% for the pigmented PET bottle and 40% for the clear PET bottle.

The shelf life of low fat pasteurized milk is approximately 5 days and the best overall protection for the product is provided by the multilayer with high light and oxygen barrier (Moyssiadi et al., 2004). Samples of green table olives fermented with lactic acid bacteria 100 g are placed in PE pouches under modified atmospheres (70% N_2, 30% CO_2) and stored at 4 and 20°C for 12 months. The result showed that after 6 months at 4°C, both starter strains are recovered in high percentages, the product is organoleptically more acceptable after 6 months than 12 months of storage at both temperatures (Argyri et al., 2015).

EVOH are semicrystalline random copolymers of ethylene and vinyl alcohol, which are widely used in the food packaging industry owing to their excellent chemical resistance, high transparency as well as excellent gas barrier properties to oxygen, organic solvents, and food aromas (Lasagabáster et al., 2009; Kucukpinar and Doruker, 2003). A major disadvantage of EVOH is that they lose oxygen barrier properties due to water absorption in humid conditions. Therefore, EVOH are usually components of multilayer structures and are sandwiched between hydrophobic polyolefins such as PP and PE (Cabedo et al., 2004). To improve the barrier property, PA/EVOH/LLDPE multilayer films is prepared and used for sausage packaging. Compared with PA/PE packages, PA/EVOH/LLDPE effectively restrains the oxidation, discoloration, and microbial growth of sausage in a 6-month storage period due to its excellent oxygen barrier (Song, 2013). The burgers are packed in 200-mm thickness trays made of PET/PE/EVOH/PE and laminated with a top film consisting of APET/PE/EVOH/PE combined with MAP (70% O_2 and 30% CO_2). The packaged burgers are stored at 4 ± 2°C. Good barrier packages combined with suitable atmosphere and bioprotective culture can improve the sensory properties of hamburger meat and potentially extend shelf life up to 12 days. The multilayer coextruded films are composed of seven layers. The medium-barrier to oxygen plain (MBP) film. To get super-high-barrier films, EVOH is composited with PA and named EHA, the barrier of PE/EVOH/PE, PE/PA/EVOH/PAEHA/PE, PA, EVOH, EHA, and PA6/PE films was assessed and used for vacuumed packaging of chilled meat. The result showed that the EHA multilayer film has excellent water vapor and oxygen barrier properties and the chilled meat packaged by it exceeded 20 days in shelf life (Hu et al., 2015; Liang et al., 2015). Meanwhile, in further work, EHA/PE high barrier composite membrane is used for preservation of fresh-cut asparagus lettuce at 5 ± 2°C. The unpackaged fresh-cut asparagus lettuce is unacceptable by the 4th day and the total numbers of colony reached 5.91 CFU, but the other group vacuum-packaged by EHA/PE is acceptable by the 16th day (as shown in Fig. 5.3), the total numbers of colony reached 5.69 CFU until the 18th day and the weight and V_c loss is smaller. The EHA/PE high barrier greatly extended the shelf life of fresh-cut asparagus lettuce (Wang et al., 2015).

Figure 5.3. The photographs of fresh-cut asparagus stalk of unpacked group and in EHA/PE packs.

Much effort has recently been made to develop biodegradable materials because of the worldwide environment and resources problems resulted from petroleum-derived plastics. Degradable organisms materials play an important role in the medical field. For example, PLA as a kind of green biodegradable material has been applied extensively in the tissue engineering scaffolds and drug-controlled release devices. Meanwhile, lots of scholars apply themselves to the further wider application of several biodegradable materials in food packaging. In the early 1970s, researcher used regenerated cellulose and biodegradable polyvinyl film combined with tripolyphosphate (Kena) and absorbent cardboard tray for the perseveration of turkey steaks under 40 + 2°F. The data indicate that cellophane is an excellent overwrapping material tested and that the shelf life of turkey steaks can be extended by using Kena (Taylor et al., 1965). In order to evaluate the effect of new multilayer packaging materials versus a traditional one on biscuit quality, three different packaging materials with different barriers were tested: metalized PLA/paper, metalized OPP with EVA prooxidant additive (EVA-POA)/paper, and metalized orientated polypropylene (OPP)/paper (control) were used for biscuit packaging. The quality of biscuit is studied during accelerated storage at 25, 35, 45°C and 50% RH for 92 days. The results demonstrate that the new biodegradable PLA/paper and EVA-POA/paper have similar performance and preservation effects on the overall product quality to those of the standard one and both are suitable for commercial biscuit storage (Romani et al., 2014). PVA is one of the few completely biodegradable synthetic polymers. It is a synthetic polyhydroxy polymer, having very good water absorption and high gas barrier properties. Poly(propylene carbonate) (PPC) is one of the most promising biodegradable materials in the future. Biodegradable high oxygen barrier PPC/PVA/PPC three-layer films are prepared by a solution casting method and used for chilled meat packaging. Compared with neat PPC, with a 20 wt.% PVA middle layer in PPC/PVA/PPC film, the oxygen barrier of film at 0 and 50 RH% greatly improved by about 500 times. The chilled meat packaged by PPC/PVA/PPC film is acceptable by 19th at 4°C (Dong et al., 2015b). Traces of trehalose (TH) is further added to PPC/PVA/PPC films, and vacuum packaging of chilled meat compared with commercial. A number of studies on the antimicrobial characteristics of films made from biodegradable materials has been carried out (Dong et al., 2014b) and are summarized in Table 5.3.

PA/PE (high barrier) multilayer film and PE packaging films are always used as chilled meat packs. A recently study found that the shelf life of chilled meat in a innovative biodegradable PPC/PVA/PPC-TH multilayer film packs achieved 28~32 days, but samples packaged with PPC/PVA/PPC and traditional PA/PE multilayer

Table 5.3 The Antimicrobial Characteristics of Films Made From Biodegradable Materials

Polymer Formulation	Substrate	Prepared Methods	Applications Comments	References
PLA/CS	Mince pork meat	Extrusion	PLA/CS showed an antimicrobial activity against microorganisms as compared to noncoated samples at 2 log CFU/mL	Bonilla et al. (2013)
PCL/MMT/CS	Pork meat	Solvent casting	The oxygen and water vapor barrier of PCL was improved by MMT and the chilled meat packaged by PCL/MMT/CS reached 23 days	Dong et al. (2014a)
PLA/Nisin	Skim milk liquid egg white	Solvent casting	Glass jar coated with PLA/250 mg nisin inactivated cell of *L. monocytogenes* in liquid egg white and skim milk at storage temperature of 4 and 10°C	Jin (2010)
PPC/PVA/PPC-TH	Pork meat	Solvent casting	The results showed that the shelf life of chilled meat packaged with PPC/PVA/PPC-TH multilayer film could achieve 28~32 days	Zhang et al. (2015b)

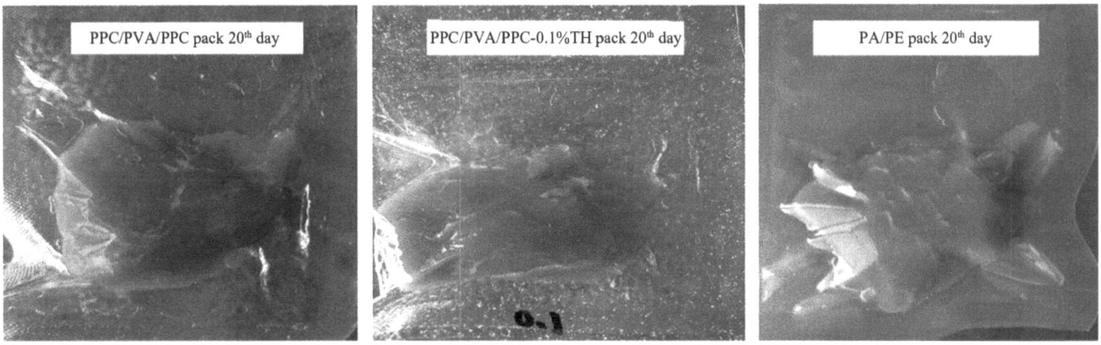

Figure 5.4. Meat preserved in PPC/PVA/PPC, PPC/PVA/PPC-TH, and PE/PA packs on 20th day.

film just could achieve 20 days (as shown in Fig. 5.4), PE film only 12 days. Biodegradable high barrier PPC/PVA/PPC with trace TH, leading to an extension of the shelf life, could replace traditional oil-based plastics for food packaging (Zhang et al., 2015b).

PLA can be synthesized from renewable bio-derived monomers and, as such, it is an alternative to conventional petroleum-based polymers. Since PLA is a relatively new polymer, much effort has been directed toward its development in order to make it an acceptable and effective option to the more traditional petroleum-based polymers. Commercially, PLA has received considerable attention in food packaging applications with a focus on films and coatings that are suitable for short shelf life and ready-to-eat food products (Tawakkal et al., 2014). PLA also been used in antimicrobial systems for food packaging applications. For example, biodegradable PLA polymer incorporated with nisin is evaluated for its application as a material for antimicrobial food packaging. The results demonstrated PLA/nisin film greatly restrained the growth of *L. monocytogenes* in liquid egg white and the culture medium. It also decreased the cell population of *Escherichia coli* in orange juice from 7.5 to 3.5 log at 72 hours. A 2 log reduction of *S. Enteritidis* in liquid egg white packed by PLA/nisin film was found at 24°C. The nisin additive into the PLA polymer presented good antimicrobial effectiveness against foodborne pathogens after 21 days at 4°C. It has a great potential for use in antimicrobial food packaging (Jin and Zhang, 2008).

PLA/CS film is prepared by extrusion and used for antimicrobial packaging, the results showed that PLA/CS showed an antimicrobial activity against microorganisms as compared to noncoated samples at 2 log CFU/mL (Bonilla et al., 2013). Poly(ε-caprolactone) (PCL) is a biodegradable linear aliphatic polyester and has been used in medicinal suture. PCL/MMT/CS and PCL CS antimicrobial films are prepared and used for meat packaging. The oxygen and water vapor barrier of PCL is improved by MMT and the chilled meat packaged by PCL/MMT/CS reached 23 days shelf life (Dutta et al., 2009).

6 Conclusions and Outlooks

Plastic films with good barrier properties used for food packaging play a key role in the preservation of meat products, chilled meat, cheese, salad sauce, seafood, agricultural products, and so on. High barrier packaging materials combined with packaging technology such as vacuum and MAP can insulate external environment and kept a suitable internal atmosphere in packs to inhibit the growth of bacteria and reduce the loss of nutrients in food. Thus the extension of the shelf life of food is expected to be achieved. To get a flexible film with excellent barrier to meet food-packaging requirements, technologies including orientation, coating, nanocompositing, blending, and layer-by-layer nanoassembiles are usually used in researches. But most of these technologies have not been fully developed and widely applied. Simplifying and further improving these fabrication technologies to realize industrial production is the key for the wide acceptance and application of these methods. Besides, the modification and application of biodegradable packaging materials with excellent properties will be a very important part of food packaging industry.

References

Abdollahi, M., Alboofetileh, M., Rezaei, M., Behrooz, R., 2013. Comparing physico-mechanical and thermal properties of alginate nanocomposite films reinforced with organic and/or inorganic nanofillers. Food Hydrocoll. 32, 416–424.

Airado-Rodríguez, D., Intawiwat, N., Skaret, J., Wold, J.P., 2011. Effect of naturally occurring tetrapyrroles on photooxidation in cow's milk. J. Agr. Food Chem. 59, 3905–3914.

Alexandre, B., Langevin, D., Médéric, P., Aubry, T., Couderc, H., Nguyen, Q.T., Saitera, A., Maraisa, S., 2009. Water barrier properties of polyamide 12/ montmorillonite nanocomposite membranes: structure and volume fraction effects. J. Membrane Sci. 328, 186–204.

Amato, L., Ritschard, J.S., Sprecher, I., Lacroix, C., Schuppler, M., Meile, L., 2015. Effect of packaging materials, environmental factors, and RNA transcriptional activity of surface microflora on red-smear cheese defect development. Int. Dairy J., 50–57.

Andiç, S., Tunçtürk, Y., Gençcelep, H., 2011. The effect of different packaging methods on the formation of biogenic amines and organic acids in kashar cheese. J. Dairy Sci. 94, 1668–1678.

Argyri, A.A., Nisiotou, A.A., Pramateftaki, P., Doulgeraki, A.I., Panagou, E.Z., Tassou, C.C., 2015. Preservation of green table olives fermented with lactic acid bacteria with probiotic potential under modified atmosphere packaging. LWT—Food Sci. Techn. 62, 783–790.

Armstrong, S., Freeman, B., Hiltner, A., Baer, E., 2012. Gas permeability of melt-processed poly(ether block amide) copolymers and the effects of orientation. Polymer 53, 1383–1392.

Arora, A., Padua, G.W., 2010. Review: nanocomposites in food packaging. J. Food. Sci. 75, R43–R49.

Asuka, K., Liu, B.P., 2006. Minoru terano, koh-hei nitta, macromol. Rapid Commun. Mass. Sp. 27, 910–913.

Aulin, C., Karabulut, E., Tran, A., Wagberg, L., Lindström, T., 2013. Transparent nanocellulosic multilayer thin films on polylactic acid with tunable gas barrier properties. ACS Appl. Mater. Interf. 5, 7352–7359.

Bang, G., Kim, S.W., 2012. Biodegradable poly(lactic acid)-based hybrid coating materials for food packaging films with gas barrier properties. J. Ind. Eng. Chem. 18, 1063–1068.

Bhatia, A., Gupta, R.K., Bhattacharya, S.N., Choi, H.J., 2012. Analysis of gas permeability characteristics of poly(lactic acid)/poly(butylene succinate) nanocomposites. J. Nanomater. 2012, 5833–5845.

Bonilla, J., Fortunati, E., Vargas, M., Chiralt, A., Kenny, J.M., 2013. Effects of chitosan on the physicochemical and antimicrobial properties of PLA films. J. Food Eng. 119, 236–243.

Brody, A.L., Bugusu, B., Han, J.H., Sand, C.K., Mchugh, T.H., 2008. Scientific status summary: innovative food packaging solutions. J. Food. Sci. 73, R107–R116.

Cabedo, L., Giménez, E., Lagaron, J.M., Gavara, R., Saura, J.J., 2004. Development of evoh-kaolinite nanocomposites. Polymer 45, 5233–5238.

Cardoso, D.R., Libardi, S.H., Skibsted, L.H., 2012. Riboflavin as a photosensitizer: effects on human health and food quality. Food Funct. 3, 487–502.

Carosio, F., Colonna, S., Fina, A., Rydzek, G., Hemmerle, J., Jierry, L., Schaaf, P., Boulmedais, F., 2014. Efficient gas and water vapor barrier properties of thin poly(lactic acid) packaging films: functionalization with moisture resistant nafion and clay multilayers. Chem. Mater. 26 (19), 5459–5466, 2014.

Carrera, M.C., Erdmann, E., Destéfanis, Hugo, A., 2013. Barrier properties and structural study of nanocomposite of HDPE/montmorillonite modified with polyvinylalcohol. J. Chem. 2013, 1–7.

Chandra, R., Rustgi, R., 1998. Biodegradable polymers. Prog. Polym. Sci. 23, 1273–1335.

Chen, T.N., Wuu, D.S., Wu, C.C., Chiang, C.C., Chen, Y.P., Horng, R.H., 2006a. High-performance transparent barrier films of SiOx/SiNx stacks on flexible polymer substrates. J. Electrochem. Soc. 153, F244–F248.

Chen, T.N., Wuu, D.S., Wu, C.C., Chiang, C.C., Lin, H.B., Chen, Y.P., Horng, R.H., 2006b. Effects of plasma pretreatment on silicon nitride barrier films on polycarbonate substrates. Thin Solid Films 514, 188–192.

Cho, S.W., Gallstedt, M., Hedenqvist, M.S., 2010. Properties of wheat gluten/poly(lactic acid) laminates. J. Agric. Food Chem. 58, 7344–7350.

Choe, E., Min, D.B., 2006. Mechanisms and factors for edible oil oxidation. Compr. Rev. Food Sci. F. 5, 169–186.

Choi, J., Rubner, M.F., 2005. Influence of the degree of ionization on weak polyelectrolyte multilayer assembly. Macromolecules 38, 116–124.

Chu, F., Kimura, Y., 1996. Structure and gas permeability of microporous films prepared by biaxial drawing of β-form polypropylene. Polymer 37, 573–579.

Colomines, G., Ducruet, V., Courgneau, C., Guinault, A., Domenek, S., 2010. Barrier properties of poly(lactic acid) and its morphological changes induced by aroma compound sorption. Polym. Int. 59, 818–826.

Dadbin, S., Noferesti, M., Frounchi, M., 2008. Oxygen barrier ldpe/lldpe/organoclay nano-composite films for food packaging. Macromol. Symp. 274, 22–27.
Dan, C.H., Lee, M.H., Kim, Y.D., Min, B.H., Kim, J.H., 2006. Effect of clay modifiers on the morphology and physical properties of thermoplastic polyurethane/clay nanocomposites. Polymer 47, 6718–6730.
Decher, G., 1997. Fuzzy nanoassemblies: toward layered polymeric multicomposites. Science 277, 1232–1237.
Decher, G., Hong, J.D., 1991. Buildup of ultrathin multilayer films by a self-assembly process: II. Consecutive adsorption of anionic and cationic bipolar amphiphiles and polyelectrolytes on charged surfaces. Berichte Der Bunsengesellschaft Für Physikalische Chemie 95, 1430–1434.
Dennler, G., Lungenschmied, C., Neugebauer, H., Sariciftci, N.S., Latrèche, M., Czeremuszkin, G., Wertheimerc, M.R., 2006. A new encapsulation solution for flexible organic solar cells. Thin Solid Films 511, 349–353.
Denoya, G.I., Vaudagna, S.R., Polenta, G., 2015. Effect of high pressure processing and vacuum packaging on the preservation of fresh-cut peaches. LWT—Food Sci. Technol. 62, 801–806.
Dong, T., Hu, H., Liang, X.H., Wang, Y., Renqing, K., Ding, C.M., Jin, Y., 2014a. Preparation of poly(ε-caprolactone)/montmorillonite/chitosan layer film and its application. Package Eng. 35, 1–5.
Dong, T., Yun, X., Shi, C., Sun, W., Fan, G., Jin, Y., 2014b. Improved mechanical and barrier properties of PPC multilayer film through interlayer hydrogen bonding interaction. Polym. Sci. Ser. A+ 56, 830–836.
Dong, T., Yu, Z., Wu, J., Zhao, Z., Yun, X., Wang, Y., Jin, Y., Yang, J., 2015a. Thermal and barrier properties of stretched and annealed polylactide films. Polym. Sci. Ser. A+ 57 (6), 738–746.
Dong, T., Yun, X., Li, M., Sun, W., Duan, Y., Jin, Y., 2015b. Biodegradable high oxygen barrier membrane for chilled meat packaging. J. Appl. Polym. Sci. 132, 41871.
Dutta, P.K., Tripathi, S., Mehrotra, G.K., Dutta, J., 2009. Perspectives for chitosan-based antimicrobial films in food applications. Food Chem. 114, 1173–1182.
Erlat, A.G., Spontak, R.J., Clarke, R.P., Robinson, T.C., Haaland, P.D., Tropsha, Y., Harvey, N.G., Vogler, E.A., 1999. SiOx gas barrier coatings on polymer substrates: morphology and gas transport considerations. J. Phys. Chem. B 103, 6047–6055.
Farris, S., Introzzi, L., Fuentes-Alventosa, J.M., Santo, N., Rocca, R., Piergiovanni, L., 2012. Self-assembled pullulan-silica oxygen barrier hybrid coatings for food packaging applications. J. Agric. Food Chem. 60, 782–790.
Fereydoon, M., Ebnesajjad, S., 2013. 5-development of high-barrier film for food packaging. Plastic Films Food Packag., 71–92.
Fereydoon, M., Tabatabaei, S.H., Ajji, A., 2015. Effect of uniaxial stretching on thermal, oxygen barrier, and mechanical properties of polyamide 6 and poly(m-xylene adipamide) nanocomposite films. Polym. Eng. Sci. 55, 1113–1127.
Fukuzumi, H., Saito, T., Iwata, T., Kumamoto, Y., Isogai, A., 2009. Transparent and high gas barrier films of cellulose nanofibers prepared by tempo-mediated oxidation. Biomacromolecules 10, 162–165.
Gain, O., Espuche, E., Pollet, E., Alexandre, M., Dubois, P.H., 2005. Gas barrier properties of poly(ε-caprolactone)/clay nanocomposites: Influence of the morphology and polymer/clay interactions. J Polym. Sci. Pol. Phys. 43, 205–214.
Giannelis, E.P., 1996. Polymer layered silicate nanocomposites. Adv. Mater. 8, 29–35.
Gorrasi, G., Tortora, M., Vittoria, V., Pollet, E., Lepoittevin, B., Alexandre, M., Philippe, D., 2003. Vapor barrier properties of polycaprolactone montmorillonite nanocomposites: effect of clay dispersion. Polymer 44, 2271–2279.

Groner, M.D., George, S.M., Mclean, R.S., Carcia, P.F., 2006. Gas diffusion barriers on polymers using Al_2O_3 atomic layer deposition. Appl. Phys. Lett. 88, 051907.
Grüniger, A., Rohr, P.R.V., 2003. Deposition of SiO_2 -like diffusion barriers on PET and paper by pecvd. Surf. Coat. Tech. 174, 1043–1047.
Gruniger, A., Bieder, A., Sonnenfeld, A., Rohr, P.R.V., Müller, U., Hauert, R., 2006. Influence of film structure and composition on diffusion barrier performance of SiOx thin films deposited by pecvd. Surf. Coat. Technol. 200, 4564–4571.
Gu, C.H., Wang, J.J., Yu, Y., Sun, H., Shuai, N., Wei, B., 2013. Biodegradable multilayer barrier films based on alginate/polyethyleneimine and biaxially oriented poly(lactic acid). Carbohydr. Polym. 92, 1579–1585.
Hagen, D.A., Saucier, L., Grunlan, J.C., 2014a. Controlling effective aspect ratio and packing of clay with pH for improved gas barrier in nanobrick wall thin films. ACS Appl. Mater. Interfaces 6, 22914–22919.
Hagen, D.A., Foster, B., Stevens, B., Grunlan, J.C., 2014b. Shift-time polyelectrolyte multilayer assembly: fast film growth and high gas barrier with fewer layers by adjusting deposition time. ACS Macro. Lett. 3, 663–666.
Hailu, M., Workneh, T.S., Belew, D., 2014. Effect of packaging materials on shelf life and quality of banana cultivars (*Musa spp.*). J. Food Sci. Tech. 51, 2947–2963.
Halasz, K., Hosakun, Y., Csoka, L., 2015. Reducing water vapor permeability of poly(lactic acid) film and bottle through layer-by-layer deposition of green-processed cellulose nanocrystals and chitosan. Int. J. Polym. Sci. 2015, 1–6.
Ham, M., Kim, J.C., Chang, J.H., 2013. High gas barrier properties of poly(vinyl alcohol) nanocomposite films with various equibiaxial stretching ratios. Macromol. Res. 21, 1289–1292.
Henry, B.M., Erlat, A.G., Mcguigan, A., Crm, G., Gad, B., Tsukahara, Y., Miyamoto, T., Noguchi, N., Niijima, T., 2001. Characterization of transparent aluminium oxide and indium tin oxide layers on polymer substrates. Thin Solid Films 382, 194–201.
Henry, K.G., Savell, J.W., Smith, G.C., And, J.G.E., Vanderzant, C., 2006. Physical, sensory, and microbiological characteristics of lamb retail cuts vacuum packaged in high oxygen-barrier film. J. Food Sci. 48, 1735–1740.
Herrera, M.A., Mathew, A.P., Oksman, K., 2014. Gas permeability and selectivity of cellulose nanocrystals films (layers) deposited by spin coating. Carbohyd. Polym. 112, 494–501.
Herrera-Alonso, J.M., Marand, E., Little, J.C., Cox, S.S., 2009. Transport properties in polyurethane/clay nanocomposites as barrier materials: effect of processing conditions. J. Membrane Sci. 337, 208–214.
Herrera-Alonso, J.M., Sedlakova, Z., Marand, E., 2010. Gas barrier properties of nanocomposites based on in situ polymerized poly(n-butyl methacrylate) in the presence of surface modified montmorillonite. J. Membrane Sci. 349, 251–257.
Holder, K.M., Priolo, M.A., Secrist, K.E., Greenlee, S.M., Nolte, A.J., Grunlan, J.C., 2012. Humidity-responsive gas barrier of hydrogen-bonded polymer-clay multilayer thin films. J. Phys. Chem. C 116, 19851–19856.
Holsti-Meittinen, R.M., Perttila, K.R., Seppala, J.V., Heino, M.T., 1995. Oxygen barrier properties of polypropylene/polyamide 6 blends. J. Appl. Polym. Sci. 58, 1551–1560.
Hosseini, S.H., Mirdamadi, S., Rastegari, S., 2015. Investigating efficiency of α-Al_2O_3 diffusion barrier layer in oxidation of EB-PVD NiCrAlY coatings. Surf. Eng. 31, 146–155.
Hu, Z., Yi, C., Fan, H., Shi, H., Luo, Z., Bi, S., 2008. The polyurethane/SiO_2 nano-hybrid membrane with temperature sensitivity for water vapor permeation. J. Membrane Sci. 318, 71–78.

Hu, H., Wang, Y., Liang, X.H., Jin, Y., Dong, T., 2015. Preparation and application of nylon composite films. China Plastics Ind. 43, 65–69.
Huang, H.D., Ren, P.G., Chen, J., Zhang, W.Q., Xu, J., Li, Z.M., 2012. High barrier graphene oxide nanosheet/poly(vinyl alcohol) nanocomposite films. J. Membrane Sci. 409, 156–163.
Hutchings, J.B., 1999. Food Color and Appearance. Aspen Publishers, Gaithersburg, MD.
Hwang, T., Pu, L., Kim, S.W., Oh, Y.S., Nam, J.D., 2009. Synthesis and barrier properties of poly(vinylidene chloride- *co* -acrylonitrile)/SiO_2 hybrid composites by sol-gel process. J. Membrane. Sci. 345, 90–96.
Inagaki, N., Tasaka, S., Makino, M., 1997. J. Appl. Polym. Sci. 64, 1031–1039.
Incarnato, L., Acierno, D., Russo, P., Malinconico, M., Laurienzo, P., 1999. Influence of Composition on Properties of Nylon6/EVOH Blends. J. Polym. Sci. B. 37, 2445–2455.
Irwandi, J., Man, Y.B.C., Yusof, S., Jinap, S., Sugisawa, H., 1998. Effects of type of packaging materials on physicochemical, microbiological and sensory characteristics of durian fruit leather during storage. J. Sci. Food Agr. 76, 427–434.
Iwamori, S., Gotoh, Y., Moorthi, K., 2003a. Characterization of silicon oxynitride gas barrier films. Vacuum 68, 113–117.
Iwamori, S., Gotoh, Y., Moorthi, K., 2003b. Silicon oxide gas barrier films deposited by reactive sputtering. Surf. Coat. Tech. 166, 24–30.
Izquierdo, A., Ono, S., Voegel, J.C., Schaaf, P., Decher, G., 2005. Dipping versus spraying: exploring the deposition conditions for speeding up layer-by-layer assembly. Langmuir. 21, 7558–7567.
Jang, W.S., Rawson, I., Grunlan, J.C., 2008. Layer-by-layer assembly of thin film oxygen barrier. Thin Solid Films 516, 4819–4825.
Jarus, D., Hiltner, A., Baer, E., 2002. Barrier properties of Polypropylene/polyamide blends produced by microlayer coextrusion. Polymer 43, 2401–2408.
Jin, T., Zhang, H., 2008. Biodegradable polylactic acid polymer with nisin for use in antimicrobial food packaging. J. Food Sci. 73, M127–M134.
Jin, T., 2010. Inactivation of listeria monocytogenes in skim milk and liquid egg white by antimicrobial bottle coating with polylactic acid and nisin. J. Food Sci. 75, M83–M88.
Jin, H.C., Park, Y.W., Park, T.H., Song, E.H., Lee, H.J., Kim, H., Shin, S.J., Fai, V.L.C., Ju, B.K., 2012. Fuzzy nanoassembly of polyelectrolyte and layered clay multicomposite toward a reliable gas barrier. Langmuir 28, 6826–6831.
Joardder, M.U.H., Karim, A., Kumar, C., Brown, R.J., 2014. Determination of effective moisture diffusivity of banana using thermogravimetric analysis. Procedia Eng. 90, 538–543.
Johnson, D.S., 2013. Protective effects of titanium dioxide packaging modification on sensory and oxidative changes in milk over 35-day shelf-life. Master of Science in Life Science. https://vtechworks.lib.vt.edu/handle/10919/49695.
Jongwhan, R., Hong, S.I., Changsik, H., 2009. Tensile, water vapor barrier, and antimicrobial properties of pla/nanoclay composite films. LWT—Food Sci. Technol. 42, 612–617.
Jung, M.Y., Yoon, S.H., Lee, H.O., Min, D.B., 1998. Singlet oxygen and ascorbic acid effects on dimethyl disulfide and off-flavor in skim milk exposed to light. J. Food Sci. 63, 408–412.
Kamal, M.R., Jinnah, I.A., Utracki, L.A., 1984. Permeability of oxygen and water vapor through polyethylene/polyamide films. Polym. Eng. Sci. 24 (17), 1337–1347.
Kasahara, I., Carrasco, V., Aguilar, L., 2015. Inactivation of *Escherichia coli* in goat milk using pulsed ultraviolet light. J. Food Eng. 152, 43–49.

Kasanen, J., Suvanto, M., Pakkanen, T.T., 2011. Improved adhesion of TiO_2-based multilayer coating on HDPE and characterization of photocatalysis. J. Appl. Polym. Sci. 119, 2235–2245.
Kasirajan, S., Ngouajio, M., 2012. Polyethylene and biodegradable mulches for agricultural applications: a review. Agron. Sustain. Dev. 32, 501–529.
Ke, Y.C., Stroeve, P., 2005. Background on polymer-layered silicate and silica nanocomposites. Polymer-Layered Silicate and Silica Nanocomposites, 1–67.
Ke, Y.C., Wang, Y.G., Yang, L., 2010. Improving the hydrophobic, water barrier and crystallization properties of poly(ethylene terephthalate) by incorporating monodisperse SiO_2 particles. Polym. Int. 59, 1350–1359.
Keiji, A., Kiyoharu, T., Tsutomu, M., 1999. Water and oxygen permeability of silica thin films containing organic polymers coated on poly(ethylene terephtalate) substrate by the sol-gel method. J. Ceram. Soc. Jpn. 107, 293.
Khan, A., Huq, T., Khan, R.A., Riedl, B., Lacroix, M., 2014. Nanocellulose-based composites and bioactive agents for food packaging. Crit. Rev. Food Sci. 54, 163–174.
Kim, S.W., 2008. Preparation and barrier property of poly(vinyl alcohol)/SiO_2 hybrid coating films. Korean J. Chem. Eng. 25, 1195–1200.
Kim, S.W., Chun, Y.H., 1999. Barrier property by controlled laminar morphology of LLDPE/EVOH blends. Korean J. Chem. Eng. 16, 511–517.
Kim, D., Kim, S.W., 2003. Barrier property and morphology of polypropylene/polyamide blend film. Korean J. Chem. Eng. 20, 776–782.
Kim, J.K., Hu, C., Woo, R.S.C., Sham, M.L., 2005. Moisture barrier characteristics of organoclay-epoxy nanocomposites. Compos. Sci. Technol. 65, 805–813.
Kim, H., Miura, Y., Macosko, C.W., 2010. Graphene/polyurethane nanocomposites for improved gas barrier and electrical conductivity. Chem. Mater. 22, 3441–3450.
Kim, S.G., Park, O.K., Lee, J.H., Ku, B.C., 2013. Layer-by-layer assembled graphene oxide films and barrier properties of thermally reduced graphene oxide membranes. Carbon Lett. 14, 247–250.
Kolasinska, M., Krastev, R., Gutberlet, T., Warszynski, P., 2008. Layer-by-layer deposition of polyelectrolytes: dipping versus spraying. Langmuir 25, 1224–1232.
Kontturi, E., Johansson, L., Kontturi, K., Ahonen, P., Thune, P., 2007. Cellulose nanocrystal submonolayers by spin coating. Langmuir 23, 9674–9680.
Kucukpinar, E., Doruker, P., 2003. Molecular simulations of small gas diffusion and solubility in copolymers of styrene. Polymer 44, 3607–3620.
Kwak, S., Jun, J., Jung, E.S., 2009. Gas barrier film with a compositional gradient interface prepared by plasma modification of an organic/inorganic hybrid sol-gel coat. Langmuir 25, 8051–8055.
Lange, J., Wyser, Y., 2003. Recent innovations in barrier technologies for plastic packaging: a review. Packag. Technol. Sci. 16, 149–158.
Lange, S., Arroval, T., Saar, R., Kink, I., Aarik, J., Krumme, A., 2015. Oxygen barrier properties of Al_2O_3- and TiO_2-coated LDPE films. Polym.-Plast. Technol. 54, 301–304.
Lasagabáster, A., Abad, M.J., Barral, L., Ares, A., Bouza, R., 2009. Application of FTIR spectroscopy to determine transport properties and water–polymer interactions in polypropylene (PP)/poly(ethylene-*co*-vinyl alcohol) (EVOH) blend films: effect of poly(ethylene-*co*-vinyl alcohol) content and water activity. Polymer 50 (13), 2981–2989.
Lau, O.W., Wong, S.K., 2000. Contamination in food from packaging materials. J. Chrom. A 882, 255–270.
Lee, S.Y., Kim, S.C., 1997. Laminar morphology development and oxygen permeability of LDPE/EVOH blends. Polym. Eng. Sci. 37, 463–475.

Lee, K.H., Hong, J.H., Kwak, S.J., Min, P., Son, J.G., 2014. Spin self-assembly of highly ordered multilayers of graphene-oxide sheets for improving oxygen barrier performance of polyolefin films. Carbon 83, 40–47.
Leterrier, Y., 2003. Durability of nanosized oxygen-barrier coatings on polymers. Prog. Mater. Sci. 48, 1–55.
Liang, X.H., Hu, H., Wang, Y., Jin, Y., Dong, T., 2015. The impact of high barrier composite film on the shelf life of chilled meat. Food Sci. Technol. 40, 150–153.
Lin, Y.J., Dias, P., Chen, H.Y., Chum, S., Hiltner, A., Baer, E., 2008. Oxygen permeability of biaxially oriented polypropylene films. Polym. Eng. Sci. 48, 642–648.
Lin, Y., Hiltner, A., Baer, E., 2010. Nanolayer enhancement of biaxially oriented polypropylene film for increased gas barrier. Polymer 51, 5807–5814.
Liu, R.Y.F., Schiraldi, D.A., Hiltner, A., Baer, E., 2002. Oxygen barrier properties of cold-drawn polyesters. J. Polym. Sci. Pol. Phys. 40, 862–877.
Lutkenhaus, J.L., Hrabak, K.D., McEnnis, K., Hammond, P.T.J., 2005. Elastomeric flexible free-standing hydrogen-bonded nanoscale assemblies. Am. Chem. Soc. 127, 17228–17234.
Magniez, K., Fox, B.L., Looney, M.G., 2009. Nonisothermal crystallization behavior of poly(m-xylene adipamide)/montmorillonite nanocomposites. J. Polym. Sci. Pol. Phys. 47, 1300–1312.
Maikap, A.N., Jain, S., Singh, H.K., Gupta, A., Behl, S., 2015. Effect of primers and surface treatments on the gas barrier, mechanical, optical, printing, and flex-crack properties of the polyvinylidene chloride coated biaxially oriented polyethylene terephthalate films. J. Plast. Film Sheet 31, 158–185.
Marsh, K., Bugusu, B., 2007. Food packaging-roles, materials, and environmental issues. J. Food. Sci. 72, R39–R55.
Martucci, J.F., Ruseckaite, R.A., 2010. Three-layer sheets based on gelatin and poly(lactic acid), Part 1: Preparation and properties. J. Appl. Polym. Sci. 118, 3102–3110.
Masuko, T., Tanaka, H., kajima, S., 1970. Studies on biaxial stretching of polypropylene film. IV. Overall orientation during one-step biaxial stretching and its application to the estimation of the principal refractive indices of isotactic polypropylene. J. Polym. Sci. Pol. Phys. 8, 1565–1574.
Mikkonen, K.S., Tenkanen, M., 2012. Sustainable food-packaging materials based on future biorefinery products: xylans and mannans. Trends Food Sci. Tech. 28, 90–102.
Moyssiadi, T., Badeka, A., Kondyli, E., Vakirtzi, T., Savvaidis, I., Kontominas, M.G., 2004. Effect of light transmittance and oxygen permeability of various packaging materials on keeping quality of low-fat pasteurized milk: chemical and sensorial aspects. Int. Dairy J. 14, 429–436.
Nam, J.D., Kim, S.W., 2010. Phase compatibility and barrier properties of ethylene/vinyl alcohol copolymer–based hybrid materials. J. Appl. Polym. Sci. 115, 1663–1670.
Nazarenko, S., Meneghetti, P., Julmon, P., Olson, B.G., Qutubuddin, S., 2007. Gas barrier of polystyrene montmorillonite clay nanocomposites: effect of mineral layer aggregation. J. Polym. Sci. Pol. Phys. 45, 1733–1753.
Nir, M.M., Miltz, J., Ram, A., 1993. Update on plastics and the environment: progress and trends. Plast. Eng. 49, 75–93.
Osman, M.A., Mittal, V., Morbidelli, M., Suter, U.W., 2003. Polyurethane adhesive nanocomposites as gas permeation barrier. Macromolecules 36, 9851–9858.
Park, S.H., Lee, H.S., Choi, J.C., Jeong, C.M., Sung, M.H., Park, H.J., 2012. Improvements in barrier properties of poly(lactic acid) films coated with chitosan or chitosan/clay nanocomposite. J. Appl. Polym. Sci. 125, E675–E680.

Picard, E., Vermogen, A., Gérard, J.F., Espuche, E., 2007. Barrier properties of nylon 6-montmorillonite nanocomposite membranes prepared by melt blending: influence of the clay content and dispersion state. J. Membrane Sci. 292, 133–144.
Ploehn, H.J., Liu, C., 2006. Quantitative analysis of montmorillonite platelet size by atomic force microscopy. Ind. Eng. Chem. Res. 45, 7025–7034.
Podsiadlo, P., Michel, M., Lee, J., Verploegen, E., Kam, N.W.S., Ball, V., Lee, J., Qi, Y., Hart, A.J., Hammond, P.T., Kotov, N.A., 2008. Exponential growth of LBL films with incorporated inorganic sheets. Nano Lett. 8 (6), 1762–1770.
Prins, W., Hermans, J.J., 1959. Theory of permeation through metal-coated polymer films. J. Phys. Chem. 63, 716–722.
Priolo, M.A., Gamboa, D., Grunlan, J.C., 2010a. Transparent clay-polymer nano brick wall assemblies with tailorable oxygen barrier. ACS Appl. Mater. Inter. 2, 312–320.
Priolo, M.A., Gamboa, D., Holder, K.M., Grunlan, J.C., 2010b. Super gas barrier of transparent polymer-clay multilayer ultrathin films. Nano Lett. 10, 4970–4974.
Priolo, M.A., Holder, K.M., Gamboa, D., Grunlan, J.C., 2011. Influence of clay concentration on the gas barrier of clay-polymer nanobrick wall thin film assemblies. Langmuir 27, 12106–12114.
Priolo, M.A., Holder, K.M., Guin, T., Grunlan, J.C., 2015. Recent advances in gas barrier thin films via layer-by-layer assembly of polymers and platelets. Macromol. Rapid Comm. 36, 866–879.
Ramanathan, T., Abdala, A.A., Stankovich, S., Dikin, D.A., Herrera-Alonso, M., Piner, R.D., Adamson, D.H., Schniepp, H.C., Chen, X., Ruoff, R.S., Nguyen, S.T., Aksay, I.A., Prud'-Homme, R.K., Brinson, L.C., 2008. Functionalized graphene sheets for polymer nanocomposites. Nat. Nanotechnol. 3, 327–331.
Ray, S.S., Maiti, P., Okamoto, M., Yamada, K., Ueda, K., 2002. New polylactide/layered silicate nanocomposites. 1. Preparation, characterization, and properties. Macromolecules 35, 3104–3110.
Razavi, S.M., Dadbin, S., Frounchi, M., 2012. Oxygen-barrier properties of poly(lactic acid)/poly(vinyl acetate-*co*-vinyl alcohol) blends as biodegradable films. J. Appl. Polym. Sci. 125, E20–E26.
Reddy, N.R., Schreiber, C.L., Buzard, K.S., Skinner, G.E., Armstrong, D.J., 1994. Shelf life of fresh tilapia fillets packaged in high-barrier film with modified atmospheres. J. Food Sci. 59, 260–264.
Reddy, M.M., Vivekanandhan, S., Misra, M., Bhatia, S.K., Mohanty, A.K., 2013. Biobased plastics and bionanocomposites: current status and future opportunities. Prog. Polym. Sci. 38, 1653–1689.
Rhim, J.W., Lee, J.H., Ng, P.K.W., 2007. Mechanical and barrier properties of biodegradable soy protein isolate-based films coated with polylactic acid. LWT—Food Sci. Technol. 40, 232–238.
Risch, S.J., 2009. Food packaging history and innovations. J. Agr. Food Chem. 57, 8089–8092.
Robertson, G.L., 2010. Food packaging and shelf life. In: Robertson, G.L. (Ed.), Food Packaging and Shelf Life: A Practical Guide. CRC Press, Boca Raton, FL.
Romani, S., Tappi, S., Balestra, F., Rodriguez Estrada, M.T., Siracusa, V., Rocculi, P., et al., 2014. Effect of different new packaging materials on biscuit quality during accelerated storage. J. Sci. Food Agr. 95, 1736–1746.
Ryu, J.G., Kim, H., Lee, J.W., 2004. Characteristics of polystyrene/polyethylene/clay nanocomposites prepared by ultrasound-assisted mixing process. Polym. Eng. Sci. 44, 1198–1204.
Sanchez-Garcia, M.D., Nordqvist, D., Hedenqvist, M., Lagaron, J.M., 2011. Incorporating amylopectin in poly(lactic acid) by melt blending using poly(ethylene-*co*-vinyl alcohol) as a thermoplastic carrier. II. Physical properties. J. Appl. Polym. Sci. 119, 3708–3716.

Sattar, A., Deman, J.M., Alexander, J.C., 1977. Wavelength effect on light-induced decomposition of vitamin a and β-carotene in solutions and milk fat. Can. I. Food Sci. Technol. J. 10, 56–60.

Schauberger, J.G., Riess, G., Kern, W., 2013. Preparation of UV reactive montmorillonite and characterization of its nanocomposites with poly(vinyl alcohol). J. Appl. Polym. Sci. 130, 665–672.

Semagoto, H.M., Liu, D., Koboyatau, K., Hu, J.H., Lu, N.Y., Liu, X.M., Regenstein, J.M., Zhou, P., 2014. Effects of UV induced photo-oxidation on the physicochemical properties of milk protein concentrate. Food Res. Int. 62, 580–588.

Shi, C., Zhang, S., Li, M., Sun, W., Fan, G., Jin, Y., Yang, J., Dong, T., 2013. Barrier and mechanical properties of biodegradable poly(ε-caprolactone)/cellophane multilayer film. J. Appl. Polym. Sci. 130, 1805–1811.

Shim, J., Yoon, H.G., Na, S.H., Kim, I., Kwak, S., 2008. Silicon oxynitride gas barrier coatings on poly(ether sulfone) by plasma-enhanced chemical vapor deposition. Surf. Coat. Tech. 202, 2844–2849.

Shiratori, S.S., Rubner, M.F., 2000. pH-dependent thickness behavior of sequentially adsorbed layers of weak polyelectrolytes. Macromolecules 33, 4213–4219.

Shukla, A., Avadhany, S.N., Fang, J.C., Hammond, P.T., 2010. Tunable vancomycin releasing surfaces for biomedical applications. Small 6, 2392.

Silva Sobrinho, A.S.D., Latreche, M., Czeremuszkin, G., Klemberg-Sapieha, J.E., Wertheimer, M.R., 1998. Transparent barrier coatings on polyethylene terephthalate by single- and dual-frequency plasma-enhanced chemical vapor deposition. J. Vac. Sci. Technol. A 16, 3190–3198.

Sinha Ray, S., Yamada, K., Okamoto, M., Fujimoto, Y., Ogami, A., Ueda, K., 2003. New polylactide/layered silicate nanocomposites. 5. Designing of materials with desired properties. Polymer 44, 6633–6646.

Somlai, L.S., Liu, R.Y.F., Landoll, L.M., Hiltner, A., Baer, E., 2005. Effect of orientation on the free volume and oxygen transport of a polypropylene copolymer. Polym. Sci. Pol. Phys. 43, 1230–1243.

Song, Y., 2013. Effect of EVOH high barrier packaging in preserving chilled distributed processed meat quality. Plast. Packag. 24, 17–23.

Song, E.H., Kang, B.H., Kim, T.Y., Lee, H.J., Park, Y.W., Kim, Y.C., Ju, B.K., 2015. Highly oriented gold/nanoclay-polymer nanocomposites for flexible gas barrier films. ACS Appl. Mater. Interf. 7, 4778–4783.

Souza, A.C., Benze, R., Ferrao, E.S., Ditchfield, C., Coelho, A.C.V., Tadini, C.C., 2012. Cassava starch biodegradable films: influence of glycerol and clay nanoparticles content on tensile and barrier properties and glass transition temperature. LWT—Food Sci. Technol. 46, 110–117.

Stankovich, S., Dikin, D.A., Dommett, G.H.B., Kohlhaas, K.M., Zimney, E.J., Stach, E.A., Piner, R.D., Nguyen, S.T., Ruoff, R.S., 2006. Graphene-based composite materials. Nature. 442, 282–286.

Stevens, B., Dessiatova, E., Hagen, D.A., Todd, A.D., Bielawski, C.W., Grunlan, J.C., 2014a. Low-temperature thermal reduction of graphene oxide nanobrick walls: unique combination of high gas barrier and low resistivity in fully organic polyelectrolyte multilayer thin films. ACS Appl. Mater. Interf. 6, 9942–9945.

Stevens, B.E., Odenborg, P.K., Priolo, M.A., Grunlan, J.C., 2014b. Hydrophobically modified polyelectrolyte for improved oxygen barrier in nanobrick wall multilayer thin films. J. Polym. Sci. Pol. Phys. 52, 1153–1156.

Su, C.C., Wan, G.H., Im, S.S., Oh, S.G., 2002. Poly(ethylene terephthalate)(PET) nanocomposites filled with fumed silicas by melt compounding. Macromol. Res. 10, 221–229.

Subramanian, R.M., 1985. Permeability barriers by controlled morphology of polymer blends. Polym. Eng. Sci. 25, 483–487.
Subramanian, P.M., Mehra, M., 1987. Laminar morphology in polymer blends: structure and properties. Polym. Eng. Sci. 27, 663–668.
Tabatabaei, S.H., Carreau, P.J., Ajji, A., 2008. Microporous membranes obtained from polypropylene blend films by stretching. J. Membrane Sci. 325, 772–782.
Tabatabaei, S.H., Carreau, P.J., Ajji, A., 2009. Effect of processing on the crystalline orientation, morphology, and mechanical properties of polypropylene cast films and microporous membrane formation. Polymer 50, 4228–4240.
Taraiya, A.K., Orchard, G.A.J., Ward, I.M., 1993. The effect of amorphous orientation on the oxygen permeability of polypropylene films. J. Appl. Polym. Sci. 31, 641–645.
Tawakkal, I.S., Cran, M.J., Miltz, J., Bigger, S.W., 2014. A Review of poly(Lactic acid)-based materials for antimicrobial packaging. J. Food Sci. 79, R1477–R1490.
Taylor, M.H., Smith, L.T., Jd, M., 1965. The effect of packaging materials and a tripolyphosphate (KENA) on the shelf life of turkey steaks. Poultry Sci. 44, 297–298.
Tsai, J.L., Hsiao, H., Cheng, Y.L., 2010. Investigating mechanical behaviors of silica nanoparticle reinforced composites. J. Compos. Mater. 44, 505–524.
Tseng, W.J., Chang, J.H., 2014. Preparation of K-Al_2O_3/resin composite coating on polyethylene terephthalate foil for gas-permeation barrier. Ceram. Int. 40, 16779–16784.
Tzeng, P., Lugo, E.L., Mai, G.D., Wilhite, B.A., Grunlan, J.C., 2014. Super hydrogen and helium barrier with polyelectolyte nanobrick wall thin film. Macromol. Rapid Comm. 36, 96–101.
Vaia, R.A., Ishii, H., Giannelis, E.P., 1993. Synthesis and properties of two-dimensional nanostructures by direct intercalation of polymer melts in layered silicates. Chem. Mater. 5, 1694–1696.
Vaia, R.A., Jandt, K.D., Kramer, E.J., Giannelis, E.P., 1995. Kinetics of polymer melt intercalation. Macromolecules 28, 8080–8085.
Walther, A., Bjurhager, I., Malho, J.M., Pere, J., Ruokolainen, J., Berglund, L.A., Ikkala, O., 2010. Large-area, lightweight and thick biomimetic composites with superior material properties via fast, economic, and green pathways. Nano Lett. 10, 2742–2748.
Wang, D., Jin, Z., Qiang, Y., Wilkie, C.A., 2002. A comparison of various methods for the preparation of polystyrene and poly(methyl methacrylate) clay nanocomposites. Chem. Mater. 14, 3837–3843.
Wang, H., Keum, J.K., Hiltner, A., Baer, E., 2009. Confined crystallization of peo in nanolayered films impacting structure and oxygen permeability. Macromolecules 42, 7055–7066.
Wang, Y., Yun, X., Zhang, X., Li, J., Hu, X., Liang, X., Hu, H., Dong, T., 2015. Effect of EHA/PE high barrier composite membrane on preservation of fresh-cut asparagus lettuce. Available from: http://www.cnki.net/kcms/detail/11.1759.TS.20150603.0944.001.html
Weiss, J., Leppin, C., Mader, W., Salzberger, U., 1989. Aluminium metallization of polyester and polypropylene films: properties and transmission electron microscopy microstructure investigations. Thin Solid Films (Part 1) 174, 155–158.
Wendy, A., Allan, A., Brian, T., 1998. A review of biodegradable polymers: uses, current developments in the synthesis and characterization of biodegradable polyesters, blends of biodegradable polymers and recent advances in biodegradation studies. Polym. Int. 47, 89–144.
Xia, H., Shaw, S.J., Song, M., 2005. Relationship between mechanical properties and exfoliation degree of clay in polyurethane nanocomposites. Polym. Int. 54, 1392–1400.

Xiang, F., Ward, S.M., Givens, T.M., Grunlan, J.C., 2014a. Super stretchy polymer multilayer thin film with high gas barrier. ACS Macro Lett. 3, 1055–1058.
Xiang, F., Tzeng, P., Sawyer, T.S., Regev, O., Grunlan, J.C., 2014b. Improving the gas barrier property of clay-polymer multilayer thin films using shorter deposition times. ACS Appl. Mater. Interf, 6, 6040–6048.
Yang, Q., Fukuzumi, H., Saito, T., Isogai, A., Zhang, L., 2011a. Transparent cellulose films with high gas barrier properties fabricated from aqueous alkali/urea solutions. Biomacromolecules 12, 2766–2771.
Yang, Y.H., Haile, M., Yong, T.P., Malek, F.A., Grunlan, J.C., 2011b. Super gas barrier of all-polymer multilayer thin films. Macromolecules 44, 1450–1459.
Yang, Y.H., Bolling, L., Priolo, M.A., Grunlan, J.C., 2013. Super gas barrier and selectivity of graphene oxide-polymer multilayer thin films. Adv. Mater. 25, 503–508.
Yeh, J.M., Huang, H.Y., Chen, C.L., Su, W.F., Yu, Y.H., 2006. Siloxane-modified epoxy resin-clay nanocomposite coatings with advanced anticorrosive properties prepared by a solution dispersion approach. Surf. Coat. Tech. 200, 2753–2763.
Yeo, J.H., Lee, C.H., Park, C.S., Lee, K.J., Nam, J.D., Kim, S.W., 2001. Rheological, morphological, mechanical, and barrier properties of PP/EVOH blends. Adv. Polym. Tech. 20, 191–201.
Yoo, D., Shiratori, S.S., Rubner, M.F., 1998. Controlling bilayer composition and surface wettability of sequentially adsorbed multilayers of weak polyelectrolytes. Macromolecules 31, 4309–4318.
Yoo, P.J., Nam, K.T., Qi, J.F., Lee, S.K., Park, J., Belcher, A.M., Hammond, P.T., 2006. Spontaneous assembly of viruses on multilayered polymer surfaces. Nat. Mater. 5, 234–240.
Yu, H.Y., Qin, Z.Y., Sun, B., Yang, X.G., Yao, J.M., 2014. Reinforcement of transparent poly (3-hydroxybutyrate-*co*-3-hydroxyvalerate) by incorporation of functionalized carbon nanotubes as a novel bionanocomposite for food packaging. Compos. Sci. Technol. 94, 96–104.
Yun, X., Zhao, S., Liang, X., Zhang, X., Jin, Y., Dong, T., 2015. Performance of PCL/PPC blend film. Package Eng. 36, 46–50.
Zhang, S., Xue, W., Yu, Z., 2015a. Moisture barrier evaluation of SiOx/SiNx stacks on polyimide substrates using electrical calcium test. Thin Solid Films 580, 101–105.
Zhang, X.Y., Yun, X.Y., Liang, M., Wang, Y., Jin, Y., Zhao, L., Dong, T., 2015b. Influence of biodegradable film containing trehalose on fresh-keeping and color protection of chilled meat. Sci. Technol. Food Ind. 8, 298–304.
Zhang, L., Zhou, Y., Huang, M., Sha, Y., Tian, J., Ye, Q., 2014. Effect of nanoparticle surface modification on charge transport characteristics in XLPE/SiO_2 nanocomposites. Dielectrics and Electrical Insulation, IEEE Trans. 21, 424–433.
Zhou, X.X., 2001. Production and function of high-resistant film. China Packag. Ind. 8, 29–36.
Zhou, J.W., Zhang, G.H., Hu, H.Y., Wang, Z.L., 2012. Effect of packaging materials on the shelf life of bamboo shoots. App. Mech. Mater. 200, 540–544.

6

BIODEGRADABLE FOOD PACKAGING NANOCOMPOSITES BASED ON ZnO-REINFORCED POLYHYDROXYALKANOATES

Ana M. Díez-Pascual
Alcalá University, Analytical Chemistry, Physical Chemistry and Chemical Engineering Department, Faculty of Biology, Environmental Sciences and Chemistry, Madrid, Spain

1 Introduction

The foremost purpose of food packaging is to preserve the quality and safety of food products throughout storage and transportation, and to expand their shelf life by averting unfavorable conditions like spoilage microorganisms, contaminants, light, oxygen, and moisture (Rhim et al., 2013). The food package ought to hamper the increase or loss of moisture, avoid microorganism infection, and act as a barrier toward permeation of water vapor, oxygen, carbon dioxide, and other volatile compounds (Marsh and Bugusu, 2007). The four basic packaging materials employed over the past decades were paper and paperboard, plastics, glasses, and metals. Among these packaging materials, petroleum-based plastics have been extensively used over the 20th century, owing to their low cost, convenience of use, easy processability, and exceptional physicochemical properties. However, these cause a serious environmental problem since they are not easily degraded in the environment subsequent to use. Current concerns about environmental pollution caused by accumulation of nondestructible solid waste has stimulated the design of biobased biodegradable polymers and composite materials as an alternative to synthetic plastic packaging materials, particularly for use in short-term packaging and disposable applications, such as, disposable cutlery, plates, cups, and utensils, fast-food containers, and beverage containers (Siracusa et al., 2008). The best biodegradable packaging

Food Packaging. http://dx.doi.org/10.1016/B978-0-12-804302-8.00006-6

materials are biopolymers in which at least one step in the degradation process is through metabolism of naturally occurring organisms. Biopolymers can be divided into three different groups (Clarinval and Halleux, 2005) depending on the origin of the raw materials and their manufacturing processes: (1) those obtained from renewable resources, such as, carbohydrates (starch, cellulose, alginate, chitosan, agar) and animal or plant-derived proteins (soy protein, wheat gluten, gelatin, collagen, whey protein, corn zein, casein, etc.); (2) those synthesized microbially such as polyhydroxyalkanoates (PHAs) and microbial polysaccharides like pullulan and curdlan; (3) synthetic biodegradable polymers synthesized from biomass such as poly(l-lactide) (PLA), or from petroleum-based chemicals like poly(glycolicacid), poly(butylene succinate), poly(ε-caprolactone) (PCL), and poly(vinyl alcohol). Nonetheless, biopolymers exhibit relatively poor mechanical and barrier properties, which limit their practical applications. Their shortcomings are threefold: price, performance, and processing. More specifically, their brittleness, high gas and vapor permeability, low heat distortion temperature, and poor resistance to prolonged processing operations have strongly restricted their industrial uses (Vroman and Tighzert, 2009). To overwhelm these shortcomings in order to expand their range of commercial uses, novel strategies are required, such as, blending with other polymers (ie, PCL, PLA, Sudesh et al., 2000) or filling with nanofillers including carbon nanotubes, montmorillonite, layered silicates, and nanohydroxyapatite (Sadat-Shojai et al., 2013). The extraordinary success of nanocomposites based on synthetic polymers has encouraged new research on nanocomposites with biodegradable polymers as matrix.

Polymer composites are mixtures of polymers with inorganic or organic fillers with certain geometries (fibers, flakes, spheres, particulates). The use of fillers that have at least one dimension in the nanometric range (nanoparticles) produces polymer nanocomposites. Three types of fillers can be distinguished, depending on how many dimensions are in the nanometer scale: 1-dimensional like layered silicates, 2-dimensional like nanotubes or whiskers, and 3-dimensional like spherical silica particles (Ghanbarzadeh et al., 2014). Inorganic nanoparticles are much cheaper than carbon based nanofillers, such as, nanotubes (CNTs) or nanofibers (CNFs), which is of a great interest from an application viewpoint. Further, they display higher specific surface area compared to microparticles, hence larger interface and stronger interaction with the host matrix; therefore lower nanoparticle concentration (1–10 vol.%) is necessary to achieve properties that are comparable or even better than those attained by conventional loadings (15–40 vol.%) of microfillers (Naffakh et al., 2013). This makes the processing easier and reduces composite weight. A number of methods have been described for the preparation of polymer/nanoparticle nanocomposites, including solution mixing, melt-compounding and in situ polymerization (Zou et al., 2008). In the solution processing route, the nanoparticle powder is normally dispersed in a liquid medium by stirring and/or sonication and then mixed with the polymer solution, followed by evaporation of the solvent. Melt-processing is the most widespread technique for the manufacture of nanoparticle-reinforced thermoplastic composites, and is also appropriate for polymers with low solubility in common solvents. It consists in the blending of the polymer melt with the nanomaterial by application of intense shear forces, and is frequently combined with a shape-forming step like injection molding, spinning, or hot compression. The in situ polymerization of monomers in the presence of nanoparticles has been extensively explored for the preparation of polymer-grafted nanoparticle composites. This technique is especially important for insoluble and thermally unstable matrices, and produces homogenous composites with high nanofiller weight fractions.

The nanofiller effect on the composite performance is highly influenced by its shape, size, surface characteristics, degree of dispersion, and interfacial adhesion with the host matrix. A regular distribution of the nanofiller within the polymer matrix is essential to manufacture composites with superior properties, especially mechanical and tribological; nevertheless, this is a hard task owing to the strong tendency to agglomerate nanoparticles. In addition, the strength of interactions between the organic and inorganic phases is another key factor that conditions the final properties. Surface modification of the nanoparticles either by chemical or physical methods is a common strategy to improve the compatibility between the hydrophobic polymer and the hydrophilic nanofillers (Kickelbick, 2003). Chemical methods involve modification with coupling agents like silanes; the coupling process is normally carried out via reaction between the trialkoxy groups of silane molecules and the hydroxyl groups on the nanoparticles (Diez-Pascual et al., 2014), while other functional groups of the coupling agent (ie, ethylene, amine, epoxy, phenyl) are suitable to interact with the polymer matrix. On the other hand, physical routes utilize surfactants or polar molecules like stearic acid or sodium stearate that adsorb onto the particle surface by means of electrostatic interactions. Both types of surface modifiers lessen the interactions among the inorganic particles, thereby inducing their disaggregation, and also strengthen the nanofiller-matrix interfacial adhesion.

The present chapter deals with the preparation and characterization of novel PHA-based bionanocomposites reinforced with zinc oxide (ZnO) nanoparticles with a view to use them as food packaging materials. In this regard, different amounts of ZnO have been incorporated in poly(3-hydroxybutyrate) (PHB) and poly(3-hydroxybutyrate-*co*-3-hydroxyvalerate) (PHBV) by a solution casting method without using surfactants or coupling agents, and their morphology, thermal, mechanical, barrier, migration, and antibacterial properties have been carefully analyzed through a variety of techniques. The desired properties for the nanocomposites are to impede gain of moisture, avoid microbial contamination as well as permeation of oxygen and water vapor, have good thermal, chemical and dimensional stability, and high mechanical strength. The following sections describe in detail the influence of the nanoparticle concentration on the different properties of each type of nanocomposite systems, analyzing their potential specific applications as food packaging materials. Finally, conclusions and future perspectives are summarized.

2 Polyhydroxyalkanoates (PHAs): Synthesis, Structure, Properties, and Applications

2.1 Synthesis

PHAs are a family of water insoluble, biodegradable, and biocompatible polyesters produced as reserve materials by a large number of bacteria that have recently attracted much attention as an alternative to petroleum-based polymeric materials. More than 150 different types of PHAs consisting of various nonomers have been reported, thus the properties of this family are very broad. PHAs can be divided into two main groups: short-chain PHAs that contain monomer units with 3–5 carbon atoms and medium-chain PHAs that contain monomer units of 6–18 carbon atoms. To synthesize PHAs, a culture of a microorganism is located in an appropriate medium and fed suitable nutrients to make it multiply quickly. Once the population has reached a significant level, the nutrient composition is modified to compel the microorganism to produce the PHA. The synthesis is typically induced when the microorganisms experience a metabolic stress like limitation by an essential nutrient (eg, lack of trace elements or oxygen) and the simultaneous excess supply of carbon sources. (Jacquel et al., 2008). Several companies are currently producing PHAs at a large scale: Metabolix (USA), Biocycle (United Kingdom), Biomer (Germany), and so on. In the industrial production, the polyester is obtained from the bacteria by improving the fermentation conditions of sugar or glucose. Carbohydrates are frequently employed as raw materials for the fermentation, albeit glycerine or vegetable oil can also be used. At the end of the 20th century, the price of a kilogram of PHAs was approximately 16 times higher than a kilogram of polypropylene (PP). To diminish the whole cost, PHAs with elevated productivity and high yield have to be produced, hence alternative production paths have to be considered. Researchers in industry all over the world are now working on methods to develop more efficient and economical processes for PHA production. For instance, techniques using transgenic crops as well as methods for producing PHA from wastewater are being developed (Sudesh et al., 2000).

2.2 Properties

PHA polymers are typically thermoplastic materials with melting points ranging from 40 to 180°C, and can be processed by conventional techniques used for synthetic thermoplastics like injection molding, extrusion, blown film extrusion, and thermoforming to form containers, bottles, razors, and materials for food packaging. Their properties strongly depend on their composition (homo- or copolyester, length of hydroxy fatty acids, side chains). By controlling the monomer composition, the physical properties of the polymer can be regulated to a great extent. Furthermore, the rate of degradation of PHA in various environments can be controlled by precisely altering its monomer compositions. They have high ultraviolet (UV) stability and display low water permeability. Their crystallinity is in the range of 5–70%, and they can be brittle or more ductile depending on the percentage of comonomers. Further, they have low thermal stability and water solubility albeit they are soluble in halogenated solvents such as chloroform, dichloromethane, or dichloroethane (Jacquel et al., 2007).

2.3 Structure

The homopolymer, PHB is an isotactic linear thermoplastic comprising 3-hydroxybutyrate (HB) units with a chemical formula of $[\text{-O-CH-(CH}_3\text{)-CH}_2\text{-CO-}]_n$ (Fig. 6.1) that was synthesized for the first time in the 1920s by the French bacteriologist Lemoigne at the Lille branch of the Pasteur Institute from *Bacillus megaterium* (Laycock et al., 2013) and it is now produced at an industrial scale. PHB does not contain any residues of catalysts like other synthetic polymers, is perfectly isotactic and does not include any chain branching, therefore it flows easily during processing. This fully biodegradable and biocompatible polyester exhibits optical activity, piezoelectricity, and very good barrier properties. It has

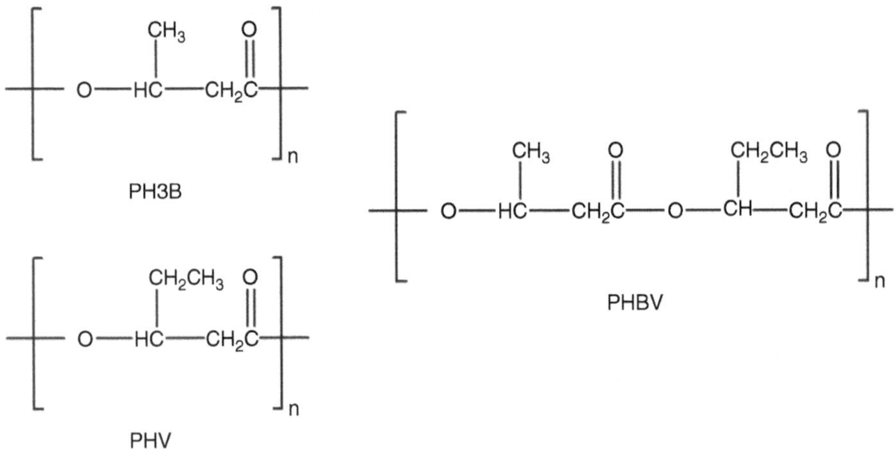

Figure 6.1. Chemical structure of poly(hydroxybutyrate) (PHB), poly(hydroxyvalerate) (PHV), and poly(3-hydroxybutyrate-*co*-3-hydroxyvalerate) (PHBV).

better oxygen, aroma, and water vapor barrier properties than PP, and is low permeable to O_2, H_2O, and CO_2, hence has already been used in small disposable products like diaper sheets and packaging materials. Nevertheless, it has certain limitations for practical applications like elevated cost, high crystallinity (>50%), poor thermal stability, brittleness and low impact resistance, together with a rather high melting temperature (c. 175°C) that leads to a small processing window (Savenkova et al., 2000).

PHA copolymers are composed primarily of HB units with a fraction of longer chains monomers, such as, 3-hydroxyvalerate (HV), 3-hydroxyhexanoate (HHx), or 3-hydroxypropionate (HP). Thus when HV units are incorporated into PHB to form PHBV, with a chemical formula of $[COCH_2CH(CH_3)O]_m[COCH_2CH(C_2H_5)O]_n$ (Fig. 6.1) the level of crystallinity, glass transition, and melting temperature drop while the processing window becomes somewhat wider since these monomers have superior melt stability at low processing temperatures. Depending on the HV content, PHBV copolymers can display physical properties and processing behavior resembling polyethylene (PE) or PP, and can range from brittle plastics to elastomers. PHBV copolymers with different content of hydroxyvalerate have been used recently as matrices in ecocomposites (Yu et al., 2012; Jiang et al., 2008).

2.4 Applications

The versatility of PHAs has made them excellent candidates as polymer materials in a variety of areas ranging from biomedical/

medical fields to food packaging, textiles, and household materials (Laycock et al., 2013). Perspective areas of PHA application are (Bonartsev et al., 2007): (1) in packaging for deep drawing articles in the food industry like bottles, laminated foils, disposable packages, fast food, one-way cups; (2) in agriculture, systems for prolonged release of fertilizers and agrochemicals; (3) in the field of textiles for clothes and shoes, sport's bags, and equipment; (4) in medicine, for medical devices, tissue engineering, and systems of sustained drug delivery (Philip et al., 2007). PHB is compatible with the blood and tissues of mammals. The monomer of PHB is a normal metabolic in the human blood. Over the past years, PHAs, particularly PHB, have been used to fabricate devices such as surgical sutures, screws, and plates for bone fixation, patches, pins, scaffolds, membranes for periodontal treatment, surgical meshes for hernioplastic surgery, and wound dressings (Chen and Wu, 2005). Moreover, the capability of PHB for encapsulation and controlled release of different drugs allows applying PHB as material for cell and tablet packaging and for development of therapeutic systems of sustained drug delivery [systems for delivery of antiproliferative drugs like dipyridamole, antibiotics (rifampicin, metronidazole, ciprofloxacin, levofloxacin), antiinflammatory drugs (indomethacin, flurbiprofen, dexamethasone, prednisolone), and antitumor drugs (paclitaxel)]. Therapeutic systems based on microspheres and microcapsules from biodegradable polymers are a new and promising trend in the modern pharmacology (Zhou, 2004). For applications in the biomedical field, PHAs of high purity are required, hence biologically active contaminants such as proteins and lipopolysaccharides must be decreased to a very low level. In particular, the US Food and Drug Administration (FDA) stipulates that the endotoxin content of medical devices ought to not exceed 20 US Pharmacopeia endotoxin units. Consequently, PHAs should be purified using standard techniques like redissolution and precipitation, washing with a nonsolvent, purification by chromatography, treatments with chemical agents and filtration as well as aqueous digestion.

3 ZnO Nanoparticles: Structure, Properties, Synthesis Methods, and Applications

3.1 Properties

In recent times, ZnO nanostructures have acquired great interest within the scientific society as "future materials" due to their cheap price, high availability, possibility of carrying out surface modifications with several functional groups and biocompatibility. With a wide bandgap of 3.4 eV and a large exciton binding energy of 60 meV at room temperature, they are widely used for optoelectronic and photocatalytic applications (Coleman and Jagadish, 2006). They also display UV-blocking capacity, and hence provide effective protection of skin from UV-A radiation. These environmentally friendly materials have large surface-area-to-volume ratio, elevated catalytic activity, highly crystalline structure, radiation resistance, enhanced mechanical properties (Kucheyev et al., 2002), including a Young's modulus of 111 GPa and hardness of ~5 GPa, low coefficient of thermal expansion ($\sim 2.5 \times 10^{-6}$ K^{-1} at 300 K) and high thermal conductivity (~1 W cm^{-1} K^{-1}) (Adachi, 2005). They also display antimicrobial activity at pH 7–8 even without illumination, and have been broadly applied as antimicrobial agents (Yamamoto, 2001). Thus ZnO has been reported to exhibit significantly stronger biocide action on *Staphylococcus aureus* than other metal oxide nanoparticles like TiO_2, MgO, Al_2O_3. Besides, ZnO nanofillers have a broad spectrum of antibacterial activities on both Gram-positive and Gram-negative bacteria, including *Escherichia coli, Listeria monocytogenes, Salmonella enteritidis, Bacillus subtilis, Pseudomonas aeruginosa*, and *Enterococcus faecalis.* In addition, these nanofillers are regarded as "nontoxic" and considered by the FDA as "generally recognized as safe" (GRAS) compounds, and it has been demonstrated that they do not induce any damage to the human DNA (Yamada et al., 2007).

The antimicrobial action of ZnO nanoparticles has been ascribed to several mechanisms (Hajipour et al., 2012), as depicted in Fig. 6.2: (1) induction of oxidative stress due to reactive oxygen species (ROS) generation, especially H_2O_2 that interacts with proteins, DNA, and lipids of the microorganism causing death; (2) membrane disorganization due to accumulation of ZnO nanoparticles in the bacterial membrane and also their cellular internalization; (3) release of Zn^{2+} ions that bind to the membrane of the microorganisms.

However, the toxicity of ZnO nanoparticles is not directly related to their penetration into the bacteria cell, but with the fact that their intimate contact with the cell causes either increase in metal solubilization or ROS generation, that finally result in cell membrane damage. Moreover, the toxicity of ZnO nanoparticles is not only affected by the light via ROS production, but may also happen in the dark although the mechanism is not fully understood yet. The generation of ROS can be explained as: when ZnO defects absorb UV and visible light, electron-hole pairs (e^-h^+) can be created. The holes split H_2O molecules into OH^- and H^+. Dissolved oxygen molecules are transformed to superoxide radical

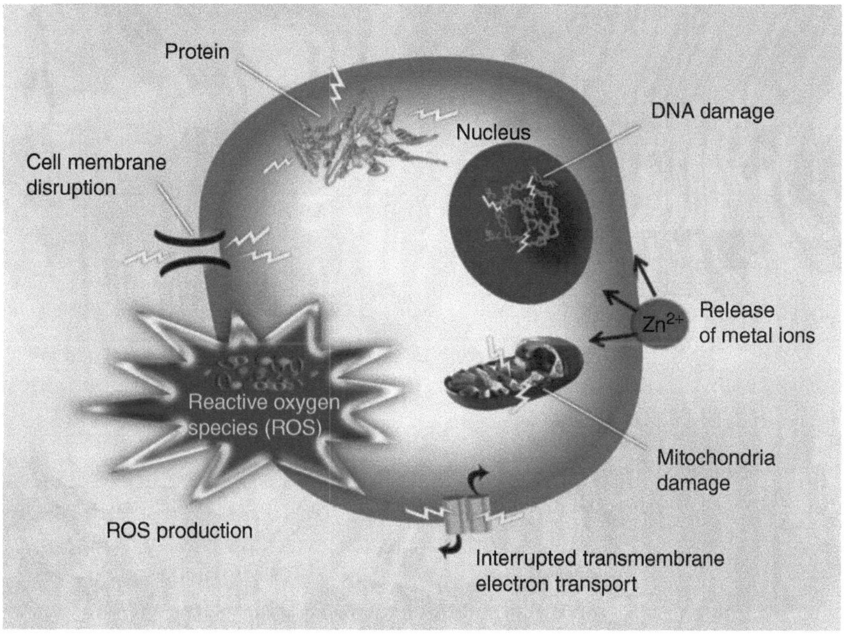

Figure 6.2. Mechanisms of toxicity of ZnO nanoparticles against bacteria. Adapted from Hajipour et al. (2012), with permission from Elsevier.

anions (O_2^-), which react with H^+ to generate ($HO_2\cdot$) radicals that subsequently collision with electrons producing hydrogen peroxide anions (HO_2^-). These react with hydrogen ions to produce H_2O_2 molecules that can penetrate the cell membrane and kill the bacteria. The concentration of H_2O_2 generated increases with decreasing particle size, due to the increased number of particles per unit volume.

3.2 Structure

At ambient pressure and temperature, ZnO crystallizes in the wurtzite (B4 type) structure, as shown in Fig. 6.3. It is a hexagonal lattice that belongs to the spatial group $P6_3mc$, and is characterized by two interconnecting sublattices of Zn^{2+} and O^{2-}, so that each Zn ion is surrounded by a tetrahedra of O ions and vice versa, hence exhibits the sp^3 covalent bonding (Coleman and Jagadish, 2006). This coordination results in polar symmetry along the hexagonal axis, which is responsible for the piezoelectricity and spontaneous polarization of this oxide. Nevertheless, the Zn—O bond presents very strong ionic character, hence ZnO lies on the borderline between being classified as a covalent and ionic compound, with an ionicity of 0.616 on the Phillips ionicity scale.

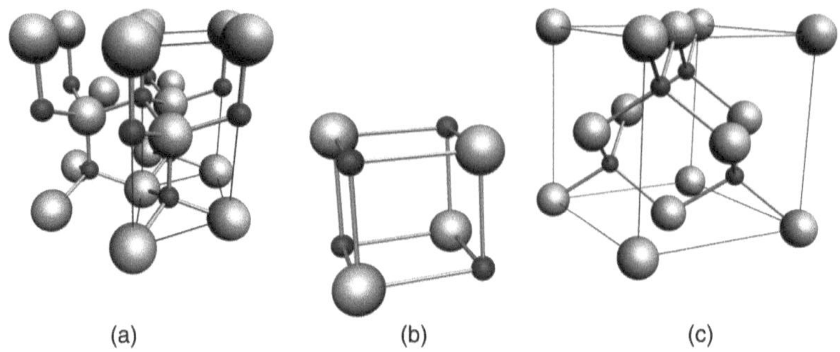

Figure 6.3. Hexagonal wurtzite structure (a), rock salt (b), and zincblende (c) phases of ZnO. The *big white spheres* correspond to O atoms and the *small black ones* to Zn atoms.

The lattice parameters of the hexagonal unit cell are a = 3.2495Å and c = 5.2069 Å, the density is 5.605 g cm^{-3} and the melting point 1975°C. In addition to the wurtzite phase, ZnO can crystallize in the cubic zincblende (Fig. 6.3) by growth on cubic structures and in the rocksalt (NaCl) structure at ~10 GPa.

3.3 Synthesis Methods

Zinc oxide can occur in 1-dimensional, 2-dimensional, and 3-dimensional structures. One-dimensional structures include nanorods, helixes, springs, tubes, wires, and needles. Two-dimensional structures are nanoplates, nanosheets, and nanopellets. Three-dimensional structures include flowers, snowflakes, and spherical particles. The use of ZnO depends mainly on its microstructure, namely crystallite size, particle size, orientation, and shape. To obtain a proper microstructure, the most challenging is to synthesize small-sized particles with a few agglomerates and a narrow size distribution. ZnO nanoparticles with different sizes and shapes can be prepared by various methods including laser ablation, hydrothermal methods, electrochemical depositions, sol-gel method, chemical vapor deposition, thermal decomposition, and combustion (Raoufi, 2013; Diez-Pascual et al., 2014). Among them, the most employed are the sol-gel and the hydrothermal method (Kołodziejczak-Radzimska and Jesionowski, 2014). The former process involves conversion of monomers into a colloidal solution (*sol*) that acts as the precursor for the discrete particles. Typical precursors are zinc nitrate and acetate. The latter does not necessitate the use of organic solvents or further processing of the product (grinding and calcination), which makes it a straightforward and environmentally friendly technique. The synthesis occurs in an autoclave, where the mixture of substrates is heated steadily to a temperature of 100–300°C and keeps for several days. This process leads to products of high degree of crystallinity and high purity. The mechanochemical process is an inexpensive and easy method of obtaining ZnO nanoparticles on a large scale (Ao et al., 2006). It employs anhydrous $ZnCl_2$ and or Na_2CO_3 as starting materials, and requires high-energy dry milling, which starts a reaction through ball–powder impact in a ball mill at low temperature. A "thinner" is added to the system in the form of a solid (usually NaCl), which acts as a reaction medium and separates the nanoparticles being formed. The advantages of this technique are the high homogeneity of the crystalline structure and morphology, low production costs, small particle sizes, and restricted tendency of the particles to agglomerate. A shortcoming of this method is that it requires long milling times to attain a uniform grinding of the powder, which leads to a greater amount of impurities. Recently, ZnO nanoparticles have been prepared by sonochemical process, microwave heating, anodization, and electrophoretic deposition (Kołodziejczak-Radzimska and Jesionowski, 2014). In the sonochemical method, an aqueous solution of the precursor (typically zinc nitrate) is added to an amine and exposed to ultrasound for several minutes (c. 30) under ambient conditions. In the microwave heating, the solution is heated using microwaves at 90°C for a few minutes. The precipitate obtained is then centrifuged, washed with water and ethanol, and finally dried. These methods of obtaining ZnO enable high yields, in excess of 90%, do not require surfactants, can be applied on a large scale with low production costs, and are simple and energy efficient.

3.4 Applications

Owing to their varied properties, both chemical and physical, ZnO nanoparticles are broadly used in many areas, ranging from tires to ceramics, from pharmaceuticals to agriculture, and from paints to chemicals.

3.4.1 Rubber Industry

Worldwide production of ZnO is approximately 10^5 tons per year, and a main part is consumed by the rubber industry to manufacture various different cross-linked rubber products. The thermal conductivity of pure silicone rubber is rather low; nonetheless, it can be increased by the incorporation of low amounts of nanoscale fillers like ZnO nanoparticles. However, the nanoparticles tend to aggregate together to form larger particles due to intense hydrogen bonding interactions. To overcome this drawback, surface modification either by chemical or physical methods is required to enhance the compatibility between the hydrophobic polymer and the hydrophilic nanofillers (Zou et al., 2008). Chemical methods involve modification with coupling agents like silanes; the coupling process is usually accomplished via reaction between the trialkoxy groups of silane molecules and the hydroxyl groups on the nanoparticles (Grasset et al., 2003) while other functional groups of the coupling agent (ie, ethylene, amine, epoxy, phenyl) are suitable to interact with the polymer matrix. On the other hand, physical routes use surfactants or molecules with polar groups like stearic acid or sodium stearate that adsorb onto the particle surface via electrostatic interactions. Both types of surface modifiers decrease the interaction among the inorganic particles, hence inducing their disaggregation, and also strengthening the nanofiller-matrix interfacial adhesion. ZnO nanoparticles have been used as crosslinking agents for carboxylated elastomers (Przybyszewska and Zaborski, 2009), and they produced vulcanizates with high-tensile strength, tear resistance, hardness, and hysteresis.

3.4.2 Pharmaceutical and Cosmetic Products

Due to its antibacterial, disinfecting, and drying properties, ZnO is widely used in the production of various kinds of medicines like those for epilepsy and diarrhea. It accelerates wound healing, hence it is employed in dermatological products against inflammation and itching, and at high concentrations it has a peeling effect. Further, it is used in dentistry as a component of dental pastes, and also for temporary fillings. In addition, ZnO is a very good media in sun creams, since it absorbs UV radiation, does not irritate the skin, and is easily absorbed into the skin (Cross et al., 2007).

3.5 Textile Industry

ZnO nanostructures have become very attractive as UV-protective textile coatings. Their self-cleaning and water repellent properties make them very helpful for preventing unwelcome stains on clothes. Different methods have been reported for the production of UV-protecting textiles using ZnO nanoparticles, including hydrothermal grown on SiO_2-coated cotton fabrics or deposition on cotton and wool fabrics (Becheri et al., 2008). Therefore ZnO nanoparticles could serve as ideal multifunctional coatings for textiles.

3.6 Electronics

ZnO is an important semiconductor with a wide range of applications in electronics. It can be used in photoelectronic and electronic equipments, field emitters, sensors, UV lasers, and solar cells (Przybyszewska and Zaborski, 2009). ZnO also exhibits the phenomenon of photoluminescence, that is, emission of light under exposure to electromagnetic radiation, hence it is employed in field emission display equipments like televisions. Thin films made of ZnO nanoparticles exhibit high conductivity and excellent permeability by visible rays, and thus have potential uses in solar batteries, transparent electrodes in photovoltaic and electroluminescent equipments, and materials for UV-emitting devices (Wang and Wang, 2008). ZnO nanoparticles are also employed in gas sensors (Lin et al., 1998), particularly for the detection of CO and CO_2 (in mines and in alarm equipments), but can also be used for the detection of other gases (H_2, SF_6, C_4H_{10}, C_2H_5OH). The sensitivity of such devices depends on the porosity and particle size: sensitivity increases as the size decreases. One of the foremost applications of ZnO is the production of varistors, which are resistors with a nonlinear current-voltage characteristic. They are used as lightning protectors, to protect high-voltage lines, and in electrical equipment providing protection against atmospheric and network voltage surges (Tsonos et al., 2011).

3.7 Bactericidal Agents

ZnO nanoparticles can be effectively used as antimicrobial agents due to their photocatalytic activity under UV light. Their extremely high specific surface area allows a large area of contact with the microorganisms, hence lead to strong mechanical damage to the cell membrane. Consequently, these inorganic nanoparticles are more efficient for bacteria killing than the corresponding microparticles. ZnO nanoparticles have already been used as active components in antibacterial cream, lotions, and ointments. Applications in mouthwashes and paints as an antimicrobial agent and in surface coatings to prevent biofilm growth have also been investigated (Jones et al., 2008). Most importantly, the antimicrobial effect of ZnO nanoparticles against foodborne pathogens makes them ideal candidates for application in food packaging and preservation process. The advantage of using them as antimicrobial agents is that they contain mineral elements essential to humans and exhibit strong activity even when administered in small amounts. Consequently, they can also be applied for various biological applications like biological sensing, biological labeling, drug and gene delivery, and nanomedicine (Visaria et al., 2006).

3.8 Other Applications

ZnO nanoparticles are also used as additives in lubricating oils for vehicles, decreasing consumption, and oxygen corrosion.

Due to their nontoxicity, low cost, and chemical stability under air conditions, they can be employed to develop novel ecofriendly substances for cell marking (Tang et al., 2009). In addition, electrochemical biosensors based on ZnO nanoparticles have been successfully employed for the determination of proteins (Topoglidis et al., 2005) or glucose (Wang et al., 2006), ascribed to their remarkable properties such as biosafety, superior analytical performance, high sensitivity, and ease of manufacture. ZnO has high biocompatibility and fast electron transfer kinetics, and such characteristics support the use of this nanomaterial as a biomimic membrane to immobilize and modify biomolecules. Due to its high hydrophilicity, it is also employed for self-cleaning surfaces like self-cleaning windows.

4 Preparation of PHB and PHBV-Based Bionanocomposites

PHB and PHBV nanocomposites can be manufactured by several techniques, including melt-extrusion, solution intercalation, and solution casting processes (Pandey and Reddy, 2014). These thermoplastics display a low thermal stability and a relatively narrow melt processing window (185–200°C), and even certain degradation can start at around 180°C; consequently, they are more safely processed in solution using *N,N*-dimethylformamide or chloroform as solvents. In this regard, Diez-Pascual and Diez-Vicente developed a series of nanocomposite films with ZnO contents in the range of 1.0–10.0 wt.% via a solution casting technique using chloroform. The ZnO nanopowder and the polymers were dispersed in the solvent via ultrasonication and then casted onto a glass petri dish to evaporate the chloroform at room temperature (Diez-Pascual and Diez-Vicente, 2014a). Recently, other environmentally friendly and cost-effective processing methods including wet electrospinning (Chan et al., 2009) and spray drying (Yun et al., 2010) have been employed to develop PHA/CNT nanocomposites.

5 Characterization of PHB and PHBV-Based Bionanocomposites

5.1 Morphology and Structure

The surface morphology of the nanocomposites is frequently examined by scanning electron microscopy (SEM). Representative SEM images of PHBV nanocomposites with 1.0 and 8.0 wt.%

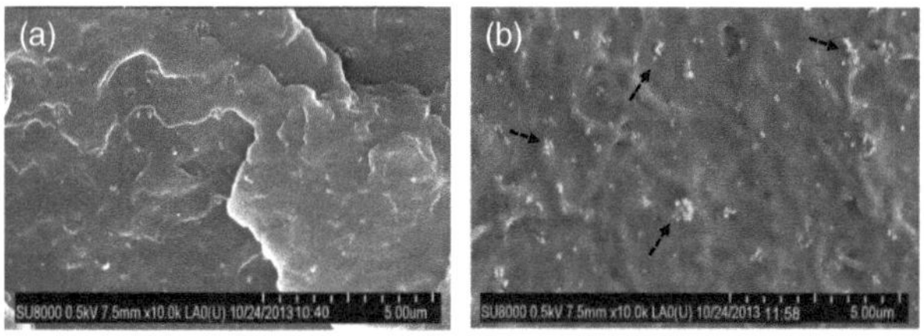

Figure 6.4. SEM images from fractured surfaces of PHBV/ZnO nanocomposites with (a) 1.0 wt.% and (b) 8.0 wt.% nanoparticle content. Reprinted from Diez-Pascual and Diez-Vicente (2014b), with permission from the American Chemical Society.

ZnO are shown in Fig. 6.4 (Diez-Pascual and Diez-Vicente, 2014b). A random and very uniform nanofiller dispersion can be observed in the nanocomposite with the lowest ZnO content, without any agglomerates. The interactions between the hydroxyl moieties of the nanoparticles and the polar groups of PHBV likely impede ZnO agglomeration and improve the filler-matrix compatibility. Nonetheless, the nanocomposite with 8.0 wt.% ZnO shows poorer nanofiller dispersion, with small clusters composed of a few particles. Thus it was not feasible to individually disperse such high ZnO loading, albeit the nanoparticles were reasonably well distributed throughout the matrix. PHBV/ZnO (1.0 wt.%) has a relatively smooth morphology, whereas the composite with 8.0 wt.% loading displays a rougher surface, suggesting a modification in the crystallinity of the polymer.

Similar behavior was found for PHB/ZnO nanocomposites. Uniform nanoparticle dispersion was detected for all the samples except that with 10.0 wt.% ZnO loading that exhibited not only discrete particles but also a few small clusters comprising several particles. Analogously, Yu et al. reported that PHBV-grafted-multiwall carbon nanotubes (MWCNTs) were well dispersed within PHBV matrix for reinforcement contents ≤7.0 wt.%, and that existed strong interfacial interaction between the nanotube and the polymer phases. Nonetheless, in the sample with 10.0 wt.% PHBV-g-MWCNTs, many dots with an average diameter of 180 nm were detected, indicating that slight agglomeration took place, albeit no microscale aggregates were observed (Yu et al., 2014). Comparable conclusions can be drawn from the morphological study of PHB and PHBV/fibrous clay nanocomposites prepared via solution intercalation (Thire et al., 2011). The composites with low clay loadings displayed the nanofibrils homogenously distributed within polymer matrix, whereas those with contents ≥5.0 wt.% showed micrometric agglomerates nonhomogenously dispersed within the PHA matrices. In some regions, voids around the agglomerates were observed, indicating poor nanofiller-matrix interface. In contrast, optimal nanofiller dispersion was found for single wall carbon nanotube (SWCNT)-reinforced PHB composites prepared via ultrasonication followed by spray drying (Yun et al., 2010), which corroborates that the processing route strongly conditions the composite morphology, and consequently, its macroscopic properties.

The morphology of the nanocomposites can also evaluated by transmission electron microscopy (TEM). Statistical analysis is carried out from the TEM micrographs to gain information about the particle size distribution and particle-particle distance in the nanocomposites. Typical images of the nanocomposites with 1.0 and 10.0 wt.% loading are shown in Fig. 6.5 (Diez-Pascual and Diez-Vicente, 2014a). In the nanocomposites with 1.0 wt.% ZnO, the nanoparticles are regularly and well distributed, without any agglomerate. They show a narrow size distribution, with diameters between 40 and 100 nm, an average value of 80 nm and a mean particle-particle distance of 180 nm. On the other hand, in the nanocomposite with 10.0 wt.% loading the nanoparticle dispersion is uneven, the size distribution is wider (diameters in the range of 40–150 nm), the mean diameter larger (~98 nm), and the average interparticle distance smaller (75 nm). On the whole, statistical analysis corroborates that ZnO is dispersed in PHB as discrete particles at low loadings, while it appears both as individual particles and small clusters at higher contents. It is important to note that ZnO dispersion within PAHs was attained without using surfactants or coupling agents, making the manufacture process of these nanocomposites simple, fast, and cheap.

X-ray diffraction (XRD) is one of the most important techniques to characterize polymeric materials, since it gives information about the size and the shape of their unit cell from the peak positions and information on electron density inside the unit cell, namely where the atoms are located, from peak intensities. The XRD patterns of PHBV, ZnO nanoparticles, and the nanocomposites are displayed in Fig. 6.6 (Diez-Pascual and Diez-Vicente, 2014b). The nanoparticles show typical peaks at 2θ ~31.8°, 34.4°, and 36.2°, corresponding to the crystalline reflection of the (100), (002), and (101) planes, respectively, of the hexagonal wurtzite structure of ZnO (JCPDS 36-1451). The sharp diffraction peaks are indicative of the high purity and quality of the nanoparticles. PHBV exhibits most important peaks at 2θ values of 13.6°, 17.1°, 20.3°, 21.7°, 22.8°, 25.7°, 27.4°, and 30.8°, which arise from the (020), (110), (021), (101), (111), (121), (040), and (200) crystalline planes, respectively, of the orthorhombic unit cell, space group $P2_12_12_1$ ($D2^4$) (Scandola et al., 1992). These reflections are also visible in the diffractograms of the nanocomposites, simultaneously

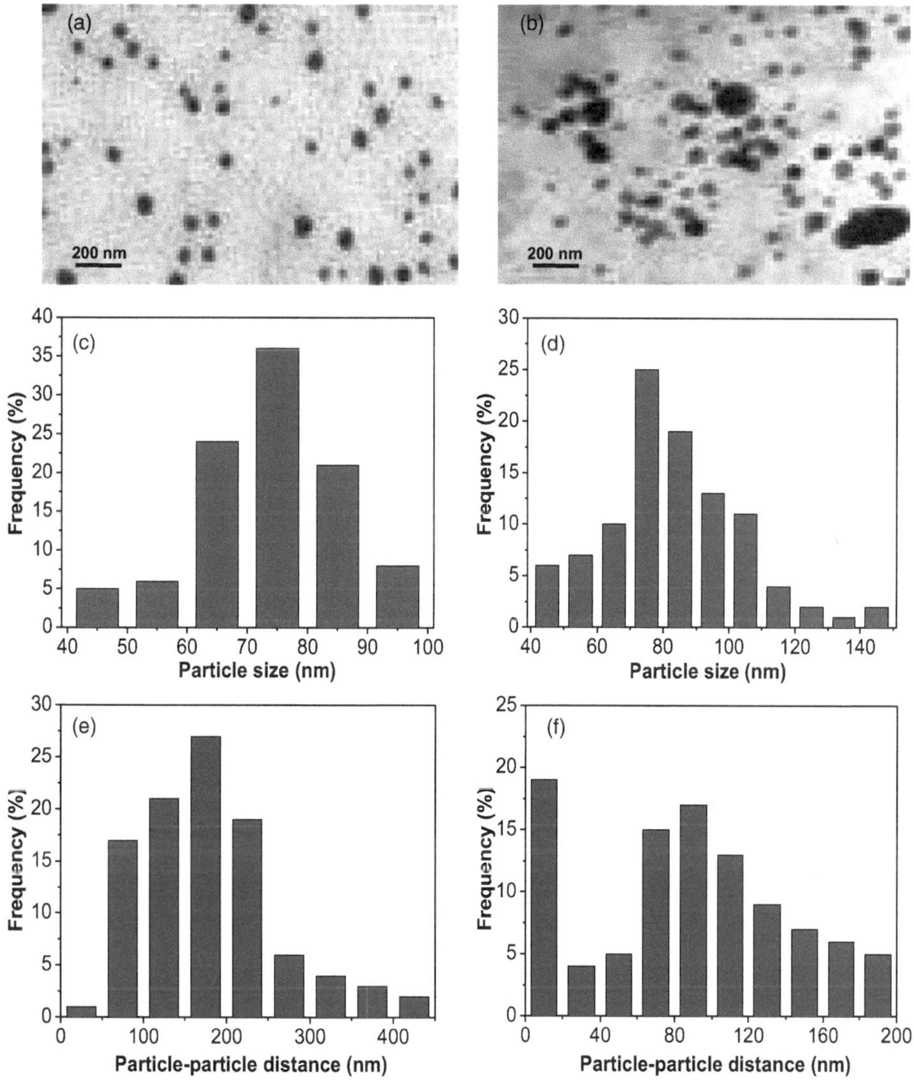

Figure 6.5. TEM micrographs of PHB/ZnO nanocomposites with (a) 1.0 wt.% and (b) 10.0 wt.% ZnO. Particle size and particle–particle distance distribution in nanocomposites with 1.0 wt.% (c and e) and 10.0 wt.% (d and f) ZnO. Reprinted from Diez-Pascual and Diez-Vicente (2014a), with permission from MDPI.

with those characteristic of ZnO, hinting that the crystalline structure of both constituents is preserved. The peaks in the nanocomposites are thinner and more intense than in the pure PHA, suggesting that ZnO nanoparticles behave as nucleating agents for PHBV crystallization, resulting in smaller crystals. For example, ~30% increase in the full width at half maximum (FWHM) of the (100) reflection is observed in the composite with 4.0 wt.%

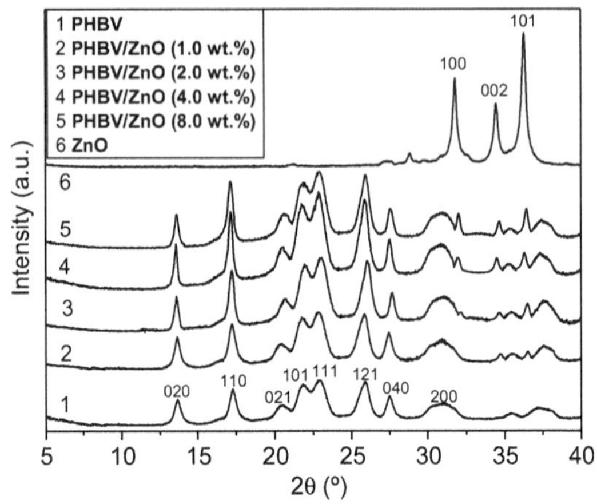

Figure 6.6. XRD patterns of PHBV, ZnO, and the indicated nanocomposites.
Adapted from Diez-Pascual and Diez-Vicente (2014b), with permission from the American Chemical Society.

ZnO, suggesting a drop in the average crystallite size, given that the crystal dimension is inversely proportional to the peak width following the Scherrer equation (Alexander and Krieger, 1969). Nonetheless, the crystallite size of the sample with 8.0 wt.% ZnO is a little bigger than that of the composite with 4.0 wt.% loading, ascribed to the stronger ZnO-PHBV hydrogen bonding interactions that could limit the crystal growth. Similar phenomenon has been observed for PHBV composites reinforced with graphene (Sridhar et al., 2013) and PHBV-g-MWCNTs (Yu et al., 2014), where an increase in FWHM was found upon raising nanofiller concentration. Further, in the former composites, the level of crystallinity derived from the diffractograms rose progressively with increasing graphene concentration, whereas in the latter the crystallinity raised from about 58% to a maximum value of 63% for the nanocomposite with 7.0 wt.% PHBV-g-MWCNTs and then slightly decreased at higher loadings, ascribed to the chain entanglement effect that restricted the mobility of the crystalline PHBV chains.

From the earlier conclusions, it is important to analyze the interactions between the nanofiller and the biopolymer matrix. The interactions between PHB and ZnO nanoparticles have been investigated by attenuated total reflectance Fourier transformed infrared (ATR-FTIR) spectroscopy (Diez-Pascual and Diez-Vicente, 2014a), and the obtained spectra are displayed in Fig. 6.7. The ATR-FTIR spectrum of ZnO displays a wide peak at about 3450 cm^{-1} related to the OH stretching of hydrogen-bonded surface hydroxyl groups. A strong peak is also detected at 440 cm^{-1} assigned to the Zn—O stretching (Abdolmaleki et al., 2012). PHB

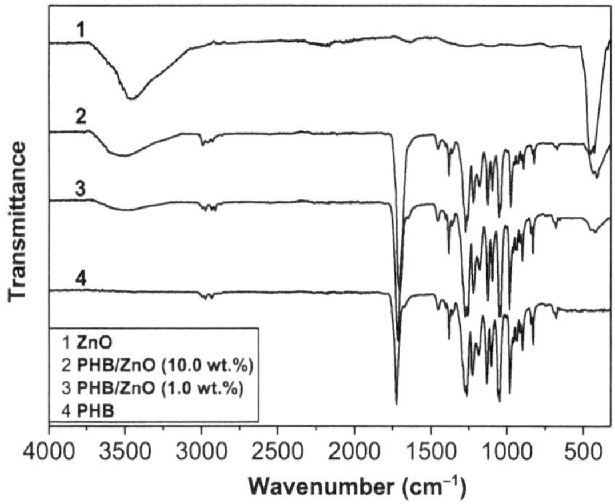

Figure 6.7. FT-IR spectra of ZnO, PHB, and the nanocomposites with 1.0 and 10.0 wt.% ZnO. Adapted from Diez-Pascual and Diez-Vicente (2014a), with permission from MDPI.

shows a strong band at 1723 cm^{-1} corresponding to the C═O stretching of the ester linkage, and peaks between 1280 and 1097 cm^{-1} assigned to C—O—C stretching vibrations. The CH_3 asymmetric bending occurs at 1457 cm^{-1}, and the H—C—O in-plane bending at 1380 cm^{-1}. The peaks characteristics of both PHB and ZnO can be found in the spectra of the nanocomposites, albeit a widening and upshift of the OH stretching band is found, which likely indicates a change from intramolecular H-bonding between OH groups of the ZnO surface to intermolecular OH—OH interactions with the C═O of the ester group of PHB. Further, the C═O stretching is broader, stronger, and shifted to lower wave number, also pointing toward the formation of hydrogen bonds. Similar downshift phenomenon is observed for the peak related to Zn—O stretching. With increasing concentration, the shifts are more pronounced, suggesting stronger interactions.

5.2 Thermal Properties

One of the foremost disadvantages of PHAs is their low thermal stability, since thermal degradation can take place during the polymer melt processing. Accordingly, it is important to examine the effect of nanofillers on the decomposition of these biopolymers. Fig. 6.8 shows as an example the thermogravimetric analysis curves under a nitrogen atmosphere of PHVB/ZnO nanocomposites at a heating rate of 10°C/min (Diez-Pascual and Diez-Vicente, 2014b). The pristine nanoparticles show a little weight loss (~2.5 wt.%) below 300°C ascribed to the elimination

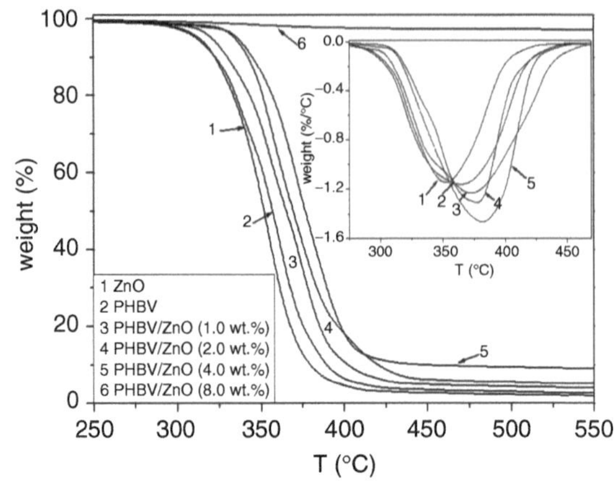

Figure 6.8. TGA curves under nitrogen atmosphere for neat PHBV, ZnO, and the nanocomposites with different nanoparticle loadings. The inset shows derivative (DTG) curves. Adapted from Diez-Pascual and Diez-Vicente (2014b), with permission from the American Chemical Society.

of surface adsorbed water. PHBV has only one decomposition stage with an initial degradation temperature (T_i) at 2% weight loss of 298°C and a maximum rate of weight loss (T_{max}) of 350°C. The degradation takes place by a random *cis*-elimination reaction (Yu et al., 2014), producing unsaturated carboxyl acid and ester products. A single stage decomposition process is also found for the nanocomposites, though the curves are shifted to higher temperatures, demonstrating the remarkable thermal stabilization induced by the presence of the nanoparticles. Accordingly, T_i, T_{10}, and T_{max} show maximum increments of 22, 25, and 31°C, respectively, for the nanocomposite with 8.0 wt.% loading. Very similar results were obtained for PHB/ZnO composites (Diez-Pascual and Diez-Vicente, 2014a). The homogenous dispersion of ZnO within the matrix and the strong interactions between the two composite components via hydrogen bonding induce a barrier effect versus the transfer of degradation products from the inner part of the composite to the outer region. Besides, the elevated thermal conductivity of ZnO likely promotes heat dissipation along the sample, resulting in superior thermal stability. A steady rise in the char residue is found with increasing ZnO concentration, suggesting that a higher fraction of material did not volatilize upon thermal degradation. Similar behavior was reported for PHBV-*g*-MWCNT-reinforced PHBV composites, attributed to the shielding effect of the nanofillers that prevented the evaporation of the polymer (Yu et al., 2014). Noteworthy, the increments in T_i and T_{max} obtained in ZnO-reinforced PHB and PHBV composites are comparable to those reported for PHVB/cellulose nanocrystals (Yu et al., 2012), PHBV/PHBV-g-MWCNT (Yu et al., 2014), or PHB/MWCNT nanocomposites (Xu and Qiu, 2011) and higher than those observed for PHVB/SiO_2 (Ma et al., 2008) or PHB/clay nanocomposites (Yeo et al., 2010). Overall, it is found that the nanocomposites can be processed in a larger processing window than neat PHAs, which is highly interesting from a practical viewpoint.

The analysis of the crystallization and melting behavior is crucial since these strongly affect the final properties of the material. To assess these effects, Differential Scanning Calorimetry (DSC) analysis under a nitrogen atmosphere was performed on PHB and its ZnO-reinforced nanocomposites, and their first cooling and second heating nonisothermal scans at a rate of 10°C/min are compared in Fig. 6.9 (Diez-Pascual and Diez-Vicente, 2014a).

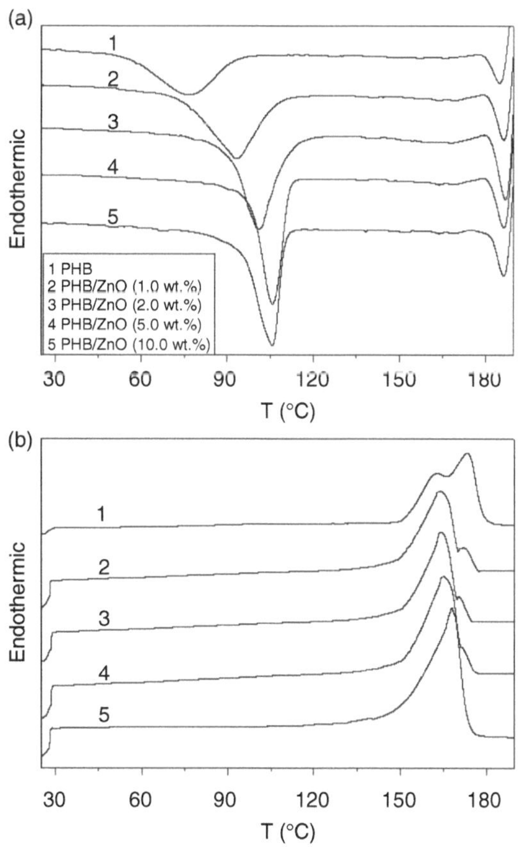

Figure 6.9. DSC cooling (a) and heating (b) thermograms of PHB/ZnO nanocomposites at a rate of 10°C/min. Adapted from Diez-Pascual and Diez-Vicente (2014a), with permission from MDPI.

The crystallization peak temperature (T_p) of PHB is about 76°C (Fig. 6.9a), and increases gradually with increasing ZnO nanoparticles. This rise is more prominent at low ZnO contents (approximately 17°C rise at only 1.0 wt.%), while it becomes nearly constant at higher loadings (≥5.0 wt.%), the maximum rise in T_p being approximately 30°C. An analogous behavior was observed for PHBV/ZnO nanocomposites (Diez-Pascual and Diez-Vicente, 2014b), pointing out that ZnO nanoparticles speed up the crystallization of PHAs via heterogeneous nucleation. A similar increasing tendency is detected for the level of crystallinity, estimated as: $X_m = \Delta H_{m,PHB} / (\Delta H^{\circ}_{m,PHB} \times w_{PHB})$, being $\Delta H_{m,PHB}$ the apparent melting enthalpy of PHB, w_{PHB} the weight fraction of PHB in the composites and $\Delta H^{\circ}_{m,PHB}$ is the theoretical enthalpy of a 100% crystalline polymer (146 J/g) (Gunaratne and Shanks, 2005). Thus X_m grew from ~53% for the neat matrix to 57, 59, 63, and 62% for the composites with 1.0, 2.0, 5.0, 10.0 wt.%, respectively. Diverse nucleating agents such as orotic acid, α-cyclodextrin, boron nitride, talc, terbium oxide, lanthanum oxide, saccharin, and phthalimide have been tested for enhancing the crystallization of PHAs (Liu et al., 2002), and it was found that boron nitride was one of the most effective, strongly increasing in the overall crystallization rate of PHBV, although it did not influence the mechanism of nucleation and growth of the polymer crystals.

As can be observed in the heating scans (Fig. 6.9b), PHB shows a bimodal melting peak ascribed to the formation of two crystalline phases with different sizes of lamellae of varying thickness (Sridhar et al., 2013). Further, this behavior has been interpreted in terms of the melting, recrystallization, and remelting model (Gunaratne and Shanks, 2005). The first peak appearing at ~163°C (T_{m1}) is believed to correspond to the melting of the crystals formed via crystallization from the melt, while that at about 174°C (T_{m2}) is attributed to the melting of crystals developed by recrystallization and reorganization during the heating. The composites with low nanoparticle cocentration display two melting peaks too, an intense one at lower temperature and a shoulder at higher temperature. Analogous double-melting behavior was observed for PHVB composites filled with graphene (Sridhar et al., 2013), cellulose nanowhiskers (Jiang et al., 2008), as well as ZnO (Diez-Pascual and Diez-Vicente, 2014b). Fig. 6.9 reveals a small increase in T_{m1} with increasing nanofiller loading, and the peak becomes narrower. The reduction in the area of T_{m2} in these nanocomposites compared to PHB suggests that these fillers restrain the recrystallization of the polymer crystals. However, the sample with 10.0 wt.% ZnO exhibits only one peak at ~169°C. The suppression of two melting peaks corroborates the extremely efficient nucleation activity of these nanoparticles. Therefore the crystals in the nanocomposites are more perfect and stable than those of neat PHB. Similarly, PHBV crystals with greater perfection were achieved by the addition of PHBV-*g*-MWCNTs (Yu et al., 2014). Further, the mobility restrictions due to the strong nanoparticle-matrix interactions also contribute to hinder the recrystallization of the matrix in the nanocomposites.

5.3 Mechanical Properties

Packaging materials are commercially used below room temperature (ie, at frozen conditions between −30 and −18°C or refrigeration temperatures in the range of 3–6°C), therefore it is important to evaluate their mechanical performance under such circumstances. Inorganic nanofillers are frequently incorporated into biopolymers to improve the mechanical properties of the resulting composites. In particular, food packaging materials should have adequate rigidity and strength to be self-supporting and withstand handling damage. The evolution of the storage modulus (E') and tan δ for neat PHBV and the composites with various ZnO loadings was investigated by dynamic mechanical analysis (DMA), at a heating rate of 2° C/min and frequency of 1 Hz, and representative curves are displayed in Fig. 6.10a and b, respectively (Diez-Pascual and Diez-Vicente, 2014b). The storage modulus indicates the capacity of a material to accumulate mechanical energy without loss; the higher E', the more rigid the material is. At temperatures below the glass transition (T_g), a significant improvement in E' is detected with increasing ZnO concentration, the highest increment being about ~44% for the nanocomposite with 4.0 wt.% ZnO at –100°C. A similar trend is found for PHB/ZnO composites, with a maximum increment of 51% at 10.0 wt.% ZnO. These enhancements are ascribed to the rise in the PHA crystallinity, together with a strong ZnO-PHA interfacial adhesion via H-bonding interactions.

The efficiency of fillers on the moduli of composites can be represented by a coefficient C obtained through the equation: $C = (E'_g/E'_r)_{comp}/(E'_g/E'_r)_{polym}$, where the *subindexes* "g" and "r" denote the glassy and rubbery regions, respectively. This coefficient indicates the capacity to perturb polymer chain motion, and depends on factors like the polymer nature, the filler content and dispersion as well as the filler–filler and filler–matrix interactions. The lower the coefficient, the higher is the reinforcement effect. The coefficient was determined at 35, 50, 65, and 80°C using E'_g at −50°C, and the data obtained are shown in Fig. 6.10d. C falls with increasing temperature, demonstrating better reinforcement

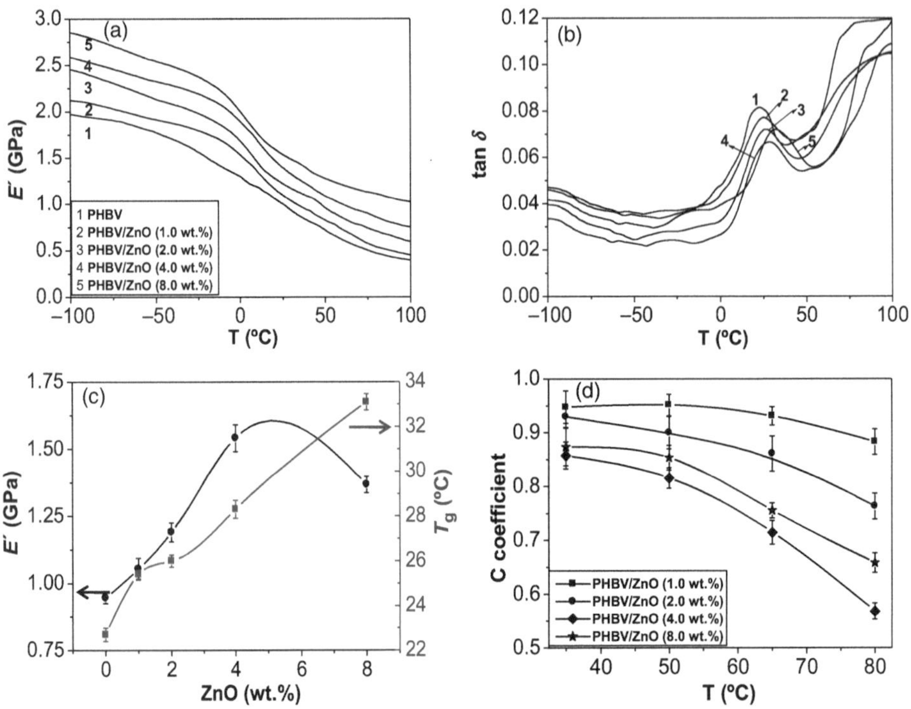

Figure 6.10. DMA measurements for PHBV/ZnO nanocomposites. (a) Storage modulus and (b) tan δ as a function of temperature; (c) E' and T_g values versus ZnO content; (d) C coefficient at different temperatures. Adapted from Diez-Pascual and Diez-Vicente (2014b), with permission from the American Chemical Society.

at high temperatures. Moreover, the nanocomposite with 4.0 wt.% ZnO has the lowest C values, corroborating that this is the most favorable nanoparticle content for enhancing PHA mechanical performance.

The plot of tan δ as a function of temperature for both PHB and PHBV reveals an intense peak (α relaxation) related to the T_g at 18 and 23°C, respectively, and a low temperature relaxation (γ) at about −100°C, ascribed to water absorbed in the amorphous regions (Fig. 6.10b). The addition of ZnO provokes an upshift of this transition temperature, indicating restrained mobility of the polymer matrix. Thus the T_g raises by ~10°C and 15°C for the nanocomposites with 8.0 and 10.0 wt.% ZnO content in comparison to that of PHBV and PHB, respectively. Further, the nanocomposites display reduced height of the tan δ peak than the pure PHAs. The incorporation of the nanoparticles causes a reduction in chain mobility, therefore lowering the peak maximum. Moreover, the lower tan δ in the nanocomposites hints that when the tension is eliminated, the stored energy is recuperated faster than in the pure polymers. A broadening of tan δ is also found on increasing nanoparticle content, explained by the bigger volume at the interface. Accordingly, an increase in the nanoparticle content would lead to stronger nanoparticle-matrix interactions and increased interfacial area, thereby contributing to the damping of the material. DMA results demonstrate that these PHA/ZnO nanocomposites exhibit adequate stiffness under subzero and refrigeration conditions.

The static mechanical performance of ZnO-reinforced PAH nanocomposites has been evaluated by tensile tests carried out at 23°C and 50% relative humidity (RH). The values of Young's modulus (E), tensile strength (σ_y), and elongation at break (ε_b), and toughness (T) obtained from the stress-strain curves of PHB/ZnO samples are plotted as a function of ZnO concentration in Fig. 6.11 (Diez-Pascual and Diez-Vicente, 2014a). The addition of ZnO nanoparticles increases E of PHB (~1.2 GPa), leading to an utmost increment of 43% at 10.0 wt.%. This outstanding E improvement is ascribed to the homogenous nanofiller dispersion, strong ZnO-PHA interfacial adhesion and the nanoparticles nucleating effect. Analogous tendency is detected for the tensile strength, where the highest rise (~32%) is found at 5.0 wt.%. Almost the same results were obtained for PHBV/ZnO composites (Diez-Pascual and Diez-Vicente, 2014b) demonstrating that ZnO nanoparticles are suitable for enhancing the mechanical performance of biodegradable PHAs. The reinforcing effect of ZnO is more effective than that reported for similar contents of other nanofillers such as layered nanoclays (Prakalathan et al., 2014) or graphene (Sridhar et al., 2013), albeit smaller than the attained with PHBV-*g*-MWCNTs, where the tensile strength and modulus improved by 88% and 172%, respectively, at 7.0 wt.% loading, likely because the grafting of the CNTs to the polymer chains enables a better stress transfer. Moreover, the Young's modulus of MWCNTs is in the range of 1.7–2.4 TPa (Ruoff et al., 2003), about 10 times higher than that of ZnO (Coleman and Jagadish, 2006).

The theoretical moduli for the nanocomposites can be calculated using the Krenchel's rule of mixtures (Krenchel, 1964) considering the Young's modulus of the nanofiller and matrix phases and their volume fractions, and the results are plotted in Fig. 6.11a as a dotted line (Diez-Pascual and Diez-Vicente, 2014a). Experimental Young's moduli of PHB/ZnO nanocomposites are thoroughly lower than the theoretical values, which indicates that this model is not appropriate for these nanocomposites. This discrepancy should be related to the fact that the model assumes perfect filler-matrix adhesion and optimal distribution of individual particles. A better approximation can be attained using

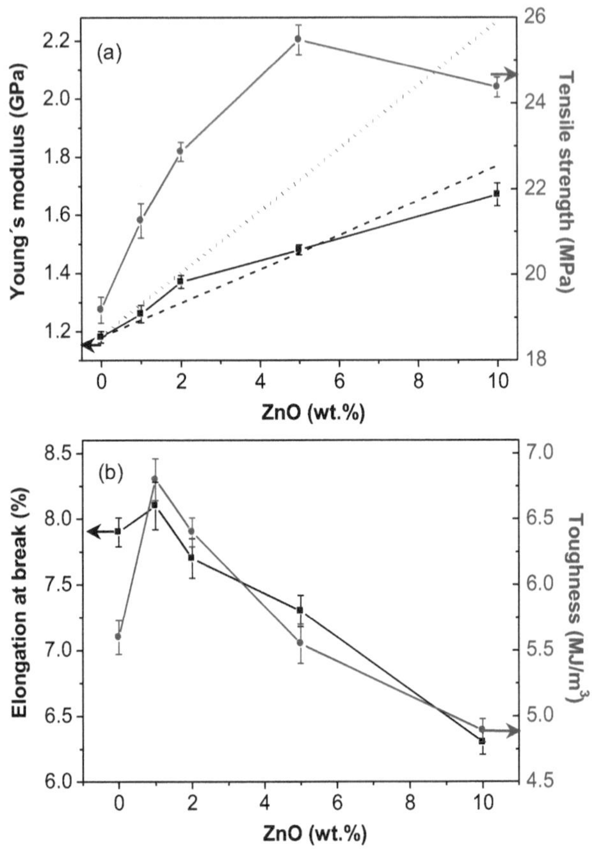

Figure 6.11. Tensile properties of PHB and PHB/ZnO nanocomposites with different nanoparticle loadings. (a) Young's modulus and tensile strength; (b) strain at break and toughness. The *dotted* and *dashed lines* in Figure 6.11A correspond to the predictions according to the Krenchel's rule of mixtures and Einstein's equation, respectively. Adapted from Diez-Pascual and Diez-Vicente (2014a), with permission from MDPI.

the Einstein's equation (Fu et al., 2008), which is only suitable for composites with low amounts of rigid and individually dispersed particles. Data drawn from this model are plotted as a dashed line in Fig. 6.11, and the differences with the experimental values are quite low (<7%).

The elongation at break hardly changes up to 2.0 wt.% ZnO and then decreases, by ~20% at 10.0 wt.% ZnO. This is the typical behavior expected for filler-reinforced composites, given that the fillers confine the ductile flow of the polymeric segments. The toughness of the nanocomposites calculated as the area under the stress-strain curve rises with the addition of nanoparticles up to

2.0 wt.%, and the highest enhancement (~21%) is found for the sample with 1.0 wt.% ZnO, while the composite with 5.0 wt.% presents comparable toughness to that of PHB, and that with 10.0 wt.% displays a decrease of approximately 12%. Consequently, ZnO contents between 1.0 and 5.0 wt.% seem to be appropriate for improving the impact resistance of PAH matrices.

5.4 Antibacterial Properties

The incorporation of antimicrobial additives into food packaging materials has lately attracted a lot of interest. Nanocomposite antimicrobial systems are successful due to the high specific surface area and surface reactivity of the nanometer-sized antimicrobial agents, which increases their effectiveness against the expansion of pathogen microorganisms in comparison to composites filled with their micro- or macroscale counterparts. The antimicrobials most frequently used are metal oxides (TiO_2, ZnO, MgO), metallic nanoparticles (Ag, Cu, Au), and carbon nanotubes. Despite that silver has been widely employed as antimicrobial additive for preservation of long-term food and beverages, ZnO nanoparticles are expected to offer a safer and cheaper alternative in the near future. Accordingly, the antibacterial action of PHBV/ZnO nanocomposites has been evaluated against *S. aureus* (*S. aureus*, ATCC 12600) and Gram-negative *E. coli* (*E. coli*, ATCC 25922), which represent one of the leading causes of death in Europe and the USA. The survival ratio (SR) was determined as: $SR = (N/N_0) \times 100$, being N_0 and N the mean number of bacteria on the neat polymer and the nanocomposites, respectively. The tests were performed in dark conditions, and the results are shown in Fig. 6.12 (Diez-Pascual and Diez-Vicente, 2014b). The survival ratios for both bacteria fall exponentially upon raising ZnO concentration, and the highest biocide action is found for the nanocomposite with 5.0 wt.% content (Fig. 6.12a). The nanocomposite with 8.0 wt.% loading has lower bactericide effect than those with ZnO loadings of 5.0 or 5.5 wt.%, attributed to its less homogenous nanoparticle dispersion. In fact, to take complete advantage of the antimicrobial properties of nanoparticles, these must be individually dispersed without forming any aggregates. In the case of PHB-based composites, the highest antibacterial activity (~97 and 94% growth inhibition for *E. coli* and *S. aureus*, respectively) was achieved in the composite with 10.0 wt.% loading (Diez-Pascual and Diez-Vicente, 2014a). Outstanding antibacterial activity toward these two bacteria combined with in vitro cell compatibility has also been reported PHBV scaffolds with Ag nanoparticle contents ≤1.0 wt.% (Xing et al., 2010).

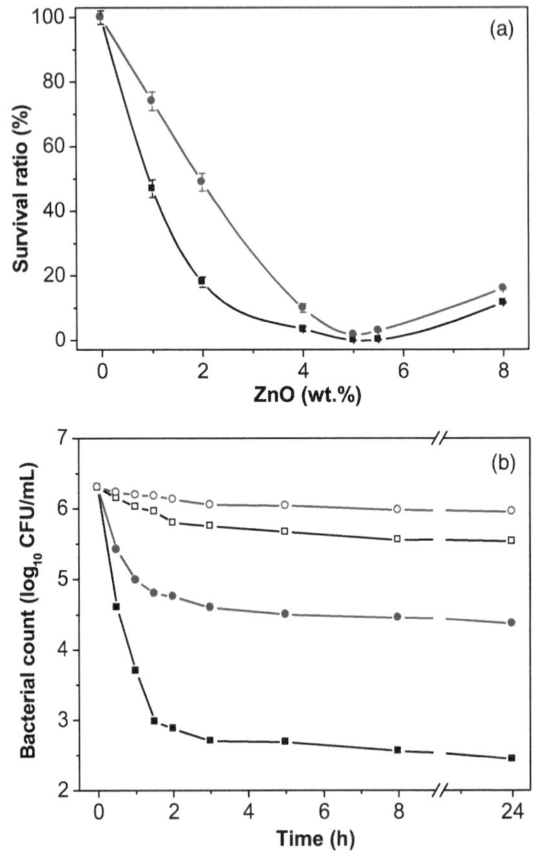

Figure 6.12. (a) Effect of PHBV/ZnO nanocomposites on the survival ratio of *E. coli (squares)* and *S. aureus (circles)*. (a) Kinetics of the antibacterial activity of PHBV/ZnO nanocomposites with 2.0 and 5.0 wt.% loading (*open and solid symbols*, respectively) against *E. coli (squares)* and *S. aureus (circles)*. Taken from Diez-Pascual and Diez-Vicente (2014b), with permission from the American Chemical Society.

Interestingly, the biocide effect of both PHB and PHBV nanocomposites on *E. coli* is systematically stronger than on *S. aureus*, in agreement with previous studies (Liu and Yang, 2003) that reported higher efficacy of ZnO nanoparticles versus Gram-negative species, and that composites with elevated ZnO concentrations can radically restrain the growth of the microorganisms. The different action versus these two types of microorganisms has been ascribed to differences in their structure and chemical composition (Brayner et al., 2006). Gram-positive bacteria only possess a cytoplasmic membrane with a thick wall comprising peptidoglycan layers, whereas Gram-negative bacteria present a complex wall structure, with a peptidoglycan layer stuck between the outer and the cytoplasmic membranes. For both bacteria, the antimicrobial activity has been attributed to different mechanisms: the release of Zn^{2+} ions as a result of ZnO decomposition; nanoparticle adsorption onto the bacteria surface, leading to physicochemical interactions that cause cellular damage; internalization of the nanoparticles in the cell and destruction of the bacteria from the interior; phagocytosis of the nanoparticles by the microorganisms leading to the cell death. Even though the precise mechanism of action is unclear yet, the generation of H_2O_2, a strong oxidizing compound, from the ZnO surface has been considered as the foremost factor responsible for the antibacterial action of ZnO-reinforced nanocomposites (Tam et al., 2008), and the different activity to *E. coli* and *S. aureus* most likely originates from the different sensitivities toward the H_2O_2 produced.

To get an insight into the kinetics of the antibacterial activity, residual colony-forming units (CFUs) were counted at different periods of time, and the data for PHBV nanocomposites with 2.0 and 5.0 wt.% ZnO are plotted in Fig. 6.12b (Diez-Pascual and Diez-Vicente, 2014b). The nanocomposite with 5.0 wt.% ZnO displays intense and rapid bactericidal action toward *E. coli*, resulting in nearly 4-log reduction in 3 h. In contrast, for the same bacteria, the nanocomposite with 2.0 wt.% exhibits slower biocide effect, causing less than 1-log fall, which corroborates the strong dependence of the bactericide activity on the ZnO concentration. An analogous tendency is found versus *S. aureus*, notwithstanding with smaller reductions in the number of CFUs. Once more, the antibacterial effect of the nanocomposites on *E. coli* is always more intense than on *S. aureus*. On the whole, experimental data reveal the great potential of PHA/ZnO nanocomposites to prevent microbial proliferation in packaged food.

5.5 Barrier and Migration Properties

A decisive concern in food packaging is that of migration and permeability. Excessive moisture and oxygen are responsible for food spoilage, hence these are the two foremost permeants investigated in packaging applications. Undesirable components can permeate through the package from the ambient atmosphere and reduce the product quality and shelf life. However, the particular barrier necessities depend on the product characteristics and the intended end-use application. For instance, containers for carbonated drinks require a high oxygen barrier to avoid oxidation of the beverage, whereas packages for fresh fruits and vegetables should allow oxygen diffusion.

The results of water uptake and water vapor permeability (WVP) for PHB/ZnO nanocomposites are displayed in Fig. 6.13a

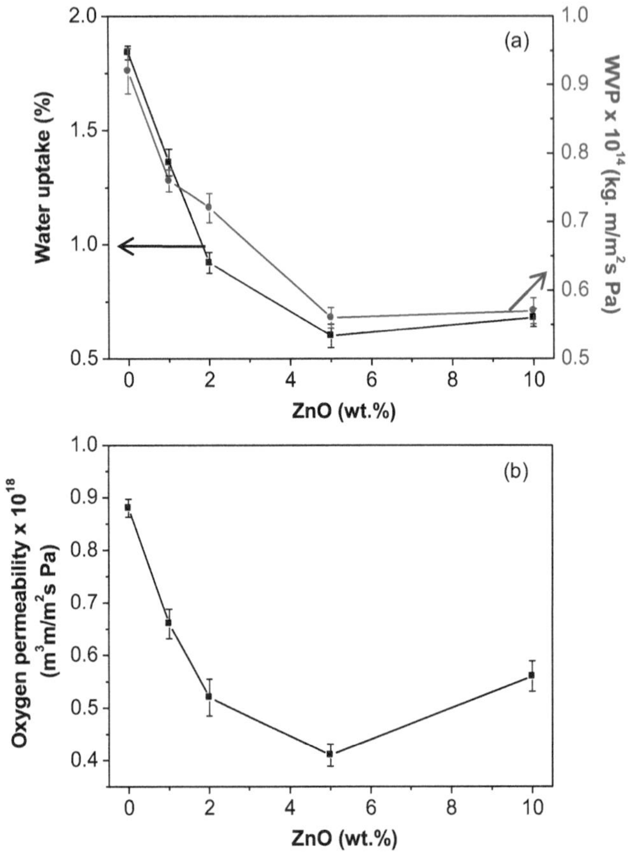

Figure 6.13. Barrier properties of PHB/ZnO nanocomposites as a function of ZnO content. (a) Water uptake and water vapor permeability (WVP); (b) oxygen permeability. Adapted from Diez-Pascual and Diez-Vicente (2014a), with permission from MDPI.

(Diez-Pascual and Diez-Vicente, 2014a). Both parameters drop upon increasing ZnO content, by up to 66% and 38% at 5.0 wt.% nanoparticle content, respectively, demonstrating that the nanocomposites have superior barrier properties to water and water vapor than PHB, in agreement with the higher degree of crystallinity of the composites, as revealed by DSC data, since the water molecules cannot permeate the matrix crystals. However, the water uptake and WVP slightly increase upon addition of 10.0 wt.% ZnO. Analogous trend was found for PHBV/ZnO nanocomposites (Diez-Pascual and Diez-Vicente, 2014b), with minimum values (~46 and 69% decrease in WVP and water uptake, respectively) at a critical loading of 4.0 wt.%. This behavior is probably the result of various effects, namely water affinity, level of crystallinity, and tortuosity. First, PHBV is more hydrophobic than ZnO, hence the hydrophilicity of the nanocomposites should rise on increasing nanoparticle concentration. Second, the crystallinity raises with increasing ZnO content, and water molecules cannot permeate the polymer crystals. Third, the higher crystallinity boosts the tortuosity of the transport path. It seems that the greater tortuosity and crystallinity outweigh the increase in hydrophilicity, and the outcome is a reduction in the water uptake and WVP. In the nanocomposite with 8.0 wt.% ZnO, the diminution in both parameters is smaller, probably due to the presence of small nanoparticle clusters, which might result in preferential penetrant pathways.

Fig. 6.13b shows the oxygen permeability (OP) versus ZnO content for PHB-based nanocomposites (Diez-Pascual and Diez-Vicente, 2014a). Similar behavior was found for PHBV/ZnO samples. A strong reduction in OP is observed with increasing ZnO content (Fig. 6.13b), showing a 53% decrement as compared to PHB at a concentration of 5.0 wt.%. This better barrier performance is related to the uniform nanoparticle distribution, its strong adhesion with the matrix that limits chain mobility combined with the increase in crystallinity. However, the OP of the composite with 10.0 wt.% loading is higher than that with 5.0 wt.% ZnO, given that nanoparticle agglomerates begin to occur, leading to preferential paths for the permeants. Interestingly, the improvements in barrier properties found for PHA/ZnO composites are higher than those reported for MWCNT (Yu et al., 2014) or carbon nanofiber-reinforced PHBV (Sanchez-Garcia et al., 2010), and comparable to those found for composites incorporating nanoclays (Sanchez-Garcia and Langaron, 2010).

It is well known that chemical components from packaging can migrate into foodstuff, which is a key issue given that many substances represent a hazard to human health and/or alter the food composition. Overall migration tests are normally carried in nonpolar and polar simulants such as ethanol 10% (v/v) and isooctane, as indicated in the EN-1186-1 standard. Fig. 6.14 presents overall migration data obtained for PHB-based nanocomposites (Diez-Pascual and Diez-Vicente, 2014a). Qualitatively similar results were obtained for PHBV-based samples (Diez-Pascual and Diez-Vicente, 2014b). For both PHB and PHBV composites and both simulants, the maximum migration values are noticeably below the legal limits stabilized for food contact materials (60 mg/kg or 10 mg/dm^2 of stimulant, Commission Directive 2002/72/EC). Increasing the ZnO content remarkably reduces the migration in ethanol 10% (v/v), by up to 54% at 5.0 wt.% loading. Analogous trend is found for migration in isooctane, the biggest drop being ~29% at the same ZnO concentration. The different

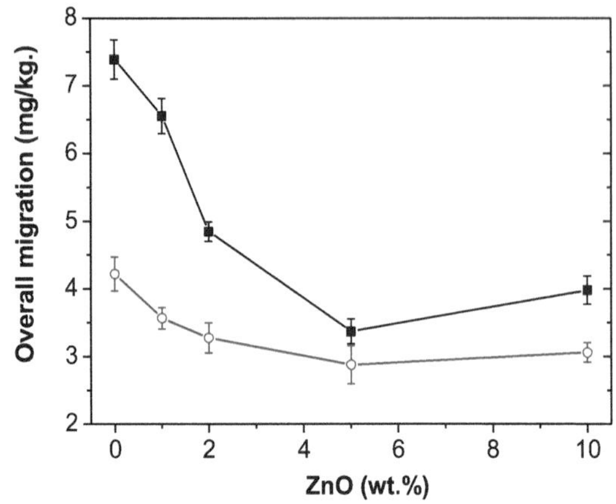

Figure 6.14. Overall migration data in ethanol 10% (v/v) (*solid squares*) and isooctane (*open circles*) for PHB and its bionanocomposites. Taken from Diez-Pascual and Diez-Vicente (2014a), with permission from MDPI.

migration values obtained for both simulants could be associated to polarity and solubility differences between the polymer, nanofiller, migrants, and the food stimulant. Very similar behavior has been reported for PHBV/PHBV-g-MWCNT composites (Yu et al., 2014), with reductions in the migration in ethanol by up to 80% at 7.0 wt.% loading. Results confirm that the incorporation of ZnO to PHA matrices has a beneficial effect on the barrier and migration properties, rendering better materials for packaging food and beverage products under different storage conditions.

6 Conclusions and Future Perspectives

In this chapter, we described the preparation and characterization of biodegradable nanocomposites by adding ZnO nanoparticles to bacterial polyester PHB homopolymer and PHBV copolymer via a solution casting technique. The nanoparticles were uniformly dispersed within the biopolymers without the aid of coupling agents, and acted effectively as nucleating agents, raising the crystallization temperature and the level of crystallinity of the matrix. The nanocomposites displayed superior mechanical properties while reduced water uptake, oxygen, and WVP compared to the neat matrices, related to the strong matrix-nanofiller interfacial adhesion attained via hydrogen bonding interactions. At an optimal concentration of 4.0–5.0 wt.% ZnO, the tensile strength, Young's, and storage moduli showed a maximum that coincided with the highest crystallinity and the best barrier properties. The nanocomposites also showed antibacterial activity against human pathogen bacteria, and the overall migration levels of the nanocomposites in both nonpolar and polar well below the limits required by the current normative for food packaging materials. The performance of these novel biocomposites has been compared to that of composites incorporating similar amounts of other nanofillers, such as, CNTs or nanoclays. Despite PHA/CNT nanocomposites show comparable or even better properties than ZnO-reinforced ones, the toxic effects of carbon nanotubes remain still unclear. Further, the elevated price of CNTs makes those composites not cost-effective. Therefore sustainable PHA/ZnO biocomposites are expected to offer a safer and cheaper alternative to synthetic plastic packaging materials. In particular, they are very promising to be used as packaging materials for cheese, cereals, bakery, and meat products as well as for the manufacture of thermoformed containers for fruit juices and dairy products, bottles for water, beer, or carbonated drinks. They can also be suitable for disposable applications like cutlery, cups and utensils, blister packages, trash bags, overwrap and lamination films.

In summary, significant progress in the development of ZnO-reinforced PHA-based composites has produced very promising results thus far. However, further research is still required to fully exploit the potential technological applications of these nanocomposites in order to compete with petroleum-based plastics. Thus in-depth studies on structure–property relationships as well as the manufacturing and characterization of the bionanocomposites at a large scale are required prior to be used in commercial applications. The development of the optimum formulation for each polymeric system and processing method to meet specific requirements and cost reduction of the bionanocomposites are also desired. In addition, there are some safety concerns about the use of these nanocomposites as food contact materials. Additional research is still required to assure their nontoxicity as well as the environmental safety of their use. Nonetheless, these bionanocomposites appear to have a very bright future for a wide range of applications in the food industry as food preservation materials.

Acknowledgment

Dr. Ana Diez-Pascual wishes to acknowledge the Ministerio de Economía y Competitividad (MINECO) for a "Ramón y Cajal" Senior Research Fellowship cofinanced by the EU.

References

Abdolmaleki, A., Mallakpour, S., Borandeh, S., 2012. Effect of silane-modified ZnO on morphology and properties of bionanocomposites based on poly(ester-amide) containing tyrosine linkages. Polym. Bull. 69, 15–28.

Adachi, S., 2005. Properties of Group-IV, III-V and II-VI Semiconductors. John Wiley & Sons, England.

Alexander, L.E., Krieger, R.E., 1969. X-ray Diffraction Methods in Polymer Science. Wiley, New York.

Ao, W., Li, J., Yang, H., Zeng, X., Ma, X., 2006. Mechanochemical synthesis of zinc oxide nanocrystalline. Powder Technol. 168, 128–151.

Becheri, A., Maximilian, D., Lo Nostro, P., Baglioni, P., 2008. Synthesis and characterization of zinc oxide nanoparticles: application to textiles as UV-absorbers. J. Nanopart. Res. 10, 679–689.

Bonartsev, A.P., Myshkina, V.L., Nikolaeva, D.A., Furina, E.K., Makhina, T.A., Livshits, V.A., Boskhomdzhiev, A.P., Ivanov, E.A., Iordanskii, A.L., Bonartseva, G.A., 2007. Biosynthesis, biodegradation, and application of poly(3-hydroxybutyrate) and its copolymers -natural polyesters produced by diazotrophic bacteria. In: Méndez-Vilas, A. (Ed.), Communicating Current Research and Educational Topics and Trends in Applied Microbiology. Formatex, Spain, pp. 295–307.

Brayner, R., Ferrari-Iliou, R., Brivois, N., Djediat, S., Benedetti, M.F., Fievet, F., 2006. Toxicological impact studies based on *Escherichia coli* bacteria in ultrafine ZnO nanoparticles colloidal medium. Nano Lett. 6, 866–870.

Chan, K.H.K., Wong, S.Y., Tiju, W.C., Li, X., Kotaki, M., He, C.B., 2009. Morphologies and electrical properties of electrospun poly[(*R*)-3-hydroxybutyrate-*co*-(*R*)-3-hydroxyvalerate]/multiwalled carbon nanotubes fibers. J. Appl. Polym. Sci. 116, 1030–1035.

Chen, G.Q., Wu, Q., 2005. The application of polyhydroxyalkanoates as tissue engineering materials. Biomaterials 26, 6565–6578.

Clarinval, A.-M., Halleux, J., 2005. Classification of biodegradable polymers. In: Smith, R. (Ed.), Biodegradable Polymers for Industrial Applications. Woodhead Publishing Ltd, Cambridge, UK, pp. 3–31.

Coleman, A., Jagadish, C., 2006. Basic properties and applications of ZnO. In: Jagadish, C., Pearton, S. (Eds.), Zinc Oxide Bulk, Thin Films and Nanostructures. Elsevier, Amsterdam.

Cross, S.E., Innes, B., Roberts, M.S., Tsuzuki, T., Robertson, T.A., McCormick, P., 2007. Human skin penetration of sunscreen nanoparticles: in vitro assessment of novel micronized zinc oxide formulation. Skin Pharmacol. 20, 148–154.

Diez-Pascual, A.M., Diez-Vicente, A.L., 2014a. Poly(3-hydroxybutyrate)/ZnO bionanocomposites with improved mechanical, barrier and antibacterial properties. Int. J. Mol. Sci. 15, 10950–10973.

Diez-Pascual, A.M., Diez-Vicente, A.L., 2014b. ZnO-reinforced poly(3-hydroxybutyrate-*co*-3-hydroxyvalerate) bionanocomposites with antimicrobial function for food packaging. ACS Appl. Mater. Interf. 6, 9822–9834.

Diez-Pascual, A.M., Xu, C.P., Luque, R., 2014. Development and characterization of novel poly(ether ether ketone)/ZnO bionanocomposites. J. Mater. Chem. B 2, 3065–3078.

Fu, S.-Y., Feng, X.-Q., Lauke, B., Mai, Y.-W., 2008. Effects of particle size, particle/matrix interface adhesion and particle loading on mechanical properties of particulate–polymer composites. Compos. Part B 39, 933–961.

Ghanbarzadeh, B., Oleyaei, S.A., Almasi, H., 2014. Nano-structured materials utilized in biopolymer based plastics for food packaging applications. Crit. Rev. Food Sci. Nutr. 54, 299–313.

Grasset, F., Saito, N., Li, D., Park, D., Sakaguchi, I., Ohashi, N., Haneda, H., Roisnel, T., Mornet, S., Duguet, E., 2003. Surface modification of zinc oxide nanoparticles by aminopropyltrietoxi silane. J. Alloys Comp. 360, 298–311.
Gunaratne, L.M.W.K., Shanks, R.A., 2005. Multiple melting behaviour of poly(3-hydroxybutyrate-*co*-hydroxyvalerate) using step-scan DSC. Eur. Polym. J. 41, 2980–2988.
Hajipour, M.J., Fromm, K.M., Ashkarran, A.A., Jimenez de Aberasturi, D., Ruiz de Larramendi, I., Rojo, T., Serpooshan, V., Parak, W.J., Mahmoudi, M., 2012. Antibacterial properties of nanoparticles. Trends Biotechnol. 30, 499–511.
Jacquel, N., Lo, C.-W., Wei, Y.-H., Wu, H.-S., Wang, S.S., 2007. Solubility of polyhydroxyalkanoates by experiment and thermodynamic correlations. AIChE J. 53, 2704–2714.
Jacquel, N., Lo, C.-W., Wei, Y.-H., Wu, H.-S., Wang, S.S., 2008. Isolation and purification of bacterial poly(3-hydroxyalkanoates). Biochem. Eng. J. 39, 15–27.
Jiang, L., Morelius, E., Zhang, J., Wolcott, M., Holbery, J., 2008. Study of the poly(3hydroxybutyrate-*co*-3-hydroxyvalerate)/cellulose nanowhisker composites prepared by solution casting and melt processing. J. Compos. Mater. 42, 2629–2645.
Jones, N., Ray, B., Ranjit, K.T., Manna, A.C., 2008. Antibacterial activity of ZnO nanoparticle suspensions on a broad spectrum of microorganisms. FEMS Microbiol. Lett. 279, 71–77.
Kickelbick, G., 2003. Concepts for the incorporation of inorganic building blocks into organic polymers on a nanoscale. Prog. Polym. Sci. 28, 83–114.
Kołodziejczak-Radzimska, A., Jesionowski, T., 2014. Zinc oxide—from synthesis to application: a review. Materials 7, 2833–2881.
Krenchel, H., 1964. Fibre Reinforcement. Akademisk Forlag, Copenhagen.
Kucheyev, O., Bradby, J.E., Williams, J.S., Jagadish, C., Swain, M.V., 2002. Mechanical deformation of single crystal ZnO. Appl. Phys. Lett. 80, 956–958.
Laycock, B., Halley, P., Pratt, S., Werker, A., Lant, P., 2013. The chemomechanical properties of microbial polyhydroxyalkanoate. Prog. Polym. Sci. 38, 536–583.
Lin, H.M., Tzeng, S.J., Hsiau, P.J., Tsai, W.L., 1998. Electrode effects on gas sensing propertiesof nanocrystalline zinc oxide. Nanostruct. Mater. 10, 465–477.
Liu, H.L., Yang, T.C.K., 2003. Photocatalytic inactivation of *Escherichia coli* and *Lactobacillus helveticus* by ZnO and TiO_2 activated with ultraviolet light. Process Biochem. 39, 475–481.
Liu, W., Yang, H.L., Wang, Z., Dong, L.S., Liu, J.J., 2002. Effect of nucleating agents on the crystallization of poly(3-hydroxybutyrate-*co*-3-hydroxyvalerate). J. Appl. Polym. Sci. 86, 2145–2152.
Ma, P.M., Wang, R., Wang, S.F., Zhang, Y., Zhang, Y.X., Hristova, D., 2008. Effect of fumed silica on the crystallization behavior and thermal properties of poly(hydroxybutyrate-*co*-hydroxyvalerate. J. Appl. Polym. Sci. 108, 1770–1777.
Marsh, K., Bugusu, B., 2007. Food packaging: roles, materials, and environmental issues. J. Food Sci. 72, R39–R55.
Naffakh, M., Díez-Pascual, A.M., Marco, C., Ellis, G.J., Gómez-Fatou, M.A., 2013. Opportunities and challenges in the use of inorganic fullerene-like nanoparticles to produce advanced polymer nanocomposites. Prog. Polym. Sci. 38, 1163–1231.
Pandey, J.K., Reddy, K.R., 2014. Handbook of Polymer Nanocomposites: Processing, Performance and Application. Springer, New York, p. 5.
Philip, S., Keshavarz, T., Roy, I., 2007. Polyhydroxyalkanoates: biodegradable polymers with a range of applications. J. Chem. Technol. Biotechnol. 82, 233–247.
Prakalathan, K., Mohanty, S., Nayak, S.K., 2014. Reinforcing effect and isothermal crystallization kinetics of poly(3-hydroxybutyrate) nanocomposites blended with organically modified montmorillonite. Polym. Compos. 35, 999–1012.

Przybyszewska, M., Zaborski, M., 2009. The effect of zinc oxide nanoparticle morphology on activity in crosslinking of carboxylated nitrile elastomer. Express Polym. Lett. 3, 542–552.
Raoufi, D., 2013. Synthesis and microstructural properties of ZnO nanoparticles prepared by precipitation method. Renew. Energ. 50, 932–937.
Rhim, J.-W., Park, H.-M., Ha, C.-S., 2013. Bio-nanocomposites for food packaging applications. Prog. Polym. Sci. 38, 1629–1652.
Ruoff, S., Qian, D., Liu, W.K., 2003. Mechanical properties of carbon nanotubes: theoretical predictions and experimental measurements. C. R. Physique 4, 993–1008.
Sadat-Shojai, M., Khorasani, M.-T., Jamshidi, A., Irani, S., 2013. Nano-hydroxyapatite reinforced polyhydroxybutyrate composites: a comprehensive study on the structural and in vitro biological properties. Mater. Sci. Eng. C 33, 2776–2787.
Sanchez-Garcia, M.D., Langaron, J.M., 2010. Novel clay-based nanobiocomposites of biopolyesters with synergistic barrier to UV light, gas, and vapour. J. Appl. Polym. Sci. 118, 188–199.
Sanchez-Garcia, M.D., Langaron, J.M., Hoa, S.V., 2010. Effect of addition of carbon nanofibers and carbon nanotubes on properties of thermoplastic biopolymers. Compos. Sci. Technol. 70, 1095–1105.
Savenkova, L., Gercberga, Z., Nikolaeva, V., Dzene, A., Bibers, I., Kalnin, M., 2000. Mechanical properties and biodegradation characteristics of PHB-based films. Process Biochem. 35, 573–579.
Scandola, M., Ceccorulli, G., Pizzoli, M., Gazzano, M., 1992. Study of the crystal phase and crystallization rate of bacterial poly(3-hydroxybutyrate-*co*-3hydroxyvalerate). Macromolecules 25, 1405–1410.
Siracusa, V., Rocculi, P., Romani, S., Dalla Rosa, M., 2008. Biodegradable polymers for food packaging: a review. Trends Food Sci. Technol. 19, 634–643.
Sridhar, V., Lee, I., Chun, H.H., Park, H., 2013. Graphene reinforced biodegradable poly(3-hydroxybutyrate-*co*-4-hydroxybutyrate) nano-composites. eXPRESS Polym. Lett. 7, 320–328.
Sudesh, K., Abe, H., Doi, Y., 2000. Synthesis, structure and properties of polyhydroxyalkanoates: biological polyesters. Prog. Polym. Sci. 2000 (25), 1503–1555.
Tam, K.H., Djurišiæ, A.B., Chan, C.M.N., Xi, Y.Y., Tse, C.W., Leung, Y.H., Chan, W.K., Leung, F.C.C., Au, D.W.T., 2008. Antibacterial activity of ZnO nanorods prepared by a hydrothermal method. Thin Sol. Films 516, 6167–6174.
Tang, X., Choo, E.S.G., Li, L., Ding, J., Xue, J., 2009. One-pot synthesis of water-stable ZnO nanoparticles via a polyol hydrolysis route and their cell labeling applications. Langmuir 25, 5271–5275.
Thire, R.M.S.M., Arruda, L.C., Barreto, L.S., 2011. Morphology and thermal properties of poly(3-hydroxybutyrate-*co*-3-hydroxyvalerate)/attapulgite nanocomposites. Mater. Res. 14, 340–344.
Topoglidis, E., Palomares, E., Astuti, Y., Green, A., Campbell, C.J., Durrant, J.R., 2005. Immobilization and electrochemistry of negatively charged proteins on modified nanocrystalline metal oxide electrodes. Electroanalysis 17, 1035–1041.
Tsonos, C., Kanapitsas, A., Triantis, D., Anastasiadis, C., Stavrakas, I., Pissis, P., Neagu, E., 2011. Interface states and MWS polarization contributions to the dielectric response of low voltage ZnO varistor. Ceram. Int. 37, 207–214.
Visaria, R.K., Griffin, R.J., Williams, B.W., Ebbini, E.S., Paciotti, G.F., Song, C.W., Bischof, J.C., 2006. Enhancement of tumor thermal therapy using gold nanoparticle-assisted tumor necrosis factor-alpha delivery. Mol. Cancer Ther. 5, 1014–1020.

Vroman, I., Tighzert, L., 2009. Biodegradable olymers. Materials 2, 307–344.
Wang, M., Wang, X., 2008. Electrodeposition zinc-oxide inverse opal and its application in hybrid photovoltaics. Sol. Energy Mater. Sol. Cells 92, 357–362.
Wang, J.X., Sun, X.W., Wei, A., Lei, Y., Cai, X.P., Li, C.M., Dong, Z.L., 2006. Zinc oxide nanocomb biosensor for glucose detection. Appl. Phys. Lett. 88, 233106–233108.
Xing, Z-C., Chae, W-P., Baek, J-Y., Choi, M-J., Jung, Y., Kang, I.-K., 2010. In vitro assessment of antibacterial activity and cytocompatibility of silver-containing PHBV nanofibrous scaffolds for tissue engineering. Biomacromolecules 11, 1248–1253.
Xu, C., Qiu, Z., 2011. Crystallization behavior and thermal property of biodegradable poly(3-hydroxybutyrate)/multi-walled carbon nanotubes nanocomposite. Polym. Adv. Technol. 22, 538–544.
Yamada, H., Suzuki, K., Koizumi, S., 2007. Gene expression profile in human cells exposed to zinc. J. Toxicol. Sci. 32, 193–196.
Yamamoto, O., 2001. Influence of particle size on the antibacterial activity of zinc oxide. Int. J. Inorg. Mater. 3, 643–646.
Yeo, S.Y., Tan, W.L., Bakar, M.A., Ismail, J., 2010. Silver sulfide/poly(3-hydroxybutyrate) nanocomposites: thermal stability and kinetic analysis of thermal degradation. Polym. Degrad. Stab. 95, 1299–1304.
Yu, H.-Y., Qin, Z.-Y., Liu, Y.-N., Chen, L., Liu, N., Zhou, Z., 2012. Simultaneous improvement of mechanical properties and thermal stability of bacterial polyester by cellulose nanocrystals. Carbohyd. Polym. 89, 971–978.
Yu, H.-Y., Qin, Z.-Y., Sun, B., Yang, X.-G., Yao, J.-M., 2014. Reinforcement of transparent poly(3-hydroxybutyrate-*co*-3-hydroxyvalerate) by incorporation of functionalized carbon nanotubes as a novel bionanocomposite for food packaging. Compos. Sci. Technol. 94, 96–104.
Yun, S.I., Lo, V., Noorman, J., Davis, J., Russell, R.A., Holden, P.J., 2010. Morphology of composite particles of single wall carbon nanotubes/biodegradable polyhydroxyalkanoates prepared by spray drying. Polym. Bull. 64, 99–106.
Zhou, S., 2004. Synthesis and characterization of biodegradable low molecular weight aliphatic polyesters and their use in protein-delivery systems. J. Appl. Polym. Sci. 91, 1848–1856.
Zou, H., Wu, S., Shen, J., 2008. Polymer/silica nanocomposites: preparation, characterization, properties, and applications. Chem. Rev. 108, 3893–3957.

7

BIOPLASTICS FROM AGRO-WASTES FOR FOOD PACKAGING APPLICATIONS

Isabel Gonçalves de Moura, Arsénio Vasconcelos de Sá, Ana Sofia Lemos Machado Abreu, Ana Vera Alves Machado

University of Minho, Institute of Polymers and Composites (IPC) and Institute of Nanostructures, Nanomodelling, and Nanofabrication, Guimarães, Portugal

1 Introduction

For many years, synthetic plastic materials produced from petrochemicals, have been used in a variety of applications in everyday life, such as, in packaging, automotive and health-care applications, and communication and electronic industries (Ohtake et al., 1998). They became essential due to their versatility, aesthetic qualities, durability, and low cost (Bayer et al., 2014). However, due to their excessive use, several drawbacks, such as, its resistance to decomposition, toxicity after incineration, and accumulation in the environment, induce negative ecological impact in landfills and water contamination. Moreover, the growing ecological responsiveness together with the novel environmental laws, are pushing the industries to search for more ecofriendly materials, specifically for food packaging, where the material has a very short lifetime (Ohtake et al., 1998; Scott, 2000).

Thus, minimizing their unfavorable ecological impact became a challenge. A possible solution to solve this problem could be to replace commodity-synthetic polymers by biodegradable ones, with good performance. The latter, after usage, would be prone to degradation and digestion by microorganisms, which would minimize their unfavorable ecological impact (Mohanty et al., 2000; Scott, 2000).

Simultaneously, the high amount of agro-wastes generated by the food industry has been a growing concern, which around 1.3 billion tons left-over, which becomes an environmental and financial issue (Bayer et al., 2014). Among these agro-wastes,

Food Packaging. http://dx.doi.org/10.1016/B978-0-12-804302-8.00007-8

food waste stands out, halum, and stems of vegetables/fruits, grains and seeds, from which it is possible to obtain natural polymers (Shah et al., 2008).

Thus, in this chapter the discussion is on how to directly extract natural polymers, namely cellulose, from vegetable agro-wastes by applying green chemistry methodology, and its modification in order to develop new biopolymers with potential application in food packaging.

2 Synthetic Plastics

During the past decades, petroleum-based synthetic polymers, commonly known as plastics, have gradually expanded and become one of the most appealing types of materials. The worldwide production is estimated to be around 140 million tons per year (Shah et al., 2008). This achievement is mostly associated with their properties, such as, low price, good processability, low density, good aesthetic qualities, and resistance to degradation.

The most widely-used plastics are polyethylene (PE, 29%), polypropylene (PP, 19%), polyvinyl chloride (PVC, 13%), polystyrene (PS, 7%), poly(ethylene terephthalate) (PET, 7%), polyurethane (PUR, 6%), and others (Fig. 7.1).

Polyolefins are the synthetic polymers with the highest commercial success, accounting for more than 47% of Western Europe's total consumption, 24.1 million tons per year. They present a combination of physical properties (flexibility, strength, lightness, stability, and impermeability) that are ideally suited to a wide variety of applications, such as food packaging (Simon et al., 1996).

Owing to the growing increase in the production of commodity plastics, as well as, the deposition rate in the past two decades of the 20th century, they are currently the most common and persistent pollutants worldwide. Once such material became part of the natural ecosystem, the negative effect of it is a long-lasting contribution to environmental contamination is depicted in Fig. 7.2 (Shah et al., 2008; Thompson et al., 2009). The growing ecological responsiveness, together with the new environmental laws, are pushing the industries to search for more ecofriendly plastics, specifically in applications where they have a very short lifetime (Moura et al., 2011).

Under natural conditions, the degradation of synthetic plastics is a very slow process that involves environmental factors, followed by the action of wild microorganisms (Albertsson et al., 1994; Cruz-Pinto et al., 1994). This has raised growing concern about the development of biodegradable polymers,

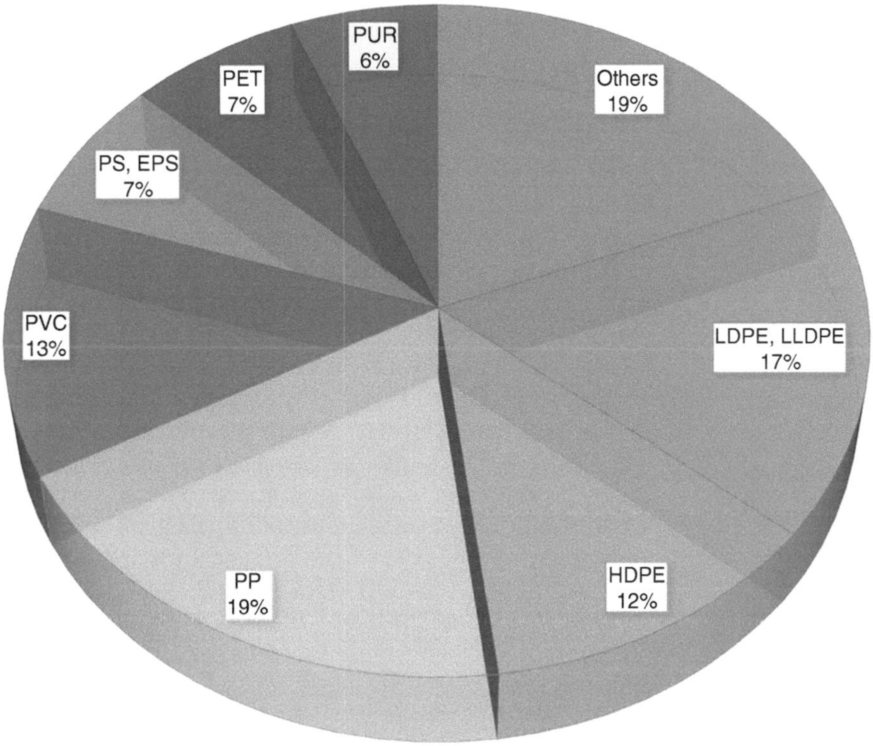

Figure 7.1. Global plastics market consumption.

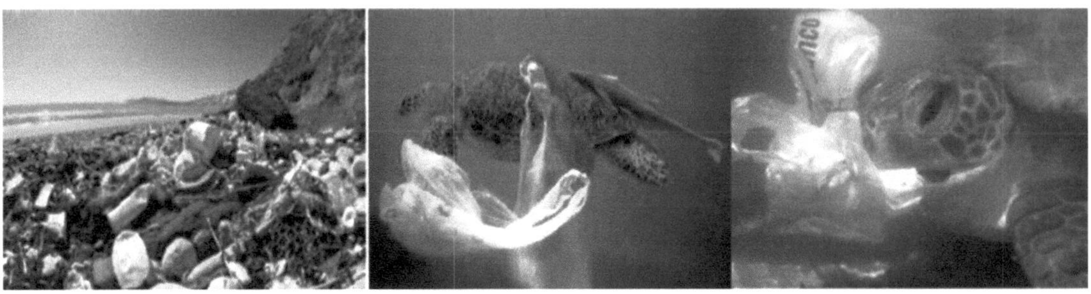

Figure 7.2. The negative impact of plastics waste disposal.

promoting research activity worldwide to either modify current products, in order to induce degradability, or to develop new alternatives that are biodegradable after use. The use of these materials, in applications with a short-life cycle, namely food packaging would be an ecological alternative to reduce the solid-plastic waste (Albertsson et al., 1994).

3 Petroleum-Based Biodegradable Polymers and Polymers Derived From Renewable Resources

Limited resources of petroleum-based polymers and increased environmental awareness on sustainability, has attracted a higher interest toward biodegradable polymers from renewable resources, for industrial application (Nabar et al., 2005). The use of these materials is an alternative to conventional nonbiodegradable plastics, which could be a more friendly solution for the environmental problem, induced by plastics wastes (Tserki et al., 2006).

The research and consumption of biodegradable polymers that are derived from renewable resources has increased in the past decades. The target market is mainly food packaging materials, hygiene products, and agricultural tools. Nevertheless, it is still a competition between commodity plastics and biodegradable ones, due the low cost of the former, functional properties, and processability (Gross and Kalra, 2002; Sudesh and Iwata, 2008).

Bioplastics can be made of biodegradable petroleum-based polymers, 100% renewable material, or a combination of renewable and fossil materials (Fig. 7.3).

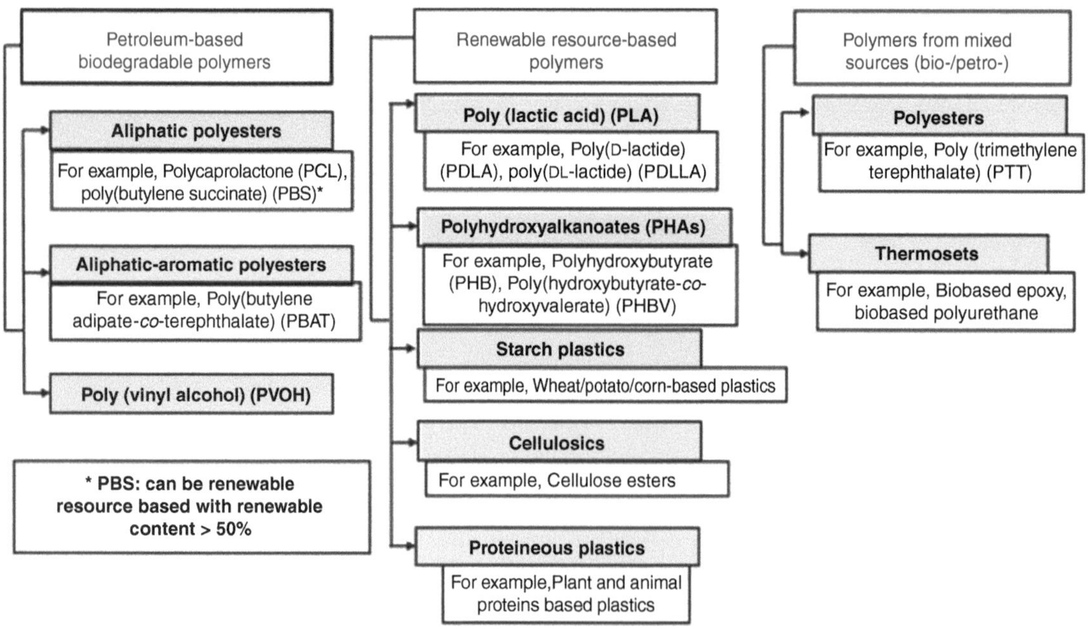

Figure 7.3. Resources of biodegradable polymers. From Reddy et al. (2013).

According to Reddy et al. (2012) biodegradable polymers can be divided as follows:

1. *Petroleum-based biodegradable polymers*: They are synthesized from petroleum resources that are biodegradable at the end of their functionality. Poly(ε-caprolactone) (PCL) and poly(butylene adipate-*co*-terephthalate) (PBAT) are some examples.
2. *Bioplastics from mixed sources (bio-/petro-)*: These biodegradable materials are developed from a combination of bio-based and petroleum monomers; they include polymers as poly(trimethylene terephthalate) (PTT), biothermosets, and biobased blends.
3. *Renewable resource-based polymers*: These biopolymers are either synthesized naturally from plants and animals, or from renewable resources. Some of the most significant examples are polylactic acid (PLA), cellulose, starch, chitosan, lignin, and proteins.

This chapter focuses on bioplastics from renewable resources and their application in the food packaging industry. The current status of production and application of the renewable resource-based bioplastics, specifically PLA, starch, and cellulose (Table 7.1) and the new trends in obtaining biopolymers directly from agro-wastes, such as cellulose, using a green chemistry methodology.

3.1 Polyethylene-Based Polymers

Synthetic polymers similar to polyolefins are resistant to biodegradation by the action of microorganisms and have a long lifetime (Sastry et al., 1998). In its original form, PE does not degrade, which is related to the high molecular weight and also to the high hydrophobic character. Thus, the use of conventional PE in food packaging has a negative environmental impact. However, research studies concerning polyolefins' biodegradation have demonstrated that these polymers have short carbon chains and can be used by the microorganisms as a carbon source, which, therefore, makes them biodegradable (Yamada-Onodera et al., 2001). Accordingly, to turn PE biodegradable, it is essential to change its features, for instance, molecular weight and crystallinity level, which enhance the resistance to degradation. Studies related with the biodegradation of PE were carried out by Wang et al. (2004) and Bonhomme et al. (2003), and it was found that its degradation occurred in two distinct steps, hydrodegradation and oxo-biodegradation (Bonhomme et al., 2003). Other researchers also observed that the oxidation products of polyolefins are biodegradable (Smith, 2005). According to Albertsson et al. (1993), the biodegradation mechanism of PE occurs in different pathways. The first step is the oxidation process, in which the PE is oxidized,

Table 7.1 Current Applications of Plastics

Packaging Applications	Biopolymer
PLA	
Coffee and tea	Cardboard cups coated with PLA
Beverages	PLA cups
Fresh salads	PLA bowls
Carbonated water, fresh juices, and dairy drinks	PLA bottles
Freshly cut fruits, whole fruit vegetables, bakery goods, and salads	Rigid PLA trays and packs
Organic pretzels and potato chips	PLA bags
Yogurt	PLA jars
Frozen fries	PLA films (Bio-Flex)
Organic fruit and vegetables	PLA packaging
Pasta	PLA packaging
Herbs	PLA packaging
Prepared sandwiches and pasta salads	PLA bowls, packaging
Bread	Paper bags with PLA window
Organic poultry	PLA bowls and absorb pads
Starch-based	
Milk chocolates	Corn-starch trays
Organic tomatoes	Corn-based packaging
Cellulose	
Kiwi	Biobased trays wrapped with cellulose film
Potato chips	Metalized cellulose film
Organic pasta	Cellulose-based packaging
Sweets	Metalized cellulose film

From Peelman et al. (2013).

resulting in the formation of hydroxyl (—OH), carbonyl (—C=O), and carboxylic acid groups (—COOH), which leads to the decrease of the polymer chain molecular weight, followed by the biodegradation due the microbial action. The oxidation process plays a crucial role in the reduction of PE's molecular weight and consequently, the formation of compounds with reduced size, which can be subsequently be assimilated by the microorganisms. It has been shown that PE molecular fragments are easily degraded. Another approach to increase the polyolefin biodegradation consists in blending them with native biodegradable polymers, such as, starch or even with other synthetic biodegradable polyesters like PLA, PCL, and poly(glycolic acid) (PGA). Iwamoto et al. (1994) evaluated the biodegradation of blends of low-density polyethylene (LDPE) and PCL and concluded that when samples are submitted to an enzymatic degradation they suffer a significant weight loss. Machado et al. (2007) evaluated the biodeterioration of high-density polyethylene (HDPE) blended with biodegradable polyesters, like Mater-Bi (thermoplastic starch with PLA or PCL), that is processed in a corotating twinscrew extruder. In a previous study, Albertsson et al. (1993) showed that starch, when blended with LDPE matrix, produces a material with higher susceptibility to biodegradation when compared with neat LDPE. Another alternative to the synthesized biodegradable PE consists is the graft copolymerization of a biodegradable polymer. Many research works reported the preparation of several grafted-polyethylene copolymers by chemical modification of natural polymers (eg, starch) for numerous potential applications, including food packaging. Maharana and Singh (2006) studied the graft copolymerization of starch onto LDPE, using an initiator. The soil-burial tests of PE-*g*-starch showed that by grafting, starch moieties onto synthetic PE, it is possible to enhance the percentage of biodegradation, which is correlated with the graft yield. Another study, carried out by Gautam (2010), reported the biodegradation of graft copolymers based on starch and PE. The biodegradability tests performed by soil-burial tests showed a growth in the number of microorganisms. Additionally, the products resulting from the biodegraded sample hydrolyze faster. Kaur et al. (2008) described the modification of polyethylene by the formation of graft copolymers based on gelatin (PE-g-gelatin). The biodegradation of these samples was monitored by soil-burial assays, by determining their weight loss. According to the analysis, it was observed that a growth in the number of microorganisms on the grafted material, as well as, an increase in the percentage weight loss occurs. Thus, the rational design of biodegradable polyolefins with the required features for a specific application, and with the capacity of being degradable/biodegradable after use, has been the subject of many works searching for valid alternatives to conventional nondegradable plastics.

3.2 Poly(ε-Caprolactone)

PCL is a biodegradable polyester, obtained from the ring-opening polymerization (ROP) of the corresponding monomer, ε-caprolactone (ε-CL). This thermoplastic is quite flexible, having low viscosity and a melting point around 58–60°C, presenting good chemical resistance, and being easily processable (Calil et al., 2006). Nevertheless, it presents some disadvantages associated with the low mechanical performance and high price. Thus, it is necessary to blend it with other polymers to get biodegradable materials at lower cost for some applications. Many studies have been carried out to develop fully biodegradable polymer blends containing PCL (Koenig and Huang, 1995; Pranamuda et al., 1996). The research findings demonstrated that the mechanical performance of the resulting blends (PCL/starch) became poorer with increasing amount of starch. The results are attributed to the poor compatibility between both components, due to their distinct hydrophilic character, since PCL is hydrophobic and starch hydrophilic. Yang et al. (2001) prepared biodegradable composites based on PCL and chitosan by melt mixing. It was shown that increasing the amount of chitosan did not affect the melting temperature, which is explained by the immiscibility of the components. Other studies using PCL and biodegradable thermoplastic starch were performed in order to adjust the rheological behavior of the melt (Matzinos et al., 2002). They used several starch grades produced by Novamont and verified that the degradation of PCL was faster for higher starch content. Besides that, the mechanical behavior of PCL/starch blends was also evaluated. The increase in the amount of modified starch, induces a decrease in the elongation at break and tensile strength and an increase in the Young's modulus compared to neat PCL (Yavuz and Babaç, 2003). One approach to enhance the mechanical performance of PCL/starch blends is by using synthetic polymers, such as LDPE. Also, Bastioli et al. (1995) showed that the biodegradation of PCL can be significantly accelerated by the presence of starch.

Nevertheless, by simple bending PCL with other biodegradable polymers fails in several applications, owing to the incompatibility between the components. Thus, one possible way to overcome this issue is by the use of a compatibilizer, which induces an enhancement of the interfacial adhesion between both components. Choi et al. (1999), prepared PCL and starch blends using starch-*g*-PCL graft copolymer as compatibilizer. The mechanical performance of the resulting blends correlated with the effect of the compatibilizer addition. According to the reported results, it was observed that the starch-*g*-PCL graft copolymer induced a remarkable increase in the elongation and toughness.

3.3 Poly(Butylene Adipate-*co*-Terephthalate)

PBAT is a commercially well-known biodegradable copolymer, which is obtained by polycondensation of adipic acid,

1,4-butanediol, and dimethylterephthalate along with a catalyzer, for example, tetrabutylorthotitanate (Jiang et al., 2009). This polymer shows good mechanical properties (high flexibility, elasticity, and elongation at break) and suitable thermal properties for concentration of terephtalic acid above 35%. It has resistance to wear and fracture, good hydrophilic- and processing properties. Nevertheless, the high cost associated with its production, as well as, the fragile Young's modulus, circumscribe the application of this polymer. For values higher than 55% of terephalic acid, its biodegradability decreases (Witt et al., 2001). Several biodegradable copolyesters resulting from the modification of PET with different content of terephtalic acid were prepared (Mohanty et al., 2000). Nevertheless, PBAT is a polymer whose production is dependent on petroleum-based materials and is associated with a high production cost, limiting its application in industry. Thus, the development of blends based on PBAT and other biodegradable polymers can impart the properties of the blend components to the final polymer. Among the biodegradable polyesters, PLA is a strong candidate to be blended with PBAT, because it is possible to reduce the overall cost and enhance the mechanical performance of the final PLA/PBAT blend (Shahlari and Lee, 2012). Shahlari and Lee (2012) blended PBAT and PLA and achieved good elongation at break, an enhanced Young's modulus, and reduced the overall cost of the produced material. Additionally, Zarrinbakhsh et al. (2013) added inexpensive biomass in poly(hydroxybutyrate-*co*-valerate) (PHBV)/PBAT to develop biodegradable composites with improved mechanical properties. Remarkably, the incorporation of some types of stearic, caproic, and lauric saturated fatty acids enhanced the degree of crystallinity of the biodegradable polymers prepared with PBAT, starch, and glycerol. Also, agar biopolymer has been used to get reinforced PBAT. The resulting biodegradable composites developed by injection molding and extrusion were characterized in terms of their thermal, physicochemical, and morphological properties. Another approach consisted in the preparation of PBAT-biodegradable nanocomposites by using organically modified silicates . The results showed that by the incorporation of the nanofiller, the thermal stability, and stiffness of PBAT increased, producing a more sustainable polymeric material (Someya et al., 2005).

3.4 Polylactic Acid

PLA is a biodegradable thermoplastic polyester produced both from fossil and renewable resources, that is nowadays one of the most promising polymers for commercially replacing LDPE,

HDPE, PET, and PS (Peelman et al., 2013). The production of PLA presents advantages over other synthetic materials: (1) PLA can be obtained from the processing of renewable agricultural sources, like corn, or other carbohydrate sources, into dextrose, followed by a fermentation step, into lactic acid (LA); (2) its production consumes CO_2, providing significant energy savings; and (3) PLA is recyclable and compostable (Drumright et al., 2000; Whiteman, 2000; Massardier-Nageotte et al., 2006).

Early economic studies have shown that PLA is an economically feasible material that can be used as a packaging material (Datta et al., 1995). PLA properties are determined by both the polymer architecture (stereochemical makeup of the backbone) and the molecular weight, the latter being controlled by the addition of hydroxylic compounds. PLA can exist as a racemic mixture, denominated as dl, or as two stereoisomers, designated as L and D. While the DL form is optically inactive, L and D forms are active. Both, poly(D-lactic acid) (PDLA) and poly(L-lactic acid) (PLLA) are semicrystalline, whereas poly(DL-lactic acid) (PDLLA) is completely amorphous (Vasanthan and Ly, 2009). The control of the polymer stereochemical architecture allows a precise control over the crystallization rate, the crystallinity degree, mechanical properties, and processing temperature (Auras et al., 2004). PLA is a biodegradable polyester with one of the highest melting temperatures, around 160–190°C.

Bacterial fermentation is used to produce LA from corn or cane sugar. However, LA cannot be polymerized as a useful product, because during the polymerization reaction, water molecules are generated, and its presence degrades the formed polymer chain. Thus, PLA of high molecular weight is produced by ring-open polymerization of LA using a catalyst, by solvent-free continuous process, and distillation method (Shah et al., 2008). This mechanism does not generate additional water, and thus, a wide range of molecular weights is achievable.

PLA is currently used in industrial packaging and biomedical applications, because the processing possibilities of this transparent material are very wide, ranging from injection molding to blow molding and thermoforming (Nair and Laurencin, 2006; Siracusa et al., 2008).

For commercial applications, due to the cost and brittleness of PLA, it is essential to blend it with other biodegradable polymers, to improve several properties and reduce its cost. Thus, some research studies were carried out on binary polymeric systems. As an example, Nijenhuis et al. (1996) blended PLA with up to 50% of polyethylene oxide (PEO), which results in blends with a single glass transition temperature (T_g), suggesting a compatibility of both components in the amorphous phase. Similarly, by blending PLA with polyvinyl alcohol (PVA) a single *Tg* was observed, which indicates that the PLA/PVA system is compatible (Gajria et al., 1996). Other experiments performed by Sheth et al. (1997) have shown that by blending PLA with polyethylene glycol (PEG) it is possible to obtain PLA/PEG-miscible blends, at a defined concentration of PEG. The biodegradability results, achieved from an enzymatic degradation, indicate significant decrease in the weight loss when compared with neat PLA. For a PEG content of 30%, a lower weight loss of the samples is essentially due to the enzymatic degradation of the PLA. However, for a content above 30% of PEG, the weight loss is mainly due to the PEG dissolution. A combination of both, degradation of PLA and PEG dissolution, occurs for PLA/PEG blends up to 30% of PEG. Zhang et al. (1995) also studied the miscibility of PLA and PCL blends. They observed that to a certain extent of miscibility it is possible to get fine morphology and also fast hydrolysis. At the same time, some works devoted to the synthesis of PLA-based biodegradable copolymers have been carried out. Cohn and Younes (1988) synthesized block copolymers by using PLA and PEO aiming to develop bioadsorbable polymeric materials to overcome the mechanical limitations and degradation most of the biodegradable polymers. It was demonstrated that the incorporation of PEO into the PLA resulted in materials with a high hydrophilic character, suitable equilibrium water contents around 60%. Poly(lactic-*co*-glycolic) acid (PLGA) is also a PLA-based biodegradable copolymer that, due its biodegradability and biocompatibility, has been one of the most attractive polymers for tissue engineering and drug delivery applications.

3.5 Starch

Starch is a carbohydrate that consists of a large number of glucose units joined together by glyosidic bonds containing, usually, 20–25% amylose and 75–80% amylopectin. It occurs widely in plants, like rice, corn, cassava, and potatoes. In all of these plants, starch is produced in the form of granules, varying in size and in composition, according to the plant used. Starch granules are hydrophilic and the water content of starch varies with relative humidity. While the branched amylopectin component contains crystalline areas, the linear amylose is mostly amorphous. Starch granules can be gelatinized in water at lower temperatures in alkaline solution and can be used as a thickening, stiffening, and gluing agent, giving wheat paste (Postek and Brown, 2009).

There are several degradable plastics made from starch (Narayan, 1994; Chandra and Rustgi, 1998; Lawton, 1996). For instance, a fully biodegradable starch-based polymer is prepared from corn or potato starch, along with smaller amounts of food-grade additives. The resin is suitable for manufacturing injection molded pieces, films, and starch-based loose-fill packaging material (Briassoulis, 2004). These pieces degrade in an active biological environment.

Starch has a wide variety of applications, including adhesives and industrial emulsions, construction, glass fiber, medical gloves, personal care, packaging, and in agriculture. The interest in this biopolymer has been recently renewed due to its abundance, low cost, availability, biodegradability, possibility of blending with conventional polymers, and the possibility to be processed using conventional polymer processing equipment, such as, extrusion and injection molding (Yew et al., 2005). However, starch by itself presents some limitations, owing to its hydrophilic character. In this context several studies have been carried out in order to associate thermoplastic starch with other biodegradable polymers to overcome the starch limitations. Among them, the most common are aliphatic polyesters, such as PLA, PCL, PHBV, poly(ester amide) (PEA), and poly(butylene succinate adipate) (PBSA), or even an aromatic polyester like PBAT (Schwach and Avérous, 2004). There are some starch-based blends, such as Mater-Bi (from Novamont, Italy), that are presently commercialized. The incorporation of the above biodegradable polymers improves the mechanical performance of plasticized starch. Even though blending is inexpensive and easier, its main disadvantage is the immiscibility of most of the polymers because of their distinct chemical structures. Therefore, another approach involves grafting other acrylic and vinyl monomers onto the starch polymer chain, increasing its hydrophilicity or hydrophobicity, depending on the conditions used.

Maharana and Singh (2006) have successfully prepared biodegradable PE by graft copolymerization of starch, and the biodegradability was evaluated. They proved that this approach allows high biodegradability, which increases with the amount of grafted starch. In other study Halley et al. (2001) synthesized starch-based biodegradable films based on starch-based biodegradable polymers mixed with high performance polyesters. The obtained films displayed excellent tensile strength and biodegradability. The biodegradation of the mentioned polymeric materials can occur by different pathways; when disposed in landfills, they undergo photodegradation by the action of catalysts, oxygen, water, microorganisms, sunlight, and enzymes of the soil (Saroja et al., 2000). Exceptional susceptibility to microorganisms' growth has been perceived in a humid environment for starch-graft copolymers. Thus, starch is a potential polymeric material that allows development a wide variety of products. It can be integrated in the polymer, thermoplastic starch (TPS), graft copolymers, among others. Since it is an inexpensive material, ecofriendly, and biodegradable, starch is a good candidate to replace conventional petroleum-based nondegradable plastics.

3.6 Cellulose

Cellulose is one of the most abundant polymer presents in nature, and is found as a structural component in the cell wall of plants and algae, and it can be also produced by various species of bacteria (Bishop, 2011). It is well known that cellulose is a very important and attractive biopolymer and an almost inexhaustible and renewable raw material, because it presents unique advantages and properties, including biodegradability, recyclability, renewability, biocompatibility, relatively high resistance, and stiffness (Dai et al., 2013). Even though wood is the most available raw source of natural cellulose, other sources have also been explored (eg, sisal, cotton, ramie) that can be used as raw materials for several products in a wide range of industries. These include coatings, laminates, optical films, and sorption media, additives to develop construction materials, composites, nanocomposites, and pharmaceuticals (Klemm et al., 2005; Moon et al., 2011).

The particular structure of cellulose provides interesting properties. The repeating units of glucose that constitute this macromolecule can be organized in different ways, resulting in its specificity, reactivity, and diversity of architectures (Moon et al., 2011). Its properties and specificity for a given reaction depend on several factors, such as, the process used to isolate the cellulose, chain lengths, its distribution, number of inter- and intramolecular hydrogen bonds, distribution of functional groups within the repeating units, crystallinity, and the degree of polymerization, which render cellulose as a unique material (Poletto et al., 2013; Åkerholm et al., 2004; Oh et al., 2005).

As an example, cellulose derived from wood pulp has molecules with 300–1700 repeating units, whereas cellulose from other plant fibers, cotton, as well as from bacteria, has polymer chains with 800–10,000-repeating units. Despite its hydrophilicity, the major drawback is related to its water-insolubility, as well as, in most of the organic solvents (Pingali et al., 2010; Onda et al., 2008; Nishiyama et al., 2002; Zhao et al., 2007). Only treatment with concentrated acids, at high temperatures, or even using Schweitzer reagents breaks down cellulose into its glucose units (Stenius, 2000). These solvents are used in the production of regenerated cellulose from dissolving pulp. Also, the dissolution process of cellulose can be significantly improved with many kinds of ionic liquids (Wang et al., 2012). For this reason, only a very small fraction of cellulose is used for the production of commodity materials and chemicals.

Cellulose can easily react with other synthetic polymers, because it has many hydroxyl groups. Examples of some studied cellulose-based blends are cellulose with PCL, which are partially compatible (Nishio and Manley, 1990), and blends of poly(4-vinylpyrrolidone) with cellulose in different ratios (Masson and Manley, 1991). Other authors reported the study of blends based on PVA and cellulose (Masson and Manley, 1991). The properties of resulting blends of cellulose and PE were also studied. It was demonstrated that the biodegradability ratio increases (Kaczmarek and Ołdak, 2006). According to the results, it was concluded that the LDPE/cellulose blends are less biodegradable, due to the fact that the film's degradation occurs on the surface, under environmental conditions. Rogovina et al. (2013) developed LDPE/cellulose and LDPE/cellulose/PEO (or chitosan, chitin) binary and ternary blends, respectively. It was demonstrated that the incorporation of PEO or the polysaccharides (chitin or chitosan) as a third constituent enhances the biodegradability.

3.6.1 Cellulose From Plants and Bacterial Cellulose

Even though cellulose is the most abundant naturally occurring polymer of glucose, found as the main constituent of plants and natural fibers, it can be also synthesized by specific bacteria, such as, those of the genera *Gluconacetober*, *Agrobacterium*, *Aerobacter*, *Azotobacter*, *Rhizobium*, *Salmonella*, and *Acetobacterxylinum* (Ross et al., 1991). Both, bacterial cellulose (BC) and plant cellulose (PC) are chemically similar, though their macromolecular structure and physical properties are different (Czaja et al., 2007) (Scheme 7.1).

For both BC and PC, the units of glucose units are kept together by 1,4-β-glucosidic linkages, which determines the crystallinity degree of cellulose, being above 60% for BC and 40–60% for PC (Sannino et al., 2009) that contributes to its insolubility in common solvents and water, as previously referred. The fibers produced have reduced size when compared with BC are smaller than PC fibers and displays a fiber network with high tensile strength and water-retention capability compared to PC. Additionally, BC is purely contrasting with PC that is it is generally related with other components, including pectin and lignin (Keshk, 2014). Thus, while BC can be used without any further purification after synthesis by bacteria, PC requires additional modification and purification. Concerning its chemical modification, it is usually accomplished by etherification or esterification of the hydroxyl groups, aiming to

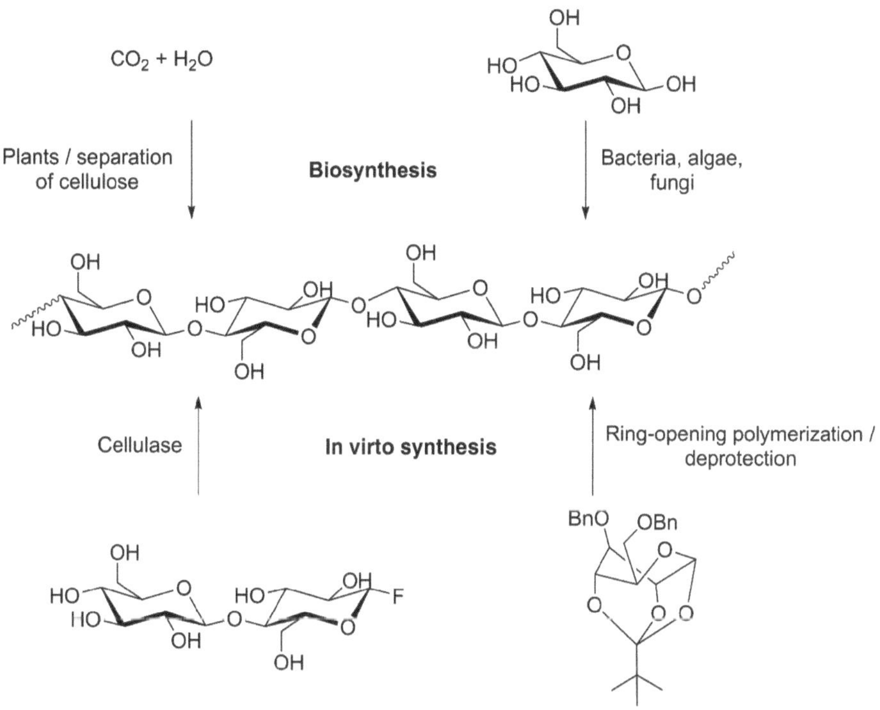

Scheme 7.1. Pathways to cellulose formation.

produce cellulose derivatives. These derivatives are more processable and find large applications at the industrial level due to the fact that cellulose and its derivatives are biodegradable and eco-friendly as are degraded by many microorganisms present in the environment (Tomšič et al., 2007). Several studies were carried out to evaluate their biodegradation. It was found that during this process a significant decrease in the molecular weight occurs, leading to the deterioration of the mechanical performance. Also, higher biodegradability of cellulose is confirmed by a decrease of the crystallinity degree and enhanced solubility in water (Miyamoto et al., 1989). Despite the advantages of BC compared to PC, it is well known that the latter has been used as a major source of raw materials and products.

3.6.2 Cellulose From Agro-Wastes

Cellulose from vegetable agro-wastes is an example of plant-derived polymers that shall be exploited for new bioplastics production. These materials should have unique properties of considerable interest to replace conventional plastics. Herein, the direct synthesis of new bioplastics, from vegetable agro-wastes and tailoring their properties by a grafting approach using controlled radical polymerization (CRP) chemistry, is a challenge.

Along with the environmental pressures and carbon footprints issues that surround traditional petroleum-based synthetic polymers, there is a possibility to replace some of these by biopolymers directly by synthesized agro-wastes.

This chapter provides an overview on the development of biodegradable materials based on cellulose, focusing on cellulose derived from vegetable agro-wastes. Green chemistry aspires to design chemical products that are fully effective, yet have little or no toxicity, which break down into innocuous substances if released into the environment after use, and/or that are based on renewable feedstocks, such as agricultural wastes. An overview of the green chemistry methodologies that have been used to extract natural polymers namely cellulose, from vegetable agro-wastes will be pointed out.

4 Extraction of Cellulose From Food Agro-Wastes

The high amount of agro-wastes generated from the food industry has been a growing concern, because it generates enormous amounts of inedible residues, originating from processed edible vegetables and cereals, all over the world. In Europe, more than 24 million tons of processed vegetable waste are generated and worldwide around 1.3 billion tons are left over per annum (Gaiker, 2004), which becomes an environmental and financial issue (Bayer et al., 2014).

Cellulose exhibits unique properties and can be produced from plants and agro-wastes, but its production with high purity from vegetable agro-wastes is still a challenge. Owing to its highly structured intermolecular hydrogen-bonding network, this polymer cannot be melted or dissolved by standard processes, such as thermoforming or dissolution (Pingali et al., 2010; Onda et al., 2008; Nishiyama et al., 2002; Zhao et al., 2007). For this reason, cellulose is mainly used in industrial applications in the form of cellulose derivatives, including cellulose esters or ethers. Conventional processing methods cannot be used, leaving industry to resort to cellulose derivatives, consuming extra time and costly chemical-purification steps to regenerate it (Kraemer, 1938). Trifluoroacetic acid (TFA), a natural organic acid with low toxicity, is biodegradable by microbial action and is capable for extraction of cellulose bioplastic (Kim et al., 2000). Bayer et al. (2014), used TFA as a unique solvent to cosolubilize cellulose with other

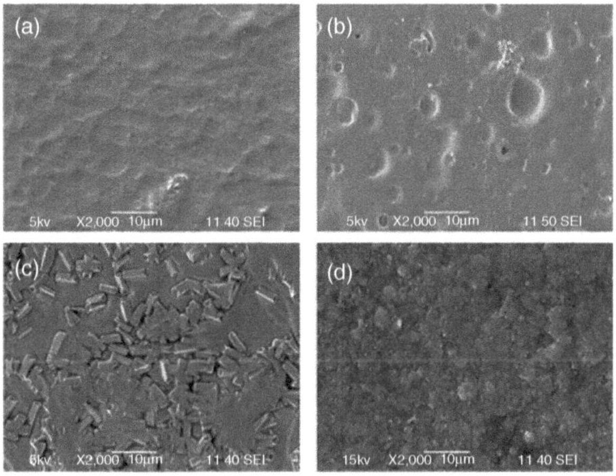

Figure 7.4. SEM images of the surfaces of (a) parsley, (b) rice, (c) spinach, and (d) cocoa samples, respectively. From Bayer et al. (2014).

organic-contained matter to obtain biopolymers from lignocellulosic biomass. Using this methodology, cellulose-based bioplastics from wastes of various edible vegetables and cereals (parsley and spinach stems, cocoa pod husks, and rice hulls) with diverse bioorigins, were prepared without requirement the of cellulose regeneration.

SEM images (Fig. 7.4) show different morphologies for all cereals thin films. For parsley and rice samples (Fig. 7.4a and b) small islands of inorganic aggregates are detected, while for spinach (Fig. 7.4c) film self-assembled embedded crystallites are evident. The new biopolymers exhibit broadly tunable-mechanical properties that are dependent on the biosource.

Bioplastic films, in comparison with commonly engineering polymers (Fig. 7.5), such as, PP, PE, polyester, elastomers (eg, silicone and PU elastomers), starch-based polymers, starch-polymer blends, and biodegradable polyesters (eg, PCL and PLA), have a mechanical performance between synthetic polymers and biopolymers.

Fig. 7.5b presents a comparison of ultimate tensile strength (UTS) versus Young's modulus. Amorphous neat cellulose films exhibit high UTS at high Young's moduli comparable to PET, while rice, spinach, and parsley biodegradable films have a similar behavior to elastomers and some thermoplastics, such as, LDPE. Bioplastics produced from cocoa-pod husk are comparable with petroleum-based thermoplastics, for instance, HDPE, and PP. The production of biopolymers with properties among those of PP and PET is also feasible. For example, blending microcrystalline

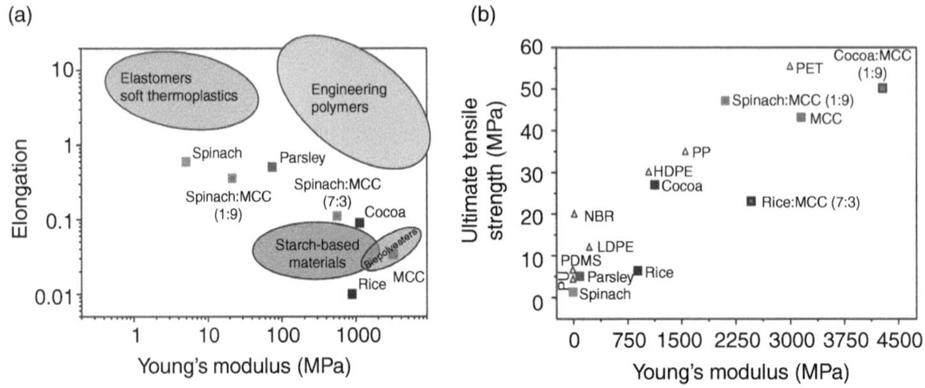

Figure 7.5. (a) Elongation versus Young's modulus data where the property domains of elastomers/soft thermoplastics (PU, rubbers, and silicone polymers), engineering polymers, (PP, PE, PS, polycarbonate, etc.), biopolyesters, and starch-based materials [PLA, poly(3-hydroxybutyrate), PCL, modified starches, etc.] are marked for comparisons, and (b) ultimate tensile strength versus Young's modulus data. From Bayer et al. (2014).

cellulose with spinach produces new materials with properties ranging between both components. This demonstrates that an extensive variety of materials with specific mechanical properties can be developed, by blending vegetable waste bioplastics that, most of the time, cannot be accomplished by solution- or even thermoform blending of conventional polymers, owing to its incompatibility.

Macedo et al. (2015) investigated the transformation of different agro-wastes, such as, raw parsley, carrot, and cabbage, by using TFA as a solvent to obtain biodegradable materials (Fig. 7.6).

Concerning the lignin component of the bioplastics, the FTIR spectra of the derived plastics from the earlier mentioned agro-wastes, indicates the existence of characteristic lignin bands in all cases, as shown in Fig. 7.7. The signal at 1740 cm^{-1} is attributed to C=O stretching of acetyl or carboxylic acid, which are also present in the structure of lignin (Dong et al., 2009). The signals at 1610–1595 cm^{-1} are attributed to the C=C aromatic stretching bands that are common to lignin. The peak that appears at 1465 cm^{-1} can be attributed

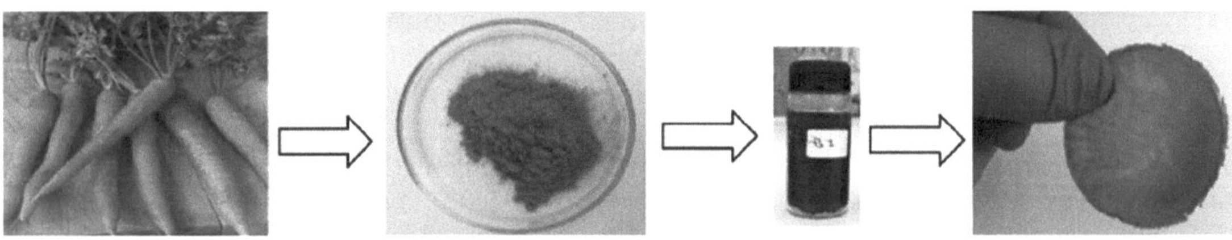

Figure 7.6. Biopolymer film production from raw of parsley wastes.

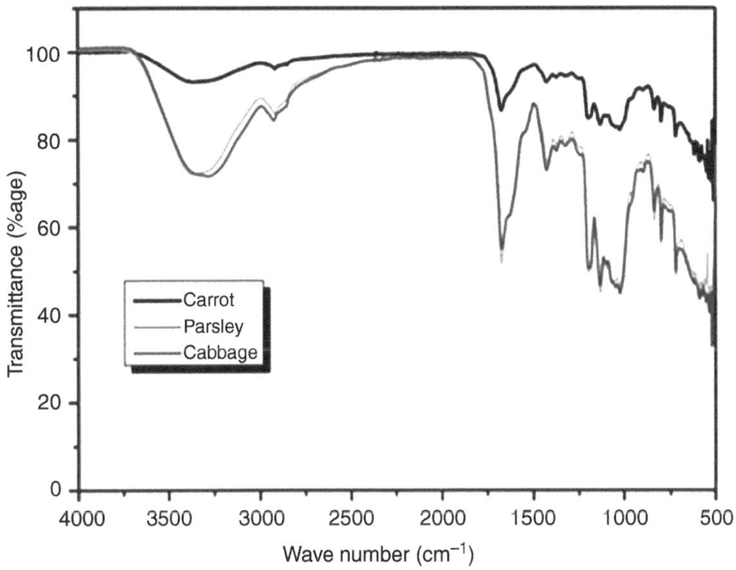

Figure 7.7. FTIR spectra of bioplastics after TFA digestion. From Macedo et al. (2015).

to the asymmetric bending of lignin CH_3 bonds. The results demonstrate that TFA can partially dissociate or break down lignin, which can be dispersed into the bioplastic matrix in its new forms.

The mechanical properties of the produced bioplastics were comparable to many synthetic polymers, in view of the substitution part of the latter, with degradable bioplastics having matching properties.

All of the earlier results indicate that agro-wastes rich in cellulose can be transformed into bioplastics by simply aging them in TFA solutions regardless of their bioorigin. Nevertheless, depending on the plant species, biopolymers were found to display diverse mechanical properties, ranging from brittle and rigid to soft and stretchable. Blending these vegetable waste solutions with TFA solutions of pure cellulose, all-natural plasticization of amorphous cellulose is achieved. These new bioplastics can replace many nondegradable plastics, preserving the environment.

Even though TFA demonstrated promising results, organic salts so-called ionic liquids (ILs), have received attention as promising green solvents for lignocellulose pretreatment or fractionation. If these solvents are efficiently recycled, the overall waste production of the process decreases, making this a green feature. However, some ILs are still synthesized using conventional organic solvents, which detracts from their overall green chemistry value. This is an issue that needs to be addressed in research. The

Table 7.2 Solubility of Cellulose in Different Ionic Liquids (ILs) Using Conventional Heating or Microwave Irradiation

Ionic Liquid	Heating Method	Solubility
[BMIM]Cl	Heating at 100°C	10%
[BMIM]Cl	Heating at 70°C	3%
[BMIM]Cl	Heating at 80°C and sonication	5%
[C6MIM]Cl	Heating at 100°C	5%
[C8MIM]Cl	Heating at 100°C	Poor
[BMIM]Cl	Microwave irradiation with pulses of 3–5 s	25%
[BMIM]Br	Microwave irradiation	5–7%
[BMIM]SCN	Microwave irradiation	5–7%
[BMIM]PF4	Microwave irradiation	None
[BMIM]PF6	Microwave irradiation	None

From Swatloski et al. (2002).

study by Swatloski et al. (2002) can be seen as a cornerstone for the dissolution of cellulose by ILs. They demonstrated that cellulose (Table 7.2) could be dissolved at high concentrations in imidazolium-based ILs.

Swatloski et al. (2002) suggested that the major contributor for the solubility of cellulose in 1-butyl-3-methyl-imidazolium chloride ([BMIM]Cl) over conventional solvents is the high chloride concentration that allows rapid and efficient dissolution, by breaking the prevailing hydrogen-bond network. When the length of the alkyl chain increases from C6 to C8, as in [C6MIM]Cl and [C8MIM]Cl, the solubility of cellulose decreased.

Most of the research studies covered in this chapter are concentrated purely on the synthesis of cellulose from plants using ILs. The first study on cellulose dissolution in ILs was published by Swatloski et al. (2002). ILs based on BMIM cation with diverse anions was tested as a suitable solvent for cellulose. The results showed higher efficiency of cellulose dissolution when chloride anion was present when compared with noncoordinating anions. Other studies regarding the solubility of cellulose using ILs, such [BMIM]Cl containing halide counter ions or even counter anions, such as, acetate, formate, and acetate, have been described in the literature (Swatloski et al., 2002; Gericke et al., 2012). However, ILs containing halide counter anions present some drawbacks associated with their high viscosity, which reduces cellulose dissolution. Contrarily, ILs liquids with other anions, such as, phosphate, acetate, and formate show reduced viscosity compared to the former, which makes them more suitable to be used in several applications (Gericke et al., 2012).

Nowadays, commercial cellulose is produced using 1-ethyl-3-methylimidazolium acetate ([EMIM]OAc) IL, which exhibits reduced viscosity and enhanced cellulose dissolution capability. In several studies published on ILs, it was assumed that the anion used has a crucial role, being responsible for the cellulose dissolution, while the cation has a minor role. Nevertheless, other studies have demonstrated that the dissolution process is affected by both, the anion and cation structure. Accordingly, acidic protons present in the heterocyclic rings enhance the solubility considerably, which is explained by the formation of hydrogen bonds with cellulose, through its hydroxyl and ether oxygens (Andanson et al., 2014). Although ([EMIM]OAc) IL exhibited a solubility of 23 g mol^{-1} at 40°C, modifying its structure to 1-methoxyethyl-3-methylimidazolium resulted in a substantial decrease of the solubility (8 g mol^{-1}) at the same temperature. Currently, cellulose solubility has reached 14.5 wt.% using 1-allyl-3-methylimidazolium chloride at a temperature of 80°C (Zhang et al., 2005) and 16 wt.% applying 1-ethyl-3-methylimidazolium acetate IL at 90°C (Zavrel et al., 2009). However, the solubility can be improved till 25 wt.% by microwave heating. All these studies demonstrate that the effective cellulose dissolution using ionic liquids depends on both, that is, ILs should comprise cations including resilient acidic protons without atoms with high electronegativity and anions with resilient-hydrogen bond acceptability.

Wang et al. (2011) dedicated their study to the cellulose extraction from pine, poplar, Chinese parasol, and catalpa wood chips in an ionic liquid 1-allyl-3-methylimidazolium chloride ([AMIM]Cl). The results show that pine is the most suitable wood species for cellulose extraction with ILs, and the extraction rate reached 62% under optimized conditions and its cellulose content was 85% when DMSO/water was used as the precipitant (Wang et al., 2011). Teck Nam et al. (2011) studied the efficiency of ionic liquids in the dissolution of rice husk. They noticed that the ILs that dissolved cellulose provided an option to pretreat lignocellulosic biomass. The ionic liquids 1-ethyl-3-methylimidazolium diethyl phosphate ([EMIM]DEP), [BMIM]Cl, and [EMIM]OAc, were able to dissolve rice husk cellulose to varying extends, with the highest dissolution (36.7%)

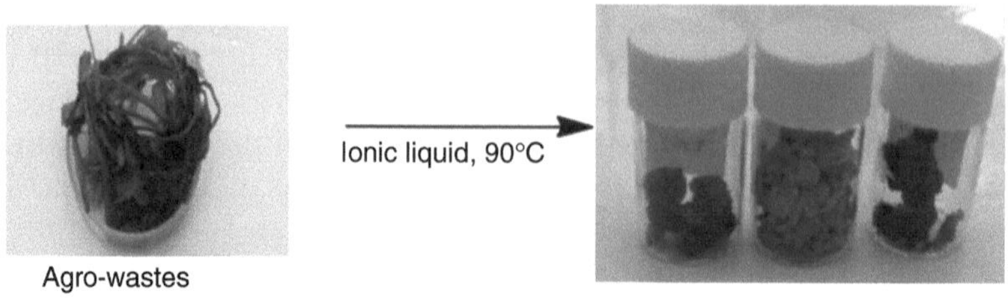

Figure 7.8. Cellulose (and other products) extracted from different food agro-residues: raw products of carrot, parsley, and cabbage. From Macedo et al. (2015).

given by [EMIM]OAc. It was observed that the regenerated cellulose was more amorphous when compared to untreated-rice husk.

Abushammala et al. (2015) reported the direct extraction of cellulose nanocrystals from wood by means of a [EMIM]OAc treatment. A native cellulosic product was recovered with 44% yield with respect to wood cellulose content. A cosolvent, DMSO, was also included to increase cellulose dissolution. It was demonstrated that the addition of an organic solvent increases the potential of the IL, reducing the time necessary for dissolution, independently of the used temperature (Andanson et al., 2014; Ries et al., 2014). Owing to the capabilities demonstrated by ILs to dissolve cellulose, several ILs have been studied not only to dissolve the cellulose but also to functionalize it.

As mentioned, ILs are used to make cellulose prone to hydrolysis (FitzPatrick et al., 2010), however, with exception of Macedo et al. (2015) no one has attempted to obtain biopolymers from lignocellulosic biomass from food agro-wastes through ILs reaction as illustrated in Fig. 7.8.

According SEM micrographs (Fig. 7.9), it is possible to observe the existence of different shapes and size distribution of the cellulose microfibers. Regarding Fig. 7.9b and d, cellulosic microfibers are clearly composed by multiple fibers. These results indicate that cellulose was successfully extracted from both, raw of carrot, parsley using [BMIM]Cl ionic liquid (Macedo et al., 2015), even though some contamination is caused by the ionic liquid inducing some agglomeration of the microfibers.

5 Cellulose Modification

5.1 Preparation of Blends of Cellulose and Biodegradable Polymers

As stated before, the most abundant biomacromolecule, cellulose, has attracted considerable attention in recent years due to

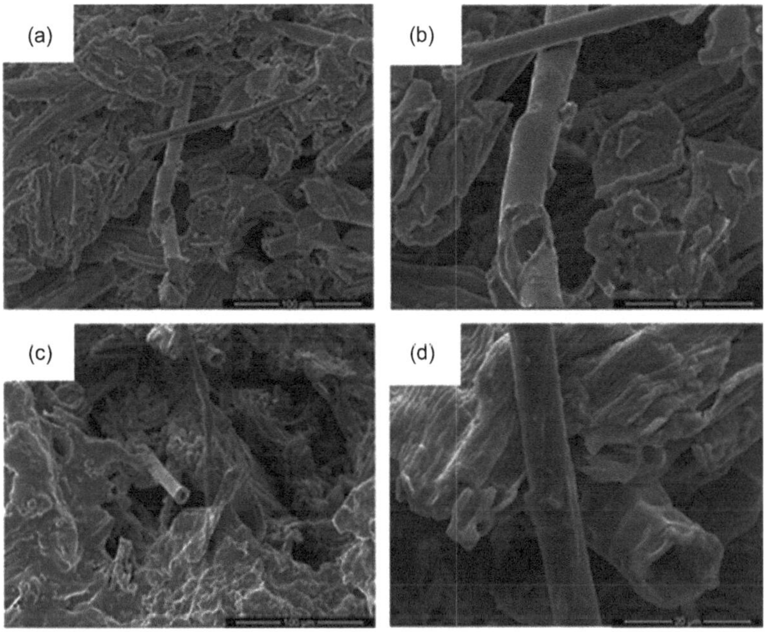

Figure 7.9. SEM micrographs of cellulose (and other products) extracted from (a) and (b) raw of carrot and (c,d) raw of parsley.

its biodegradability, biocompatibility, and renewable properties (Meng et al., 2009). Several studies reported the addition of cellulose into a biodegradable polymer matrix as a way to improve the final properties of the developed material. The most pronounced effect of this consists of the reduction of water vapor permeability, but also several mechanical properties, like tensile strength and Young's modulus, were significantly affected, mainly the elongation at break. Dias et al. (2011), observed that the addition of cellulose fibers were well covered in the plasticized starch-polymer matrix, which is attributed to the chemical resemblance between starch and natural fibers that provides a strong interaction (Avérous et al., 2001). The effect of the incorporation of cellulose fibers on the water permeability of starch films is related to the highly crystalline and hydrophobic character of the fibers. The addition of cellulose fibers decreases the water vapor transmission by increasing the diffusion path length through the film. However, this effect can be reversed by the addition of excess fiber, causing congregation (Dias et al., 2011). Müller et al. (2009) also found reinforcement of starch films lowered water vapor permeability after incorporation of cellulose fibers. Also, application of carboxymethyl cellulose (CMC) in starch-based films showed a decrease of the water vapor permeability with increasing CMC content

(Ghanbarzadeh et al., 2011). Gáspár et al. (2005) also tested the effect of different natural additives, such as, cellulose and hemicellulose, on the water resistance of a starch-based film (70% corn starch and 30% glycerol). Hemicellulose filled starch-based film showed better tensile strength and modulus than the pure film, but elongation of all four filled films was lower than the pure TPS. Water resistance of cellulose and hemicellulose films was better than pure TPS (Sanchez-Garcia et al., 2008). Authors have studied the effect of cellulose fibers in PHBV films. Water permeability decreased by 52% for 10 wt.% fiber content and by 71% for 1 wt.% fiber content. Petersson and Oksman (2006) devoted their work to the incorporation of microcrystalline cellulose. The results showed no reduction in the oxygen permeability of PLA. Conversely, Sanchez-Garcia and Lagaron (2010) found a 90% oxygen-permeability reduction for PLA after addition of 1, 2, 3, or 5 wt.% cellulose nanowhiskers. Thus, several studies on cellulose-based films showed that they could be used as an alternative for packaging several food products. Popa and Belc (2007) stated that paper and board, based on cellulose, are the most widely used renewable packaging materials nowadays.

Similarly, a lot of cellulose derivatives are produced commercially, on which cellulose acetate (CA) is the most commonly used for food packaging (fresh produce, baked goods). Makino and Hirata (1997), found that a laminate of chitosan–cellulose and PCL had a similar film permeability as LDPE, which makes this laminate a possible material for packaging for fresh produce. Popa and Belc (2007) found that coating of cellophane with nitrocellulose or PVDC (polyvinylidine chloride) improved barrier properties and this film could be used for packaging of candies, processed meat, cheese, and baked goods. Furthermore, Rhim and Kim (2009) stated that paperboard coated with PLA could be used as a substitute for PE-coated paperboard in the manufacture of paper cups or containers for high moisture foods (beverage cartons and ice cream containers).

Although many studies reported the development of cellulose derivatives and their potential applications, more research is required to explore more environmentally friendly methodologies to extract cellulose from renewable sources and to overcome the inherent difficulty in the modification of its properties associated with the reduced availability of common solvents to dissolve cellulose (Swatloski et al., 2003). Traditional cellulose-dissolution processes are often expensive and require the use of unusual solvents, typically with high ionic strength and relatively harsh conditions (Wu et al., 2004). Moreover, these processes sometimes cause severe environmental problems because they cannot be recovered and reused (Zhang et al., 2005). In recent years, the rational green-comprehensive utilization of cellulose resources has received great attention (Shengdong et al., 2005). The common methodologies used for cellulose extraction and dissolution processes are fronting challenges due to energy and environmental problems (Shengdong et al., 2005; Zhu et al., 2005). Therefore, to make cellulose resources useful, it is necessary to develop green cellulose extraction, suitable cellulose dissolution approaches, and novel modification methodologies.

5.2 Cellulose Modification Using Ionic Liquids

Cellulose-based materials are currently a large renewable resource to reduce society's dependence on nonrenewable petroleum-based feedstocks (Zhu et al., 2006a). Therefore, to improve the properties of biodegradable/biobased materials, the addition of cellulose has been the focus of many research studies. The most investigated effect was the development of plastics with lower water vapor permeability, as well as, the enhancement of mechanical properties of the industrialized materials.

Homogeneous-cellulose derivatizations have been successfully obtained by using common solvent systems, such as reaction media, which are used to process cellulose at an industrial scale (Barthel and Heinze, 2006; Heinze et al., 2005; Köhler et al., 2008; Edgar et al., 1995; Liebert and Heinze, 2001; Heinze and Schaller, 2000; Regiani et al., 1999). However, due to their toxicity, together with the partial recovery, they have limited the use of conventional solvents for this purpose. Owing to their complex structure, to use cellulose as a green material, many research efforts have been made (Zhu et al., 2006b). As previously mentioned, cellulose can be dissolved in ILs and it can be easily also be regenerated from them. Because they are chemically and thermally stable, nonflammable, and have low vapor pressure, they are called green solvents. This has been providing a new platform for the green-comprehensive utilization of cellulose resources. Thus, the dissolution of cellulose with ILs is reviewed.

Besides cellulose dissolution, ILs can also be used for cellulose functionalization and its chemical modification. Though specific attention has been given to the acetylation of cellulose due to its wide variety of applications, other functionalizations to produce other cellulose derivatives (cellulose butyrate, cellulose phthalate, and cellulose benzoate) have been performed (Liu et al., 2007; Zhang et al., 2009; Schenzel et al., 2014).

In addition to esterification and etherification reactions, cellulose derivatization with various functional groups, including

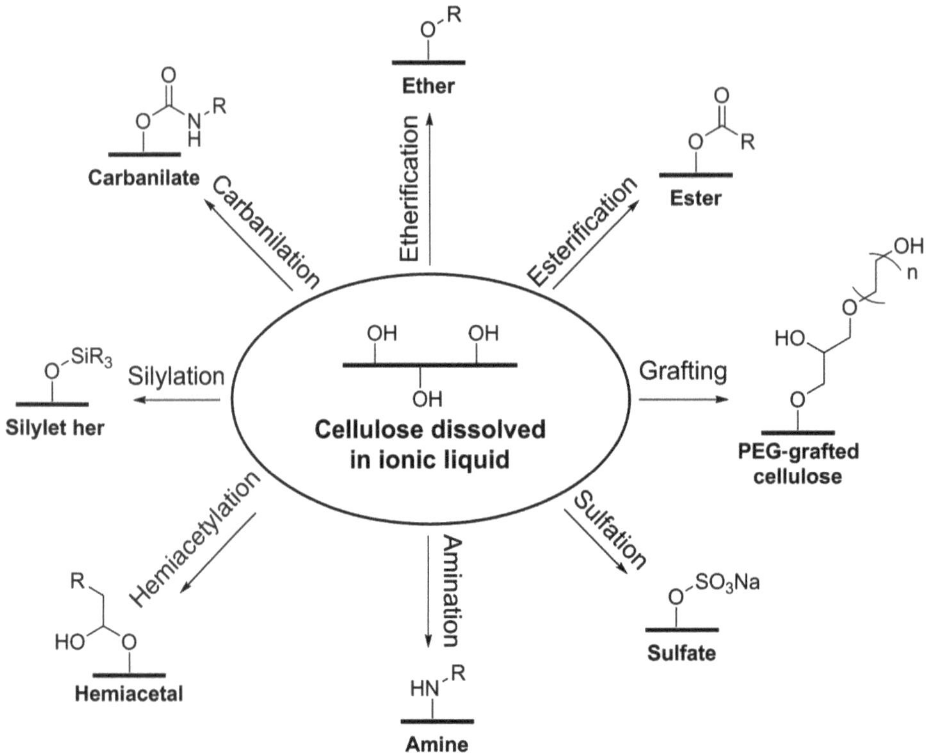

Figure 7.10. Chemical modifications of cellulose using ILs. From Isik et al. (2014).

amines, silylates, sulfonates, sulfates, and carbanilates has been also effectively accomplished by modification of cellulose hydroxyl groups (Fig. 7.10) (Erdmenger et al., 2007; Köhler et al., 2008; Liebert et al., 2006; Song et al., 2008). Tome et al. (2011) used tetradecyltrihexylphosphoniumbis(trifluoromethylsulfonyl)imide ([TDTHP][NTF_2]) to solubilize and functionalize cellulose with anhydrides. Using this IL, they were able to catalyze the esterification reaction. As a result, depending on the anhydride applied in the modification process the grafted fibers showed greater surface hydrophobicity. Moreover, the grafted-fibers characterization demonstrated that the esterification occurs at the fibers' peripheral layers, corroborating that the ultrastructure was not affected during the process.

5.3 Cellulose Modification Using by Grafting via Controlled Living Radical Polymerization

Another fascinating way to modify cellulose consists in the grafting of polymers by branching into cellulose, in order to impart new properties, retaining its intrinsic properties (Semsarilar et al., 2010). Grafting reactions of cellulose and its derivatives has been widely studied to develop new materials, allowing the combination of properties of one or more polymers in one new material. Thus, it is possible to attain properties, such as, temperature responsiveness (Chauhan et al., 2000), hydrophobicity and oleophobicity, flexibility (Guo et al., 2013), sorbancy (Chauhan et al., 2000; Gupta and Khandekar, 2003; Waly et al., 1998; El-Salmawi et al., 2001), ion-exchange capability (Waly et al., 1998), and polarization (O-Rak et al., 2013). These materials are promising candidates to replace the conventional petroleum-based materials that are currently available in the market. Some examples of their applications include their use in drug delivery, as sorption agents, in coatings, and in membranes (Nishio, 2006).

Several methodologies are used for graft copolymerization of several monomers on the cellulose backbone, including free radical polymerization (ROP), nitroxide-mediated polymerization, reversible addition–fragmentation chain transfer (RAFT) polymerization, and atom transfer radical polymerization (ATRP) (Meng et al., 2009).

Cellulose was chemically modified by ROP. Usually, cyclic monomers (eg, epoxides, lactones, and anhydrides) are used along with cellulose, which are dissolved in an appropriate IL. The hydroxyl groups of cellulose are subsequently used to accomplish ROP of the monomers to graft polymers onto the cellulose polymer chains (Guo et al., 2013). Hufendiek et al. (2014) converted cellulose into a macro-RAFT agent in [BMIM]Cl. Cellulose-*g*-polyacrylamide copolymers with temperature-responsive properties was prepared using the previous marco-RAFT and polyacrylamide. ATRP was also used to graft, by polymerization of vinyl monomers onto cellulose and cellulose derivatives in a living/controllable way. Synthesis of cellulose-based graft copolymers by ATRP has been carried out in either heterogeneous- or homogenous-reaction medium. The heterogeneous ATRP was mainly used for surface modification of solid-cellulose materials, such as, paper or film (Carlmark and Malmström, 2002; Lindqvist and Malmström, 2006; Lee et al., 2004; Nystrom et al., 2006), fibers (Carlmark and Malmström, 2003; Plackett et al., 2005), and particles (Lindqvist and Malmström, 2006; Coskun and Temüz, 2005). The homogenous reactions, graft polymerization via ATRP, were mostly performed on cellulose derivatives (Chang et al., 2008; Shen and Huang, 2004), such as CA (Vlček et al., 2006), ethyl cellulose (EC) (Tang et al., 2007), and hydroxypropyl cellulose (HPC) (Östmark et al., 2007). However, a few studies were also found on homogeneous graft polymerization via ATRP on underivatized cellulose backbone without any protection of the reactive groups. Direct grafting of polymers, for example, synthetic ones, on the underivatized-cellulose backbone, in a homogenous reaction medium, is expected to generate a new kind of polymer that combines the advantages of natural- and synthetic-polymers properties. Graft polymerization via ATRP in homogenous reaction media requires the synthesis of macroinitiator based on cellulose. Nevertheless, for underivatized cellulose this has been restricted by the lack of effective solvents. Bontempo et al. (2006) have reported a versatile ATRP method for polysaccharide grafting, in homogeneous conditions, without using protecting group chemistry. *N,N*-dimethylformamide (DMF) was used as a reaction medium for the synthesis of polysaccharide-based macroinitiators and water/DMF mixtures with different compositions as the polymerization medium for ATRP. However, this method cannot be directly applied for underivatized cellulose because it is insoluble in this solvent system and other common solvents. Wu et al. (2004) found out that the IL, [AMIM]Cl, at room temperature, was a powerful nonderivatizing solvent for cellulose and it was found as a desirable medium for the cellulose acylation. Meng et al. (2009) synthesized a macroinitiator, cellulose 2-bromoisobutyrylate, through homogeneous acylation of underivatized cellulose with 2-bromoisobutyryl bromide in ionic liquid [AMIM]Cl. Then, the cellulose-based macroinitiators were used for graft copolymerization of methyl methacrylate (MMA) and styrene.

Morphological studies of cellulose-g-PMMA copolymer (Fig. 7.11) show that the average diameters of sphere-like particulates were roughly 50 nm. Using this new method for graft modification of cellulose directly onto its backbone under a homogeneous condition has been successfully carried out. The

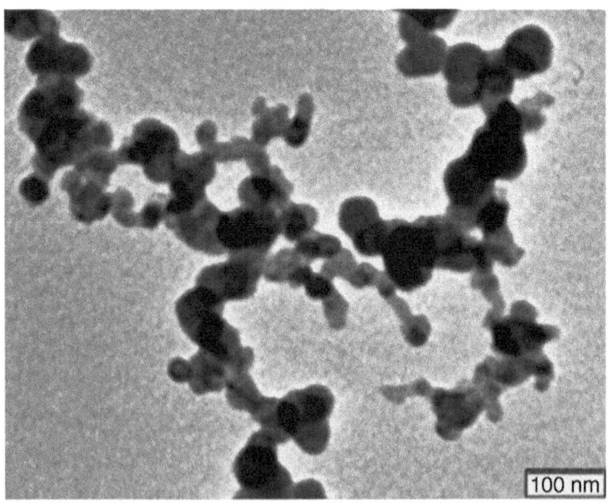

Figure 7.11. TEM pictures for cellulose-*g*-PMMA. From Meng et al. (2009).

macroinitiator for ATRP was synthesized by direct homogenous acylation of cellulose in [AMIM]Cl.

Moreover, RAFT polymerization has been employed to prepare cellulose-*g*-copolymers using (soluble) cellulose derivatives (eg, hydroxyisopropylcellulose and ethyl hydroxylethyl cellulose), which are commercially available, for subsequent modification with variable polymer side chains [eg, PS, poly(vinyl acetate), poly(ethyl acrylate), PNIPAM, polyacrylamide] using RAFT polymerization (Stenzel et al., 2003; Hernández-Guerrero et al., 2005; Fleet et al., 2008; Hiltunen et al., 2009).

Recently Hufendiek et al. (2014) synthesized well-defined temperature responsive cellulose-g-polyacrylamide copolymers in a grafting RAFT approach in an IL (Scheme 7.2).

The method of synthesis of cellulose-*g*-polyacrylamides, employing RAFT polymerization, takes place in a controlled fashion and may be transferred to the synthesis of a variety of other

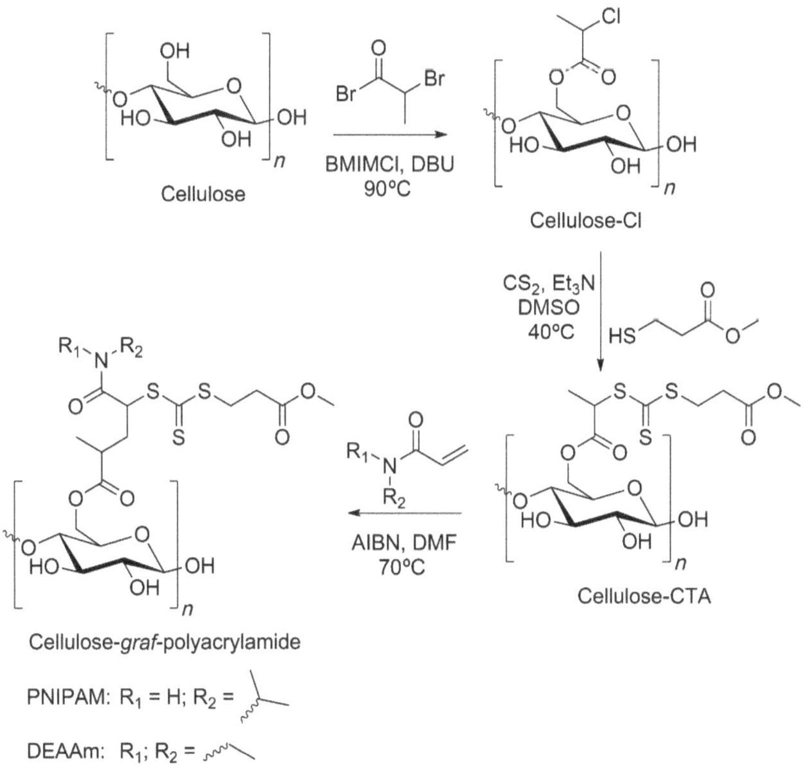

Scheme 7.2. Synthetic pathway for cellulose-*g*-polyacrylamide copolymers via RAFT-polymerization. DBU, 1,8-Diazabicycloundec-7-ene; AIBN, Azobisisobutyronitrile; PNIPAM, Poly(*N*-isopropylacrylamide); CS2, Carbon disulfide; Et3N, Triethylamine. From Hufendiek et al. (2014).

cellulose-*g*-copolymers. The result provides a benign and efficient access route to produce cellulose-hybrid materials, introducing a synthetic-platform technology that may be exploited for the employment of cellulose materials in applications where a stimuli–response is of importance.

These new copolymers developed using different routes are at this time new candidates for a wide field of applications, such as drug delivery. Moreover, from the mentioned examples, the dissolution of cellulose with an IL in moderate conditions enables the accomplishment of homogeneous reactions, resulting in different materials with several properties, applying a green-chemistry methodology.

Agriculture crops have been used as alternative sources of biodegradable polymers, but the agricultural potential of Earth is limited, therefore, attention is necessary to avoid competition with food crops. In this context, agro-food waste industry appears as a sustainable and continuous source of bioplastics. Nevertheless, there no scientific investigation reporting the extraction of cellulose from agro-wastes using ILs and the further chemical modification by grafting via CRP in ILs extraction mediums to achieve cellulose biopolymers.

Cellulose fibers have been much explored as a reinforcing material in biopolymer-based matrix material, such as PLA (Siqueira et al., 2010). The challenge lies in dispersing these fibers uniformly in the PLA matrix. However, due to the presence of hydroxyl groups on their surface, cellulose fibers have the tendency to stick together or agglomerate resulting in crack formation and failure of the composite. To overcome this issue, some scientific research has been made using variations of the dispersion methods (Eichhorn et al., 2010). Several methodologies have been explored to improve the cellulose fibers-dispersion mechanism, including the use of surfactants (Oksman et al., 2006), acetylation (Braun and Dorgan, 2009), silylation (Pei et al., 2010), and grafting (Pracella et al., 2010).

Recently, cellulose extracted from food agro-wastes was grafted into PLA as the base matrix via esterification reaction, according to Scheme 7.3.

SEM images were used to analyze the bulk morphology of the developed bioplastics (Fig. 7.12). Cellulose-*g*-PLA from raw of parsley and cellulose-*g*-PLA of carrot have a heterogeneous bulk structure, whereas films have some cracks in the bulk, indicating a highly brittle material in both cases. Nevertheless, it is possible to observe that the fibers of parsley and raw of carrot residues are entangled in a PLA matrix. In addition, a noncompact structure can be observed, suggesting that the reactions were not complete.

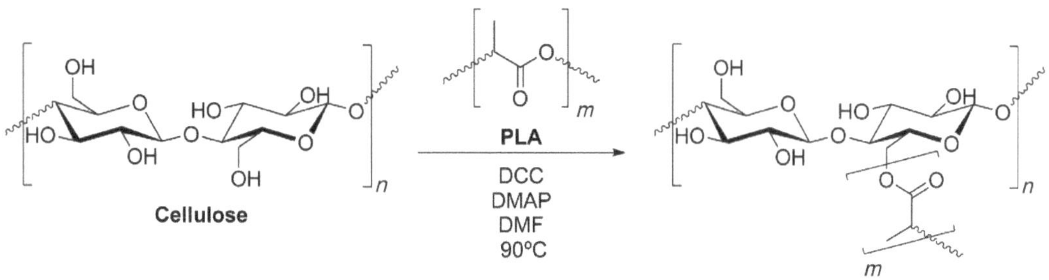

Scheme 7.3. Esterification reaction between cellulose extracted from food agro-wastes and polylactic acid (PLA). DCC, *N,N′*-Dicyclohexylcarbodiimide; DMAP, *N,N*-Dimethylpyridin-4-amine. From Macedo et al. (2015).

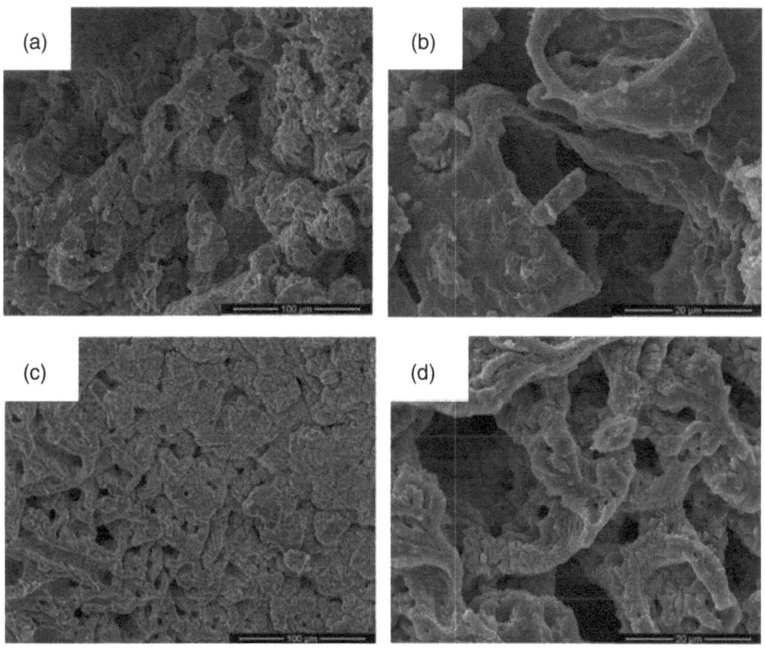

Figure 7.12. SEM micrographs resulting from the esterification reaction between cellulose extracted from (a, b) raw of parsley (c, d) raw of carrot and PLA. From Macedo et al. (2015).

Thus, even the similarity among both films, depending on the batch of waste used, different forms of self-assembled crystal structures emerged, similar to the naturally synthesized minerals of plant leaves (Ball, 2012).

Moreover, the samples prepared by transesterification-reaction polymerization show differences in biodegradability behavior, cellulose-g-PLA from raw of carrot being the more biodegradable sample (Fig. 7.13).

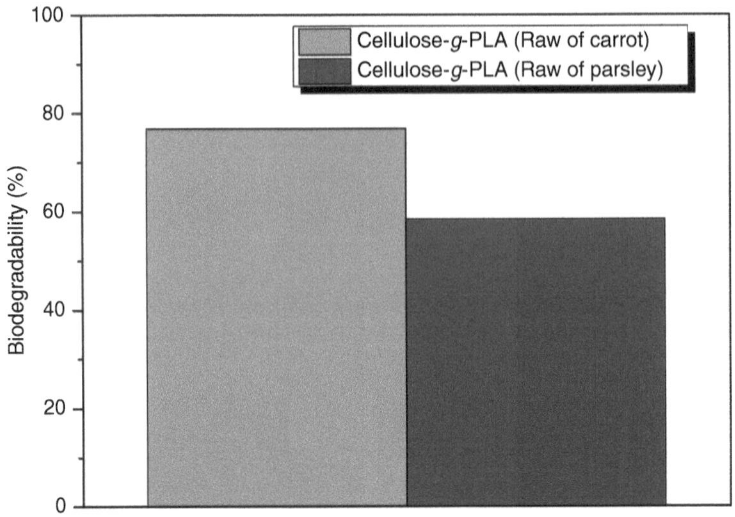

Figure 7.13. Biodegradability of food agro-waste residues according to ISO 14851(1999).

6 Conclusions

This chapter focused on the cellulose derived from food agro-wastes dissolution, using ILs to build up a new platform to get green cellulose and to improve the cellulose-bioplastics properties, by grafting via CRP in TFA and ILs extraction mediums. It is possible to adapt the chemistry, by the right selection of the IL and the cellulose resource from food agro-wastes, as well as, the synthesis parameters to use CRP methodologies for (1) innovation in production from lignocellulosic materials and reduce society's dependence on nonrenewable petroleum-based feed-stocks, in order to solve the environmental and energy issues keeping social sustainable developments; and (2) providing the possibility to develop various advanced materials, including cellulose derivatives and new cellulose composites, to replace conventional polymers with a slow biodegradation rate, mainly for products with short life times, as food packaging.

There is no doubt that the examples presented in this chapter are only some illustrations of the huge potential of the synthesis of cellulose bioplastics by a direct digesting and valorization of vegetable agro-wastes, including its regeneration and tailoring of properties by chemical modification.

Acknowledgments

The authors would like acknowledge the n-STeP—Nanostructured systems for Tailored Performance, with reference NORTE-07-0124-FEDER-000039, supported by the Programa Operacional Regional do Norte (ON.2), PEst-C/CTM/LA0025/2013 (Strategic Project—LA 25—2013-2014).

References

Abushammala, H., Krossing, I. Laborie, M.P., 2015. Ionic liquid-mediated technology to produce cellulose nanocrystals directly from wood. Carbohydr. Polym. 134, 609–616.

Åkerholm, M., Hinterstoisser, B., Salmén, L., 2004. Characterization of the crystalline structure of cellulose using static and dynamic FT-IR spectroscopy. Carbohydr. Res. 339, 569–578.

Albertsson, A.A., Barenstedt, C., Karlsson, S., 1993. Increased biodegradation of a low-density polyethylene (LDPE) matrix in starch-filled LDPE materials. J. Environ. Polym. Degrad. 1, 241–245.

Albertsson, A.C., Barenstedt, C., Karlsson, S., 1994. Degradation of enhanced environmentally degradable polyethylene in biological aqueous media: mechanisms during the first stages. J. Appl. Polym. Sci. 51, 1097–1105.

Andanson, J.M., Bordes, E., Devemy, J., Leroux, F., Padua, A.A.H., Gomes, M.F.C., 2014. Understanding the role of co-solvents in the dissolution of cellulose in ionic liquids. Green Chem. 16, 2528–2538.

Auras, R., Harte, B., Selke, S., 2004. An overview of polylactides as packaging materials. Macromol. Biosci. 4, 835–864.

Avérous, L., Fringant, C., Moro, L., 2001. Plasticized starch—cellulose interactions in polysaccharide composites. Polymer 42, 6565–6572.

Ball, P., 2012. Material witness: why leaves have stones. Nat. Mater. 11, 271–1271.

Barthel, S., Heinze, T., 2006. Acylation and carbanilation of cellulose in ionic liquids. Green Chem. 8, 301–306.

Bastioli, C., Cerutti, A., Guanella, I., Romano, G.C., Tosin, M., 1995. Physical state and biodegradation behavior of starch-polycaprolactone systems. J. Environ. Polym. Degrad. 3, 81–95.

Bayer, I.S., Guzman-Puyol, S., Heredia-Guerrero, J.A., Ceseracciu, L., Pignatelli, F., Ruffilli, R., Cingolani, R., Athanassiou, A., 2014. Direct transformation of edible vegetable waste into bioplastics. Macromolecules 47, 5135–5143.

Bishop, C.A., 2011. Vacuum Deposition onto Webs, Films and Foils, second ed. William Andrew Publishing, Oxford.

Bonhomme, S., Cuer, A., Delort, A.M., Lemaire, J., Sancelme, M., Scott, G., 2003. Environmental biodegradation of polyethylene. Polym. Degrad. Stabil. 81, 441–452.

Bontempo, D., Masci, G., De Leonardis, P., Mannina, L., Capitani, D., Crescenzi, V., 2006. Versatile grafting of polysaccharides in homogeous mikle conditions by using atom transfer radical polymerization. Biomacromolecules 7, 2154–2161.

Braun, B., Dorgan, J.R., 2009. Single-step method for the isolation and surface functionalization of cellulosic nanowhiskers. Biomacromolecules 10, 334–341.

Briassoulis, D., 2004. An overview on the mechanical behaviour of biodegradable agricultural films. J. Polym. Environ. 12, 65–81.

Calil, M.R., Gaboardi, F., Guedes, C.G.F., Rosa, D.S., 2006. Comparison of the biodegradation of poly(ε-caprolactone), cellulose acetate and their blends by the Sturm test and selected cultured fungi. Polym. Test. 25, 597–604.

Carlmark, A., Malmström, E., 2002. Atom transfer radical polymerization from cellulose fibers at ambient temperature. J. Am. Chem. Soc. 124, 900–901.
Carlmark, A., Malmström, E.E., 2003. ATRP grafting from cellulose fibers to create block-copolymer grafts. Biomacromolecules 4, 1740–1745.
Chandra, R., Rustgi, R., 1998. Biodegradable polymers. Prog. Polym. Sci. 23, 1273–1335.
Chang, F., Yamabuki, K., Onimura, K., Oishi, T., 2008. Modification of cellulose by using atom transfer radical polymerization and ring-opening polymerization. Polym. J. 40, 1170–1179.
Chauhan, G.S., Mahajan, S., Guleria, L.K., 2000. Polymers from renewable resources: sorption of Cu^{2}+ ions by cellulose graft copolymers. Desalination 130, 85–88.
Choi, E.J., Kim, C.H., Park, J.K., 1999. Structure–property relationship in PCL/starch blend compatibilized with starch-*g*-PCL copolymer. J. Polym. Sci. Part B 37, 2430–2438.
Cohn, D., Younes, H., 1988. Biodegradable PEO/PLA block copolymers. J. Biomed. Mater. Res. 22, 993–1009.
Coskun, M., Temüz, M.M., 2005. Grafting studies onto cellulose by atom-transfer radical polymerization. Polym. Int. 54, 342–347.
Cruz-Pinto, J.J.C., Carvalho, M.E.S., Ferreira, J.F.A., 1994. The kinetics and mechanism of polyethylene photo-oxidation. Die Angew. Makromol. Chem. 216, 113–133.
Czaja, W.K., Young, D.J., Kawecki, M., Brown, R.M., 2007. The future prospects of microbial cellulose in biomedical applications. Biomacromolecules 8, 1–12.
Dai, D., Fan, M., Collins, P., 2013. Fabrication of nanocelluloses from hemp fibers and their application for the reinforcement of hemp fibers. Ind. Crop. Prod. 44, 192–199.
Datta, R., Tsai, S.P., Bonsignore, P., Moon, S.H., Frank, J.R., 1995. Technological and economic potential of poly(lactic acid) and lactic acid derivatives. FEMS Microbiol. Rev. 16, 221–231.
Dias, A.B., Müller, C.M.O., Larotonda, F.D.S., Laurindo, J.B., 2011. Mechanical and barrier properties of composite films based on rice flour and cellulose fibers. LWT Food Sci. Technol. 44, 535–542.
Dong, D., Sun, J., Huang, F., Gao, Q., Wang, Y., Li, R., 2009. Using trifluoroacetic acid to pretreat lignocellulosic biomass. Biomass Bioenerg. 33, 1719–1723.
Drumright, R.E., Gruber, P.R., Henton, D.E., 2000. Polylactic acid technology. Adv. Mater. 12, 1841–1846.
Edgar, K.J., Arnold, K.M., Blount, W.W., Lawniczak, J.E., Lowman, D.W., 1995. Synthesis and properties of cellulose acetoacetates. Macromolecules 28, 4122–4128.
Eichhorn, S.J., Dufresne, A., Aranguren, M., Marcovich, N.E., Capadona, J.R., Rowan, S.J., Weder, C., Thielemans, W., Roman, M., Renneckar, S., Gindl, W., Veigel, S., Keckes, J., Yano, H., Abe, K., Nogi, M., Nakagaito, A.N., Mangalam, A., Simonsen, J., Benight, A.S., Bismarck, A., Berglund, L.A., Peijs, T., 2010. Review: current international research into cellulose nanofibres and nanocomposites. J. Mater. Sci. 45, 1–33.
El-Salmawi, K.M., Zaid, M.M.A., Ibraheim, S.M., El-Naggar, A.M., Zahran, A.H., 2001. Sorption of dye wastes by poly(vinyl alcohol)/poly(carboxymethyl cellulose) blend grafted through a radiation method. J. Appl. Polym. Sci. 82, 136–142.
Erdmenger, T., Haensch, C., Hoogenboom, R., Schubert, U.S., 2007. Homogeneous tritylation of cellulose in 1-butyl-3-methylimidazolium chloride. Macromol. Biosci. 7, 440–445.
FitzPatrick, M., Champagne, P., Cunningham, M.F., Whitney, R.A., 2010. A biorefinery processing perspective: treatment of lignocellulosic materials for the production of value-added products. Bioresour. Technol. 101, 8915–8922.

Fleet, R., Mcleary, J.B., Grumel, V., Weber, W.G., Matahwa, H., Sanderson, R.D., 2008. RAFT mediated polysaccharide copolymers. Eur. Polym. J. 44, 2899–2911.
Gaiker, C.T., 2004. Handbook for the Prevention and Minimization of Waste and Valorisation of By-Products in European Agro-Food Industries. AWARENET, Beirut, Lebanon.
Gajria, A.M., Davé, V., Gross, R.A., Mccarthy, P.S., 1996. Miscibility and biodegradability of blends of poly(lactic acid) and poly(vinyl acetate). Polymer 37, 437–444.
Gáspár, M., Benkő, Z., Dogossy, G., Réczey, K., Czigány, T., 2005. Reducing water absorption in compostable starch-based plastics. Polym. Degrad. Stabil. 90, 563–569.
Gautam, I.K.A.N., 2010. Starch grafted polyethylene evincing biodegradation behaviour. Malay. Polym. J. 5, 26–38.
Gericke, M., Fardim, P., Heinze, T., 2012. Ionic liquids—promising but challenging solvents for homogeneous derivatization of cellulose. Molecules 17, 7458.
Ghanbarzadeh, B., Almasi, H., Entezami, A.A., 2011. Improving the barrier and mechanical properties of corn starch-based edible films: effect of citric acid and carboxymethyl cellulose. Ind. Crop. Prod. 33, 229–235.
Gross, R.A., Kalra, B., 2002. Biodegradable polymers for the environment. Science 297, 803–807.
Guo, Y., Wang, X., Shen, Z., Shu, X., Sun, R., 2013. Preparation of cellulose-graft-poly(ε-caprolactone) nanomicelles by homogeneous ROP in ionic liquid. Carbohydr. Polym. 92, 77–83.
Gupta, K.C., Khandekar, K., 2003. Temperature-responsive cellulose by ceric(iv) ion-initiated graft copolymerization of n-isopropylacrylamide. Biomacromolecules 4, 758–765.
Halley, P., Rutgers, R., Coombs, S., Kettels, J., Gralton, J., Christie, G., Jenkins, M., Beh, H., Griffin, K., Jayasekara, R., Lonergan, G., 2001. Developing biodegradable mulch films from starch-based polymers. Starch Stärke 53, 362–367.
Heinze, T., Schaller, J., 2000. New water soluble cellulose esters synthesized by an effective acylation procedure. Macromol. Chem. Phys. 201, 1214–1218.
Heinze, T., Schwikal, K., Barthel, S., 2005. Ionic liquids as reaction medium in cellulose functionalization. Macromol. Biosci. 5, 520–525.
Hernández-Guerrero, M., Davis, T.P., Barner-Kowollik, C., Stenzel, M.H., 2005. Polystyrene comb polymers built on cellulose or poly(styrene-*co*-2-hydroxyethylmethacrylate) backbones as substrates for the preparation of structured honeycomb films. Eur. Polym. J. 41, 2264–2277.
Hiltunen, M., Riihelä, S., Maunu, S.L., 2009. New associative EHEC-*g*-PAam copolymers: their syntheses, characterization, and rheological behavior. J. Polym. Sci. Part B 47, 1869–1879.
Hufendiek, A., Trouillet, V., Meier, M.A.R., Barner-Kowollik, C., 2014. Temperature responsive cellulose-graft-copolymers via cellulose functionalization in an ionic liquid and RAFT polymerization. Biomacromolecules 15, 2563–2572.
Isik, M., Sardon, H., Mecerreyes, D., 2014. Ionic liquids and cellulose: dissolution, chemical modification and preparation of new cellulosic materials. Int. J. Mol. Sci. 15, 11922.
Jiang, L., Liu, B., Zhang, J., 2009. Properties of poly(lactic acid)/poly(butylene adipate-*co*-terephthalate)/nanoparticle ternary composites. Ind. Eng. Chem. Res. 48, 7594–7602.
Kaczmarek, H., Ołdak, D., 2006. The effect of UV-irradiation on composting of polyethylene modified by cellulose. Polym. Degrad. Stabil. 91, 2282–2291.
Kaur, I., Bhalla, T.C., Deepika, N., Gautam, N., 2008. Biodegradation and swelling studies of gelatin-grafted polyethylene. J. Appl. Polym. Sci. 107, 3878–3884.
Keshk, S.M.A.S., 2014. Bacterial cellulose production and its industrial applications. J. Bioprocess. Biotech 4, 2.

Kim, B.R., Suidan, M.T., Wallington, T.J., Du, X., 2000. Biodegradability of trifluoroacetic acid. Environ. Eng. Sci. 17, 337–342.
Klemm, D., Heublein, B., Fink, H.P., Bohn, A., 2005. Cellulose: fascinating biopolymer and sustainable raw material. Angew. Chem. Int. Ed. Engl. 44, 3358–3393.
Koenig, M.F., Huang, S.J., 1995. Biodegradable blends and composites of polycaprolactone and starch derivatives. Polymer 36, 1877–1882.
Köhler, S., Liebert, T., Heinze, T., 2008. Interactions of ionic liquids with polysaccharides. VI. Pure cellulose nanoparticles from trimethylsilyl cellulose synthesized in ionic liquids. J. Polym. Sci. Part A 46, 4070–4080.
Kraemer, E.O., 1938. Molecular weights of celluloses and cellulose derivates. Ind. Eng. Chem. 30, 1200–1203.
Lawton, J.W., 1996. Effect of starch type on the properties of starch containing films. Carbohydr. Polym. 29, 203–208.
Lee, S.B., Koepsel, R.R., Morley, S.W., Matyjaszewski, K., Sun, Y., Russell, A.J., 2004. Permanent, nonleaching antibacterial surfaces. 1. Synthesis by atom transfer radical polymerization. Biomacromolecules 5, 877–882.
Liebert, T.F., Heinze, T.J., 2001. Exploitation of reactivity and selectivity in cellulose functionalization using unconventional media for the design of products showing new superstructures. Biomacromolecules 2, 1124–1132.
Liebert, T., Hänsch, C., Heinze, T., 2006. Click chemistry with polysaccharides. Macromol. Rapid Commun. 27, 208–213.
Lindqvist, J., Malmström, E., 2006. Surface modification of natural substrates by atom transfer radical polymerization. J. Appl. Polym. Sci. 100, 4155–4162.
Liu, C.F., Sun, R.C., Zhang, A.P., Ren, J.L., 2007. Preparation of sugarcane bagasse cellulosic phthalate using an ionic liquid as reaction medium. Carbohydr. Polym. 68, 17–25.
Macedo, M.J., Moura, I., Oliveira, M., Machado, A.V., 2015. Development of bioplastics from agro-wastes. Abstract in MATERIAIS 2015. Porto, Portugal.
Machado, A.V., Moura, I., Duarte, F.M., Botelho, G., Nogueira, R., Brito, A.G., 2007. Evaluation of properties and biodeterioration potential of polyethylene and aliphatic polyester blends. Int. Polym. Process. 22, 512–518.
Maharana, T., Singh, B.C., 2006. Synthesis and characterization of biodegradable polyethylene by graft copolymerization of starch using glucose–Ce(IV) redox system. J. Appl. Polym. Sci. 100, 3229–3239.
Iwamoto, A., Tokiwa, Y., 1994. Environmentally degradable polymers enzymatic degradation of plastics containing polycaprolactone. Polym. Degrad. Stab. 45, 205–213.
Makino, Y., Hirata, T., 1997. Modified atmosphere packaging of fresh produce with a biodegradable laminate of chitosan-cellulose and polycaprolactone. Postharvest Biol. Technol. 10, 247–254.
Massardier-Nageotte, V., Pestre, C., Cruard-Pradet, T., Bayard, R., 2006. Aerobic and anaerobic biodegradability of polymer films and physico-chemical characterization. Polym. Degrad. Stabil. 91, 620–627.
Masson, J.F., Manley, R.S.J., 1991. Cellulose/poly(4-vinylpyridine) blends. Macromolecules 24, 5914–5921.
Matzinos, P., Tserki, V., Kontoyiannis, A., Panayiotou, C., 2002. Processing and characterization of starch/polycaprolactone products. Polym. Degrad. Stabil. 77, 17–24.
Meng, T., Gao, X., Zhang, J., Yuan, J., Zhang, Y., He, J., 2009. Graft copolymers prepared by atom transfer radical polymerization (ATRP) from cellulose. Polymer 50, 447–454.
Miyamoto, T., Takahashi, S.I., Ito, H., Inagaki, H., Noishiki, Y., 1989. Tissue biocompatibility of cellulose and its derivatives. J. Biomed. Mater. Res. 23, 125–133.

Mohanty, A.K., Misra, M., Hinrichsen, G., 2000. Biofibres, biodegradable polymers and biocomposites: an overview. Macromol. Mater. Eng. 276-277, 1–24.
Moon, R.J., Martini, A., Nairn, J., Simonsen, J., Youngblood, J., 2011. Cellulose nanomaterials review: structure, properties and nanocomposites. Chem. Soc. Rev. 40, 3941–3994.
Moura, I., Machado, A.V., Duarte, F.M., Nogueira, R., 2011. Biodegradability assessment of aliphatic polyesters-based blends using standard methods. J. Appl. Polym. Sci. 119, 3338–3346.
Müller, C.M.O., Laurindo, J.B., Yamashita, F., 2009. Effect of cellulose fibers addition on the mechanical properties and water vapor barrier of starch-based films. Food Hydrocoll. 23, 1328–1333.
Nabar, Y., Raquez, J.M., Dubois, P., Narayan, R., 2005. Production of starch foams by twin-screw extrusion: effect of maleated poly(butylene adipate-co-terephthalate) as a compatibilizer. Biomacromolecules 6, 807–817.
Nair, L.S., Laurencin, C.T., 2006. Polymers as biomaterials for tissue engineering and controlled drug delivery. Adv. Biochem. Eng. Biotechnol. 102, 47–90.
Narayan, R., 1994. Impact of governmental policies, regulations, and standards activities on an emerging plastics industry. In: Doi, Y., Fukuda, K. (Eds.), Biodegradable Plastics and Polymers. Elsevier, Amsterdam, Netherlands.
Nijenhuis, A.J., Colstee, E., Grijpma, D.W., Pennings, A.J., 1996. High molecular weight poly(L-lactide) and poly(ethylene oxide) blends: thermal characterization and physical properties. Polymer 37, 5849–5857.
Nishio, Y., 2006. Material functionalization of cellulose and related polysaccharides via diverse microcompositions. Adv. Polym. Sci. 205, 97–151.
Nishio, Y., Manley, R.S.J., 1990. Blends of cellulose with nylon 6 and poly(epsillon-caprolactone) prepared by a solution-coagulation method. Polym. Eng. Sci. 30, 71–82.
Nishiyama, Y., Langan, P., Chanzy, H., 2002. Crystal structure and hydrogen-bonding system in cellulose from synchrotron x-ray and neutron fiber diffraction. J. Am. Chem. Soc. 124, 9074–9082.
Nystrom, D., Lindqvist, J., Ostmark, E., Hult, A., Malmstrom, E., 2006. Superhydrophobic bio-fibre surfaces via tailored grafting architecture. Chem. Commun., 3594–3596.
Oh, S.Y., Yoo, D.I., Shin, Y., Kim, H.C., Kim, H.Y., Chung, Y.S., Park, W.H., Youk, J.H., 2005. Crystalline structure analysis of cellulose treated with sodium hydroxide and carbon dioxide by means of X-ray diffraction and FTIR spectroscopy. Carbohydr. Res. 340, 2376–2391.
Ohtake, Y., Kobayashi, T., Asabe, H., Murakami, N., 1998. Studies on biodegradation of LDPE—observation of LDPE films scattered in agricultural fields or in garden soil. Polym. Degrad. Stabil. 60, 79–84.
Oksman, K., Mathew, A.P., Bondeson, D., Kvien, I., 2006. Manufacturing process of cellulose whiskers/polylactic acid nanocomposites. Compos. Sci. Technol. 66, 2776–2784.
Onda, A., Ochi, T., Yanagisawa, K., 2008. Selective hydrolysis of cellulose into glucose over solid acid catalysts. Green Chem. 10, 1033–1037.
O-Rak, K., Ummartyotin, S., Sain, M., Manuspiya, H., 2013. Covalently grafted carbon nanotube on bacterial cellulose composite for flexible touch screen application. Mater. Lett. 107, 247–250.
Östmark, E., Harrisson, S., Wooley, K.L., Malmström, E.E., 2007. Comb polymers prepared by ATRP from hydroxypropyl cellulose. Biomacromolecules 8, 1138–1148.
Peelman, N., Ragaert, P., De Meulenaer, B., Adons, D., Peeters, R., Cardon, L., Van Impe, F., Devlieghere, F., 2013. Application of bioplastics for food packaging. Trends Food Sci. Technol. 32, 128–141.

Pei, A., Zhou, Q., Berglund, L.A., 2010. Functionalized cellulose nanocrystals as biobased nucleation agents in poly(L-lactide) (PLLA) – Crystallization and mechanical property effects. Compos. Sci. Technol. 70, 815–821.
Petersson, L., Oksman, K., 2006. Biopolymer based nanocomposites: comparing layered silicates and microcrystalline cellulose as nanoreinforcement. Compos. Sci. Technol. 66, 2187–2196.
Pingali, S.V., Urban, V.S., Heller, W.T., Mcgaughey, J., O'neill, H., Foston, M., Myles, D.A., Ragauskas, A., Evans, B.R., 2010. Breakdown of cell wall nanostructure in dilute acid pretreated biomass. Biomacromolecules 11, 2329–2335.
Plackett, D., Jankova, K., Egsgaard, H., Hvilsted, S., 2005. Modification of jute fibers with polystyrene via atom transfer radical polymerization. Biomacromolecules 6, 2474–2484.
Poletto, M., Zattera, A.J., Pistor, V., 2013. Structural characteristics and thermal properties of native cellulose. In: van de Van, T., Godbout, L. (Eds.), Cellulose Fundamental Aspects. InTech.
Popa, M., Belc, N., 2007. Packaging. In: McElhatton, A., Marshall, R. (Eds.). Food Safety: A Practical and Case Study Approach. Springer, New York.
Postek, M., Brown, E., 2009. Sustainable, renewable nanomaterials may replace carbon nanotubes. SPIE Newsroom 17.
Pracella, M., Haque, M.M.-U., Alvarez, V., 2010. Functionalization, compatibilization and properties of polyolefin composites with natural fibers. Polymers 2, 554.
Pranamuda, H., Tokiwa, Y., Tanaka, H., 1996. Physical properties and biodegradability of blends containing poly(ε-caprolactone) and tropical starches. J. Environ. Polym. Degrad. 4, 1–7.
Reddy, M.M., Misra, M., Mohanty, A.K., 2012. Bio-based materials in the new bio-economy. Chem. Eng. Prog. 108, 37–42.
Reddy, M.M., Vivekanandhan, S., Misra, M., Bhatia, S.K., Mohanty, A.K., 2013. Biobased plastics and bionanocomposites: current status and future opportunities. Prog. Polym. Sci. 38, 1653–1689.
Regiani, A.M., Frollini, E., Marson, G.A., Arantes, G.M., El Seoud, O.A., 1999. Some aspectsof acylation of cellulose under homogeneous solution conditions. J. Polym. Sci. Part A 37, 1357–1363.
Rhim, J.W., Kim, J.H., 2009. Properties of poly(lactide)-coated paperboard for the use of 1-way paper cup. J. Food Sci. 74, E105–E111.
Ries, M.E., Radhi, A., Keating, A.S., Parker, O., Budtova, T., 2014. Diffusion of L-ethyl-3-methyl-imidazolium acetate in glucose, cellobiose, and cellulose solutions. Biomacromolecules 15, 609–617.
Rogovina, S., Aleksanyan, K., Prut, E., Gorenberg, A., 2013. Biodegradable blends of cellulose with synthetic polymers and some other polysaccharides. Eur. Polym. J. 49, 194–202.
Ross, P., Mayer, R., Benziman, M., 1991. Cellulose biosynthesis and function in bacteria. Microbiol. Rev. 55, 35–58.
Sanchez-Garcia, M., Lagaron, J., 2010. On the use of plant cellulose nanowhiskers to enhance the barrier properties of polylactic acid. Cellulose 17, 987–1004.
Sanchez-Garcia, M.D., Gimenez, E., Lagaron, J.M., 2008. Morphology and barrier properties of solvent cast composites of thermoplastic biopolymers and purified cellulose fibers. Carbohydr. Polym. 71, 235–244.
Sannino, A., Demitri, C., Madaghiele, M., 2009. Biodegradable cellulose-based hydrogels: design and applications. Materials 2, 353–373.
Saroja, N., Gowda, L.R., Tharanathan, R.N., 2000. Chromatographic determination of residual monomers in starch-*g*-polyacrylonitrile and starch-*g*-polyacrylate. Chromatographia 51, 345–348.
Sastry, P.K., Satyanarayana, D., Mohan Rao, D.V., 1998. Accelerated and environmental weathering studies on polyethylene–starch blend films. J. Appl. Polym. Sci. 70, 2251–2257.

Schenzel, A., Hufendiek, A., Barner-Kowollik, C., Meier, M.A.R., 2014. Catalytic transesterification of cellulose in ionic liquids: sustainable access to cellulose esters. Green Chem. 16, 3266–3271.

Schwach, E., Avérous, L., 2004. Starch-based biodegradable blends: morphology and interface properties. Polym. Int. 53, 2115–2124.

Scott, G., 2000. "Green" polymers. Polym. Degrad. Stabil. 68, 1–7.

Semsarilar, M., Ladmiral, V., Perrier, S., 2010. Synthesis of a cellulose supported chain transfer agent and its application to RAFT polymerization. J. Polym. Sci. Part A 48, 4361–4365.

Shah, A.A., Hasan, F., Hameed, A., Ahmed, S., 2008. Biological degradation of plastics: a comprehensive review. Biotechnol. Adv. 26, 246–265.

Shahlari, M., Lee, S., 2012. Mechanical and morphological properties of poly(butylene adipate-*co*-terephthalate) and poly(lactic acid) blended with organically modified silicate layers. Polym. Eng. Sci. 52, 1420–1428.

Shen, D., Huang, Y., 2004. The synthesis of CDA-*g*-PMMA copolymers through atom transfer radical polymerization. Polymer 45, 7091–7097.

Shengdong, Z., Ziniu, Y., Yuanxin, W., Xia, Z., Hui, L., Ming, G., 2005. Enhancing enzymatic hydrolysis of rice straw by microwave pretreatment. Chem. Eng. Commun. 192, 1559–1566.

Sheth, M., Kumar, R.A., Davé, V., Gross, R.A., Mccarthy, S.P., 1997. Biodegradable polymer blends of poly(lactic acid) and poly(ethylene glycol). J. Appl. Polym. Sci. 66, 1495–1505.

Simon, C.M., Kaminsky, W., Schlesselmann, B., 1996. Pyrolysis of polyolefins with steam to yield olefins. J. Anal. Appl. Pyrol. 38, 75–87.

Siqueira, G., Bras, J., Dufresne, A., 2010. Cellulosic bionanocomposites: a review of preparation. Polymers 2, 728–765.

Siracusa, V., Rocculi, P., Romani, S., Rosa, M.D., 2008. Biodegradable polymers for food packaging: a review. Trends Food Sci. Technol. 19, 634–643.

Smith, R., 2005. Biodegradable Polymers for Industrial Applications. CRC Press, New York.

Someya, Y., Sugahara, Y., Shibata, M., 2005. Nanocomposites based on poly(butylene adipate-*co*-terephthalate) and montmorillonite. J. Appl. Polym. Sci. 95, 386–392.

Song, Y., Sun, Y., Zhang, X., Zhou, J., Zhang, L., 2008. Homogeneous quaternization of cellulose in NaOH/urea aqueous solutions as gene carriers. Biomacromolecules 9, 2259–2264.

Stenius, P., 2000. Papermaking Science and Technology: Forest Products Chemistry Book 3. Fapet, Helsinki, Finland.

Stenzel, M.H., Davis, T.P., Fane, A.G., 2003. Honeycomb structured porous films prepared from carbohydrate based polymers synthesized via the RAFT process. J. Mater. Chem. 13, 2090–2097.

Sudesh, K., Iwata, T., 2008. Sustainability of biobased and biodegradable plastics. Clean Soil Air Water 36, 433–442.

Swatloski, R.P., Spear, S.K., Holbrey, J.D., Rogers, R.D., 2002. Dissolution of cellulose with ionic liquids. J. Am. Chem. Soc. 124, 4974–4975.

Swatloski, R.P., Holbrey, J.D., Rogers, R.D., 2003. Ionic liquids are not always green: hydrolysis of 1-butyl-3-methylimidazolium hexafluorophosphate. Green Chem. 5 (4), 361–363.

Tang, X., Gao, L., Fan, X., Zhou, Q., 2007. Controlled grafting of ethyl cellulose with azobenzene-containing polymethacrylates via atom transfer radical polymerization. J. Polym. Sci. Part A 45, 1653–1660.

Teck Nam, A., Yoon, L.W., Lee, K.M., Ngoh, G.C., Chua, A.S.M., Lee, M.G., 2011. Efficiency of ionic liquids in the dissolution of rice husk. Bioresources 6, 4790–4800.

Thompson, R.C., Moore, C.J., Vom Saal, F.S., Swan, S.H., 2009. Plastics, the environment and human health: current consensus and future trends. Phil. Trans. R. Soc. B 1, 1–14.

Tome, L.C., Freire, M.G., Rebelo, L.P.N., Silvestre, A.J.D., Pascoal Neto, C., Marrucho, I.M., Freire, C.S.R., 2011. Surface hydrophobization of bacterial and vegetable cellulose fibers using ionic liquids as solvent media and catalysts. Green Chem. 13, 2464–2470.

Tomšič, B., Simončič, B., Orel, B., Vilčnik, A., Spreizer, H., 2007. Biodegradability of cellulose fabric modified by imidazolidinone. Carbohydr. Polym. 69, 478–488.

Tserki, V., Matzinos, P., Pavlidou, E., Vachliotis, D., Panayiotou, C., 2006. Biodegradable aliphatic polyesters. Part I. Properties and biodegradation of poly(butylene succinate-*co*-butylene adipate). Polym. Degrad. Stabil. 91, 367–376.

Vasanthan, N., Ly, O., 2009. Effect of microstructure on hydrolytic degradation studies of poly (L-lactic acid) by FTIR spectroscopy and differential scanning calorimetry. Polym. Degrad. Stabil. 94, 1364–1372.

Vlček, P., Janata, M., Látalová, P., Kríž, J., Cˇadová, E., Toman, L., 2006. Controlled grafting of cellulose diacetate. Polymer 47, 2587–2595.

Waly, A., Abdel-Mohdy, F.A., Aly, A.S., Hebeish, A., 1998. Synthesis and characterization of cellulose ion exchanger. II. Pilot scale and utilization in dye heavy metal removal. J. Appl. Polym. Sci. 68, 2151–2157.

Wang, Y.-Z., Yang, K.-K., Wang, X.-L., Zhou, Q., Zheng, C.-Y., Chen, Z.-F., 2004. Agricultural application and environmental degradation of photo-biodegradable polyethylene mulching films. J. Polym. Environ. 12, 7–10.

Wang, X., Li, H., Cao, Y., Tang, Q., 2011. Cellulose extraction from wood chip in an ionic liquid L-allyl-3-methylimidazolium chloride (AmimCl). Bioresource Technol. 102, 7959–7965.

Wang, H., Gurau, G., Rogers, R.D., 2012. Ionic liquid processing of cellulose. Chem. Soc. Rev. 41, 1519–1537.

Whiteman, N.F., 2000. New packaging thermoplastic, polylactide. Polymers, Laminations and Coatings Conference, Chicago, USA.

Witt, U., Einig, T., Yamamoto, M., Kleeberg, I., Deckwer, W.D., Müller, R.J., 2001. Biodegradation of aliphatic–aromatic copolyesters: evaluation of the final biodegradability and ecotoxicological impact of degradation intermediates. Chemosphere 44, 289–299.

Wu, J., Zhang, J., Zhang, H., He, J., Ren, Q., Guo, M., 2004. Homogeneous acetylation of cellulose in a new ionic liquid. Biomacromolecules 5, 266–268.

Yamada-Onodera, K., Mukumoto, H., Katsuyaya, Y., Saiganji, A., Tani, Y., 2001. Degradation of polyethylene by a fungus, *Penicillium simplicissimum*. Polym. Degrad. Stabil. 72, 323–327.

Yang, A., Wu, R., Zhu, P., 2001. Thermal analysis and miscibility of chitin/polycaprolactone blends. J. Appl. Polym. Sci. 81, 3117–3123.

Yavuz, H., Babaç, C., 2003. Preparation and biodegradation of starch/polycaprolactone films. J. Polym. Environ. 11, 107–113.

Yew, G.H., Yusof, A.M.M., Ishak, Z.A.M., Ishiaku, U.S., 2005. Natural weathering effects on mechanical properties of polylactic acid/rice starch composites. 8th Polymers for Advanced Technologies International Symposium, Budapest, Hungary.

Zarrinbakhsh, N., Mohanty, A., Misra, M., 2013. Improving the interfacial adhesion in a new renewable resource-based biocomposites from biofuel coproduct and biodegradable plastic. J. Mater. Sci. 48, 6025–6038.

Zavrel, M., Bross, D., Funke, M., Büchs, J., Spiess, A.C., 2009. High-throughput screening for ionic liquids dissolving (ligno-)cellulose. Bioresour. Technol. 100, 2580–2587.

Zhang, L., Xiong, C., Deng, X., 1995. Biodegradable polyester blends for biomedical application. J. Appl. Polym. Sci. 56, 103–112.

Zhang, H., Wu, J., Zhang, J., He, J., 2005. 1-Allyl-3-methylimidazolium chloride room temperature ionic liquid: a new and powerful nonderivatizing solvent for cellulose. Macromolecules 38, 8272–8277.
Zhang, J., Wu, J., Cao, Y., Sang, S., Zhang, J., He, J., 2009. Synthesis of cellulose benzoates under homogeneous conditions in an ionic liquid. Cellulose 16, 299–308.
Zhao, H., Holladay, J.E., Kwak, J.H., Zhang, Z.C., 2007. Inverse temperature-dependent pathway of cellulose decrystallization in trifluoroacetic acid. J. Phys. Chem. B 111, 5295–5300.
Zhu, S., Wu, Y., Yu, Z., Zhang, X., Wang, C., Yu, F., Jin, S., Zhao, Y., Tu., S., Xue, Y., 2005. Simultaneous saccharification and fermentation of microwave/alkali pre-treated rice straw to ethanol. Biosyst. Eng. 92, 229–235.
Zhu, S., Wu, Y., Yu, Z., Zhang, X., Wang, C., Yu, F., Jin, S., 2006a. Production of ethanol from microwave-assisted alkali pretreated wheat straw. Process Biochem. 41, 869–873.
Zhu, S.D., Wu, Y.X., Chen, Q.M., Yu, Z.N., Wang, C.W., Jin, S.W., Ding, Y.G., Wu, G., 2006b. Dissolution of cellulose with ionic liquids and its application: a mini-review. Green Chem. 8, 325–327.

STUDY OF THE STRUCTURE/ PROPERTY RELATIONSHIP OF NANOMATERIALS FOR DEVELOPMENT OF NOVEL FOOD PACKAGING

Agustín González*, Cintia B. Contreras, Cecilia I. Alvarez Igarzabal**, Miriam C. Strumia****

**Universidad Nacional de Córdoba, Juan Filloy s/n, Ciudad Universitaria, Instituto de Ciencia y Tecnología de los Alimentos Córdoba (ICYTAC), CONICET, Facultad de Ciencias Químicas, Córdoba, Argentina; **Universidad Nacional de Córdoba, Medina Allende and Haya de la Torre, Ciudad Universitaria, Instituto Multidisciplinario de Biología Vegetal (IMBIV), CONICET, Facultad de Ciencias Químicas, Córdoba, Argentina*

1 Introduction

The importance of packaging in the food supply chain is fully recognized (Brody et al., 2008). It is known that the goal of food packaging is to maintain food safety and to minimize environmental impact. One of the major concerns for the industry is food safety and quality. For this, several years ago, researchers studied ways of trying to extend the life of food and retain its nutritional properties and sensory characteristics. The new generation of packaging provides additional functions such as gas (O_2, CO_2), vapor and aroma barrier or mechanical, optical, and thermal properties, Fig. 8.1.

Over the past decades, the use of polymers as a food packaging material has increased enormously due to their advantage over other traditional materials (Sinha Ray and Okamoto, 2003; Silvestre et al., 2011). Plastics are the materials typically used for packaging because their versatility to adapt to different demands and requirements. They exhibit many advantageous characteristics such as transparency, softness, heat seal ability, and good strength to weight ratio and flexibility. In addition, they are generally

Food Packaging. http://dx.doi.org/10.1016/B978-0-12-804302-8.00008-X

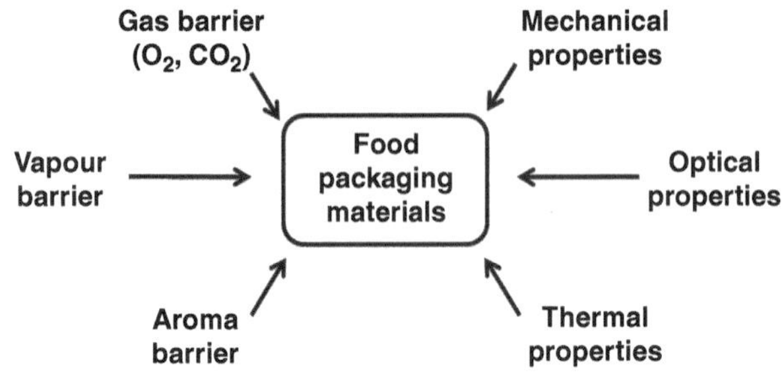

Figure 8.1. Additional properties required for new generation of food packaging materials.

inexpensive, show efficient mechanical properties including tear and tensile strength, good barriers to oxygen, moisture and heat, and resistance to food component attack. In addition to these functions, it provides essential product information to consumers to facilitate the promotion and advertisement of the product (Robertson, 2008). In spite of this, conventional packaging polymers have begun to be questioned due to increasing environmental issues.

Today, the preparation and development of economic packaging that can replace traditional materials derived from petroleum is an important research goal. Desirably, they have to be biodegradable or environmentally friendly and they have to preserve the properties of synthetic polymers currently used (González and Alvarez Igarzabal, 2015; González and Alvarez Igarzabal, 2013; Siracusa and Dalla, 2008; Schmid et al., 2012; Rhim et al., 2013b; Jiménez et al., 2012; González et al., 2015b). Therefore, the use of natural polymers or biopolymers as packaging materials may relieve the environmental impact caused by excessive use of conventional polymeric materials and thus reduce the increased production of plastics (Da-Silva et al., 2013; Jiménez et al., 2012). However, the materials extracted from natural sources are associated with poor properties and often exhibit inferior properties compared to those of traditional polymers. Unfortunately, currently used bioplastics still do not meet all the requirements for use in packaging. Specifically, as regards its mechanical and barrier properties (Hendrix et al., 2012a).

Polymer nanotechnology is a broad interdisciplinary area of research, development, and industrial activity that involves the design, manufacture, processing, and application of polymer materials filled with devices that have one or more dimensions of the order of 100 nanometers (nm) or less (Paul and Robeson, 2008; Utracki, 2004; Sinha Ray and Okamoto, 2003).

These nanomaterials exploit novel properties and functions that occur in matter at this scale, because the high surface area-to-volume ratio enhances surface reactivity of nanosized systems. Nanocomposite systems are those in which the dispersed phase is nanostructured; when used for the preparation of packaging they give rise to a new class of materials called nanopackaging.

Nanosystems could be classified by their composition or by their shape. The first type may include any of the following nanomaterials: metallic nanoparticles (gold, silver), metal oxide nanoparticles (TiO_2, ZnO, MgO), natural nanoparticles (chitosan, alginate, starch, cellulose), nanotubes/fullerenes, nanoclays, and so on (Cushen et al., 2012; Azeredo, 2013).

Depending on their size and shape, nanomaterials can be classified in three types. 3-dimensional nanoreinforcements in the nanoscale are particulated (nanoparticles and nanocrystals); 2-dimensional nanoreinforcements in the nanoscale are fibrillated (nanofibers, nanotubes); and 1-dimensional nanoreinforcements in the nanometric scale are laminated (nanoclays). The classification of nanoreinforcement according to their shape is shown in Fig. 8.2. The nanosystems can be used to reinforce different classes of materials for improving their properties (such as stiffness, flexibility, barrier properties). Thus, combining two or more different materials, a new material with the desired combination of properties can be achieved.

The application of polymer nanotechnology can indeed extend and implement all the main functions of packaging, such as containment, protection, and preservation, marketing and communication (Chaudhry et al., 2008). It can also provide new food packaging materials with improved properties, such as barrier protection to light and gases, in addition to adding strength, stiffness, dimensional stability, and heat resistance. Further advantages include: the reduction to zero of any critical interaction with food matrices and with human health: reduction of energy-inputs for the production and volume of waste material to be disposed of in landfills, transported and stored; and increase

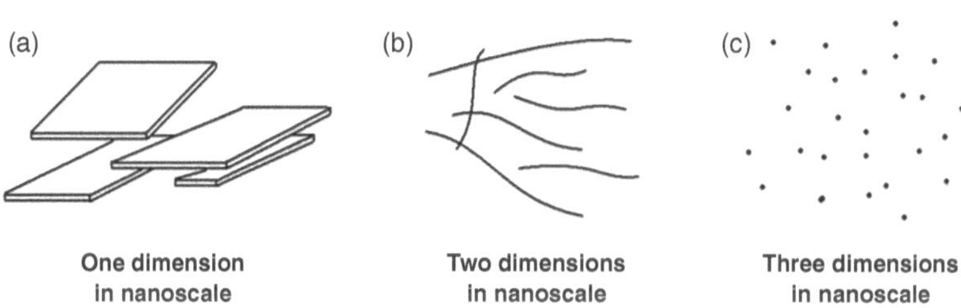

Figure 8.2. Classification of nanoreinforcement according to their shape. Laminated (a), fibrillated (b), and particulated (c) reinforcement.

of biodegradability and contribution to decreasing CO_2 emissions (Arvanitoyannis, 2005; Brody, 2008; Lagaron et al., 2005; Arora and Padua, 2010). A further interesting application is the possibility of using nanosensors for tracing and monitoring the condition of food during transport and storage. All these advantages open up significant challenges for packaging technology based on the inclusion of nanosystems in the preparation of packages with new or improved properties as compared to those already known.

The main uses of nanosystems for food packaging preparation (Silvestre et al., 2011) are oriented to three types of technological advantages, such as reinforcement effect, preparation of active materials and preparation of intelligent packaging. They are schematized in Fig. 8.3 and discussed in three sections as follows:

1. The development of materials with *improved properties* through the presence of nanonanosystems in the polymer matrix, which confer stiffness, flexibility, better barrier properties or temperature/moisture stability;
2. The development of *active packaging* through the presence of nanosystems, which allows packages to interact with food and the environment, playing a dynamic role in food preservation.
3. The development of *intelligent packaging* through the presence of nanodevices in the polymer matrix to monitor the condition of packaged food or the environment surrounding the food, acting as a guard against fraudulent imitation.

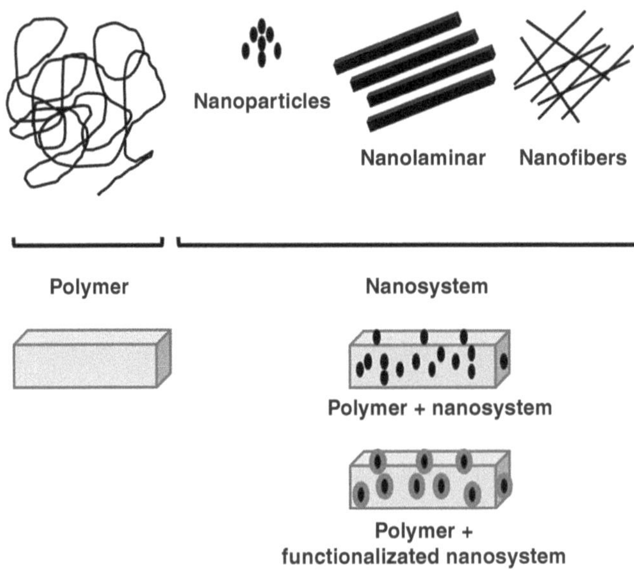

Figure 8.3. Schematic illustration of nanosystems used to combine with polymers to yield nanopackaging with improved properties.

The main advances and improvements developed in these three areas of research involved in food packaging field are summarized in the following sections. The nanosystems classified as nanocomposites have had a rapid development due to the significant improvement in the properties of their container and their easy processing by incorporation into the plastic materials used for the manufacture of packaging. Likewise, it is important to notice that this field is the most developed in packaging nanotechnology due to its low cost and excellent properties.

2 Development of Films With Improved Properties

As was previously commented, biobased materials must exhibit several properties such containment and protection of food quality to act as selective barriers to moisture, oxygen uptake, oxidation of lipids and loss of volatile aromas and flavors, while maintaining properties as sensorial quality and food security (Roy et al., 2012).

The materials prepared from natural and renewable polymers are often associated with poor mechanical properties. Many bioplastics currently available do not satisfy prerequisites for food packaging (Hendrix et al., 2012b). Therefore, numerous modification strategies have been developed in order to achieve optimum properties for the specific uses. As one of the most important characteristics, the biopolymers possess suitable functional groups available to be chemically modified.

The possibility of the incorporation of active ingredients, the adhesion to the food surfaces, the stability under storage conditions, the water resistance, the barrier effect to the gases and the mechanical properties to avoid rupture of the material, are the most important properties of the developed packaging. Improvements in these properties are usually achieved through different modification strategies of the starting material.

Many nanocomposites have been used in the development of films with good properties as coating material. In general, a synergistic effect through the combination of the individual properties of both, the matrix precursor of the film and the reinforcement, is achieved. In this case, an interface between the nanoreinforcement and the matrix with strong interactions is created, improving the physical properties, especially by the high surface/volume. The incorporation of nanomaterials in polymeric matrices produces composites with excellent properties without impairing the density, transparency and processability of the film (Gao et al., 2011).

In the literature, several examples of nanomaterials used as reinforcement in natural matrices have been reported (Kristo and Biliaderis, 2007).

2.1 Different Nanocrystals as Film Reinforcement

The most used nanoreinforcements are those nanocrystals prepared from semicrystalline polysaccharides such as cellulose, chitin, chitosan, and starch. These resulted in ideal for the development materials with high performance even in low concentrations (Dufresne, 2010; Chen et al., 2012; Flauzino Neto et al., 2013; Lin et al., 2011). These developed products showed very good mechanical properties, biodegradability, and surface area. Reports on the improvement in films yielded with starch nanocrystals (SNC) are extensively described in the literature (García et al., 2011; Chen et al., 2008; Li et al., 2015a; González et al., 2015a; González and Alvarez Igarzabal, 2015; Kristo and Biliaderis, 2007).

Native starch granules are formed by amylose and amylopectin and have an annular alternating crystalline and semicrystalline layer structure. Amylose is composed of long linear chains and can acquire a 3-dimensional helical conformation, wherein each helical turn consists of six glucose molecules. The inside of the helix contains only hydrogen atoms, and is thus lipophilic, while the outside of the helix contains/holds hydroxyl groups. Most starches contain about 25% amylose. Moreover, amylopectin contains branches, which are attached to the main chain (similar to amylose) every 15–25 glucose units. Amylopectin constitutes about 75% of the most common starches (García et al., 2011). Due to its structural architecture, this component has high crystallinity, it being the cause of the crystalline regions of the starch granule.

To obtain nanocrystals from native starch, the isolation of the crystalline region is needed. This can be accomplished by acid hydrolysis using hydrochloric or sulfuric acid. At temperatures below gelatinization temperature, acid molecules attack the amorphous regions of the granule (Jenkins and Donald, 1997), thus these regions are hydrolyzed faster than crystalline regions (Wang et al., 2003). SNC obtained from corn starch through 3.16 M sulfuric acid hydrolysis by 5 days at 40°C are ranged between 30 and 60 nm in diameter determined by dynamic light scattering. Scanning electron microscopy (SEM) images show the spherical morphology of the nanocrystals (González and Alvarez Igarzabal, 2015). The crystalline nature of the SNC could be determined by X-ray diffractometry (XRD). A typical semicrystalline A-type polymorphic structure characteristic of the starch from cereals is identified in the diffractogram of the native corn starch sample (Kristo and Biliaderis, 2007). On the other hand, SNCs show a pattern similar to that of native starch although the presence of more intense crystalline peaks indicates greater crystallinity of SNC compared to that of native starch, due to a higher amylopectin proportion.

In order to study the effect of the addition of nanocrystals to biodegradable films as compared to the addition of native starch, Chen and coworkers prepared pea SNC. The tensile, moisture uptake, and light transmittance effects of the addition of nanocrystals and native pea starch on the structure and properties of poly(vinyl alcohol) films were comparatively investigated. The authors found that nanocomposite films containing 5 and 10% (w/w) of SNC exhibited improved physical properties over the unreinforced film. In contrast, light transmittance, tensile strength, and elongation at break of the films containing native starch were lower than those of poly(vinyl alcohol) (PVA) film, and decreased when the native starch content was higher (Chen et al., 2008). In addition, when pea starch films were reinforced with waxy maize SNC, the mechanical, water vapor barrier, and morphological properties of composite films improved. In this sense, Li and coworkers founded that the addition of SNC increased the tensile strength of the films when the starch nanocrystal content was 5% (w/w). The moisture content and water vapor permeability (WVP) of the composite films significantly decreased as SNC content increased (Li et al., 2015a).

The use of nanocrystals derived from hydrophilic sources such as starch to reinforce materials of similar chemical structure, such as, polysaccharides or proteins, promotes good miscibility and interfacial adhesion between the components (Lin et al., 2011). In line with this analysis, several works reported protein films being reinforced with SNC. Condés and coworkers studied the effect of SNC prepared from waxy and normal corn starch to the general properties of amaranth protein films. The authors found that the addition of maize SNC to protein formulations improved the physical behavior of protein films since nanocomposites showed improved WVP, water uptake, surface hydrophobicity, and mechanical behavior than those of neat protein films. Moreover, the botanical origin of starch influenced the reinforcement effect, since more significant effect was observed for normal maize nanocrystals compared to waxy starch, which, however, exhibited lower crystallinity (Condes et al. 2015).

Soy protein films were reinforced with 2, 5, 10, 20, and 40% (w/w) of corn SNC.

The major amount of nanocrystals produced an increase in the opacity and the degree of crystallinity of the films. A marked effect on the moisture content, total soluble matter, swelling in water properties was observed when the nanocrystals of starch were added, since the films showed lower affinity for water. Additionally, WVP decreased for reinforced films. This optimization of permeability properties is generally described for reinforced films since the permeation process is controlled by the diffusion of water through the film and not by the solubility of water in the matrix (Slavutsky and Bertuzzi, 2014). According to Kristo and Biliaderis, the addition of rigid crystalline structures produces a tortuous path that hinders the passage of water molecules through the film matrix decreasing the permeation of water vapor or oxygen through the matrix (Kristo and Biliaderis, 2007).

Mechanical properties of soy protein/SNC films were determined showing that reinforced films became more resistant and less elongable as the amount of nanocrystals increased (González and Alvarez Igarzabal, 2015). The substantial improvement of mechanical performance is generally attributed to the enduring stress of rigid nanocrystals and the stress transfer mediated by strong interactions between the filler and the matrix. With the incorporation of a considerable amount of reinforcement, a marked variation was observed in these properties.

As natural matrices, synthetic films can also be reinforced with SNC. Nanocomposite materials were prepared by casting with waterborne polyurethane as the matrix and SNC as the filler. The strength and Young's modulus of the nanocomposites were enhanced while elongation was maintained. Herein, composites containing 10% (w/w) of nanocrystals showed the maximum tensile strength and the higher Young's modulus (Zou et al., 2011).

Several reports describing cellulose nanocrystals can be found in the literature. Probably their extensive study resides in their low cost because they can be extracted from any natural source of biomass. Cellulose nanocrystals can be obtained from acid hydrolysis of rice husk (Johar et al., 2012), pineapple leaf (Dos Santos et al., 2013), potato peel waste (Chen et al., 2012), soy hulls (Flauzino Neto et al., 2013), sugarcane bagasse (Slavutsky and Bertuzzi, 2014), and so on.

Cellulose nanocrystals were obtained by Slavutsky and Bertuzzi in order to reinforce starch films. The authors found that film solubility, contact angle, and water sorption isotherm indicated that reinforced films had lower affinity to water molecules than starch films, because the similar chemical structure of the polymer matrix and the filler led to strong adhesion between them through hydrogen bonding (Slavutsky and Bertuzzi, 2014). A further example of reinforcement was described for PVA and thermoplastic starch (TPS) films reinforced with low amounts of cellulose nanocrystals (1 and 2% w/w). The results indicated that an increase of 19 and 33% for starch composite and 38 and 49% for PVA composite in tensile modulus was observed when 1 and 2% of reinforcement was used, respectively. In addition, water vapor transmission measurements showed a decrease in water permeability for PVA composite (Chen et al., 2012).

The optical properties of reinforced films were studied when cellulose nanocrystals were used. The opacity of reinforced films is defined as the degree to which light is allowed to pass through the material. An increase in opacity means an increase in the amount of light that can pass through the material. In addition, transparency is defined as the opposite of opacity. In this sense, no significant changes in the transparency of the films were reported from the addition of cellulose nanocrystals to biodegradable films. El-Miri and coworkers developed carboxymethyl cellulose/starch blend films reinforced with cellulose nanocrystals extracted from sugarcane bagasse. Hence, the optical transparency of the films was performed by acquiring transmittance spectra in the wavelength region of 200–800 nm. The authors found that all studied films exhibited optical transmittance between 95 and 80% under the visible light region (El-Miri et al., 2015). The transparency level of carboxymethyl cellulose films was not largely affected by its blending with starch, and bionanocomposite films maintained the transparency properties of the blended matrix, confirming the dispersion at the nanoscale of the nanocrystals in bionanocomposite films (Fortunati et al., 2015).

Likewise, Pereda and coworkers developed chitosan films plasticized with glycerol and reinforced with cellulose nanocrystals. The authors observed that the increase in the amount of nanocrystals (up to 7% w/w) led to a decrease in the opacity values of the films. With higher amounts of nanocrystals, opacity increases again, since a clear aggregation of nanocrystals takes place. A plausible explanation for this relates to the fact that when adding nanocrystals to chitosan/glycerol solution, cellulose nanocrystals interact mainly with glycerol molecules because of their hydroxyl groups, thus leaving less free glycerol in the glycerol-rich phase and consequently reducing the size of these domains, which leads to reduced opacity values (Pereda et al., 2014).

According to the reinforcement of chitosan films with cellulose nanocrystals, a chemical strategy was developed by de Mesquita and coworkers in which biobased nanocomposites were obtained through covalent bonds between matrix and filler. The strategy consists in a first functionalization of nanocrystals with methyl adipoyl chloride and the reaction onto the surface of the nanocrystals with the amino groups of chitosan. A replace of up to 8% of the hydroxyl groups was observed in the functionalized nanocrystals through different characterization techniques.

The chemically bonded nanocomposites demonstrated significant improvement in mechanical performance (increase in tensile strength of up to 150%) compared to that of neat chitosan. In addition, results for water uptake measurements showed a remarkable decrease in the hydrophilicity of chitosan (De Mesquita et al., 2012).

Other nanoparticles are also being studied. Chitosan offers real potential for applications in the food industry due to its particular physicochemical properties, short-time biodegradability, biofunctional, biocompatible, antimicrobial, and antifungal activity, and nontoxicity (Aider, 2010). The preparation of chitosan nanoparticles by ionic gelation between chitosan and sodium tripolyphosphate is reported in the literature. Its use as a filler in a fish gelatin matrix to produce bionanocomposite films was studied by Hosseini and coworkers. The obtained particles exhibited spherical shape with a size range of 40–80 nm, and a positively charged surface with a zeta potential value of +10 mV. XRD results confirmed the cross-linking reaction between chitosan and sodium tripolyphosphate. SEM images showed that nanoparticles could be well dispersed in gelatin polymer matrix at low content, while higher loadings (8% w/w) led to the aggregation of particles in the composites. The evaluation of the mechanical properties showed important improvements since tensile strength increased from 7.44 to 11.28 MPa and elastic modulus increased approximately 60% for a composite with 8% (w/w) of nanoparticles with respect to unreinforced films. Furthermore, the addition of chitosan nanoparticles contributed to the significant decrease in WVP, leading to a 50% decline at 6% (w/w) filler.

Another example of the influence of chitosan nanocrystals was developed through electrospun chitosan-based randomly oriented fibrous mats containing chitin nanocrystals (ChNC) as a reinforcement and cross-linked with genipin. Results showed that electrospun porous random mats comprising ChNC were free of any defects due to the homogeneous dispersion of ChNC in the chitosan matrix, indicating good chemical compatibility between matrix and chitin. The addition of ChNC improved the moisture stability of the as-spun mats, facilitating water-mediated cross-linking processes. The cross-linked nanocomposite fiber mats with 50 wt.% ChNC showed high tensile and flexible strength (Naseri et al., 2014).

In addition to nanocrystals, biopolymer matrices may be reinforced with laminar or fiber-type nanoreinforcement.

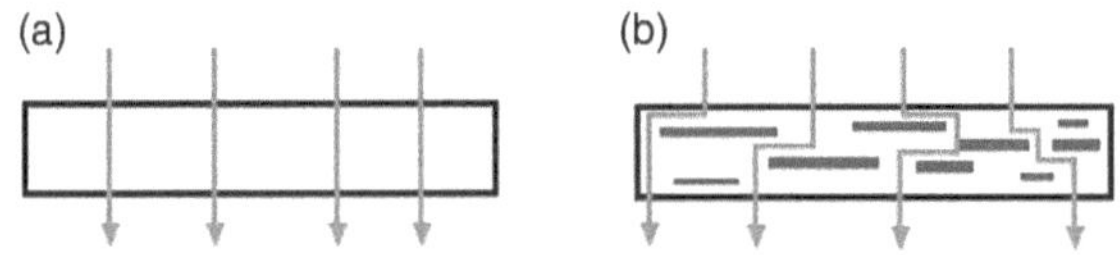

Figure 8.4. (a) Gas diffusion across a film without reinforcement and (b) tortuous way in presence of laminar reinforcement.

2.2 Inorganic Laminar Fillers as Film Reinforcement

Within the laminar nanoreinforcement, the packaging industry focuses its attention on layers of inorganic solids such as clays due to availability, low cost, and enhancements offered to the films in terms of permeability to gases and water vapor and mechanical properties (Chung et al., 2010; Arora and Padua, 2010). The theories most commonly used to explain the improvements of nanocomposite materials in the barrier properties are based on the formation of a tortuous path formed by the presence of the reinforcement, forcing the permeable gas to pass a long way to diffuse through the film. The presence of clays in the biopolymer increases the diffusion distance of the molecule to penetrate, conferring excellent barrier properties as seen in Fig. 8.4.

Within clay types, the most investigated are phyllosilicates such as montmorillonite (MMT). This inorganic material is particularly effective as reinforcement filler due to its high surface area. The dispersion of clays in biopolymers is usually inhomogeneous because of their hydrophilic surface.

There are reports of the use of these clays as reinforcement of films formed from natural sources. In films prepared from soy protein, the use of exfoliated MMT produced between protein chains increases the breaking strength and elasticity module, reducing moisture content, solubility, and vapor permeability water (Echeverría et al., 2014). In addition, the increased thermal stability of the films (Kumar et al., 2010) is produced. For films formed by polysaccharides such as starch, improvements in thermal stability, reduced water absorption and increased Young's modulus (Cyras et al., 2008; Schlemmer et al., 2010) were reported. As a result of the optimizations of the final properties of these films, they prove useful materials for rigid or flexible food packaging.

2.3 Cellulose Nanofibers as Film Reinforcement

Natural fibers stand out among nanofibrillar reinforcements. At present, biofibers as fillers and reinforcements of natural films stand out due to their flexibility during processing, high longitudinal stiffness, and low cost. In general, organic waste is an important source of fibers, conferring high added value. Thus, the preparation of nanocellulose fibers from plant residues such as soybean pods (Flauzino Neto et al., 2013), pineapple plant leaves (Dos Santos et al., 2013), or rice hulls (Johar et al., 2012) has been reported. Some natural nanofibers most commonly used derived from tow (jute, flax), leaf (sisal, pineapple), seed (coconut, cotton), core (jute), sugarcane (wheat, corn, rice, and sugarcane), and from wood and roots (Faruk et al., 2012). With the utilization of these nanofibers, some improvements were obtained in mechanical properties of several films derived from the great tensile strength that this reinforcements exhibit (Lodha and Netravali, 2005).

Nanocellulose fibers are inexpensive, available, strong, and rigid, lightweight, biodegradable, and low power consumption is required for their manufacture Therefore, nanomaterials are attractive for the production of inexpensive, lightweight, and high strength nanobiocomposites. In various reports (Azeredo, 2009; Abdul Khalil et al., 2012, 2014; Siró and Plackett, 2010; Ja et al., 2014; Abdul Sirviö et al., 2014; Lavoine et al., 2012; Xu et al., 2015), nanocellulose-based reinforcements have been extensively described and reviewed. These include the structure of nanofibers and whiskers (crystalline rod-like nanoparticles obtained by acid hydrolysis of cellulose fibers), the methods of preparation, the ways of modification, their properties, the factors that affect the behavior of these reinforcements, the applications and the effects in polymer matrices, and the improvement in compatibility between nanocellulose fibers and hydrophobic matrices.

Nanocellulose fibers have been used to reinforce starch films showing improved mechanical properties, increased barrier properties to water vapor, and increased resistance in high humidity environments (Müller et al., 2009a, 2009b; Savadekar and Mhaske, 2012). Other studies have demonstrated the potential of nanocellulose fibers as reinforcing elements in various polymers including chitosan, poly(lactic acid) (PLA), PVA, and polypropylene (Yuwawech et al., 2015; Fernandes et al., 2010; Prakobna et al., 2015; Iwamoto et al., 2014; Okubo et al., 2009; Nakagaito et al., 2009) to improve the mechanical properties, the behavior as barrier and heat sealability, as well as other functional properties such as transparency, biodegradability, and specific surface properties.

Nanopaper has been produced even from nanofibrillated cellulose, which has proved a very interesting packaging material due to its high hardness and high oxygen-barrier effectiveness (Henriksson et al., 2008; Nogi et al., 2009).

As a specific example, the effect of different concentrations (up to 36 g/100 g) of cellulose nanofibers (CNFs) added as a nanoreinforcing component of mango puree edible films was evaluated on tensile properties, WVP, and glass transition temperature (*Tg*) (Azeredo et al., 2009). The cellulose reinforcement was well dispersed into the mango puree matrix and the performance of the final films was improved by CNF reinforcement. CNFs were effective in increasing tensile strength, and its effect on Young's modulus (YM) was even more noticeable, especially at higher concentrations, suggesting the formation of a fibrillar network within the matrix. The WVP was significantly decreased when CNFs were incorporated at loadings of at least 10%. The effect of the filler on *Tg* was low but significant. The study demonstrated that the properties of mango puree edible films can be significantly improved through CNF reinforcement. In other studies, Azeredo et al. (2010) evaluated the effect of different concentrations of CNF as a nanoreinforcement and plasticizer (glycerol) on the same properties, but on chitosan edible films. The experiment was conducted according to a central composite design, with two variables: CNF (0–20 g/100 g) and glycerol (0–30 g/100 g) concentrations in the film produced by casting. Most responses (except by elongation at break, EB) were favored by high CNF concentrations and low glycerol content. The optimization was based on maximizing tensile strength, YM and *Tg*, and decreasing WVP, while maintaining a minimum acceptable EB of 10%. The authors defined the optimum conditions of glycerol and CNF concentration for chitosan films and demonstrated how CNF can improve their mechanical and water vapor barrier properties.

Minelli et al. (2010) conducted research on the mass transport properties of water and oxygen in four different film materials obtained from two types of microfibrillated cellulose (MFC), with and without glycerol as a plasticizer, in order to understand particular structure–property relationships in the films. Two different types of MFC, G1 and G2, were used. MFC G1, a commercial bleached sulfite softwood pulp consisting of 40% pine and 60% spruce with hemicellulose content of 13.8% and lignin content of 1%, and MFC G2 prepared from sulfite softwood pulp with hemicellulose content of 4.5% and lignin content of 0.6%. In general MFC G1 showed a more clearly defined fibrous structure compared to that of MFC G2, which, with higher charge density and lower nanofibril dimensions, leads to a more homogeneous and less porous film structure. The addition of plasticizer decreased the porosity of the materials in both cases and in MFC G2 it led to a substantially solid-like structure. Films prepared from MFC G2 exhibited higher water sorption and lower diffusivity than those prepared from MFC G1. The addition of glycerol increased diffusion coefficient and influenced water sorption, leading to decreased water uptake at low water pressure while water uptake was enhanced at high activity. It was suggested that the differences in the sorption of the two types of materials could be related to the higher charge density of MFC G2 and to its greater tendency to absorb water in the regions between the fibril microstructure. All films displayed excellent oxygen-barrier properties under dry conditions possibly resulting from a very dense lamellar structure with no porosity.

Belbekhouche and coworkers prepared cellulosic membranes using nanoparticles (whiskers vs. MFC). The rod-like straight cellulose whiskers and the long flexible entangled MFC were extracted from sisal fibers. The water vapor and gas (carbon dioxide, nitrogen, and oxygen) permeability of the transparent films was analyzed by comparison between both nanoparticles. It was concluded that the diffusion of water was controlled by the surface rather than by the core, probably due to the barrier effect related to the presence of water at the surface during the sorption kinetics. The cellulose whisker film was found to be much more permeable to gases than MFC. This difference ascribed to the higher porosity of the former and to the possibility of entanglement of MFC and different surface chemistry. MFC contains residual lignin, extractive substances, and fatty acids at the surface. In its film, gas molecules penetrate much more slowly by a longer diffusion path. In addition, the entanglements of these long flexible nanoparticles and the lower porosity of the films represent barrier domains leading to a higher tortuosity of the diffusion pathway (Belbekhouche et al., 2011).

Li et al. prepared nanosized chitin whiskers by hydrolyzing α-chitin sample and used them at different ratios to improve the strength properties of paper sheets. The purpose was to develop a convenient approach to enhance these paper properties through the electrostatic assembly between chitin whisker and cellulosic fiber in an aqueous medium. Thus, eucalyptus cellulose fibers were first treated with carboxymethylcellulose to increase the negative charge density of the fiber surface and form paper sheets. Fibers were then further dip-coated with different concentrations of chitin whisker suspension. Transmission electron microscopy, field emission scanning electron microscopy, and zeta potential were used to analyze the composite material. The authors suggested that electrostatic attraction between cellulose fibers and chitin whiskers can aid retention, allowing hydrogen bonds to be formed between the solid surfaces during drying, in such a way as to increase connectivity within cellulose nanostructures. Therefore, coating with chitin whiskers was shown to increase tear strength, burst strength, and wet and dry tensile strength (Li et al., 2015b).

2.4 Cellulose Nanowhiskers as Film Reinforcement

Azeredo et al. (2012b) evaluated the tensile properties and the water vapor barrier of alginate-acerola puree films plasticized with corn syrup, and studied the influence of cellulose whiskers from different origins (from cotton or from coconut husk fibers submitted to different bleaching levels) on film properties. The whiskers improved the tensile properties and water vapor barrier of the films proving their use as edible packaging. In another report (Azeredo et al., 2012a), the combination between acerola puree and alginate was reinforced with cellulose whiskers (CW) or MMT to form nanocomposite edible films (casted on glass plates) and edible coatings (applied to acerola fruit surfaces). Both nanofillers (CW and MMT) reduced WVP of films. When applied to fresh acerolas, coatings decreased fruit weight loss, decay incidence, and ripening rates. Moreover, MMT maintained the red color of the coating anthocyanins, making the visual acceptance of coated acerolas be similar to that of uncoated fruits.

In another report, microcrystalline cellulose from cotton linter was first prepared by acid hydrolysis, followed by a process to prepare cotton linter nanowhiskers (CLNW) (Savadekar et al., 2015). These were used as reinforcing agents in TPS to prepare nanocomposite films and to improve the performance properties. TPS/CLNW nanocomposite films were characterized for tensile, thermal, morphological, light transmittance, WVP, and oxygen permeability properties exhibiting excellent performance. The improvements were proposed mostly due to the hydrophilic nature of both materials, TPS and CLNW, showing better interaction between them.

Cellulose nanowhisker (CNW) was also used to reinforce polylactic acid (PLA) composites using whole milk casein protein (CS) for being an environmentally compatible dispersant (Gu and Catchmark, 2013). The presence of whole milk casein improved the dispersion of CNWs in the PLA matrix. The evaluation of the mechanical properties of pure PLA and its composites shows the reinforcement effect of CNWs. The improved dispersion of CNWs provided higher modulus of PLA composites at a higher reinforcement loading and maintained the strain and stress at break of composites at a relatively low reinforcement loading. The affinity of casein as a dispersant to PLA proved important for the ultimate strength and stiffness of the composites.

2.5 Carbon Nanotubes as Film Reinforcement

Other (inorganic) fibrillar materials are carbon nanotubes (CNTs). CNTs consist of a material of 1-atom thick and single wall, or a series of concentric tubes called multiple nanotubes that present high modulus of elasticity. To enhance their interaction with other biopolymers, they can be modified by introducing carboxyl groups on their surface. CNTs have been used to improve thermal stability and tensile strength (George et al., 2015).

Reinforced composites formed by soy proteins and CNTs showed an increased resistance of the material to water and a particularly marked improvement in its mechanical properties, despite being dependent on the amount and size of nanotube added (Zheng et al., 2007).

Multifunctional polymeric nanocomposites containing chitosan–starch with uniformly distributed decorated magnetite—CNTs were also described. The most homogeneous particle distribution was obtained from 0.1% magnetite-decorated nanotubes. Higher concentrations of magnetite-decorated nanotubes showed formation of agglomerate resulting in a decrease in storage modulus. The magnetic and electric properties also depend on the content of magnetite-decorated multiwall CNTs. The inclusion of CNTs yields a conducting polymer with a potential application as a biosensor in biomedicine. The magnetic response of these multifunctional nanocomposites provides the opportunity to control the bioactive molecule release (Castrejón-Parga et al., 2014).

The study of graphene/CNT (G/CNTs) hybrid fillers was described by Ma and coworkers. XRD and SEM were used to characterize the obtained hybrid fillers. The solution blending of G/CNTs with PVA matrix was used to prepare PVA composite films with enhanced mechanical properties and thermal stability. The Young's modulus and tensile strength of PVA composite films containing 1 wt.% G/CNTs, were 33.6 and 56% higher than PVA pure, respectively. This behavior was possibly due at the interconnected 3-dimensional structure of G/CNTs with the homogeneous dispersion and the strong interaction of hydrogen bond between G/CNTs PVA and macromolecular chains (Ma et al., 2016).

Recently, other synthetic nanoreinforcements have been developed in order to yield films with improved properties. Thermocompressed films of thermoplastic corn starch reinforced with talc nanoparticles in 0, 1, 3, and 5% w/w were developed. The mechanical properties of films were studied, revealing starch films reinforcement by the addition of talc. The mechanical improvement of TPS films with a talc concentration higher than 3% w/w was reflected in the increase in strength properties, since tensile strength increased 68% with no detriment in elongation at break. Additionally, barrier properties were determined; it was concluded that water vapor and oxygen permeability were reduced in a 54 and 26% proportion, respectively, by addition of talc (3% w/w) to the TPS (Castillo et al., 2013).

3 Active Packaging

Active packaging focused on providing food protection and preservation through some mechanism activated by intrinsic and/or extrinsic factors, Fig. 8.5 (Lim, 2011).

The incorporation of nanoscale active compounds (antimicrobials, antioxidants, preservatives, flavors, etc.) in food packaging materials that are capable of improving shelf life, food quality, and/or safety are termed active packaging. Active compounds are more suitable when covalently bonded to the package material than when physically incorporated since they can migrate to the food. In the first case, the compounds produce an active effect but are not ingested with the food while in the last case, are considered as food additives and could be used only under regulation.

The possibility of incorporating or binding active compounds in the different nanosystems prior to their incorporation into packaging material has been described. However, there is another way to prepare active packaging through nanostructured coating in food packaging. In nanocoating, the nanomaterial can be applied to the package surface or can be incorporated as a laminate composite layer. Recently, much attention has been directed to the use of nanostructured coating due to its inherent advantages over multilayer films to improve the barrier performance and antimicrobial aspects of the package.

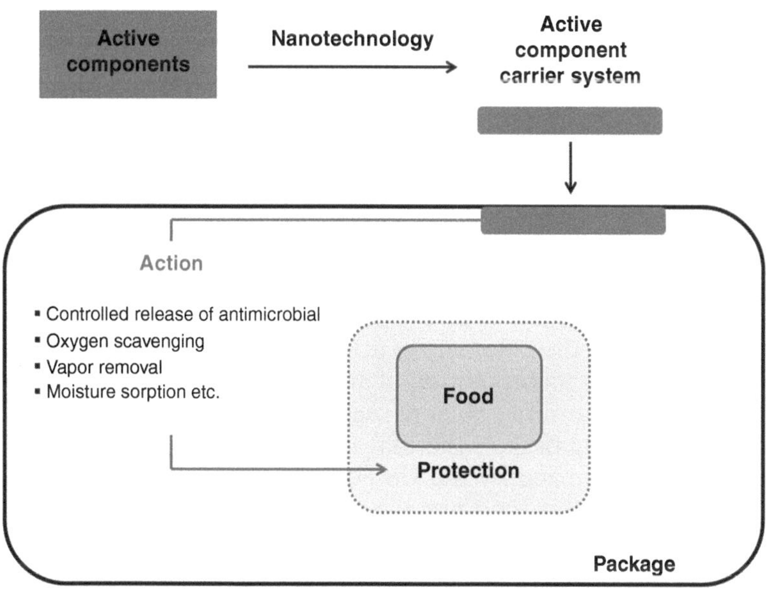

Figure 8.5. Schematic diagram showing the active packaging concept and its nanotechnology involvement. Adapted from Mihindukulasuriya and Lim (2014).

The effective antimicrobial properties of calcium silicate and titanium dioxide (TiO_2) nanostructured surfaces were described (Johnston et al., 2008; Fu et al., 2005), respectively. Recent advances in the packaging industry showed the potential use of nanostructured surface coating for the preparation of water-repellent papers using TiO_2 nanoparticles (Teisala et al., 2010). This effect was attributed to its superhydrophobic nanoscale surface (Contreras et al., 2014).

At the moment, active packaging has been mainly developed for antimicrobial packaging applications. Other promising applications comprise oxygen scavengers, ethylene removers, and carbon dioxide absorbers/emitters.

3.1 Antimicrobial Packaging

Antimicrobial packaging is an area with high potential for applying nanotechnology to the control of undesirable microorganisms on food by means of the incorporation of active molecules such as antimicrobial compounds in coating into packaging materials (Sozer and Kokini, 2009). Nanoparticles or nanocomposite materials have been investigated for antimicrobial activity such as growth inhibitors, antimicrobial agents, antimicrobial carriers, or antimicrobial packaging films. The main potential food applications for antimicrobial nanosystem films include meat, fish, poultry, bread, cheese, fruits, and vegetables (Rhim et al., 2013a).

Metal and metal oxide nanoparticles and CNTs are the nanosystems most widely used to develop antimicrobial active packaging. These particles function in direct contact; however, they can also migrate slowly and react preferentially with organic compounds in the food.

Several studies have demonstrated the antimicrobial activity in chitosan of silver nanoparticles included in PVA with cellulose nanocrystals, or polyethylene (PE) films that are effective against both Gram-positive and Gram-negative bacteria, fungi, protozoa, and certain viruses (Costa et al., 2011, 2012; Gammariello et al., 2011; Cruz-Romero et al., 2013; Mihindukulasuriya and Lim, 2014). The incorporation of silver (Ag) nanoparticles in the preparation of antimicrobial packaging, offers important advantages, conferring besides thermal stability. This nanocomponent has been recognized by the US Food and Drug Administration (FDA) for its use as food preservative, although it is still considered as an additive, with general restrictions, by the European Food Safety Authority (EFSA). Materials used for packaging containing Ag nanoparticles are widely used for packaging of vegetables, fruits, and fruit juices. (Costa et al., 2011). Azlin-Hasim and coworkers have studied the antimicrobial effect of low-density polyethylene (LDPE) nanocomposite films containing different concentrations of Ag nanoparticles. The nanocomposite films were obtained by extrusion and the antimicrobial assay was performed on chicken breast fillets. The authors demonstrated that Ag/LDPE nanocomposite films extended significantly the shelf life of chicken breast fillets and these nanocomposite films have the potential to be used as antimicrobial packaging of food products (Shafrina Azlin-Hasim et al., 2015). Gammariello and coworkers evaluated the effects of an alginate-based coating containing Ag-MMT nanoparticles on the shelf life of *Fior di latte* cheese packaged with modified atmosphere packaging (Gammariello et al., 2011).

TiO_2, zinc oxide (ZnO), silicon oxide (SiO_2), and magnesium oxide (MgO) are the oxide nanoparticles most widely studied for their ability to be UV blockers and photo-catalytic disinfecting agents (Fujishima et al., 2000). It is well known that photo-catalyst titanium dioxide (TiO_2), a wide band gap (3.2 eV) semiconductor under UV illumination, generates energy-rich electron hole pairs that can be transferred to the surface of TiO_2, and promotes reactivity with the surface-absorbed molecules leading to the production of active radicals. These active radicals oxidize C—H bonds of polyunsaturated phospholipids of cell membranes of microorganisms resulting in their degradation. The photo-catalyst reaction of TiO_2 leads to generation of hydroxyl radicals (·OH) and reactive oxygen species (ROS) on the TiO_2 surface, resulting in the oxidation of polyunsaturated phospholipids of the cell membranes of microorganisms. As a consequence, the microorganism is inactivated. TiO_2 has been approved by the FDA as a nontoxic material for human food, drugs, cosmetics, and use in the food industry. Currently, there is considerable interest in the self-disinfecting property of TiO_2 for meeting hygienic design requirements in food processing and packaging surfaces. TiO_2 photo-catalyst has been used to degrade organic pollutants and inactivate a wide spectrum of microorganisms. Some interesting studies have demonstrated the efficiency of TiO_2 nanoparticles. Bodaghi et al. prepared a LDPE-TiO_2 nanocomposite by melt blending method and studied in vitro and in vivo the photo-catalytic antimicrobial effect of nanocomposites. The in vitro test was performed with *Pseudomonas* spp. and *R. musilaginosa* and the in vivo test was carried out on fresh pears. The authors observed that the film caused inactivation of *Pseudomonas* spp., *R. mucilaginosa*, and mesophilic bacteria in saline solution and on pears when exposed to UV light. The number of microorganisms on LDPE-TiO_2 nanocomposite film plus UV light was lower than that of nanocomposite film without UV light and LDPE film exposed to UV light. These results are of industrial relevance because an effective technique for developing a novel photo-catalyst thin film as an active packaging system (Bodaghi et al., 2013) was presented.

More recently, the antimicrobial properties of nano-ZnO and MgO have been reported (Silvestre et al., 2011). These nanoparticles are expected to provide a more affordable and safe food packaging solution in the future. It was demonstrated that ZnO nanoparticles exhibit better antibacterial activity when particle size decreases (Yamamoto, 2001), and their activity does not to require the presence of UV light but the presence of visible light (Jones et al., 2008). This class of nanoparticles has been incorporated in a number of different polymers (Azeredo, 2013; Silvestre et al., 2011). Premanathan et al. (2011) reported antimicrobial activity of ZnO nanoparticles against *Listeria monocytogenes*, *Salmonella enteritidis*, and *Escherichia coli*, indicating that it was essential for nanoparticles to contact or penetrate into microbial cells. They were bacteriostatic or bactericidal depending on the concentration.

On the other hand, the bactericidal mechanism of MgO nanoparticles was related to the production of high concentrations of superoxide anions (O_2^-) on MgO surface, which can react with the carbonyl groups of peptide linkages in bacterial cell walls and destroy them.

The effectiveness of MgO nanowires as antimicrobial agents has been demonstrated by Al-Hazmi et al. They showed increasing activity against *E. coli* and *Bacillus* sp., with increasing concentration of MgO nanowires (Al-Hazmi et al., 2012).

CNTs could be used for improving the properties of polymer matrix and for their antibacterial properties. Several studies suggest that CNTs are cytotoxic to human cells, at least when in contact with skin (Warheit et al., 2004), therefore their application was stopped.

Biopolymer nanoparticles or whiskers are usually used for their intrinsic antimicrobial properties, as chitosan or as nanocarriers of antimicrobial compounds. Chitosan nanoparticles (CSNP) and their derivatives have wide antimicrobial effects. The most accepted antimicrobial mechanism for chitosan is related to interactions of cationic chitosan with anionic cell membranes, increasing membrane permeability and eventually resulting in rupture and leakage of the intracellular material (Zhao et al., 2011).

Composite chitosan-ZnO nanofibers were obtained using PVA as a support. The presence of nano-ZnO in the nanofibers was demonstrated to contribute to the antimicrobial activity of chitosan. The coantimicrobial effect of chitosan and nano-ZnO was attributed to a sequence of effects, including: adherence of nanofibers to cell membranes through electrostatic attraction, leading to changes in microbial cell membranes (denaturation of proteins and permeability changes); disruption of cell membranes by ROS produced by nano-ZnO, causing leakage of intercellular contents; interference in cell metabolism by nano-ZnO and chitosan combination with DNA and RNA, blocking genome replication (Cruz-Romero et al., 2013).

3.2 Antioxidant Packaging

Waste products from fruit and vegetable processing provide a practical and economical source of potent antioxidants/antimicrobials that could replace synthetic preservatives. These preservatives could be included in nanocarriers. As an example, functionalized nanoclays were used as carriers of the natural extract obtained from a brewery waste stream and incorporated into a polymer film to develop an active packaging film with both antioxidant and antimicrobial properties (Barbosa-pereira et al., 2014). The suitability of the natural extract for incorporating into a polymeric matrix by extrusion was assessed by testing thermal stability. Two polymeric matrices (LDPE and PEVA) were compared and the optimal concentration of the natural extract, the overall migration, and the antimicrobial and antioxidant activities of the active films were tested. In addition, the use of functionalized nanoclays (prepared with the natural extract) improved the effectiveness of the active packaging and minimized the amount of natural extract required. Barbosa-pereira et al. (2014) demonstrated the extension of the shelf life of meat products.

A further important type of active packaging is that which contributes to decreasing the deterioration of many foods, either directly or indirectly by oxidation with nano-O_2 scavengers. Several nanoparticles, including TiO_2, were used to produce oxygen-scavenger films. Others based on Ag nanoparticles are able to absorb and decompose ethylene (Li et al., 2009). Removing ethylene from a package environment helps to extend the shelf-life of fresh produce like fruits and vegetables.

4 Intelligent Packaging

Intelligent packaging systems are those with enhanced function in communication and marketing of traditional packaging. They have been produced and manufactured as an indicator system that monitors the physicochemical properties of the packaged food or the environment during transport or storage properties.

Therefore, intelligent packaging provides the correct information on food conditions, its environment and packaging integrity (Vanderroost et al., 2014).

Since the beginning of nanotechnology, food packaging innovation activities have gradually expanded toward the development of intelligent packaging. The key elements of nanotechnology-based intelligent packaging systems are showed in Fig. 8.6.

At this point, intelligent packaging should not be confused with an active packaging, because they have different functions. Active packaging improves and extends the functions of the traditional packaging and is designed to interact with food and its environment to extend food quality and shelf life, whereas intelligent packaging is defined as an extension of the communication function of traditional food packaging. However, Vanderroost et al. defined both kinds of packaging as not mutually exclusive, which can work synergistically to provide so-called smart packaging. They form a total packaging solution since they could monitor changes in the product or environment and act upon these changes (Vanderroost et al., 2014).

So far, there are three systems that act as intelligent packaging: sensors, indicators and radio frequency identification (RFID) systems.

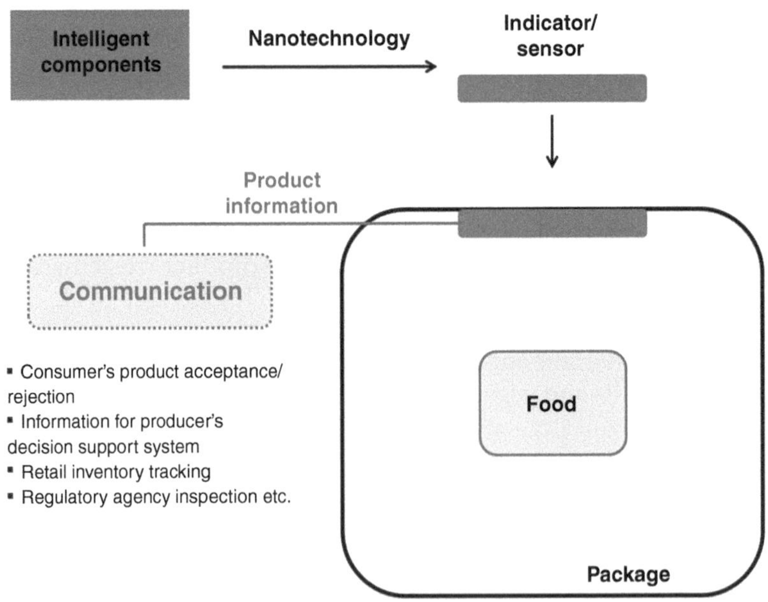

Figure 8.6. Schematic diagram showing the intelligent packaging concept and its nanotechnology involvement. Adapted from Mihindukulasuriya and Lim (2014).

The nanotechnologies most often developed for the preparation of intelligent packaging are involved in sensors/indicators of compounds such as oxygen, detection of spoilage and pathogenic microorganisms, time, temperature, humidity, and freshness indicators (Mihindukulasuriya and Lim, 2014; Cushen et al., 2012).

Time-temperature indicators are useful to track thermal history during food storage, handling, and distribution. They are simple, effective, and easy-to-use devices for monitoring and accumulatively indicating the overall influence of temperature on quality, from manufacturing to the end consumer. These types of intelligent packaging have been widely applied to monitor and evaluate the storage shelf-life of many chilled and frozen foods, such as fish and seafood, vegetables, and dairy products (Pereira et al., 2015).

Nanoparticles of TiO_2 are used as a UV-activated oxygen indicator. They are typically encapsulated in a polymer carrier, as a coating or film (Mills and Hazafy, 2008; Mills and Hazafy, 2009). Mihindukulasuriya and Lim prepared a membrane as an indicator of oxygen using glycerol, methylene blue, and TiO_2 nanoparticles encapsulated in poly(ethylene oxide) fibers obtained by electrospin technique. The sensitivity of this membrane was higher than that of cast membrane carrier (Mihindukulasuriya and Lim, 2014).

Detection of humidity is an indicator of the integrity of the package and the quality and safety of the food product. Zhou studied the use of nanocrystalline cellulose film made by casting to form a thick iridescent film in which chiral nematic textures are developed by the self-assembly of rigid rod crystallites. This film was capable of changing from blue-green to red-orange in less than 2 s by exposure to high humidity or water (Zhou, 2013).

The function of the freshness indicators is to provide information for the producer, retailer, and/or consumer on the actual product quality during storage and distribution. Smolander et al. invented a coating of Ag and/or Cu nanoparticles of 1–10 nm thick on plastic film or paper packaging. After reacting with sulfide volatile compounds produced from the spoilage of fresh meats, the thin layer becomes dark (Hurme et al., 2004). Other researchers described the use of CNTs for the development of a sensor to detect CO_2, volatile compounds and ethylene during the repining of fruits (Esser et al., 2012), or the use of a cross-reactive conjugated poly(thiophene) indicator in the detection of amine profiles in tuna fish (Maynor et al., 2007).

There are important research studies that are not currently applied to technologies for food packaging; yet, it is desirable that these works can be technologically applied, especially in the development of advanced intelligent labeling for the detection of pathogens and their toxins in food. For example, antibodies conjugated to nanomaterials, such as, quantum dots, have been developed to detect bacteria. They exhibited blue-shifted fluorescence emission when bound to target bacteria surfaces (Dwarakanath et al., 2004). Yang and Li (2006) investigated the use of quantum dots for simultaneous detection of *E. coli* O157:H7 and *Salmonella typhimurium* (Yang and Li, 2006). Quantum dots are being only used today in labeling imaging and therapy areas. A further interesting example is the use of triangular Ag nanoplates as colorimetric indicators for time-temperature history (Zeng et al., 2010). Ag nanoplates have sharp corners and display strong in-plane dipole resonance mode in visible radiation spectrum. As the corner of the nanoplate increases, it becomes more rounds during storage, and a gradual blue shift results for the resonance peak position, changing from cyan to blue with time; the rate of this color transformation is temperature-dependent. The main challenges of applying time-temperature indicators (TTIs) to reflect food quality relates to the fact that the kinetics of the TTIs response must match those of the key degradation reactions in food (Mihindukulasuriya and Lim, 2014).

5 Conclusions

The various nanosystems that have been discussed in this chapter show the major advantages of their uses for the development of a new generation of packaging with improved performance. This advancement will positively affect the shelf life, quality, safety, and security of food, which will ultimately benefit both producers and consumers.

Although there are several publications of the use of nanosystems for the preparation of nanomaterials for food packaging, there are few studies into the structure/property relationship. By structure we not only mean their chemical structure but also their shape. For instance, it is known that nanolaminates improve barrier properties, while fibers improve mechanical properties; cellulose nanocrystals or nanofibers are more efficient in biopolymer-derived materials due to their chemical affinity; Ag nanoparticles incorporated in polymers convert packagings into active or intelligent. Therefore, the understanding of this relationship is key for the selection of nanosystems required to provide specific properties in a nanopackaging.

For an extensive use of nanotechnology in food packaging, it is necessary to know the risk assessment and effect on the environment and health. However, precautionary principles should be applied and further research would be needed. Today, it is difficult to determine how many of these products can find suitable application. It is clear that these new technologies, if managed and regulated correctly, can play a central role in improving product and process development.

References

Abdul Khalil, H.P.S., et al., 2014. Production and modification of nanofibrillated cellulose using various mechanical processes: a review. Carbohyd. Polym. 99, 649–665.

Abdul Khalil, H.P.S., Bhat, a.H., Ireana Yusra, a.F., 2012. Green composites from sustainable cellulose nanofibrils: a review. Carbohyd. Polym. 87 (2), 963–979.

Aider, M., 2010. Chitosan application for active bio-based films production and potential in the food industry: Review. LWT—Food Sci. Technol. 43 (6), 837–842, http://www.sciencedirect.com/science/article/pii/S002364381000040X

Al-Hazmi, F., et al., 2012. A new large-scale synthesis of magnesium oxide nanowires: structural and antibacterial properties. Superlat. Microstruct. 52 (2), 200–209, http://www.ncbi.nlm.nih.gov/pmc/articles/PMC3990453/

Arora, A., Padua, G.W., 2010. Review: nanocomposites in food packaging. J. Food Sci. 75 (3), 43–49.

Arvanitoyannis, I.S., 2005. Food packaging technology.In: Coles, R., McDowell, D., Kirwan, M.J. (Eds.). Blackwell Publishing, CRC Press, Oxford, 2003. Journal of the Science of Food and Agriculture, 85(6), p.1072.

Azeredo, H.M.C., 2013. Antimicrobial nanostructures in food packaging. Trends Food Sci. Technol. 30, 56–69.

Azeredo, H.M.C., Miranda, K.W.E., Ribeiro, H.L., et al., 2012a. Author's personal copy Nanoreinforced alginate—acerola puree coatings on acerola fruits. J. Food Eng. 113, 505–510.

Azeredo, H.M.C., Miranda, K.W.E., Rosa, M.F., et al., 2012b. Edible films from alginate-acerola puree reinforced with cellulose whiskers. LWT—Food Sci. Technol. 46 (1), 294–297.

Azeredo, H.M.C., et al., 2010. Nanocellulose reinforced chitosan composite films as affected by nanofiller loading and plasticizer content. J. Food Sci. 75 (1), 1–7.

Azeredo, H.M.C., et al., 2009. Nanocomposite edible films from mango puree reinforced with cellulose nanofibers. J. Food Sci. 74 (5), 31–35.

De Azeredo, H.M.C., 2009. Nanocomposites for food packaging applications. Food Res. Int. 42 (9), 1240–1253.

Barbosa-pereira, L., et al., 2014. Development of new active packaging films containing bioactive nanocomposites. Innov. Food Sci. Emerg. Technol. 26, 310–318.

Belbekhouche, S., et al., 2011. Water sorption behavior and gas barrier properties of cellulose whiskers and microfibrils films. Carbohyd. Polym. 83 (4), 1740–1748.

Bodaghi, H., et al., 2013. Evaluation of the photocatalytic antimicrobial effects of a TiO_2 nanocomposite food packaging film by in vitro and in vivo tests. LWT - Food Sci. Technol. 50, 702–706.

Brody, A.L., et al., 2008. Innovative food packaging solutions. J. Food Sci. 73 (8), R107–R116.

Brody, A.L., 2008. Packaging by the numbers. Food Technol. 62, 89–91.

Castillo, L., et al., 2013. Thermoplastic starch films reinforced with talc nanoparticles. Carbohyd. Polym. 95 (2), 664–674.

Castrejón-Parga, K.Y., et al., 2014. Chitosan–starch film reinforced with magnetite-decorated carbon nanotubes. J. Alloys Comp. 615, S505–S510.

Chaudhry, Q., et al., 2008. Applications and implications of nanotechnologies for the food sector. Food Addit. Contam. Part A 25 (3), 241–258.

Chen, D., et al., 2012. Biocomposites reinforced with cellulose nanocrystals derived from potato peel waste. Carbohyd. Polym. 90 (1), 709–716.

Chen, Y., et al., 2008. Comparative study on the films of poly (vinyl alcohol)/ pea starch nanocrystals and poly (vinyl alcohol)/ native pea starch. Carbohyd. Polym. 73, 8–17.

Chung, Y.L., et al., 2010. Preparation and properties of biodegradable starch–clay nanocomposites. Carbohyd. Polym. 79 (2), 391–396.

Condes, M.C., et al., 2015. Amaranth protein films reinforced with maize starch nanocrystals. Food Hydrocoll. 47, 146–157.

Contreras, C.B., et al., 2014. Permanent superhydrophobic polypropylene nanocomposite coatings by a simple one-step dipping process. Appl. Surf. Sci. 307, 234–240.

Costa, C., et al., 2011. Antimicrobial silver-montmorillonite nanoparticles to prolong the shelf life of fresh fruit salad. Int. J. Food Microbiol. 148 (3), 164–167.

Costa, C., et al., 2012. Calcium-alginate coating loaded with silver-montmorillonite nanoparticles to prolong the shelf-life of fresh-cut carrots. Food Res. Int. 48 (1), 164–169.

Cruz-Romero, M.C., et al., 2013. Antimicrobial activity of chitosan, organic acids and nanosized solubilisates for potential use in smart antimicrobially-active packaging for potential food applications. Food Cont. 34, 393–397.

Cushen, M., et al., 2012. Nanotechnologies in the food industry: recent developments, risks and regulation. Trends Food Sci. Technol. 24, 30–46.

Cyras, V.P., et al., 2008. Physical and mechanical properties of thermoplastic starch/montmorillonite nanocomposite films. Carbohyd. Polym. 73 (1), 55–63.

Da-Silva, M.A., et al., 2013. Evaluation of the antimicrobial potential of alginate and alginate/chitosan films containing potassium sorbate and natamycin. Pack. Technol. Sci. 26 (8), 479–492.

Dos Santos, R.M., et al., 2013. Cellulose nanocrystals from pineapple leaf, a new approach for the reuse of this agro-waste. Ind. Crops Prod. 50, 707–714.

Dufresne, A., 2010. Processing of polymer nanocomposites reinforced with polysaccharide nanocrystals. Molecules 15, 4111–4128.

Dwarakanath, S., et al., 2004. Quantum dot-antibody and aptamer conjugates shift fluorescence upon binding bacteria. Biochem. Biophys. Res. Comm. 325 (3), 739–743.

Echeverría, I., Eisenberg, P., Mauri, A.N., 2014. Nanocomposites films based on soy proteins and montmorillonite processed by casting. J. Memb. Sci. 449, 15–26.

El-Miri, N., et al., 2015. Bio-nanocomposite films reinforced with cellulose nanocrystals: rheology of film-forming solutions, transparency, water vapor barrier and tensile properties of films. Carbohyd. Polym. 129, 156–167.

Esser, B., Schnorr, J.M., Swager, T.M., 2012. Selective detection of ethylene gas using carbon nanotube-based devices: utility in determination of fruit ripeness. Angew. Chem. Int. Ed. 51 (23), 5752–5756.

Faruk, O., et al., 2012. Biocomposites reinforced with natural fibers: 2000–2010. Prog. Polym. Sci. 37 (11), 1552–1596.

Fernandes, S.C.M., et al., 2010. Transparent chitosan films reinforced with a high content of nanofibrillated cellulose. Carbohyd. Polym. 81 (2), 394–401.

Flauzino Neto, W.P., et al., 2013. Extraction and characterization of cellulose nanocrystals from agro-industrial residue: soy hulls. Ind. Crops Prod. 42 (1), 480–488.
Fortunati, E., et al., 2015. Processing of PLA nanocomposites with cellulose nanocrystals extracted from Posidoniaoceanica waste: innovative reuse of coastal plant. Ind. Crops Prod. 67, 439–447.
Fu, G., Vary, P.S., Lin, C.-T., 2005. Anatase TiO_2 nanocomposites for antimicrobial coatings. J. Phys. Chem. B 109 (18), 8889–8898.
Fujishima, A., Rao, T.N., Tryk, D.A., 2000. Titanium dioxide photocatalysis. J. Photochem. Photobiol. C 1, 1–21.
Gammariello, D., et al., 2011. Bio-based nanocomposite coating to preserve quality of Fior di latte cheese. J. Dairy Sci. 94 (11), 5298–5304.
Gao, W., et al., 2011. Effects of clays with various hydrophilicities on properties of starch–clay nanocomposites by film blowing. Carbohyd. Polym. 88 (1), 321–328.
García, N.L., et al., 2011. Effect of glycerol on the morphology of nanocomposites made from thermoplastic starch and starch nanocrystals. Carbohyd. Polym. 84, 203–210.
George, N., et al., 2015. High performance natural rubber composites with conductive segregated network of multiwalled carbon nanotubes. Compos. Sci. Technol. 116, 33–40.
González, A., Alvarez Igarzabal, C.I., 2015. Nanocrystal-reinforced soy protein films and their application as active packaging. Food Hydrocoll. 43, 777–784.
González, A., Alvarez Igarzabal, C.I., 2013. Soy protein—poly (lactic acid) bilayer films as biodegradable material for active food packaging. Food Hydrocoll. 33 (2), 289–296.
González, K., et al., 2015a. Starch and cellulose nanocrystals together into thermoplastic starch bionanocomposites. Carbohyd. Polym. 117, 83–90.
González, Agustín; Strumia, Miriam; Alvarez Igarzabal, C., 2015. Modification strategies of proteins for food packaging applications. In: Giuseppe Cirillo, F.I., Spizzirri, U.G. (Eds.). Functional Polymers in Food Science: From Technology to Biology, vol. 1, Food Packaging. Wiley Online Publication, pp. 127–147.
Gu, J., Catchmark, J.M., 2013. Polylactic acid composites incorporating casein functionalized cellulose nanowhiskers. J. Biol. Eng. 7, 31.
Hendrix, K.M., et al., 2012a. Defatted mustard seed meal-based biopolymer film development. Food Hydrocoll. 26, 118–125.
Hendrix, K.M., et al., 2012b. Defatted mustard seed meal-based biopolymer film development. Food Hydrocoll. 26, 118–125.
Henriksson, M., et al., 2008. Cellulose nanopaper structures of high toughness. Biomacromolecules 9 (6), 1579–1585.
Hurme, E. et al., 2004. United States Patent US WO2004,102,185A1. Indicator. Available from: <http://www.google.com/patents/WO2004102185A1?cl=en>
Iwamoto, S., et al., 2014. Mechanical properties of polypropylene composites reinforced by surface-coated microfibrillated cellulose. Compos. Part A 59, 26–29.
Ja, S., et al., 2014. Strong, self standing oxygen barrier films from nanocelluloses modified with regioselective oxidative treatments. ACS Appl. Mater. Interfaces 6 (16), 14384–14390.
Jenkins, P.J., Donald, A.M., 1997. The effect of acid hydrolyis on native starch granule structure. Starch 8, 262–267.
Jiménez, A., et al., 2012. Edible and biodegradable starch films: a review. Food Biopro. Technol. 5, 2058–2076.
Johar, N., Ahmad, I., Dufresne, A., 2012. Extraction, preparation and characterization of cellulose fibres and nanocrystals from rice husk. Ind. Crops Prod. 37 (1), 93–99.

Johnston, J.H., et al., 2008. Nanostructured composite calcium silicate and some novel applications. Curr. Appl. Phys. 8 (3–4), 504–507.
Jones, N., et al., 2008. Antibacterial activity of ZnO nanoparticle suspensions on a broad spectrum of microorganisms. FEMS Microbiol. Lett. 279 (1), 71–76.
Kristo, E., Biliaderis, C.G., 2007. Physical properties of starch nanocrystal-reinforced pullulan films. Carbohyd. Polym. 68, 146–158.
Kumar, P., et al., 2010. Preparation and characterization of bio-nanocomposite films based on soy protein isolate and montmorillonite using melt extrusion. J. Food Eng. 100 (3), 480–489.
Lagaron, J.M., et al., 2005. Improving packaged food quality and safety. Part 2: Nanocomposites. Food Addit. Contam. 22 (10), 994–998.
Lavoine, N., et al., 2012. Microfibrillated cellulose—its barrier properties and applications in cellulosic materials: a review. Carbohyd. Polym. 90 (2), 735–764.
Li, H., et al., 2009. Effect of nanopacking on preservation quality of Chinese jujube (*Ziziphus jujuba* Mill. var. inermis (Bunge) Rehd). Food Chem. 114 (2), 547–552.
Li, X., et al., 2015a. Mechanical, barrier and morphological properties of starch nanocrystals-reinforced pea starch films. Carbohyd. Polym. 121, 155–162.
Li, Z. et al., 2015b. Structure and properties of chitin whisker reinforced papers for food packaging application, *Bioresources*, 10 (2), 2995–3004.
Lim, L.T., 2011. Active and intelligent packaging materials. Comprehensive Biotechnology, second ed., Elsevier, Amsterdam, The Netherlands, pp. 629–644.
Lin, N., et al., 2011. Preparation, modification, and application of starch nanocrystals in nanomaterials: a review. J. Nanomat. 2011, 1–13.
Lodha, P., Netravali, A.N., 2005. Characterization of stearic acid modified soy protein isolate resin and ramie fiber reinforced green composites. Compos. Sci. Technol. 65 (7–8), 1211–1225.
Ma, H.-L., et al., 2016. Radiation preparation of graphene/carbon nanotubes hybrid fillers for mechanical reinforcement of poly(vinyl alcohol) films. Radiat. Phys. Chem. 118, 21–26.
Maynor, M.S., et al., 2007. A food freshness sensor using the multistate response from analyte-induced aggregation of a cross-reactive poly(thiophene). Organ. Lett. 9 (17), 3217–3220.
De Mesquita, J.P., et al., 2012. Bio-based nanocomposites obtained through covalent linkage between chitosan and cellulose nanocrystals. Carbohyd. Polym. 90 (1), 210–217.
Mihindukulasuriya, S.D.F., Lim, L., 2014. Nanotechnology development in food packaging: a review. Trends Food Sci. Technol. 40(2), 149–167.
Mills, A., Hazafy, D., 2008. A solvent-based intelligence ink for oxygen. Analyst 133 (2), 213–218.
Mills, A., Hazafy, D., 2009. Nanocrystalline SnO_2-based, UVB-activated, colourimetric oxygen indicator. Sens. Actuat. B: Chem. 136 (2), 344–349.
Minelli, M., et al., 2010. Investigation of mass transport properties of microfibrillated cellulose (MFC) films. J. Memb. Sci. 358 (1–2), 67–75.
Müller, C.M.O., Laurindo, J.B., Yamashita, F., 2009a. Effect of cellulose fibers addition on the mechanical properties and water vapor barrier of starch-based films. Food Hydrocoll. 23 (5), 1328–1333.
Müller, C.M.O., Laurindo, J.B., Yamashita, F., 2009b. Effect of cellulose fibers on the crystallinity and mechanical properties of starch-based films at different relative humidity values. Carbohyd. Polym. 77 (2), 293–299.
Nakagaito, A.N., et al., 2009. Production of microfibrillated cellulose (MFC)-reinforced polylactic acid (PLA) nanocomposites from sheets obtained by a papermaking-like process. Compos. Sci. Technol. 69 (7–8), 1293–1297.

Naseri, N., et al., 2014. Electrospun chitosan-based nanocomposite mats reinforced with chitin nanocrystals for wound dressing. Carbohyd. Polym. 109, 7–15.

Nogi, M., et al., 2009. Optically transparent nanofiber paper. Adv. Mater. 21 (16), 1595–1598.

Okubo, K., Fujii, T., Thostenson, E.T., 2009. Multi-scale hybrid biocomposite: processing and mechanical characterization of bamboo fiber reinforced PLA with microfibrillated cellulose. Compos. Part A 40 (4), 469–475.

Paul, D.R., Robeson, L.M., 2008. Polymer nanotechnology: nanocomposites. Polymer 49, 3187–3204.

Pereda, M., et al., 2014. Polyelectrolyte films based on chitosan/olive oil and reinforced with cellulose nanocrystals. Carbohyd. Polym. 101, 1018–1026.

Pereira, V.A., de Arruda, I.N.Q., Stefani, R., 2015. Active chitosan/PVA films with anthocyanins from *Brassica oleraceae* (red cabbage) as time–temperature indicators for application in intelligent food packaging. Food Hydrocoll. 43, 180–188.

Prakobna, K., et al., 2015. Core-shell cellulose nanofibers for biocomposites: nanostructural effects in hydrated state. Carbohydr. Polym. 125, 92–102.

Premanathan, M., et al., 2011. Selective toxicity of ZnO nanoparticles toward Gram-positive bacteria and cancer cells by apoptosis through lipid peroxidation. Nanomed.: Nanotechnol., Biol., Med. 7 (2), 184–192.

Rhim, J.-W., Park, H.-M., Ha, C.-S., 2013a. Bio-nanocomposites for foodpackaging applications. Prog. Polym. Sci. 38 (10–11), 1629–1652.

Rhim, J., Park, H., Ha, C., 2013b. Progress in polymer science bio-nanocomposites for food packaging applications. Prog. Polym. 38, 1629–1652.

Robertson, G., 2008. Environmentally Compatible Food Packaging. Elsevier, Amsterdam, Netherlands.

Roy, N., et al., 2012. Biodegradation of PVP–CMC hydrogel film: a useful food packaging material. Carbohyd. Polym. 89, 346–353.

Savadekar, N.R., et al., 2015. Preparation of cotton linter nanowhiskers by high-pressure homogenization process and its application in thermoplastic starch. Appl. Nanosci. 5 (3), 281–290.

Savadekar, N.R., Mhaske, S.T., 2012. Synthesis of nano cellulose fibers and effect on thermoplastics starch based films. Carbohyd. Polym. 89 (1), 146–151.

Schlemmer, D., Angélica, R.S., Sales, M.J.a, 2010. Morphological and thermomechanical characterization of thermoplastic starch/montmorillonite nanocomposites. Compos. Struct. 92 (9), 2066–2070.

Schmid, M., et al., 2012. Properties of whey-protein-coated films and laminates as novel recyclable food packaging materials with excellent barrier properties. Int. J. Polym. Sci. 2012, 1–7.

Shafrina, Azlin-Hasim, Malco, C., Cruz-Romero,Morris, M.A., Cummins, Enda, Kerry, J.P., 2015. Effects of a combination of antimicrobial silver low density polyethylene nanocomposite films and modified atmosphere packaging on the shelf life of chicken breast fillets. Food Packaging and shelf life, 4, pp. 26–35.

Silvestre, C., Duraccio, D., Cimmino, S., 2011. Food packaging based on polymer nanomaterials. Prog. Polym. Sci. (Oxford) 36, 1766–1782.

Sinha Ray, S., Okamoto, M., 2003. Polymer/layered silicate nanocomposites: a review from preparation to processing. Prog. Polym. Sci. (Oxford) 28, 1539–1641.

Siracusa, V., Dalla, M., 2008. Biodegradable polymers for food packaging: a review. Trends Food Sci. Technol. 19 (12), 634–643.

Siró, I., Plackett, D., 2010. Microfibrillated cellulose and new nanocomposite materials: a review. Cellulose 17 (3), 459–494.

Sirviö, J.A., et al., 2014. Biocomposite cellulose-alginate films: promising packaging materials. Food Chem. 151, 343–351.
Slavutsky, A.M., Bertuzzi, M.A., 2014. Water barrier properties of starch films reinforced with cellulose nanocrystals obtained from sugarcane bagasse. Carbohyd. Polym. 110, 53–61.
Sozer, N., Kokini, J.L., 2009. Nanotechnology and its applications in the food sector. Trends Biotechnol. 27, 82–89.
Teisala, H., et al., 2010. Development of superhydrophobic coating on paperboard surface using the liquid flame spray. Surf. Coat. Technol. 205 (2), 436–445.
Utracki, L.A., 2004. Clay-Containing Polymeric Nanocomposites Vol. 1. Rapra Technology Ltd, Shawbury, Shrewsbury.
Vanderroost, M., et al., 2014. Intelligent food packaging: the next generation. Trends Food Sci. Technol. 39, 47–62.
Wang, Y.-J., Truong, V.-D., Wang, L., 2003. Structures and rheological properties of corn starch as affected by acid hydrolysis. Carbohyd. Polym. 52, 327–333.
Warheit, D.B., et al., 2004. Comparative pulmonary toxicity assessment of single-wall carbon nanotubes in rats. Toxicol. Sci. 77 (1), 117–125.
Xu, C., et al., 2015. Isolation and properties of cellulose nanofibrils from coconut palm petioles by different mechanical process. PLoS One 10 (4), e0122123.
Yamamoto, O., 2001. Influence of particle size on the antibacterial activity of zinc oxide. Int. J. Inorgan. Mater. 3 (7), 643–646.
Yang, L., Li, Y., 2006. Simultaneous detection of *Escherichia coli* O157:H7 and *Salmonella typhimurium* using quantum dots as fluorescence labels. Analyst 131 (3), 394–401.
Yuwawech, K., Wootthikanokkhan, J., Tanpichai, S., 2015. Effects of two different cellulose nanofiber types on properties of poly (vinyl alcohol) composite films. J. Nanomat. 2015, 1–10.
Zeng, J., Roberts, S., Xia, Y., 2010. Nanocrystal-based time–temperature indicators. Chemistry 16 (42), 12559–12563.
Zhao, L.M., et al., 2011. Preparation and application of chitosan nanoparticles and nanofibers. Br. J. Chem. Eng. 28 (03), 353–362.
Zheng, H., et al., 2007. Thermoplastic soy protein nanocomposites reinforced by carbon nanotubes. Macromol. Mater. Eng. 292, 780–788.
Zhou, C., 2013. Theoretical analysis of double-microfluidic-channels photonic crystal fiber sensor based on silver nanowires. Optics Comm. 288, 42–46.
Zou, J., et al., 2011. Effects of starch nanocrystals on structure and properties of waterborne polyurethane-based composites. Carbohyd. Polym. 85 (4), 824–831.

9

BIOACTIVE FOOD PACKAGING WITH NANODIAMOND PARTICLES MANUFACTURED BY DETONATION AND PLASMA-CHEMICAL METHODS

Katarzyna Mitura*, Konrad Wyrębski, Paweł K. Zarycki[†]**
**Koszalin University of Technology, Department of Biomedical Engineering, Koszalin, Poland; **PLASTMOROZ Sp. z o.o. Sp., Białogard, Poland; [†]Koszalin University of Technology, Faculty of Civil Engineering, Environmental and Geodetic Sciences, Department of Environmental Technologies and Bioanalytics, Koszalin, Poland*

1 Introduction

Diamonds have been arousing human interest for centuries. The "magical properties" of the diamond have been known since the ancient times. In ancient Greece, it was named "adamas", meaning indestructible or invincible. The first data and information about its origin, properties, and application can be found in Pliny the Elder's "Natural History."

In the diamond lattice, each carbon atom is connected to four other ones. All valence electrons are involved in creating chemical bonds. It should be noticed that four outer shell electrons may determine the number of carbon structures in chemical compounds as well as pure carbon modifications. Diamonds, in the form of microcrystals, have been extensively used in the industry for the past half century. The diamond synthesis in the metastability field demands a long time because it is highly improbable, mostly due to extremely high diamond surface energy. More feasible should be the epitaxial regrowth of the ready seed crystal. Furthermore, exactly in this direction the first independent efforts were done by the regrowth of diamond seed crystals at isothermal decomposition of simple carbon precursors

Food Packaging. http://dx.doi.org/10.1016/B978-0-12-804302-8.00009-1

(Spitsyn and Derjaguin, 1956; Eversole, 1958). The first diamond self-nucleation was realized in Poland in what has been a recently found explanation due to the following factors: the chemisorptions on critical nucleus with strong C—H bonds formation, the atomic hydrogen recombination and, in some cases, the surface charge. All valence electrons are involved in creating chemical bonds. Diamond nanoparticles have been found both, in plasma-chemical reaction products and in some meteorites about 20 years ago (Mitura, 1987, 2009).

Recently, the number of new materials has been invented on the basis of small diamond particles. Diamond nanoparticles were manufactured using Danilenko detonation method in 1963 in the Soviet Union from graphite in presence oxygen-deficient TNT (2-methyl-1,3,5-trinitrobenzene)/hexogen composition in inert media. A negative oxygen balance is the necessary condition for producing of ultrananocrystalline diamond particles. Detonation nanodiamonds consist of nanodiamond particles grain size from 2 to 10 nm, but the nanodiamond particles tend to aggregate. This spontaneous agglomeration of nanoparticles allows the use of nanodiamonds depending on the size of the agglomerates and their surface properties in various biological systems (Danilenko, 2004; Dolmatov and Yu, 2008; Krüger et al., 2005; Krüger, 2008; Larionova et al., 2006; Shenderova and McGuire, 2006). Ultrananocrystalline diamond particles produced by detonation of carbon explosive materials (size grain from 4 to 6 nm) have chemically active surface depends on the size of grains and free active bonds on the nanodiamond surface ready for chemical and plasma-chemical functionalization (Ray et al., 2006; Shenderova and McGuire, 2006; Mitura et al., 2013; Adach, 2013; Ceynowa et al., 2014). Detonation nanodiamond particles are hydrophilic and the nanodiamond surface has many dangling bonds, which are chemically reactive and ready for chemical functionalization. The optimum grain size for the chemical functionalization is from 5 to 60 nm. Single nanodiamonds forming agglomerates of nanoparticles reaching such proportions have developed a surface reaction and are able to join functional groups (Jee and Ah- Lee, 2009).

Nanodiamond particles obtained by detonation method are described in the literature as the so-called onion-like carbon (OLC) (Larionova et al., 2006; Shenderova and Gruen, 2006; Shenderova and Ciftan Hens, 2010). Due to the nanoscale size of, for example, detonation nanodiamond particles, such material can be used for biological studies (Puzyr et al., 2007; Mochalin et al., 2012). The series of research began by checking the biological properties of nanodiamond particles, which differ in content of a diamond phase and a grain size (Baidakova et al., 2013). The in vitro and in vivo studies provide valuable information regarding the effects on living organism. In vitro, in vivo and clinical studies on nanodiamond particles are the most spectacular in dermatology and cosmetology, and thus they are used as cosmetics and dermocosmetics (Batory and Mitura, 2009; Batory et al., 2012; Mihranyan et al., 2012). The diamond phase determines the biological activity of the powder. However, the presence of other carbon allotropic forms (graphite and carbynes), in diamond powder mixtures, stabilizes the powder biological activity. The grain size is of high importance when it comes to contact with biological material. Therefore nanodiamond powders are more biologically active and cause no mechanical damage to cells and microbes.

Eco Management and Auditing Scheme and Cleaner Production, which were established by European Union Council Regulations in 1993, actively promote the principles of sustainable development in production and consumption, and create a framework for effective and continuous improvement of business impacts on the environment. These systems and the related ordinance will force the use of new, biodegradable materials, and adopt the principle of zero waste technologies. Generally, traditional food packaging involves nondegradable materials that generate a number of environment pollution problems. Degradation of packaging materials may be improved by incorporation of inorganic nanoparticles as the commissures to the biopolymer matrix (Sozer and Kokini, 2009). Newly invented biodegradable food packaging layers containing nanodiamonds are characterized by improved durability and flexibility as well as temperature and humidity resistance (Rossi et al., 2014). Incorporation of functionalized nanodiamonds to the polymer matrix may significantly change the oxygen-barrier and antimicrobial properties of new food packaging materials. It should be noted that nanodiamond modification of food packaging layers cannot generate problems with waste utilization because such additives are considered as nontoxic. This is especially important for transferring of this eco-friendly and nonexpensive technology for developing countries. It should be noted that the global production of plastics exceed 200 billion tons/year and the negative impact of resulting plastic waste on the environment is obvious. Therefore researchers are seeking innovative technologies that are predominantly based on biodegradable polymers originated, for example, from vegetable polymers. Biodegradable materials, resulting from processing of plant sugars, are used for the manufacturing of food packaging products. Biodegradable polymers involving natural components are also used in the electronics industry. These materials are considered as the polymeric materials of the 21st century and currently they are produced at amount of 400,000 tons/year. It should be noted that there is potential risk associated with direct contact of food products with nanoparticles emitted from newly invented nanomaterial-based packaging. Therefore it is important to conduct appropriate research involving tissues, isolated cells, and nanoparticles in order to obtain full information about their toxicity after penetration to food products. Several published studies described that new invented packaging polymers may exhibit low toxicity (Chaudhry and Castle, 2011).

2 Bioactivity of Nanodiamond Powder Particles

2.1 In Vitro Studies

The main objective of the work was the biological research of primary vascular endothelial cells (HUVECs) isolated from the umbilical cord of healthy women at the High Risk Pregnancy Clinic of Medical University in Łódź—the Consent of Bioethics Committee of Medical University in Łódź (No. RNN/8/05/KE-2005).

Cell proliferation was evaluated in MTT cytotoxicity test by measuring the ability of living cells to metabolize the tetrazolium salt of a white or yellow color [3-(4.5-dimethylthiazol-2-yl)-2.5-diphenyltetrazolium bromide] to blue formazan. The crystals of formazan were dissolved in DMSO. The absorbance was measured at a wavelength of 580 nm. The impact of nanodiamond particles and graphite powder on the viability of endothelial cells was evaluated by flow cytometry. Tested powders were added to a cell culture, which was placed on gelatin plates, hung in the medium, and incubated for 16 h before the powders were added. After 24 h of incubation, the powders were analyzed by the method of flow cytofluorometry. The cells showing the green fluorescence (positive for calcein) were living cells and the cells showing the red fluorescence (positive for ethidium) were dead cells.

Findings of the research on the viability of endothelial cells in MTT cytotoxicity test by the flow cytofluorometry method showed that at the concentration of nanodiamond particles below 50 μg/mL there is no statistically significant impact on the viability of endothelial cells, while above 50 μg/mL the powders cause statistically significant reduction in the viability of endothelial cells. The graphite powder has the most cytotoxic effect on the viability of endothelial cells.

The bioactivity of nanodiamond particles depends on the content of the diamond phase, that is, the sp^3 bonds, in comparison with the graphite phase of sp^2, which does not affect the biological activity of the powders. During this research, the influence of nanodiamond particles and control graphite powder on the level of reactive oxygen species (ROS) and reactive nitrogen species (RNS) was also examined. Nanodiamond particles exhibit antioxidant activity—they stop the production of ROS. Graphite powder has a pro-oxidative effect—it enhances an oxidative stress. The enhancement of the oxidative stress of cells (changes in an oxidation–reduction potential) causes vascular endothelial cell dysfunction, such as, a change in permeability of a cell membrane. Nanodiamond particles do not influence the level of RNS. The grain size and the shape of the powders do not affect their biological properties in a significant way (Czerniak-Reczulska et al., 2010).

The second series of research described the interaction of diamond nanopowder obtained by the detonation method, with microorganisms: *Staphylococcus aureus* ATCC 25923, and thrush fungus (yeast fungus), *Candida albicans* ATCC 24433, and the metal nanopowders: silver nanopowder, gold nanopowder, and platinum nanopowder. Hydrocolloids from nanodiamond particles and metal nanopowders were added separately to a suspension of microorganisms and the observation was carried out using transmission electron microscope (TEM)—JEM-2000EX. In the microscopic examination, nanodiamond particles joined nonspecifically to the wall of bacteria and showed relation with a peptidoglycan layer. In a reaction with *C. albicans*, nanodiamond particles surrounded the fungus cells very tightly, probably collecting cytoplasm. Metal nanopowders present a different morphological picture in an interaction with microorganisms. The silver nanopowder adhered to the bacterial and fungal cell at specific places, thus causing damage and disintegration of the cell wall and the cytoplasmic membrane. The gold nanopowder did not surround bacterial cells but; however, it was trapped inside the biofilm produced by the bacterial cell. It damaged the cell wall and the cytoplasmatic membrane of the fungal cell yet it was not visible around the fungal cells. The gold nanopowder adhered to the thread-like substance produced by a torn fungal cell. The platinum nanopowder caused the bacterial cell produced substances, which were damaging and disintegrating the cell wall and the cytoplasmatic membrane, like the rest of nanopowders. The gold nanopowder stimulated the fungal cell to secrete substances that leaked from the cell.

The research results indicate a direct link between the electronic structure of nanoparticles and the zeta potential with the type of interaction with microorganisms. The nanodiamond particles are dielectric and have the positive zeta potential, which differs substantially from a membrane potential of microorganisms and, therefore the diamond nanopowder uniformly surrounded tested microorganisms without causing visible damage and destruction of the cells. Metal nanopowders have the negative zeta potential of less value than the zeta potential of bacterial cells and, therefore cause damage to bacterial cells. It is known from literature that the silver nanopowder has antibacterial properties. Ions silver and silver compounds have a strong bactericidal effect. Silver nanoparticles inactivate Gram-negative bacteria, for example: *Salmonella, Acinetobacter, Escherichia, Pseudomonas, Vibrio,* and Gram-positive bacteria, for example: *Streptococcus, Bacillus, Listeria, Clostridium,* and resistant strains of *S. aureus,* causing a purulent infection of skin, subcutaneous tissues and soft tissue, as well as an infection of the respiratory system, nervous system, urinary tract, digestive system, circulatory system, heart muscle, and an infection of hematopoietic system, very often being the reason of sepsis. On the basis of the findings of this research, it was found that an antifungal activity of silver nanopowder is possible. Metal nanopowders caused damage to the tested cells of *S. aureus,* as well as to fungal cells of thrush fungus. The nanodiamonds joined closely to the surface of tested microorganism, did not cause visible damage of tested bacterial and fungal cells observed in TEM. However, it indicates a good ability to self-organization, which leads to the conclusion-tested pathogenic microorganisms: *S. aureus* and thrush fungus (*C. albicans*) can be potentially be used as biological carriers of nanodiamond, which deactivates the activity of microorganisms without causing direct damage to bacterial and fungal cells (Chwalibóg et al., 2010; Pineda et al., 2012).

2.2 In Vivo Studies

The examinations concern the effects of nanodiamond on phagocytic activity (evaluation of neutrophils isolated from the whole blood of rats) and oxidation–reduction potential as well as on immunological parameters of Wistar rats. Biological tests were carried out on 21 male individuals, divided into 3 groups and staying in 3 separate cages for 10 days in standard conditions, such as, temperature: 22°C, humidity: 50–70%, the cycle of light/darkness: 12 h/12 h. The laboratory animals were given unlimited access to water and food. Rats received a suspension of diamond nanopowder (the nanopowder was dissolved in deionized water at a concentration of 500 mg/L). The first group of animals received the suspension of nanodiamond particles intravenously. Group 2 received the nanodiamond intraperitoneally, group 3 (a control group) received the deionized water intravenously. When the experiment was finished, the rats were put to sleep by 5% ketamine hydrochloride. A blood sample was taken from rats' hearts, put into heparinized tubes, and cooled down to 4°C. The samples of whole blood were prepared for the determination of glutathione peroxidase (GPx). Plasma and erythrocytes from the venous whole blood of rats were separated.

The erythrocytes were prepared for the measurement of the activity of superoxide dismutase (SOD). The plasma was prepared for the measurement of total antioxidant capacity, activity of glutathione reductase (GR) and the concentration of thiobarbituric acid reactive substances (TBARS—formed as a product of lipid peroxidation due to the degradation of fatty acids). The research aimed at the determination of the enzymatic activity, affecting oxidative stress in cells, in the presence of nanodiamond particles on an animal model. These are the tests that allow for an evaluation of the extended surface, that is, nanodiamond particles, in contact with biological material at the micro- and nanolevel. The results of this research can provide valuable information that may be used in the application of medical implants coated with the nanodiamond layer. The presence of diamond nanoparticles in the living body does not pose a threat to the patient. Carbon is biocompatible with living organisms and, at the same time, one should be aware of diamond nanoparticles' bioactivity in contact with biological material. The results of this research indicate a protective antioxidant action of diamond nanoparticles by increasing the activity of SOD in erythrocyte, sweeping free radicals, and inhibiting oxidative stress of cells. Whereas, there has been no impact noticed on the processes of lipid peroxidation, that is, the oxidative stress, associated with the oxidation of fatty acids in cell membranes (TBARS measurement). The nanodiamond particles suspension decreases an activity of GPx and GR, and, also in this case, it does not provide antioxidant activity. The oxidative stress, which is associated with neutrophils (white blood cells), was measured after a stimulation of *Escherichia coli* by formyl peptide (a factor stimulating phagocytosis of neutrophils). The findings indicate an increase in the number of phagocytic cells after intravenous and subperitoneal administration, whereas the intensity of fluorescence decreases in groups of rats in which the nanodiamond was administered, in comparison with the control group.

Summing up the results of research, it indicates that nanodiamond particles affect the antioxidant activity of SOD, which, in addition to the role of sweeping free radicals, is an antiinflammatory factor. This is the effect of its action at the molecular level, involving slowing down of immunological and inflammatory process, driven by a tumor necrosis factor-alpha (TNF-α). The oxidative stress induces an inflammatory process that stimulates the production of pro-inflammatory interleukins. The diamond nanopowder weakens the process of phagocytosis of neutrophils. It is likely that the mechanism of an interaction between diamond nanoparticles and the cellular membrane has a receptor nature and is associated with the presence of a large amount of sp^3 free bonds on the extended surface of diamond nanopowder, to which antibodies may join by Fc regions. Nanodiamonds obtained by the detonation method, change the oxidation–reduction potential of erythrocyte membrane cell and it acts as an immunosuppressant on innate immune cells (Niemiec et al., 2011).

2.3 Interaction Between Bacteria and Nanodiamond Particles

The next series concern the interaction of nanodiamond particles with Gram-negative (−) *Salmonella enteritidis* bacteria and Gram-positive (+) *Listeria monocytogenes* bacteria. As a comparison, the silver nanopowder was also examined. Such a choice was made because of the proven antibacterial properties of silver. The aim of the study was to demonstrate the interaction between nanodiamond particles and to describe the morphology of this interaction by means of TEM and the possibility of creating the complex: "nanodiamond-bacteria" in order to neutralize it.

S. enteritidis is a bacteria that causes immunological and inflammatory reaction in the human body. It results in an infection of, for example, intestines, blood, and lymphatic system. *L. monocytogenes* is a relatively anaerobic and intracellular bacteria, which is an etiological agent of listeriosis. It is one of the most virulent pathogens transmitted by food (20–30% of infections end in death).

Hydrocolloidal samples of nanodiamond particles were prepared in quantity of 200 μL and they were added to *S. enteritidis* and *L. monocytogenes* bacterial suspensions (each bacterial suspension -200 μL). The same was done with the silver nanopowder. The TEM picture of an interaction between diamond nanopowder and *S. enteritidis* indicates harmful action of nanodiamond particles on the morphology of bacterial cells. Nanodiamonds form aggregates, which partly join to the flagella and partly adhere nonspecifically to a bacterial wall. The silver nanopowder penetrates into *S. enteritidis* bacterial cells, causing a damage to its walls and the cell membrane. The cytoplasm and the elements of bacterial walls are removed from the cells or leak from the bacterial cell and the silver nanopowder is absorbed by these components, creating at the same time packages adjacent to the silver nanopowder. The silver nanopowder surrounds the *L. monocytogenes* bacterial cell closely, creating the rim of the cell membrane.

Nanodiamond particles adhere to the bacterial flagella that demonstrate the immunogenic structure (H antigen), contain a few amino acids, simultaneously almost do not contain amino acids with a sulfur group, and also have high levels of aspartic and glutamic acid. The most spectacular effect of interaction between nanodiamonds and *S. enteritidis* is creating strong connections between the packages of nanoparticles and bacterial flagella. This ability of diamond and silver nanopowder to penetrate into the center of the bacteria can be used to design "nanobacterial vehicles," which support active substances adherent to nanopowders (Sawosz et al., 2011).

2.4 High Resolution Transmission Electron Microscope (HR-TEM) Visualization of Different Types of Nanodiamond Particles Using In Vitro Study

The nanodiamond particles produced by RF PACVD method (radio frequency plasma-assisted chemical vapor deposition), by MW/RF PACVD method (microwave and radio frequency plasma-assisted chemical vapor deposition method), and detonation method and graphite powder as a control nanodiamond particles on proliferation HUVEC-ST cells in MTT test (cytotoxic test) were investigated by HR-TEM. Nanodiamonds produced by RF PACVD and MW/RF PACVD contained amorphous phase on the surface and the grain size was in the range of 150–800 nm. The graphite powder has the patch structure and the grain size was less than 1 μm. The detonation nanodiamond had a structure that was characteristic for the nanodiamond and a grain size of 2–10 nm (Mitura, 2009).

The findings reported that nanodiamond particles manufactured by RF PACVD and MW/RF PACVD methods have the low cytotoxicity and low biological activity in comparison with detonation nanodiamonds. RF PACVD and MW/RF PACVD powders have the amorphous structure—more sp^2 bonds on the diamond surface (proved by Raman spectroscopy) and higher than the medium grain size (proved by HR TEM). The detonation nanodiamonds revealed higher biological activity, more cytotoxicity to HUVEC-ST cells that is connected with a vast amount of sp^3 phase in nanopowder, the presence of free reactive bonds on the surface, and the lower grain size (Mitura, 2009).

2.5 Apoptotic Properties of Nanodiamond Particles

Nanodiamond particles can induce apoptosis, that is, programmed cell death. There was included a material characteristics of diamond nanoparticles (grain size, crystallographic structure) in order to systematize the results of the biological research. The study was carried out on human umbilical vein endothelial cells (HUVEC-ST)—immortalized cell lines. Apoptotic, dead and live cells were quantitatively double-stained with acridine orange and ethidium bromide, and measured using a flow cytofluorometry. In order to confirm the induction of apoptosis, alexin was added to diamond nanopowders. The fluorescence measurement was performed on a cytometer.

The research results indicate that the diamond nanoparticles can induce apoptosis in the examined vascular endothelial cell lines; however, this effect depends on the structure and size of the grains, which are dependent on the manufacturing methods of nanodiamond particles. Apoptosis and necrosis are induced only by nanodiamond detonation after 24 h of incubation, depending on the concentration of the diamond nanoparticles (the higher the concentration, the greater the number of apoptotic and necrotic cells). Detonation nanodiamond increases the production of ROS and RNS in the cells and induces apoptosis in HUVEC-ST cells, that is, slows down a cancerous process at the in vitro cellular level.

Detonation nanodiamond shows the greatest biological activity, conditioned by small grain size (2–10 nm) and a very extended surface of reactive sp^3 free bonds on the surface of powder (Solarska et al., 2012).

2.6 Surface Modification of Detonation Nanodiamond Particles

Chemical modification of detonation nanodiamond particles was done with the use of Fenton's reaction, which is carried out by placing hydroxyl groups —OH into the diamond nanoparticles. The aim of the study was to compare the effect of diamond nanopowder, which was unmodified or modified by Fenton's reaction (Martin et al., 2009), on the production of ROS and RNS (the study of the activity of antioxidant enzymes: SOD, catalase, GPx, GR, glutathione transferase) and glutathione content in the cell; by vascular endothelial cells (HUVEC-ST). It was important to carry out the research of modified and unmodified nanodiamond particles by using an infrared spectroscopy (FTIR, Fourier transform infrared) in order to confirm the presence of hydroxyl groups (hydroxylation of the diamond nanopowder surface). Comparatively, the graphite powder was examined. The findings of FTIR studies reported the presence of the hydroxyl groups on the nanodiamond particles, after Fenton's reaction, in the stretched strand —OH from 3600 to 3300 cm^{-1}. A similar effect was obtained for the graphite powder.

The research results show that modification of the nanodiamond surface using Fenton's reaction (surface hydroxylation) results in a change of surface properties of the nanopowder and the reduction of the grain size. Modified and unmodified nanopowder show similar antioxidant effects; however, the modified powder exhibits more spectacular antioxidant effects on cells of the vascular endothelium (HUVEC-ST). The modified nanodiamond reduces the production of NO in vascular endothelial cells (Solarska-Ściuk et al., 2013).

2.7 Antioxidant Properties of Nanodiamonds

From the beginning, there was a focus on the lipid peroxidation (affecting oxidative stress, and, therefore the aging processes) and the factors that influence it, for example, vegetable flavonoids, that is, rutin, hesperidin, and quercetin. Inhibition of the lipid peroxidation, by the aforementioned flavonoids, was a promising effect that encouraged to carry out further research in this area, for example, the research regarding the impact of flavonoids on inhibition of peroxidation of linolenic acid induced with hydrogen peroxide. The research findings appeal for antioxidative activity of flavonoids (Pietras et al., 1999).

2.8 Nanocrystalline Diamond Coatings (NCD) in Biomedical Application

The next task was to carry out research on the impact of metal implants on the emergence of the so-called contact allergy that causes degradation and damage to the implant. These studies were conducted in cooperation with the Centre of Occupational Allergy and Environmental Health of J. Nofer Institute of Occupational Medicine in Łódź. The studies reported that patients had a massive allergic reaction to nickel, chromium, cobalt, and palladium. Before that, the patch tests had been done on patients who experienced complications due to the presence of a metal implant. The test results showed that the majority of patients with poor tolerance on metal implant in patch tests had a contact allergic reaction to metal (Kręcisz et al., 2006).

Further studies were carried out regarding the interaction between diamond coatings and the surroundings of living organisms. The research characterized diamond coatings [diamond like carbon

(DLC) and NCD] in atomic force microscope (AFM)—the grain size, the structure of the surface, the types of bonds characteristic for a diamond, graphite, and carbynes. The next step was to carry out the corrosion tests in order to describe the NCD coating on the surface of medical steel, AISI 316L, which is used for the production of some medical implants. The research results showed that DLC and NCD coatings differ in the size of the grain. The grain in the NCD coating is smaller than 10 nm, while grain in the DLC coating has a size of about 50 nm. The surface of NCD coating is more developed than the DLC coating. Moreover, an amorphous structure of DLC coating contains less diamond (about 60%) in comparison with the nanocrystalline NCD coatings containing about 95% of the diamond. The corrosion potential of NCD coating is 2.5 times larger than the potential of AISI 316L steel, which clearly argues for greater resistance to corrosion of the implant, with a NCD coating than without it, in the aggressive environment of body fluids of a living organism. This publication also describes the research on microbiological growth of *E. coli* bacteria on such substrates as: AISI 316L medical steel, titanium, and a nanocrystalline coating of the NCD diamond. The research results reported an antibacterial nature of NCD diamond coating in comparison with titanium and AISI 316L medical steel, on which the growth of pathogenic strains of *E. coli* was the largest (Mitura et al., 2006b; Mitura, 2007).

Nanodiamond coatings in terms of diffusion barriers are formed by diamond coatings on medical implant between the biomaterial and the environment of a living organism. The research was focused mainly on orthopedic implants. There were tests carried out involving Schanz screws (tap screws) and wires, which were covered with NCD coatings. Then, the screw and the wire, with the diamond coating, were implanted to a femoral and humeral bone. An early general inflammatory reaction did not occur. After 18 months, the Schanz screw with the diamond coating was removed from the patient's body. At the macroscopic level, there were no changes, such as, engorgements, inflammatory reactions, excessive proliferation of fibrous connective tissue or organ reconstruction. Studies show that NCD coatings on selected orthopedic implants constitute an important "diffusion barrier" for metal ions, protecting the organism from metallosis (Grabarczyk et al., 2007).

Material analysis of jewelry covered with an amorphous carbon coatings (DLC) and NCD coatings was examined (Bociaga and Mitura, 2008). The essence of this research was to demonstrate the changes in living organisms (in vivo studies), which are caused by metal with regard to the use of metal implants (metallosis), and the need for using carbon and diamond coatings as a diffusion barrier.

2.9 Clinical Research With Nanodiamond Creams

The limited opportunities for the studies on diamond coatings in contact with biological material encouraged researchers to carry out the research in the field of cosmetology with the use of nanodiamond particles. The videodermatoscopy research contains characteristics of nanodiamond creams and clinical research on female patients in order to evaluate how the cream with nanodiamond particles acts. Test results showed that the creams with diamond powders enhanced the protection of hydrolipid coat of patients' skin, while, at the same time, maintaining an appropriate level of hydration and lubrication of the skin, and thus inhibiting the aging process of the skin while maintaining homeostasis. Researchers also participated in the research concerning in vivo studies assessing the impact of diamond powders on mice ear skin. The studies shown that diamond powders do not cause allergies (Batory and Mitura, 2009; Batory et al., 2012).

3 Nanotechnology in Food

Nanotechnology may provide new ways and tools not only for controlling properties and structuring foods, but also for introducing new features and creating added values. Nanotechnology, as a technique, involves the manipulation of the matter at a very small scale—generally ranging from 1 to 100 nm. Nanomaterials and nanoparticles may consist of any of the following nanoforms: nanoparticles, nanotubes, fullerenes, nanofibers, nanowhiskers, nanoshits. A nanoparticle is defined as a discrete entity of dimensions of about 100 nm or less (Rossi et al., 2014; Cushena et al., 2012; Som et al., 2010; Siegrist et al., 2007). Whereas, a nanomaterial is determined as an "insoluble or biopersistent and intentionally manufactured material with one or more external dimensions, or an internal structure, on the scale from 1 to 100 nm" as was expanded in the recent EC Cosmetics Regulation (EC No. 1223/2009). The regulation (EC) No. 1223/2009, also referred to as the Cosmetics Regulation, concerns in particular nanomaterials and offers the first legal definition (Article 2, paragraph 1, section k), which was described as subject to change according to the developing research (Article 3, paragraph 3). The offered definition notably excludes soluble or physiologically unstable nanoscale systems from the specific notification requirements of the nanomaterial. Nanomaterials present in cosmetics must be safe and reported to the commission (Article 16). The associated foreseeable exposure conditions must be notified as well. Specific information on such products must be reported stated 6 months prior to the day they are being put on the market and include such data as: physicochemical specifications, estimated volume (to be made available on the market per year), toxicological profile, safety assessment, and exposure conditions. The ingredients in nanoforms must be indicated on the list of ingredients by the presence of the word "nano" in brackets (Article 19). Although this regulation is for cosmetics, not for food, it sets regulation of nanomaterials in consumer products and outlines a need for toxicological data and safety assessment to be conducted before such products are brought to the market (Mihranyan et al., 2012).

The impact of nanosciences in many fields related to health and well-being is nowadays playing a crucial role and the nanotechnological applications are expected to bring considerable benefits to the food and nutrition sector. The phenomena that take place at the nanometer scale and have the potential to have substantial influence on the food industry worldwide, as nanotechnologies, can be applied to the whole food chain, including such processes as: production to processing, safety, packaging, transportation, storage, and delivery. The impacts of and needs for nanotechnological applications in food and related sectors are presented in Fig. 9.1 (Rossi et al., 2014).

Functionalities of such materials (eg, particle size, size distribution, potential agglomeration, and surface charge) can be affected by the biological matrix, in which they are held, such as, the food composition. The complete characteristics of nanoparticles

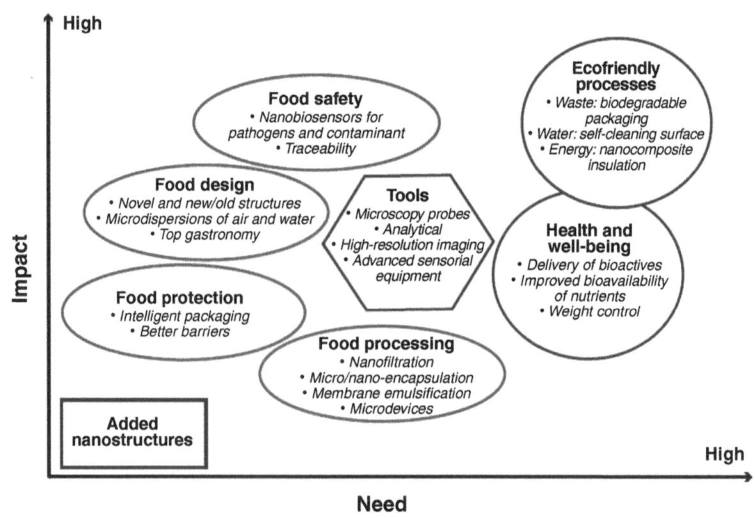

Figure 9.1. Impacts and needs of nanotechnological applications in foods and food processing. Reproduced from Rossi et al. (2014), with permission.

includes such features as size and size distribution, shape, and other morphological features (eg, crystallinity, porosity, surface roughness), chemistry of the material, solubility, surface area, state of dispersion, surface chemistry, and other physicochemical properties. The geometric surface area is a function of the number and shape of particles and square of the particle diameter. As the size of particles decreases, the specific surface area (usually expressed as square meters per gram) increases rapidly. Such situations mean having geometric specific surface areas in the hundreds of square meters per gram, and are not unusual for nanomaterials. The generation of ROS, rate of dissolution, equilibrium solubility, and so on, can be dramatically enhanced by such vast surface areas. The surface chemistry consists of a wide variety of properties that govern the way particles interact with their environment, in this case, the biological environment. Surface chemistry includes the number of studies like solubility, equilibrium, catalytic properties, surface charge, surface adsorption and desorption of molecules from solution. Most of these properties are functions of the atomic or molecular composition of the surface and the physical surface structure. It is noteworthy to mention that the surface chemistry of particles can change in a variety of ways, particularly through the adsorption, coating of the particles with proteins, or other species from the biological fluid surrounding (Powers et al., 2006).

Nanotechnologies will likely be diversified so as to defy the generic classification when it comes to evaluating potential health impact. It is therefore crucial to be able to define the criteria distinguishing between technologies and products, more or less likely to present a health risk, if we are to avoid inappropriate, and possibly deleterious, sweeping conclusions regarding the potential impact. Other studies show that healthy, intact skin presents a good barrier against nanostructured particles. However, there are indications that hair follicles could act as a repository of nanometer-diameter particles and that the chemistry of carrier liquids may affect the penetration potential (Maynard, 2006).

Even if nanoscale particles are able to penetrate through the outer layers of the skin, there is very little information on the hazard they might carry. The research conducted with the use of subcutaneously introduced nanoscale particles suggests that they can be transported within the lymphatic system, thus raising questions about how they might influence the immune responses. Moreover, there are certain indications that neuronal uptake and transportation may occur (Oberdörster et al., 2005).

At some point, there has been expressed a concern that the photogeneration of hydroxyl radicals by nanosized particles,

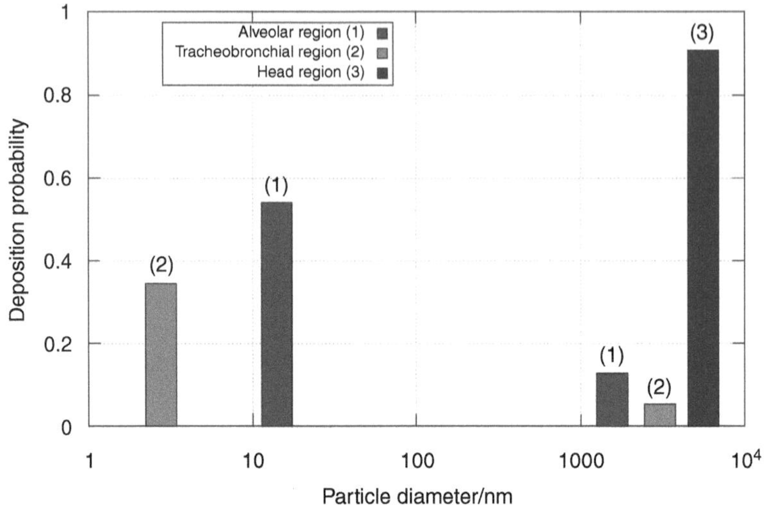

Figure 9.2. Deposition probability of macro- and nanoparticles fractions that can be retained within the respiratory track including (1) alveolar, (2) tracheobronchial, and (3) head regions. Based on data from Maynard (2006), ICRP (1994).

made of materials like TiO_2 and zinc oxide, may lead to an oxidative damage in the skin, although the use of surface modification in such nanoparticles has shown to suppress free-radical generation (Lademann et al., 2000; Wakefield et al., 2004). Most airborne particles, smaller than a few tens of micrometers in diameter, can be inhaled. Once in the respiratory system, particles will deposit in different regions according to their shape, diameter, and density. Diffusion-based deposition mechanisms result in relatively high probability of particle deposition in the alveolar region of the lungs, for particles smaller than approximately 300 nm (although particles smaller than 4 μm have a greater than 50% probability of penetrating this region) (ISO, 1995). Below approximately 30 nm, diffusion leads to high deposition probabilities throughout the respiratory system, including the upper airways. At smaller diameters, deposition in the upper airways, especially within the nasal area, begins to dominate deposition in the alveolar region (Fig. 9.2).

3.1 Nanotoxicology of Food Packaging

Interactions between nanomaterials and biological organisms typically take place at the particle's surface. Therefore the surface area is of primary importance in determining possible toxic effects (Brown et al., 2001). In order to assign proper mechanisms or reasons for toxic effects of nanoscale materials, their properties and characteristics, both outside and within the biological environment, need to be well understood. As toxicologists began to examine nanomaterials in a systematic fashion, there is a consensus that a series of guidelines or recommended practices is necessary for basic characterization of nanomaterials. These recommended practices should be developed jointly by physical scientists, skilled in nanocharacterization, and biological scientists, experienced in toxicology research (Powers et al., 2006).

Evaluations of the health risk associated with aerosol exposure are generally based on the assumption that toxicity is related to the mass and the chemical composition of the inhaled material. This mass-based approach has been very effective in the past and led to substantial reductions in the respiratory disease with reduced exposures. The observed surface-associated material toxicity can be seen as an extension of macroscopic properties, rather than something unique to nanoscale materials. In comparison, a number of studies have shown that the particle diameter may have a previously unrecognized role in determining the fate and the potential impact of nanometer-diameter particles in the lungs. Particles, smaller than a few hundred nanometers in diameter, may be able to enter the lung interstitium following the deposition in the lungs, and there is an increasing evidence that nanometer-diameter particles can pass from the lungs into the bloodstream. Particle penetration into cells has been observed, and in vitro studies using ambient ultrafine particles have shown the evidence of particle localization in mitochondria, where they induce major structural damage. The 2004 report on nanotechnology, from the Royal Society and Royal Academy of Engineering, highlighted nanotechnologies associated with unbound sub-100 nm diameter particles as being of particular interest to human health. The Oberdörster team supports this emphasis on sub-100 nm diameter particles in a discussion on the emerging field of nanotoxicology (Oberdörster et al., 2005). However, it is clear from published toxicity studies that particle size individually is not a good criteria for differentiating between more or less hazardous materials and technologies. The dependence of engineered nanomaterials' behavior on physical and chemical structure significantly increases the difficulty in developing a sound understanding of material toxicity (Maynard, 2006; Maynard and Baron, 2004; Ferin and Oberdörster, 1992; Oberdörster et al., 2002; Li et al., 2003).

Particles that enter the bloodstream may affect the blood vessel lining or function and promote blood clot formation (Nemmar et al., 2004). They may also be associated with cardiovascular effects linked to inhaling ambient ultrafine particles (Pekkanen et al., 2002; Wichmann and Peters, 2000).

3.2 Other Allotropic Forms of Carbon in Food Packaging

Carbon nanotubes are already used in commercial products (predominantly encapsulated in composite materials); relatively few studies have investigated their potential toxicity. In a recently published in vitro study, carbon nanotubes were found to penetrate into human epidermal keratinocyte cells and to elicit production of an inflammatory cytokine (Monteiro-Riviere et al., 2005). The Bottini group has shown that carbon nanotubes are cytotoxic to human T-cells, with oxidized nanotubes being significantly more potent than pristine nanotubes (Bottini et al., 2006).

In studies on food packaging, the role of the antibacterial activity of the packaging itself should be underlined . It has been reported that carbon (ie, nanotubes and fullerenes) and diamond nanoparticles possess antimicrobial properties. The new allotrope of carbon, graphene, has antibacterial activity. The interactions between bacterial cells and graphene materials depend on the site of interaction: its surface or edges. Fig. 9.3 demonstrated that foodborne bacterial pathogens, *L. monocytogenes*, were arranged at the edges of pristine graphene and graphene oxide, while with graphene oxide they adhered to

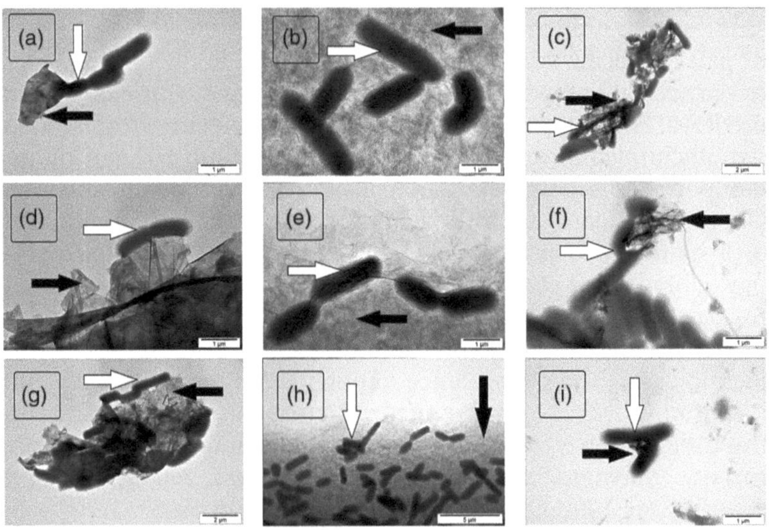

Figure 9.3. TEM visualization of the interaction of graphene with *Listeria monocytogenes* (foodborne bacterial pathogens) with pristine graphene (a, d, g), graphene oxide (b, e, h), and reduced graphene oxide (c, f, i). *Black arrows* indicate the graphene and *white arrows* the bacterial cells. Reproduced from Kurantowicz et al. (2015), with permission.

the nanoparticle surface. Graphene oxide was found to have the highest antibacterial activity (Kurantowicz et al., 2015; Jastrzębska et al., 2012).

The evaluation of the potential cytotoxicity of the graphene surface is crucial for its application in medicine as bioactive molecules, drug carriers, or biosensors. There are several reasons that chicken embryos are often considered to build a useful toxicity model of the organism compartments including the brain: its fast growth rate, external treatments sensitivity and, especially, lack of the blood/brain barrier at the early stage of embryogenesis.

Toxic effects of pristine grapheme administered in ovo to chicken embryos were proved. It has been found that albumins from egg consists of a number of nutrients and they may be attached to the graphene surface during the self-assembly process. Proteins have hydrophobic regions, which can interfere with nonpolar graphene molecules. Importantly, depending on the graphene/protein balance, the graphene structure may surround or be surrounded by proteins (Sawosz et al., 2014; Prasek et al., 2013).

Carbon nanotubes, graphene, and fullerenes next to silver are the most common nanomaterials used in cosmetics packaging. Pristine fullerenes have low toxicity and no risks would be expected for humans exposed to fullerenes under good hygiene conditions (Mihranyan et al., 2012).

Graphite nanoparticles influence the development of chicken embryos, as well as vascularization of the chorioallantoic membrane and heart, at the morphological and molecular level and does not affect either body/heart weight or serum indices of the embryos' health. The application of graphite nanoparticles significantly downregulated gene and protein expression of the proangiogenic basic fibroblast growth factor (bFGF), indicating that graphite nanoparticles inhibit angiogenesis (Wierzbicki et al., 2013).

Carbon atoms, located at the edge of graphene platelets, have special chemical reactivity. In vitro results indicate that graphene oxide platelets is less toxic to glioma cells than reduced graphene oxide platelets, which induced cell death mostly through the apoptosis pathway. The in vivo results demonstrated that both graphene platelets injected into glioblastoma tumors decreased the volume and the weight of tumors. Interaction between two types of surface graphene platelets (graphene oxide or reduced graphene oxide) and glioma cells were caused by the reactive free bonds on the surface or their saturation (Jaworski et al., 2015; Uo et al., 2011).

4 Bioactive Food Packaging With Nonmodified and Modified Nanodiamond Particles by Plasma-Chemical and Chemical Method

4.1 Nonmodified Nanodiamond Particles in Nanonutrition

Food packaging materials that are capable of releasing nanoscale antimicrobial compounds, antioxidants and/or flavors, which would improve the shelf life or sensory characteristics of food, are termed "active." Bioactive food packaging seems to be a logical following from nanofiller technology and intelligent packaging, a type of bioactive packaging, seems to be where research has found a new focus. The incorporation of active compounds into food packaging materials, where they are bound rather than designed to migrate, are more common than packaging designed to release particulate nanomaterials into foods. Nanosensors can be incorporated into food packaging matrices that have the ability to identify specific microbial and chemical contaminants, or environmental conditions. The connection of food packaging materials and bioactive substances is a method to control surface microbial contamination of foods (Cushena et al., 2012).

Diamond nanoparticles have high biocompatibility and low toxicity in the process of forming capillary blood vessels from preexisting vessels in chicken embryos model. Vascularization of the heart and the density of branched vessels were significantly reduced after the treatment with diamond nanoparticles and, to a lesser extent, graphite nanoparticles. Diamond nanoparticles inhibit angiogenesis through the downregulation of proangiogenic basic fibroblast growth factor (Wierzbicki et al., 2013).

The use of antibiotics in the food industry can reduce the sensitivity and effectiveness of antimicrobial treatment. They appear as strains resistant to standard antimicrobial treatment. Therefore researchers should look for alternative methods of preventing microbial contamination of food (Pineda et al., 2012). Silver nanoparticles influence tissue, cell, and molecular levels in ovo application in broiler chickens. Examined silver nanoparticles inhibit bacterial strains and do not interfere with immune balance, and decrease body weight of about 100 g in comparison with the control group without silver nanoparticles. This research does not contain any information on adverse reactions to the presence of metal in a living organism. For this reason, studies are made on the diamond nanoparticles that do not cause metallosis and contain biologically safe and active diamond phase hybridized sp^3. Modern broiler lines are intensively selected for a higher growth rate and increased size of muscles, including pectoral muscles. Diamond nanoparticles are highly biocompatible nontoxic carbonic structures. Biocomplexes of diamond nanoparticles with L-glutamine may accelerate growth and maturation of muscle cells (Pineda et al., 2012; Czerniak-Reczulska et al., 2010; Grodzik et al., 2013; Petracci and Cavani, 2012).

Nanodiamond particles can protect antioxidant activity from environmental stress (Bakowicz, 2003). Nanodiamond particles colloid increase the activity of the antioxidants (eg, ascorbic acid, vitamin C). The results of these research evaluate the effect of intravenous and intraperitoneal administration of nanodiamond particles to male Wistar rats on biochemical and hematological parameters. It was interpreted that the results of blood chemistry values may provide the precise picture of the animal health status at the time of sampling (ie, nutritional status, disease condition, and stress after environmental changes). Intravenous and intraperitoneal injection of detonation nanodiamond particles increased platelet counts on rats. Additionally, the intravenous injection of nanoparticles decreased serum level of glucose and total protein. Nanodiamond particles stimulate the activity of pro-inflammatory cytokines. All results indicate the need to assess the minimum toxic dose for a living body (Niemiec et al., 2013; Szmidt et al., 2013).

4.2 Genotoxic, Mutagenic, and Anticancerogenic Activities of Bioactive Nanodiamond Particles

The food contact packaging is crucial issue for mutagenicity and genotoxicity of biomaterials that are in contact with food. This is due to interactions at the molecular level and has a direct relationship with the surface reaction of the nanoparticles. Nanodiamond particles reveal high biocompatibility with selected in vitro cellular systems. The potential genotoxic and mutagenic activities were evaluated with detonation nanodiamond particle (particle size <10 nm) with human peripheral lymphocytes in vitro. The results indicate that detonation nanodimaond particles were able to provoke cytotoxicity at higher concentration, while at lower concentration genotoxic and aneugenic effects were stimulated (Dworak et al., 2014).

Nanodiamond particles have antioxidant and anticancerogenic properties in living organisms. The present results indicate that the diamond is not inert at the molecular level (Bakowicz-Mitura et al., 2007).The results show that the extended surface of nanodiamond particles has a very high beneficial bioactivity on the molecular level (ie, via gene expression). The results also indicate that nanodiamond particles influence gene expression and inhibit oxidative stress, cellular stress, and genotoxic stress in human genes. The effect possibly depends on the extension of the diamond surface.

4.3 Chemical Modification of Detonation Nanodiamond Particles

Functionalization of detonation nanodiamond particles included the processes of incorporation of the hydroxyl, amino, and carboxyl groups on the surface of nanodiamond. The hydroxyl-modified nanodiamond particles were obtained by Fenton's reaction (Martin et al., 2009), amine-functionalized NDPs by chemical reduction of the nitro-functionalized surface, carboxyl-modified NDPs by oxidation with the use of H_2O_2 under acidic conditions (Satyanarayana et al., 2012).

Obtained functionalized by hydroxyl and amine groups, and amino groups of nanodiamond particles were used for covalent binding of nonsteroidal antiinflammatory pharmaceuticals (aspirin, ketoprofen, ibuprofen, naproxen) via ester or amide bond. These studies were to verify the activity of the conjugates of active substance-NDP and study the rate of the release of active substance from the NDPs surface. In addition, the HO—NDPs and H_2N—NDPs were modified by incorporation of β-alanine residue as a linker for better separation of active pharmaceuticals from the nanodiamond particles surface, expecting that it will affect the activity and the rate of release of the therapeutic ingredient. In parallel, the conjugates of NDPs were obtained with the active substance bounded by ionic bond between amine function of H_2N—NDPs and carboxylic function of nonsteroidal antiinflammatory pharmaceuticals. It was expected that the presence of weak ionic bond affects the increase of the rate of releasing of the active substance from the NDPs surface. Nanodiamond particles, with carboxylic groups on the surface, were applied to preparation of the conjugates with Paclitaxel (anticancer drug). Two synthetic approaches were used: one carboxyl groups of HOOC—NDPs were coupled with Paclitaxel, while the second approach relied on incorporation of β-alanine as a linker between nanodiamond particles and Paclitaxel molecule. 4-(4,6-Dimethoxy-1,3,5-triazin-2-yl)-4-methylmorpholinium toluene-4-sulfonate was used as a coupling reagent for amide and ester preparation (Mitura et al., 2013; Kolesinska et al., 2015).

Nanodiamond particles modified by Fenton's reaction exhibit the phenomenon of fluorescence in the presence of a strain of *Pseudomonas aeruginosa* ATCC 9027 that can be used in microbiological diagnostics packages containing diamond nanoparticles (Mitura et al., 2013).

Detonation nanodiamond particles functionalized with hydroxyl groups were verified in studies evaluating their antioxidant potential. Detonation nanodiamond particles delayed soybean oil from going rancid, which is important in the potential for using them in bioactive packaging extending the period of own consumption of the food (Adach, 2013).

Functionalization of pristine nanodiamond influences on fragmentation of nanoparticles, their wettability and catalytic properties. Oxidation of the particle surface results in the possibility of selective attachment of functional groups and molecules. Further chemical modifications (eg, through an amide bonds) can engage the intrinsic carboxylic groups that are present on the nanodiamond surface. In this way the surface is modified with attached amines. Chemical modification of nanodiamonds by amide functionalization increases their surface polarity. Polar nanodiamond surface, measured by dispersibility test in water and alcohol, is necessary for particular chemical modifications for given medical applications (Basiuk et al., 2013).

4.4 Characterization of Antibacterial and Other Biological Properties of Nanodiamond Particles as Extended Surface of Nanocrystalline Diamond Coatings (NCD)

NCDs onto medical implants are fully biocompatible with living organism tissues at the macroscopic level in postmortem histopathological examinations. Experiments involving the test layer at the cellular level is usually difficult (mainly due to cells' poor adhesion and lack of adequate amounts of free bonds on the surface of the diamond). Also, research on the molecular level (gene expression) requires large diamond surface reactivity, which causes exposition on a large number of free reactive bonds. Therefore in order to determine the properties of the NCD testing is performed diamond nanoparticles as so-called extended surface, reactive biologically, chemically and physically. NCD coatings protect against metallosis and thus the use of diamond films for medical implants form a protective coating against the penetration of metal ions into the surrounding tissues, cells, and body fluids (Bociaga and Mitura, 2008; Mitura et al., 2006a,b, 2013, 2015; Bakowicz-Mitura et al., 2007; Adach, 2013).

Examples of current or intended uses of nanotechnology in packaging, in general, include antibacterial coatings (eg, nanosilver, NCD coatings) and detonation nanodiamond particles to increase the shelf life of the product (Mitura et al., 2006a; Grabarczyk et al., 2007; Mihranyan et al., 2012). The National Institutes of Health USA reported that more than 60% of all microbial infections are caused by biofilms. The results indicate that antibacterial and antiadhesive properties of NCDsurfaces allow for the use of hard antibacterial coating for medical implants (Medina et al., 2012).

L. monocytogenes (G+) reveal the possibilities of constructing nanoparticle-bacteria vehicles. The interaction between nanodiamond particles and *L. monocytogenes* results in the presence of aggregates of nanodiamonds on the bacterial wall, moreover, a large amount of diamond nanoparticles surrounding the cell leads to the loss of cytoplasm. Diamond nanoparticles present a potential package to inactivate pathogenic *L. monocytogenes* liveliness and thus may prevent infection (Sawosz et al., 2011).

It has been found that the presence of metal implants in the body may generate a number of decomposition products causing the spread of inflammation. This phenomenon may generate a number of problems including a loss of biomechanical stability or mobility of orthopedic implants (eg, hip replacement) and increasing of pain intensity, which creates serious mechanical damage of the implant at macroscopic scale. It is connected with the cascade of inflammation and the induction of macrophages activity and pro-inflammatory cytokines. Therefore the use of biocompatible diamond coatings for long-term metal implants significantly decreases the degradation process at molecular levels by the presence of protective diffusion barrier (Thomas et al., 2012; Mitura et al., 2006a,b). The biological properties of diamond coatings on orthopedic implants were tested using various nanodiamond particles (including detonation nanodiamonds) characterized by different grain size. The effects on macrophage activity and basic inflammatory cells in the environment of the implant were investigated. Assessment of the inflammatory process also includes evaluation of the pro-inflammatory cytokines and growth factors: tumor necrosis factor-α (TNF-α) and vascular endothelial growth factor (VEGF). The results show that increasing concentrations of nanodiamond particles in phosphate-buffered saline is associated with inhibition of macrophage proliferation and decreasing of their biological activity. The bigger grain size of nanodiamond particles causes the decrease in cell proliferation and release of their metabolism. The presence of suspension of nanodiamond particles inhibits the gene expression of TNF-α, VEGF, and pro-inflammatory cytokines and significantly reduces the inflammatory process at the molecular level (Thomas et al., 2012; Ingham and Fisher, 2005; Merkel et al., 1999).

Similar to ultrananocrystalline diamonds particles, ultrananocrystalline diamond coatings have grain sizes from 2 to 5 nm and contain the big amount of sp^3-hybridized carbon and the small amount of sp^2 carbon phase (Shenderova and McGuire, 2006). In vitro examinations with influence dexamethasone–nanodiamond complex influence on murine macrophages proved the antiinflammatory properties of nanodiamond coatings measured by decreasing of pro-inflammatory cytokine levels (Narayan et al., 2011; Lam and Ho, 2009). Ultrananocrystalline diamond implants can be considered as the hermetic coatings. Using this technology the retinal implants in the rabbit eyes were manufactured. Such implants are characterized by the absence of intraocular inflammation. The damage of this layer causes the acute tissue reaction on silicon degradation products from background (Xiao et al., 2006). Nanodiamond particles with grain sizes from 2 to 8 nm connected with doxorubicin cause the apoptosis in murine macrophages (Huang et al., 2007). It was reported that a complex of nanodiamond with Paclitaxel drug was blocked the tumor growth and lung cancer cells. Nanodiamonds modified with Paclitaxel involving ester bonds are sensitive for esterase enzyme resulting with strong bioactivity of this complex (Liu et al., 2010).

NCD-coated borosilicate glass substrates and nanoparticulate diamond powder (DP)-coated borosilicate glass substrates were investigated involving in vitro studies (Lechleitner et al., 2008). Particularly, proliferation of human renal epithelial proximal tubular cells (Hk-2 cell line) with biomaterials contact, was examined. The effects of surface structure and the type of biomaterial on cell adhesions were determined. The results have revealed that hydrogenated NCD-coated borosilicate glass surface inhibits the cell adhesion, which can be linked with the absence of polar groups. However, oxygen-terminated NCD and DP surfaces reveal good adhesion of the cells investigated. Oxygenated NCD significantly increases the cell proliferation. The authors have demonstrated that increase of the free bonds density on the nanodiamond surface significantly change the cells viability.

4.5 Plasma-Chemical Modification of Detonation Nanodiamond Particles

Functionalization of nanodiamond particles can be performed by using a cold plasma discharge generated with fluorine containing gas. Fluorination of nanodiamonds allowed colloidal suspension in anisole that was previously impossible (Ray et al., 2006).

Surface modification of detonation nanodiamonds affects (increase) nucleation of diamond on NCD films. FTIR spectroscopy showed that plasma treated diamond nanoparticles have number of functional groups on the surface and such nanostructures can be easily dispersed in polar liquids (eg, alcohols). For this reason, NCD films can be applied in solar cells (Lin et al., 2014).

Physical and physicochemical properties of nanodiamonds that are synthesized using detonation protocols determine their applications in biology, medicine, pharmacy, and the food industry. Such particles may be applied as efficient adsorbents for a number of biochemical procedures including proteins extraction and/or purification from, for example, *E. coli* bacterial cells (Puzyr et al., 2007).

Nanodiamond particles at grain sizes of 100 nm can be modified by physical adsorption. For example, diamond nanopowder was treated in oxidative acid conditions at elevated temperature. The nanodiamond was then functionalized by immobilization of enzymes and lysozyme (antimicrobial protein). It has been documented that such new hybrid organic–nonorganic materials possess similar properties as the initial active organic molecules (Nguyen et al., 2007).

Detonation nanodiamond particles (the grain sizes from 2 to 5 nm) were plasmochemically modified in the rotary reactor chamber by microwave plasma activated chemical vapor deposition (MW PACVD). In order to modify the nanodiamond from all sides, 5 g nanopowder was placed in a rotary drum. Before plasmochemical modification, after pumping, the powder was annealed at 200°C for about 3 days. It was successfully managed to carry out the project, from concept through preliminary design to the implementation of the microwave modification system of detonation nanodiamond particles. Objectives of the project have been completed, and to the surprise of skeptics, managed to generate stable microwave plasma. The implementation of rotation within the vacuum chamber caused a lot of problems, but in the end, thanks to an innovative solution, worked out. Currently the results of the first modification of the nanodiamond particles are awaited (Ceynowa et al., 2014; Mitura et al., 2015). Data visualized in Fig. 9.4 presents hemocompatibility of detonation nanodiamond modified by plasma-chemical method with human erythrocytes cells (Mitura et al., 2015). Interestingly, the human red blood cells that are interacting with pristine graphene may agglomerate causing to decrease hemolytic properties (Liao et al., 2011).

The surface energy of biomaterial affects the cell-material interactions. In hemocompatibility test, the silicon-doped diamond-like carbon were examined (Major et al., 2014). The results revealed

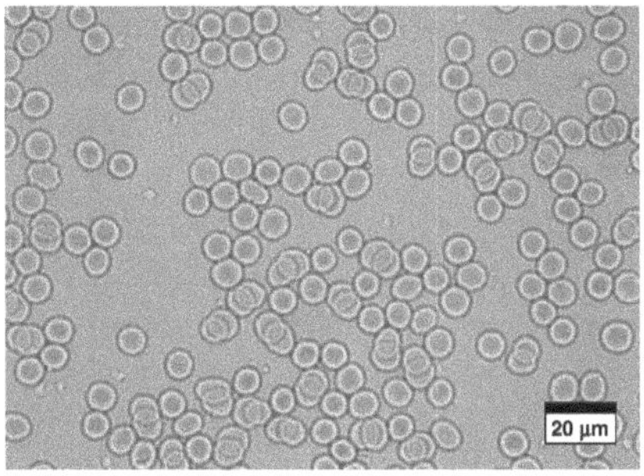

Figure 9.4. The normal human erythrocytes with detonation nanodiamond particles modified by plasma-chemical method (MW/PACVD) in rotary chamber, optical microscope Olympus CX31, 400 times of magnification. Reproduced from Mitura et al. (2015), with permission.

good hemocompatibility of silicon modified by diamond-like carbon with platelets. Such surface functionalization influences endothelial cell behavior and reduces the growth of smooth muscle cells. Modification of silicon by diamond-like carbon layer allows preparation of hemocompatible and antithrombogenic biomaterial's surface.

5 Conclusions

The bioactivity of nanodiamond particles depends on the content of the diamond phase, the sp^3 bonds. The graphite phase that is based on sp^2 bonds does not affect the biological activity of the powders. Recently, it has been found that the surface of detonation nanodiamonds can be biologically active. The number of research communications indicates the critical role of free bonds affecting high reactivity with different biological materials. However, the research results are sometimes surprising and have a prospective nature. Controlled surface modification of nanodiamonds allows them to obtain biologically active nanodiamond particles, which can be used in the future technologies, especially in food packaging.

A number of studies revealed the fact that nanodiamond particles affect the antioxidant activity of key enzymes like SOD. Moreover, the diamond nanopowders weaken the process of phagocytosis of neutrophils. It is likely that mechanism of an interaction between diamond nanoparticles and the cellular membrane has a receptor nature and is associated with the presence of a large amount of sp^3 free bonds on the extended surface of diamond nanoparticles. Detonation nanodiamond shows the greatest biological activity including cytotoxicity to given cells, conditioned by small grain size (2–10 nm) and a very extended surface of reactive sp^3 free bonds on the surface of powder.

Published data indicate that modification of the nanodiamond surface using Fenton's reaction (surface hydroxylation) results in a change of surface properties of the nanopowder and the reduction of the grain size. Modified and unmodified nanopowder show similar antioxidant effects; however, the modified powder exhibits more spectacular antioxidant effects on cells. There are some evidences indicating that detonation nanodiamonds functionalized with hydroxyl groups delayed soybean oil from going rancid, which is important in the potential for using them in bioactive packaging extending the period of consumption of the food. Moreover, nanodiamond particles modified by Fenton's reaction exhibit the phenomenon of fluorescence in the presence of given microorganisms that can be used in microbiological diagnostics packages containing diamond nanoparticles.

Several papers reported application of nanodiamonds as nanobiosensors identifying the pathogenic microorganisms, demonstrating endemic antiinflammatory and antibacterial activity as well as inactivates the foodborne bacterial pathogens. It can be concluded that the future packaging technology will demand smart biosensors that will protect food, pharmaceutics, and related products against micropathogens. It is hoped that the main benefit of such approaches will be the expiration date extension and the protection of edible products from being spoiled by toxic molecules. Nanoparticles based on carbon materials, also using selected microbes, seem to be an interesting proposal for such nanobiosensor development. On the other hand, it is well documented that the interaction between nanodiamond particles and microorganisms may result in inactivating foodborne bacterial pathogens. Such nanodiamond food additives may prevent serious infections.

There is potential risk associated with direct contact of food products with nanoparticles emitted from newly invented nanodiamond-based packaging materials. Nanoparticles spatial properties (size, 3-dimensional shape), and surface chemical functionalization may strongly influence their migration to and within living organism compartments. Therefore it is important to conduct appropriate research involving tissues, isolated cells, and nanoparticles in order to obtain full information about their toxicity after penetration of functionalized nanodiamonds particles to food products. Newly invented food packaging layers containing nanodiamonds are characterized by improved durability, flexibility as well as temperature and humidity resistance. Moreover, incorporation of functionalized nanodiamonds to the polymer matrix of food packaging layers may significantly change the oxygen-barrier and antimicrobial properties of new packaging materials. However, a number of biodegradation aspects must be investigated in terms of toxicity of such food packaging products as well as nanoparticles bioconcentration in various environment ecosystems.

References

Adach, K., 2013. Chemical modification of diamond nanoparticles produced by detonation method. PhD Thesis. Technical University of Łódź.

Baidakova, M.V., Kukushkina, Y.A., Sitnikova, A.A., Yagovkina, M.A., Kirilenko, D.A., Sokolov, V.V., Shestakov, M.S., Vul, Ya. A., Zousman, B., Levinson, O., 2013. Structure of nanodiamonds prepared by laser synthesis. Phys. Sol. State 55, 1747–1753.

Bakowicz, K., 2003. Bioactivity of diamond. PhD Thesis, Technical University of Łódź.

Bakowicz-Mitura, K., Bartosz, G., Mitura, S., 2007. Influence of diamond powder particles on human gene expression. Surf. Coat. Technol. 201, 6131–6135.

Basiuk, E.V., Santamaria-Bonfil, A., Meza-Laguna, V., Gromovoy, T.Y., Alvares-Zauco, E., Contreras-Torres, F.F., Rizo, J., Zavala, G., Basiuk, V.A., 2013. Solvent-free covalent functionalization of nanodiamond with amines. Appl. Surf. Sci. 275, 324–334.

Batory, M., Mitura, K., 2009. Application of diamond powders in cosmetics. Pol. J. Cosmetol. 12, 238–250.

Batory, M., Batory, D., Grabarczyk, J., Kaczorowski, W., Kupcewicz, B., Mitura, K., Nasti, T.H., Yusuf, N., Niedzielski, P., 2012. Biological properties of carbon powders synthesized using chemical vapour deposition. J. Nanosci. Nanotechnol. 12, 9037–9046.

Bociaga, D., Mitura, K., 2008. Biomedical effect of tissue contact with metallic material used for body piercing modified by DLC coatings. Diam. Relat. Mater. 17, 1410–1415.

Bottini, M., Bruckner, S., Nika, K., Bottini, N., Bellucci, S., Magrini, A., Bergamaschi, A., Mustelin, T., 2006. Multi-walled carbon nanotubes induce T lymphocyte apoptosis. Toxicol. Lett. 160, 121–126.

Brown, D.M., Wilson, M.R., MacNee, W., Stone, V., Donaldson, K., 2001. Size-dependent proinflammatory effects of ultrafine polystyrene particles: a role for surface area and oxidative stress in the enhanced activity of ultrafines. Toxicol. Appl. Pharmacol. 175, 191–199.

Ceynowa, P., Mitura, K., Zinka, W., Mitura, S., 2014. System modyfikacji nanoproszków diamentowych (DPP) w rotacyjnej komorze reaktora plazmo-chemicznego (MW PACVD). Elektronika 10, 47–50.

Chaudhry, Q., Castle, L., 2011. Food applications of nanotechnologies: an overview of opportunities and challenges for developing countries. Trends Food Sci. Technol. 22, 595–603.

Chwalibóg, A., Sawosz, E., Hotowy, A., Szeliga, J., Mitura, S., Mitura, K., Grodzik, M., Orłowski, P., Sokołowska, A., 2010. Visualisation of interaction between inorganic nanoparticles and bacteria or fungi. Int. J. Nanomed. 5, 1085–1094.

Cushena, M., Kerryb, J., Morrisc, M., Cruz-Romeroband, M., Cumminsa, E., 2012. Nanotechnologies in the food industry: recent, developments, risks and regulation. Trends Food Sci. Technol. 24, 30–46.

Czerniak-Reczulska, M., Niedzielski, P., Balcerczyk, A., Bartosz, G., Karowicz-Bilińska, A., Mitura, K., 2010. Biological properties of different type carbon

particles in vitro study on primary culture on endothelial cells. J. Nanosci. Nanotechnol. 10, 1065–1071.
Danilenko, V.V., 2004. On the history of the discovery of nanodiamond synthesis. Phys. Sol. State 46, 595–599.
Dolmatov, V.Yu., 2008. Modern industrial technology of reception detonation nanodiamonds and the basic areas of their use. Nanotechnics 1, 56–78.
Dworak, N., Wnuk, M., Zebrowski, J., Bartosz, G., Lewińska, A., 2014. Genotoxic and mutagenic activity of diamond nanoparticles in human peripheral lymphocytes in vitro. Carbon 68, 763–776.
Eversole, W.G., 1958. US Patent No. 3,050,187.
Ferin, J., Oberdörster, G., 1992. Translocation of particles from pulmonary alveoli into the interstitium. J. Aerosol Med. 5, 179–187.
Grabarczyk, J., Batory, D., Louda, P., Couvrat, P., Kotela, I., Bąkowicz-Mitura, K., 2007. Carbon coatings for medical implants. JAMME 20, 107–110.
Grodzik, M., Sawosz, F., Sawosz, E., Hotowy, A., Wierzbicki, M., Kutwin, M., Jaworski, S., Chwalibog, A., 2013. Nano-nutrition of chicken embryos: the effect of in ovo administration of diamond nanoparticles and l-glutamine on molecular responses in chicken embryo pectoral muscles. Int. J. Mol. Sci. 14, 23033–23044.
Huang, H., Pierstorff, Osawa, E., Ho, D., 2007. Active nanodiamonds hydrogels for chemotherapeutic delivery. Nano Lett. 7, 3305–3314.
International Commission on Radiological Protection (ICRP), 1994. Human respiratory tract model for radiological protection, publication66, Elsevier, Oxford.
Ingham, E., Fisher, J., 2005. The role of macrophages in osteolysis of total joint replacement. Biomaterials 26, 1271–1286.
International Standards Organisation (ISO), 1995. Air quality—particle size fraction definitions for health-related sampling, Geneva, ISO 7708:1995.
Jastrzębska, A.M., Kurtycz, P., Olszyna, A.R., 2012. Recent advances in graphene family materials toxicity investigations. J. Nanopart. Res. 14, 1320.
Jaworski, S., Sawosz, E., Kutwin, M., Wierzbicki, M., Hinzman, M., Grodzik, M., Winnicka, A., Lipińska, L., Włodyga, K., Chwalibog, A., 2015. In vitro and in vivo effects of graphene oxide and reduced graphene oxide on glioblastoma. Int. J. Nanomed. 10, 1585–1596.
Jee, Ah-Y., Lee, M., 2009. Surface functionalization and physicochemical characterization of diamond nanoparticles. Curr. Appl. Phys. 9, 144–147.
Kolesinska, B., Rozniakowski, K.K., Fraczyk, J., Relich, I., Papini, A.M., Kamiński, Z.J., 2015. The effect of counterion and tertiary amine on the efficiency of n-triazinylammonium sulfonates in solution and solid-phase peptide synthesis. Eur. J. Org. Chem. 2, 401–408.
Kręcisz, B., Kieć-Świerczyńska, M., Bąkowicz-Mitura, K., 2006. Allergy to metals as a cause of orthopedic implant failure. Int. J. Occup. Med. Environ. Health 19, 178–180.
Krüger, A., 2008. The structure and reactivity of nanoscale diamond. J. Mater. Chem. 18, 1485–1792.
Krüger, A., Kataoka, F., Ozawa, M., Fujino, T., Suzuki, Y., Aleksenskii, A.E., Vul, A. Ya., Osawa, E., 2005. Unusually tight aggregation in detonation nanodiamond: identification and disintegration. Carbon 43, 1722–1730.
Kurantowicz, N., Sawosz, E., Jaworski, S., Kutwin, M., Strojny, B., Wierzbicki, M., Szeliga, J., Hotowy, A., Lipińska, L., Koziński, R., Jagiełło, J., Chwalibog, A., 2015. Interaction of graphene family materials with *Listeria monocytogenes* and *Salmonella enterica*. Nanoscale Res. Lett. 10, 1–12.
Lademann, J., Weighmann, H.-J., Shäfer, H., Müller, G., Sterry, W., 2000. Investigation of the stability of coated titanium microparticles used in sunscreens. Skin Pharmacol. Appl. Skin Physiol. 13 (258–264).

Lam, R., Ho, D., 2009. Nanodiamonds as vehicles for systemic and localized drug delivery. Exp. Opin. Drug Del. 6, 883–895.
Larionova, I., Kuznetsov, V., Frolov, A., Shenderova, O., Moseenkov, S., Mazov, I., 2006. Properties of individual fractions of detonation nanodiamond. Diam. Relat. Mater. 15, 1804–1808.
Lechleitner, T., Klauser, F., Seppi, T., lechner, J., Jennings, P., Perco, P., Mayer, B., Steinmuller-Nethl, D., Preiner, J., Hinterdorfer, P., Hermann, M., Bertel, E., Pfaller, K., Phaller, W., 2008. The surface properties of nanocrystalline diamond and nanoparticulate diamond powder and their suitability as cell growth support surfaces. Biomaterials 29, 4275–4284.
Li, N., Ning Li, Sioutas, C., Cho, A., Schmitz, D., Misra, Ch., Sempf, J., Wang, M., Oberley, T., Froines, J., Nel, A., 2003. Ultrafine Particulate Pollutants Induce Oxidative Stress and Mitochondrial Damage. Environ. Health Perspect. 111, 455–460.
Liao, K.H., Lin, Y.S., Macosko, C.W., Haynes, C.L., 2011. Cytotoxicity of graphene oxide and graphene in human erythrocytes and skin fibroblasts. ACS App. Mater. Interf. 3, 2607–2615.
Lin, Ch.-R., Wei, D.-H., BenDao, M.-K., Chang, Ch.-M., Chen, W.-E., Lee, J.A., 2014. Effects of surface modification of nanodiamond particles for nucleation enhancement during its film growth by microwave plasma jet chemical vapour deposition technique. Adv. Mater. Sci. Eng. 2014, 1–5.
Liu, K.K., Zheng, W.W., Wang, C.C., Chiu, Y.C., Cheng, C.L., Lo, Y.S., Chen, C., Chao, J.I., 2010. Covalent linkage of nanodiamond-paclitaxel for drug delivery and cancer therapy. Nanotechnology 21, 315106.
Major, R., Bruckert, F., Lackner, J.M., Marczak, J.M., Major, J.B., 2014. Surface treatment of thin-film materials to allow dialogue between endothelial and smooth muscle cells and the effective inhibition of platelet activation. Royal Soc. Chem. 4, 9491–9502.
Martin, R., Heydron, P., Alvaro, M., Garcia, H., 2009. General strategy for high–density covalent functionalization of diamond nanoparticles using fenton chemistry. Chem. Mater. 21, 4505–4514.
Maynard, A.D., 2006. Nanotechnology: assessing the risks. Nano Today 1, 22–33.
Maynard, A.D., Baron, P.A., 2004. Aerosol measurement in the workplace. In: Ruzer, L.S., Harley, N.H. (Eds.), Aerosols Handbook. Measurement, Dosimetry and Health Effects. CRC Press, Boca Raton, p. 225.
Medina, O., Nocua, J., Mendoza, F., Gómez-Moreno, R., Ávalos, J., Rodríguez, C., Morell, G., 2012. Bactericide and bacterial anti-adhesive properties of the nanocrystalline diamond surface. Diam. Relat. Mater. 22, 77–81.
Merkel, K.D., Erdman, J.M., McHugh, K.P., Abu-Amer, Y., Ross, F.P., Teitelbaum, S.L., 1999. Tumor necrosis factor—a mediates orthopedic implant osteolysis. Am. J. Pathol. 154, 203–210.
Mihranyan, A., Ferraz, N., Strømme, M., 2012. Current status and future prospects of nanotechnology in cosmetics. Prog. Mater. Sci. 57, 875–910.
Mitura, S., 1987. Nucleation of diamond powder particles in a RF methane plasma. J. Cryst. Growth 80, 417–424.
Mitura, S., 2007. Nanodiamonds. JAMME 24, 166–171.
Mitura, K., 2009. HR TEM examinations of nanodiamond particles for biomedical application. JAMME 2, 317–322.
Mitura, S., Mitura, K., Niedzielski, P., Louda, P., Danilenko, V.V., 2006a. Nanocrystalline diamond, its synthesis, properties and applications. JAMME 16, 9–16.
Mitura, K., Niedzielski, P., Bartosz, G., Moll, J., Walkowiak, B., Pawłowska, Z., Louda, P., Kieć-Świerczyńska, M., Mitura, S., 2006b. Interactions between carbon coatings and tissue. Surf. Coat. Technol. 201, 2117–2123.

Mitura, K., Gisterek, I., Ceynowa, P., Wachowicz, A., Pich, G., Woś, R., Adach, K., 2013. Badania mikrobiologiczne bioaktywnych proszków nanodiamentowych modyfikowanych chemicznie. In: Leniowska, L., Nawrat, Z. (Eds.), Postępy Inz˙ ynierii Biomedycznej. Uniwersytet Rzeszowski, Rzeszów, pp. 157–167.
Mitura, K., Jędrzejewska-Szczerska, M., Ceynowa, P., Dudek, M., Cicha, M., Kotela, I., Mitura, S., 2015. Haemocompatibility of non-functionalized and plasma-chemical functionalized detonation nanodiamond particles. Arch. Metal. Mater. 60.
Mochalin, V.N., Shenderova, O., Ho, D., Gogotsi, Y., 2012. The properties and applications of nanodiamonds. Nat. Nanotechnol. 7, 11–23.
Monteiro-Riviere, N.A., Nemanich, R.J., Inman, A.O., Wang, Y.Y., Rivier, e J.E., 2005. Multi-walled carbon nanotube interactions with human epidermal keratinocytes. Toxicol. Lett. 155, 377–384.
Narayan, R.J., Boehm, R.D., Sumant, A.V., 2011. Medical applications of diamond particles & surfaces. Mater. Today 14, 154–163.
Nemmar, A., Hoylaerts, M.F, Hoet, P.H., Nemery, B., 2004. Possible mechanisms of the cardiovascular effects of inhaled particles: systemic translocation and prothrombotic effects. Toxicol. Lett. 149, 243–253.
Nguyen, T.T.-B., Chang, H.Ch., Wu, V.W.-K., 2007. Adsorption and hydrolytic activity of lysozyme on diamond nanocrystallites. Diam. Relat. Mater. 16, 872–876.
Niemiec, T., Szmidt, M., Sawosz, E., Grodzik, M., Mitura, K., 2011. The effect of diamonds nanoparticles on redox and immune parameters in rats. J. Nanosci. Nanotechnol. 11, 1–6.
Niemiec, T., Szmidt, M., Kruk-Roszkowska, A., Sawosz-Chwalibóg, E., Mitura, K., 2013. Carbon synthetized by RF PACVD method enhances the activity of antioxidants. Anim. Sci. 52, 151–156.
No. 1223/2009 (2009). Regulation (EC) No. 1223/2009 of the European Parliament and of the Council of 30 November 2009 on cosmetic products. Official Journal of the European Union.
Oberdörster, G., Sharp, Z., Atudorei, V., Elder, A., Gelein, R., Lunts, A., Kreyling, W., Cox, Ch., 2002. Extrapulmonary translocation of ultrafine carbon particle following whole body inhalation exposure of rats. J. Toxicol. Environ. Health A 65, 1531–1543.
Oberdörster, G., Oberdörster, E., Oberdörster, J., 2005. Nanotoxicology: an emerging discipline involving from studies of ultrafine particles. Environ. Health Perspect. 113, 823–839.
Pekkanen, J., Peters, A., Hoek, G., Tiittanen, P., Brunekreef, B., de Hartog, J., Heinrich, J., Ibald-Mulli, A., Kreyling, W.G., Lanki, T., Timonen, K.L., Vanninen, E., 2002. Particulate air pollution and risk of ST-segment depression during repeated submaximal exercise tests among subjects with coronary heart disease: the Exposure and Risk Assessment for Fine and Ultrafine Particles in Ambient Air (ULTRA) study. Circulation 106, 933–938.
Petracci, M., Cavani, C., 2012. Muscle growth and poultry meat quality issues. Nutrients 4, 1–12.
Pietras, T., Bakowicz, K., Kędziora, P., Nowak, D., Glinka, 1999. The influence of rutine, hesperidine and quercitine on LPS—induced lipid peroxydation in murine lungs. Eur. Respir. J. 14, 231.
Pineda, L., Sawosz, E., Lauridsen, Ch., Enberg, R.M., Elnif, J., Hotowy, A., Sawosz, F., Chwalibóg, A., 2012. Influence of in ovo injection and subsequent provision of silver nanoparticles on growth performance, microbial profile, and immune status of broiler chicken. Anim. Physiol. 4, 1–8.
Powers, K.W., Brown, S.C., Krishna, V.B., Wasdo, S.C., Moudgil, B.M., Roberts, S.M, 2006. Research strategies for safety evaluation of nanomaterials. In: Rai, M.,

Ribeiro C., Mattoso, L., Duran, N. (Eds.), Part VI. Characterization of Nanoscale Particles for Toxicological Evaluation. Toxicol. Sci. 90, 296–303.

Prasek, M., Sawosz, E., Jaworski, S., 2013. Influence of nanoparticles of platinum on chicken embryo development and brain morphology. Nanoscale Res. Lett. 8, 251–260.

Puzyr, A.P., Baron, A.V., Purtov, K.V., Bortnikov, E.V., Skobelev, N.N., Mogilnaya, O.A., Bondar, V.S., 2007. Nanodiamonds with novel properties: a biological study. Diam. Relat. Mater. 16, 2124–2128.

Ray, M.A., Shenderova, O., Hook, W., Martin, A., Grishko, V., Tyler, T., Cunningham, G.B., McGuire, G., 2006. Cold plasma functionalization of nanodiamond particles. Diam. Relat. Mater. 15, 1809–1812.

Rossi, M., Cubaddac, F., Dinid, L., Terranovae, M.L., Aurelic, F., Sorbof, A., Passeri, D., 2014. Scientific basis of nanotechnology, implications for the food sector and future trends. Trends Food Sci. Technol. 40, 127–148.

Satyanarayana, M., Vitali, F., Frost, J.R., Fasan, R., 2012. Diverse organo-peptide macrocycles via a fast and catalyst-free oxime/intein-mediated dual ligation. Chem. Comm. 48, 1461–1463.

Sawosz, E., Chwalibóg, Mitura, K., Mitura, S., Szeliga, J., Niemiec, T., Rupiewicz, M., Grodzik, M., Sokołowska, A., 2011. Visualisation of morphological interaction of diamond and silver nanoparticles with *Salmonella enteritidis* and *Listeria monocytogenes*. J. Nanosci. Nanotechnol. 11, 1–7.

Sawosz, E., Jaworski, S., Kutwin, M., Hotowy, A., Wierzbicki, M., Grodzik, M., Kurantowicz, N., Strojny, B., Lipińska, L., Chwalibóg, A., 2014. Toxicity of pristine graphene in experiments in a chicken embryo model. Int. J. Nanomed. 9, 3913–3922.

Shenderova, O.A., Ciftan Hens, S.A., 2010. Detonation nanodiamond particles processing, modification and bioapplications. In: Ho, D. (Ed.), Nanodiamonds. Application in Biology and Nanoscale Medicine. Springer, New York.

Shenderova, O.A., Gruen, D.M., 2006. Ultrananocrystalline Diamond: Synthesis, Properties, and Applications. William Andrew Publishing, Norwich, New York.

Shenderova, O.A., McGuire, G., 2006. Nanocrystalline diamond. In: Gogotsi, Y. (Ed.), Nanomaterials Handbook. CRC Press, Taylor and Francis Group, Boca Raton, pp. 203–237.

Siegrist, M., Cousin, M.E., Kastenholz, H., Wiek, A., 2007. Public acceptance of nanotechnology foods and food packaging: the influence of affect and trust. Appetite 49, 459–466.

Solarska, K., Gajewska, A., Bartosz, G., Mitura, K., 2012. Induction of apoptosis in human endothelial cells by nanodiamond particles. J. Nanosci. Nanotechnol. 12, 1–5.

Solarska-Ściuk, K., Gajewska, A., Skolimowski, J., Mitura, K., Bartosz, G., 2013. Stimulation of production of reactive oxygen and nitrogen species in endothelial cells by unmodified and Fenton-modified ultradisperse detonation diamond. Biotechnol. Appl. Biochem. 60, 259–265.

Som, C., Berges, M., Chaudhry, Q., Dusinska, M., Fernandes, T.F., Olsen, S.I., Nowack, B., 2010. The importance of life cycle concepts for the development of safe nanoproducts. Toxicology 269, 160–169.

Sozer, N., Kokini, J.L., 2009. Nnaotechnology and its applications in food sector. Trends Biotechnol. 27, 82–89.

Spitsyn, B.V., Derjaguin, B.V., 1956. USSR Author's Certificate No 339134. Appl. 964957/716358.

Szmidt, M., Niemiec, T., Sawosz- Chwalibóg, E., Mitura, K., 2013. The influence of nanodiamond particles on rat health status. Anim. Sci. 52, 195–201.

Thomas, V., Halloran, B., Ambalavanan, N., Catledge, S.A., Vohra, Y.K., 2012. In vitro studies on the effect of particle size on macrophage responses to nanodiamond wear debris. Acta Biomater. 8, 1939–1947.

Uo, M., Akasaka, T., Watari, F., Sato, Z., Tohji, K., 2011. Toxicity evaluations of various carbon nanomaterials. Dent. Mater. J. 30, 245–263.

Wakefield, W., Green, S., Lipscomb, S., Flutterr, B., 2004. Modified titania nanomaterials for sunscreen applications—reducing free radical generation and DNA damage. Mater. Sci. Technol. 20, 985–988.

Wichmann, H., Peters, A., 2000. Epidemiological evidence of the effects of ultrafine particle exposure. Philos. Trans. R. Soc. Lond. Ser. A. 358, 2751–2769.

Wierzbicki, M., Sawosz, E., Grodzik, M., Hotowy, A., Prasek, M., Jaworski, S., Sawosz, F., Chwalibóg, A., 2013. Carbon nanoparticles downregulate expression of basic fibroblast growth factor in the heart during embryogenesis. Int. J. Nanomed. 8, 3427–3435.

Xiao, X., Wang, J., Liu, C., Carlisle, J.A., Mech, B.V., Greenberg, R.J., Guven, D., Freda, R., Humayun, M.S., Weiland, J.D., Auciello, O., 2006. In vitro and in vivo evaluation of ultrananocrystalline diamond for coating of implantable retinal microchips. J. Biomed. Mat. Res. Part B 77, 273–281.

10

BIODEGRADABLE POLYMERNANOCOMPOSITES FOR PACKAGING APPLICATIONS

Ana Sofia Lemos Machado Abreu, Isabel Gonçalves de Moura, Arsénio Vasconcelos de Sá, Ana Vera Alves Machado
University of Minho, Institute of Polymers and Composites (IPC) and Institute of Nanostructures, Nanomodelling, and Nanofabrication, Guimarães, Portugal

1 Introduction

The packaging material evolution is due to the new challenges that arise mainly from the advancements of materials science and technology, a customer's lifestyle, and social needs. Nowadays, the packaging industry not only is essential for the protection and preservation of the product, from its production until consumption, but also to provide information on the product quality and especially in the case of active packaging, to interact with the product and to increase its shelf life. Therefore, in developed countries, packaging represents about 2% of the gross national product (GNP) and is considered as the third largest industry in the world.

Packaging must be developed from specific materials into specific shapes and texture, scientifically designed to suit the product being packed, reduce vulnerabilities during the transit journey and storage, maximize the shelf life of the product, and ultimately, positively influence the consumer's decision. In the recent years, the production and use of nonbiodegradable materials or plastics for food packaging application have significantly increased. To meet the growing demand for sustainability and environmental safety, a growing number of studies have been focused on the development of food packaging materials that could rapidly degrade and completely mineralize in the environment, after its lifetime (Othman, 2014; Majeed et al., 2013). Biodegradable polymers have been extensively investigated as promising alternatives to be exploited and developed into ecofriendly food packaging materials

Food Packaging. http://dx.doi.org/10.1016/B978-0-12-804302-8.00010-8

(Yates and Barlow, 2013). These environmental-friendly packages have the capability of being degraded, or broken down, in nature due to the action of microorganisms that are able to convert them, for instance, into carbon dioxide (CO_2) and water (H_2O) (Araujo et al., 2014). This degradation is either induced by hydrolysis or chain scission depending on the polymer's chemical structure and environmental conditions (Sinha Ray and Bousmina, 2005).

Most of the biodegradable polymers exhibit excellent properties, similar to several nonbiodegradable commodity polymers obtained from nonrenewable sources. Due to this, they are attracting high interest in a wide range of applications (Sinha Ray and Bousmina, 2005). Therefore, the competition between biodegradable and nondegradable polymers has increased and the annual sales of polymers from renewable sources has presented a huge expansion (Mohanty et al., 2002). Nevertheless, the use of biodegradable polymers has limitations in a wide range of applications due to some of their inherent properties, such as brittleness, low melt viscosity, low heat distortion temperature, and high gas and vapor permeability (Sinha Ray and Bousmina, 2005), which conventional nonbiodegradable materials made from petroleum can overcome. Biodegradable polymers can be categorized into three major groups, depending on their origin, such as, natural biopolymers extracted from renewable resources (starch, chitin, cellulose, casein, zein, and soy protein), synthetic polymers based on raw materials obtained from biomass [eg, polylactic acid (PLA)] or from petroleum derivatives [eg, poly(ε-caprolactone) (PCL)], and produced by microorganisms [eg, polyhydroxyalkanoates (PHAs)] (Othman, 2014) (Fig. 10.1). Carbohydrates, such as, starch, cellulose and chitin, and protein-based polymers, namely zein or

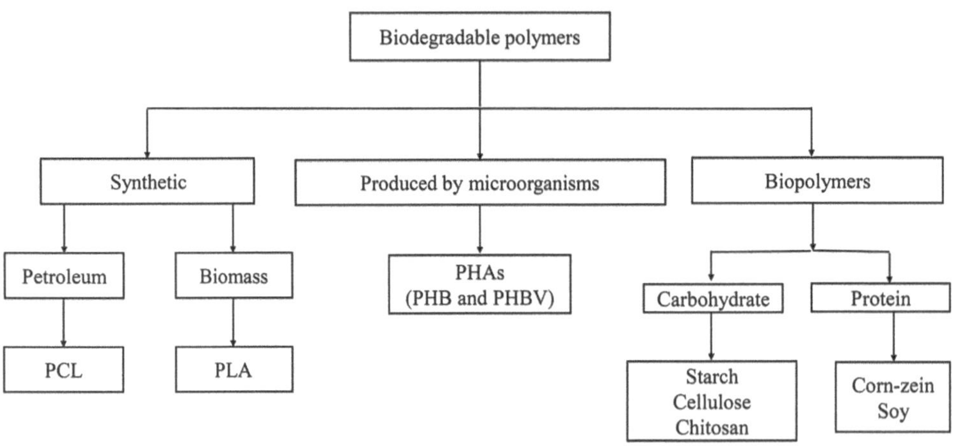

Figure 10.1. Classification of biodegradable polymers discussed in this chapter.

soy protein, among others, are the most common type of biodegradable polymers for packaging application (Rhim et al., 2013). Therefore, using innovative processing technologies for the modification of biodegradable polymers is still a main task for polymer chemists and engineers. Due to this, a lot of effort has been made to improve the biopolymers' properties.

2 Nanocomposites

In the past decades, in polymer science, the addition of an inorganic phase to an organic one became a very attractive subject. This is due to the improvement in properties obtained when a small amount of the inorganic phase can be dispersed at the molecular level in the polymer matrix (Uskoković, 2007). In the 70's, this new class of materials was classified as nanocomposites. The properties' enhancement is associated with the high surface area of the nanosized molecules and the particles interface which acts as a key point for the macroscopic properties.

Typically, nanofillers can be classified into three main classes, such as, particles, platelets, and fibers (Thostenson et al., 2005). Table 10.1 has several examples of organic- or inorganic nanofillers that have been studied in the development of biodegradable food packaging materials.

Although nanocomposites preparation has been investigated for a few years and a lot of results have been published, it is still a challenge to achieve a homogeneous dispersion of the nanoparticles in polymeric matrices. Frequently the van der Waals attraction between nanoparticles is higher than that between the nanoparticle and polymer molecules, which results in the formation of nanoparticles agglomerates. Moreover, surface properties play an important role; hydrophobic polymers are not compatible with hydrophilic nanoparticles. Consequently, these materials exhibited poor interfacial adhesion, bad dispersion, and poor properties. Thus, occasionally nanocomposites exhibited worse properties than the polymer matrices, limiting their efficient application.

Aiming to overcome these difficulties, researchers have been focusing on particles modification, processing methods, as well as, on the development of new methods where chemical reactions can take place.

Nanocomposites have been produced by different methods, such as, in situ nanoparticles formation, in situ intercalative polymerization, solution intercalation, and melt intercalation (Rhim and Kim, 2014). Among these methods, solution intercalation and melt intercalation have been most regularly used to

Table 10.1 General Fillers That Have Been Used for Food Packaging Applications

Classification	Type of Filler	Example	Reference
Organic	Natural biopolymers	Cellulose	Abdollahi et al. (2013)
		Chitosan	Shen et al. (2010)
	Natural compounds	D-limonene	Arrieta et al. (2014)
Inorganic	Metal ions	Silver	Abreu et al. (2015a)
		Copper	Conte et al. (2013)
		Gold	Youssef et al. (2014)
	Metal oxide	ZnO	De Silva et al. (2015)
		MgO	Sanuja et al. (2014)
		TiO_2	El-Wakil et al. (2015)
		Ag_2O	Tripathi et al. (2011)
	Clay	MTT	Echeverría et al. (2014)
		Closite–Na^+	Maisanaba et al. (2013)
		Closite–10A	Molinaro et al. (2013)
		Closite–20A	Rhim et al. (2006)
		Closite–30B	Araújo et al. (2014)
		Dellite–43B	Araújo et al. (2014)

manufacture bionanocomposites (Ojijo and Sinha Ray, 2013). In solution intercalation, clays (nanofiller) and the polymer (intercalant) are dissolved either in the same solvent or in different solvents. Therefore, the polymer chains intercalate and displace the solvent within the interlayer of the nanoparticles. This method is suitable for the intercalation of polymers, with low or no polarity, into layered structures, which produces thin films of polymer-oriented clay intercalated layers. Thus, it has been widely used in the preparation of water-soluble biopolymers nanocomposites (Shipp, 2011). In the melt intercalation, the clay, or other nanoparticle, and the polymer are mixed above the melting temperature of the polymer. After this, they may be kept at this temperature for a controlled period of time or put under shear, or other conditions, to stimulate the intercalation or exfoliation, in the case of the clay (Fig. 10.2). The melt-intercalation method is the most attractive approach because of its versatility, compatibility with current

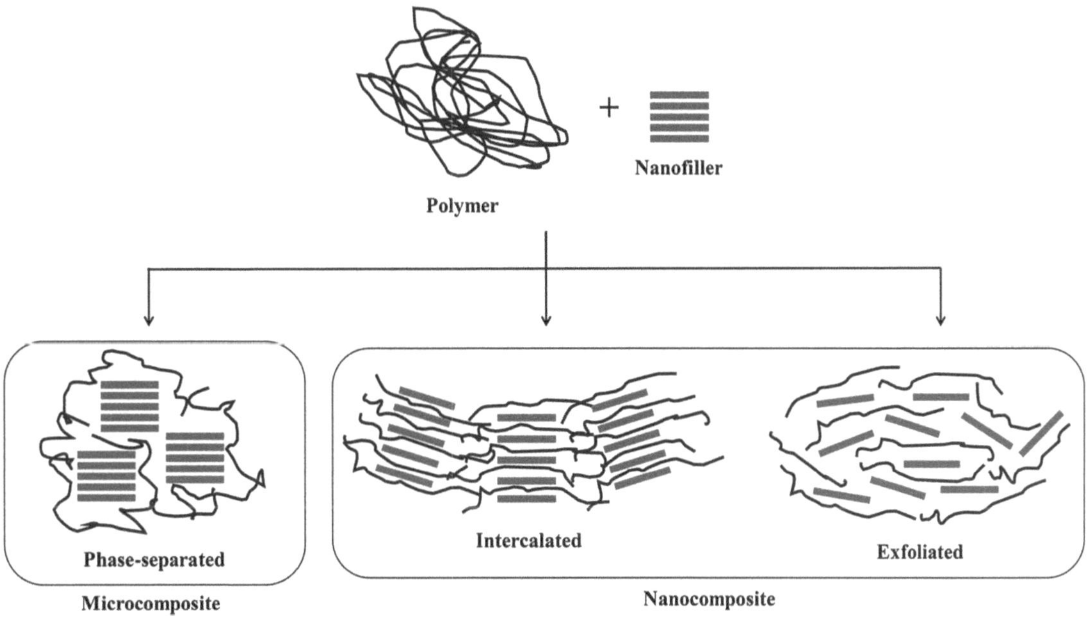

Figure 10.2. Schematic representation of the possible structures achieved during preparation of polymer nanocomposites.

polymer processing equipment, such as, extrusion and injection molding, and its environmental-friendly character, due to the absence of solvents. Moreover, this process can easily be used at the industrial scale, which would be most appropriate to produce new packaging materials.

Consequently, using different methods for the processing of bionanocomposites, it has been possible to develop advanced structures for active- and intelligent-packaging applications (Yam et al., 2005).

This chapter contains a state-of-the-art overview of a new class of materials, biodegradable polymer-based nanocomposites, with potential applications in the food-packaging industry. These materials can be used to ensure safety and quality standards in the development of active and smart/intelligent packaging materials.

3 Biopolymers

3.1 Polymers From Natural Resources

Biodegradable polymers obtained from renewable resources have attracted high attention in recent years. This new interest results from global environmental-respect awareness and the fossil-depletion problem. Biopolymers research and development, as well as, their production, have been the fastest for several years. Main agro-polymers presented in this chapter are the proteins and the polysaccharides. They are often used in the production of multiphase materials.

3.1.1 Corn-Zein

Corn-zein is obtained from the corn gluten, which is a by-product of corn-wet milling industry, has great potential to be used in the food packaging application (Corradini et al., 2014). Corn grains usually contain between 7% and 11% protein. Commercially available corn protein, known as zein, consists mainly of the prolamin fraction of corn and has a protein content greater than 90%. Proteins are mostly hydrophilic, being very sensitive to humidity and frequently soluble in water, establishing an important limitation for their use in food-packaging applications. Zein, on the other hand, is not soluble in water. However, its properties are dependent on relative humidity (Shukla and Cheryan, 2001). The hydrophobicity of corn-zein is related to its high content in nonpolar amino acids, such as, leucine, proline, and alanine, which represent approximately 35% of weight (Graciela and Qin, 2002). This property allows corn-zein to provide fairly good moisture-barrier properties when compared with other proteins. Moreover, materials produced from this protein also display excellent oxygen-barrier properties. However, poor mechanical and flexibility properties have limited the use of corn-zein in food-packaging materials.

Films of corn-zein are normally produced by the casting-solution method, using aqueous solutions of aliphatic alcohols. The solvent evaporation promotes the formation a film that, upon complete evaporation of the solvent, remains characteristically hard and glossy. Several chemical compounds, such as formaldehyde, glutaraldehyde, epichlorohydrin, citric acid, 1,2,3,4-butanetetracarboxylic acid, polymeric dialdehyde starch, 1,2-epoxy-3-chloropropane, and dialcohols can be used to induce crosslinking between zein molecules. This structural alteration results in the enhancement of the tensile strength of the material (Parris and Coffin, 1997; Yamada et al., 1995). Additionally, the use of plasticizers is an approach that enhances the flexibility of corn-zein films (Hernandez-Izquierdo and Krochta, 2008; Lawton, 2004), enhancing its sustainability and excellent oxygen-barrier properties for the production of food-packaging materials.

The development of materials for food packaging, based on corn-zein, involved different approaches, such as, lamination of zein with other proteins and lipids (Hernandez-Izquierdo and

Krochta, 2008), blending the protein with synthetic polymers or biopolymers (Corradini et al., 2004; Liu et al., 2012; Ozcalik and Tihminlioglu, 2013), and coating polyolefin surfaces with corn zein (Shin et al., 2002; Yu et al., 2006). In recycling facilities, identification and separation of layers of synthetic films is problematic, and the use of multilayer structures for barrier applications results in the accumulation of high amounts of waste (Han, 2014). Due to this, biopolymer coatings, such as corn-zein, may be advantageous from the ecofriendly point of view.

Recently, the use of lignin as an additive in thermoplasticized zein was studied (Oliviero et al., 2011). The results revealed that this interactive additive allowed a supramolecular control over the zein structure and produced a true bionanocomposite suitable for application in food packaging.

The introduction of clay nanoplatelets, intercalated or exfoliated, into zein matrices is well reviewed in the literature, where zein nanocomposites have shown improved thermal stability and hydrophilicity (Park et al., 2013; Sanchez-Garcia et al., 2010; Corradini et al., 2014).

In summary, corn-zein bionanocomposites or zein blends are excellent candidates for the production of materials that can be used in the food-packaging manufacture (Sanchez-Garcia et al., 2010).

3.1.2 Soy

Soy protein is extracted from soybean seeds, widely grown in the world. These seeds, used in the production of soy oil, can be used in the preparation of materials for food packaging because soy meets food-grade standards (Garrido et al., 2013). Soy proteins are composed of a mixture of albumin and globulin proteins with globular structure. The subunits of globulins are associated via hydrophobic, hydrogen bonding, and disulfide bonds (Ciannamea et al., 2014). These proteins, obtained from renewable resource, have been used as base material in the production of edible films and coatings with potential application in the food industry. Their good biodegradability, biocompatibility, and the fact that they allow the preparation of green composites have attracted the interest of researchers (Koshy et al., 2015). Studies have shown that films based on soy protein are good oxygen- and lipid barriers. However, due to their hydrophilic nature, they have limited water-barrier properties (Kurt and Kilincceker, 2011).

Different approaches to improve the mechanical properties and to decrease the water sensibility of materials based on soy proteins have been reported in literature (Song et al., 2011). Blends of soy protein with organic- and inorganic fillers have been prepared by solution mixing and melting compression (Ai et al., 2007; Ciannamea et al., 2014). Other blends of soy protein with biodegradable hydrophobic polymers, such as, PCL (Zhong and Sun, 2001), PLA (González and Alvarez Igarzabal, 2013), polyurethane (Madbouly and Lendlein, 2012), or waterborne polyurethane (Zhang et al., 2012) can be found in literature. These materials are examples of efforts to enhance water resistance and mechanical performance without compromising biodegradability of soy protein-based composites.

Nanocomposite films based on these proteins have been recently processed by solvent-casting and reinforced with the natural clay montmorillonite (MMT). Soy–MMT films display highly intercalated–exfoliated structures with enhanced mechanical properties and susceptibility to water (Echeverria et al., 2014).

Studies of reinforcement of soy with the addition of chitin whiskers (Lu et al., 2004) revealed that strong interactions between both components have an important contribution in supporting the composites without interfering in their biodegradability. Using banana fibers, a waste product of banana cultivation, Kumar et al. (2008) developed 100% biodegradable soy-protein composites.

3.1.3 Starch

Starch is a polymer obtained from agro-resources, which is known for its complete biodegradation in soil and water. In addition to its degradability, its low cost attracts considerable attention in the food-packaging industry (Avella et al., 2005). Starch is a promising raw material because of its availability from many plants (Chen and Evans, 2005). This polysaccharide can promote the biodegradability of nonbiodegradable plastics (Abreu et al., 2015b). It can also be used in the production of low-cost polymer blends with fully biodegradable-synthetic materials (Wang et al., 2003). Starch is a mixture of two polymers, amylose, a linear 1,4′-α-D-glucan, and amylopectin, a highly branched molecule formed by short chains of 1,4′-α-D-glucose with 1,6′-α-crosslinking. The polymeric chains of amylose commonly have more than 6,000-D-glucopyranose units and their molecular weight ranges between 150,000 and 600,000 Da (Fig. 10.3).

Starch remains in granular form in the plastic matrix and thus may act as a filler. One major problem with granular starch composites is their limited processability, due to the big particle sizes (5–100 μm). Therefore, it is commonly pretreated with a plasticizer in order to acquire thermoplastic properties to allow starch to be processed in the melt. Thermoplastic starch (TPS) has been developed by gelatinizing starch with 6–10% moisture, heat, and pressure (George et al., 1994). However, TPS suffers from several

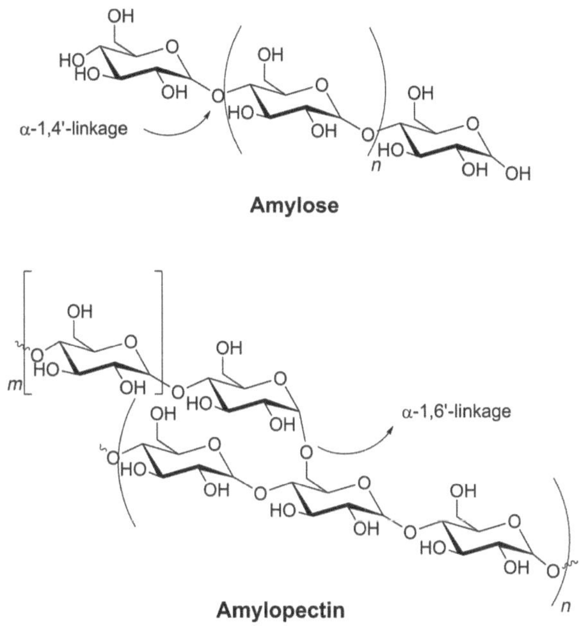

Figure 10.3. Chemical structure of starch—amylose and amylopectin.

limitations, such as, poor water resistance, low strength, and poor mechanical properties, and a large number of research works have been published on this subject.

Dispersion of several nanosized fillers into a starch biopolymer matrix have been studied by Xie et al. (2013). Several types of nanofillers have been used with plasticized starch, including polysaccharide nanofillers (nanowhiskers, cellulose-based nanoparticles, and chitosan) (Nasseri and Mohammadi, 2014; Liu et al., 2013) and phyllosilicates (MMT, hectorite, sepiolite, etc.) (Chung et al., 2010; Avérous and Pollet, 2014; Chivrac et al., 2010). Carbon nanotubes, graphite oxide, and carbon black have also been used within starch-polymer matrix. However, the main advantages of using these nanofillers are that they can provide new functionalities to starch-based materials, in addition to the general reinforcement, but compromise their completely biodegradability (Cheng et al., 2013). Phyllosilicates (particularly MMT) have been the most-used nanofillers due to their advantages, such as, wide availability, low cost, high-aspect ratio, and their swelling nature. Table 10.2 lists some of the most commonly used nanofillers in the preparation of starch-based bionanocomposites for food packaging, through solvent casting or by melt processing, found in the literature.

Table 10.2 Starch/Plasticizer/Clay Prepared Through Solvent Casting (SC) or by Melt Processing (MP)

Plasticizer	Clay	Process	Reference
Glycerol	Hectorite–Ca^{2+}	SC	Wilhelm et al. (2003)
Glycerol/water	MMT with CEC[a]	MP	Avella et al. (2005)
Water	MMT–Na^{+}	MP	Dean and Wu (2007)
Glycerol/urea/formamide	MMT	MP	Tang et al. (2008)
Water	C25A	MP	Chaudhary (2008)
Glycerol/urea/ethanolamine	MMT–Na^{+}/C30B	SC	Zeppa et al. (2009)
Pequi oil	MMT (K10)	SC	Schlemmer et al. (2010)

[a] CEC, Cation-exchange capacity

Gycerol-plasticized starch/clay-biocomposite films, obtained by solvent casting method, incorporating small-layered crystals (hectorite–Ca^{2+}) revealed significantly increased storage modulus value and mechanical properties, revealing the clay-reinforcing effect on the starch matrix (Wilhelm et al., 2003). Other authors developed novel biodegradable-starch/clay nanocomposite films to be used as food-packaging material, using glycerol and water as plasticizer, and dispersing MMT via polymer-melt processing. The results revealed a good intercalation of the polymeric phase into clay-interlayer galleries, together with an enhancement of the mechanical parameters, such as, modulus and tensile strength. The conformity of the developed bionanocomposites were verified by migration tests and by placing the films in contact with vegetables and simulants, according to regulations and European directives, demonstrating low overall-migration limits (Avella et al., 2005).

Intercalated and exfoliated structures of gelatinized-starch/clay nanocomposites have been developed. Nanoclay dispersions [sodium MMT (Na–MMT) and fluorohectorite (Na–FHT)] using different plasticizer amounts were prepared, either by standard mixing or by the use of ultrasonics. The authors showed that enhanced properties were obtained from a gelatinized starch film with high-exfoliation level after optimization of the amount of both plasticizer and nanoclay. They also found out that the clay dispersion at medium clay concentrations, in the Na–MMT nanocomposites, and at higher clay concentrations, in the Na–FHT, was improved using ultrasonics. Nevertheless, the same level of exfoliation, and consequently, similar mechanical properties were achieved via standard mixing, when the amount of clay, water, and starch were optimized (Dean and Wu, 2007). Tang et al. (2008) have successfully produced biodegradable nanocomposites from corn starch and MMT by melt-extrusion processing. The influence of glycerol content was investigated and it was observed that by decreasing it from 20% to 5%, the degree of clay exfoliation increased. The authors have also tested urea and formamide as alternative plasticizers, observing an increasing degree of clay exfoliation. The formamide plasticized-starch/clay nanocomposite films exhibited lower water-vapor permeability, higher glass-transition temperature, and higher tensile strength when compared to those films containing glycerol and urea at the same amount. The amylose–water interaction process within starch matrix has been studied (Chaudhary, 2008), being observed with slow recrystallization of amylose chains and growth of anisotropic properties in the starch matrix. The authors have found that increasing nanoclay content from 2% to 4% by weight, the composite modulus jumped to 100%, demonstrating the excellent interaction between clay nanocomposites and starch polymer. Zeppa et al. (2009) investigated nanocomposites based on a plasticized natural-biodegradable matrix. Glycerol and a urea/ethanolamine mixture with a natural and an organically modified MMT were used as plasticizers. The use of the natural MMT provided a significant reduction of oxygen permeability, which is related to the high dispersion state of this clay. For urea–ethanolamine composites, specific compatibilizer/clay interactions also led to an improvement of the barrier properties. More recently, TPS/MMT nanocomposites were prepared using starch and pequi (*Caryocar brasiliense*) oil as plasticizer (Schlemmer et al., 2010). Exfoliated- and intercalated nanocomposites were found to be dependent on the MMT content. When MMT content is superior to 5 wt.% intercalation was observed while exfoliation predominates when the content reaches a maximum of 5 wt.%. The introduction of low content of MMT improves the thermal stability and the stiffness of the materials.

Nevertheless, polysaccharide nanofillers are the second most-used fillers due to their abundance in nature and biological sources. Cellulose nanofibers (CNFs) have been recently used as reinforcement of starch nanocomposites, enhancing mechanical properties and preserving their biodegradability (Babaee et al., 2015). A comparative study between cellulose whiskers (CNWs) and starch nanoparticles in the production of starch nanocomposites were performed. The results revealed that starch nanoparticles shows higher reinforcement in dynamic mechanical tests compared to CNWs nanocomposites (Nasseri and Mohammadi, 2014).

Starch-based nanocomposite blends with biodegradable polymers have also been the focus of several studies, including the use of PCL (Pérez et al., 2013), natural rubber (Mondragón et al., 2009), PLA (Ayana et al., 2014), polyvinyl alcohol (PVA) (Tang and Alavi, 2011), and poly(butylene succinate adipate) (Yu et al., 2007; Bocchini et al., 2010).

3.1.4 Cellulose

Cellulose is the most abundant biopolymer in nature, which means that it is an attractive material that can be obtained from natural resources. It as a linear polymeric chain of 1,4′-β-D-glucose (Fig. 10.4), which is responsible for the formation of cellulose microfibrils through inter- and intramolecular hydrogen bonds.

Cellulose-based films have been investigated for controlling the migration of moisture, gas, and hydrocarbons in several types of foods, such as, meat products, fruits, and vegetables. These studies showed that cellulose-based films are very efficient oxygen and hydrocarbon barriers, and their water vapor-barrier properties may be improved by the addition of lipids (Rhim and Ng, 2007).

Cellulose bionanocomposites based on particles, such as clay, have been processed using melt compounding or solution-casting techniques (Park et al., 2006; Delhom et al., 2010; Rodríguez et al., 2012). Cellulose has been recently used as reinforcement in nanocomposites (eg, nanowhiskers and CNFs) and is a new field of interest (Abdul Khalil et al., 2012; Eichhorn, 2011). There are several aspects and properties of cellulose that make its usage advantageous, including its low price and being a biodegradable material obtained from renewable sources, as well as, its low density, low-energy consumption, low abrasivity during processing, wide variety of available fillers, reactive surface, and high specific properties (Azizi Samir et al., 2005). CNFs and CNWs, derived from renewable biomass, have attracted interests during the past years due to their appealing intrinsic properties, including nanoscale dimensions, high surface area, low density, high

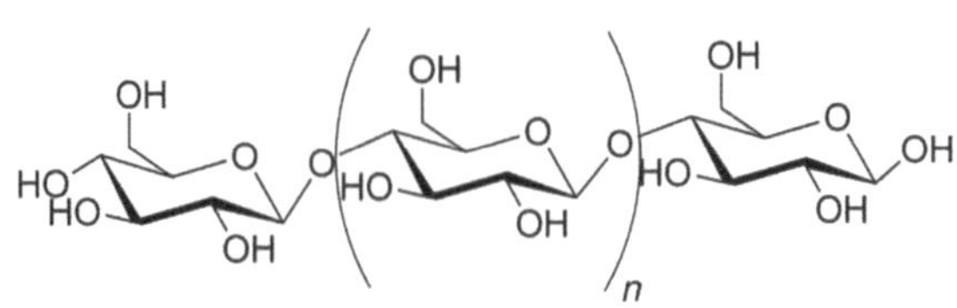

Figure 10.4. Chemical structure of cellulose.

Table 10.3 Biodegradable Nanocomposites With Cellulose Nanoparticles

Biopolymer	Cellulose Nanoparticles	Process	Reference
Cellulose	CNWs	SC	Qi et al. (2009)
Chitosan	CNWs	SC	Li et al. (2009)
PHA	CNFs	SC	Bhardwaj et al. (2014)
Alginate	CNFs	SC	Sirviö et al. (2014)
Starch	CNFs	SC	Babaee et al. (2015)
PLA	CNFs	SC	Fortunati et al. (2015)
Wheat gluten	CNFs	SC	El-Wakil et al. (2015)

mechanical strength, renewability, and biodegradability (Abdul Khalil et al., 2012; Eichhorn, 2011).

In recent studies focused on the development of new materials for food packaging, several researchers have used cellulose nanoparticles to reinforce bionanocomposites using biodegradable polymers (Table 10.3).

Qi et al. (2009) prepared bionanocomposite films from native CNWs and cellulose matrix regenerated from an aqueous NaOH/urea-solvent system. The bionanocomposite films were transparent to visible light and showed good mechanical and thermal properties, as a result of the reinforcement by the whiskers (5–10 wt.% CNWs). The well-dispersed CNWs played an important role in the improvement of the properties of the cellulose materials. The biggest advantage of these new bionanocomposite films is the fact that they are fully biobased, easily recyclable, and biodegradable.

Bionanocomposites of chitosan reinforced with CNWs were successfully developed by Li et al. (2009). The results showed that the bionanocomposites have good miscibility, excellent thermal stability, water resistance, strong interactions between the whiskers and the chitosan matrix, and enhanced tensile strength. Nanocellulose was used to improve the properties of biodegradable polymers obtained from microbial fermentation, especially PHAs, ensuring that the biodegradable nature of the material is retained. The authors have observed that the optimum amount of CNFs to improve biopolymer properties was approximately 2% (Bhardwaj et al., 2014). Biocomposite films based on cellulose and alginate have been produced using CNFs, which had fiber widths between 19.0 μm and 25 nm. The biocomposite films showed excellent grease-barrier properties and reduced water-vapor permeability (WVP) after the addition of cellulose fibers (Sirviö et al., 2014). TPS nanocomposite with CNFs were successfully prepared by using the solution-casting method. Mechanical studies showed that the tensile strength and modulus of nanocomposites increased when compared to the neat matrix (Babaee et al., 2015). Fortunati et al. (2015) produced, by solvent casting, PLA-nanocomposite films reinforced with CNFs (1–3 wt.%), using a commercial surfactant to increase the dispersion of CNFs in the biodegradable matrix. The migration levels for all the studied nanocomposites that were well below the legislative limits required for their use as food packaging materials. Wheat gluten-based films reinforced with CNFs, and incorporated with TiO_2 nanoparticles and antimicrobial agents, were developed to be a material for active-food packaging (El-Wakil et al., 2015).

In general, the addition of cellulose nanoparticles improved the degradation rate of the bionanocomposites, the physical barrier, thermal and mechanical properties of the biopolymers. The research works with cellulose nanoparticles demonstrate that biopolymer films based on these materials have a high potential to be used as high strength-packaging materials.

3.1.5 Chitosan

Chitosan is a polysaccharide prepared by the deacetylation of chitin, which is one of the most abundant natural polymers found in living organisms, such as, crustacea, insects, and fungi (Van Den Broek et al., 2015). Its polymeric chain is formed by 1,4′-β-glucosamine units and it can contain random *N*-acetylated-glucosamine units (Fig. 10.5).

The interest in chitosan is supported by its unique properties, such as, being renewable, biocompatible, biodegradable, nontoxic, and having excellent-adsorption properties (Barikani et al., 2014). Chitosan also has interesting properties, such as, excellent film-forming capacity, and gas- and aroma-barrier properties at dry conditions, which makes it a suitable material for

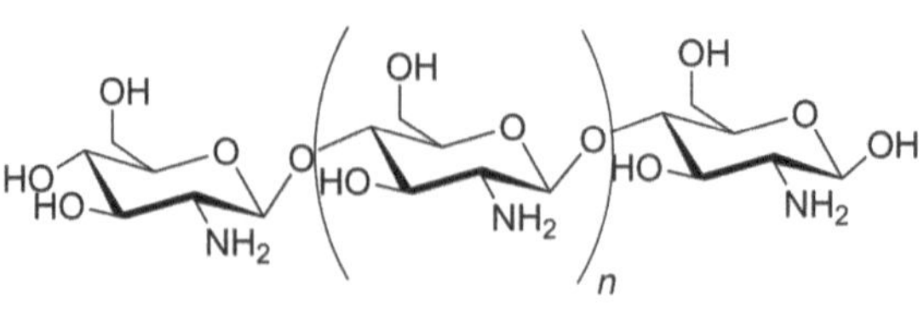

Figure 10.5. Chemical structure of chitosan.

designing food coatings and packaging structures (Caner, 2005). All the aforementioned properties also make chitosan a very good candidate for the development of novel antimicrobial-active packaging technologies that can improve the quality and safety and also extend the shelf life of perishable foods (Zou et al., 2016). Chitosan is soluble in several acidic solvents and has high antimicrobial activity against many pathogenic and spoilage microorganisms, including Gram-positive and Gram-negative bacteria.

In the production of chitosan films there are several parameters, such as, solvent pH, ionic strength, type of solvent (acid), and annealing treatment, which are manipulated to alter the mechanical properties and membrane porosity. Ionic strength or pH can be controlled in order to reduce intermolecular- and intramolecular-electrostatic repulsion between chitosan chains, thus allowing the chains to approach each other and enhance the formation of intermolecular- and intramolecular hydrogen bonds. Interestingly, some antibacterial and antifungal activities have been associated with chitosan and modified chitosan (Van Den Broek et al., 2015). The potential of chitosan as an antimicrobial preservative is appealing, especially with regard to limiting the development of certain bacteria, such as *Listeria monocytogenes*, which are unacceptable in the food industry because of their pathogenicity.

The good film-forming properties of chitosan allows the production of films to be used as coating material or in the production of membranes semipermeable to gases. For instance, it is known that these films possess low-oxygen permeability (Aider, 2010). Chitosan films can be divided into edible films or coatings (<30 μm) used directly on food in order to improve food safety and shelf life, films (>30 μm) and blends, which can be homogenous or heterogeneous mixtures.

Chitosan nanocomposites are mostly clay-based. The solution casting technique has been the method of choice in the preparation of chitosan/clay nanocomposites. Due to the polycationic nature of chitosan in acidic media, it appears as a natural candidate for intercalation in MMT–Na^+ through cationic-exchange process in the solvent-casting technique. Generally, intercalated- and exfoliated structures have been obtained, which is of high importance for food-packaging applications (Giannakas et al., 2014; Darder et al., 2003).

Rao et al. (2010) prepared films performing the casting technique, using chitosan and guar gum, a polysaccharide, in different ratios. The chitosan and guar gum can be used to make a biocomposite film that has desirable packaging characteristics. Furthermore, the antimicrobial activity of films containing 15% (v/v) guar gum was comparable to chitosan films against *Escherichia coli* and *Staphylococcus aureus*. Composite films attained from chitosan and guar gum might decrease environmental problems related to synthetic packaging.

The potential use of antimicrobial agents in materials for packaging applications have also been investigated. Chitosan–silver- and chitosan–gold-nanocomposites films were synthesized by a simple chemical method and the resulting films evaluated against Gram-positive (*S. aureus*) and Gram-negative bacteria (*Pseudomonas aerugenosa*), fungi (*Aspergillus niger*), and yeast (*Candida albicans*), showing great antimicrobial activity (Youssef et al., 2014).

Recently, chitosan-based bionanocomposites reinforced with CNFs have been studied by de Mesquita et al. (2012). Biobased nanocomposites were obtained through covalent linkage between CNFs and chitosan. The prepared materials demonstrated a significant improvement in the mechanical performance and a considerable decrease in the hydrophilicity when compared with chitosan itself.

3.1.6 Others

Polysaccharides, such as, konjac glucomannan, pullulan, alginate, carrageenan, and pectin have been used for the preparation of edible films. Among them, a few carbohydrates are used for commercial edible films as food packaging. Konjac glucomannan is an essential polysaccharide, which is the main component of the roots and tubers of the *Amorphophallus konjac* plant and a good source for film-forming materials (Zhang et al., 2005). Pectin, a complex anionic polysaccharide, is a linear polymer of D-galacturonic-acid units with their methyl esters connected through 1,4′-α-glycosidic bonds. This biopolymer is a potential film-forming material that needs to be blended with other polymers (Thakur et al., 1997; Bierhalz et al., 2012), because alone they do not provide satisfactory functionality for most applications (Bierhalz et al., 2012). Polysaccharides, such as, pectin, alginate, and carrageenan, have been blended with other compounds to make edible films, such as, starch, polyvinyl alcohol, soy protein, gelatin, chitosan, hydroxypropyl methylcellulose, and ethylcellulose for food-packaging applications (Sanchez-García, 2011; Bierhalz et al., 2012; González and Alvarez Igarzabal, 2013; Van Den Broek et al., 2015).

3.2 Biodegradable Polyesters

Biodegradable synthetic polymers offer a number of advantages over other materials for developing biodegradable packaging. Among the families of biodegradable synthetic polymers, polyesters have been attractive specifically in applications with short lifetime, such as for packaging applications.

The key advantages of biodegradable polyesters include the ability to be tailored to have different mechanical properties and degradation kinetics. Also, because of the ease of degradation of the ester linkage by hydrolysis, the degradation products can be resorbed through the metabolic pathways. Biodegradable synthetic polyesters are also attractive due to the fact that they can be fabricated into various shapes with desired morphologic features. Thus, the use of these materials, instead of nonbiodegradable-synthetic polymers, becomes an ecological alternative to reduce the solid-plastic waste.

3.2.1 Fossil-Based Polyesters

Biodegradable polymers can be obtained from fossil and natural sources, or even be produced by bacteria (Sinha Ray and Bousmina, 2005).

Aliphatic polyesters are the most well-known fossil source-derived biodegradable polymers, due to their synthetic versatility and variety. Recently, a growing interest has been demonstrated in the synthesis of completely biodegradable polyesters (Edlund and Albertsson, 2003). Thus, several chemical strategies allow the development of new biodegradable synthetic polyesters. Some examples of these are the ring-opening polymerization (ROP) methodology, which enables the synthesis of polymers with high molecular weight, or the polycondensation route, from which low molecular-weight polymers are synthesized (Lofgren et al., 1995). The most common synthetic biodegradable polyesters are PCL and PLA, prepared by using ROP of the respective cyclic monomers, LA and ε-CL. The major peculiarity of these polymers is related to the existence of hydrolysable-ester bonds, making them compatible with the natural environment and susceptible to hydrolytic- and biological degradation (Duda and Penczek, 2005). Nevertheless, their biodegradation rate is dependent on several factors, including their hydrolysable-ester bonds in the main chain, the chains' chemical structure, degree of crystallinity, molecular weight, and their morphology (Kim et al., 2001). Both synthetic approaches allowed the synthesis of biodegradable-synthetic polyesters that have been used in a number of food-packaging applications (Byun and Kim, 2014).

3.2.1.1 Polylactic Acid

PLA is renewable, biocompatible, and a biodegradable polymer that has good mechanical- and optical properties, and is one of the most widely used bioplastics (Fig. 10.6).

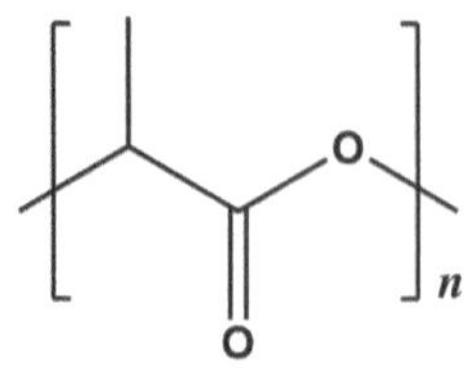

Figure 10.6. Chemical structure of PLA.

As previously mentioned, PLA is obtained either by ROP or by direct polycondensation of LA monomer. The monomer is a chiral molecule, existing as D- and L-LA isomers, and can be easily biologically or chemically obtained. The large-scale use of PLA as a packaging material is still disadvantageous due to its high cost and low performance when compared to other polymers. The most important limitation for the application of PLA in food packaging is its low gas-barrier properties. The properties of PLA, such as, thermal stability and impact resistance, are lower when compared to other conventional polymers used for thermoplastic applications.

Nanocomposites technology has the potential to improve polymer properties and expand the applications of PLA (Rhim et al., 2007). Of particular interest, clay-containing bionanocomposites that consist of PLA and organically modified clay (organoclay); often exhibit improved properties when compared with those of neat PLA or PLA-based conventional composites. For example, reduced amounts of organoclay particles added (approximately 5%) show flexural moduli enhancements with respect to neat PLA, comparable to that achieved by using approximately 15% loading of conventional fillers (Araújo et al., 2014). The lower loadings also simplify processing and reduce component weight. Nevertheless, the most important feature of PLA clay-based nanocomposites is the unique value-added properties, not usually possible with traditional fillers, such as, improved heat-distortion temperature and flexural properties, reduced permeability, and optical clarity. Also, clay particles are environmentally friendly and do not significantly affect the inherent compostability of the PLA after the formation of the nanocomposite (Pandey et al., 2005).

Many researchers had been focused on developing PLA bionanocomposites for food-packing applications. Besides the use of clays, another nanofiller, such as CNFs, as biobased nanofillers, have been investigated.

Other authors have investigated the development of blends of biopolymers with nanocomposites. In Table 10.4 summarizes some representable biodegradable-PLA nanocomposites with a high potential for packaging application.

Table 10.4 Biodegradable PLA Nanocomposites

Biopolymer	Nanofillers	Process	Reference
PLA–PCL blends	MMT	Melting	Cabedo et al. (2006)
PLA	C30B or C20A	Extrusion	Molinaro et al. (2013)
TPS–PLA blend	MMT–Na^+	SC	Ayana et al. (2014)
PLA–PHB blend	CNFs	Microextrusion	Arrieta et al. (2015)

Cabedo et al. (2006) developed, by melt, biodegradable-nanocomposites blends of amorphous PLA and PCL. The bionanocomposites revealed a high degree of interaction between the nanoclays and the PLA, which led to an enhancement of the oxygen-barrier properties of about 50%. In this study, the effect of plasticizers on oxygen permeability of PLA was offset by the formation of the kaolinite nanocomposite. Molinaro et al. (2013) demonstrated that the nanoclay-type and the L-isomer content affected the oxygen-transmission rate (OTR) and water vapor-transmission rate (WVTR). The incorporation of C30B improved the OTR by 50% and C20A improved the WVTR by 53%. These nanocomposites were shown to have improved functional properties and consequently may have the potential for application within the food-manufacturing sector. Using the melt-mixing method, known to be a green approach, bionanocomposites of TPS/PLA/MMT–Na^+ with a high degree of exfoliation were prepared by dispersion of clay in gelatinized starch, followed by the addition and blending with PLA (Ayana et al., 2014). This led to the development of a new green material in the field of packaging. The presence of the clay had a major effect on the mechanical and thermomechanical properties than the addition of PLA. While the addition of PLA decreased the water absorption (approximately 8%), the effect of both PLA and clay (8–12%) was higher (Ayana et al., 2014).

Arrieta et al. (2015) prepared, by melt blending, optically transparent plasticized PLA-based bionanocomposite films for food packaging. The materials were plasticized with 15 wt.% of acetyl(tributyl citrate) (ATBC) to improve the material processability and to obtain flexible films. Polyhydroxybutyrate (PHB) was used to increase PLA's crystallinity and thermal stability was improved by the addition of 5 wt.% of CNFs. The combination of ATBC and CNFs, mainly the better dispersed CNFs, improved the interaction between PLA and PHB. Consequently, an improvement in the oxygen barrier and stretch ability was achieved, and all bionanocomposite films presented appropriate disintegration in compost, suggesting their possible application as biodegradable-packaging materials (Arrieta et al., 2015). The same authors blended PLA and PHB using a natural plasticizer, terpene D-limonene, with the dual objective of increasing PLA's crystallinity and obtaining flexible films projected for food-packaging applications. The addition of D-limonene affected the mechanical properties of the developed material, which presented higher elongation at break. Disintegrability under composting conditions showed that, while PHB delays the PLA disintegrability, D-limonene increases it (Arrieta et al., 2014).

3.2.1.2 Poly(ε-Caprolactone)

PCL is a biodegradable polyester that can be gradually consumed by microorganisms. The hydrophobic linear chains are polymerized using monomers from petroleum-based sources. It is partially crystalline and has a low melting point of around 60°C and a glass-transition temperature of about −60°C (Fig. 10.7).

The addition of nanoparticles in a PCL matrix improves significantly the thermal and mechanical performance, but also decreases gas permeability and electric conductivity. Therefore, these new materials can be good candidates to replace conventional nonbiodegradable polymers in specific areas, such as, medicine, agriculture, and packaging applications (Vroman and Tighzert, 2009).

Solvent-casting method was used by Jimenez et al. (1997) to prepare PCL-based biodegradable nanocomposites, using MMT modified with distearyldimethylammonium cations. PCL was dissolved in hot chloroform and a given amount of the modified clay was added. Homogeneous nanocomposites films were obtained after solvent evaporation. Nevertheless, after the films analysis it was found that under those conditions, no intercalation of the clay occurred in the polymer matrix. In order to improve dispersion, a different method was used; the layered silicates were mixed, in the molten state, with the polymer matrix. Since this process involves temperature and shear, the polymeric chains could penetrate into the clay platelets, increasing the interlayer space that resulted both in intercalated- and exfoliated structures. A study on PCL/clay-nanocomposites

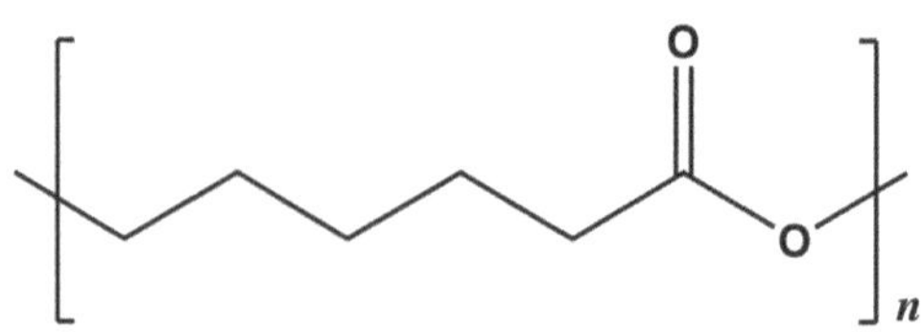

Figure 10.7. Chemical structure of PCL.

preparation, following this approach, demonstrated that the clays were exfoliated in the polymeric matrix and that the vapor-barrier properties were improved when compared to neat PCL. Similar results were obtained by Gorrasi et al. (2003) using the same system, but in their study, intercalated/exfoliated structures could be noticed among the morphologies achieved.

Correlation between morphology and vapor-barrier properties of PCL/MMT composites has been studied by Gorrasi et al. (2002). They prepared different compositions of PCL/organically modified clay nanocomposites by melt blending or catalyzed ROP of CL. Exfoliated nanocomposites were obtained by in situ ROP of CL with an organically modified clay, by using dibutylindimethoxide as an initiator/catalyst.

Melt-mixed PLA/PCL blend bionanocomposites with small amounts of TiO_2 nanoparticles were investigated by Mofokeng and Luyt (2015). Blending PLA and PCL resulted in the reduction of the thermal stabilities of both polymers. However, incorporation of TiO_2 nanoparticles enhanced their thermal stability.

3.2.2 Bacterial-Based Polyesters

In the recent years, biodegradable polymers obtained from bacteria have attracted high attention due to the fossil depletion problem.

Bacteria are able to synthesize several types of biodegradable polyesters, with specific properties that are suitable alternatives to oil-based materials in several industrial and medical applications. The bacterial-based biopolymers can be divided into semi-synthetic or microbial polymers, depending on their origin. Thus, the former are generated by polymerization of monomers resulting from the fermentation process, and the latter are produced by a wide range of microorganisms under different environmental conditions (Madison and Huisman, 1999). An example of a semi-synthetic polymer is PLA, obtained from the respective monomer (LA), which in turn, is produced through starch fermentation from lactic bacteria (Hartmann, 1998).

Microbial polymers integrate PHAs, which include a variety of natural polyesters produced by several bacteria. Although, carbohydrate polymers, such as pullulan, generated by the fermentation process of sugar feedstock with bacteria or fungi, have been also subject of studies (Piermaria et al., 2009).

3.2.2.1 Polyhydroxyalkanoates

PHA (Fig. 10.8) is the family of biopolyesters that are synthesized by the bacterial fermentation of several substrates that act as carbon sources. The PHAs are stored within the microorganisms,

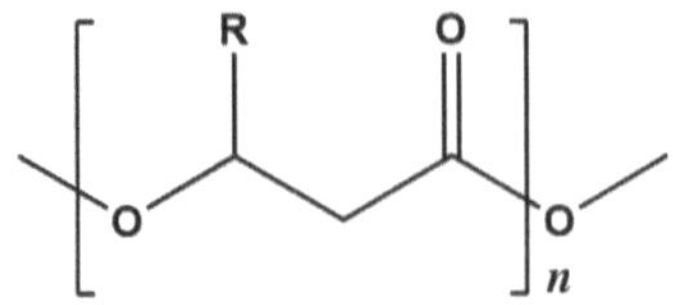

Figure 10.8. General chemical structure of PHAs.

as intracellular reserves of carbon and energy. This group of biopolymers has diverse structures and display properties according to the varying the size of the alkyl chain, hence the PHAs' properties are different. Besides that, their features can be modified by incorporating inorganic materials, blending with other polymers, and performing surface modification, which qualify these materials for several applications. In terms of production, they scarcely differ from plastics based on petrochemical raw materials. Thus, facilities specialized in plastics technology are also suitable for the fabrication of PHAs.

Moreover, PHAs are completely biodegradable, biocompatible, and nontoxic. Their biodegradability depends, mainly, on crystallinity and on the polymer type; copolymers degrade faster than homopolymers. Therefore, because of their biodegradability and biotechnical production from biogenic raw materials, PHAs are considered alternatives to nondegradable polymers based on fossil-derived raw materials (Plackett and Siró, 2011). More than 150 different types of PHAs, including both homopolymers and copolymers, can be synthesized by employing different bacterial species and growth conditions.

PHB and poly(hydroxybutyrate-*co*-hydroxyvalerate) (PHBV) are the most well-known members of the PHA family. They are produced when bacteria are exposed to carbon sources while all other necessary nutrients become limited (Chen, 2009).

PHB is a biodegradable polyester synthesized by biotechnological approaches. It shows a high crystallinity degree, being the neat homopolymer, very brittle, and the mechanical performance is quite similar to other synthetic biodegradable polyesters, such as PLA (Savenkova et al., 2000). Even though its susceptibility to the biodegradation by the action of several microorganisms (fungi, algae, and bacteria), PHB has some drawbacks, namely its low thermal stability, brittleness, and difficulty in processing. Nevertheless, these problems can be overcome by performing several modifications, including grafting different monomers onto PHB's backbone or by blending it with other polymers. Accordingly, it was demonstrated that the grafting degree depends on the monomer used (Hsieh et al., 2009). Also, it was shown that it is possible to successfully obtain PHB-based polymeric systems using radical polymerization or by melt mixing (Avella et al., 1996). Moreover, it should be mentioned that, besides the bacterial synthetic route, other chemical methodologies could be used for the synthesis of PHB, for instance, ROP of β-butyrolactone (Juzwa and Jedlinski, 2006).

PHBV, the major biodegradable microbial copolymer, is synthesized by the addition of propionic acid to nutrient feedstock provided by the bacteria. It has high crystallinity, being the neat copolymer, is quite brittle, but less than PHB, and the mechanical performance can be improved by changing the content of the hydroxyvalerate unit. Also, it was shown that its degradation rate is affected by the crystallinity, processing conditions, and structure (Parikh et al., 1998).

Several companies are already stimulating the introduction of PHAs, such as, PHB or PHBV, in the production of commercial commodities for several applications, including food packaging.

In order to produce PHAs films with better mechanical and barrier properties, a number of studies have been conducted on PHAs/organomodified MMT or PHAs/layered-double-hydroxides nanocomposites, either by solution casting or by melt processing (Erceg et al., 2010; Botana et al., 2010; Cabedo et al., 2009). One of the drawbacks of PHB as a candidate material for food packaging is its lack of flexibility. For this reason, there has been a particular interest in PHB's copolymers (eg, PHBV), which display better properties in this respect. Another disadvantage of PHAs is their relative heat sensitivity, consequently requiring some attention when processing PHAs nanocomposites (Cabedo et al., 2009). PHAs generally have relatively good gas- and water vapor–barrier properties and probably, as a result of this, there have been a rather limited number of studies on PHA nanocomposites for packaging uses. Cyras et al. (2007) developed double-layer biocomposites based on PHB and cellulose paper. The authors observed a decrease in the moisture and water absorption, as well as in water-vapor permeation and surface roughness in the double-layer films, when compared to cellulose paper. The elastic modulus and mechanical strength of the biocomposites increased with the PHB content. This method offers one interesting alternative to the production of less expensive biodegradable materials, with good properties, using cellulose paper coated by a thin layer of PHB.

However, the commercialization of some PHAs is limited by their insufficient amounts, which restricts their study. Also, another disadvantage of PHAs is that they are costly when used individually. Therefore, they are often mixed with other less expensive materials, having complementary features.

In this context, researchers have been working on the improvement of PHAs' mechanical, biodegradation, and morphological properties to broaden their applicability in different industries.

Blends of PHB and PHBV with starch were prepared by Verhoogt et al. (1994). It was demonstrated that these PHAs are able to form miscible blends with polymers containing the desirable functional groups. Also, studies regarding the mixture of PHB with starch revealed that the properties of the final blend were not affected by the amount of starch, with exception of an enhancement of the mechanical performance and a reduction in the overall cost (Kotnis et al., 1995).

Blends of PHBV with biodegradable polymers, such as PCL (Chun and Kim, 2000), PLA (Nanda et al., 2011), PBAT (Javadi et al., 2010a), starch (Avella and Errico, 2000), and corn starch (Gordon et al., 2000) have been studied. The studies of blends of starch with the aliphatic polyester PHBV showed that only low amounts of starch are suitable.

Furthermore, some scientific works have been done using several natural fibers, such as wood fiber (Singh and Mohanty, 2007), bamboo fiber (Jiang et al., 2008), jute (Bledzki and Jaszkiewicz, 2010), wheat straw (Avella et al., 2000), and coir fiber (Javadi et al., 2010b) in the production of biocomposites. With the continuous development of new PHBV-based blends, biocomposites, and new processing technologies, an even broader range of applications are anticipated for biobased and biodegradable PHBV. Globally, more than 20 companies have been established to commercialize these developed products.

Since PHB and PHBV polymers can be produced in processing lines used for the production of common synthetic-polymer materials, they offer the possibility to obtain several products by different techniques, including thermoforming, molding, injection, as well as, reactive blown-film extrusion.

4 Biodegradation

The use of biodegradable polymers appears as a solution in the management of polymer wastes in the environment. Unlike conventional polymers, these biodegradable polymers are degraded and assimilated by microorganisms, such as bacteria, fungi, and algae, which are able to chemically modify them (Gautam et al., 2007). Thus, biodegradable polymers have a life cycle similar to biomass, as depicted in Fig. 10.9.

As result of the biodegradation process, the polymers are converted into carbon dioxide (CO_2), water (H_2O), methane (CH_4), and biomass. Nevertheless, the simultaneous formation of CO_2 and CH_4 only occurs in rare occasions. Their formation is dependent on the presence of oxygen during the degradation process. Therefore, the biodegradation of polymers can be according to the following equations, aerobic or anaerobic (Kyrikou and Briassoulis, 2007).

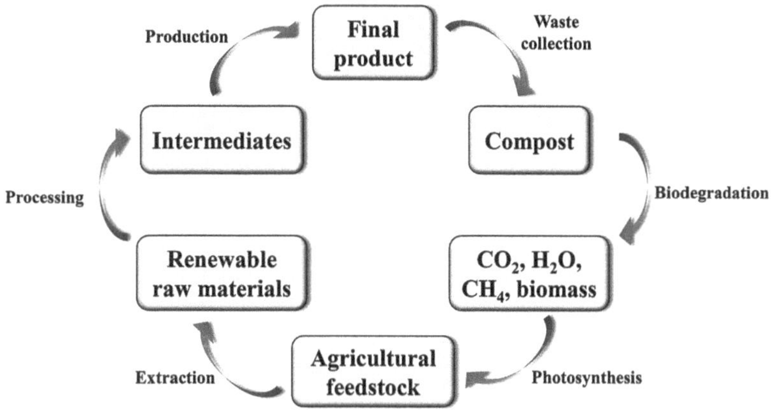

Figure 10.9. Life cycle of a biodegradable polymer.

$$\text{Polymer} + O_2 \xrightarrow{\text{microorganisms}} CO_2 + H_2O + \text{biomass} \tag{10.1}$$

$$\text{Polymer} \xrightarrow{\text{microorganisms}} CH_4 + H_2O + \text{biomass} \tag{10.2}$$

The degradation process of polymers in nature can be divided into two stages. In the first stage, the material may suffer mechanical, chemical, or thermal degradation, which results in its fragmentation. This increases the surface area of the material that will be exposed to microorganisms. In the second stage, these microorganisms attack the polymer structure transforming it in the final products already described (Kyrikou and Briassoulis, 2007).

4.1 Limitations to Biodegradation

The biodegradation of polymeric materials is dependent on several aspects, most of them related to the chemical structure of the material. Properties, such as, surface area, hydrophilicity, molecular weight, and its distribution, glass transition, melting temperatures, crystallinity, and elasticity are several examples of properties influencing polymer biodegradation (Tokiwa et al., 2009). Polymers, for instance, polyesters with side chains are less assimilated by microorganisms, thus having slower degradations rates. The same aspect is observed when the molecular weight of a polymer increases. Enzymes preferably attack amorphous regions in the polymer structure, which means that higher crystallinity will contribute to reduced biodegradability (Tokiwa et al., 2009).

However, the environment conditions also have impact on the degradability of polymers, because there are several aspects, such as, moisture content, temperature, pH, available O_2, porosity,

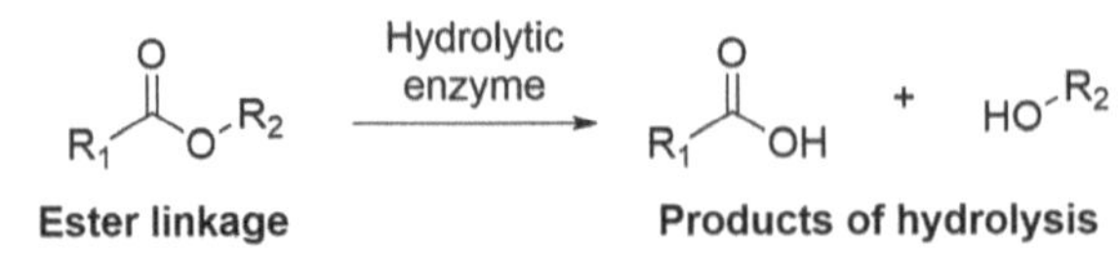

Figure 10.10. General scheme for the hydrolysis of polymers' ester linkages by hydrolytic enzymes.

and type of microorganisms that can influence the process rate (Kyrikou and Briassoulis, 2007).

4.2 Biodegradation of Polyesters

Although biodegradation of polymers is associated with microorganisms, the process is mostly due to enzymes produced by them. Water-insoluble polymers, which have a high molar mass, do not cross cell membranes. For that reason, microorganisms produce extracellular enzymes that, in the first stage, are able to convert polymers into smaller water-soluble compounds. Due to their size, extracellular enzymes only attack the polymer surface, and for that reason, the degradation of polymers occurs as a typical surface erosion. These resulting intermediates of the degradation stage can cross cell membranes and be part of the microorganisms' metabolic pathways, where they are converted into the final products of the degradation of a given polymer (Mueller, 2006). Therefore, it is important to highlight the enzymatic degradation mechanism of polymers. Biodegradable polymers, like PCL, PLA, and PHAs are polyesters, which mean that they are hydrolyzed by esterases, lipases, and proteases. These enzymes hydrolyze the ester-functional groups that form the polymer chains, resulting in the formation of a carboxylic acid and an alcohol functional group per ester linkage (Fig. 10.10).

5 Production of Food Packaging

The use of biodegradable polymers in the production of food packages is related to the requirement of a sustainable management of the environmental impact of plastics. These materials are already being used in disposable packaging with short life and for single-use, as well as, in the production of service ware items and disposable nonwoven fabrics, among others (Weber, 2000). This means that these type of polymers should allow the same industrial production techniques of packaging items than the conventional polymers, such as coextrusion, film casting, thermoforming, blown film extrusion, injection blow molding, extrusion coating, and production of fibers and nonwovens.

In the food-packaging field, there are several examples of commercially available products based on biodegradable polymers. Cellophane, a biodegradable cellulose-based material, is used in the food packaging industry, because its films are prepared by extrusion. Cereplast Compostable 3000 is a film used for food packaging that is prepared by blown film extrusion of the PLA/PBAT blend, registered by BASF as Ecoflex. Clamshells made of PLA (Ingeo, NatureWorks) is another example of commercially biodegradable-food packages.

6 Conclusions

This chapter illustrated the potential of bionanocomposites in food packaging. The main focus of this field is the need of better barrier properties provided by biodegradable polymeric materials. Better barrier properties against the migration of oxygen, carbon dioxide, and water vapor would have a major impact on the shelf life of fresh and processed foods. In the case of biopolymers, improving water- and gas-barrier properties is a critical issue. Nanocomposite technology is still in its early stages. MMT and kaolinite clays have shown good potential in the improvement of the properties of biopolymeric materials. Best results are generally observed for lower contents (approximately 5 wt.%). A good compatibility between filler and the biopolymer is essential, thus chemically modified clays are important. With respect to the processing, although exfoliation is recognized as a processing goal, an orderly array of platelets in the biopolymer matrix, which would maximize effectiveness, is still largely unachieved. Improvements in barrier properties are often reported at 50% of that of the neat polymer. Thus, biopolymers with intrinsically good barrier properties will render even better products as bionanocomposites. Further improvements could be expected from the development of more compatible filler-biopolymer systems, better processing technologies, and a systems approach to the design of biopolymer–plasticizer–filler.

In addition, there are some safety concerns related to the use of nanocomposites as food contact materials. There is limited scientific data about migration of nanostructures from the packaging materials into the food. Significant research is still required to evaluate the potential toxicity of nanocomposite materials, as well as, the environmental safety of their use. Nevertheless, bionanocomposite packaging materials seem to have a very bright future in a wide range of applications in the food industry, including innovative active- and intelligent-food packaging with high barrier-packaging properties, nanosensors, freshness indicators, self-cleaning, and nanocoatings with biofunctional properties.

References

Abdollahi, M., Alboofetileh, M., Rezaei, M., Behrooz, R., 2013. Comparing physico-mechanical and thermal properties of alginate nanocomposite films reinforced with organic and/or inorganic nanofillers. Food Hydrocoll. 32, 416–424.

Abdul Khalil, H.P.S., Bhat, A.H., Ireana Yusra, A.F., 2012. Green composites from sustainable cellulose nanofibrils: a review. Carbohydr. Polym. 87, 963–979.

Abreu, A.S., Oliveira, M., De Sá, A., Rodrigues, R.M., Cerqueira, M.A., Vicente, A.A., Machado, A.V., 2015a. Antimicrobial nanostructured starch based films for packaging. Carbohydr. Polym. 129, 127–134.

Abreu, A.S., Oliveira, M., Machado, A.V., 2015b. Effect of clay mineral addition on properties of bio-based polymer blends. Appl. Clay Sci. 104, 277–285.

Ai, F., Zheng, H., Wei, M., Huang, J., 2007. Soy protein plastics reinforced and toughened by SiO_2 nanoparticles. J. Appl. Polym. Sci. 105, 1597–1604.

Aider, M., 2010. Chitosan application for active bio-based films production and potential in the food industry: review. LWT Food Sci. Technol. 43, 837–842.

Araujo, A., Oliveira, M., Oliveira, R., Botelho, G., Machado, A.V., 2014. Biodegradation assessment of PLA and its nanocomposites. Envir. Sci. Pollut. Res. Int. 21, 9477–9486.

Araújo, A., Botelho, G., Oliveira, M., Machado, A.V., 2014. Influence of clay organic modifier on the thermal-stability of PLA based nanocomposites. Appl. Clay Sci. 88-89, 144–150.

Arrieta, M.P., López, J., Hernández, A., Rayón, E., 2014. Ternary PLA–PHB–limonene blends intended for biodegradable food packaging applications. Eur. Polym. J. 50, 255–270.

Arrieta, M.P., Fortunati, E., Dominici, F., López, J., Kenny, J.M., 2015. Bionanocomposite films based on plasticized PLA–PHB/cellulose nanocrystal blends. Carbohydr. Polym. 121, 265–275.

Avella, M., Errico, M.E., 2000. Preparation of PHBV/starch blends by reactive blending and their characterization. J. Appl. Polym. Sci. 77, 232–236.

Avella, M., Immirzi, B., Malinconico, M., Martuscelli, E., Volpe, M.G., 1996. Reactive blending methodologies for Biopol. Polym. Int. 39, 191–204.

Avella, M., Rota, G.L., Martuscelli, E., Raimo, M., Sadocco, P., Elegir, G., Riva, R., 2000. Poly(3-hydroxybutyrate-co-3-hydroxyvalerate) and wheat straw fibre composites: thermal, mechanical properties and biodegradation behaviour. J. Mater. Sci. 35, 829–836.

Avella, M., De Vlieger, J.J., Errico, M.E., Fischer, S., Vacca, P., Volpe, M.G., 2005. Biodegradable starch/clay nanocomposite films for food packaging applications. Food Chem. 93, 467–474.

Avérous, L., Pollet, E., 2014. Nanobiocomposites based on plasticized starch. In: Avérous, P.J.H. (Ed.), Starch Polymers. Elsevier, Amsterdam.

Ayana, B., Suin, S., Khatua, B.B., 2014. Highly exfoliated eco-friendly thermoplastic starch (TPS)/poly (lactic acid)(PLA)/clay nanocomposites using unmodified nanoclay. Carbohydr. Polym. 110, 430–439.

Azizi Samir, M.A.S., Alloin, F., Dufresne, A., 2005. Review of recent research into cellulosic whiskers: their properties and their application in nanocomposite field. Biomacromolecules 6, 612–626.

Babaee, M., Jonoobi, M., Hamzeh, Y., Ashori, A., 2015. Biodegradability and mechanical properties of reinforced starch nanocomposites using cellulose nanofibers. Carbohydr. Polym. 132, 1–8.

Barikani, M., Oliaei, E., Seddiqi, H., Honarkar, H., 2014. Preparation and application of chitin and its derivatives: a review. Iran. Polym. J. 23, 307–326.

Bhardwaj, U., Dhar, P., Kumar, A., Katiyar, V., 2014. Polyhydroxyalkanoates (PHA)-cellulose based nanobiocomposites for food packaging applications. Food Additives and Packaging. ACS Symposium Series, vol. 1162. ACS Publications, pp. 275–314.
Bierhalz, A.C.K., Da Silva, M.A., Kieckbusch, T.G., 2012. Natamycin release from alginate/pectin films for food packaging applications. J. Food Eng. 110, 18–25.
Bledzki, A.K., Jaszkiewicz, A., 2010. Mechanical performance of biocomposites based on PLA and PHBV reinforced with natural fibres: a comparative study to PP. Compos. Sci. Technol. 70, 1687–1696.
Bocchini, S., Battegazzore, D., Frache, A., 2010. Poly (butylensuccinate-*co*-adipate)-thermoplastic starch nanocomposite blends. Carbohyd. Polym. 82, 802–808.
Botana, A., Mollo, M., Eisenberg, P., Torres Sanchez, R.M., 2010. Effect of modified montmorillonite on biodegradable PHB nanocomposites. Appl. Clay Sci. 47, 263–270.
Byun, Y., Kim, Y.T., 2014. Utilization of bioplastics for food packaging industry. In: Han, J.H. (Ed.), Innovations in Food Packaging. second ed. Academic Press, San Diego.
Cabedo, L., Luis Feijoo, J., Pilar Villanueva, M., Lagarón, J.M., Giménez, E., 2006. Optimization of biodegradable nanocomposites based on APLA/PCL blends for food packaging applications. Macromol. Symp. 233, 191–197.
Cabedo, L., Plackett, D., Giménez, E., Lagarón, J.M., 2009. Studying the degradation of polyhydroxybutyrate-co-valerate during processing with clay-based nanofillers. J. Appl. Polym. Sci. 112, 3669–3676.
Caner, C., 2005. The effect of edible eggshell coatings on egg quality and consumer perception. J. Sci. Food Agric. 85, 1897–1902.
Chaudhary, D.S., 2008. Understanding amylose crystallinity in starch–clay nanocomposites. J. Polym. Sci. Part B 46, 979–987.
Chen, G.Q., 2009. A microbial polyhydroxyalkanoates (PHA) based bio- and materials industry. Chem. Soc. Rev. 38, 2434–2446.
Chen, B., Evans, J.R.G., 2005. Thermoplastic starch–clay nanocomposites and their characteristics. Carbohydr. Polym. 61, 455–463.
Cheng, J., Zheng, P., Zhao, F., Ma, X., 2013. The composites based on plasticized starch and carbon nanotubes. Int. J. Biol. Macromol. 59, 13–19.
Chivrac, F., Pollet, E., Schmutz, M., Avérous, L., 2010. Starch nano-biocomposites based on needle-like sepiolite clays. Carbohydr. Polym. 80, 145–153.
Chun, Y.S., Kim, W.N., 2000. Thermal properties of poly(hydroxybutyrate-co-hydroxyvalerate) and poly(ε-caprolactone) blends. Polymer 41, 2305–2308.
Chung, Y.L., Ansari, S., Estevez, L., Hayrapetyan, S., Giannelis, E.P., Lai, H.M., 2010. Preparation and properties of biodegradable starch–clay nanocomposites. Carbohydr. Polym. 79, 391–396.
Ciannamea, E.M., Stefani, P.M., Ruseckaite, R.A., 2014. Physical and mechanical properties of compression molded and solution casting soybean protein concentrate based films. Food Hydrocoll. 38, 193–204.
Conte, A., Longano, D., Costa, C., Ditaranto, N., Ancona, A., Cioffi, N., Scrocco, C., Sabbatini, L., Contò, F., Del Nobile, M.A., 2013. A novel preservation technique applied to fiordilatte cheese. Innov. Food Sci. Emerg. Technol. 19, 158–165.
Corradini, E., Mattoso, L.H.C., Guedes, C.G.F., Rosa, D.S., 2004. Mechanical, thermal and morphological properties of poly(ε-caprolactone)/zein blends. Polym. Adv. Technol. 15, 340–345.
Corradini, E., Curti, P.S., Meniqueti, A.B., Martins, A.F., Rubira, A.F., Muniz, E.C., 2014. Recent advances in food-packing, pharmaceutical and biomedical applications of zein and zein-based materials. Int. J. Mol. Sci. 15, 22438–22470.
Cyras, V.P., Commisso, M.S., Mauri, A.N., Vázquez, A., 2007. Biodegradable double-layer films based on biological resources: polyhydroxybutyrate and cellulose. J. Appl. Polym. Sci. 106, 749–756.

Darder, M., Colilla, M., Ruiz-Hitzky, E., 2003. Biopolymer—clay nanocomposites based on chitosan intercalated in montmorillonite. Chem. Mater. 15, 3774–3780.

de Mesquita, J.P., Donnici, C.L., Teixeira, I.F., Pereira, F.V., 2012. Bio-based nanocomposites obtained through covalent linkage between chitosan and cellulose nanocrystals. Carbohydr. Polym. 90, 210–217.

De Silva, R.T., Pasbakhsh, P., Lee, S.M., Kit, A.Y., 2015. ZnO deposited/encapsulated halloysite–poly (lactic acid) (PLA) nanocomposites for high performance packaging films with improved mechanical and antimicrobial properties. Appl. Clay Sci. 111, 10–20.

Dean, K., Yu, L., Wu, D.Y., 2007. Preparation and characterization of melt-extruded thermoplastic starch/clay nanocomposites. Compos. Sci. Technol. 67, 413–421.

Delhom, C.D., White-Ghoorahoo, L.A., Pang, S.S., 2010. Development and characterization of cellulose/clay nanocomposites. Compos. Part B 41, 475–481.

Duda, A., Penczek, S., 2005. Aliphatic polyester formation, mechanisms of polyesters. Biopolymers Online. Wiley-VCH Verlag GmbH & Co. KGaA, Germany.

Echeverria, I., Eisenberg, P., Mauri, A.N., 2014. Nanocomposites films based on soy proteins and montmorillonite processed by casting. J. Memb. Sci. 449, 15–26.

Echeverría, I., Eisenberg, P., Mauri, A.N., 2014. Nanocomposites films based on soy proteins and montmorillonite processed by casting. J. Memb. Sci. 449, 15–26.

Edlund, U., Albertsson, A.C., 2003. Polyesters based on diacid monomers. Adv. Drug Del. Rev. 55, 585–609.

Eichhorn, S.J., 2011. Cellulose nanowhiskers: promising materials for advanced applications. Soft Matter 7, 303–315.

El-Wakil, N.A., Hassan, E.A., Abou-Zeid, R.E., Dufresne, A., 2015. Development of wheat gluten/nanocellulose/titanium dioxide nanocomposites for active food packaging. Carbohydr. Polym. 124, 337–346.

Erceg, M., Kovačić, T., Perinović, S., 2010. Isothermal degradation of poly(3-hydroxybutyrate)/organically modified montmorillonite nanocomposites. Polym. Compos. 31, 272–278.

Fortunati, E., Luzi, F., Puglia, D., Petrucci, R., Kenny, J.M., Torre, L., 2015. Processing of PLA nanocomposites with cellulose nanocrystals extracted from *Posidonia oceanica* waste: innovative reuse of coastal plant. Ind. Crops Prod. 67, 439–447.

Garrido, T., Etxabide, A., Peñalba, M., De La Caba, K., Guerrero, P., 2013. Preparation and characterization of soy protein thin films: processing–properties correlation. Mater. Lett. 105, 110–112.

Gautam, R., Bassi, A.S., Yanful, E.K., 2007. A review of biodegradation of synthetic plastic and foams. Appl. Biochem. Biotechnol. 141, 85–108.

George, E.R., Sullivan, T.M., Park, E.H., 1994. Thermoplastic starch blends with a poly(ethylene-*co*-vinyl alcohol): processability and physical properties. Polym. Eng. Sci. 34, 17–23.

Giannakas, A., Grigoriadi, K., Leontiou, A., Barkoula, N.M., Ladavos, A., 2014. Preparation, characterization, mechanical and barrier properties investigation of chitosan–clay nanocomposites. Carbohydr. Polym. 108, 103–111.

González, A., Alvarez Igarzabal, C.I., 2013. Soy protein—poly (lactic acid) bilayer films as biodegradable material for active food packaging. Food Hydrocoll. 33, 289–296.

Gordon, S.H., Imam, S.H., Shogren, R.L., Govind, N.S., Greene, R.V., 2000. A semiempirical model for predicting biodegradation profiles of individual polymers in starch–poly- (β-hydroxybutyrate-co-β-hydroxyvalerate) bioplastic. J. Appl. Polym. Sci. 76, 1767–1776.

Gorrasi, G., Tortora, M., Vittoria, V., Galli, G., Chiellini, E., 2002. Transport and mechanical properties of blends of poly(ε-caprolactone) and a modified montmorillonite- poly(ε-caprolactone) nanocomposite. J. Polym. Sci. Part B 40, 1118–1124.

Gorrasi, G., Tortora, M., Vittoria, V., Pollet, E., Lepoittevin, B., Alexandre, M., Dubois, P., 2003. Vapor barrier properties of polycaprolactone montmorillonite nanocomposites: effect of clay dispersion. Polymer 44, 2271–2279.

Graciela, W. P., Qin, W., 2002. Formation and properties of corn zein films and coatings. In: Gennadios, A. (Ed.), Protein-Based Films and Coatings. CRC Press, Boca Raton, USA.

Han, J.H., 2014. A review of food packaging technologies and innovations. In: Han, J.H. (Ed.), Innovations in Food Packaging. second ed. Academic Press, San Diego.

Hartmann, M.H., 1998. High molecular weight polylactic acid polymers. In: Kaplan, D. (Ed.), Biopolymers from Renewable Resources. Springer, Berlin Heidelberg.

Hernandez-Izquierdo, V.M., Krochta, J.M., 2008. Thermoplastic processing of proteins for film formation: a review. J. Food Sci. 73, R30–R39.

Hsieh, W.C., Wada, Y., Chang, C.P., 2009. Fermentation, biodegradation and tensile strength of poly(3-hydroxybutyrate-co-4-hydroxybutyrate) synthesized by *Delftia acidovorans*. J. Taiwan Inst. Chem. Eng. 40, 143–147.

Javadi, A., Kramschuster, A.J., Pilla, S., Lee, J., Gong, S., Turng, L.-S., 2010a. Processing and characterization of microcellular PHBV/PBAT blends. Polym. Eng. Sci. 50, 1440–1448.

Javadi, A., Srithep, Y., Pilla, S., Lee, J., Gong, S., Turng, L.S., 2010b. Processing and characterization of solid and microcellular PHBV/coir fiber composites. Mater. Sci. Eng. C 30, 749–757.

Jiang, L., Huang, J., Qian, J., Chen, F., Zhang, J., Wolcott, M., Zhu, Y., 2008. Study of poly(3-hydroxybutyrate-*co*-3-hydroxyvalerate) (PHBV)/bamboo pulp fiber composites: effects of nucleation agent and compatibilizer. J. Polym. Environ. 16, 83–93.

Jimenez, G., Ogata, N., Kawai, H., Ogihara, T., 1997. Structure and thermal/mechanical properties of poly (ε-caprolactone)-clay blend. J. Appl. Polym. Sci. 64, 2211–2220.

Juzwa, M., Jedlinski, Z., 2006. Novel synthesis of poly(3-hydroxybutyrate). Macromolecules 39, 4627–4630.

Kim, M.N., Kim, K.H., Jin, H.J., Park, J.K., Yoon, J.S., 2001. Biodegradability of ethyl and n-octyl branched poly(ethylene adipate) and poly(butylene succinate). Eur. Polym. J. 37, 1843–1847.

Koshy, R.R., Mary, S.K., Thomas, S., Pothan, L.A., 2015. Environment friendly green composites based on soy protein isolate: a review. Food Hydrocoll. 50, 174–192.

Kotnis, M.A., Obrien, G.S., Willett, J.L., 1995. Processing and mechanical-properties of biodegradable poly(hydroxybutyrate-*co*-valerate)-starch compositions. J. Environ. Polym. Degrad. 3, 97–105.

Kumar, R., Choudhary, V., Mishra, S., Varma, I., 2008. Banana fiber-reinforced biodegradable soy protein composites. Front. Chem. China 3, 243–250.

Kurt, S., Kilincceker, O., 2011. Performance optimization of soy and whey protein isolates as coating materials on chicken meat. Poult. Sci. 90, 195–200.

Kyrikou, I., Briassoulis, D., 2007. Biodegradation of agricultural plastic films: a critical review. J. Polym. Environ. 15, 125–150.

Lawton, J.W., 2004. Plasticizers for zein: their effect on tensile properties and water absorption of zein films. Cer. Chem. J. 81, 1–5.

Li, Q., Zhou, J., Zhang, L., 2009. Structure and properties of the nanocomposite films of chitosan reinforced with cellulose whiskers. J. Polym. Sci. Part B 47, 1069–1077.

Liu, L.S., Kost, J., Yan, F., Spiro, R.C., 2012. Hydrogels from biopolymer hybrid for biomedical, food, and functional food applications. Polymers 4, 997.

Liu, H., Adhikari, R., Guo, Q., Adhikari, B., 2013. Preparation and characterization of glycerol plasticized (high-amylose) starch–chitosan films. J. Food Eng. 116, 588–597.

Lofgren, A., Albertsson, A.C., Dubois, P., Jerome, R., 1995. Recent advances in ring-opening polymerization of lactones and related-compounds. J. Macromol. Sci. C35, 379–418.

Lu, Y., Weng, L., Zhang, L., 2004. Morphology and properties of soy protein isolate thermoplastics reinforced with chitin whiskers. Biomacromolecules 5, 1046–1051.

Madbouly, S.A., Lendlein, A., 2012. Degradable polyurethane/soy protein shape-memory polymer blends prepared via environmentally-friendly aqueous dispersions. Macromol. Mater. Eng. 297, 1213–1224.

Madison, L.L., Huisman, G.W., 1999. Metabolic engineering of poly(3-hydroxyalkanoates): from DNA to plastic. Microbiol. Mol. Biol. Rev. 63, 21–53.

Maisanaba, S., Puerto, M., Pichardo, S., Jordá, M., Moreno, F.J., Aucejo, S., Jos, Á., 2013. In vitro toxicological assessment of clays for their use in food packaging applications. Food Chem. Toxicol. 57, 266–275.

Majeed, K., Jawaid, M., Hassan, A., Abu Bakar, A., Abdul Khalil, H.P.S., Salema, A.A., Inuwa, I., 2013. Potential materials for food packaging from nanoclay/natural fibres filled hybrid composites. Mater. Des. 46, 391–410.

Mofokeng, J.P., Luyt, A.S., 2015. Morphology and thermal degradation studies of melt-mixed poly(lactic acid) (PLA)/poly(ε-caprolactone) (PCL) biodegradable polymer blend nanocomposites with TiO_2 as filler. Polym. Test. 45, 93–100.

Mohanty, A.K., Misra, M., Drzal, L.T., 2002. Sustainable bio-composites from renewable resources: opportunities and challenges in the green materials world. J. Polym. Environ. 10, 19–26.

Molinaro, S., Cruz Romero, M., Boaro, M., Sensidoni, A., Lagazio, C., Morris, M., Kerry, J., 2013. Effect of nanoclay-type and PLA optical purity on the characteristics of PLA-based nanocomposite films. J. Food Eng. 117, 113–123.

Mondragón, M., Hernández, E.M., Rivera-Armenta, J.L., Rodríguez-González, F.J., 2009. Injection molded thermoplastic starch/natural rubber/clay nanocomposites: morphology and mechanical properties. Carbohydr. Polym. 77, 80–86.

Mueller, R.J., 2006. Biological degradation of synthetic polyesters—enzymes as potential catalysts for polyester recycling. Process Biochem. 41, 2124–2128.

Nanda, M.R., Misra, M., Mohanty, A.K., 2011. The effects of process engineering on the performance of PLA and PHBV blends. Macromol. Mater. Eng. 296, 719–728.

Nasseri, R., Mohammadi, N., 2014. Starch-based nanocomposites: a comparative performance study of cellulose whiskers and starch nanoparticles. Carbohydr. Polym. 106, 432–439.

Ojijo, V., Sinha Ray, S., 2013. Processing strategies in bionanocomposites. Prog. Polym. Sci. 38, 1543–1589.

Oliviero, M., Verdolotti, L., Di Maio, E., Aurilia, M., Iannace, S., 2011. Effect of supramolecular structures on thermoplastic zein-lignin bionanocomposites. J. Agric. Food Chem. 59, 10062–10070.

Othman, S.H., 2014. Bio-nanocomposite materials for food packaging applications: types of biopolymer and nano-sized filler. Agric. Agric. Sci. Procedia 2, 296–303.

Ozcalik, O., Tihminlioglu, F., 2013. Barrier properties of corn zein nanocomposite coated polypropylene films for food packaging applications. J. Food Eng. 114, 505–513.
Pandey, J.K., Kumar, A.P., Misra, M., Mohanty, A.K., Drzal, L.T., Palsingh, R., 2005. Recent advances in biodegradable nanocomposites. J. Nanosci. Nanotechnol. 5, 497–526.
Parikh, M., Gross, R.A., Maccarthy, S.P., 1998. The influence of injection molding conditions on biodegradable polymers. J. Injection Molding Technol. 2, 30.
Park, H.M., Mohanty, A., Drzal, L., Lee, E., Mielewski, D., Misra, M., 2006. Effect of sequential mixing and compounding conditions on cellulose acetate/layered silicate nanocomposites. J. Polym. Environ. 14, 27–35.
Park, J.H., Park, S.M., Kim, Y.H., Oh, W., Lee, G.W., Karim, M.R., Park, J.H., Yeum, J.H., 2013. Effect of montmorillonite on wettability and microstructure properties of zein/montmorillonite nanocomposite nanofiber mats. J. Compos. Mater. 47, 251–257.
Parris, N., Coffin, D.R., 1997. Composition factors affecting the water vapor permeability and tensile properties of hydrophilic zein films. J. Agric. Food Chem. 45, 1596–1599.
Pérez, E., Pérez, C.J., Alvarez, V.A., Bernal, C., 2013. Fracture behavior of a commercial starch/polycaprolactone blend reinforced with different layered silicates. Carbohydr. Polym. 97, 269–276.
Piermaria, J.A., Pinotti, A., Garcia, M.A., Abraham, A.G., 2009. Films based on kefiran, an exopolysaccharide obtained from kefir grain: development and characterization. Food Hydrocoll. 23, 684–690.
Plackett, D., Siró, I., 2011. Polyhydroxyalkanoates (PHAs) for food packaging. In: Lagarón, J.M. (Ed.), Multifunctional and Nanoreinforced Polymers for Food Packaging. Woodhead Publishing, Cambridge, UK.
Qi, H., Cai, J., Zhang, L., Kuga, S., 2009. Properties of films composed of cellulose nanowhiskers and a cellulose matrix regenerated from alkali/urea solution. Biomacromolecules 10, 1597–1602.
Rao, M.S., Kanatt, S.R., Chawla, S.P., Sharma, A., 2010. Chitosan and guar gum composite films: preparation, physical, mechanical and antimicrobial properties. Carbohydr. Polym. 82, 1243–1247.
Rhim, J.W., Kim, Y.T., 2014. Biopolymer-based composite packaging materials with nanoparticles. In: Han, J.H. (Ed.), Innovations in Food Packaging, second ed. Academic Press, San Diego.
Rhim, J.W., Ng, P.K.W., 2007. Natural biopolymer-based nanocomposite films for packaging applications. Crit. Rev. Food Sci. Nutr. 47, 411–433.
Rhim, J.W., Hong, S.I., Park, H.M., Ng, P.K.W., 2006. Preparation and characterization of chitosan-based nanocomposite films with antimicrobial activity. J. Agric. Food Chem. 54, 5814–5822.
Rhim, J.W., Lee, J.H., Hong, S.I., 2007. Increase in water resistance of paperboard by coating with poly(lactide). Pack. Technol. Sci. 20, 393–402.
Rhim, J.W., Park, H.M., Ha, C.S., 2013. Bio-nanocomposites for food packaging applications. Prog. Polym. Sci. 38, 1629–1652.
Rodríguez, F.J., Galotto, M.J., Guarda, A., Bruna, J.E., 2012. Modification of cellulose acetate films using nanofillers based on organoclays. J. Food Eng. 110, 262–268.
Sanchez-García, M.D., 2011. Carrageenan polysaccharides for food packaging. In: Lagarón, J.M. (Ed.), Multifunctional and Nanoreinforced Polymers for Food Packaging. Woodhead Publishing, Cambridge, UK.
Sanchez-Garcia, M.D., Hilliou, L., Lagaron, J.M., 2010. Nanobiocomposites of carrageenan, zein, and mica of interest in food packaging and coating applications. J. Agric. Food Chem. 58, 6884–6894.

Sanuja, S., Agalya, A., Umapathy, M.J., 2014. Studies on magnesium oxide reinforced chitosan bionanocomposite incorporated with clove oil for active food packaging application. Int. J. Polym. Mater. Polym. Biomater. 63, 733–740.
Savenkova, L., Gercberga, Z., Nikolaeva, V., Dzene, A., Bibers, I., Kalnin, M., 2000. Mechanical properties and biodegradation characteristics of PHB-based films. Proc. Biochem. 35, 573–579.
Schlemmer, D., Angélica, R.S., Sales, M.J.A., 2010. Morphological and thermomechanical characterization of thermoplastic starch/montmorillonite nanocomposites. Compos. Struct. 92, 2066–2070.
Shen, X.L., Wu, J.M., Chen, Y., Zhao, G., 2010. Antimicrobial and physical properties of sweet potato starch films incorporated with potassium sorbate or chitosan. Food Hydrocoll. 24, 285–290.
Shin, G.H., Lee, Y.H., Lee, J.S., Kim, Y.S., Choi, W.S., Park, H.J., 2002. Preparation of plastic and biopolymer multilayer films by plasma source ion implantation. J. Agric. Food Chem. 50, 4608–4614.
Shipp, D. A., 2011. Polymer-layered silicate nanocomposites. In: Wiederrecht, D.L.A.D.S.P. (Ed.), Comprehensive Nanoscience and Technology. Academic Press, Amsterdam.
Shukla, R., Cheryan, M., 2001. Zein: the industrial protein from corn. Ind. Crops Prod. 13, 171–192.
Singh, S., Mohanty, A.K., 2007. Wood fiber reinforced bacterial bioplastic composites: fabrication and performance evaluation. Compos. Sci. Technol. 67, 1753–1763.
Sinha Ray, S., Bousmina, M., 2005. Biodegradable polymers and their layered silicate nanocomposites: in greening the 21st century materials world. Prog. Mater. Sci. 50, 962–1079.
Sirviö, J.A., Kolehmainen, A., Liimatainen, H., Niinimäki, J., Hormi, O.E.O., 2014. Biocomposite cellulose-alginate films: promising packaging materials. Food Chem. 151, 343–351.
Song, F., Tang, D.L., Wang, X.L., Wang, Y.Z., 2011. Biodegradable soy protein isolate-based materials: a review. Biomacromolecules 12, 3369–3380.
Tang, X., Alavi, S., 2011. Recent advances in starch, polyvinyl alcohol based polymer blends, nanocomposites and their biodegradability. Carbohydr. Polym. 85, 7–16.
Tang, X., Alavi, S., Herald, T.J., 2008. Effects of plasticizers on the structure and properties of starch–clay nanocomposite films. Carbohydr. Polym. 74, 552–558.
Thakur, B.R., Singh, R.K., Handa, A.K., 1997. Chemistry and uses of pectin: a review. Crit. Rev. Food Sci. Nutr. 37, 47–73.
Thostenson, E.T., Li, C., Chou, T.W., 2005. Nanocomposites in context. Compos. Sci. Technol. 65, 491–516.
Tokiwa, Y., Calabia, B.P., Ugwu, C.U., Aiba, S., 2009. Biodegradability of Plastics. Int. J. Mol. Sci. 10, 3722–3742.
Tripathi, S., Mehrotra, G.K., Dutta, P.K., 2011. Chitosan–silver oxide nanocomposite film: preparation and antimicrobial activity. Bull. Mater. Sci. 34, 29–35.
Uskoković, V., 2007. Nanotechnologies: what we do not know. Technol. Soc. 29, 43–61.
Van Den Broek, L.A.M., Knoop, R.J.I., Kappen, F.H.J., Boeriu, C.G., 2015. Chitosan films and blends for packaging material. Carbohydr. Polym. 116, 237–242.
Verhoogt, H., Ramsay, B.A., Favis, B.D., 1994. Polymer blends containing poly(3-hydroxyalkanoate)s. Polymer 35, 5155–5169.
Vroman, I., Tighzert, L., 2009. Biodegrad. Polym. Materials 2, 307–344.
Wang, X.L., Yang, K.K., Wang, Y.Z., 2003. Properties of starch blends with biodegradable polymers. J. Macromol. Sci. Part C 43, 385–409.

Weber, C.J., 2000. Biobased packaging materials for the food industry: status and perspectives, a European concerted action. Report for the European Union Food Biopack Project. Available from: http://www.bc.bangor.ac.uk/_includes/docs/pdf/biobased%20packaging%20materials.pdf
Wilhelm, H.M., Sierakowski, M.R., Souza, G.P., Wypych, F., 2003. Starch films reinforced with mineral clay. Carbohydr. Polym. 52, 101–110.
Xie, F., Pollet, E., Halley, P.J., Avérous, L., 2013. Starch-based nano-biocomposites. Prog. Polym. Sci. 38, 1590–1628.
Yam, K.L., Takhistov, P.T., Miltz, J., 2005. Intelligent packaging: concepts and applications. J. Food Sci. 70, R1–R10.
Yamada, K., Takahashi, H., Noguchi, A., 1995. Improved water resistance in edible zein films and composites for biodegradable food packaging. Int. J. Food Sci. Technol. 30, 599–608.
Yates, M.R., Barlow, C.Y., 2013. Life cycle assessments of biodegradable, commercial biopolymers: a critical review. Res. Cons. Recycl. 78, 54–66.
Youssef, A.M., Abdel-Aziz, M.S., El-Sayed, S.M., 2014. Chitosan nanocomposite films based on Ag-NP and Au-NP biosynthesis by *Bacillus subtilis* as packaging materials. Int. J. Biol. Macromol. 69, 185–191.
Yu, L., Dean, K., Li, L., 2006. Polymer blends and composites from renewable resources. Prog. Polym. Sci. 31, 576–602.
Yu, L., Dean, K., Yuan, Q., Chen, L., Zhang, X., 2007. Effect of compatibilizer distribution on the blends of starch/biodegradable polyesters. J. Appl. Polym. Sci. 103, 812–818.
Zeppa, C., Gouanvé, F., Espuche, E., 2009. Effect of a plasticizer on the structure of biodegradable starch/clay nanocomposites: thermal, water-sorption, and oxygen-barrier properties. J. Appl. Polym. Sci. 112, 2044–2056.
Zhang, Y.Q., Xie, B.J., Gan, X., 2005. Advance in the applications of konjac glucomannan and its derivatives. Carbohydr. Polym. 60, 27–31.
Zhang, M., Song, F., Wang, X.L., Wang, Y.Z., 2012. Development of soy protein isolate/waterborne polyurethane blend films with improved properties. Colloids Surf. B 100, 16–21.
Zhong, Z., Sun, X.S., 2001. Properties of soy protein isolate/polycaprolactone blends compatibilized by methylene diphenyl diisocyanate. Polymer 42, 6961–6969.
Zou, P., Yang, X., Wang, J., Li, Y., Yu, H., Zhang, Y., Liu, G., 2016. Advances in characterisation and biological activities of chitosan and chitosan oligosaccharides. Food Chem. 190, 1174–1181.

11

FLEXIBLE PACKAGING FOR NONTHERMAL DECONTAMINATION BY HIGH HYDROSTATIC PRESSURE

Mythili Prakasam, Alain Largeteau
University of Bordeaux, CNRS, Pessac, France

1 Introduction

In the fast pace of life, modern technologies help us in various domains to live at ease and benefit from the riches and diversities of the world resources. In ancient times due to the lack of preservation knowledge of food, much of the food was being wasted and some of the foods were regional/seasonal making few of the food products exotic to a part of the population. Even in modern times, most of the food products are wasted due to the lack of knowledge on food preservation technologies. With the advancement of food engineering, it is possible to make ready-to-eat food, conservation of vegetables, dairy products, animal products, fruits, and medicine by prolonging their shelf lives. Thanks to the food engineering and packaging technologies that made the preservation technologies (Balasubramaniam et al., 2004; Butz et al., 1996; Devlieghere et al., 2004; Hayashi, 1991; Hotchkiss, 1995; Masuda et al., 1992; Ozen and Floros, 2001) easier. Foods are a complex blend of a huge number of different compounds. Based on the chemical and physical properties of food, it is prone to degradation. There are various factors that influence the degradation of foods. The physical degradation of food being induced by bruising and mechanical deterioration due to the handling process of food products, temperature-induced texture changes due to the phase changes, moisture gain or loss, flavor changes due to the absorption of flavor due to the contact of the carton layer with the surrounding package containing the food are a few to mention. The chemical reactions of the food can alter the food chemical composition (Butz and Taucher, 2002), which can lead to degradation in

Food Packaging. http://dx.doi.org/10.1016/B978-0-12-804302-8.00011-X

terms of nutritive values. Further during the processing of foods, the damage caused to the tissue causes the release of various chemical constituents into the cellular fluid, which interacts with the environment and degrades the surface. Food items containing unsaturated fats are influenced by direct interaction with oxygen through an autocatalytic free radical mechanism causing rancid flavors in the food, which is undesirable to eat (Butz et al., 1992). Also, any fine trace metals that are present in the food could accelerate the reaction rate manifold. There are other fruit/food items that can undergo browning, which is caused by the reduction of sugars and amino groups in the food, which causes sour tastes or goes stale. The moisture changes also can trigger the undesired chemical changes in foods. Microorganisms play a major role in the loss of quality, which prevail on the food and decreases the shelf life (Alderton and Snell, 1969; Alderton et al., 1976; Caner et al., 2004; Hoover, 1993; Mackey et al., 1995; Smelt, 1998). There are various methods that retard/repress them in many ways, such as low temperature to slow growth, raise temperature to kill them, remove or bind water to slow or prevent growth, lower pH, control oxygen/carbon dioxide (O_2/CO_2) level to control microorganism population or add preservatives.

Food-borne pathogens lead to various illness such as infection, intoxication, and toxin-mediated infection. There are various types of food-borne pathogens (Ballestra et al., 1996; Berlin et al., 1999; Butz et al., 1996, 1990; Ishiguro et al., 1993; Kakugawa et al., 1996; Kalchayanand et al., 1994; Miyakawa et al., 1994; Zobell, 1970) such as bacteria (eg, *Salmonella, Escherichia coli, Listeria, Camphylobacter, Shigella, Closridium botulinum, Staphylococcus aureus*), viruses (eg, hepatitis A, norovirus), parasites (eg, cyclospora, giardia, trichinella), and fungi (eg, *Aspergillus* spp.). In order to avoid potential problems in food, it is very important to control/protect the food product. One of the measures to protect the food for safe consumption is through food packaging, which also preserves the food for long shelf life. Food packaging ensures that the food does not come in contact with the pathogens. Food packaging materials should adhere to the safety norms and ensure no interaction with the food, which could alter its taste, smell, texture, or nutritional property. Different types of food packing are used for different food products such as paper bags, carton boxes, metal cans, or plastic bags. Currently, plastic bags are the most widely used form for the food protection as bags, films, containers, and boxes. Plastic bags also allow the printing, holding, mechanical strength, perforation, and see-through of the packaging without opening.

Among the various food conservation techniques (Benito et al., 1999; Bridgman, 1912; Kata Galic et al., 2011; Meyer, 2000;

Min and Zhang, 2005; Yuste et al., 2002), thermal processes such as cooking, blanching, pasteurization, and sterilization, to inactivate microorganisms, enzymes, and other chemical reactions in food materials are widely employed. Also, cooking raw foods by extending the period of desirable quality and safety level is prevalent. Recently other than the thermal processing of the food, nonthermal processing is used as the food preservation technology (Arroyo et al., 1997; Hite et al., 1914; Hite, 1899; Hoover, 1997; Horie et al., 1991; Lelieveld, 2006; Mertens 1993; Zook et al., 1999). Nonthermal process helps in inactivating spoilage and pathogenic microorganisms at temperatures well below those that are used in thermal processing methods. Nonthermal processing is applied when the product is altered too much by the temperature and generally does not cause any significant changes to the flavor, color, taste, nutrients, and functionalities. Currently, used nonthermal processing for food conservation and microorganism inactivation are microwave, radiofrequency, ohmic, and inductive heating, high pressure processing, pulsed electric field, high voltage arc discharge, oscillating magnetic fields, ultraviolet light, ultrasound, and X-rays. Of the nonthermal processes, we focus on high pressure processing/high hydrostatic pressure processing (HHP) as an alternative technique used for preservation of food and inactivation of food-borne pathogens. Nonthermal processes can be employed for both prepackaged and unpackaged food products. But here we narrow down our discussion to the prepackage made of flexible polymer containing food. This chapter explains in the following sections about the nonthermal processing, HHP processing, currently used flexible polymers for food packaging and the influence of HHP processing on them.

2 High Hydrostatic Pressure Processing and the Related Parameters for Food Products

Various microbiological factors influence the preservation of food and also influence the process by which the inactivation of the microbiological organism takes place. Different microorganisms have different resistance (Timson and Short, 1965; Metrick et al., 1989; Johnson and Campbell, 1945; Dring, 1976; Butz et al., 1996) to the inactivation process such as thermal inactivation, irradiation, and high-pressure processing. Generally the processing parameters that are used for food processing are classified by thermal, electrothermal, and nonthermal food processing. Among the various nonthermal treatments, we focus on high-pressure processing as an alternative technique for the preservation of food and inactivation of food-borne pathogens. Earlier reports on usage of HHP technique for processing food dates back to 1914, when Hite et al. (1914) showed the possibility of increasing the shelf life of food and milk products by using high-pressure. Various industrial applications and advancements experienced in the ceramics and metallurgical industries by the use of pressure led to the usage of HHP to the food technology domain. Japan was the first country to commercialize food cold decontaminated by HHP. HHP process has stringent regulatory measures such as microbiological and toxicological safety of food products.

HHP, or sometimes called ultrahigh pressure (UHP) processing, is a process where sample/product either in the solid or liquid form with or without the packaging is subjected to pressure (100–800 MPa). Usual temperature limits during the HHP are in the range below 0°C to above 100°C. Vessels capable of sustaining high pressure exposure are employed. The material subjected to the HHP will preserve the chemical structure of the different constituents (flavor, color, vitamins, etc.) due to the little/no effect on the covalent bonds. In fact, HHP transmits a weak energy in comparison with temperature, the energy of one liter of water at ambient pressure (0.1 MPa) from 20 to 25°C is the same at ambient temperature (20°C) from 0.1 to 400 MPa. This low energy cannot destroy the vitamins as is the case by thermal treatment as sterilization. The real effect of pressure on the microorganisms is not well known but probably of the modifications of the physical and chemical properties of the internal membranes of the microorganisms, which is not capable of surviving and then cannot proliferate. HHP in conjunction with heat can lead to inactivate microbes and enzymes. Due to the instantaneous and even distribution of pressure on the food product, package size, shape, and their composition is not a decisive factor for employing HHP on food products. The adiabatic increase of heat due to the increase of pressure is ~3°C per 100 MPa. This increase in temperature is influenced by factors such as the fat content in the product subjected to HHP (Alpas et al., 1994; Hayashi and Balny, 2005). Further temperature increases during the pressure application/compression, the influence of the temperature distribution also depends on the factors such as dwell time, heat transfer to the product from the autoclave walls. In order to obtain the isothermal condition during compression, the vessel should be held at the desired final temperature of the product after HHP. If temperature is used in conjunction with high pressure for pathogen inactivation, it is essential to optimize the temperature cycle for microbial inactivation and process reproducibility. It is noted that in a few cases, the high-pressure level could lead to change in texture and color of the product treated by

HHP. HHP processing can cause denaturation of the food product rich in protein and cause changes in structurally weak foods such as fruits (berries, grapes) and leafy salads. Currently, HHP is used for food items (Aleman et al., 1994, 1996; El Moueffak et al., 1995; Linton et al., 1999; Naik et al., 2013) (such as jellies, jams, fruit juices, purees, liquid salad dressings, sea foods, foie gras, ham, deshelling, and pasteurizing of crustaceans, and mollusks). Under the influence of HHP, the product tends to decrease in volume as a function of applied pressure. Decompression also leads to increase of the volume, hence the package holding the product should be capable of accommodating up to the changes in the volume and yet retain seal integrity and barrier properties.

3 Food Packaging and Nonthermal Processing

3.1 Nonthermal Processing and Food Packaging: Outline

Packaged food is designed in such a way for long-term storage, preservation, and protection. Other essential roles of food packaging is better quality maintenance and easy food handling. These primary food packagings also help in efficient food marketing, distribution, and stock management. Currently, nonthermal processes are also used as a mode for food preservation. Nonthermal processing, as the name indicates, is a food preservation procedure that involves no or less amounts of thermal energy for processing. Though nonthermal processing has lesser success in microbiological sterility, it is efficient in diminishing the microorganism population. When athermal processing is used with other techniques involving less heat, the shelf life of the food product can be increased by preserving nutritional properties. The oldest known nonthermal processing used for food preservation is by ionizing radiation. The characteristics pertaining to thermal processes and nonthermal processes are significantly different. Pulsed electric fields and HHP are the most widely used nonthermal treatment techniques in comparison to their counterparts such as irradiation and chemical treatments. It has to be noted that these athermal processing sometimes becomes inefficient to eliminate the spores, though is effective in destroying the microorganisms. But due to the minimal processing conditions that are required in the case of nonthermal processing, the freshness and nutritional aspects of the food products will be retained. The degree of good processing by thermal treatment is usually evaluated in terms of its temperature and the time required to cause death time for a microorganism. But it is difficult to evaluate the same for the nonthermal processing due to the diversified resistance of various microorganisms to different athermal treatments. It is difficult to standardize a particular technique with prior processing parameters to inactivate microorganisms by nonthermal treatment. Athermal process can be a part of the multiple process constituting of two or more processes of nonthermal treatments for inactivating microorganisms. However, such a combinatorial process will be difficult to be used to target on a particular microorganism, because the procedure will disturb the resistance of the microorganism to be treated.

Progress of athermal processing systems has become a driving force that has been changing the functions of food packaging. The various food packagings that are used are designed for fresh vegetables/fruits, fresh meats or heat-treated foods. Nonthermal processing food requires additional properties such as inhibition of migration of the package material into food and any other possible reaction between the food contact and the package material. Food products and packaging materials interact with each other. The interactions include the absorption of flavors and colors into packaging structures and oxidative reactions. Few of the food products also necessitate the usage of high temperatures to pasteurize the food products with or without the package structure. Under such situations, the physical and chemical properties of the package materials (Caner et al., 2000, 2003; Dobiás et al., 2004; Galatto et al., 2008; Kübel et al., 1996; Lambert et al., 2000a, 2000b) are required in addition to the phase transitions, mechanical properties, barrier properties of the packaging material. The processing conditions also depend on the temperature at which the microorganism thermal death takes place, hence the properties of the packaged material could change depending on the processed temperature. Thermal processing of packaged food could also alter some of the characteristics such as adhesion, sealing, gas permeability, and moisture condensation inside the packages. Due to the extensive research by various research groups, it is now possible to foretell almost all the changes of the food packages by thermal packages, which could guide as a precautionary report for processing.

When the nonthermal process is required to be used in conjunction with the sterility processes, it is necessary that the packaging material should be resistant to the processing conditions of the nonthermal process. Some of the examples of athermal processes that are currently used are pulsed light emission, irradiation, batch high pressure processes. Nonthermal processes help in destroying/suppressing the microorganisms in the food product through irreversible internal membrane modification. The nonthermal processes could also induce some changes to the cells and tissues of the animal products leading to some changes in its original form. However, on the contrary, in the case of vegetables it can cause desirable changes, which could positively influence by safeguarding the nutrients in them. Besides the aforesaid characteristics of the packaging materials for nonthermal treatment by HHP, it is necessary to study the headspace gas, oxygen content in food and the total volume of the package.

Protection from the package plays a significant role in the conservation of the food. Packaging protects the food from the environmental influences such as humidity, water, gases, odors, dust, vibrations, compressive forces, and pathogens. For example, sterile packaged milk and fruit juices are safe only until the package is intact. Similarly, vacuum-sealed meat products are safe only if the package can avoid O_2 diffusion into the meat. Based on the processing method used for the preservation of the food products, the package material should be compatible with the processing parameters of the technique used. For example, as in the case of HHP, if the material is not compatible or resistant to the processing parameters, the packaging materials (Calvert and Billingham, 1979; López-Rubio et al., 2005; Schauwecker et al., 2002; Wessling et al., 1999) are susceptible to failure and modifications in the physical and chemical structure of the packaging polymer material in addition to its barrier properties. Fleckenstein has explained in detail about the direct effects of the high pressure and change in properties due to the compressed food products and headspace gas phases. There are other modifications in the package material such as microscopic structural changes, which could lead to volume change or irreversible crystallization process, changes in permeability through package walls, development of defects in the material layer, and delamination of multilayer structures. Hence, the choice of any package materials should be evaluated based on its mechanical properties such as mechanical integrity, delamination of multilayer structures, and barrier properties.

For the consumer health and environmental health protection from the potential hazards of packaging materials, European Union (EU) legislation has a set of rules that are to be taken into consideration to ensure the safety of food packaging materials. According to the EU with regulation (EC) 1935/2004 of the European Parliament and of the Council of October 27, 2004, on materials and articles intended to come into contact with food are being regulated. EU Directive 2002/72/EC has limited the concentration of certain substances in packaging or of migrants in foodstuffs or corresponding food simulants. These regulations are based on the toxicological data of substances. EU Directive 2002/72/EC has approved global migration from the package to the food stuff to a maximum of 10 mg/dm^2 or 60 mg/kg.

3.2 Packaging—Fabrication and Their Diverse Roles

Packaging is being employed in many domains, but food has a major share in terms of utilization (Fig. 11.1). Based on the consumer requirements and safety norms, most often single monolayer material fails to satisfy the packaging design requirements. Under such cases, multilayers are used, which essentially combine various required properties to meet the package design requirements. Barrier to moisture and gases such as oxygen is required in most of the cases and the permeabilities to water vapor and oxygen is also required. Depending on the package requirements, the thickness and the grade of barrier and permeability layers are chosen. In order to design the package material of desired properties, multilayer materials are used. The multilayer package should satisfy some of the requirements such as strength, sealing and dead folding. For the fabrication of multilayer materials, there are various methodologies that are employed, which are explained here:

1. Lamination or adhesive lamination: A process that assembles multilayers of different layers of materials with adhesives and possible combination of partial extrusion and coating. The overall thickness of the multilayer can be up to 400–500 μm.

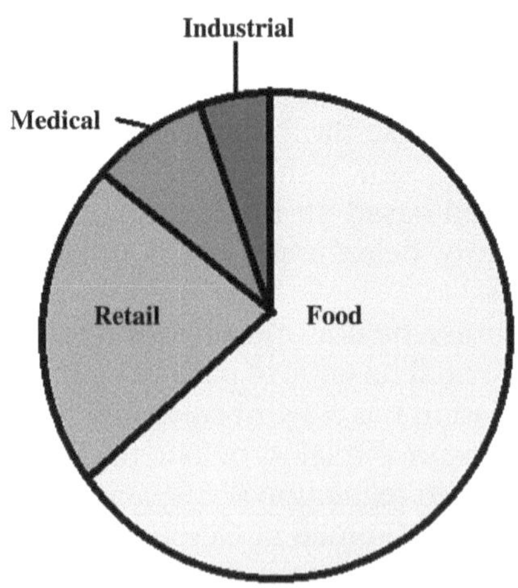

Figure 11.1. Packaging market based on its utilization in various domain.

Lamination packages offer good gas barriers in comparison to the coextruded packages. This is the only process that enables the possibility to use nonplastic materials, paper, and aluminium foil.

2. Coextrusion: A process of extrusion of multiple structure with several layers of resins passed through a single die simultaneously to form a single integral structure. This technique permits to bond various plastics and eliminates the inconveniences such as delamination and air entrapment. This technique is economical in comparison to lamination technique.
3. Extrusion lamination: This process helps in extruding a resin that acts as a bond between two layers.
4. Coating: This process can be used in conjunction with coextrusion and lamination techniques for a broad range of multilayer materials. One of the examples of the multilayer film made by this technique is oriented polypropylene (OPP) film.
5. Metallizing: A process that is mainly used to deposit a thin layer of aluminum (Al) instead of the aluminum foil used on plastics or papers; this technique, however, leads to poor adhesion. In a few instances, materials with excellent barrier properties can be relatively fragile. Under such conditions, it is necessary to protect the package with more robust materials to ward off mechanical damage, consequent cracking, and pinholes: for example, polyethylene (PE) on either side of Al foil in a beverage carton. Other vital properties of the packaging material that need to be taken in to account are moisture control, oxygen control, and carbon dioxide control.

3.2.1 Moisture Control

One of the very essential functions of the package material is moisture control and is largely used in wide variety of applications ranging from food, electronic devices, and so on. The water vapor permeable plastic may be spun-bonded polyolefin/porous paper. The essential task is to remove the bulk of excess liquid purge, drip, or condensation of the product.

3.2.2 Oxygen Control

Oxygen is one of the prime agents leading to biochemical reactions toward product degradation and acts as a stimulant for the microbiological growth in food products. The O_2 is to be avoided by gas barrier packaging in food preservation. The removal of O_2 can be done by vacuum, inert gas injection, or by steaming. Other means to remove O_2 also is by absorbers such as ferrous iron, unsaturated hydrocarbons, nylons, and benzyl acrylate copolymers. However, in the case of red meat products and fresh produce (Rivas-Cañedo et al., 2009), O_2 is required to maintain its freshness. In such cases, O_2 is added to the interior of the package by using high gas transmission package material such as percarbonates.

3.2.3 Carbon Dioxide Control

Carbon dioxide (CO_2), an antimicrobial agent alters the pH (reducing) of the food product. CO_2 may be added to complement existing procedure. With the headspace gas, the ability to pass sufficient gas in or out to accomplish increased shelf life and food preservation.

4 Polymers Employed for Food Packaging

Currently polymers in the form of plastic bags are the most commonly used material for packaging. Various forms of packages are used, which can be used as flexible or rigid package. Generally, packaging materials are made of plastic, which needs to abide by the following criteria such as flexibility, package design, integrity, barrier properties, minimum headspace, compatibility of packaging materials, and high pressure. Fig. 11.2 shows the wide range of consumer products that employs packaging for food preservation. The packaging material has to be flexible and should be capable of sustaining the compression forces by retaining its physical integrity and avoiding irreversible deformations. Usually high barrier flexible packages made of polymers are employed for processing food products in solid or liquid state. The various polymers that are used in food packaging is given in Table 11.1 and

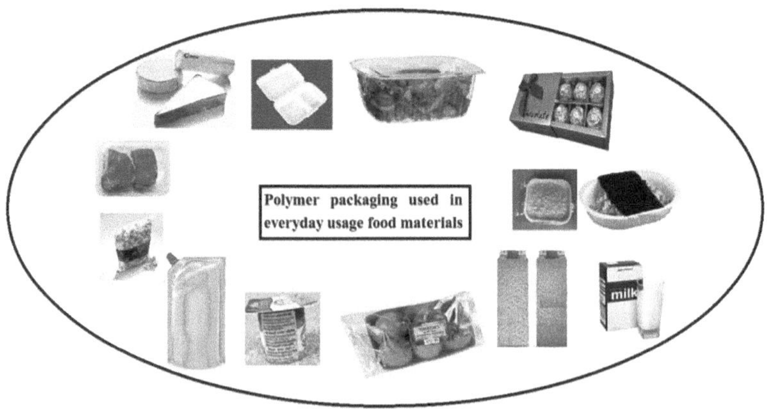

Figure 11.2. Few examples of polymer packaging used in everyday food materials.

Table 11.1 Flexible Packaging Laminates Used in Food Sector

Abbreviation	Polymer
PET	Polyester
MPET	Metallized polyester
PE	Polyethylene
OPP	Oriented polypropylene
MOPP	Metallized oriented polypropylene
BOPP	Bioriented polypropylene
AL	Aluminium
CPP	Cast polypropylene
OPA	Oriented polyamide
MBOPP	Metallized bioriented polypropylene

Table 11.2 and list the different types of flexible polymers that are used for different daily use food products. Of the different types of plastic forms are polymers, polyethylenes (PE), polypropylenes (PP), poly(ethylene vinyl acetates) (EVA), polystyrenes (PS), polyvinyl chlorides (PVC), polyethylene terephthalates (PET), polyvinyl acetates (PVAc), polycarbonates (PC), polyamides (PA), polyvinyl alcohols (PVOH), and polyvinylidene chlorides (PVDC) are used for food packaging.

Among PE is the most commonly used form of polymer with more than 80 million tons manufactured each year. The primary use of PE is plastic bags, plastic films, containers including bottles. Low density PE (LDPE), medium density PE (MDPE), and high density PE(HDPE) have excellent chemical resistance, light weight, low moisture absorption rates, resistant to staining, good fatigue and wear resistance, good resistance to organic solvents with high mechanical strength in comparison to other polymers. Individually quick frozen films are used for packaged food that requires low temperature stability and noncracking when cold. Stand-up pouch structures are designed for a wide range of products. Stand-up pouches consist of a lamination and a sealant layer. Stand-up pouches consist of two structures, such as high barrier films and low-to-medium barrier films. PP is a polymer used for a wide variety of applications, including packaging, which has excellent resistance to stress, high resistance to cracking, high operational

Table 11.2 Different Food Laminates Used in Food Sector

Product	Polymer Used as Package
Coffee	PET/AL/PE; PAPER/AL/PE; PET/MPET/PE; PET/MOPP/PE; OPP/AL/PE
Rice/pasta	PE/PE; PAPER/CPP; OPP/CPP; OPP/PE; OPP
Frozen food	PET/AL/PE; PET/PE; PE; MPET/PE; OPP/MPET/PE
Snacks/chips	OPP/OPP; OPP/OPP barrier; OPP/PE; OPP/MPET/PE; PAPER/MOPP/PE; PAPER /MOPP
Snacks/dried fruits	OPA/PE; OPA/PE barrier; OPA barrier/PE; OPA/OPA/ PE
Biscuits/chocolates	OPP; OPP treated; OPP/MOPP; PET/MOPP
Breads/Cackes	CPP bag; OPP; OPP treated; OPP/MOPP; PET/MOPP
Ice cream	MOPP/cold seal; MBOPP white; BOPP white; two sides acrylic lacquered white MOPP; two sides acrylic lacquered white OPP
Ham/cheese	• Lids: PVDC/PET/PE • Bottom Film (tray): PE/PA • Bottom Film (tray): LLDPE; EVOH; LLDPE/PA • Lids for ALU trays: PET; PET peelable
Soups/sauces/spices	PAPER/PE; PET/EVOH; PAPER/PVDC; PAPER/white MOPP; PAPER/AL/PE; PET/AL/PE; PAPER; OPP; PET/MOPP/PE
Tea	Crimped PAPER; white OPP; PAPER/PE; PAPER/PVDC; PAPER/EVOH; PAPER/MOPP; PAPER/PE/PET SiO_x/PE; PET/AL/PE
Dessert mixes	PAPER/PE; PAPER/PE; PVDC; PAPER/PVDC; PAPER/PVDC/PE; PAPER/AL/PE; MPET/PE; PAPER/MOPP/PE

temperatures, nontoxic, high resistance to most alkalis and acid, organic solvents, degreasing agents, and electrolytic attacks. PP-contained laminated films, PP food containers, PP peel system are a few types of the PP polymers. PET is another polymer that is used in beverage, food, and other light containers, thermoforming applications and as engineering resins. PET is hard, strong with superior dimensional stability, light-weight, high gas barrier properties, low water absorbability, good chemical resistance (except to alkalis, which hydrolyze it), and excellent wear resistance and is resistant to microbial and nonbiodegradable. According to the food safety element requirements, PET should contain low hazard, Bisphenol A free and extremely low toxicity. PS is a synthetic aromatic polymer, which can be rigid or foamed and used in protective packaging, containers, lids, and bottles. PS has poor barriers to oxygen and water vapor with relatively low melting point.

PET is transparent in addition to having good tensile strength and yielding strength properties, and is being used for cold beverages with no chemical interactions. Opaque/clouded containers for milk is made of HDPE (Miltz et al., 1988). PVC types of materials are used as plastic wrappers and are easily stretchable with the capability to be stretchable. LDPE is known for its excellent barrier properties and is used for food storage bags due to its low cost and stretching capacity. PP is used in rigid containers such as cups and bowls, but is slightly more expensive. PS is commonly used in Styrofoam food containers and cups as well as meat and egg trays to be heat resistant. In some cases, the leaching effect could cause contamination due to the loss of mechanical and barrier properties in long duration storage cases. Fillers (to lower cost), plasticizers (strengthen flexibility, ductility, and toughness of polymers) and stabilizers (prevent deterioration of mechanical) have the possibility to interact with the food product stored and cause leakage leading to the development of food-borne pathogens by interaction with external environment. Upon using the techniques such as UV radiation and microwave radiation treatment, it is possible to heat the matrix that could loosen or distort the structure leading to loss of mechanical property. In the following section, we study the influence of nonthermal processing by HHP on various properties of the flexible polymer packaging materials.

5 Nonthermal Processing by HHP and its Influence on Various Factors

5.1 HHP Processing: General Introduction

HHP process aids in microbiological reduction and texturation. HHP process also assists in food processing in the case of seafoods such crustaceans. Nonthermal food processing technology is successfully employed to obtain high quality food product such as fruit juices, ready to eat meals, smoothies, salsa, and guacamole to be served as cool products. Until now wider commercial applications of this process are limited, however, if it is combined with other preservation technologies broader commercial applications can be expected. As explained in the previous section, food products are subjected to high pressures in the range of 100–800 MPa for short duration in the temperature ranges from 0°C to 100°C. Water is most generally used as the pressure transmitting medium. The food products are usually packaged in the flexible pouches or containers. Currently available commercial HHP equipments are capable of processing various quantities based on the volume of the vessel used for processing and other practical parameters related to package and experiment such as pressure, duration of HHP treatment, decompression and compression rate, product temperature (initial and final), pH, composition, type of pressurizing medium, type of microorganism to be deactivated and state of food product to be treated (water percent content). Few of the disadvantages of HHP process do persist such as follows: the HHP process is said to cause deprotonation of charged groups, rupturing salt bridges and hydrophobic bonds in cell membranes of microorganisms. In general, bacterial with spore form is more pressure resistant (baroresistant) than yeasts and molds. In a few other cases, HHP is used in combination with other techniques to effectively reduce the spores in the product. There are reports on the recurrence of pressure resistant vegetative bacteria (less baroresistant than spore) after HHP treatments. Currently, HHP processed foods are available for commercial sales in European, Japanese, and American markets. Some of the commercialized HHP food products are ready-to-eat meals, fruit juices, smoothies, jams, jellies, salsa, guacamole, oysters, ham, salad dressing, chicken, and rice-based cakes. However, there are different types of interactions and influences on the food product and packaging during HHP treatment, which are discussed in the following sections.

5.2 HHP Flexible Package Prerequisites

For successful execution of the HHP treatment on the food product, packaging is one of the essential factors. In order to evaluate the effectiveness of the package, the overall properties including the interaction of the food product with the package material under HHP has to be understood, as the food product is usually packaged before HHP treatment. It is important to have a reversible response of the package containing food to compression for evaluation of successful HHP treatment. Usually flexible packaging materials are used for HHP treatment, because of its capability to withstand the compression forces without compromise to physical integrity. Various flexible systems, including PP, polyester tubes, PE pouches, and nylon cast PP pouches are currently used in HHP treatments. HHP processing is compatible with most of the flexible packaging systems. However, Al metal layer present in the multilayered packaging material negatively impacts its barrier properties, due to the plasticity of the metallic layer such as an Al layer in a multilayered packaging material negatively affected its barrier properties. In the presence of aluminium trioxide and silicon oxide coating in multilayered packaging material positive influences the mechanical and barrier properties. Choice of the packaging material for HHP treatment is also dependent on the initial processing steps to which the food product will be subjected. If high temperature is employed in the initial processing steps could lead to many changes in the structural, physical, chemical, and barrier properties. It has been concluded that the high pressure individually is not capable of inducing any negative effects on the packaging material. However, the problems related to the adhesion in the multilayer structure leading to loss of structural integrity are possible. All the physical processes that the food/drug product undergoes during HHP is experienced by the packaging material as well. For example, compression heating. PP and PE undergoes compression heating under HHP in all the temperature ranges from 10°C up to 90°C. As the volume filled by the packaging material in the vessel is significant, the heat transfer from the package material containing the product leads to the temperature nonequilibrium in the vessel. In the case of bigger headspace due to the high compressible nature of the air by HHP treatment could lead to deformation strains in the packaging materials.

Other than the aspects discussed earlier, with the prime objective of HHP treatment being the long shelf life and the preservation of the properties of the food product, it is necessary to study barrier, mechanical, and mass transfer properties such as sorp tion and migration properties of the packaging. The material used for packaging is important in addition to sealing process, which determines the sealing strength in order to avoid leakage. In order to avoid the undesirable reactions between the oxygen and the product content, the food product is usually vacuum packed. Vacuum packaging also improves the loading factor and increases the number of products to be processed at a time. Flexible packaging materials show good compatibility with most of the food products used. Packaging materials are made with one- or multilayers such as PET, PE, PP, and ethylene vinyl alcohol (EVOH) films are used for pasteurization. Coextruded films with a combination of polymeric barrier layers on polymer or inorganic layer such as aluminium foil.

5.3 Influence of HHP on Microorganisms

With the present state of the art, the complete knowledge about the microorganism inactivation by nonthermal processing is not understood completely. It has been observed that Gram-positive bacteria is high pressure resistive in comparison to Gram-negative bacteria, yeasts, and molds. Spores were also reported to be more pressure resistant than vegetative cells, which is due to the calcium rich dipicolinic acid preventing the spore from excess ionization. In a similar fashion, the microbes that were resistive to heat were also exhibiting a similar resistance toward high pressure. Depending on the growth phase of the cell either in the exponential or in the stationary phase, the sensitivity to the pressure varied. In the former phase, the cell was found to be more pressure sensitive than the latter. Population of yeasts, psychrotrophs, molds, and coliforms diminished with the increase of pressure, but there was little or no influence of pressure on the heat/acid-resistant bacteria and proteolytic microorganisms. In most of the cases, the vegetative bacterial cells are inactivated by high pressures in the range of 400–600 MPa. Regarding the virus deactivation, it depends on the denaturation of capsid proteins necessary for host cell attachment. Studies have indicated that HHP treatment does not influence the nutritional aspects of orange juice, tomato puree, and milk. During HHP treatment, some of the effects due to high pressure is seen in the form of reduced turbidity, casein dissociation from colloidal to soluble phase, increased pH, and reduction in whiteness. Recently, HHP processing is applied in medicine also for the conservation of colostrum and human milk efficiently by retaining their bioactive properties.

5.4 Influence of HHP Processing on Physical, Chemical, and Mechanical Properties of Packaging

Based on the processing parameters, the influence of HHP on physical, chemical, and mechanical properties on the food packages have been studied. Effect on the tensile properties and barrier properties for the multilayer plastic films such as linear low-density polyethylene (LLDPE), EVA, EVOH, and ethylene-vinyl acetate (EVAC) after HHP processing at 400 MPa at 60°C for 30 min have been reported by Mertens (1993). It was inferred that HHP processing did not have any significant changes in the mechanical and barrier properties. On the other Lambert et al. observed delamination in the case of the coextruded polyamide (PA) and PE films which were filled with food simulants, however, no changes in the case of other types of multilayer packages were observed. The HHP processing conditions used were in the range of 200–500MPa at ambient temperature with a dwell time of 30 min. There were also no other changes in the barrier properties, physical integrity/mechanical changes in the films. In other research work from (Lambert et al., 2000a,b) have reported on the tensile strength of various multilayer films such as PA/PE, PET/PVDC/PE,

Table 11.3 Mechanical Properties of Common Food Plastics

Material Characteristics	PET	HDPE	PVC	LDPE	PP	PS
Density (g/cm^3)	1.35	0.959	1.3–1.58	0.925	0.905	1.05
Yield strength (MPa)	59.3	26.2–33.1	40.7–44.8	9–14.5	31–37.2	—
Modulus of elasticity (GPa)	2.76–4.14	1.08	2.41–4.14	0.172–0.282	1.14–1.55	2.28–3.28
Tensile strength (MPa)	48.3–72.4	22.1–31	40.7–51.7	8.3–31.4	31–41.4	35.9–51.7
Percent elongation (%)	30–300	10–1200	40–80	100–650	100–600	1.2–2.5
Fracture toughness (MPa.m^2)	5	—	2–4	—	3–4.5	0.7–1.1

and PA/PP/PE. It was observed that after HHP processing, the multilayer flexible packages became rigid and less flexible. Table 11.3 shows the mechanical properties of the commonly used plastics in food packaging.

Caner et al. (2000) studied the effect of mechanical and physical characteristics of the different types of multilayer films after HHP processing on PET/SiO_x/LDPE, PET/Al_2O_3/LDPE, PET/PVDC/nylon/HDPE/PP, PE/nylon/EVOH/PE, PE/nylon/PE, and PET/PVDC/EVA. The aforesaid films were made into pouches and filled with distilled water, sealed, and subjected to pressure in the domain of 600–800 MPa for 5, 10, and 20 min at ambient temperature. It was observed that the metallized polyester was affected significantly after HHP but all the other types of multilayer high barrier films were found to be apt for the usage of HHP process without any significant damage.

The effect of pressure in the range of 600 MPa for 60 min was analyzed by (Dobiás et al., 2004) and offered an insight on the HHP effects for long dwell times. Seven different monomaterials and laminated films were studied. The effect on the mechanical properties (tensile strength and seal strength), transparency (absorbance at 450 and 600 nm), water vapor permeability, overall migration into fat food simulants (95% ethanol and iso-octane), and sorption of food simulants (water and olive oil) into packaging materials were studied, which evidenced that variation/loss of sealability and changes in migration levels, depending on the studied materials. Le-Bail et al. (2006) studied multilayer package materials with internal films of PE in direct contact with food. These packaging materials were filled with water as a food simulant and subjected to HHP at 10°C for 10 min in the pressure range of 200–600 MPa. Depressurization rate was used as the modulating factor by employing either rapid depressurization rate (pressure drop in <10 s) or slow (20 MPa/min). Maximal stress, rupture stress and strain at rupture were evaluated with nontreated and treated samples. Results showed that HPP minimally affects the mechanical strength of packaging material. The depressurization rate did not have any significant influence on the studied conditions. Water vapor permeability was not significantly affected and was even slightly enhanced for LDPE, which is commonly used in HHP applications as a food contact material.

Schauwecker et al. (2002) reported that there were no differences in the properties after thermal treatment at 690 MPa at 95°C for 10 min in the case of PA/EVOH/PE. Morphological properties and oxygen transmission rates of EVOH-based materials subjected to HHP at 400 and 800 MPa, for 5 and 10 min at 40 and 75.8°C has been analyzed by (López-Rubio et al., 2005), and results were compared with those obtained after a sterilization process. It was inferred that HHP do not affect the packaging material in comparison to the consequences leading to deterioration of properties caused by retorting. No significant changes in the structure of the copolymers containing high ethylene content (EVOH) was observed and on a positive note HHP lead to improvement in the crystalline morphology of EVOH, which gave improved barrier properties than in the case of nontreated material.

Galatto et al. (2008) investigated the effect of HHP at 400 MPa for 30 min at 20 and 60°C (pressure/temperature combined treatment) on the mechanical and thermal properties for complex films of PE/EVOH/PE (53/12/53 μm) with 44 mol% of ethylene content, PETmet/PE (75/19 μm), PET/PE (93/17 μm), and $PPSiO_x$ (21 ± 1 μm) as a function of the contact food simulant. None of the aforesaid materials showed significant changes in their mechanical properties, and film orientation did not affect the suitability of packaging materials, even when many of them showed superficial damages. The materials with aluminum or SiO_x cover presented disruption of the recoverage, delamination, and wrinkles, which did not affect mechanical properties, since the metal and polymer layers have different compressibility factors. Only $PPSiO_x$ mechanical properties were affected by HHP treatment. There is no influence of packaging phase contact such as aqueous, olive oil, or empty space. For PE/EVOH/PE, thermal properties were affected by the simulant in contact, as shown by a decrease in the melting point in those pouches filled with water, while other materials were not affected. This study demonstrated that those materials selected for their high-barrier qualities could be affected by the treatment, depending on the film and the processing conditions (high-pressure treatment, temperature, and time).

HHP influences the shift of pH of the food as a function of applied pressure. A decrease of pH in apple juice by about 0.2 units/100 MPa has been reported with the increase of pressure. The shift of pH cannot be predetermined to generalize for all the food products as each product will have the shift depending on its chemical constituent and pH value. pH value and water activity have remarkable effects on the inactivation of pathogens. Though covalent bonds are not disturbed by HHP treatments, it can influence the ionic bonds. Influence of HHP on proteins vary due to the hydrophobic interactions, which lead to volume increase and beyond this pressure range a volume decrease is associated with hydrophobic interactions and the pressure tends to stabilize these reactions.

5.5 Influence of HHP on Migration Properties

The selection of the apt packaging material is vital. The three main interactions occur between the food and the packaging material such as sorption, permeation, and migration. Flexible plastic structures, which bounce back to their initial position, are most sought out for HHP processing. By high pressure compression, the volume of the plastic material decreases and hence loses its capacity to absorb flavor from the food. When the pressure is released, the package material should regain its original dimension, where the sorption and diffusion characteristics should be similar to that of those expected at the ambient pressure. In a few cases, the package materials fail to regain their original volume after depressurization.(Masuda et al.,1992) reported on the decrease of sorption of D-limonone into LDPE and EVA films at 400 MPa treated for 10 min. Similarly (Kübel et al., 1996) also reported on the effects of sorption of aroma compounds such as p-cymene and actetophenone into plastic films. It was observed that the sorption of aroma compounds was lower in the case of HHP treated films than in the nontreated film. Glassy state transition occurring at high pressure is attributed to causing the diminution in the sorption of aroma compounds. Due to the case of delamination, appearance of the pouch after processing and the reduction in mechanical properties paves the way to choose polymeric multilayer films for HHP treatment. Largeteau et al. reported on the type of polymer without EVOH constituting of PA and PE assembled in two layer films PA/PE. Largeteau et al. (2010) claimed that RILTHENE obtained by lamination fabrication process is the best candidate to be used at high pressure at negative temperatures for inactivation of resistant microorganisms such as *S. aureus*. VIROFLEXAL obtained by co-extrusion process is suited for applications involving high pressure at ambient temperatures. However, no reports or data on the extreme pressure conditions have been reported for VIROFLEXAL films.

Polymer films containing antioxidants such as BHT and Irganox are reported to inhibit the deterioration of the food products by oxidation by acting as a functional barrier to oxygen and any auto-oxidation of the product. In order to diminish the degradation due to oxidation, antioxidants in the percentage of 0.001–1 wt.% are added to the polymer. It is also reported that it helps in the protection against the thermal degradation of the packaging film. However, the probability of antioxidants migrating into the food products is high due to its low molecular weight. The risk of migration of the antioxidants from the polymer in to the food stored depends on the state (solid/liquid) and nature of the food (acidic/alcoholic/fatty). Food simulants used in the polymers could also positively/negatively influence the migration process. Caner et al. (2004) reported the effect of HHP process to study the effect of migration of Irganox from PP film to food-simulating liquids (FSL) with variable processing parameters such as antioxidant levels, temperature, type of food simulant, and time of exposure. It was concluded that the additive migration from plastics into food simulants and food takes place in case of long contact time. However, no significant differences in the migration levels of Irganox were observed in the case of HHP treatments in comparison to the other reference materials. But the level of migration from PP film into 95% ethanol FSL was higher in comparison to that in 10% ethanol FSL. Based on the contact phase properties of food nature, migration level was highly influenced. When the HHP processing temperature was increased or when the storage time after HHP processing is increased, there was a significant increase in the migration level.

5.6 Influence of HHP on Barrier Properties

HHP processing also has a significant influence on barrier properties of packaging materials. Table 11.4 shows the barrier properties of different packaging materials used for food products. (Masuda et al., 1992) reported that the packaging materials constituting EVOH and PVOH are compatible with HHP and no barrier properties were changed. Similarly Ozen and Floros (2001) have also reported that water vapor and oxygen permeabilities of several laminated plastic films (PP/EVOH/PP), oriented polypropylene (OPPY) PVOH/PE, KOP/cast polypropylene (CPP), polyethylene

Table 11.4 Barrier Properties of Currently Used Packaged Materials

Material	O_2 Gas Transport	CO_2 Gas Transport	Water Vapor	High Temperature Limit (°C)	Chemical Resistance	Mechanical Resistance
PET	13.5	71	3.45	65	Good	Good
HDPE	185	580	0.3	120	Moderate	Moderate
PVC	17	27	3	70	Moderate	Good
LDPE	300	2700	1.3	100	Moderate	Moderate
PP	135	390	0.3	125	Moderate	Slightly weak
PS	330	1160	8.5	100	Moderate	Slightly weak

terephthalate (PETYAI/CPP) were not affected in the high pressure domain between 400 and 600 MPa. Lambert et al. (2000a,b) reported on the oxygen-barrier properties of PE coated with SiO_x and PP materials that were not changed by a treatment at 800 MPa (2 min, 25°C). On the other hand, Caner et al. (2000, 2003, 2004) reported significant increase in the permeability of oxygen, carbon dioxide, and water vapor metallized PET film after HHP processing, as its permeance values for oxygen, carbon dioxide, and water vapor increased as much as 1500/0. Caner et al. (2000, 2003, 2004) reported that the effect of high pressure treatment such as 800 MPa with increased processing time has significant influence on the permeance change on inorganic layers and in some films when the pressure is greater than 600 MPa and shorter processing time permeance changes occurred. Metallized PET undergoes changes in permeance significantly more than in other materials. Considering the little changes in permeance in the packaging films in comparison to metallized PET, it can be concluded that there are very few changes on the films used to fabricate the package during HHP processing.

5.7 Delamination and Sealing in HHP Processed Films

During HHP delamination is usually caused by various factors such as the type of glue used, presence of air pockets, and variation in compression of the multilayered material used for package. Schauwecker et al. (2002) reported on the delamination of PE/nylon/Al/PP laminate after HHP processing. Sealing integrity of various films have to be investigated as it could play a major role in the conservation of the food package after HHP processing. Any leakage during/after the HHP processing presents greater risk for pathogen development. However, there are no extensive reports on the sealing integrity after HHP treatment. Based on the available reports by Lambert et al. (2000a,b) and Dobiás et al. (2004) sealing has a positive influence due to the HHP processing.

6 Conclusions

Rapid developments in the food engineering and the improvements in the food conservation processes have enabled food security, distribution, storage, shelf life, and consumption a lot easier than since the early 20th century. Food packaging and food conservation is a field of huge progress in the past few decades. Owing to the cost-saving high-quality products that the combination of food processing and packaging offers, it has been a field of interest to many researchers and countries in the viewpoint of socioeconomic benefits. Recently, nonthermal food processing is used as a viable technique to the conventionally used thermal processing techniques. In order to sustain and respect the process conditions, sometimes that could be extreme such as high pressure, it is necessary to understand the characteristics of various packaging materials. With the wide variety of applications that the polymers could offer, the food packaging is one of them. Among the various nonthermal processes that are used high HHP technique is one of the techniques that is employed for the conservation of food to increase shelf life. Packaging materials used for food are usually multifunctional, such as for marketing, distribution, storage, and safety. Depending on the food products that should be packed and on the type of processing method that the food product will undergo, the package material is chosen based on its structural, physical, chemical, mechanical, and thermal characteristics.

Currently, polymers are largely used as the packaging material. Depending on the characteristics and the basic functions of the polymers, they are chosen as the packaging material. The ingredients in the package material could interact with the food in contact, hence it is necessary to study the sorption, permeability, and barrier properties of the polymers. For the materials that will be processed by HHP, the package polymer material is preferred to be flexible. Often the food product in the packaging also undergoes the nonthermal processing, under such circumstances there is a high probability to have changes in the overall properties and performance of the package. The active packaging components such as oxygen control, moisture control, and CO_2 control judges the freshness of the product. High pressure processing, in the case of a few of polymer packagings, could lead to undesirable changes. Current state of the art allows certifying the HHP method as a means for enhancing the shelf life and deactivation of the pathogens. However, the complete knowledge on pressure resistant pathogens, glues, and sealants is lacking. Synergies between the flexible polymers with intelligent packaging processed by HHP help in the future for more food products with extended shelf life and security. More research focus has to be invested toward the HHP processing and suitable packaging materials for a successful implementation.

References

Alderton, G., Snell, N., 1969. Bacterial spores: chemical sensitization to heat. Science 163, 1212–1213.

Alderton, G., Ito, K.A., Chen, J.K., 1976. Chemical manipulation of the heat resistance of *Clostridium botulinum* spores. Appl. Environ. Microbiol. 31, 492–498.

Aleman, G.D., Farkas, D.F., McIntyre, S., Torres, J.A., Wilhelmsen, E., 1994. Ultra-high pressure pasteurization of fresh cut pineapple. J. Food Protect. 57 (10), 931–934.

Aleman, G.D., Ting, E.Y., Mordre, S.C., Hawes, A.C.O., Walker, M., Farkas, D.F., Torres, J.A., 1996. Pulsed ultra high pressure treatments for pasteurization of pineapple juice. J. Food Sci. 61 (2), 388–390.

Alpas, H., Kalchayandand, N., Sikes, T., Dunne, C.P., Ray, B., 1994. Hydrostatic pressure among strains of foodborne pathogens. Appl. Environ. Microbiol. 65 (9), 4248–4251.

Arroyo, G., Sanz, P.D., Prestamo, G., 1997. Effect of high pressure on the reduction of microbial populations in vegetables. J. Appl. Microbiol. 82, 735–742.

Balasubramaniam, V.M., Ting, E.Y., Stewart, C.M., Robbins, J.A., 2004. Recommended laboratory practices for conducting high-pressure microbial inactivation experiments. Innov. Food Sci. Emerg. Technol. 5, 299–306.

Ballestra, P., Da Silva, A.A., Cuq, J.L., 1996. Inactivation of *Escherichia coli* by carbon dioxide under pressure. J. Food Sci. 61 (4), 829–836.

Benito, A., Ventoura, G., Casadei, M., Robinson, T., Mackey, B., 1999. Variation in resistance of natural isolates of *Escherichia coli* O157 to high hydrostatic pressure, mild heat, and other stresses. Appl. Environ. Microbiol. 65 (4), 1564–1569.

Berlin, D.L., Herson, D.S., Hicks, D.T., Hoover, D.G., 1999. Response of pathogenic vibrio species to high hydrostatic pressure. Appl. Environ. Microbiol. 65 (6), 2776–2780.

Bridgman, P.W., 1912. Watern in the liquid and five solid forms under pressure. Proc. Amer. Acad. Art Sci. 47, 441–558.

Butz, P., Tauscher, B., 2002. Emerging technologies: chemical aspects. Food Res. Int. 35 (2/3), 279–284.

Butz, P., Trangott, V., Ludwig, H., Ries, J., Weber, H., 1990. The high pressure inactivation of bacteria and bacterial spores. Die Pharm. Ind. 52, 487–491.

Butz, P., Habison, G., Ludwig, H., 1992. Influence of high pressure on a lipid-coated virus. In: Hayashi, R., Heremans, K., Masson, P. (Eds.), High Pressure and Biotechnology. John Libby & Co., Ltd, London, pp. 61–64.

Butz, P., Funtenberger, S., Haberditzl, T., Tauscher, B., 1996. High pressure inactivation of *Byssochlamys nivea* ascospores and other heat-resistant moulds. Lebensm. Wiss. Technol. 29, 404–410.

Calvert, P.D., Billingham, N.C., 1979. Loss of additives from polymer: a theoretical model. J. Appl. Polym. Sci 24, 357–370.

Caner, C., Hernandez, R.J., Pascall, M.A., 2000. Effect of high-pressure processing on the permeance of selected high-barrier laminated films. Pack. Technol. Sci. 13, 183–195.

Caner, C., Hernandez, R.J., Pascall, M.A., Riemer, J., 2003. The use of mechanical analyses, scanning electron microscopy and ultrasonic imaging to study the effects of high-pressure processing on multilayer films. J. Sci. Agri. 83, 1095–1103.

Caner, C., Hernandez, R.J., Pascall, M., Balasubramaniam, V.M., Harte, B.R., 2004. The effect of high-pressure food processing on the sorption behaviour of selected packaging materials. Packag. Technol. Sci. 17, 139–153.

Devlieghere, F., Vermeiren, L., Debevere, J., 2004. New preservation technologies: possibilities and limitations. Int. Dairy J. 14 (4), 273–285.

Dobiás, J., Voldrich, M., Marek, M., Chudáckará, K., 2004. Changes of properties of polymer packaging films during high-pressure treatment. J. Food Eng. 61, 545–549, 2004.

Dring, G.J., 1976. Some aspects of the effects of hydrostatic pressure on microorganisms. In: Skinner, S.A., Hugo, V. (Eds.), Inhibition and Inactivation of Vegetative Microbes. Academic Press Inc, New York.

El Moueffak, A.C., Antoine, M., Cruz, C., Demazeau, G., Largeteau, A., Montury, M., Roy, B., Zuber, F., 1995. High-pressure and pasteurization effect on duck foie gras. Int. J. Food Sci. Technol. 30 (6), 737–743.

Galatto, M.J., Ulloa, P.A., Hernández, D., Fernández-Martin, F., Gavara, R., Guarda, A., 2008. Mechanical and thermal behavior of flexible food packaging polymeric film materials under high pressure/temperature treatments. Pack. Technol. Sci. 21, 297–308.

Hayashi, R., 1991. Pressure-Processed Food: Research and Development. San-Ei Publishing, Kyoto, Japan.

Hayashi and Balny, C., 2005. High pressure science and biotechnology. Amsterdam, Elsevier Science, B.V. 171–174, 253–259.

Hite, B.H., 1899. The Effects of Pressure in the Preservation of Milk, 58, West Virginia University. Agricultural Experiment Station, Morgantown, Morgantown, pp. 15–35.

Hite, B.H., Giddings, N.J., Weakly, C.E., 1914. The effects of pressure on certain microorganisms encountered in the preservation of fruits and vegetables, 146, West Virginia University. Agricultural Experiment Station, Morgantown, pp. 1–67.

Hoover, D.G., 1993. Pressure effects on biological systems. Food Technol. 47 (6), 150–155.

Hoover, D.H., 1997. Minimally processed fruits and vegetables: reducing microbial load by nonthermal physical treatments. Food Technol. 51, 66–71.

Horie, Y., Kimura, K., Ida, M., Yosida, Y., Ohki, K., 1991. Jam preparation by pressurization. Nipp. Nogeikag. Kaishi. 65 (6), 975–980.

Hotchkiss, J.H., 1995. An overview of food and packaging interactions. In: Ackermann, P., Jagerstad, M., Ohlsson, T. (Eds.), Food and Packaging Interactions. The Royal Society of Chemistry, London, pp. 3–11.

Ishiguro, Y., Sato, T., Okamoto, T., Okamoto, K., Sakamoto, H., Inakuma, T., Sonoda, Y., 1993. Effects of hydrostatic pressure and antimicrobial substances on the sterilization of tomato juice. Nipp. Nogeikag. Kaishi. 67 (12), 1707–1711.

Johnson, F.H., Campbell, D.H., 1945. The retardation of protein denaturation by hydrostatic pressure. J. Cell Comp. Physiol. 26, 43–46.
Kakugawa, K., Okazaki, T., Yamauchi, S., Morimoto, K., Yoneda, T., Suzuki, K., 1996. Thermal inactivating behavior of *Bacillus stearothermophilus* under high pressure. In: Hayashi, R., Balny, C. (Eds.), High Pressure Bioscience and Biotechnology. Elsevier Science B.V, Amsterdam, pp. 171–174.
Kalchayanand, N., Sikes, T., Dunne, C.P., Ray, B., 1994. Hydrostatic pressure and electroporation have increase bactericidal efficiency in combination with bacteriocins. Appl. Environ. Microbiol. 60, 4174–4177.
Kata Galic, Mario Scetar, Mia Kurek, 2011. The benefits of processing and packaging. Trends Food Sci. Technol. 22, 127–137.
Kübel, J., Ludwig, H., Marx, H., Taucher, B., 1996. Diffusion of aroma compounds into packaging films under high pressure. Pack. Technol. Sci. 9, 143–152.
Lambert, Y., Demazeau, G., Largeteau, A., Bouvier, J.M., 2000a. Packaging for high-pressure treatments in the food industry. Packag. Technol. Sci. 13, 63–71.
Lambert, Y., Demazeau, G., Largeteau, A., et al., 2000b. New packaging solutions for high pressure treatments of food. High Press. Res. 19, 207–212.
Largeteau, A., Angulo, I., Coulet, J.P., Demazeau, G., 2010. Evaluation of films for packaging applications in high pressure processing. J. Phys.: Conf. Ser. 215, 11 pp.
Le-Bail, A., Hamadami, N., Bahuaud, S., 2006. Effect of high pressure processing on the mechanical and barrier properties of selected packagings. Pack. Technol. Sci. 19, 237–243.
Lelieveld, H.L.M., 2006. Pulsed electric field pasteurization of foods. New Food 9 (1), 31–33.
Linton, M., McClements, J.M.J., Patterson, M.F., 1999. Inactivation of *Escherichia coli* O157:H7 in orange juice using a combination of high pressure and mild heat. J. Food Protect. 62 (3), 277–279.
López-Rubio, A., Lagaron, J.M., Hernánez-Muñoz, P., et al., 2005. Effect of high pressure treatments on the properties of EVOH-based food packaging materials. Innovat. Food Sci. Emerg. Technol. 6, 51–58.
Mackey, B.M., Forestiere, K., Isaacs, N., 1995. Factors affecting the resistance of *Listeria monocytogenes* to high hydrostatic pressure. Food Biotechnol. 9, 1–11.
Masuda, M., Saito, Y., Iwanami, T., Hirai, Y., 1992. Effects of hydrostatic pressure on packaging materials for food. High Pressure and Biotechnology, vol. 224. Colloque INSERM/John Libby Eurotext Ltd., London, pp. 545–547.
Mertens, B., 1993. Packaging aspects of high-pressure food processing technology. Pack. Technol. Sci. 6, 31–36.
Metrick, C., Hoover, D.G., Farkas, D.F., 1989. Effects of high hydrostatic pressure on heat-resistant and heat-sensitive strains of *Salmonella*. J. Food Sci. 54, 1547–1564.
Meyer, P., 2000. US Patent No. 6,017,572. Ultra high pressure, high temperature, food preservation process.
Miltz, J., Hoojjat, P., Han, J.K., Giacin, J.R., Harte, B.R., Gray, I.J., 1988. Loss of antioxidants from high-density polyethylene: its effect on oatmeal cereal oxidation. In: Hotchkiss, J.H. (Ed.), Food and Packaging Interactions. American Chemical Society, Washington, DC, pp. 83–93.
Min, S., Zhang, Q.H., 2005. Packaging for non-thermal food processing. In: Han, J.H. (Ed.), Innovations in Food Packaging. Elsevier Academic Press, Oxford, pp. 482–500.
Miyakawa, H., Anjitsu, K., Ishibashi, N., Shimanura, S., 1994. Effects of pressure on enzyme activities of *Lactobacillus helveticus* LHE-511. Biosci. Biotechnol. Biochem. 58 (3), 606–607.

Naik, L., Sharma, R., Rajput, Y.S., Manju, G., 2013. Application of high pressure processing technology for dairy food preservation—future perspective: a review. J. Anim. Prod. Adv. 3 (8), 232–241.

Ozen, B.F., Floros, J.D., 2001. Effects of emerging food processing techniques on the packaging materals. Trends Food Sci. Technol. 12, 60–67.

Rivas-Cañedo, A., Fernández-García, E., Nuñez, M., 2009. Volatile compounds in fresh meats subjected to high pressure processing: effect of the packaging material. Meat Sci. 81, 321–328.

Schauwecker, A., Balasubramanian, V.M., Sadler, G., Pascall, A., Adhikari, C., 2002. Influence of high-pressure processing on selected polymeric materials and on the migration of a pressure-transmitting fluid. Pack. Technol. Sci. 15, 255–262.

Smelt, J.P.P., 1998. Recent advances in the microbiology of high pressure processing. Trends Food Sci. Technol. 9, 152–158.

Timson, W.J., Short, A.J., 1965. Resistance of microorganisms to hydrostatic pressure. Biotechnol. Bioeng. 7 (1), 139–159.

Wessling, C., Nielsen, T., Leufven, A., Jagerstad, M., 1999. Retention of alpha-tocopherol in low-density polyethylene (LDPE) and polypropylene (PP) in contact with foodstuffs and food simulating liquids. J. Sci. Food Agri. 79, 1635–1641.

Yuste, J., Capellas, M., Pla, R., Fung, D.Y.C., Mor-Mur, M., 2002. High pressure processing for food safety and preservation: a review. J. Rap. Meth. Auto. Microbiol. 9, 1–10.

ZoBell, C.E., 1970. Pressure effects on morphology and life process of bacteria. In: Zimmermann, A.M. (Ed.), High Pressure Effects on Cellular Processes. Academic Press Inc, New York, pp. 85–130.

Zook, C.D., Parish, M.E., Braddock, R.J., Balaban, M.O., 1999. High pressure inactivation kinetics of *Saccharomyces cerevisiae* ascospores in orange and apple juice. J. Food Sci. 64 (3), 533–535.

12

AN OVERVIEW OF NANOPARTICLES ROLE IN THE IMPROVEMENT OF BARRIER PROPERTIES OF BIOPLASTICS FOR FOOD PACKAGING APPLICATIONS

Marina Patricia Arrieta*, Laura Peponi*, Daniel López*, Juan López, José María Kenny***

**Institute of Polymer Science and Technology, Madrid, Spain; **Polytechnical University of Valencia, Instituto de Tecnología de Materiales, Alcoy, Spain*

1 Introduction

The packaging field is the largest application sector for the plastics industry. In Europe it represents about 40% of the total plastics demand (Plastics Europe, 2015). Plastics packaging has important economical and practical advantages that make it well suited for food packaging applications, such as, their low cost, lightness, flexibility, easy to handling in integrated production lines, excellent printability, and their higher resistance with respect to other food packaging materials (ie, glass, metal, ceramics, or cardboard). Besides containment and information, the packaging should protect packaged foods from not only contamination but also the loss of food quality (Arrieta et al., 2015a). Despite the many advantages of plastics for packaging applications, their inherent permeability to small molecules such as gases, water, organic vapors, and liquids is perhaps the foremost drawback that could lead to the loss of some food components (ie, aroma, flavor, and fatty compounds) as well as to the transference of gases or moisture from the surrounding to the food (Lagaron and Lopez-Rubio, 2011; Auras, 2007). In this context, the loss of specific chemical compounds, such as aroma, flavor, and fatty compounds, as

Food Packaging. http://dx.doi.org/10.1016/B978-0-12-804302-8.00012-1

well as the gain of off-odors, residues, and additives of the packaging material as a consequence of the mass transfer processes, can lead to a reduction of food product quality, resulting in a reduction in shelf life of foodstuffs (Auras, 2007). Thus, the barrier properties, such as, water vapor permeability, gases transfer (oxygen, carbon dioxide, anhydride, etc.), aroma, and flavor scalping are of fundamental importance for materials intended for food packaging applications. Mass transfer processes between the polymer, a permeant, and/or the environment include sorption, permeation, and migration processes (Auras, 2007).

The most commonly used polymers in the field of food packaging are synthetic polymers such as low-density polyethylene (LDPE), high-density polyethylene (HDPE), polystyrene (PS), polypropylene (PP), polyethylene terephthalate (PET), polyvinyl chloride (PVC), and nylon. All these traditional plastics are derived from nonrenewable petroleum resources and give nonbiodegradable plastic materials after their useful life. In order to recover the material, and valorizing the plastics wastes a commonly used practice, is to recycle them (Samper et al., 2014, 2010; Garcia et al., 2007; Navarro et al., 2008; Parres et al., 2008). However, the valorization of plastics coming from the food packaging sector shows a rather negative ecobalance due to the need to wash the plastic wastes and to the energy consumption during the process phases, such as, waste grinding, separation, and plastic reprocessing (Avérous, 2004). For this, short-term food packaging materials bring two serious global environmental problems: (1) a high consumption of nonrenewable petrochemical sources, and (2) the high amount of plastic waste generated after their useful life (Ghanbarzadeh et al., 2014; Lagaron and Lopez-Rubio, 2011). This is why the research and industry sectors are focusing their attention on the development of new biobased and biodegradable materials for sustainable food packaging applications.

The most commonly biobased and biodegradable plastics studied for food packaging applications are classified in three types: (1) polymers conventionally and chemically synthesized, (2) agro-polymers extracted from biomass, and (3) polymers synthesized by microorganisms (Chivrac et al., 2009). The monomers of polymers conventionally and chemically synthesized are obtained from agro-resources, such as, the polylactic acid (PLA). Agro-polymers are extracted from biomass, such as plasticized starch, which are also commonly called thermoplastic starches (TPS), cellulose, proteins, and chitin (Avérous, 2004). Polymers obtained by microbial production, such as, the family of polyhydroxyalkanoates (PHA), in which the most common representatives used for food packaging are poly(3-hydroxybutyrate) (PHB) and its copolymerization with hydroxyvalerate (HV) leading to the biopolymer poly(3-hydroxybutyrate-*co*-3-hydroxyvalerate) (PHBV) (Arrieta et al., 2014e; Bittmann et al., 2013; Xavier et al., 2015). Unfortunately, so far the biobased and/or biodegradable polymers present some drawbacks with respect to the traditional plastic frequently used in the food packaging field, such as the narrow processing window, low heat deflection temperatures, hydrophilicity, or poor barrier performance (Reddy et al., 2013). One critical key point is that the material should be durable enough for food protection but still biodegradable. Thus, the structural and mechanical properties of the materials are essential requirements for food protection. It is worth noting that maintaining the physical packaging integrity is fundamental to guaranteeing its barrier performance (Berthet et al., 2015). Only a few scientific articles reported in the literature focused on the topic of the structural and mechanical properties after migration studies of biodegradable packaging materials.

Moreover, an ideal packaging material requires positive effect on food preservation, water resistance, oxygen barrier, and lower overall migration in food simulants (Yu et al., 2014). On the other hand, plasticizers are additives frequently incorporated into packaging materials intended for film applications to overcome ductility, to improve polymers processability, and to get the flexibility required for film manufacturing (Arrieta et al., 2014a). However, the plasticizers act by reducing the polymer chain interactions and consequently increasing the polymer chains mobility and thus decreasing the material barrier properties (Arrieta et al., 2014g, 2013; Burgos et al., 2013; Martino et al., 2009b).

Initially, the materials for food applications were semirigid or flexible monolayer systems. Subsequently, to improve the barrier properties they were replaced by the more complex multilayer systems widely used today (Lagaron and Lopez-Rubio, 2011). However, its difficulty of recycling and the previously mentioned cost limitation of the recycling processes as well as the worldwide trend to reduce the polymer consumption per package unit, demand the development of simplest packaging formulations, particularly focused on melt blending strategies (Arrieta et al., 2014c).

Moreover, the introduction of the biobased and biodegradable polymers into the food packaging market chain, which shows lower performance than traditional plastics, has also led to the need of new strategies to improve their properties, particularly focused on the barrier performance. The most frequently used strategies to enhance polymers barrier properties are the use of polymers blends, coatings, multilayer systems, natural biopolymers with chemical modifications and composites (Ghanbarzadeh et al., 2014; Arrieta et al., 2013, 2014f; Peponi et al., 2012). Polymer composites typically consist of a mixture of a polymer matrix as a continuous phase with inorganic or organic fillers as discontinuous phase. Several composites have been developed by adding reinforcing compounds to polymer matrices to enhance their thermal, mechanical, and barrier performance. However, most of these reinforcing materials present poor adhesion at the interface resulting in lower interaction with polymer matrix (Azeredo, 2009; Ghanbarzadeh et al., 2014). Macroscopic reinforcing materials usually contain defects, which become less important as the particles of the reinforcing component are smaller (Azeredo, 2009).

Recently, an increasing number of government, scientific, and institutional reports have concluded that nanotechnology could make a significant contribution to achieving the Millennium Development Goals, but sure with a caution on the potential risks of nanotechnology (Mukhopadhyay, 2014). In this sense, nanotechnology has gained considerable attention in the food packaging field since it offers innovative solutions to improving not only the polymers and biopolymers performance for the intended use, but also the cost-price efficiency. In fact, nanotechnology can provide new lightweight materials with stronger packaging barriers, and to better protect food quality during transport and storage (Bumbudsanpharoke and Ko, 2015). Moreover, the resultant nanomaterials are more efficient in regard to manufacturing, cost, and recycling than other currently used food packaging systems, such as, multilayer packaging (Mitrano et al., 2015). Accordingly, the addition of nanoparticles allows tuning the polymers barrier properties through a relatively simple approach.

Polymer nanocomposites refer to composites where at least one of the constituent phases, commonly the nanofiller, has one dimension in the nanoscale range (below 100 nm) (Raquez et al., 2013). Thus, for the food packaging field, nanocomposites represent a new alternative additive source for improving polymeric performances of monolayer packaging materials, such as, barrier, mechanical, and heat resistance (Bumbudsanpharoke and Ko, 2015). The main safety concern around the use of nanocomposites to enhanced food storage is the migration onto/into the food for consumption (Mitrano et al., 2015) because it is known that materials that are 300 nm or less in diameter can be taken up by individual cells (Bumbudsanpharoke and Ko, 2015). Therefore biocompatible nanoparticles are preferred for food packaging purposes (Mitrano et al., 2015). The introduction of bionanocomposites to the food packaging market offers the opportunity to optimize functional performance of nanocomposites and reduce the environmental damages due to the use of renewable and biodegradable materials, but at the same time, they should ensure consumer safety.

This chapter aims to explain the role of nanoparticles on the polymers permeation and diffusion processes, mainly moisture and oxygen permeation, focusing on food safety and environmental concerns. Primarily, it describes the main interactions of polymers with low molecular weight molecules (gases, vapors, and organic compounds). Second, it explains the nanoparticles properties and merits on the improvement of polymer barrier performances. Finally, it describes the recent approaches on bionanocomposites for the improvement of biopolymer barrier properties for sustainable food packaging applications.

2 Mass Transfer Processes

In food packaging applications, as a consequence of that polymeric materials are not totally inert, mass transfer interactions take place between them and the food or the atmosphere. In this sense, when polymers are in contact with liquids, such as foods, or chemical compounds, generally two mass transfers take place quite simultaneously: the liquid enters the polymer, while the additives of the polymer diffuse out of it. These two transfers are controlled by transient diffusion, and they are connected with each other (Vergnaud, 1998). Hence, permeation, sorption, and migration processes are typical mass transfer phenomena occurring in food packaging systems (Auras, 2007). All these mass transfer processes are schematically represented in Fig. 12.1. In food packaging materials, mass transfer processes take place for many types of molecules, all of them having an impact on foodstuff quality and

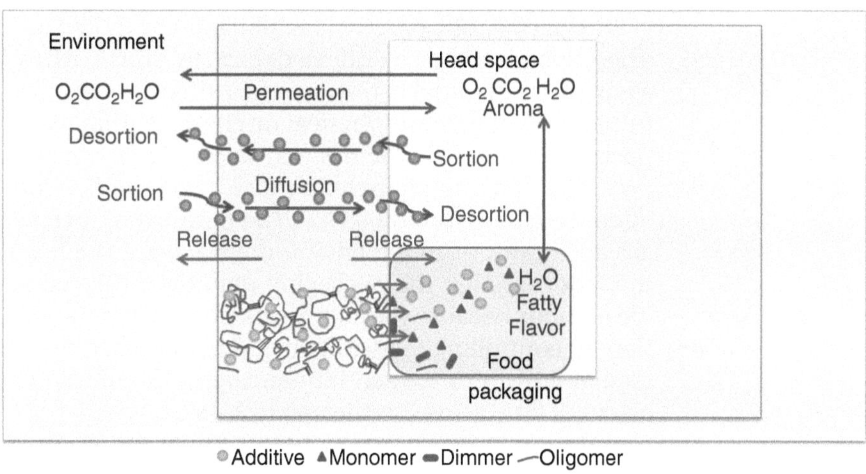

Figure 12.1. Schematic representation of mass transfer processes. Adapted from Arrieta (2014).

safety (Berthet et al., 2015). As a result of the mass transfer processes, specific chemical compounds can be lost (aroma, flavor, and fatty compounds) or gained (off-odors and additives from the packaging material) and, thus, the quality of packed food can be deteriorated caused by chemical, nutritional, and sensory changes (Catalá et al., 2015; Auras, 2007).

The mass transfer processes are generally considered negative aspects of plastic materials because they can compromise the quality of food and consequently the safety of consumers. Nevertheless, another rising tendency in food packaging research is to take advantage of the interactions between materials and foodstuff for developing active packaging systems (Arrieta et al., 2013; Xavier et al., 2015; Sánchez-González et al., 2011). However, there is an inevitable potential for conflict between innovation in food packaging, through nanotechnology approaches, which benefit the consumer by opening new applications and the restrictions of regulatory control required to ensure satisfactory consumer safety protection (Vergnaud, 1998). There are several factors affecting mass transfer processes in semicrystalline polymers, some of them are related to the polymer nature, such as, the crystallinity, free volume in the amorphous phase, orientation, cohesive and adhesive forces. A further factor is the low molecular weight compound itself (ie, molecular size). Some environmental factors, such as the temperature, pH, and relative humidity (RH%), could also affect the mass transfer processes (Auras, 2007).

2.1 Permeation, Migration, and Barrier Functions

2.1.1 Permeation

The permeation process is a diffusion-controlled process where a permeant is sorbed by, diffused through, and then desorbed from a polymeric material in response to a difference in partial pressure (Auras, 2007), without physical or chemical modifications of the packaging material. The gas and water vapor transmission rates (WVTR) of packaging materials give a good indication of permeation (Han and Scanlon, 2005). The permeation of environmental gas and vapor from the environment toward the headspace and the food, through the packaging materials, will reduce the shelf life of foods because they are key factors of food quality preservation by controlling foodstuffs degradation reactions, such as microbial growth or oxygen influencing oxidation of vitamins, and essential fatty acids (Berthet et al., 2015).

Sorption is a term that embraces both adsorption and absorption phenomena, which sometimes are difficult to distinguish from each other (Kadam et al., 2015). Adsorption is the process in which the adsorbate molecules from the food, such as, aroma, flavor, odor, or coloring compounds are transferred to the polymer packaging material (Auras, 2007; Kadam et al., 2015), for instance, aroma compounds of coffee or carbon dioxide of carbonated beverages (Berthet et al., 2015). When the adsorbate penetrates the surface layer and enters the structure of the bulk polymer, it is termed absorption (Kadam et al., 2015). Sorption and its counterpart desorption measure the affinity of a given substance for two media with which it comes into contact. The affinity of a substance for a material can be expressed using the solubility (S) or partition (K) coefficient (Han and Scanlon, 2005). Since flavor scalping of D-limonene from orange juice to a flexible packaging material is a good example of absorption process in materials (Han and Scanlon, 2005), the barrier properties to organic vapors is frequently measured by means of D-limonene (Auras, 2007).

Diffusion is the movement of substances in a medium caused by a concentration difference acting as a driving force (Han and Scanlon, 2005). It is commonly the slowest process and therefore the substance mass transport process is controlled by the diffusivity (diffusion coefficient, D) value (Gómez-Estaca et al., 2014). Diffusivity is a measure of how well the compound diffuses in the medium (Han and Scanlon, 2005). Diffusion is crucial for gas permeability whereas both sorption and diffusion are essential for moisture transfer (Guilbert et al., 1997). While water vapor permeation of polymers is directed by vapor pressures and concentration gradients on both polymer surfaces, the gas permeability is generally the combined effect of diffusion and solubility through the voids present between the segments of the polymer chains (Pankaj et al., 2014a).

2.1.2 Migration

Migration is the transfer of substances originally present in the plastic material to the packaged product (Auras, 2007). In fact, migration is a diffusion of substances from a zone of higher concentration, as the food-contact packaging layer, to one of a lower concentration, usually the food surface (Bhunia et al., 2013). Nevertheless, migration phenomenon could be bidirectional, that is, from packaging to food or from food to packaging (Dhar et al., 2015). Therefore the migration of undesirable molecules potentially toxic for human health in long-term exposure conditions, such as, chemical additives or residual monomers regulated to a maximum limit from the packaging toward the food should be controlled (Berthet et al., 2015). But it should be highlighted that the migration phenomena can also be a benefit for the development of active materials, which are the new generation of food packaging. The strategy of active packaging is based on the inclusion of additives in the packaging material instead of the direct addition of active compounds (antimicrobials, antioxidants, vitamins, etc.) into the foodstuff, so that when active compounds migrate from the packaging to the food they slow down food deterioration processes. For active packaging, the major complexity emerges for migration/diffusion processes that includes both: no release or slow release of package components itself (ie, plasticizers, oligomers), as well as the effective incorporation of the active agent with the subsequently ability to be released. In both contexts, the evaluation of material compliance with regulations includes migration monitoring for package component (ie, monomer) and additives (ie, active agent) (Jamshidian et al., 2010). It is extremely difficult to measure the nanoparticles migration directly into a food, as most foodstuffs have complicated compositions and a complete extraction of nanoparticles from food matrices is required to detect, characterize and quantify their migration in food products. Thus, the number of studies directly related to detection and quantification of nanoparticles in real food products is really small (Singh et al., 2014). Conversely, the migration behavior can be easily measured in so-called food simulants (Avella et al., 2005) and this is why the literature mainly reviews the evaluation of nanoparticles migration in food simulants instead of in real food products. A intensive review on sampling and extraction methods of nanoparticles from food products as well as the available analytical methods for nanoparticles detection has been recently published by Singh et al. They concluded that an appropriate approach to quantify nanoparticles in food products may include the complete release of nanoparticles from food matrices (simulating gastric juice, organic solvents, chemical oxidants, or enzymes), the sample clean-up (ie, filtration, centrifugation, and phase separation/extraction processes), followed by the sample preparation to be detected and finally their quantified by specific analytic techniques. Most studies use imaging analysis typically by transmission electron microscopy (TEM) or scanning electron microscopy (SEM) and, to quantify inorganic particles some combination of imaging analysis and elemental analysis almost always by inductively coupled plasma mass spectrometer (ICP-MS). More details on the current analytical techniques used for nanoparticles quantification are well described elsewhere (Singh et al., 2014).

Transfer of chemical compounds from plastics into food has raised concerns about the potentially adverse effects of food products on human health (Bhunia et al., 2013). To ensure the protection of the health of the consumers, two types of migration limits have been established in the area of plastic materials intended to be in contact with food: an overall migration limit that applies to all substances that can migrate from the food contact material to the foodstuff and a specific migration limit, which applies to individual authorized substances and is fixed on the basis of the toxicological evaluation of the substance (Avella et al., 2005). The legislation establishes overall migration limits at 60 mg kg^{-1} of food or food simulant and then establishes also specific migration limits for every substance listed in a positive appropriate list (European Commission, 2011).

2.1.3 Barrier Properties

The specific barrier requirement of the packaging material is related to the food product characteristics and the intended end-use application (Siracusa et al., 2008). Generally, plastics are relatively permeable to small molecules and they present a broad range of mass transfer characteristics, ranging from excellent to low barrier values. This broad range of mass transfer properties is essential in the case of foods that display, depending on their nature, different sensitivity to moisture and/or oxygen (Berthet et al., 2015).

Packaging plays a key role for protecting foodstuff from oxygen and as a result, oxygen permeability is one of the most important properties of a packaging material for deciding its suitability for their application in different food products (Pankaj et al., 2014a). Oxygen is the main factor of organoleptic and nutritional quality degradation of food throughout the oxidation of vitamins, aroma compounds, pigments, lipids and proteins compounds, and so on. It is also involved in microbial development and maturation rate of fresh living products, such as, fresh fruits and vegetables or fermented cheeses (Berthet et al., 2015). Moisture barrier property is another important basic property when packing many foods, whether dry or moist, to preserve the texture (crispiness, softness, firmness, etc.) and control microbial development of aerobic spoilage and pathogen species (Galic et al., 2009). Thus, materials with substantial gas- and moisture-barrier properties are required for many food packaging applications to control gas exchange for fresh food, to reduce moisture exchange with the external atmosphere (Guilbert et al., 1997; Pankaj et al., 2014b).

The oxygen and water moisture are two of the main barrier properties studied in food packaging applications, because they may transfer from the internal or external environment through the polymer package wall, resulting in a continuous change in product quality and shelf life (Siracusa et al., 2008). Oxygen diffusion process could be assessed by the determination of oxygen transmission rate per packaging material thickness (OTR·e, cm^3 mm m^{-2} day^{-1}), which indicates the oxygen volumetric flow rate per unit area of the membrane and per time and membrane thickness (Burgos et al., 2013) or by the oxygen permeability coefficients (OPC, kg mm^{-2} s^{-1} Pa^{-1}), which indicates the amount of oxygen that permeates per unit of area and time in a packaging material (Siracusa et al., 2008). The water vapor transfer of food packaging must be as low as possible to efficiently limit the moisture when it is in contact with food systems (Cano et al., 2015). Meanwhile, moisture barrier property is usually expressed by the WVTR, which is the amount of water vapor that permeates per unit of area and time through the packaging materials or more adequately by the water vapor permeability (WVP, m^3 m m^{-2} s^{-1} Pa^{-1}), which takes into account the influence of material thickness and water vapor partial pressure gradient (Berthet et al., 2015).

As was already commented, the biodegradable polymers present lower oxygen barrier, generally a value one or more order of magnitude below the synthetic polymer used in the same field (Siracusa et al., 2008). It is widely known that PET shows good barrier properties to gases and moisture as well as very good barrier properties to flavors and contaminants. The good oxygen barrier properties of PET (OTR·e = 3 cm^3 mm m^{-2} day^{-1}) were reported as considerably better than those of pure LDPE (OTR·e = 160 cm^3 mm m^{-2} day^{-1}), which is currently widely used in film manufacturing (Martino et al., 2009a). The oxygen permeability is often necessary to control either materials with high barrier properties or for applications involving the use of food packaging under controlled conditions (Burgos et al., 2013). Nanocomposites are a new class of materials that have interesting barrier properties (Lavoine et al., 2012).

2.1.4 *Wettability*

Although wettability is not a mass transfer process, it determines the surface hydrophobicity/hydrophilicity and, thus, it is an essential issue in materials intended for food packaging. In effect, food packaging is generally required to protect foodstuff from humidity during transport, handling, and storage (Arrieta et al., 2015b). The hydrophilicity/hydrophobicity of packaging material surfaces is frequently investigated by water contact angle measurements (Fombuena et al., 2013; Hambleton et al., 2009). The water contact angle will decrease with increasing surface hydrophilicity and vice versa (Introzzi et al., 2012; Fombuena et al., 2013). In this sense, it is widely known that modifying the physicochemical properties of the material surface the water adsorption can be controlled (Introzzi et al., 2012; Arrieta et al., 2014f). Water contact angle depends on the cohesive molecular forces, within the water, and adhesive molecular forces between the water and the polymer surface

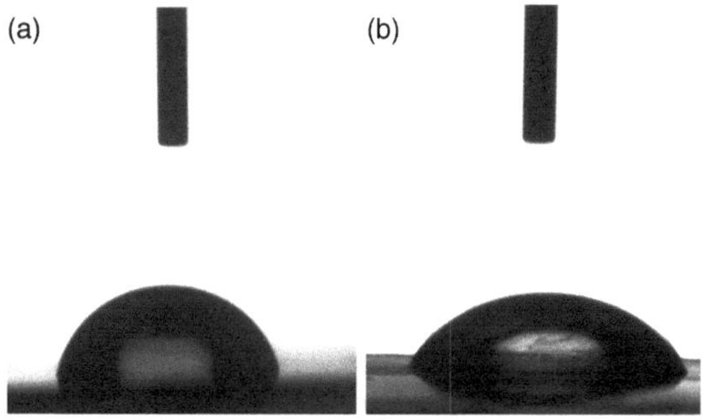

Figure 12.2. Water contact angle of (a) hydrophobic surface ($\theta° > 65°$) and (b) hydrophilic surface ($\theta° < 65°$).

(Hambleton et al., 2009). In fact, long-range attractive forces are detected only between surfaces exhibiting large water contact angles ($\theta°$), higher than 65°, with pure water adhesion. Repulsive forces are detected between surfaces exhibiting small water contact angles, lower than 65° (Vogler, 1998). Practically, a large contact angle, $\theta° > 65°$, is typical for hydrophobic surfaces (Fig. 12.2a). On the other hand, small contact angles, $\theta° < 65°$, imply hydrophilic surfaces (Fig. 12.2b) (Hambleton et al., 2009; Vogler, 1998). The decrease in the total surface free energy is due to the decrease in the polar component of surface energy (Pankaj et al., 2015).

In general, traditional plastics frequently used in the food packaging field present low wetting properties showed by their low surface energy. For example, LDPE and PP are characterized by water contact angles of about 100° (Fombuena et al., 2013; Balart et al., 2012).

Natural compounds, like cellulose and starch, are hydrophilic polymers, water wettable, or swellable, and this is why they are consequently biodegradable. But, they are not technologically useful for food packaging where water resistance is required (Siracusa et al., 2008). For instance, lignocellulosic fillers are characterized by high sensitiveness to moisture and water uptake. This can lead to some problems that can compromise the service life of composites (Revert et al., 2015). This drawback can be improved by promoting changes in surface topography (Balart et al., 2012). The wettability of a material is governed by both the chemical composition and the topography of the surfaces (Scheibe et al., 2014). For instance, nanocomposites offer the additional advantage to improve the polymer filler interaction and, thus, the nanofiller presence could produce some roughness and chemical changes on the nanocomposites surfaces, so that directly affect their wettability performance.

3 Nanoparticles and Nanocomposites Generalities

Nanoparticles are able to considerably improve the nanocomposite properties with respect to the same ones of the corresponding neat polymer, since they have a high specific surface area for a given volume, which favors the interactions between nanofiller and polymer matrix, providing better reinforcing effects and, thus, improving the performance of the resulting material (Azeredo, 2009).

Depending on the nanoparticle nature, geometry, and other characteristics, nanocomposites could exhibit drastic modifications in their properties, as improved mechanical and barrier properties (Chivrac et al., 2009). The polymer nanocomposites properties are strongly dependent on a uniform dispersion of the nanoparticles in the polymer matrix and the resultant huge interfacial area developed (Peponi et al., 2014). The large matrix/filler interfacial area leads to better interaction between composite components, which changes the molecular mobility, the relaxation behavior, and the consequent thermal and mechanical properties of the material (Azeredo, 2009). Moreover, the ability of nanoparticles to dramatically increase the surface area-to-volume ratios and the surface roughness leads to superior physiochemical properties, such as, improved water resistance owing to the topographic features influence the interaction between biomaterial and biological environment (hydrophilicity/hydrophobicity, surface energy, etc.) (Armentano et al., 2013).

Nanoparticles can be classified into their origin (natural, semisynthetic, or synthetic), depending on their synthetic pathway (Raquez et al., 2013). Nanoparticles can be also distinguished depending on their geometry as nanolayers or plate-like nanoparticles (1D), nanofibrous (2D), and nanoparticles (3D).

3.1 Nanoparticles Geometry and Properties

Nanoparticles can be distinguished depending on the geometry and how many dimensions are in the nanometric range. There are three main categories: nanolayers (1D) with only one dimension in the nanoscale range (Fig. 12.3a), 2D or nanofibrous with two dimensions in the nanometer scale (Fig. 12.3b), and 3D isodimensional nanoparticles with their three dimensions in the nanoscale range (Fig. 12.3c) (Peponi et al., 2014).

The most followed strategy to create a tortuous pathway in polymers for food packaging has been the incorporation of nanofillers and, in particular, layered silicates and cellulose nanocrystals (CNCs) because of their environmental friendliness and natural abundance (Armentano et al., 2013). It is well known that

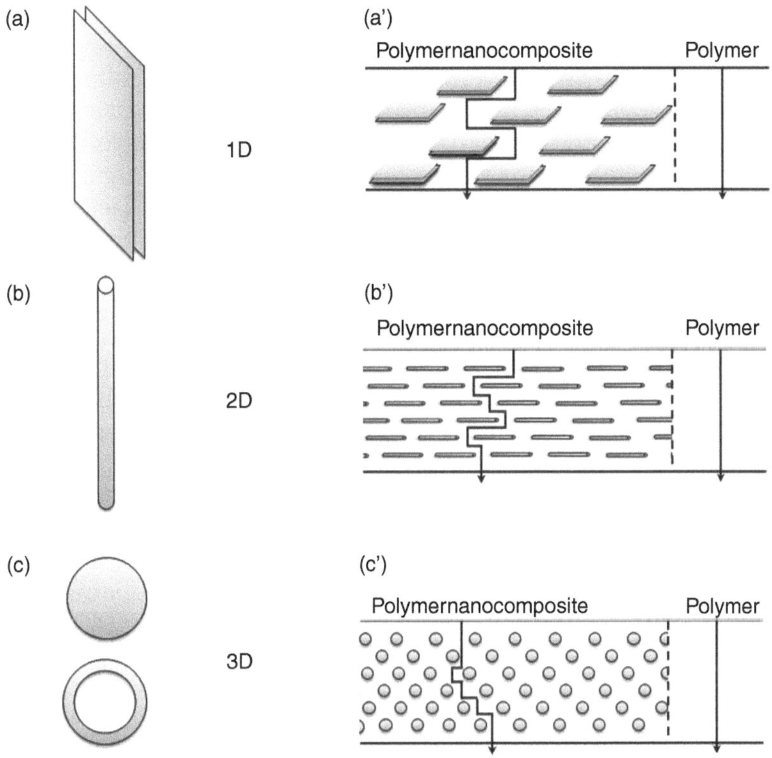

Figure 12.3. Nanoparticle geometries and its corresponding tortuosity pathway in nanocomposites. (a) 1D nanolayered, (b) 2D nanofibrous, and (c) 3D isodimensional nanoparticles.

the decrease in the permeability in nanocomposite materials is attributed to the more tortuous path created by nanoreinforcements, which increases the effective path length for vapors and gas molecules (Reddy et al., 2013). The presence of different geometries of nanofillers (Fig. 12.3a'–c') forces the low molecular weight molecules to adopt particular tortuous pathways, all of them producing a significant lag time. Higher aspect ratio of fillers forces the permeating gas molecule to follow a more tortuous path leading to improved barrier properties (Gokhale and Lee, 2014).

The most commonly used techniques to characterize the structure of nanoparticles in nanocomposites are X-ray diffraction technique, TEM, atomic force microscopy (AFM) as well as SEM (Peponi et al., 2014; Singh et al., 2014).

3.1.1 Nanolayers

Nanolayers, plate-like nanoparticles or 1D, (Fig. 12.3a) are layered particles typically with a thickness on the order of 1 nm (Raquez et al., 2013). The filler is present in the form of sheets of one to a few nanometers thick to hundreds or thousands nanometers long. The family of composites reinforced with these nanofillers is known as polymer-layered crystal nanocomposites, almost exclusively obtained by the intercalation of the polymer (or a monomer subsequently polymerized) inside the galleries of layered host crystals. There is a wide variety of both natural and synthetic crystalline fillers with the ability to intercalate a polymer, under specific conditions (Alexandre and Dubois, 2000). The most popular 1D fillers are layered silicates including smectic clays, layered double hydroxides as well as graphene sheets (Raquez et al., 2013). With the addition of small amounts of clay nanofillers it is possible to achieve considerable improvements of the mechanical properties and a decrease of the gas permeability with respect to their pure polymers or conventional composites (Peponi et al., 2014; Ramos et al., 2012). A simple permeability model for a regular arrangement of platelets has been proposed by Nielsen and is presented in Fig. 12.3a'. The nanoclays are evenly dispersed and considered to be rectangular platelets, which have perpendicular orientation to the diffusion direction. The platelets represent impermeable barriers to the diffusing molecules, forcing them to follow longer and more tortuous paths in the nanocomposite (Ghanbarzadeh et al., 2014).

Clay minerals are usually composed by phyllosilicates, which can be divided into four groups: kaolinite, montmorillonite/smectite, illite, and chlorite group that mainly differ in the crystalline structure (Peponi et al., 2014). At present, the most intensive researches in the food packaging field are focused on layered silicates, such as, montmorillonite (MMT) (Chivrac et al., 2009; Azeredo, 2009). Organoclays are cheaper than most other nanoparticles, since they come from readily available natural sources and are produced in existing, full-scale production facilities (Ghanbarzadeh et al., 2014).

3.1.1.1 Montmorillonite

Montmorillonites are clay minerals that have 2:1 sheet-structure. It means that an octahedral sheet of alumina is between two silica tetrahedral sheets and shares an oxygen atom with them (Peponi et al., 2014). While the layer thickness is about 1 nm the lateral dimensions could vary from 300 Å to several microns and even larger. The dimensions of these layers depend on the particular silicate, the clay source and the method of preparation (Alexandre and Dubois, 2000). Therefore the aspect ratio of the layers (ratio length/thickness) is particularly high, showing values higher than 1000 (Peponi et al., 2014). These types of clays are characterized by a moderate negative surface charge, which is known as the cation exchange capacity (CEC), counterbalanced by exchangeable cations placed in the galleries (Alexandre and Dubois, 2000). In pristine layered silicates, the interlayer cations are usually hydrated sodium or potassium cations that make the clay hydrophilic and consequently incompatible with most polymers. As a result, layered silicates have to be organically modified to obtain polymer-compatible clay (organoclay) (Peponi et al., 2014). The organically modified clays allow the polymer chains to penetrate within the galleries and to outdistance the single tactoids from one another. Nevertheless, the ability of organic chains to intercalate within the interlayer of the MMT mainly depends on the nature of both polymer and clay, in addition to the method of nanocomposite preparation (Peponi et al., 2014). In this sense, according to the nature of the materials used, layered silicate, organic cation, and polymer matrix, as well as the method of preparation; three main types of composites can be obtained which depends on the degree of polymer penetration into the galleries: traditional microcomposite, intercalated nanocomposite or exfoliated nanocomposite (Fig. 12.4) (Alexandre and Dubois, 2000).

If the polymer is unable to intercalate among the silicate sheets, resulting in a microphase separated final composite, the final material presents the same range of properties of traditional microcomposites (Alexandre and Dubois, 2000). In intercalated systems, the polymer is placed between clay layers, increasing

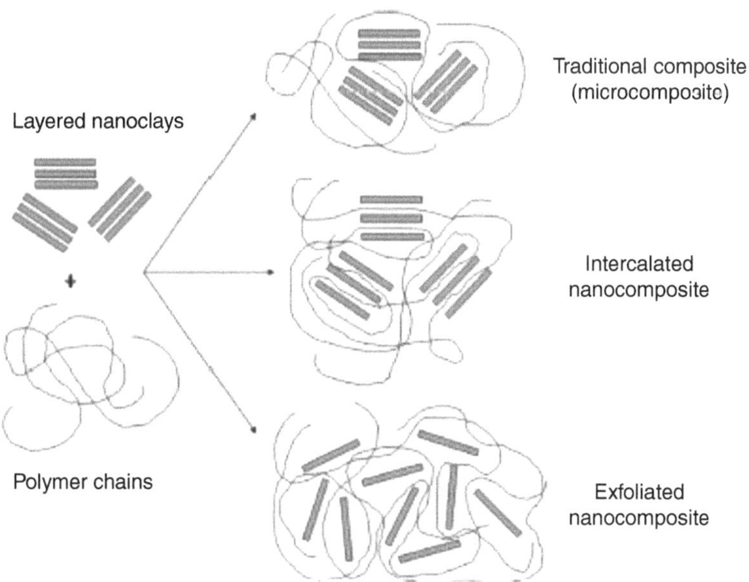

Figure 12.4. Schematic representation of types of composites based on layered clays. Reprinted with permission from Peponi et al. (2014). Copyright 2014 Elsevier Ltd.

interlayer spacing. Some degree of order is retained in parallel clay layers that are separated by alternating polymer layers with a repeated distance of few nanometers (Azeredo, 2009). Exfoliated systems (full delamination) reach complete separation of clay platelets in random arrangements (Ghanbarzadeh et al., 2014). The clay layers are individually delaminated and fully dispersed within the polymer matrix (Chivrac et al., 2010). The best combination of property enhancements is frequently presented in exfoliated nanocomposites (Peponi et al., 2014). One of the most important improvements on the polymer and biopolymers matrices for packaging proposes that provide the exfoliated structures is that they significantly enlarge the tortuous path for small diffusing molecules due to their small particle size, high aspect ratio, and exceptionally large surface areas. Therefore they are able to enhance the polymer barrier performance (Chivrac et al., 2010).

3.1.2 Nanofibrous Particles

Nanofibrous particles (Fig. 12.3b) have two dimensions in the nanometer scale with a diameter below 100 nm (Raquez et al., 2013). The third dimension is larger, forming an elongated structure (Ghanbarzadeh et al., 2014). They are characterized with an aspect ratio of at least 100 (Raquez et al., 2013). Under this category we found nanotubes and nanocellulose substrates, among others, which are widely studied as reinforcing nanofillers yielding nanocomposites with outstanding performance (Alexandre and Dubois, 2000). In fact, cellulose derivatives-based composites have many positive characteristics as reinforcements for materials intended for food packaging proposes, such as, high stiffness, excellent barrier properties, low density, low cost, abundance in nature all around the world, biodegradable under certain conditions, and generally low toxicity (Rayón et al., 2015, 2013; Sonseca et al., 2014; Torres et al., 2009). Regarding food safety concerns, potential toxicological substances able to migrate from cellulose fillers toward food, are mostly pesticide residues (including herbicides, insecticides, and fungicides) and environmental pollutants (ie, polycyclic aromatic hydrocarbons). Thus, a special attention should be paid to the decontamination efficiency during the nanocellulose extraction processes intended to be used for the development of nanocomposites for food packaging applications (Berthet et al., 2015).

3.1.2.1 Cellulose Nanocrystals (CNCs)

Cellulose is a structural material naturally organized as microfibrils linked together to form cellulose fibers. A single filament of all plant-based natural fibers consists of several cells that are formed out of cellulose-based highly ordered crystalline microfibrils linked along the disordered amorphous domains (Peponi et al., 2014; Lavoine et al., 2012). These microfibrils are built from superfine fibrils with diameters in the nanoscale level (10–30 nm) and indefinitely long containing cellulose molecules in the cross section. Each nanofibril is mainly composed by ordered crystallites (60–80%) and to a lesser part composed by disordered domains (Peponi et al., 2014). The ordered regions are cellulose chain packages that are stabilized by a strong and complex network of hydrogen bonds that resembles nanocrystalline rods. This is why the two main types of cellulose reinforcements are CNCs and cellulose microfibrils. There is yet another source of nanocellulose biosynthesized by specific bacteria known as bacterial cellulose, but their industrial use does not seem imminent (Lavoine et al., 2012). In this chapter we focus our attention on CNCs and their corresponding bionanocomposites as well as their barrier properties.

CNCs are, in general, rigid rod-like monocrystalline cellulose domains, which are between 1 and 100 nm in diameter and from 10 to hundreds of nanometers in length. CNCs are characterized by a highly crystalline structure with a very large surface area (ca. 150 $m^2\ g^{-1}$) (Fortunati et al., 2013). CNCs can be extracted by acid hydrolysis of several natural cellulose fibers such as cotton, cellulose fibers from lignocellulosic materials, and a marine animal tunicate. Therefore CNCs are commonly produced from the hydrolysis of microcrystalline cellulose (MCC) using sulfuric acid to remove the amorphous cellulose regions and form highly crystalline cellulose of colloidal dimensions with considerably smaller size than MCC (Fortunati et al., 2012a). However, the high polarity of cellulose surface, which leads to low interfacial compatibility with hydrophobic polymer matrices, makes difficult its homogenous dispersion in polymer matrices (Jonoobi et al., 2012; Aranguren et al., 2013). CNCs dispersion into polymer matrix, particularly during the extrusion process, could be problematic due to their low thermal stability (Oksman et al., 2006). It has been shown that increasing the pH of the CNCs suspensions from pH 3 to pH 9 a clear improvement of the thermal stability is achieved (Fortunati et al., 2012a). On the other hand, the homogeneous dispersion of high polarity CNCs into the hydrophobic polymers matrices can be favored modifying the nanocrystal surfaces. The surface hydroxyl groups (—OH) on the CNCs can be reacted to produce modified cellulose crystals (Aranguren et al., 2013). One strategy to favor the CNCs dispersion and therefore to increase the final properties of the final formulation is by using a surfactant, which leads to CNCs surface functionalization (Arrieta et al., 2014b,c;

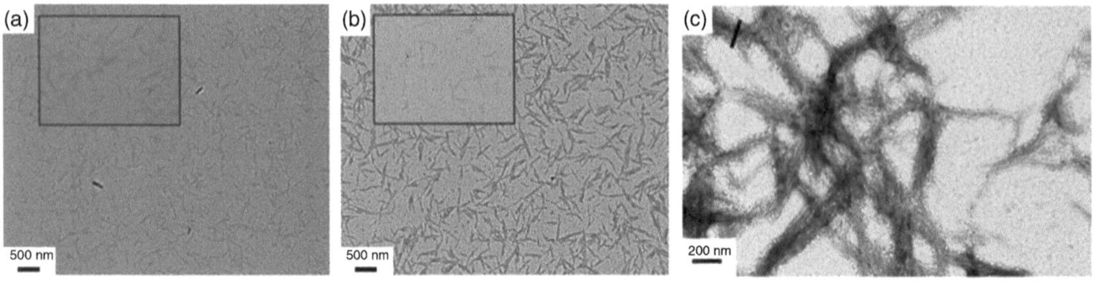

Figure 12.5. TEM micrographs of: (a) CNCs, (b) s-CNCs, and (c) CNC-*g*-PLLA. Parts (a) and (b) reprinted from Arrieta et al. (2014b), with permission. Copyright 2014 Elsevier. Part (c) reprinted from Navarro-Baena et al. (2014a), with permission. Copyright 2014 Springer.

Fortunati et al., 2012a). CNCs surface can be also modified by grafting the polymer chains onto CNCs surface, to enhance the polymer-nanofiller interaction. In fact, in this case, the superficial —OH group of the CNCs can be used as initiators for the grafting to reaction leading to the formation of functionalized CNCs. Typical CNCs are functionalized with biodegradable polymers, such as, PLA or poly(ε-caprolactone) (PCL) (Navarro-Baena et al., 2014a; Peltzer et al., 2014). Fig. 12.5 shows the TEM of CNCs (Fig. 12.5a), surface modified CNCs by means of a surfactant s-CNCs (Fig. 12.5b) (Arrieta et al., 2014b), and CNCs functionalized by grafting poly(L-lactic acid) (PLLA) chains onto their surface, CNC-*g*-PLLA (Fig. 12.5c) (Navarro-Baena et al., 2014a). The TEM images of the CNCs and s-CNCs show the typical dimensions ranging from 100 to 300 nm in length and between 5 and 10 nm in width. The unmodified CNCs appear somewhat agglomerated, while more individualized structures are obtained after the surface modification (s-CNCs) (Arrieta et al., 2014b). The CNC-*g*-PLLA shows the typical rod-like shape morphology with an average diameter of about 25 nm. The increased diameter value is due to the presence of PLLA chains grafted onto the CNCs surface (Navarro et al., 2008).

3.1.3 Isodimensional Nanoparticles

Isodimensional nanoparticles or 3D (Fig. 12.3c) are structures with three nanodimensions (below 100 nm) and are very versatile, such as, vesicles, spherical micelles, fullerenes, metal and ceramic nanoparticles. Nanoparticles of almost all metals have been produced, but the more diffused metals for polymeric nanocomposites are silver, gold, platinum, copper, iron, and so on (Peponi et al., 2008, 2011, 2014). There are also several examples of ceramic nanoparticles, such as, nanocrystalline aluminium oxide (Al_2O_3), zirconium dioxide (ZrO_2), silicon carbide (SiC), silicon nitride

(Si_3N_4), titanium dioxide (TiO_2), titanates, and ferrites (Peponi et al., 2014; Gutierrez et al., 2008). But, the most well-known 3D nanoparticles with interest in the food packaging fields, because of their larger-scale production, are undoubtedly silica particles and metal oxides like titanium or zinc oxides (TiO_2, ZnO) (Raquez et al., 2013; Peponi et al., 2014). The use of oxides, such as zinc oxide (ZnO), as nanoreinforcement, offers two advantages for food packaging proposes; they contain mineral elements, which are essential to humans and also can exhibit strong antibacterial activity (Yamamoto, 2001). Moreover, metal oxide nanoparticles are often used as photocatalysis agents to degrade organic molecules and microorganisms (Bumbudsanpharoke and Ko, 2015). Thus metal oxide nanoparticles gain considerable attention in the food packaging industry for the development of active food packaging nanocomposites. Particularly, ZnO is moderately more efficient and attractive than other nanoparticles, for example silver nanoparticles, owing to its less toxicity and cost effectiveness (Bumbudsanpharoke and Ko, 2015).

3.1.3.1 Zinc Oxide Nanoparticles

Zinc oxide isodimensional nanoparticle is a polar inorganic crystalline material and has a hexagonal quartzite structure (Ghanbarzadeh et al., 2014). Several methods have been developed to synthesize zinc oxide nanocrystals, including warm ultrasonic bath, vapor phase growth, vapor-liquid-solid process, soft chemical method, electrophoretic deposition, sol–gel process, homogeneous precipitation, and also zinc oxide can be obtained from catalyst-free combust-oxidized mesh process (Pantani et al., 2013; Ghanbarzadeh et al., 2014). Fig. 12.6 shows the field

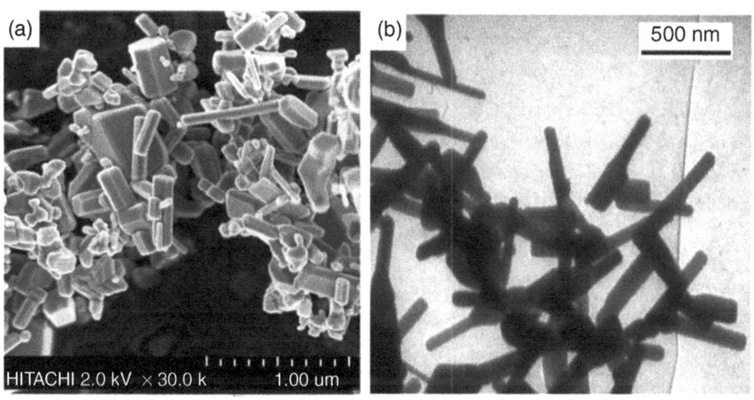

Figure 12.6. ZnO nanoparticles. (a) FESEM micrograph and (b) TEM micrograph. Reproduced with permission from Nafchi et al. (2013). Copyright 2013 Elsevier.

emission scanning electron microscopy (FESEM) and TEM micrographs of ZnO nanoparticles powder, where it is possible to see their typical structural geometry with 40–100 nm in diameter and 200–700 nm in length (Pantani et al., 2013; Nafchi et al., 2013).

The particle properties of ZnO, such as crystallinity and morphology, can be controlled by adjusting factors such as the source species, pH, reaction temperature, time, and solvent (Ghanbarzadeh et al., 2014). In the plastic industry, ZnO nanoparticles are recognized as multifunctional inorganic additives, which are biocompatible, nontoxic, and also environmentally friendly. Thus, they could be considered as multifunctional nanofillers for various polymers providing interesting properties, such as, antibacterial activity or intensive ultraviolet absorption (Pantani et al., 2013; Raquez et al., 2013). Moreover, zinc oxide is one of the five zinc compounds that are currently listed as generally recognized as safe (GRAS) by the US Food and Drug Administration (FDA) (Nafchi et al., 2013). Thus the use of ZnO nanoparticles has particular interest in the food packaging field, not only to improve the barrier properties of biopolymers, but also to develop antibacterial active packaging systems. In this regard, ZnO shows considerably higher antibacterial effect on *Staphylococcus aureus* than other metal oxides like MgO, TiO_2, Al_2O_3, CuO, or CeO_2 (Díez-Pascual and Díez-Vicente, 2014). The antimicrobial mechanism of nano-ZnO is attributed to the generation of reactive oxygen species (ROS), resulting in the oxidation of cytoplasm of bacterial cells, which leads to cell death (Bumbudsanpharoke and Ko, 2015).

4 Nanofillers Effects on the Barrier Properties of Biopolymers Matrices

In this part of the chapter, the effect of the common used nanofillers dispersed in biopolymer matrices for the improvement of food packaging barrier properties is presented. In particular, the attention is focused on MMT as 1D nanolayers, cellulose-based nanofillers as 2D and zinc oxide nanoparticles as 3D geometry.

4.1 Montmorillonites-Based Nanocomposites

MMT-based nanocomposites show great promise in providing excellent barrier performance, due to the presence of the clay layers, which interrupt the molecular pathway making the polymer diffusive path more tortuous (Gorrasi et al., 2004). As it was already commented, MMT are generally modified by cations exchange with alkylammonium, and display greatly improved properties, even at low nanofillers content (typically less than 5 wt.%).

A lot of studies have reported the effectiveness of nanoclays and particularly those organomodified nanoclays in decreasing oxygen barrier properties (Avérous, 2004). Many positive characteristics of PLA have situated it as the most used biopolyester for biodegradable food packaging industry, such as, easy processability, advanced transparency, availability in the market at competitive cost, excellent printability as well as the high rate of disintegration in compost (Arrieta et al., 2015a). However, it shows poor ductility and poor oxygen barrier properties (Martino et al., 2011). Nanocomposite films based on a PLA reinforced with organomodified MMT (O-MMT) exhibited low permeability to various substances, including water vapor, oxygen, and carbon dioxide. In this sense, the addition of 5 wt.% of O-MMT in the PLA matrix significantly improves the film barrier properties by reducing the water vapor transmission by more than 40%, oxygen by almost 40%, and of carbon dioxide by up to more than 80% (Zenkiewicz et al., 2010). Burgos et al. developed nanocomposites based on PLA reinforced with 5 wt.% O-MMT and observed a reduction of about 57% in OTR·*e* value for neat PLA from 31 cm^3 mm m^{-2} day^{-1} to 18 cm^3 mm m^{-2} day^{-1} in PLA with 5 wt.% of O-MMT (Burgos et al., 2013; Burgos, 2013). Moreover, Martino et al. reported a study on plasticized PLA by adding 15 wt.% of polyadipate Glyplast 206/7 to PLA. The addition of 5 wt.% of O-MMT to plasticized PLA produced an improvement of about 45% in oxygen barrier of nanocomposites (OTR·*e* of PLA-15% G206/7–5% C30B = 17 cm^3 mm m^{-2} day^{-1}) with respect to plasticized PLA (OTR·*e* of PLA-15% G206/7 = 31 cm^3 mm m^{-2} day^{-1}) (Martino et al., 2010). The use of essential oils for active packaging proposes frequently produces a plasticizing effect (Arrieta et al., 2013; Sánchez-González et al., 2011). In this context PLA-based nanocomposites with 5 wt.% nanoclays and 8 wt.% thymol showed 23 cm^3 mm m^{-2} day^{-1} of OTR·*e* value, near to that of the neat PLA (22 cm^3 mm m^{-2} day^{-1}) (Ramos et al., 2014, 2012). The wettability performance of PLA-O-MMT nanocomposites has been also studied and it was observed that the degree of surface hydrophobicity increased with increasing O-MMT content, reaching up to 108° of water contact angle in the case of nanocomposite with 5 wt.% O-MMT (Ocak et al., 2015).

Sanchez García et al. studied the barrier properties to oxygen, water, and limonene in several biopolymers nanocomposites, PLA, PHB, and PHBV, all of them reinforced with 5 wt.% of modified MMT were prepared by melt-compounding process. They compared their results with neat PET and PET-O-MMT nanocomposites. Comparing PET nanocomposite (PET-O-MMT 5 wt.%) with the pure sample they observed a reduction on the oxygen permeability of 55%, a reduction in D-limonene permeability of 68% and a reduction in water permeability of only 14%. The PHB nanocomposite has the lowest oxygen permeability (2.24×10^{-19} $m^3\ m\ m^{-2}\ s^{-1}\ Pa^{-1}$) value of all the biodegradable materials, closer to the PET nanocomposite ($1.69 \times 10^{-19}\ m^3\ m\ m^{-2}\ s^{-1}\ Pa^{-1}$). PHBV nanocomposite reinforced with 5 wt.% of O-MMT presents water permeability ($0.032 \times 10^{-13}\ m^3\ m\ m^{-2}\ s^{-1}\ Pa^{-1}$) and D-limonene permeability ($1.27 \times 10^{-13}\ m^3\ m\ m^{-2}\ s^{-1}\ Pa^{-1}$) similar to the neat PET (WVP = $0.03 \times 10^{-13}\ m^3\ m\ m^{-2}\ s^{-1}\ Pa^{-1}$, D-limonene permeability = $1.17 \times 10^{-13}\ m^3\ m\ m^{-2}\ s^{-1}\ Pa^{-1}$). Higher amounts of clay (10 wt.%) resulted in lower water permeability reduction ascribed to the clay content surpassing the solubility limit and, hence, resulting in detrimental agglomeration and thus lower the final performance of the nanocomposite (Sanchez-Garcia et al., 2007). As a consequence of PHB is high brittle for film for food packaging proposes, the same authors further prepare PHB blended with PCL in 80:20 proportion, respectively, and loaded with 1 wt.% and 4 wt.% of layered phyllosilicates based on commercial organomodified kaolinite and MMT (Sanchez-Garcia et al., 2008). PCL is a semicrystalline polymer, which has the melting temperature around 60°C and the glass transition temperature well below room temperature, thus at room temperature it behaves as a tough plastic (Peponi et al., 2012; Navarro-Baena et al., 2014b). Although PCL acts as a plasticizing agent for PHB promoting better dispersion of nanoplates, it is expected that it will reduce the oxygen barrier properties since neat PCL has lower oxygen transmission rate (OTR·e = 90 $cm^3\ mm\ m^{-2}\ day^{-1}$) (Valdés García et al., 2014) than PHB (OTR·e = 12 $cm^3\ mm\ m^{-2}\ day^{-1}$) (Arrieta et al., 2014d). A good dispersion of layered clays in the PCL matrix is relatively easy to achieve due to the stronger interactions between the polymer and the nanoclays, originated from the hydrogen bonding interactions between the carbonyl group of PCL and the hydroxyl groups of nanoclays organic modifier (Ramos et al., 2012). While the addition of PCL considerable improved the dispersion of the clay in the PHB-PCL blends, the addition of 4 wt.% of clay to the PHB-PCL blend clearly maintained the good oxygen barrier properties of the neat PHB in the final nanocomposite and also outperforms the barrier properties of PET at dry and at high relative humidity conditions (Sanchez-Garcia et al., 2008).

Starch-based materials generally are hydrophilic in nature and have high water transfer rates due to clustering of water molecules in their diffusion through microcavities, which act as one of the major limitations for its use in commercial food packaging applications (Pankaj et al., 2015). In the same way, plasticized starch materials are not fully suitable for packaging applications since they are highly water sensitive with poor barrier properties (Chivrac et al., 2010). While organomodified MMT is frequently used with hydrophobic polymers, the native sodium MMT is preferred for polysaccharides such as starch and plasticized starch matrices (Chivrac et al., 2010). It has been proven that cationic starch, as MMT organomodifier, can favor the MMT exfoliation process in plasticized starch matrix (Chivrac et al., 2008). In this sense, Zeppa et al. studied starch, plasticized with glycerol, reinforced with 6 wt.% of both, natural MMT and an organically modified MMT. They observed a reduction in the oxygen permeability of 14 and 44% in the presence of organically modified MMT and natural MMT, respectively (Zeppa et al., 2009). The improvement on the oxygen barrier resulted from a high dispersion state of the nanoclays within the polymeric matrix as well as from specific interactions between glycerol and the nanoclays. These interactions lead to the formation of an additional ordered phase in the material bringing a significant contribution to both water and gas barrier property improvement (Zeppa et al., 2009). The migration behavior of nanocomposites based on potato starch and purified MMT clay was studied by Avella et al. in vegetable samples, specifically lettuce and spinach, as well as in water-like simulant for vegetables products. They observed a slight increase in silica contents in the studied vegetable samples and it was attributed to the fact that clay nanoparticles are mainly composed by this element (Avella et al., 2005). Even though, they concluded that these materials can be utilized in the food packaging sector owing to their low overall migration limit (European Commission, 2011).

4.2 Nanocellulose-Based Nanocomposites

The low oxygen barrier properties as well as the water resistance of PLA have been also improved by developing PLA based nanocomposites reinforced with CNCs. Sanchez-García et al. used two methods to disperse CNCs into PLA matrix namely, freeze-dried CNCs and CNCs obtained after solvent exchange with chloroform for the development of PLA-CNCs nanocomposites by solvent casting method. The CNCs prepared by freeze drying were able to reduce the water permeability by up to 82% and the oxygen permeability by up to 90% with only 3 wt.% of CNCs content. Meanwhile, for bionanocomposites based on PLA and 3 wt.% of CNCs prepared by solvent exchange with chloroform they obtained a reduction of 49% for the water permeability. For the PLA composites containing 5 wt.% of CNCs prepared by freeze drying, the oxygen permeability was reduced of about 90% (Sanchez-Garcia and Lagaron, 2010). Fortunati et al. studied the effect of CNCs and surface modified CNCs by means of a surfactant (s-CNCs) on PLA matrix and observed a reductions of 43 and 48% in OTR·*e* values with modified and unmodified CNCs, respectively. The improvement of oxygen barrier was related with the well dispersed CNCs and surfactant modified CNCs (s-CNCs) and also with the higher levels of crystallinity of bionanocomposites in comparison with neat PLA. Meanwhile the reduction in WVP was of 4% for PLA-CNC nanocomposite and 15% in PLA-s-CNCs (Fortunati et al., 2012b). They also reported the overall migration levels in nonpolar (ethanol at 10% v/v) and polar (isooctane) food simulants and all the bionanocomposites they studied were below the overall migration limits (60 mg kg^{-1} of food simulant) (European Commission, 2011). The barrier properties were significantly enhanced with the addition of modified cellulose compared to those bionanocomposites with pristine cellulose at the same filler content (Fortunati et al., 2012b). Arrieta et al. reported the addition of CNCs and s-CNCs to PLA-PHB blends and the OTR·*e* values of ternary bionanocomposites showed a reduction of about 50% with respect to neat PLA (Arrieta et al., 2014c), closer to the neat PHB (OTR·*e* = 12 cm^3 mm m^{-2} day^{-1}) (Arrieta et al., 2014d). The same authors further plasticized these systems with acetyl tributyl citrate (ATBC) and the quaternary bionanocomposite film of PLA-PHB-s-CNC-ATBC (OTR·*e* = 23 cm^3 mm m^{-2} day^{-1}) showed a reduction of around 24% in PLA OTR·*e* value (OTR·*e* = 30 cm^3 mm m^{-2} day^{-1}) (Arrieta et al., 2015a). The good oxygen barrier performance of PLA-PHB blends reinforced with surface modified s-CNC and their plasticized system has been directly related with the increased crystallinity (Arrieta et al., 2015a, 2014c). In fact, PHB enhances the PLA crystallization as well as it maintain plasticized PLA-PHB structural and mechanical properties, even after 10 days in a fatty food simulant that represents the worst foreseeable conditions of the intended use. (Arrieta et al., 2014a). Therefore PHB ensures that the material is durable enough for food protection. In PLA-PHB-nanocellulose composites the high amount of hydroxyl groups of nanocellulose favors hydrogen bond interactions between the carbonyl groups of polymers and the hydroxyl groups of cellulose (Arrieta et al., 2015a, 2014c). The wettability performance of PLA and PLA-PHB blends reinforced with s-CNC was also measured by means of water contact angle measurements. The PLA-PHB ($\theta°$ = 89°) blend showed significantly increased water resistance in comparison with neat PLA ($\theta°$ = 83°). The addition of functionalized s-CNC in PLA and PLA-PHB also caused somewhat increase in water contact angle values, $\theta°$ of PLA-s-CNC = 85° and $\theta°$ of PLA-PHB-s-CNC = 91°, ascribed to the presence of sulfate groups on the CNC surface, which reduce the hydroxyl groups availability to interact with water and, as a result, increase the surface hydrophobicity of the final material

(Arrieta et al., 2014c). Conversely, plasticized bionanocomposites presented lower water contact angle values with respect to their unplasticized counterparts since the plasticizing effect influence the diffusion process as a consequence of the increased polymer chain mobility (Arrieta et al., 2015a). Similarly, the addition of CNCs to plasticized PLA with D-limonene provoked a reduction in the surface wettability from 54° in PLA-limonene to 75° of water contact angle in PLA-limonene blend with 3 wt.% of CNC, due to the change of both the topography and the chemical properties of the material (Fortunati et al., 2014).

Dhar et al. developed PHB-CNC based nanocomposites and observed drastic reduction in OTR values, between 46% and 65%, upon incorporation of CNCs at very low loading, 1 wt.% and 2 wt.%. Accordingly, they observed reduction in solubility and diffusion coefficients values with increasing amount of CNCs. The solubility decrease from 3.22×10^{-6} g g^{-1} Pa in neat PHB to 18.43×10^{-6} g g^{-1} Pa in PHB loaded with 5 wt.% of CNCs (PHB-CNC 5%). Meanwhile, the diffusivity changes from 10.7×10^{-12} m^2 s^{-1} in neat PHB to 2.2×10^{-12} m^2 s^{-1} in PHB-CNC 5%. They related this increased resistance to oxygen permeation to the formation of the transient percolation network of CNCs in the polymer matrix through hydrogen bonding interactions (Dhar et al., 2015). The migration studies of these films in two food simulants, polar (ethanol at 10% v/v) and nonpolar (isooctane) food simulants, showed that overall migration is within the standards limits as per the legislations (European Commission, 2011).

Nanocomposite-based packaging film based on PHBV and functionalized CNCs with methyl ester (CNC-me) were also developed (Yu et al., 2014). Compared to neat PHBV, the water uptake and WVP values were decreased by 74 and 56% by increasing the CNC-me concentration from 1 to 20 wt.%. The increased CNC-me produced more perfect crystalline structure resulting in higher tortuosity of the transport path and prevent water and water vapor molecules permeating the material (Yu et al., 2014). The nanocomposites also possess lower overall migration levels in food simulants, particularly in ethanol 10% v/v and isooctane, than the migration limits for food contact materials (European Commission, 2011). Attributed to the good interfacial adhesion or the improved interaction between CNC-me and the polymer matrix (Yu et al., 2014).

The penetration and diffusion of water into the bulk of the starch-nanocellulose nanocomposites was also studied and it was demonstrated that the movement of the starch chains is hindered due to the presence of nanocellulose, greatly reducing the water uptake (Arun et al., 2012). Cano et al. developed pea starch based materials blended with poly(vinyl alcohol) (PVA) and reinforced the starch-PVA materials with 5 wt.% of CNCs. Slight reduction of WVP values were produced due to CNCs presence, while the nanocomposites also showed lower overall migration levels in food simulants than the migration limits established (ethanol 10% v/v and isooctane) (Cano et al., 2015). El Miri et al. developed blends of carboxymethyl cellulose with starch reinforced with CNCs using the solution casting method. The WVP of bionanocomposite films with 5 wt.% of CNCs decreased about 13% with respect to the nonreinforced blend, might be due to the interaction between both biopolymers and CNCs via hydrogen bonding, which leads to the decrease of the number of free hydroxyl groups that give the hydrophilic character to biopolymers, and thus decreases WVP (El Miri et al., 2015).

4.3 Zinc Oxide Nanoparticle-Based Nanocomposites

Currently, metal oxide nanoparticles, including ZnO, have special interest in the food packaging field due to their low cost, easy availability as well as their antibacterial activity (Pantani et al., 2013). It is certainly of interest to develop PLA with antibacterial properties, thus the addition of ZnO nanoparticles has been proposed. Unfortunately, the incorporation of untreated ZnO nanoparticles into PLA matrix at melt-processing temperature results to severe degradation of the PLA matrix, attributed to *trans*-esterification reactions and unzipping depolymerization of PLA, which finally leads in a reduction of the thermo-mechanical properties (Raquez et al., 2013; Pantani et al., 2013). When ZnO are adequately surface-treated by means of silanes PLA-ZnO nanocomposites with quite good preservation of PLA-intrinsic molecular parameters as well as related physicochemical characteristics can be obtained (Pantani et al., 2013; Murariu et al., 2015). PLA-ZnO nanocomposite films were produced by melt-compounding PLA with 1 and 3 wt.% of ZnO rod-like nanoparticles by Pantani et al. and the water vapor permeability did not show significant variation at constant temperature. Whereas, the temperature increases produced an increase of the water vapor permeability (Pantani et al., 2013).

Díez-Pascual and Díez-Vicente prepared PHB/Zn bionanocomposites by solvent casting method. The nanocomposites with 5 wt.% of ZnO exhibited improved oxygen permeability (53%), reduced water uptake (66%), and water vapor permeability (38%) with respect to neat PHB. They further performed overall migration tests in nonpolar (isooctane) and polar (ethanol at 10% v/v) food simulants and the migration levels of PHB/ZnO nanocomposites decreased with increasing nanoparticle content (Díez-Pascual and Díez-Vicente, 2014), while they were below the current legislative limits for food packaging materials (European Commission, 2011).

Starch-based bionanocomposites plasticized with glycerol and sorbitol and filled with ZnO have been also reported. Increasing the level of ZnO the moisture content showed somewhat decrease, from 12% (at 58% RH) in neat starch to 10% in the nanocomposite with 5 wt.% of ZnO and ascribed this reduction to the interaction among plasticizers, biopolymer matrix, and zinc oxide nanoparticles that reduced the hydroxyl group availability to interact with water. As well, the oxygen permeability coefficient decreased by about 40% by the addition of 5 wt.% of ZnO (Nafchi et al., 2013). Similarly, Ma et al. developed ZnO nanoparticles stabilized by soluble starch as filler in a glycerol plasticized-pea starch matrix. They observed that the water vapor permeability decreased from 4.76×10^{-10} g m^{-1} s^{-1} Pa^{-1} in plasticized starch to 2.18×10^{-10} g m^{-1} s^{-1} Pa^{-1} with 4 wt.% of ZnO nanoparticles (Ma et al., 2009). The nano-ZnO based materials also showed improved absorbance of UV radiation in comparison with neat plasticized-pea starch (Ghanbarzadeh et al., 2014).

It should be highlighted that ZnO based nanocomposites have shown antibacterial effect against Gram-positive and Gram-negative indicator bacteria, *S. aureus* and *Escherichia coli* (Pantani et al., 2013; Díez-Pascual and Díez-Vicente, 2014), respectively, as well as to pathogen bacteria with particular interest in the meat food industry, such as, *E. coli* O157:H7 (Nafchi et al., 2013).

5 Conclusions

The biobased and biodegradable polymers open new windows in the food packaging industry for becoming independent from petrochemical-based polymers and also reducing the amount of nondegradable plastic waste in landfill generated after their useful life and, thus, reducing the environmental damage. However, they have lower barrier performance with respect to petrochemical traditional plastics currently used in the food packaging industry.

Nanotechnology with the ability to control the size of functional materials is a promising tool for the development of innovative biobased and biodegradable food packaging systems with improved barrier properties. In this context, to develop packaging materials with higher barrier performance, the incorporation of nanofillers is a promising strategy. New bionanocomposites have been recently developed with improved barrier properties for their application in the food packaging sector. This review aimed at presenting the current state of the art of bionanocomposites mainly those based on PLA, PHB, PHBV, and starch, as affected by different geometries nanolayers, nanofibrous, and isodimensional nanoparticles; focusing the attention on organomodified MMT as 1D nanolayer, CNCs as 2D nanofibrous and zinc oxide as 3D nanoparticles.

The nanoparticles are also able to tune the surface topography that will affect the wettability performance of the final nanocomposites. Organomodified MMT and surface modified cellulose nanocrystals show the best effectiveness in improving the oxygen and moisture barrier properties. Meanwhile, zinc oxide also shows an enhancement barrier properties effect, offering the additional antibacterial advantage for active food packaging applications.

According to its renewability, biodegradability and ability for being tailored to improve their barrier properties against gas, moisture, and aroma compounds, as well as to reduce the migration of some packaging components (ie, additives, oligomers) by means of the addition of natural based nanofillers, it is expected that the traditional petrochemical-based food packaging materials will be progressively replaced by biobased and biodegradable nanocomposites.

Acknowledgments

We are indebted to the Spanish Ministry of Economy and Competitiveness (MINECO) (MAT2013-48059-C2-1-R and MAT2014-55778-REDT) and to the Regional Government of Madrid (S2013/MIT-2862) for their economic support.

MPA and LP thank also to MINECO for the "Juan de La Cierva" (FJCI-2014-20630) and the "Ramón y Cajal" (RYC-2014-15595) contracts, respectively.

References

Alexandre, M., Dubois, P., 2000. Polymer-layered silicate nanocomposites: preparation, properties and uses of a new class of materials. Mater. Sci. Eng. R 28, 1–63.

Aranguren, M.I., Marcovich, N.E., Salgueiro, W., Somoza, A., 2013. Effect of the nano-cellulose content on the properties of reinforced polyurethanes: a study using mechanical tests and positron anihilation spectroscopy. Polym. Test. 32, 115–122.

Armentano, I., Bitinis, N., Fortunati, E., Mattioli, S., Rescignano, N., Verdejo, R., Lopez-Manchado, M.A., Kenny, J.M., 2013. Multifunctional nanostructured PLA materials for packaging and tissue engineering. Prog. Polym. Sci. 38, 1720–1747.

Arrieta, M.P. 2014. Films de PLA y PLA-PHB plastificados para su aplicación en envases de alimentos. Caracterización y análisis de los procesos de degradación. PhD Thesis, Universitat Politècnica de València.

Arrieta, M.P., López, J., Ferrándiz, S., Peltzer, M.A., 2013. Characterization of PLA-limonene blends for food packaging applications. Polym. Test. 32, 760–768.

Arrieta, M.P., Castro-López, M.D.M., Rayón, E., Barral-Losada, L.F., López-Vilariño, J.M., López, J., González-Rodríguez, M.V., 2014a. Plasticized poly(lactic acid)-poly(hydroxybutyrate) (PLA-PHB) blends incorporated with catechin intended for active food-packaging applications. J. Agr. Food Chem. 62, 10170–10180.
Arrieta, M.P., Fortunati, E., Dominici, F., Rayón, E., López, J., Kenny, J.M., 2014b. Multifunctional PLA-PHB/cellulose nanocrystal films: processing, structural and thermal properties. Carbohyd. Polym. 107, 16–24.
Arrieta, M.P., Fortunati, E., Dominici, F., Rayón, E., López, J., Kenny, J.M., 2014c. PLA-PHB/cellulose based films: mechanical, barrier and disintegration properties. Polym. Degrad. Stabil. 107, 139–149.
Arrieta, M.P., López, J., Hernández, A., Rayón, E., 2014d. Ternary PLA-PHB-Limonene blends intended for biodegradable food packaging applications. Eur. Polym. J. 50, 255–270.
Arrieta, M.P., López, J., Rayón, E., Jiménez, A., 2014e. Disintegrability under composting conditions of plasticized PLA-PHB blends. Polym. Degrad. Stabil. 108, 307–318.
Arrieta, M.P., Peltzer, M.A., López, J., Garrigós, M.D.C., Valente, A.J.M., Jiménez, A., 2014f. Functional properties of sodium and calcium caseinate antimicrobial active films containing carvacrol. J. Food Eng. 121, 94–101.
Arrieta, M.P., Samper, M.D., López, J., Jiménez, A., 2014g. Combined effect of poly(hydroxybutyrate) and plasticizers on polylactic acid properties for film intended for food packaging. J. Polym. Environ. 22, 460–470.
Arrieta, M.P., Fortunati, E., Dominici, F., López, J., Kenny, J.M., 2015a. Bionanocomposite films based on plasticized PLA-PHB/cellulose nanocrystal blends. Carbohyd. Polym.l 121, 265–275.
Arrieta, M.P., López, J., López, D., Kenny, J.M., Peponi, L., 2015b. Development of flexible materials based on plasticized electrospun PLA-PHB blends: structural, thermal, mechanical and disintegration properties. Eur. Polym. J. 73, 433–446.
Arun, S., Kumar, K.A.A., Sreekala, M.S., 2012. Fully biodegradable potato starch composites: effect of macro and nano fiber reinforcement on mechanical, thermal and water-sorption characteristics. Int. J. Plast. Technol. 16, 50–66.
Auras, R.A., 2007. Solubility of gases and vapors in polylactide polymers. Elsevier Science Ltd, United Kingdom.
Avella, M., De Vlieger, J.J., Errico, M.E., Fischer, S., Vacca, P., Volpe, M.G., 2005. Biodegradable starch/clay nanocomposite films for food packaging applications. Food Chem. 93, 467–474.
Avérous, L., 2004. Biodegradable multiphase systems based on plasticized starch: a review. J. Macromol. Sci.—Polym. Rev. 44, 231–274.
Azeredo, H.M.C.D., 2009. Nanocomposites for food packaging applications. Food Res. Int. 42, 1240–1253.
Balart, J., Fombuena, V., Boronat, T., Reig, M.J., Balart, R., 2012. Surface modification of polypropylene substrates by UV photografting of methyl methacrylate (MMA) for improved surface wettability. J. Mater. Sci. 47, 2375–2383.
Berthet, M.A., Angellier-Coussy, H., Guillard, V., Gontard, N., 2015. Vegetal fiber-based biocomposites: which stakes for food packaging applications? J. Appl. Polym. Sci. 133 (2), 42528.
Bhunia, K., Sablani, S.S., Tang, J., Rasco, B., 2013. Migration of chemical compounds from packaging polymers during microwave, conventional heat treatment, and storage. Comp. Rev. Food Sci. Food Saf. 12, 523–545.
Bittmann, B., Bouza, R., Barral, L., Diez, J., Ramirez, C., 2013. Poly(3-hydroxybutyrate-co-3-hydroxyvalerate)/clay nanocomposites for replacement of mineral oil based materials. Polym. Compos. 34, 1033–1040.

Bumbudsanpharoke, N., Ko, S., 2015. Nano-food packaging: an overview of market, migration research, and safety regulations. J. Food Sci. 80, R910–R923.
Burgos, N. 2013. Desarrollo de bionanocompuestos en base a poli(ácido láctico) y plastificantes de alta compatibilidad para el envasado de alimentos. PhD thesis, University of Alicante.
Burgos, N., Martino, V.P., Jiménez, A., 2013. Characterization and ageing study of poly(lactic acid) films plasticized with oligomeric lactic acid. Polym. Degrad. Stabil. 98, 651–658.
Cano, A., Fortunati, E., Cháfer, M., González-Martínez, C., Chiralt, A., Kenny, J.M., 2015. Effect of cellulose nanocrystals on the properties of pea starch-poly(vinyl alcohol) blend films. J. Mater. Sci. 50, 6979–6992.
Catalá, R., López Carballo, G., Hernández Muñoz, P., Gavara, R., 2015. PLA and active packaging. In: Jiménez, A., Peltzer, M.A., Ruseckaite, R.A. (Eds.), Poly(lactic acid) Science and Technology: Processing, Properties, Additives and Applications. RSC.
Chivrac, F., Pollet, E., Schmutz, M., Avérous, L., 2008. New approach to elaborate exfoliated starch-based nanobiocomposites. Biomacromolecules 9, 896–900.
Chivrac, F., Pollet, E., Avérous, L., 2009. Progress in nano-biocomposites based on polysaccharides and nanoclays. Mater. Sci. Eng. R. 67, 1–17.
Chivrac, F., Angellier-Coussy, H., Guillard, V., Pollet, E., Avérous, L., 2010. How does water diffuse in starch/montmorillonite nano-biocomposite materials? Carbohyd. Polym. 82, 128–135.
Dhar, P., Bhardwaj, U., Kumar, A., Katiyar, V., 2015. Poly (3-hydroxybutyrate)/cellulose nanocrystal films for food packaging applications: barrier and migration studies. Polym. Eng. Sci. 55 (10), 2388–2395.
Díez-Pascual, A., Díez-Vicente, A., 2014. Poly(3-hydroxybutyrate)/ZnO bionanocomposites with improved mechanical, barrier and antibacterial properties. Int. J. Mol. Sci. 15, 10950.
El Miri, N., Abdelouahdi, K., Barakat, A., Zahouily, M., Fihri, A., Solhy, A., El Achaby, M., 2015. Bio-nanocomposite films reinforced with cellulose nanocrystals: rheology of film-forming solutions, transparency, water vapor barrier and tensile properties of films. Carbohyd. Polym. 129, 156–167.
European Commission, 2011. No. 10/2011 of 14 January 2011 on plastic materials and articles intended to come into contact with food. Off. J. Eur. Comm., 1–89.
Fombuena, V., Balart, J., Boronat, T., Sánchez-Nácher, L., Garcia-Sanoguera, D., 2013. Improving mechanical performance of thermoplastic adhesion joints by atmospheric plasma. Mater. Des. 47, 49–56.
Fortunati, E., Armentano, I., Zhou, Q., Iannoni, A., Saino, E., Visai, L., Berglund, L.A., Kenny, J.M., 2012a. Multifunctional bionanocomposite films of poly(lactic acid), cellulose nanocrystals and silver nanoparticles. Carbohyd. Polym. 87, 1596–1605.
Fortunati, E., Peltzer, M., Armentano, I., Torre, L., Jiménez, A., Kenny, J.M., 2012b. Effects of modified cellulose nanocrystals on the barrier and migration properties of PLA nano-biocomposites. Carbohyd. Polym. 90, 948–956.
Fortunati, E., Puglia, D., Monti, M., Peponi, L., Santulli, C., Kenny, J.M., Torre, L., 2013. Extraction of cellulose nanocrystals from *Phormium tenax* fibres. J. Polym. Environ. 21, 319–328.
Fortunati, E., Luzi, F., Puglia, D., Dominici, F., Santulli, C., Kenny, J.M., Torre, L., 2014. Investigation of thermo-mechanical, chemical and degradative properties of PLA-limonene films reinforced with cellulose nanocrystals extracted from *Phormium tenax* leaves. Eur. Polym. J. 56, 77–91.
Galic, K., Curic, D., Gabric, D., 2009. Shelf life of packaged bakery goods: a review. Crit. Rev. Food Sci. Nutr. 49, 405–426.
Garcia, D., Balart, R., Sánchez, L., López, J., 2007. Compatibility of recycled PVC/ABS blends: effect of previous degradation. Polym. Eng. Sci. 47, 789–796.

Ghanbarzadeh, B., Oleyaei, S.A., Almasi, H., 2014. Nanostructured materials utilized in biopolymer-based plastics for food packaging applications. Crit. Rev. Food Sci. Nutr. 55, 1699–1723.
Gokhale, A.A., Lee, I., 2014. Recent advances in the fabrication of nanostructured barrier films. J. Nanosci. Nanotechnol. 14, 2157–2177.
Gómez-Estaca, J., López-De-Dicastillo, C., Hernández-Muñoz, P., Catalá, R., Gavara, R., 2014. Advances in antioxidant active food packaging. Trends Food Sci. Technol. 35, 42–51.
Gorrasi, G., Tammaro, L., Vittoria, V., Paul, M.A., Alexandre, M., Dubois, P., 2004. Transport properties of water vapor in polylactide/montmorillonite nanocomposites. J. Macromol. Sci. Phys. 43 B, 565–575.
Guilbert, S., Cuq, B., Gontard, N., 1997. Recent innovations in edible and/or biodegradable packaging materials. Food Addit. Contam. 14, 741–751.
Gutierrez, J., Tercjak, A., Garcia, I., Peponi, L., Mondragon, I., 2008. Hybrid titanium dioxide/PS-b-PEO block copolymer nanocomposites based on sol-gel synthesis. Nanotechnology, 19.
Hambleton, A., Fabra, M.-J., Debeaufort, F., Dury-Brun, C., Voilley, A., 2009. Interface and aroma barrier properties of iota-carrageenan emulsion-based films used for encapsulation of active food compounds. J. Food Eng. 93, 80–88.
Han, J.H., Scanlon, M.G., 2005. Mass transfer of gas and solute through packaging materials. Innov. Food Pack. 2, 37–49.
Introzzi, L., Fuentes-Alventosa, J.M., Cozzolino, C.A., Trabattoni, S., Tavazzi, S., Bianchi, C.L., Schiraldi, A., Piergiovanni, L., Farris, S., 2012. Wetting enhancer pullulan coating for antifog packaging applications. ACS Appl. Mater. Interf. 4, 3692–3700.
Jamshidian, M., Tehrany, E.A., Imran, M., Jacquot, M., Desobry, S., 2010. Poly-lactic acid: production, applications, nanocomposites, and release studies. Comp. Rev. Food Sci. Food Saf. 9, 552–571.
Jonoobi, M., Mathew, A., Abdi, M., Makinejad, M., Oksman, K., 2012. A comparison of modified and unmodified cellulose nanofiber reinforced polylactic acid (PLA) prepared by twin screw extrusion. J. Polym. Environ. 20, 991–997.
Kadam, A.A., Karbowiak, T., Voilley, A., Debeaufort, F., 2015. Techniques to measure sorption and migration between small molecules and packaging. J. Sci. Food Agr. 95, 1395–1407.
Lagaron, J.M., Lopez-Rubio, A., 2011. Nanotechnology for bioplastics: opportunities, challenges and strategies. Trends Food Sci. Technol. 22, 611–617.
Lavoine, N., Desloges, I., Dufresne, A., Bras, J., 2012. Microfibrillated cellulose—its barrier properties and applications in cellulosic materials: a review. Carbohyd. Polym. 90, 735–764.
Ma, X., Chang, P.R., Yang, J., Yu, J., 2009. Preparation and properties of glycerol plasticized-pea starch/zinc oxide-starch bionanocomposites. Carbohyd. Polym. 75, 472–478.
Martino, V.P., Jiménez, A., Ruseckaite, R.A., 2009a. Processing and characterization of poly(lactic acid) films plasticized with commercial adipates. J. Appl. Polym. Sci. 112, 2010–2018.
Martino, V.P., Ruseckaite, R.A., Jiménez, A., 2009b. Ageing of poly(lactic acid) films plasticized with commercial polyadipates. Polym. Int. 58, 437–444.
Martino, V.P., Ruseckaite, R.A., Jiménez, A., Averous, L., 2010. Correlation between composition, structure and properties of poly(lactic acid)/polyadipate-based nano-biocomposites. Macromol. Mater. Eng. 295, 551–558.
Martino, V.P., Jiménez, A., Ruseckaite, R.A., Avérous, L., 2011. Structure and properties of clay nano-biocomposites based on poly(lactic acid) plasticized with polyadipates. Polym. Adv. Technol. 22, 2206–2213.

Mitrano, D.M., Motellier, S., Clavaguera, S., Nowack, B., 2015. Review of nanomaterial aging and transformations through the life cycle of nano-enhanced products. Environ. Int. 77, 132–147.
Mukhopadhyay, S.S., 2014. Nanotechnology in agriculture: prospects and constraints. Nano., Sci. Appl. 7, 63–71.
Murariu, M., Paint, Y., Murariu, O., Raquez, J.-M., Bonnaud, L., Dubois, P., 2015. Current progress in the production of PLA-ZnO nanocomposites: beneficial effects of chain extender addition on key properties. J. Appl. Polym. Sci. 132 (48), 42480.
Nafchi, A.M., Nassiri, R., Sheibani, S., Ariffin, F., Karim, A.A., 2013. Preparation and characterization of bionanocomposite films filled with nanorod-rich zinc oxide. Carbohyd. Polym. 96, 233–239.
Navarro, R., Ferrándiz, S., López, J., Seguí, V.J., 2008. The influence of polyethylene in the mechanical recycling of polyethylene terephtalate. J. Mater. Process. Technol. 195, 110–116.
Navarro-Baena, I., Kenny, J.M., Peponi, L., 2014a. Thermally-activated shape memory behaviour of bionanocomposites reinforced with cellulose nanocrystals. Cellulose 21, 4231–4246.
Navarro-Baena, I., Marcos-Fernández, A., Fernández-Torres, A., Kenny, J.M., Peponi, L., 2014b. Synthesis of PLLA-b-PCL-b-PLLA linear tri-block copolymers and their corresponding poly(ester-urethane)s: effect of the molecular weight on their crystallisation and mechanical properties. RSC Adv. 4, 8510–8524.
Ocak, Y., Sofuoglu, A., Tihminlioglu, F., Böke, H., 2015. Sustainable bio-nano composite coatings for the protection of marble surfaces. J. Cult. Herit. 16, 299–306.
Oksman, K., Mathew, A.P., Bondeson, D., Kvien, I., 2006. Manufacturing process of cellulose whiskers/polylactic acid nanocomposites. Compos. Sci. Technol. 66, 2776–2784.
Pankaj, S.K., Bueno-Ferrer, C., Misra, N.N., O'Neill, L., Jiménez, A., Bourke, P., Cullen, P.J., 2014a. Characterization of polylactic acid films for food packaging as affected by dielectric barrier discharge atmospheric plasma. Innov. Food Sci. Emerg. Technol. 21, 107–113.
Pankaj, S.K., Bueno-Ferrer, C., Misra, N.N., O'Neill, L., Tiwari, B.K., Bourke, P., Cullen, P.J., 2014b. Physicochemical characterization of plasma-treated sodium caseinate film. Food Res. Int. 66, 438–444.
Pankaj, S.K., Bueno-Ferrer, C., Misra, N.N., O'Neill, L., Tiwari, B.K., Bourke, P., Cullen, P.J., 2015. Dielectric barrier discharge atmospheric air plasma treatment of high amylose corn starch films. LWT - Food Sci. Technol. 63, 1076–1082.
Pantani, R., Gorrasi, G., Vigliotta, G., Murariu, M., Dubois, P., 2013. PLA-ZnO nanocomposite films: water vapor barrier properties and specific end-use characteristics. Eur. Polym. J. 49, 3471–3482.
Parres, F., Balart, R., López, J., García, D., 2008. Changes in the mechanical and thermal properties of high impact polystyrene (HIPS) in the presence of low polypropylene (PP) contents. J. Mater. Sci. 43, 3203–3209.
Peltzer, M., Pei, A., Zhou, Q., Berglund, L., Jiménez, A., 2014. Surface modification of cellulose nanocrystals by grafting with poly(lactic acid). Polym. Int. 63, 1056–1062.
Peponi, L., Tercjak, A., Gutierrez, J., Stadler, H., Torre, L., Kenny, J.M., Mondragon, I., 2008. Self-assembling of SBS block copolymers as templates for conductive silver nanocomposites. Macromol. Mater. Eng. 293, 568–573.
Peponi, L., Tercjak, A., Martin, L., Mondragon, I., Kenny, J.M., 2011. Morphology-properties relationship on nanocomposite films based on poly(styrene-block-diene-block-styrene) copolymers and silver nanoparticles. Exp. Polym. Lett. 5, 104–118.

Peponi, L., Navarro-Baena, I., Báez, J.E., Kenny, J.M., Marcos-Fernández, A., 2012. Effect of the molecular weight on the crystallinity of PCL-b-PLLA di-block copolymers. Polymer 53, 4561–4568.
Peponi, L., Puglia, D., Torre, L., Valentini, L., Kenny, J.M., 2014. Processing of nanostructured polymers and advanced polymeric based nanocomposites. Mater. Sci. Eng. R. 85, 1–46.
Plastics Europe, 2015. Plastics—the Facts 2014/2015. An analysis of European plastics production, demand and waste data.
Ramos, M., Arrieta, M.P., Beltran, A., Garrigós, M.C., 2012. Characterization of PLA, PCL and sodium caseinate active bio-films for food packaging applications. In: Kontominas M.G. (Ed.), Food Packaging: Procedures, Management and Trends. Nova Science Publishers, Inc.
Ramos, M., Jiménez, A., Peltzer, M., Garrigós, M.C., 2014. Development of novel nano-biocomposite antioxidant films based on poly (lactic acid) and thymol for active packaging. Food Chem. 162, 149–155.
Raquez, J.M., Habibi, Y., Murariu, M., Dubois, P., 2013. Polylactide (PLA)-based nanocomposites. Prog. Polym. Sci. 38, 1504–1542.
Rayón, E., López, J., Arrieta, M.P., 2013. Mechanical characterization of microlaminar structures extracted from cellulosic materials using nanoindentation technique. Cellul. Chem. Technol. 47, 345–351.
Rayón, E., Ferrandiz, S., Rico, M.I., López, J., Arrieta, M.P., 2015. Microstructure, mechanical, and thermogravimetric characterization of cellulosic by-products obtained from biomass seeds. Int. J. Food Prop. 18, 1211–1222.
Reddy, M.M., Vivekanandhan, S., Misra, M., Bhatia, S.K., Mohanty, A.K., 2013. Biobased plastics and bionanocomposites: current status and future opportunities. Prog. Polym. Sci. 38, 1653–1689.
Revert, A., Reig, M., Seguí, V.J., Boronat, T., Fombuena, V., Balart, R., 2015. Upgrading brewer's spent grain as functional filler in polypropylene matrix. Polym. Compos..
Samper, M.D., Garcia-Sanoguera, D., Parres, F., López, J., 2010. Recycling of expanded polystyrene from packaging. Prog. Rubber, Plast. Recycl. Technol. 26, 83–92.
Samper, M.D., Arrieta, M.P., Ferrándiz, S., López, J., 2014. Influence of biodegradable materials in the recycled polystyrene. J. Appl. Polym. Sci., 131.
Sanchez-Garcia, M., Lagaron, J., 2010. On the use of plant cellulose nanowhiskers to enhance the barrier properties of polylactic acid. Cellulose 17, 987–1004.
Sanchez-Garcia, M.D., Gimenez, E., Lagaron, J.M., 2007. Novel PET nanocomposites of interest in food packaging applications and comparative barrier performance with biopolyester nanocomposites. J. Plast. Film Sheet. 23, 133–148.
Sanchez-Garcia, M.D., Gimenez, E., Lagaron, J.M., 2008. Morphology and barrier properties of nanobiocomposites of poly(3-hydroxybutyrate) and layered silicates. J. Appl. Polym. Sci. 108, 2787–2801.
Sánchez-González, L., Chiralt, A., González-Martínez, C., Cháfer, M., 2011. Effect of essential oils on properties of film forming emulsions and films based on hydroxypropylmethylcellulose and chitosan. J. Food Eng. 105, 246–253.
Scheibe, A.S., De Moraes, J.O., Laurindo, J.B., 2014. Production and characterization of bags from biocomposite films of starch-vegetal fibers prepared by tape casting. J. Food Proc. Eng. 37, 482–492.
Singh, G., Stephan, C., Westerhoff, P., Carlander, D., Duncan, T.V., 2014. Measurement methods to detect, characterize, and quantify engineered nanomaterials in foods. Comp. Rev. Food Sci. Food Saf. 13, 693–704.
Siracusa, V., Rocculi, P., Romani, S., Rosa, M.D., 2008. Biodegradable polymers for food packaging: a review. Trends Food Sci. Technol. 19, 634–643.

Sonseca, Á., Camarero-Espinosa, S., Peponi, L., Weder, C., Foster, E.J., Kenny, J.M., Giménez, E., 2014. Mechanical and shape-memory properties of poly(mannitol sebacate)/cellulose nanocrystal nanocomposites. J. Polym. Sci. A 52, 3123–3133.

Torres, F.G., Troncoso, O.P., Lopez, D., Grande, C., Gomez, C.M., 2009. Reversible stress softening and stress recovery of cellulose networks. Soft Matt. 5, 4185–4190.

Valdés García, A., Ramos Santonja, M., Sanahuja, A.B., Selva, M.D.C.G., 2014. Characterization and degradation characteristics of poly(e-caprolactone)-based composites reinforced with almond skin residues. Polym. Degrad. Stabil. 108, 269–279.

Vergnaud, J.M., 1998. Problems encountered for food safety with polymer packages: chemical exchange, recycling. Adv. Coll. Interf. Sci. 78, 267–297.

Vogler, E.A., 1998. Structure and reactivity of water at biomaterial surfaces. Adv. Coll. Interf. Sci. 74, 69–117.

Xavier, J.R., Babusha, S.T., George, J., Ramana, K.V., 2015. Material properties and antimicrobial activity of polyhydroxybutyrate (phb) films incorporated with vanillin. Appl. Biochem. Biotechnol. 176, 1498–1510.

Yamamoto, O., 2001. Influence of particle size on the antibacterial activity of zinc oxide. Int. J. Inorgan. Mater. 3, 643–646.

Yu, H., Yan, C., Yao, J., 2014. Fully biodegradable food packaging materials based on functionalized cellulose nanocrystals/poly(3-hydroxybutyrate-co-3-hydroxyvalerate) nanocomposites. RSC Adv. 4, 59792–59802.

Zenkiewicz, M., Richert, J., Rózanski, A., 2010. Effect of blow moulding ratio on barrier properties of polylactide nanocomposite films. Polym. Test. 29, 251–257.

Zeppa, C., Gouanvé, F., Espuche, E., 2009. Effect of a plasticizer on the structure of biodegradable starch/clay nanocomposites: thermal, water-sorption, and oxygen-barrier properties. J. Appl. Polym. Sci. 112, 2044–2056.

13

BIOPOLYMER/CLAY NANOCOMPOSITES AS THE HIGH BARRIER PACKAGING MATERIAL: RECENT ADVANCES

Kateryna Fatyeyeva, Corinne Chappey, Stéphane Marais

Normandie University, France; University of Rouen, Polymères, Biopolymères, Surfaces, Mont Saint Aignan, France; UMR 6270, Mont Saint Aignan, France

1 Introduction

For the past few decades, polymers have emerged as the attractive engineering materials for the different applications, such as storage tanks, toys, luggage, pipes and tubings, textile and furniture, structural sections, packaging, etc. They have replaced traditional materials (metals, glass, paper, wood) in the packaging application due to their multifunctionality, light weight, ease of processing, and low cost. In the case of the food packaging application, the synthetic polymers provide mechanical, chemical, and microbial protection from the environment and allow seeing the product. The most frequently used polymers are polyethylene terephthalate (PET), high- and low-density polyethylene, polyvinyl chloride, polypropylene, and polystyrene. Each food packaging plastic has its unique properties and, thus, is used in a certain way. PET has very good tensile strength properties and is optically transparent (Dombre et al., 2014). Besides, due to its high resistance to heat, solvents, and acids, it is generally used for the elaboration of cold mineral water bottles. For the fabrication of plastic bags and milk bottles high-density polyethylene used as a firm material is needed, but the transparency is not required. On the contrary, low-density polyethylene is used for trays and different kinds of containers because of its very low cost and a large stretch capacity. However, despite their enormous versatility, a limiting property of polymers in the food packaging is their inherent permeability to

Food Packaging. http://dx.doi.org/10.1016/B978-0-12-804302-8.00013-3

gases and vapors, including oxygen, carbon dioxide (CO_2), and organic vapors. Besides, the migration of different compounds from the polymer film into the food invokes a lot of discussions (Da Silva et al., 2011; Marcilla et al., 2008). For example, plasticizers are an important class of low molecular weight nonvolatile compounds that are widely used in polymer industry as additives (Da Silva et al., 2011). However, since the early 1980s, there has appeared profound concern, and even controversy, regarding the use of phthalates and their effects on human health and the environment. Thus, the use of plasticizers is being questioned due to their possible toxicity problems related to their migration (Marcilla et al., 2008).

Taking all the foregoing into account, increasing ecological concerns toward a reduction of CO_2 emissions during synthetic polymer production from the fossil fuels contribute to the developing of the biodegradable polymer industry. During past decade in order to decrease the environmental impact, the attention of scientists is focused on the new biodegradable materials used as food packaging (Sorrentino et al., 2007; Reddy et al., 2013; Rekha et al., 2015). At the same time, consumers demand high-quality food with the long shelf life and also they desire through the packing to see and to perceive the foodstuff quality (Rhim et al., 2013). Thus, food packaging is required not only to protect food products but also by being transparent enough to allow seeing the food through the packaging. However, biopolymer-based packaging materials are not widely used in the packaging industry, mainly because of their poor mechanical, barrier, and processing properties as well as higher production cost as compared with conventional plastic films (Othman, 2014). One of the methods to overcome such limitation of biopolymer-based packaging materials is elaborating a hybrid with other materials. Recently, biopolymer/clay nanocomposites have attracted attention as an alternative to conventional filled polymers, because of their ability for nanoscale dispersion, which brings great improvement in mechanical and physical properties compared to microscale polymer composites. This is mainly due to the strong interfacial interactions between the filler (with large specific surface area) and polymer matrix and to the formation of tortuous pathway of gas diffusion caused by the impermeable nanofillers. Several studies have reported the amelioration of mechanical properties (Bitinis et al., 2013; Youssef et al., 2015; Mallakpour and Moslemi, 2012), thermal stability (Bitinis et al., 2013; Arrieta et al., 2015), barrier properties (Crétois et al., 2014; Giannakas et al., 2014), and water solubility (Rhim et al., 2013; Crétois et al., 2014) of biopolymer films *via* the introduction of clay into the polymer matrix in the range of 1–5 wt.%. Besides, the clay introduction to biopolymers may not only improve the biopolymers mechanical and barrier properties, but also offer other functions and applications in food packaging such as antimicrobial agent, biosensors, oxygen scavengers (Rhim et al., 2013). The nanocomposite can be an active food packaging, that is the food packaging can interact with food by releasing beneficial compounds such as antimicrobial agent, antioxidant agent, or by eliminating some unfavorable elements, such as oxygen or water vapors (Othman, 2014). The nanocomposite can also be smart food packaging, which means it detects such properties of the packaged food as microbial contamination or expiry date and uses some mechanisms to register and convey information about the food quality or safety (Baeumner, 2004; Azeredo, 2009). Scientists in big name companies including Kraft, Bayer, and Kodak, as well as numerous universities and smaller companies, are developing a range of smart packaging materials that will absorb oxygen, detect food pathogens, and alert consumers of spoiled food. The smart packages, which will be able to detect public health pathogens such as *Salmonella* and *Escherichia coli*, are expected to be available within the next few years. According to industry analysts, the current US market for "active, controlled and smart" packaging for foods and beverages is estimated at $38 billion and will surpass $54 billion by 2016 (Rhim et al., 2013).

Therefore, the development of nanocomposite materials based on biopolymer and clay for food packaging is important not only to reduce the environmental problems, but also to improve the functions of the food packaging materials.

Modified clays have been the major fillers used in achieving enhancement of the barrier properties of polymer materials, owing to their rich surface chemistry and the high aspect ratio of the individual layers. However, it is critical to fully disperse the clay into the polymer matrix to realize this high aspect ratio. Different technologies have been developed to introduce clays into the polymer *via* in situ polymerization and melt compounding. Besides, for the packaging application, especially for food packaging, it is also important to have a regulatory approval for commercialization.

This chapter gives an overview of the main biopolymer/clay materials used for food packaging. Special attention is paid to the barrier properties of such nanocomposites toward oxygen and other gases.

2 Regulations

Food packaging is one of the most attractive opportunities of the nanotechnology application in the food sector. Between 400 and 500 nanopackaging products are estimated to be on the market now, while 40% of all business in the food industry in 2015 will be working with nanotechnology (LGL Bayern, 2012). The main purpose of nanopackaging is to set a longer shelf life by improving the barrier properties of food packaging to reduce gas and moisture exchange and ultraviolet (UV) light exposure (Sorrentino et al., 2007).

By 2003 more than 90% of the nanopackaging was based on nanocomposites, in which nanomaterials were used to improve the barrier properties of plastic wrapping for foods and dairy products. Nanopackaging can also be designed to release antimicrobials, antioxidants, enzymes, and flavors in order to extend a shelf life (Rhim et al., 2013).

Within the period 2012–2015, 183 patents were published that contain the keywords "nano" and "food" in the patent title (http://wokinfo.com/). Among these patents, 47 relate to packaging or coating applications. Furthermore, 19 patents concern nanoadditives, and 10 patents cover nanotechnology applications for the detection of the compounds in food.

Therefore, there is an urgent need for a regulatory system of the food regulations capable of identifying any risks associated with the nanotechnology in the food industry. However, no legal basis for regulating the nanomaterials and the application of the nanotechnology exists in Europe and the United States. The European Union (EU) regulations for food and food packaging have required specific safety standards and testing procedures for the introduction of new nanotechnology. In the United States, the safety of the food packaging is regulated by the US Food and Drug Administration (FDA) (Badgley et al., 2007), while in Australia, nanofood additives and ingredients are regulated by Food Standards Australia and New Zealand (FSANZ) under the Food Standards Code (Bowman and Hodge, 2006). In India food safety regulations are introduced but not adequate for the monitoring safety of nanoparticles. Legislation and regulation of nanomaterials were reviewed both in EU and non-EU countries on the basis of literature research and a questionnaire. Currently, only a few EU legal acts give a nanomaterial definition: No. 1169/2011 and 10/2011 on food information to consumers and on plastic food contact materials, respectively; and No. 1223/2009 and 528/2013 on cosmetic and biocidal products, respectively. Surprisingly, the definition in the two former regulations is different from the one in the two latter definitions. In 2011, the European Commission published a recommendation of the definition of a nanomaterial (2011/696/EU). A draft law on novel food (COM/2013/0435) also provides a legal basis to regulate nanomaterials or the application of nanotechnology in food products. The review of existing legislation in non-EU countries shows that these countries have limited nanomaterial specific legislation and illegally bound definition. They have adopted a broader approach, which is built mainly on the guidance for industry (Azeredo, 2009; Rhim et al., 2013).

Existing laws are inadequate to assess risks posed by the nano-based foods and packaging because:

1. Toxicity risks remain very poorly understood (because of their unique properties).
2. They are not assessed as new chemicals according to many regulations.
3. Current exposure and safety methods are not suitable for nanomaterials.
4. Many safety assessments use confidential industry studies.

We don't know whether nanoparticles in food packaging can be transferred to food and whether this is harmful to humans. Despite these matters such materials are currently being developed.

There is a necessity to better assess the risks of food packaging toxicity. Food contact packaging is rarely inert; food and packaging interactions can lead to food contamination. How to evaluate and better manage the toxic risks for the consumer? The food contact packaging materials are a subject of the constant innovation to improve the protection and conservation of food and control the conditions of its transport and storage. In addition, such materials must allow labeling and represent an important marketing tool in the food industry. The suitability of packaging materials for the contact with food is evaluated in terms of risks to the health of consumers due to possible substance transfers between packaging and food. What are the risks?

Most substances used in the food packaging are regulated by the FDA as "food contact substances" under the "food additive" provisions of the Federal Food, Drug and Cosmetic Art. The regulatory process for ensuring the safety of the food packaging is designed to perform a case-by-case assessment before new materials enter the market. The legal standard and the battery of safety tests make it difficult for the new materials to enter the food supply without FDA approval. However, the FDA approach does not take into account the nanosize of the material, but only its chemical and biological nature. It means that if a substance has already been approved for the use as food ingredients, additives, or packaging in its bulk form, it can also be used in its nanoform. On the contrary to the FDA view on the particle size, the United Kingdom's Royal Society and Royal Academy of Engineering have recommended that the nanoparticles should be treated as new chemicals and be subject to new safety assessments prior to their commercialization until testing proves their safety. Nonetheless, the European Commission intends to establish a new and more safety-oriented approach for all food packaging production companies to their products. Some European countries also have a national register of qualified Business Operator–Food Contact Expert as the professional reference able to deal with every aspect related to the food contact materials applications and safety.

However, up to now there is no international regulation of nanotechnology or nanoproducts as well as no requirement to label foods that contain nanomaterials.

3 Biopolymers

Biopolymers as an alternative to petroleum-based polymers are of great interest both in the academic world and in industry. Limited resources of crude oil, global warming, environmental problems, and new standards are some of the factors that have led to this growing interest. Biopolymers are fully renewable, do not contribute to an accumulation of CO_2 in the atmosphere after their life cycle, and are a part of the ecosystem. A number of biopolymers are also biologically degradable, thus reducing waste management problems. That's why the development of materials from natural polymers for different applications has been a hot topic for several years and will continue to play an important role in food industry. Some authors classify the polymers according to the method of production or to their source as (Bordes et al., 2009; Tang et al., 2012):

1. polymers directly extracted or removed from vegetal or animal biomass, such as polysaccharides and proteins;
2. polymers produced by classical chemical synthesis starting from renewable biobased monomers such as poly(lactic acid) (PLA);
3. polymers produced by microorganisms, such as poly(hydroxyalkanoate)s (PHAs), cellulose, xanthan, pullulan.

Polysaccharides are natural biopolymers that can be found in nature during the growth cycles of all organisms. These polymers are often chemically modified in order to change the degradation rate and to improve the mechanical properties (Vieira et al., 2011; Giannakas et al., 2014).

Starch, a completely biodegradable polysaccharide that is a result of biosynthesis of numerous plants, is one of the most abundant renewable resources known to man. In its granular form, starch is mostly composed of linear amylose and highly branched amylopectin. Thus, starch can be considered as a crystalline material. When mixed with a limited amount of water and subjected to heat and shear, starch undergoes spontaneous destruction.

A homogeneous melt known as thermoplastic starch (TPS), which possesses thermoplastic characteristics, is formed (Zdrahala, 1997). TPS can be easily used in different thermoplastification processes with standard equipment used in manufacturing synthetic polymers. These processes include injection molding, extrusion blow molding, injection compression molding, and extrusion (Averous, 2004; Mihai et al., 2007; Nafchi et al., 2013). Starch films are stronger than other biopolymer films and mechanical properties of the starch films are better than those of polysaccharides and proteins films (Nafchi et al., 2013). Starch, because of its nature, cost, and reliability of thermoplastics, is of great interest as a packaging material. However, starch alone has some disadvantages, such as hydrophilic nature and poor mechanical properties (Li et al., 2009).

Alginate is another polysaccharide, extracted from brown algae such as *Laminaria* and *Macrocystis* (Tapia et al., 2008). Alginate is a binary copolymer and it contains carboxyl groups in each constituent residue. Its structure is composed of β-D-mannuronic acid monomer linked to α-L-guluronic acid monomer, through α 1,4-glycoside linkage (George and Abraham, 2006).

Carrageenan is one of the three dominant carrageenan species, that is, κ, ι, and λ-carrageenan, which differ in their disaccharide structures. κ-Carrageenan is mainly used in food applications as a water-bonding agent (Kong and Ziegler, 2013).

Other polysaccharides, like chitin, chitosan, can also be used for the production of biobased polymers.

The film-forming abilities of some *proteins* have been known for some time. Whey proteins are by-products from the cheese industry, and consist of whey protein isolates that represent the pure form of such proteins. Whey proteins are capable to from elastic films (Wang et al., 2013), and they have been employed as a raw material for biodegradable packaging due to their good oxygen barrier and moderate moisture permeability (Ramos et al., 2012; Popović et al., 2012). Park and Chinnan report that corn zein provides a good oxygen barrier for tomatoes, resulting in doubling of their shelf life (Park and Chinnan, 1995). Barrier and mechanical properties of protein-based films are affected by various factors such as concentration and selection of plasticizer, pH of solvent, salts, and other additives. For example, Gennadios et al. report that the tensile strength of corn zein and wheat gluten (WG) films vary significantly with conditioning at different combinations of relative humidity (23–75%) and temperature (5–45°C) (Gennadios et al., 1996). Gontard et al. note that pH and ethanol concentration have strong interactive effects on film opacity, water solubility, and water vapor permeability

(Gontard et al., 1992). The concentration of plasticizer greatly affects mechanical properties of grain protein-based films. Flexibility is improved, but the puncture strength and the water vapor and oxygen barrier properties are decreased as plasticizer concentration is increased in WG films (Park et al., 1994). Thus, modification of protein-based films must be aimed primarily at improving the mechanical strength and barrier properties with respect to moisture (Bourtoom, 2009). The most commonly used plasticizers include various polyols (glycerine), oligosaccharides, and lipids that destroy the hydrogen bonds between the polymer chains and make structure more fluid, thereby increasing the elasticity. However, the barrier properties of the films are reduced (Di Pierro et al., 2006). Besides, these agents significantly increase the hydrophilicity of the film and as a result, increase its vapor permeability. Despite numerous researches carried out for these proteins, their use is very limited.

In the category of biodegradable aliphatic polyesters produced by bacterial fermentation, polylactic acid (PLA) is one of the most commercially available and exploited biopolymers. PLA is obtained from agricultural products such as corn, sugarcane, and other sources (Bang and Kim, 2012). PLA is synthesized from lactic acid produced *via* starch fermentation from lactic bacteria. Starch is converted into sugar and after that the sugar is fermented to give lactic acid. The lactic acid prepared by this biotechnological method is almost exclusively L-lactic acid (Wee et al., 2006). However, PLA can exist in three stereochemical forms: poly(L-lactide) (PLLA), poly(D-lactide) (PDLA), and poly(DL-lactide) (PDLLA) with extremely various tensile properties, depending on the degree of crystallinity (Wee et al., 2006).

The PHA family is biodegradable thermoplastic polymers produced in the microbial cells as an energy reserve through a fermentation process. There exist more than 100 types of PHAs with a large diversity of properties. Poly(3-hydroxybutyrate) (PHB) is one of the most common and well-characterized biodegradable PHA (Vieira et al., 2011). The crystallinity of PHAs can be manipulated to provide a broad range of mechanical and barrier properties, in some cases matching the performance of engineered thermoplastics. In addition, to improve the flexibility, PHB is synthesized with various copolymers such as poly-(3-hydroxyvalerate) (HV) leading to a decrease of the glass transition and melting temperatures (Modi et al., 2011; Crétois et al., 2014).

Biopolymer-based packaging materials originated form naturally renewable resources such as polysaccharides, proteins, aliphatic polyesters, or combination of these compounds have the potential to replace current synthetic polymers. Actually,

Table 13.1 Characteristics of Nanocomposites

Improved Properties	Disadvantages
Mechanical properties (tensile strength, stiffness, toughness)	Viscosity increase
Gas barrier properties	Dispersion difficulties
Dimensional stability	Optical issues
Thermal expansion	Sedimentation
Thermal conductivity	
Ablation resistance	
Chemical resistance	

biopolymers cover approximately 5–10% of the current packaging market (Vieira et al., 2011).

However, some properties like melt strength, impact strength, thermal stability, and especially permeability (water vapors and gases) of biopolymers do not meet the demands for some end-use applications and must be improved (Sorrentino et al., 2007; Reddy et al., 2013). One of the ways to improve the properties of biopolymers and greatly enhance their commercial potential is to incorporate nanosized reinforcement into the polymer (Table 13.1).

4 Different Types of Nanoclays

Different types of commercially available clays that can be incorporated into the polymer matrix to form the biopolymer nanocomposites exist actually (Sorrentino et al., 2007). Depending on the specified application, the type of clay should be determined. The most widely studied type of the clay filler is montmorillonite (MMT) (Bordes et al., 2009; Rhim et al., 2013). A hydrated alumina-silicate-layered clay consisting of two silica tetrahedral sheets fused to an edge-shared octahedral sheets of aluminium hydroxide, MMT has several advantages. Its high surface area, large aspect ratio (50–1000) and platelet thickness of 10 Å make it suitable for the nanocomposite application. However, pristine MMT is difficult to be dispersed homogeneously in an organic polymer phase due to the narrow interlayer distance (typically about 1 nm) and the hydrophilic nature of its surface (Kotov, 2003). Thus, the organophilicity of MMT must be improved by means of the surface modification. This is accomplished, among other methods, by the cation exchange with organic cations (ie, different quaternized alkyl ammonium cations) leading to organoclays with the additional increase in the distance between the silicate layers.

Silicon dioxide, or silica, has been used in food processing and food applications for many years in a form that is called synthetic amorphous silica. The bulk material is known as food additive E551 and consists of primary particles, aggregates, and agglomerates. The primary particles and part of the aggregates consist of materials with the size less than 100 nm (Peters et al., 2012).

Halloysite is a clay mineral usually described as a gibbsite octahedral sheet [$Al(OH)_3$], which is modified by siloxane groups at the outer surface and has a 1:1 Al:Si ratio and stoichiometry $Al_2(OH)_4Si_2O_5{\cdot}nH_2O$ (Guimaraes et al., 2010). According to Joussein et al., halloysite structure depends on the crystallization conditions and geological occurrences (Joussein et al., 2005) and various morphologies (ie, platy and spheroidal crystals, the scroll, glomerular, "onion-like" or hollow tubular structure) have been already reported in the literature (Joussein et al., 2005; Guimaraes et al., 2010). The most common halloysite structure is hollow tubes and the size of tubes may vary from 500 to 1000 nm in length and from 15 to 100 nm in inner diameter, depending on the substrate (Guimaraes et al., 2010). The layered halloysite exists in two different polymorphs: the hydrated form with the formula $Al_2(OH)_4Si_2O_5{\cdot}2H_2O$ and with the interlayer spacing of 10 Å, and the anhydrous form with kaolinite composition—$Al_2(OH)_4Si_2O_5$—and with the interlayer spacing of 7 Å. It was found that the intercalated water is weakly bonded and can be quickly and irreversibly removed (Joussein et al., 2005).

Due to the availability and vast range of applications halloysite nanotubes are attractive materials. Besides, naturally existing halloysite is an inexpensive nanoscale container. It has been used as a support for the controlled release of herbicides and fungicides (Shchukin and Mohwald, 2007) and to improve the mechanical performance of polymers (Hedicke-Höchstötter et al., 2009).

The use of synthetic clays such as Laponite, a hectorite of composition $Na_{0.7}[(Mg_{5.5}Li_{0.4})Si_8O_{20}(OH)_4]$, has many advantages over natural clays because of the absence of impurities (iron oxides, carbonates, etc.) and owing to well defined chemical formula. This synthetic clay has been employed in different fields such as surface coatings, paper and polymer films, as well as bioactive materials (Gaharwar et al., 2013). Laponite particles show a disk-like shape. Its relatively small size (usually around 25–30 nm in diameter and 1 nm in thickness), combined with its great ability to form transparent colloidal dispersions, facilitates the obtaining of the homogeneous polymer/clay nanocomposites by casting method (Nobel et al., 2006). The Laponite surface area is up to 370 m^2/g

(Herrera et al., 2004), which can endorse efficient interactions with species in its surrounding. Smectite clay primary particles are a set of negative charged layers arranged face to face and intercalated with hydrated cations to bring about a zero net electric charge. As in the case of MMT, each Laponite layer is formed by the condensation of three sheets. The denominated tetrahedral sheet presents continuous corner-shared tetrahedral (SiO_4) units arranged in hexagonal rings. Connected to two tetrahedral sheets, there is an octahedral sheet composed of bivalent or trivalent cations sharing the edges coordinated to hydroxyl groups. In the case of Laponite, octahedral sheet supports Mg^{2+} ions, which are partially substituted by Li^{+} ions (Brindley, 1980).

5 Biopolymer/Clay Nanocomposites

Several synthetic routes have been used to prepare polymer/clay nanocomposites:

- exfoliation-adsorption;
- in situ intercalative polymerization;
- melt intercalation;
- template synthesis.

In general, the structure of polymer/clay nanocomposites is classified according to the level of intercalation and exfoliation of polymer chains into the clay galleries (Tang et al., 2012). Various parameters including clay nature, organic modifier, polymer matrix, and preparation method influence the nanocomposite structure. Thus, depending on the nature of the organic and inorganic components (and also on the preparation method), and on the strength of the interfacial interactions between a polymer matrix and clay, three types of nanocomposites can be obtained: phase-separated microcomposite; intercalated nanocomposite and exfoliated nanocomposite. A microcomposite is obtained when the polymer chains are not at all intercalated between the nanoclay sheets, which testifies to a phase-separated composite formation. This kind of composite maintains the immiscibility between the macromolecular chains of polymer and inorganic clay. In this case a minimal reinforcement by the fillers will be obtained and, thus, the enhancement in the composite properties is negligible. The intercalation is the state in which the polymer chains are inserted between the clay layers in such a way that the interlayer spacing is expanded with a repeat distance of a few nanometers. Due to the penetration of the polymer chains into the interlayer galleries of the nanoclay, the distance between the silicate sheets increases. Exfoliation is a state in which the silicate layers are completely and uniformly separated and dispersed in a continuous polymer matrix. The distance between the silicate layers in this case depends on the clay content. Such a structure allows the maximum reinforcement because of the proper clay distribution resulting in a large surface area. The structure and properties of the resulting nanocomposits can be altered by controlling the polymer-clay interactions.

The main aim of the introduction of the clay minerals to the biopolymers is to improve the polymer properties and to produce bionanocomposites with required characteristics for the desired applications. This approach (bionanocomposite) is very promising as it will allow overcoming the drawbacks of biopolymer (ie, poor mechanical and barrier properties) while remaining the intrinsic advantages of polymer matrix (ie, biodegradability).

Due to the low cost, availability, and especially high aspect ratio, clays can provide improved properties at very low concentration, usually less than 10%. The nature and specific properties of individual components as well as the preparation method and conditions affect the final properties of bionanocomposite (Sorrentino et al., 2007). In this chapter only the barrier properties of the biopolymer/clay nanocomposite are discussed.

6 Barrier Properties of Bionanocomposites

Many foods require specific atmospheric conditions to sustain the freshness and overall quality during the storage period. Thus, to ensure a constant gas composition inside the package, the packaging material should have certain gas barrier properties. In most cases, the gas mixture inside the package consists of CO_2, oxygen, and nitrogen. The oxygen degradation can affect the color, flavor, and microbial stability of a variety of food. Maintaining the product quality relies on the packaging film's ability to act as a barrier, overcoming the driving force of the difference in the oxygen partial pressures inside the package (0–2%) versus outside the package (21%). The objective of this section is to describe the gas barrier properties of biopolymer/clay nanocomposites.

A lot of information on the barrier properties of the biopolymer/clay films can be found in the literature. However, the comparison between the different biopolymers and/or nanoclays is rather complicated and sometimes even not possible because of the different measurement conditions (temperature, relative humidity, gas pressure, etc.) and different types of equipment used.

The barrier properties of the polymers are necessarily associated with their inherent ability to permit exchange, to a greater or lower extent, of low molecular weight substances through the mass transport process such as permeation. The permeation is generally presented as a combination of two processes: solution and diffusion. A permeate gas is first dissolved into the upstream face of the polymer film, and then undergoes the molecular diffusion to the downstream face of the film, where it desorbed into the external phase again. A solution-diffusion mechanism is thus applied, which can be formally expressed in terms of permeability P, solubility S, and diffusion D coefficients by

$$P = D \cdot S \tag{13.1}$$

The solubility coefficient S is thermodynamic in nature, and it is defined as the ratio of the equilibrium concentration of the dissolved penetrant in the polymer to its partial pressure p in the gas phase (Henry's law). In polymers, this law is usually obeyed at low penetrant concentrations, that is, when S is independent of the concentration or of the partial pressure. The diffusion coefficient D (Fig. 13.1) characterizes the average ability of the sorbed permeate to move through the polymer chain segments, and it is determined from Fick's first law of diffusion, that is, the flux of permeant is proportional to the local concentration gradient through the thickness of the polymer film.

The temperature, relative humidity, pH of the environment, the molecular size, the functional groups and the chemical potential of the diffusing species, as well as the morphology, orientation, free volume and cohesion, and adhesion forces of the polymer affect the mass transfer through the polymer matrix.

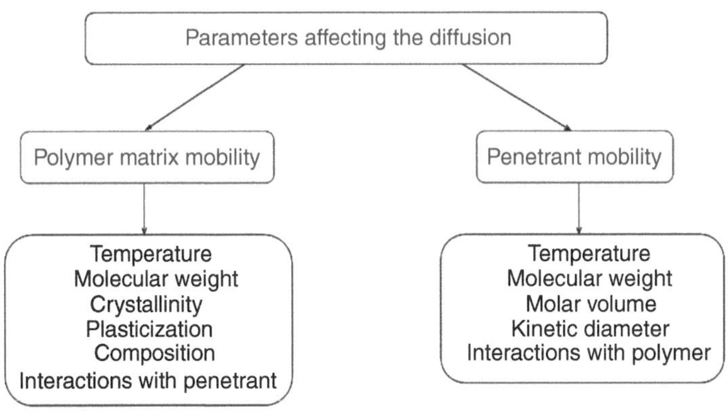

Figure 13.1. Parameters affecting the diffusion of penetrated molecules through the polymer matrix.

The addition of the clay into the polymer matrix leads to the significant improvement of the barrier properties of biopolymer. Clay nanoparticles are naturally impermeable. Therefore, the presence of clay increases of the polymer barrier properties by creating the tortuous path that retards the diffusion of the gas or water molecules through the polymer matrix. That's why the improvement of the barrier properties depends on the degree of tortuosity created by clay layers. The tortuous factor, in its turn, is determined by the aspect ratio of the clay nanoparticles dispersed in the polymer matrix. The increasing of the lateral length of clay sheet and the increasing of exfoliation or dispersion degree lead to the higher barrier improvement factor of the polymer matrix. Many research studies have reported the barrier properties of biopolymer/clay nanocomposites to various gases and vapors. Besides, simultaneous enhancement of other properties such as mechanical, thermal, and flame resistance has also been observed. All these properties are greatly improved in nanocomposites as compared to the conventional phase-separated microcomposites due to the good dispersion of the layered clays in a polymer matrix.

According to Nielsen, if the fillers are impenetrable to a diffusing gas or liquid molecule, then the diffusing molecule should go around the filler particles (Nielsen, 1967). The distance traveled by the molecule in the presence of nanofiller is called a tortuous path. Nielsen is the first to introduce the tortuosity factor τ, which is defined as a ratio of the actual distance d' that a penetrant must travel to the shortest distance d that it would have traveled in the absence of the nanoclay (Eq. 13.2). The large L/W ratio (where L is the length of the filler particle and W is the thickness of the filler plates) can dramatically reduce the permeability only if the particles are oriented in such a way that their flat surfaces are parallel to the surfaces of the film (Fig. 13.2). It should be noted that the incomplete clay dispersion, the presence of voids and/or only

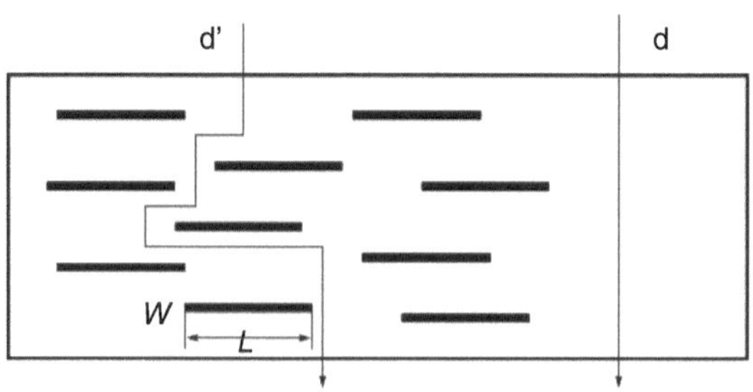

Figure 13.2. Proposed model for the tortuous pathway in polymer/clay nanocomposites.

partial orientation of the particles result in reverse effect in permeability (Chang et al., 2003).

The tortuosity factor τ is expressed as:

$$\tau = \frac{d'}{d} = 1 + \frac{L}{2W}\phi_s \qquad (13.2)$$

where ϕ_s is the volume fraction of the clay sheets.

The influence of tortuosity on the permeability is expressed as:

$$\frac{P_c}{P_p} = \frac{1-\phi_s}{\tau} \qquad (13.3)$$

where P_c and P_p represent the permeability of the nanocomposite and the pure polymer, respectively.

Taking all the foregoing into account, one can conclude that the enhancement in gas barrier properties of the nanocomposites depends on different factors, such as the volume fraction of the clay, the relative orientation of the clay sheets in the polymer matrix to the diffusion direction, the state of aggregation and dispersion (ie, exfoliated, intercalated, or mixed morphology) and the aspect ratio (L/W ratio, Fig. 13.2). With the incorporation of the nanoclays, a decrease of the solubility (Eq. 13.1) is expected in the nanocomposite because of the reduced polymer matrix volume, as well as a decrease in diffusion due to a more tortuous path for the diffusion molecules.

However, the relationship between the nanocomposite morphology and the improving of the gas barrier properties is not well understood until now. Bharadwaj proposed a simple method developed earlier by Nielsen correlating the sheet length, clay concentration, its relative orientation and the state of aggregation of the nanofiller in the polymer matrix (Bharadwaj, 2001). From Eq. 13.3, the relative permeability is a function of the clay sheet length. The permeability measurements performed for the nanocomposites with several different silicate volume fractions have shown progressively the low relative permeability values as a function of increasing the concentration and the sheet length (Nobel et al., 2006). Extremely large length-to-width ratio of the silicate sheets revealed a significant increase in tortuosity. Besides, no significant reduction in the relative permeability for the sheet length beyond $\phi_s = 0.05$ was noticed (Bharadwaj, 2001).

The arrangement of the sheets normal to the direction of the gas molecule diffusion results in the highest tortuosity. A range of different relative orientations of the sheets with respect to each other and to the plane of the film may be proposed (Bharadwaj, 2001). The two extremes, the planar and orthogonal alignments of the sheets, may be treated by simply exchanging L and W in Eq. 13.2. Bharadwaj proposed to modify the tortuosity factor in order to include the orientation order (Bharadwaj, 2001). Thus, the relative permeability is given by:

$$\frac{P_c}{P_p} = \frac{1-\phi_s}{1+(L/2W)\phi_s(2/3)(N+(1/2))} \tag{13.4}$$

where N is the type of the relative orientation of the sheets within the polymer matrix.

The aforementioned expression may be reduced to Eq. 13.3 when $N = 1$ (planar arrangement) (maximum tortuosity). In the case of $N = -1/2$ (orthogonal arrangement), Eq. 13.4 provides approximately the permeability value of the pure polymer matrix. Moreover, as the sheet length increases, the increase in the L to W ratio reduces the dependence of the permeability on the relative orientation of nanoclay sheets. In the case of the longer sheets, the barrier property decreases only in the range $-1/2 < N < 0$ at $\phi_s = 0.05$ (Bharadwaj, 2001). Besides, the introduction of sheets with $L > 500$ nm with random orientation is more beneficial for barrier properties than the case where the sheets were aligned perpendicular to the diffusing path.

Exfoliation of the nanoclays appears to be the critical factor in determining the maximum performance of the polymer nanocomposites for the barrier applications. A significant decrease of the permeability of O_2 gas through the nearly exfoliated PLA-nanocomposite compared to the pure polymer matrix was clearly observed (Chang et al., 2003) proving the direct benefit of the formation of the tortuous path.

6.1 Polysaccharide-Based Nanocomposites

The use of starches of different botanical origins for development of nanocomposite films was reported by several authors, showing the potential of this kind of polysaccharide (Slavutsky et al., 2012; Sadegh-Hassani and Nafchi, 2014; Alcantara et al., 2014; Romero-Bastida et al., 2015).

Slavutsky et al. studied the corn starch–based films with the incorporated MMT nanoparticles (Slavutsky et al., 2012). The nanoclay was incorporated using two different methodologies and the hybrid films were formed by casting. According to first method, the MMT particles were previously diluted in water in order to favor the opening of the clay pallets, and then mixed with the starch and glycerol (used as a plasticizer). In accordance with the second method, the starch, glycerol, and MMT were mixed all together in one step. The addition of the MMT particles resulted in a significant decrease in water vapor permeability (WVP) in comparison with the pure starch film. The MMT dispersion method has an important effect on the WVP value of the resulting films as well. The films obtained by the mixing of all components in one step presented a lower WVP value compared to those obtained by the first method. Such a behavior is caused by the fact that the second method (ie, the mixing of all components in one step) leads to the exfoliated nanocomposites, whereas the step-by-step addition of the components into the solution allows obtaining an intercalated morphology of starch-base films (Slavutsky et al., 2012). Therefore, the longer diffusive path that the penetrating molecules have to travel as the filler concentration has increased reduces the WVP value—from $6.4{\cdot}10^{-10}$ g/(m·Pa·s) for pure starch film up to $4.5{\cdot}10^{-10}$ g/(m·Pa·s) and $2.3{\cdot}10^{-10}$ g/(m·Pa·s) for starch-MMT (10 wt.%) films prepared by the first and the second methods, respectively.

Romero-Bastida et al. have also shown the influence of the component addition order in the case of starch/MMT nanocomposite films prepared by casting (Romero-Bastida et al., 2015). Primarily, glycerol was incorporated before the starch gelatinization and MMT addition (method 1), while in method 2 glycerol was added to the mixture after the gelatinization and MMT addition. In method 1, the WVP value was not affected neither by the MMT concentration nor the amylase content (Fig. 13.3). This behavior was caused by the major hydrogen bonds between starch and glycerol, that reduced the polar–polar attraction between starch and nanoclay and the effect of MMT on the barrier properties. On the other hand, the increase in the MMT content led to a decrease in the WVP value using method 2–$1.3{\cdot}10^{-10}$ g/(m·s·Pa) and $0.9{\cdot}10^{-10}$ g/(m·s·Pa) for the nanocomposite films with 0 and 15 wt.% MMT, respectively. During the solution preparation, the starch first interacted with MMT and only then with the plasticizer (ie, glycerol). When the higher MMT concentration is present in the nanocomposite, glycerol molecules can be accommodated between the silicate spacing, thus increasing the tortuosity and decreasing WVP.

The influence of the dilution of chitosan solution and the film preparation method on the water barrier properties of the chitosan-based nanocomposites was investigated by Giannakas et al. (2014). The nanocomposites films were obtained by the solution casting method. One part of the solution containing chitosan, clay (NaMMT), and glycerol was refluxed for 1 h before spreading onto the plastic dishes and the second part was spread directly. It was shown that the addition of NaMMT resulted in a

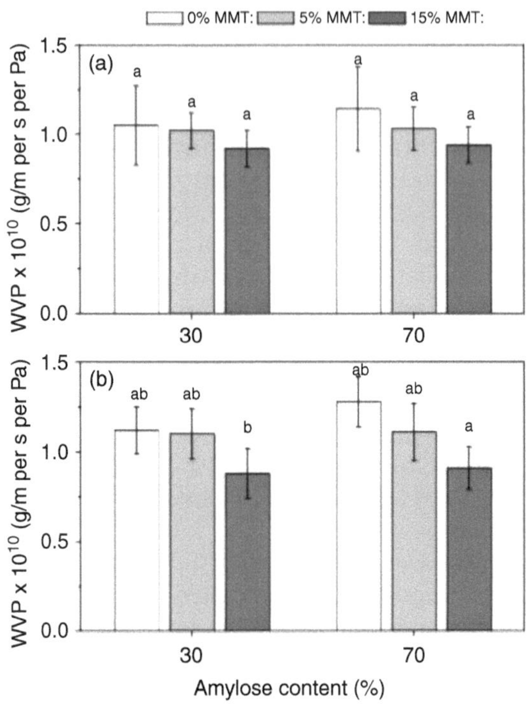

Figure 13.3. Water vapor permeability of the starch/MMT nanocomposite films prepared by method 1 (a) and method 2 (b) (Romero-Bastida et al., 2015). Reproduced with permission from Elsevier Ltd.

considerable reduction of the absorbed water—up to 75% in the case of the both preparation methods. However, the films prepared by the conventional solution method (ie, without reflux) as well as the films prepared from the diluted chitosan solutions had higher WVP values. The result can be ascribed to the higher hydrated crystallinity of the corresponding nanocomposite films revealed by X-ray diffraction analysis. Besides, the significant increasing of WVP with the addition of glycerol (relatively small hydrophilic molecule) can be explained by the possible insertion of glycerol between adjacent polymer chains, decreasing intermolecular attractions and, thus, facilitating the migration of the water vapor molecules.

Chitosan nanocomposite films were also prepared using a solvent-casting method by incorporation of an organically modified MMT (Cloisite 10A) (Oguzlu and Tihminlioglu, 2010). Since chitosan is a water sensitive polymer, the effect of the relative humidity (RH) on the water vapor permeability was studied. The sharp increase of WVP at around 50% RH was observed for the pure chitosan film. However, WVP was significantly reduced (up

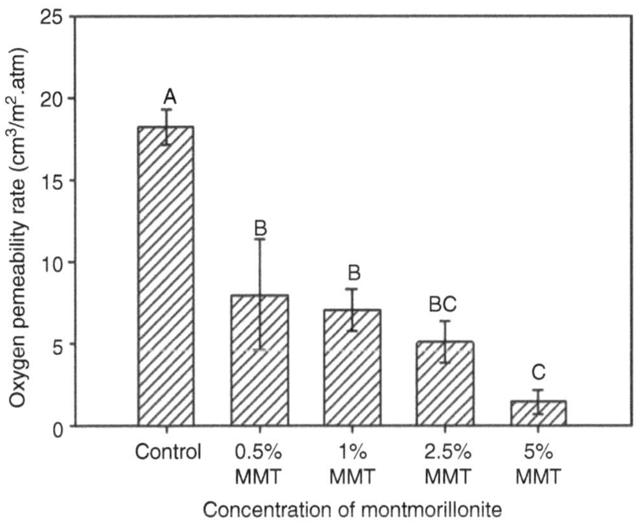

Figure 13.4. Oxygen permeability values for the chitosan/MMT nanocomposite films with different amount of MMT (Kasirga et al., 2012). Reproduced with permission from John Wiley & Sons.

to 25%) already at 2 wt.% of Cloisite 10A in the nanocomposite film. As to oxygen permeability, the permeability reduction up to 83–92% was observed with the addition of 2–10 wt.% of clay into chitosan. The authors applied the Nielsen's model (Eq. 13.2) in order to estimate the aspect ratio of the nanocomposite films. The effective aspect ratio was calculated to be between 53 and 2375 from the water vapor and oxygen permeability data, which is in good agreement with the previously reported aspect ratio range for layered silicates (10–1000) (Xu et al., 2006).

The oxygen permeability of the chitosan/MMT films ranged from 1.54 to 14.27 cm^3/m^2·day·atm, depending on the driving force used and on the nanoclay content (0.5–5 wt.%) (Kasirga et al., 2012) (Fig. 13.4).

The maximum reduction of O_2 permeability (ie, 90%) as well as of the water vapor permeability (22%) compared to the pure chitosan film was observed for 5 wt.% of MMT. Such reduction confirms that the clay platelets are well dispersed in the polymer matrix thus lengthening the pathways of the water vapor and oxygen molecules between the nanoclay lamellae nanocomposites that are not permeable to these compounds.

Souza et al. studied the influence of glycerol (plasticizer) and natural MMT on the tensile and barrier properties of the nanocomposite films based on cassava starch (Souza et al., 2012). They showed that the glycerol incorporation method did not influence the results; however, the glycerol content (from 0.17 to 0.75 g/100 g) significantly influenced the tensile and barrier properties of the films. The nanocomposite films prepared with a lower plasticizer content presented better tensile and barrier properties than the films prepared with a higher content. This fact may be explained by the antiplasticizing effect caused by the high plasticizer content (Alves et al., 2010), indicating stronger interactions between the plasticizer and the biopolymer that induce a loss of macromolecular mobility. The decreasing of the film permeability (both to water vapors and oxygen) was noted in the presence of nanoparticles, while the glass transition temperature of the nanocomposites was not influenced.

Alginate-based nanocomposites with different MMT clay loadings were produced by the solvent casting method (Alboofetileh et al., 2013). It was found that the water vapor barrier property of the composite films was significantly improved by the clay incorporation in the polymer matrix. The greatest values of WVP were obtained for the films with 1.5% alginate—$1.41{\cdot}10^{-10}$ g/ms·Pa and $2.64{\cdot}10^{-10}$ g/ms·Pa for the films with 1.0% and 1.5% of alginate, respectively. Moreover, the WVP value decreased significantly by 5–46% depending on the both alginate and clay concentrations. That decrease was explained by the presence of the water vapor-impermeable silicate layers with the large aspect ratio dispersed in the polymer matrix.

The influence of the clay content, homogenization parameters and pH on the mechanical and barrier properties of the fish gelatin/nanoclay composite films was investigated by Bae et al. (2009). Unmodified sodium MMT clay in a glycerol solution was treated with ultrasound and added to the fish gelatin solutions before casting into the films. The addition of 5 wt.% nanoclay increased the tensile strength from 2.37 to 3.30 MPa. At 9 wt.% clay loading, the gelatin films exhibited the best improvements in the oxygen and water barrier properties. The oxygen permeability decreased from 402.8 to 114.4 g·m/m^2·day·atm and WVP decreased from 31.2 to 8.1 ng·m/m^2·s·Pa.

The WVP and oxygen permeability of the potato starch–based bionanocomposite films with halloysite nanoclay are presented in Table 13.2 (Sadegh-Hassani and Nafchi, 2014). As one can see, the oxygen permeability coefficient has significantly decreased from ~78 to ~38 cm^3·μm/(m^2·day·atm) with the addition of nanoclay due to the creation of the tortuous pathway for the oxygen molecules to pass through the material. In its turn, the tortuosity factor τ calculated according to the Nielsen's model (Eq. 13.2) has increased from 1 (for the pure potato starch film) up to ~2 for the nanocomposite film with 5 wt% of nanoclay.

Table 13.2 WVP, Oxygen Permeability, and Tortuosity Factor of the Potato Starch Nanocomposites (Sadegh-Hassani and Nafchi, 2014)

Clay Content (wt.%)	WVP × 10^{11} (g/m per s per Pa)	Oxygen Permeability [cm^3/μm per (m^2·day)]	Tortuosity Factor τ
0	2.91	76.88	1
1	2.14	61.72	1.24
2	1.44	50.98	1.50
3	1.02	45.77	1.67
5	0.83	38.20	1.99

Source: Reproduced with permission from Elsevier Ltd

The influence of the talc nanoparticles' addition on the microstructure of the TPS film was evaluated (Lopez et al., 2015a). It was noted that the laminar morphology of the nanofiller led to a preferential particle orientation within the polymer matrix during the thermo-compression process. Moreover, the individual and intercalated talc platelets distributed along the TPS matrix were observed by the transmission electron microscopy (TEM) measurements. The talc addition at 1% w/w did not significantly modify the moisture vapor permeability of the TPS films (Lopez et al., 2015b). A WVP decrease of 1.4 times was observed for the nanocomposite film with 3% w/w of talc. A further addition of talc nanoparticles to TPS (up to 5% w/w) led to 54% reduction of WVP. Besides, for the film with 5% w/w of talc, an oxygen permeability decrease of about 26% was also observed. The permeability decrease in the TPS films by the talc introduction is attributed to the starch molecules immobilization because of the starch-talc particles interactions (Abreu et al., 2015). Although the talc incorporation increases the amount of rigid phase in nanocomposites, TPS stress–strain behavior is not affected (Castillo et al., 2015).

The addition of Cloisite 30B to the TPS films increases the film storage modulus (Abreu et al., 2015). TPS/Cloisite 30B (5 wt.%) nanocomposite exhibits a 28% increase compared to the pristine TPS film. A possible explanation of such increase is the formation of the 3-dimensional network in the interconnected long silicate layers, strengthening the material through the mechanical percolation (Tang et al., 2012).

In order to increase the barrier properties of the biopolymer/clay film used as the packaging material, the chemical modification of clay may be performed. For example, zein-modified fibrous clay (sepiolite) was introduced into the alginate matrix (Alcantara et al., 2015). Homogeneous and transparent films were obtained independently of the incorporated amount of the nanoclay. The permeability of the bionanocomposite films to different gases, such as He, N_2, O_2, and CO_2, was first evaluated in dry state, showing a very low permeability value, which was promising for their application. However, as a food packaging material, the composite film could be in contact with moisture, so the gas barrier properties could change. Therefore, the permeability measurements were also performed for the nanocomposite films in wet state. The higher permeability to CO_2 compared to other tested gases may be explained by higher solubility of acid gases, such as CO_2, in water, which facilitates the passage of this gas through the wetted films thus increasing its concentration inside the polymer matrix. It is found that the oxygen permeability of the elaborated films increases with the pressure and at the same time it decreases depending on the amount of the incorporated nanoclay. The film with 16 wt.% of the protein modified clay shows the higher O_2 permeability values than the alginate film with 24 wt.% of nanoclay. Although the presence of the protein limits the diffusion of the water molecules, this fact does not strongly affect the oxygen permeability values until the total zein fraction in the film reaches 10%. At higher zein content, O_2 permeability is significantly reduced indicating that the alginate films with modified clay can be an effective barrier to oxygen even under the high humidity conditions.

6.2 Protein-Based Nanocomposites

Among different proteins, WG is considered to be a valuable candidate for the packaging applications, as the WG-based films have good barrier properties under dry conditions (Gontard et al., 1992; Ture et al., 2013; Kuktaite et al., 2014; Azevedo et al., 2015). Recently, Ture et al. presented a novel approach to improve the barrier properties of a paperboard using a WG/MMT multilayer coating (Ture et al., 2013). In that multilayer structure, WG was used as the first barrier layer (on the top of the precoated paperboard) and the top layer, while MMT was deposited between the two WG layers. The idea to use a separate clay layer was caused by the wish to get locally as high concentration of clay (10 wt.%) as possible in order to effectively reduce the permeability of the composite. It was shown that the addition of single and multilayer coatings led to a decrease in the both WVP and oxygen permeability values: if in the case of the uncoated paperboard the both values were immeasurably high, then the addition of only the single WG layer was resulted in a 3.4-fold reduction in WVP. The multilayer coating (WG + clay + WG) reduced the WVP value by the factor of 6.3 compared to the value for the uncoated paperboard. In the case of the oxygen permeability measurements, the same tendency was noted, that is, the barrier improvement factor for the multilayer coating was 23 compared to the paperboard with only one WG layer. Besides, it was revealed that the multilayer coating had little, if any, mechanical effect on the paperboard properties.

The impact of the structure of the both natural and modified MMT (Closite 15A) nanoclays in the WG-urea-matrix on the barrier and mechanical properties was studied (Kuktaite et al., 2014). The decrease of the water vapor (up to 43%) and oxygen (~70%) barrier properties caused by the presence of nanoclays was found to be the same for the unmodified and modified MMT with 5 wt.%. However, the mechanical properties were different. For the nanocomposites with the unmodified MMT the Young's modulus increased and the strain at break decreased. On the contrary, a decrease in the modulus and an increase in the stain at break were observed for the nanocomposite films with Cloisite 15A. Such a result was explained by the more pronounced intercalation in the Cloisite 15A containing films due to a greater compatibility with the WG protein matrix compared to the MMT-based films.

The sodium montmorillonite, being hydrophilic, can form stable suspensions in water, which facilitate its dispersion in the water-soluble polymers such as whey protein isolate. Azevedo et al. evaluated the influence of the addition of citric acid (CA) and MMT on the thermodynamic equilibrium with the water vapors for the films based on whey protein isolate (Azevedo et al., 2015). The WVP value of the nanocomposite films decreased with the addition of MMT—$4.29 \cdot 10^{-12}$ g/(m·s·Pa) for the pure whey protein isolate and $2.11 \cdot 10^{-12}$ g/(m·s·Pa) for the film with 3 wt.% of MMT. Previously, it was reported that CA might be used as a plasticizer replacing glycerol (Olsson et al., 2013). In the presence of CA, the hydrogen bonds formed with a hydroxyl group and a carboxyl group enhance the inter- and intramolecular interactions within the polymer film. Besides, CA thanks to its hydrophilic character ameliorates the MMT dispersion that, in its turn, improves the barrier properties of the nanocomposites. Really, the addition of CA led to further decreasing of WVP, that is, $2.61 \cdot 10^{-12}$ g/(m·s·Pa) and $1.60 \cdot 10^{-12}$ g/(m·s·Pa) for the films without and with MMT,

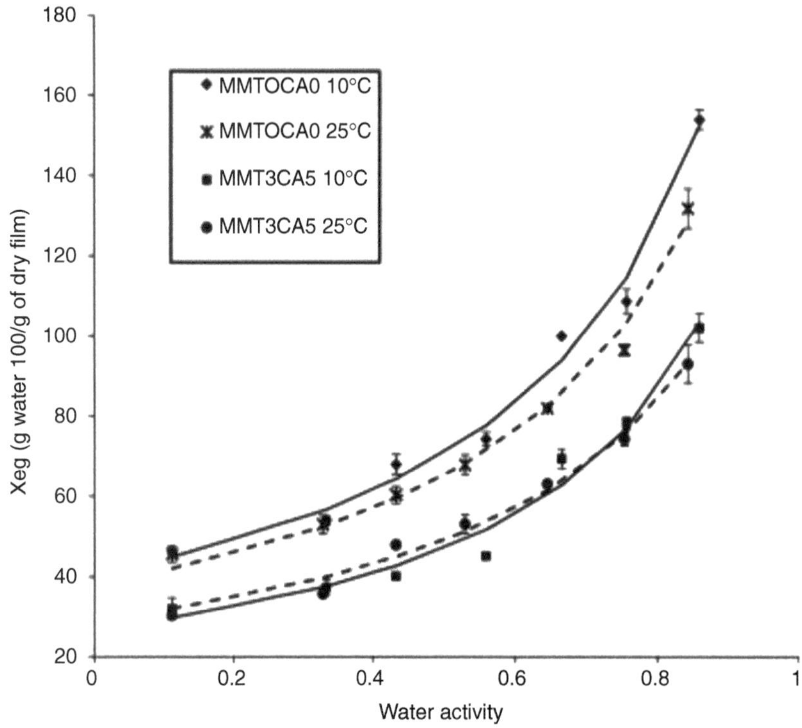

Figure 13.5. Water vapor sorption isotherms for the whey protein isolate/MMT nanocomposite films with different amounts of MMT and CA. (Azevedo et al., 2015). Reproduced with permission from Elsevier Ltd.

respectively (Azevedo et al., 2015). Moreover, the nanocomposite film that contains 3 wt.% MMT and 5 wt.% of CA acid is characterized by the lowest moisture adsorption and hence better performance and storage stability (Fig. 13.5).

The chicken feather protein (CFP) composite films with various concentrations of MMT were prepared in order to evaluate the effects of various plasticizers and nanoclay concentrations on the barrier and mechanical properties of the films (Song et al., 2013). The tensile strength of the nanocomposite films had the highest value with the addition of 7 wt.% of nanoclay, that is, the CFP film with and without nanoclay had a tensile strength of 5.95 and 4.74 MPa, respectively. Such an increase is explained by the formation of the intercalated state and the uniform dispersion of the MMT particles into the CFP matrix, indicating a strong interaction between the biopolymer and nanoclay particles. Consequently, the WVP values also decreased with the MMT content increasing—from $3.11 \cdot 10^{-9}$ g·m/(m^2·s·Pa) for the film without nanoclay particles up to $1.96 \cdot 10^{-9}$ g·m/(m^2·s·Pa) for the nanocomposite film with 7 wt.% of MMT.

6.3 Aliphatic Polymer-Based Nanocomposites

The impact of the MMT content on the water barrier properties was studied for polyhydroxybutyrate with 12 mol% of the valerate (PHBV) films (Sanchez-Garcia and Lagaron, 2010). The film containing 1 wt.% of clay shows a water permeability reduction of 61%, the film containing 5 wt.% of MMT shows a reduction of 76%, whereas for the film with 10 wt.% of clay the reduction is only 47%. Such a behavior is explained by the possible exceeding of the solubility limit resulting in the detrimental agglomeration. In addition, a reduction in the oxygen permeability of 32% was obtained for the PHBV film containing 5 wt.% of clay. Increasing nanoclay content to 10 wt.% did not result in further barrier increase. The crystalline phase is considered to be impermeable to gases and water molecules. Therefore, the highest crystallinity degree for the PHBV/clay (5 wt.%) film is responsible for this barrier enhancement indicating that 5 wt.% is the optimum clay loading for this type of the biopolymer matrix.

Miguel and Iruin carried out the studies on water, CO_2 and organic solvent transport through the films of PHB and PHBV with the 3HV molar ratios of 0, 8, 14, and 24 mol/mol (Miguel and Iruin, 1999). The films were obtained by simple casting of chloroform solutions. The authors noted that the values of the barrier performance, related to water and CO_2, of such films are similar to the common plastics (PET).

The PHB packaging for the storage of food products was investigated by Bucci et al. (2005). They produced the PHB-packaging containers by the injection molding. The packaging was evaluated by means of dimensional (volumetric capacity, mass and thickness) and mechanical (dynamic compression and impact resistance) tests, and compared to the classically used poly(propylene) (PP) packaging of the same dimension. In addition, the potential for the organoleptic contamination of some food types (cream cheese, margarine, and mayonnaise) was studied. The authors gave proofs that PHB had a significantly lower resistance to dynamic compression compared to PP. This is revealed by the deformation value of about 500% lower than that reported for PP. Hence, PHB constitutes a more rigid and less flexible material as compared with PP. Under normal freezing and refrigeration conditions, the barrier performance of PHB turned out to be inferior to that of PP, whereas at higher temperatures, PHB is even superior to PP in the barrier performance.

Some authors found out that the incorporation of low concentration of nanoclay (5 wt.%) and cellulose whiskers (3 wt.%) into the PHBV matrix and other biodegradable polymers such as PLA and polycaprolactone resulted in the improvement in oxygen permeability that could be very useful for the food packaging (Ten et al., 2012). The authors showed that the PHBV films with 1 wt.% of the alpha cellulose fiber content had a water permeability reduction of 71% compared to the unfilled film, whereas the PHBV films with the fiber content of 10 wt.% demonstrated a water permeability decrease of around 52% because of the fiber agglomeration. However, the bionanocomposites with 5 wt.% of clay revealed the lowest water permeability coefficient value due to the good morphology of these composite films. It was also shown that the mica-based nanoclays exerted a certain UV-Vis light blocking action in the PLA and PHBV matrix (Ten et al., 2012). The blocking effect of PHBV in the UV-Vis region was higher than that of PLA as PHBV was a translucent material. Moreover, the greater reduction in the vapor permeability was attained for the PHBV nanocomposites with the clay content of 1 wt.% (Halley and Dorgan, 2011).

It was found that the barrier properties of the PHBV/clay nanocomposites were dependent not only on the dispersion and orientation of nanofiller platelets, but also on the nature of the permeate molecules (Crétois et al., 2014). The decreasing of the nitrogen permeability from 0.033 Barrer for the pure PHBV film to 0.014 Barrer for the films with 7.5 wt.% of Cloisite 30B was detected. However, further increasing of N_2 permeability was found for the nanocomposite film with 10 wt.% of Cloisite 30B (0.024 Barrer). Two explanations of this result were proposed. The first one is related to the reduction of the molecular weight of the polymer occurring during the melt blending process as the thermal degradation can take place. The second explanation is based on the compatibility of the polymer matrix and nanoclays and on the formation of the weak interfaces. Due to the lack of the free space in the polymer matrix, the nanoclay particles can aggregate thus resulting in the favoring of the gas diffusion.

Furthermore, the PHBV processing behavior could be improved with the addition of the MMT nanoclays, since with the increasing of the clay content the processing temperature range widened by the lowering of the melting temperature (Sanchez-Garcia et al., 2010). The mechanical properties of the corresponding films were improved by the incorporation of 3 wt.% of clay.

Several authors were interested in the PLA/clay nanocomposite films (Chang et al., 2003; Auras et al., 2005; Martino et al., 2010; Ahmed et al., 2010; Vargas et al., 2012; Salvatore et al., 2015). A variety of the processing strategies, including the melt extrusion, in situ polymerization and solvent casting was explored. When processing conditions are optimized, the use of nanoclay can provide a significant enhancement in the barrier properties of the bionanocomposite due to the formation of a tortuous diffusion path (Fig. 13.2).

Table 13.3 Oxygen Gas Permeability of PLA/MMT Films (Chang et al., 2003)

Clay Content (wt.%)	Oxygen Permeability (cm^3/m^2 per day) Cloisite 25 A	Dodecyltrimethyl Ammonium Bromide-MMT	Hexadecylamide-MMT
0	777	777	777
4	—	455	449
6	430	353	340
10	340	330	327

Source: Reproduced with permission from John Wiley & Sons

Chang et al. noted O_2 gas permeability of the PLA nanocomposites having different organically modified MMT (Table 13.3) (Chang et al., 2003). The nanocomposites were prepared by the melt intercalation technique. The increase of the tortuous paths in the nanocomposites in the presence of the clay content can be noticed. The permeability value of the nanocomposites decreases to half of the PLA permeability value, whatever is the MMT nature.

Martino et al. studied the influence of the plasticizer (polyadipate) for the melt-blended nanocomposite films with different amounts of 1–5 wt.% of MMT (Cloisite 30B) (Martino et al., 2010). The oxygen transmission rate was notably reduced (around 45%) with increasing the amount of nanoclay. However, for the clay concentration above 3 wt.%, a significant deterioration of the ductile properties was observed. That fact was explained by the morphology of the nanocomposite films (Fig. 13.6). Moreover, for a given nanoclay loading, the increasing of the plasticizer concentration from 15 to 20 wt.% improved the clay dispersion through the polymer matrix. As it was indicated early, the storage conditions and the time were two important parameters of the composite films designed for food packaging. The elaborated nanocomposite films were stored at 25°C and 50% RH during 2 months. It was observed that the clay addition reduced the film aging since the higher shift to the lower crystallization temperatures and more pronounced increase in the oxygen barrier property were detected for the pure PLA films. Such a behavior was explained by the presence of clay that restricted the segmental motions at the organic-inorganic interfaces when the storage temperature was below the glass transition temperature (T_g) (T_g of PLA is 50–60°C). A study by Ahmed et al. confirmed that for the PLA/MMT films T_g remained

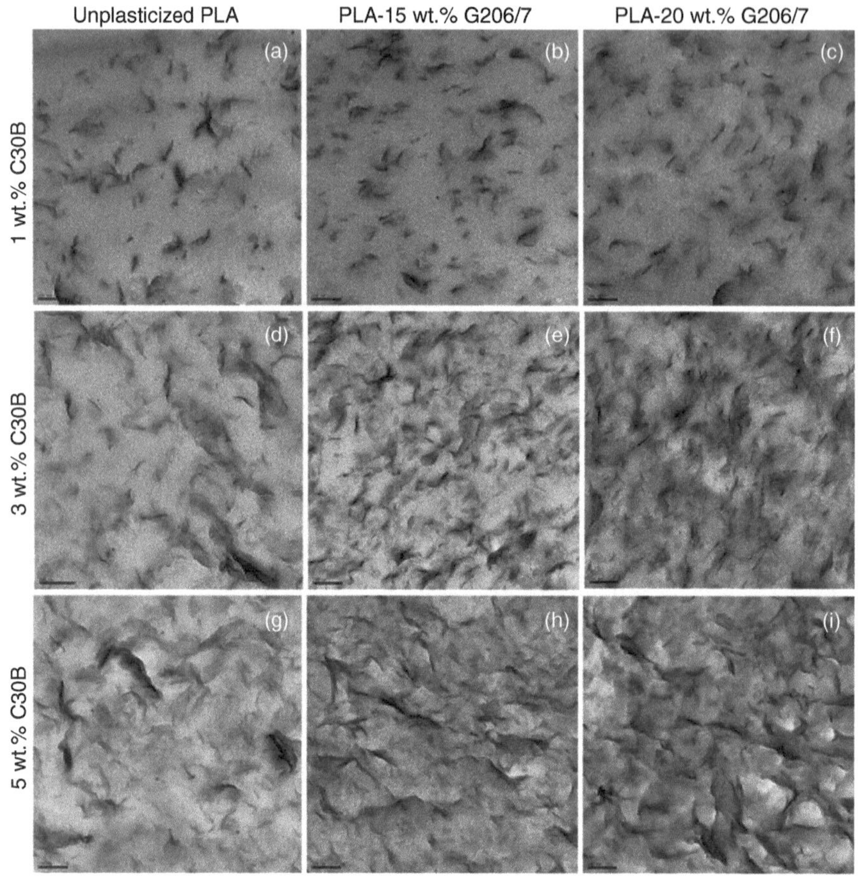

Figure 13.6. TEM images of the PLA-based nanocomposite films with different MMT and adipate contents. (Martino et al., 2010). Reproduced with permission from John Wiley & Sons.

unaffected, irrespectively on the clay concentration, due to the immobilization of the polymer chain in the clay nanocomposites (Ahmed et al., 2010).

Auras et al. have stated that the oxygen permeability of polylactide changes with the crystallinity degree (Auras et al., 2005). The higher the crystallinity degree in the polymer is, the lower permeability values are obtained as a more tortuous path is generated for the oxygen molecule permeation.

Żenkiewicz et al. studied the water vapor, oxygen and CO_2 transmission rates through the polylactide nanocomposite films with the addition of the natural and organomodified clays (Żenkiewicz et al., 2008). The transmission rate of the organomodified clay nanocomposites was reduced by 43%, 39%, and 82% for water vapors, oxygen, and CO_2, respectively. Besides, the reduction ratio of the natural clay nanocomposite was improved by 25%, 4%, and 76% for water vapors, oxygen, and CO_2 transmission rates, respectively. Moreover, the increase in the clay content resulted in the continuous decrease in the transmission rates for the all studied diffusing molecules.

In the work of Vargas et al., the nanocomposite films based on PLA, cellulose nanofibers, cellulose whiskers, or commercially available nanoclay (MMT Closite 30B) were prepared with the different content of nanofillers (1, 3, or 5 wt.%) (Vargas et al., 2012). The barrier properties of the PLA composite films were evaluated by means of the oxygen and water vapor permeability. Significantly lower oxygen permeability was found for the composite films containing nanocellulose fillers compared to the nanoclay-based composites, whereas both kind of composite films demonstrated the same behavior as to the water barrier properties. Furthermore, in the presence of more than one nanofiller inside the PLA matrix (ie, cellulose nanofibers and nanoclay or cellulose whiskers and nanoclay), an accumulative behavior in terms of the barrier properties was revealed, providing an important reduction in the permeability of the PLA films even at the low nanofiller concentration.

In order to exploit both the improvement in the biopolymer packaging performance caused by the nanotechnology and the drastic reduction of microbial contamination of food by the irradiation procedure, the PLA/MMT films were eBeam irradiated (Salvatore et al., 2015). Prior to the irradiation, the surface of the nanocomposite films was smooth and homogeneous whatever the MMT content was (Fig. 13.7).

However, after the eBeam irradiation the film surface presented the regions with higher roughness and cavities of different shapes as it was known that the surface irradiation could cause the polymer degradation and the crosslinking. The oxygen permeability of the nonirradiated film depends on the film composition, that is, the higher clay content and the higher polymer crystallinity degree lower the oxygen permeability of the nanocomposite films. For the eBeam irradiated films, the permeability for the given clay content appears to depend on the irradiation dose. For the nanocomposite films irradiated at 1 kGy, the oxygen permeability value decreases by about 4–5% compared to the untreated film at a given film composition. For example, the oxygen permeability is 385 and 358 $cm^3/m^2 \cdot day$ for the PLA/MMT (3 wt.%) film before and after the irradiation with 1 kGy, respectively. Such a decrease in permeability can be caused by the increase in the crystallinity and the formation of crosslinks, which increase the rigidity of the polymer amorphous phase. Really, the film crystallinity degree

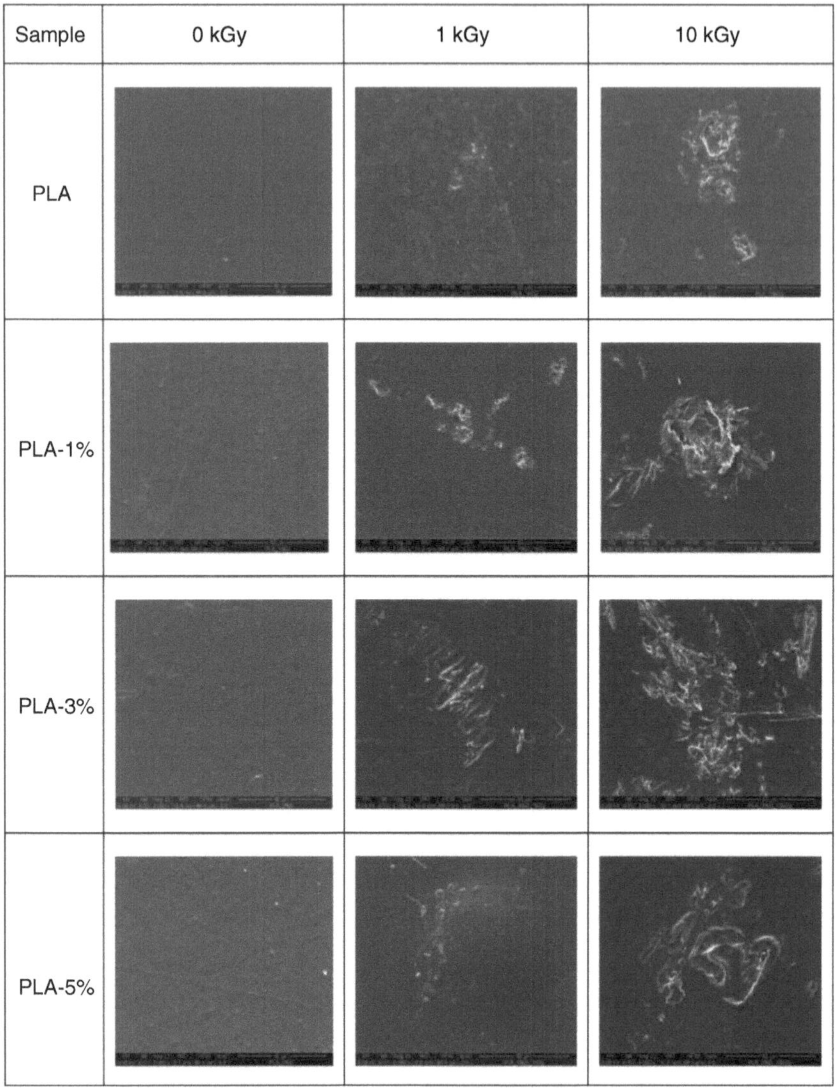

Figure 13.7. SEM images of the eBeam irradiated PLA based nanocomposites with different MMT content. (Salvatore et al., 2015). Reproduced with permission from John Wiley & Sons.

before and after irradiation is found to be 4 and 6%, respectively. When the eBeam dose is increased to 10 kGy, the permeability increases compared to the 1 kGy treated films and in some cases also with respect to those not treated. For the nanocomposite film with 3 wt.% of MMT, the oxygen permeability after irradiation at 10 kGy is found to be 390 $cm^3/m^2 \cdot day$ and the degree of crystallinity is 5%. This result is probably explained by the effect of the irradiation on the film surface as at the high irradiation deep cavities are produced thus facilitating the gas diffusion.

As noted, the complete exfoliation and orientation of nanoclays in the PLA films is a challenging task. Svagan et al. proposed an alternative strategy by using a bottom-up layer-by-layer process (Svagan et al., 2012). In that case, the free-standing films consisting of hundreds of the alternating, electrostatically self-assembled layers of clay and polymer were obtained by the consecutive adsorption of the oppositely charged clay and polymer. The novel multilayer-coated PLA films were elaborated using the polycation chitosan. Compared with uncoated PLA, the PLA films coated with 70 bilayers of MMT/chitosan had an oxygen permeability almost two orders of magnitude lower at 20% RH and more than one order of magnitude lower at 50% RH. The effective oxygen barrier action was explained by the strong MMT-chitosan molecular interaction and the possible reorganization of the chitosan macromolecular chains and the MMT sheets during the layer-by-layer assembly. The MMT/chitosan coating was revealed to be less effective at 80% RH, because of the free volume increase in the chitosan layers in the presence of the plasticizing water molecules. In contrast with the oxygen permeability, the WVP value was reduced by only 20% for the multilayer-coated PLA film.

Organically modified synthetic fluorohectorites (FHT) were explored to produce the biodegradable polybutylene succinate adipate (PBSA)/clay nanocomposites with the improved barrier properties (Dean et al., 2009). The oxygen permeability of pure PBSA was reduced by 35% by the addition of FHT-Na despite the fact that the TEM measurements showed the poor dispersion of nanoclay in the polymer matrix. The introduction of the finely dispersed organically modified FHT into the PBSA matrix led to a higher reduction of the oxygen permeability (up to 53%). This result confirms that the platelet dispersion and the clay orientation inside the polymer matrix play an important role in the gas barrier properties. Besides, the modification of nanoclay with trioctylmethylammonium leads to better compatibility of the modified clay with the PBSA matrix as it was revealed that the reduction of the oxygen permeability was not due to the increased crystallinity degree or the free volume changes.

6.4 Nanocomposites Based on the Mixture of Biopolymers

Blending with another biopolymer is also a technique to improve the barrier properties of the nanocomposites. When blending biopolymers, the compatibility of different compounds is very important. The gas barrier properties of the biodegradable films composed of the mixtures of κ-carrageenan and locust bean gun with the dispersed organically modified clay (Cloisite 30B) were studied (Martins et al., 2013). It was found that the Cloisite 30B incorporation into the biopolymer matrix decreased the water vapor permeability of the composite films—the water vapor permeability was 40% lower for the nanocomposite film with 16 wt.% of clay compared to that of the pure biopolymer matrix. Moreover, a decrease in the oxygen and CO_2 permeability through the biopolymer/clay films was observed until a critical clay concentration was attained (1 wt.%), after which an increase of permeability was detected. These results are in agreement with the results obtained by Alves et al. for κ-carrageenan/pectin blends with the mica flakes (Alves et al., 2010). This phenomenon was explained by the lack of exfoliation or the decrease of aspect ratio.

The incorporation of the nanosized sepiolite clays into the TPS/poly(butylenes adipate-*co*-terephthalate) (PBAT) films was investigated (Olivato et al., 2015). The scanning electron microscopy (SEM) images confirmed the good dispersion of the sepiolite clay with a low number of small aggregates within the polymer matrix. The addition of only 3 wt.% of sepiolite into the TPS matrix resulted in an important enhancement of the mechanical properties of TPS, reaching a 300% increase in Young's modulus and a 150% increase in tensile strength.

The mixture of TPS and MMT with the other polyester (PLA) was also intensively studied (Arroyo et al., 2010; Ayana et al., 2014). Arroyo et al. examined the properties of the TPS/PLA nanocomposites with the incorporation of MMT using a twin screw extruder (Arroyo et al., 2010). They studied the effect of water in enhancing the clay exfoliation and noticed that TPS could intercalate into the clay intragallery and that the preferential location of clay took place in the TPS phase or at the blend interface. They also noted an increase in the tensile strength and modulus simultaneously with the lessening of the fracture roughness. Ayana et al. prepared the biodegradable TPS/PLA/MMT nanocomposites using an eco-friendly approach that involved in situ gelatinization of potato starch in the presence of dispersed nanoclay followed by the melt mixing with PLA (Ayana et al., 2014). The performed morphological (TEM and SEM) analysis revealed that MMT act as the compatibilizer for the TPS/PLA films and was selectively dispersed into the TPS phase. At the same time, the water absorption of the TPS films was decreased up to 8 and 12% in the presence of PLA alone and PLA/clay, respectively. Besides, under condition of water immersion, the TPS/PLA/MMT films continued to be dimensionally stable and intact in shape, which makes them promising candidates for the food packaging application.

7 Future Developments

The biopolymer/clay nanocomposites are the subject of many scientific and research projects. When biopolymer is mixed with the nanoclay particles, the final hybrid material will exhibit improved the barrier properties and, besides, it can be used as compost. In order to obtain the optimum properties for bionanocomposites, the excellent dispersion of nanofiller and its good compatibility with the polymer matrix will be required. The packaging material made from such bionanocomposites can be elaborated at lower cost and at the same time with increased functionality and sustainability compared to the conventional films. Moreover, the obtained higher barrier properties to gases, the organic and water vapors are going to be very promising for the food packaging application. However, despite their attractive properties and industrial needs of such materials, the lack of compatibility and interfacial interactions between the biopolymer matrix and nanofiller prevent their commercialization on a large scale. Therefore, the working out of the tailored solutions for individual components (eg, the chemical modification of the nanoclay particles by grafting the different functional groups) and the optimization of the elaboration method allow developing the efficient food packaging material. The other question to which the answer should be found before the commercialization is the possibility of nanoparticles diffusion from the packaging film to the packaged food and the eventual toxicological effects if any (Huang and Zhou, 2015; Maisanaba et al., 2015). The progress in this field will lead to further development of the packaging materials and, as a result, to new commercial markets.

8 Conclusions

Nanotechnology is becoming increasingly important for the food sector. Promising results and applications have been already developed in the fields of food packaging and food safety. The introduction of nanomaterials into food packaging will improve the barrier properties of packaging materials and thereby help to reduce the use of valuable raw materials and the creation of waste products. The particularly promising application of food nanotechnology is an elaboration of nanosensors to detect microorganisms and contaminants. It must be noted, however, that no one can promise that all nanotechnology foods will be safe. Thus, each new product prepared should be tested for toxicity before marketing.

References

Abreu, A.S., Oliveira, M., Machada, A.V., 2015. Effect of clay mineral addition on properties of bio-based polymer blends. Appl. Clay Sci. 104, 277–285.
Ahmed, J., Varshney, S.K., Auras, R., 2010. Rheological and thermal properties of polylactide/silicate nanocomposites films. J. Food Sci. 75, N17–N24.
Alboofetileh, M., Rezaei, M., Hosseini, H., Abdollahi, M., 2013. Effect of montmorillonite clay and biopolymer concentration on the physical and mechanical properties of alginate nanocomposites films. J. Food Eng. 117, 26–33.
Alcantara, A.C.S., Darder, M., Aranda, P., Ruiz-Hitzky, E., 2014. Polysaccharide-fibrous clay bionanocomposites. Appl. Clay Sci. 96, 2–8.
Alcantara, A.C.S., Darder, M., Aranda, P., Ayral, A., Ruiz-Hitzky, E., 2015. Bionanocomposites based on polysaccharides and fibrous clays for packaging applications. J. Appl. Polym. 133, 42362.
Alves, V.D., Costa, N., Coelhoso, I.M., 2010. Barrier properties of biodegradable composite films based on kappa-carrageenan/pectin blends and mica flakes. Carbohyd. Polym. 79, 269–276.
Arrieta, M.P., Fortunati, E., Dominici, F., Lopez, J., Kenny, J.M., 2015. Bionanocomposite films based on plasticized PLA-PHB/cellulose nanocrystal blends. Carbohyd. Polym. 121, 265–275.
Arroyo, O., Huneault, M., Favis, B.D., Bureau, M.N., 2010. Processing and properties of PLA/thermoplastic starch/montmorillonite nanocomposites. Polym. Comp. 31, 114–127.
Auras, R.A., Singh, S.P., Singh, J.J., 2005. Evaluation of oriented poly(lactide) polymers vs. existing PET and oriented PS for fresh food service containers. Packag. Technol. Sci. 18, 207–216.
Averous, L., 2004. Biodegradable multiphase systems based on plasticized starch: a review. J. Macromol. Sci. Polym. Rev. 44, 231–274.
Ayana, B., Suin, S., Khatua, B.B., 2014. Highly exfoliated eco-friendly thermoplastic starch (TPS)/poly(lactic) acid (PLA)/clay nanocomposites using unmodified nanoclay. Carbohyd. Polym. 110, 430–439.
Azeredo, H.M.C.D., 2009. Nanocomposites for food packaging applications. Food Res. Int. 42, 1240–1253.
Azevedo, V.M., Sila, E.K., Goncalves Pereira, C.F., Gomes de Costa, J.M., Borges, S.V., 2015. Whey protein isolate biodegradable films: influence of the citric acid and montmorillonite clay nanoparticles on the physical properties. Food Hydrocoll. 43, 252–258.
Badgley, C., Moghtader, J., Quintero, E., Zakem, E., Chappel, L.M., Aviles-Vazquez, K., Salon, A., Perfecto, I., 2007. Organic agriculture and the global food supply. Renew. Agr. Food Sys. 22, 86–108.
Bae, H.J., Park, H.J., Hong, S.I., Byun, Y.J., Darby, D.O., Kimmel, R.M., Whiteside, W.S., 2009. Effect of clay content, homogenization RPM, pH, and ultrasonication on mechanical and barrier properties of fish gelatine/montmorillonite nanocomposites films. LWT Food Sci. Technol. 42, 1179–1186.
Baeumner, A., 2004. Nanosensors identify pathogens in food. Food Technol. 58, 51–55.
Bang, G., Kim, S.W., 2012. Biodegradable poly(lactic acid)-based hybrid coating materials for food packaging films with gas barrier properties. J. Ind. Eng. Chem. 18, 1063–1068.
Bharadwaj, R.K., 2001. Modeling the barrier properties of polymer-layered silicate nanocomposites. Macromolecules 34, 9189–9192.
Bitinis, N., Fornunati, E., Verdejo, R., Bras, J., Kenny, J.M., Torre, L., Lopez-Manchado, M.A., 2013. Poly(lactic acid)/natural rubber/cellulose nanocrystal

bionanocomposites. Part II: properties evaluation. Carbohyd. Polym. 96, 621–627.
Bordes, P., Pollet, E., Avérous, L., 2009. Nano-biocomposites: biodegradable polyester/nanoclay systems. Prog. Polym. Sci. 34, 125–155.
Bourtoom, T., 2009. Edible protein films: properties enhancement. Int. Food Res. J. 16, 1–9.
Bowman, D., Hodge, G., 2006. Nanotechnology: mapping the wild regulatory frontier. Futures 38, 1060–1073.
Brindley, G.W., 1980. Quantitative X-ray mineral analysis of clays. In: Brindley, G.W., Brown, G. (Eds.), Crystal Structures of Clay Minerals and Their X-Ray Identification. Mineralogical Society, London, UK, pp. 125–195.
Bucci, D.Z., Tavares, L.B.B., Sell, I., 2005. PHB packaging for the storage of food products. Polym. Test. 24, 564–571.
Castillo, L.A., Lopez, O.V., Ghilardi, J., Villar, M.A., Barbosa, S.E., Garcia, M.A., 2015. Thermoplastic starch/talc bionanocomposites: influence of particle morphology on final properties. Food Hydrocoll. 51, 432–440.
Chang, J.H., Uk-An, Y., Sur, G.S., 2003. Poly(lactic acid) nanocomposites with various organoclays. I. Thermomechanical properties, morphology, and gas permeability. J. Polym. Sci. B 41, 94–103.
Crétois, R., Follain, N., Dargent, E., Soulestin, J., Bourbigot, S., Marais, S., Lebrun, L., 2014. Microstructure and barrier properties of PHBV/organoclays bionanocomposites. J. Memb. Sci. 467, 56–66.
Da Silva, M.A., Vieira, M.G.A., Macumoto, A.C.G., Beppu, M.M., 2011. Polyvinylchrolide (PVC) and natural rubber films plasticized with a natural polymeric plasticizer obtained through polyesterification of rice fatty acid. Polym. Test. 30, 478–484.
Dean, K.M., Pas, S.J., Yu, L., Ammala, A., Hill, A.J., Wu, D.Y., 2009. Formation of highly oriented biodegradable polybutylene succinate adipate nanocomposites: effects of cation structures on morphology, free volume, and properties. J. Appl. Polym. Sci. 113, 3716–3724.
Di Pierro, P., Chico, B., Villalong, R., Mariniello, L., Damiao, A.E., Masi, P., Porta, R., 2006. Chitosan-whey protein edible films produced in the absence or presence of transglutaminase: analysis of their mechanical and barrier properties. Biomacromolecules 7, 744–749.
Dombre, C., Marais, S., Chappey, C., Lixon-Buquet, C., Chalier, P., 2014. The behaviour of wine aroma compounds related to structure and barrier properties of virgin, recylced and active PET membranes. J. Memb. Sci. 463, 215–225.
Gaharwar, A.K., Mihaila, S.M., Swami, A., Patel, A., Sant, S., Reis, R.L., Marques, A.P., Gomes, M.E., Khademhosseini, A., 2013. Bioactive silicate nanoplatelets for osteogenic differentiation of human mesenchymal stem cells. Adv. Mater. 25, 3329–3336.
Gennadios, A., Weller, C.L., Hanna, M.A., Froning, G.W., 1996. Mechanical and barrier properties of egg albumen films. J. Food Sci. 61, 585–589.
George, M., Abraham, T.E., 2006. Polyionic hydrocolloids for the intestinal delivery of protein drugs: alginate and chitosan: a review. J. Cont. Rel. 114, 1–14.
Giannakas, A., Grigoriadi, K., Leontiou, A., Barkoula, N.-M., Ladavos, A., 2014. Preparation, characterization, mechanical and barrier investigation of chitosan-clay nanocomomposites. Carbohyd. Polym. 108, 103–111.
Gontard, N., Guilbert, S., Cuq, J.-L., 1992. Edible wheat gluten films: influence of the main process variables on film properties using response surface methodology. J. Food Sci. 57, 190–195.
Guimaraes, L., Enyashin, A.N., Seifert, G., Duarte, H.A., 2010. Structural, electronic, and mechanical properties of single-walled halloysite nanotube models. J. Phys. Chem. C 114, 11358–11363.

Halley, P.J., Dorgan, J.R., 2011. Next-generation biopolymers: advanced functionality and improved sustainability. MRS Bull. 36, 687–691.
Hedicke-Höchstötter, K., Lim, G.T., Altstädt, V., 2009. Novel polyamide nanocomposites based on silicate nanotubes of the mineral halloysite. Comp. Sci. Technol. 69, 330–334.
Herrera, N.N., Letoffe, J.M., Putaux, J.L., David, L., Bourgeat-Lami, E., 2004. Aqueous dispersions of silane-functionalized Laponite clay platelets: a first step towards the elaboration of water-based polymer/clay nanocomposites. Langmuir 20, 1564–1571.
Huang, J.-Y., Zhou, W., 2015. Safety assessment of nanocomposites for food packaging application. Trends Food Sci. Technol. 45, 187–199.
Joussein, E., Petit, S., Churchman, J., Theng, B., Righi, D., Delvaux, B., 2005. Halloysite clay minerals: a review. Clay Miner. 40, 383–426.
Kasirga, Y., Oral, A., Caner, C., 2012. Preparation and characterization of chitosan/montmorillonite-K10 nanocomposites films for food packaging applications. Polym. Compos. 33, 1874–1882.
Kong, L., Ziegler, G.R., 2013. Fabrication of κ-carrageenan fibers by wet spinning: addition of ι-carrageenan. Food Hydrocoll. 30, 302–306.
Kotov, N.A., 2003. Layer-by-layer assembly of nanoparticles and nanocolloids: intermolecular interactions, structure and materials perspective. In: Decher, G., Schlenoff, J.B. (Eds.), Multilayer Thin Films: Sequential Assembly of Nanocomposite Materials. Weinheim, Germany, Wiley-VCH, pp. 207–243.
Kuktaite, R., Ture, H., Hedenqvist, M.S., Gallstedt, M., Plivelic, T.S., 2014. Gluten biopolymer and nanoclay-derived structures in wheat gluten-urea-clay composites: relation to barrier and mechanical properties. ACS Sustain. Chem. Eng. 2, 1439–1445.
LGL Bayern (Bayerisches Landesamt für Gesundheit und Lebensmittelsicherheit), 2012. Nanomaterialien in Lebensmitteln und Verbraucherprodukten. Anwendungsbereiche, Analytik, rechtliche Rahmenbedingungen. <http://www.lgl.bayern.de/publikationen/doc/nanomaterialien_lebensmittel_produkte.pdf>
Li, J.H., Hong, R.Y., Li, M.Y., Li, H.Z., Zheng, Y., Ding, J., 2009. Effects of ZnO nanoparticles on the mechanical and antibacterial properties of polyurethane coatings. Prog. Org. Coat. 64, 504–509.
Lopez, O, Castillo, L., Zaritzky, N., Barbosa, S., Villar, M., Garcia, M.A., 2015a. Talc nanoparticles influence on thermoplastic corn starch film properties. Proced. Mater. Sci. 8, 338–345.
Lopez, O.V., Castillo, L.A., Garcia, M.A., Villar, M.A., Barbosa, S.E., 2015b. Food packaging bags based on thermoplastic corn starch reinforced with talc nanoparticles. Food Hydrocoll. 43, 18–24.
Maisanaba, S., Prieto, A.I., Pichardo, S., Jorda-Neneyto, M., Aucejo, S., Jos, A., 2015. Cytotoxicity and mutagenicity assessment of organomodified clays potentially used in food packaging. Toxicol. in Vitro 29, 1222–1230.
Mallakpour, S., Moslemi, S., 2012. Dispersion of chiral amino acid organomodified Cloisite Na^+ in poly(vinyl alcohol) matrix for designing of novel bionanocomposite films. Prog. Org. Coat. 74, 8–13.
Marcilla, A., Garcia, S., Garcia-Quesada, J.S., 2008. Migrability of PVC plasticizers. Polym. Test. 27, 221–233.
Martino, V.P., Ruseckaite, R.A., Jimenez, A., Averous, L., 2010. Correlation between composition, structure and properties of poly(lactic acid)/polyadipate-based nano-biocomposites. Macromol. Mater. Eng. 295, 551–558.
Martins, J.T., Bourbon, A.I., Pinheiro, A.C., Souza, B.W.S., Cerqueira, M.A., Vicente, A.A., 2013. Biocomposite films based on κ-carrageenan/locust bean gum blends and clays: physical and antimicrobial properties. Food Bioprocess. Technol. 6, 2081–2092.

Miguel, O., Iruin, J.J., 1999. Evaluation of the transport properties of poly(3-hydroxybutyrate) and its 3-hydroxyvalerate copolymers for packaging applications. Macromol. Symp. 144, 427–438.

Mihai, M., Huneault, M.A., Favis, B.D., Li, H., 2007. Extrusion foaming of semi-crystalline PLA and PLA/thermoplastic starch blends. Macromol. Biosci. 7, 907–920.

Modi, S., Koelling, K., Vodovotz, Y., 2011. Assessment of PHB with varying hydroxyvalerate content for potential packaging applications. Eur. Polym. J. 47, 179–186.

Nafchi, A.M., Moradpour, M., Saeidi, M., Alias, A.K., 2013. Thermoplastic starches: properties, challenges, and prospects. Starch 65, 61–72.

Nielsen, L.E., 1967. Models for the permeability of filled polymers systems. J. Macromol. Sci. A 1, 929–942.

Nobel, M.L., Picken, S.J., Mendes, E., 2006. Waterborne nanocomposite resins for automotive coating applications. Prog. Org. Coat. 58, 96–104.

Oguzlu, H., Tihminlioglu, F., 2010. Preparation and barrier properties of chitosan-layered silicate nanocomposites films. Maromol. Symp. 298, 91–98.

Olivato, J.B., Marini, J., Pollet, E., Yamashita, F., Grossmann, M.V.E., Avérous, L., 2015. Elaboration, morphology and properties of starch/polyester nano-biocomposites based on sepiolite ckay. Carbohyd. Polym. 118, 250–256.

Olsson, E., Hedenqvist, M.S., Johansson, C., Järnström, L., 2013. Influence of citric acid and curing on moisture sorption, diffusion and permeability of starch films. Carbohyd. Polym. 94, 765–772.

Othman, S.H., 2014. Bio-nanocomposite materials for food packaging applications: types of biopolymer and nano-sized filler. Agr. Agr. Sci. Procedia 2, 296–303.

Park, H.J., Bunn, J.M., Weller, C.L., Vergano, P.J., Testin, R.F., 1994. Water vapor permeability and mechanical properties of grain protein-based films as effected by mixtures of polyethylene glycol and glycerin plasticizers. Am. Soc. Agric. Eng. 37, 1281–1285.

Park, H.J., Chinnan, M.S., 1995. Gas and water vapor barrier properties of edible coatings for fruit and vegetables. J. Food Eng. 25, 497–507.

Peters, R., Kramer, E., Oomen, A.G., Herrera Rivera, Z.E., Oegema, G., Tromp, P.C., Fokkink, R., Rietveld, A., Marvin, H.J.P., Weigel, S., Peijnenburg, A.A.C.M., Bouwmeester, H., 2012. Presence of nano-sized silica during in vitro digestion of foods containing silica as a food additive. Nano 6, 2441–2451.

Popović, S., Peričin, D., Vaštag, Z., Lazić, V., Popović, L., 2012. Pumpkin oil cake protein isolate films as potential gas barrier coating. J. Food Eng. 110, 374–379.

Ramos, O.L., Silva, S.I., Soares, J.C., Fernandes, J.C., Poças, M.F., Pintado, M.E., Malcata, F.X., 2012. Features and performance of edible films, obtained from whey protein isolate formulated with antimicrobial compounds. Food Res. Int. 45, 351–361.

Reddy, M.M., Vivekanandhan, S., Misra, M., Bhatia, S.K., Mohanty, A.K., 2013. Biobased plastics and bionanocomposites: current status and future opportunities. Prog. Polym. Sci. 38, 1653–1689.

Rekha, R.K., Siji, K.M., Sabu, T., Laly, A.P., 2015. Environment friendly green composites based on soy protein isolate: a review. Food Hydrocoll. 50, 174–192.

Rhim, J.-W., Park, H.-M., Ha, C.-S., 2013. Bio-nanocomposites for food packaging applications. Prog. Polym. Sci. 38, 1629–1652.

Romero-Bastida, C.A., Bello-Pérez, L.A., Velazquez, G., Alvarez-Ramirez, J., 2015. Effect of the addition order and amylose content on mechanical, barrier and structural properties of films made with starch and montmorillonite. Carbohydr. Polym. 127, 195–201.

Sadegh-Hassani, F., Nafchi, A.M., 2014. Preparation and characterization of bionanocomposite films based on potato strach/halloysite nanoclay. Int. J. Biolog. Macromol. 67, 458–462.
Salvatore, M., Marra, A., Duraccio, D., Shayanfar, S., Pillai, S.D., Cimmino, S., Silvestre, C., 2015. Effect of electron beam irradiation on the properties of polylactric acid/montmorillonite nanocomposites for food packaging applications. J. Appl. Polym. Sci. 133, 42219.
Sanchez-Garcia, M.D., Lagaron, J.M., 2010. Novel clay-based nanobiocomposites of biopolyesters with synergistic barrier to UV light, gas, and vapour. J. Appl. Polym. Sci. 118, 188–199.
Sanchez-Garcia, M.D., Lopez-Rubio, A., Lagaron, J.M., 2010. Natural micro and nanobiocomposites with enhanced barrier properties and novel functionalities for food biopackaging applications. Trends Food Sci. Technol. 21, 528–536.
Shchukin, D.G., Mohwald, H., 2007. Surface-engineered nanocontainers for entrapment of corrosion inhibitors. Adv. Funct. Mater. 17, 1451–1458.
Slavutsky, A.M., Bertuzzi, M.A., Armada, M., 2012. Water barrier properties of starch-clay nanocomposites films. Braz. J. Food Technol. 15, 208–218.
Song, N.-B., Jo, W.-S., Song, H.-Y., Chung, K.-S., Won, M., Song, K.B., 2013. Effects of plasticizers and nano-clay content on the physical properties of chicken feather protein composite films. Food Hydrocoll. 31, 340–345.
Sorrentino, A., Gorrasi, G., Vittoria, V., 2007. Potential perspectives of bio-nanocomposites for food packaging applications. Trend. Food Sci. Technol. 18, 84–95.
Souza, A.C., Benze, R., Ferrao, E.S., Ditchfield, C., Coelho, A.C.V., Tadini, C.C., 2012. Cassava starch biodegradable films: influence of glycerol and clay nanoparticles content on tensile and barrier properties and glass transition temperature. LWT Food Sci. Technol. 46, 110.
Svagan, A.J., Akesson, A., Cardenas, M., Bulut, S., Knudsen, J.C., RIsbo, J., Plackett, D., 2012. Transparent films based on PLA and montmorillonite with tunable oxygen barrier properties. Biomacromolecules 13, 397–405.
Tang, X.Z., Kumar, P., Alavi, S., Sandeep, K.P., 2012. Recent advances in biopolymers and biopolymer-based nanocomposites for food packaging materials. Crit. Rev. Food Sci. Nutr. 52, 426–442.
Tapia, M.S., Rojas-Grau, M.A., Carmonac, A., Rodrıguez, F.J., Soliva-Fortuny, R., Martin-Belloso, O., 2008. Use of alginate- and gellan-based coatings for improving barrier, texture and nutritional properties of fresh-cut papaya. Food Hydrocoll. 22, 1493–1503.
Ten, E., Bahr, D.F., Li, B., Jiang, L., Wolcott, M.P., 2012. Effects of cellulose nanowhiskers on mechanical, dielectric and rheological properties of poly(3-hydroxybutyrate-*co*-3-hydroxyvalerate)/cellulose nanowhisker. Comp. Ind. Eng. Chem. Res. 51, 2941–2951.
Ture, H., Gallstedt, M., Johansson, E., Hedenqvist, M.S., 2013. Wheat-gluten/montmorillonite clay multilayer-coated paperboards with high barrier properties. Ind. Crops Prod. 51, 1–6.
Vargas, G., Trifol, J., Algar, I., Arbelaiz, A., Mondragon, G., Fernandes, S.C.M., Mujika, F., Mondragon, I., 2012. Nanostructured composite materials reinforced with nature-based cellulose nanofibers. In: Hernandez, S., Brebbia, C.A. (Eds.), Design and Nature VI. WIT Press, UK, pp. 237–248.
Vieira, M.G.A., Altenhofen da Silva, M., Oliveira dos Santos, L., Beppu, M.M., 2011. Natural-based plasticizers and biopolymer films: a review. Eur. Polym. J. 47, 254–263.
Wang, Y., Xiong, Y.L., Rentfrow, K.G., Newman, M.C., 2013. Oxidation promotes cross-linking but impairs film-forming properties of whey proteins. J. Food Eng. 115, 11–19.

Wee, Y., Kim, J., Ryu, H., 2006. Biotechnological production of lactic acid and its recent applications. Food. Technol. Biotechnol. 44, 163–172.
Xu, B., Zheng, Q., Song, Y., Shangguan, Y., 2006. Calculating barrier properties of polymer/clay nanocomposites: effect of clay layers. Polymer 47, 2904–2910.
Youssef, A.M., El-Sayed, S.M., Salama, H.H., El-Sayed, H.S., Dufresne, A., 2015. Evaluation of bionanocomposites as packaging material on properties of soft white cheese during storage period. Carbohydr. Polym. 132, 274–285.
Zdrahala, R.J., 1997. Thermoplastic starch revisited: structure/property relationship for "dialed-in" biodegradability. Marcomol. Symp. 123, 113–121.
Żenkiewicz, M., Richert, J., 2008. Permeability of polylactide nanocomposite films for water vapour, oxygen and carbon dioxide. Polym. Test. 27, 835–840.

14

BARRIER PROPERTIES IMPROVEMENT USING ADDITIVES

J. Rodolfo Rendón-Villalobos, Javier Solorza-Feria, Francisco Rodríguez-González, Emmanuel Flores-Huicochea
Instituto Politécnico Nacional, Centro de Desarrollo de Productos Bióticos, Departamento de Desarrollo Tecnológico, San Isidro. Yautepec, Mexico

1 Introduction

In the past decades, it has increased the tendency of barrier plastic materials finding a clear substitution of chlorine-containing polymers and the rational use of these. That is to say, it is expected to get the same level with less thickness barrier of plastic materials, both laminated and coextruded packaging and, especially, with the development of biodegradable polymers from renewable natural resources and frequently used.

The barrier plastic materials are used primarily in the packaging industry, whose function is to extend the shelf life of food and beverages, because these prevent or diminish considerably the passage of gases and volatiles from or into the packaging. The presence of oxygen may accelerate loss of nutritional and sensory quality of food. Furthermore, the aroma barrier is fundamental both to prevent the egress of aromas of food as the entry of undesirable odors from the environment.

The gas permeability of the packaging materials is a gas transportation process, which occurs in several stages: adsorption and dissolution on the surface of the plastic material, subsequently, diffusion through and, finally, desorption of either the product or the environment.

Moreover, in recent years, it has also being studied the possible use of nanoscale materials that can be used to protect the food from adverse agents such as microorganisms. The use of functional substances, with possible beneficial effects on food and edible coatings deposited directly on the food itself, is also one of the trends in the packaging sector.

Food Packaging. http://dx.doi.org/10.1016/B978-0-12-804302-8.00014-5

Nanotechnology encompasses a range of techniques and sciences, in which materials, substances, and devices between 1 and 100 nm are studied, manipulated, and produced. With time and support of nanotechnology, one of the trends in the sector of plastic packaging for the food industry, is on track to produce lighter packaging, maintaining its properties.

This is how it has been working to add to nanoscale and in very small quantities; polymer particles to get improved certain properties of the packaging, such as, for example, the gas barrier.

2 Barrier Properties of Materials

There has been a renewed interest in edible and biodegradable films, which have the potential to replace conventional packaging in some applications. Polysaccharides, such as starches, cellulose derivatives, and plant gums, and proteins such as gelatin, have been used in edible films and coatings in food packaging and preservation (Debeaufort et al., 1998). An edible film can be defined as a thin layer of edible material, which can be formed on a food as a coating or preformed as a film that can be placed between food components, used as a food wrap, or formed into a pouch to contain foods. Its purpose is to inhibit migration of moisture, oxygen, carbon dioxide, aromas, and lipids; carry food ingredients and/or improve mechanical integrity or handling characteristics of foods (Krochta and de Mulder-Johnston, 1997). Most of the data on water vapor transmission (WVT) of edible films available in the literature, have been obtained by using the ASTM Standard Test Method E 96-80 (ASTM, 1989), known as the "cup method," or variations of it. According to this method, a cup with an open mouth of known area is filled with distilled water or desiccant. In a modification of the method, saturated salt solutions are also used. A film specimen is sealed on the open mouth of the cup, the assembly is weighed, and placed under controlled relative humidity and temperature conditions. A partial water vapor pressure difference ($p_{w0}-p_{w2}$), provides the driving force for water vapor flux through the film. The weight change of the cup is monitored by periodical weighing. Weight gain or loss is plotted over time and when steady state is reached the plot is usually a straight line. The slope of this line divided by the exposed film area (cup open mouth area) yields the water vapor transmission rate (WVTR) of a tested film. According to this standard method, cups should be filled with water to a level of 2 ± 0.5 cm from the film specimen. When the cups are filled with desiccant, an air space of about 0.6 cm between the underside of the film and the desiccant surface is recommended. This air gap is considered necessary to avoid film contact with water or desiccant while handling the cup. The air gap between a film and the surface of the liquid or desiccant inside the cup, is assumed to offer negligible resistance to water vapor transport (Gennadios et al., 1994).

In general, the difference on concentration of water across a film produces a chemical potential difference. It will generate a driving force to water transfer through the film known like water vapor permeability (WVP), and it turns out that these results or data can be advisable to comprehend hypothetical mass-transfer mechanisms and solute-macromolecule interactions in degradable films.

2.1 Gas Permeability

Permeability (*P*), is the product of solubility (*S*) and diffusivity (*D*), according to the following equation:

$$P = D \cdot S \tag{14.1}$$

Assuming a one direction diffusion through a flat membrane, such diffusion can be written as follows:

$$Ji = -\mathrm{Di(ci)[dci/dx]} \tag{14.2}$$

Where *J* is the flux, Di(ci) denotes that the diffusion coefficient that depends on the composition of the species that penetrate and c refers to the local gas or entrant (penetrant) concentration. The equation that describes the determination of the diffusion constant is:

$$D = d^2/6\theta \tag{14.3}$$

Where d is the thickness of the film and θ is the time lag for permeation. The lag is related to the time needed by the gas to reach an equilibrium in an initially gas-free film. Extrapolation of the pressure increase-time plot to the zero axes, will produce the time lag (θ) (Arvanitoyannis et al., 1996b). The amount of gas (*Q*) that will pass through the film, is directly proportional to the difference in pressure exerted by the gas on each side of the film ($p1-p2$) and is inversely proportional to its thickness (x). It is also directly proportional to the area exposed (*A*) and the time (*t*), during which permeation occurs. Overall, the relationship can be expressed as follows:

$$Q = [\mathrm{Pat}(p1 - p2)/x] \tag{14.4}$$

Where *P* has a constant value for a specific combination of gas and polymer at a given temperature, and is usually known as either the transmission factor or permeability factor/constant/coefficient (Van Krevelen, 1990; Arvanitoyannis et al., 1996a).

2.1.1 Water Vapor Transmission Rate (WVTR)

One of the problems encountered in the extensive application of edible films has been, and still is, their high WVTR. Usually, it is more difficult to make accurate sorption and transport measurements for water vapor than for most other penetrating species, because of the characteristic properties of water: A trend to adsorb on high energy surfaces, relatively high vaporization heat, low saturation vapor pressure, high solubility in several polymer systems, a trend to plasticize polymers and to cluster in the polymer matrix (Schult and Paul, 1996). In various publications (Gennadios et al., 1993, 1994) on WVTR of edible films and, in particular, films made from proteins or starch components such as gluten or amylose, the high WVTR values have been attributed to a "clustering" of water molecules during their diffusion through microcavities. It is not uncommon to find that an increase in the plasticizer content, either as water or as polyol (eg, glycerol, sorbitol, sucrose), results in higher WVTR values. These findings have been confirmed by theoretical calculations, using computer simulation on polysaccharides as starch and other polymers, mainly in an amorphous state (Takeuchi, 1990; Takeuchi and Okazaki, 1990; Takeuchi et al., 1990; Trommsdorf and Tomka, 1995).

2.1.2 Films Gas Permeability

Most edible films are not particularly effective as WVTR barriers, but they are quite efficient as gas barriers. A theoretical calculation of gas permeation in films has been the subject of several investigations, because it is directly related to potential applications of films (Van Krevelen, 1990; Salame, 1986). A straightforward relationship between polar groups and solubility is not that feasible, because of the complexity of interactions and inherent difficulties in evaluating affecting factors, such as accessibility of polar groups, the strength of water-water versus water-polymer interactions, crystallite sizes, and degree of crystallinity of the food matrix. The segment mobility of the amorphous component (through which permeation occurs) in a semicrystalline polymer, usually differs considerably from that of a fully amorphous polymer (Arvanitoyannis et al., 1996b; Hedenqvist and Gedd, 1996). For some penetrating molecules, their size may be even too large to go through the available interlayer separating two crystallites. The ordinarily accepted methodology for amorphous synthetic polymers to calculate gas permeability, is based in a model proposed by Salame (1986). In this model, the so called polymer permachor (π) is by definition proportional to the negative logarithm of a relative permeability. The product of $N \times \pi$ is named as molar permachor (Π), an additive molar function, according to the following equation.

$$N \times \pi = \Pi = \sum (Ni \times \pi i) \tag{14.5}$$

Where N, the number of characteristic groups per structural unit and πi, the increment of the group i.

After the numerical value of π is calculated, the permeability at ambient temperature can be estimated as:

$$P_{(298)} = P^{*}_{(298)} e^{(-s\pi)} \tag{14.6}$$

Where $P_{(298)}$, is the permeability of a standard gas (eg, N_2) in a standard polymer (eg, rubber, chosen arbitrarily) and s is the scaling factor. By substituting log P* = −12 and s = 0.122 v for nitrogen, results in the following equation (Van Krevelen, 1990):

$$\text{Log}\, P^{*}_{(298)} = -12 - 0.053\pi \tag{14.7}$$

It should be noted that the applicability of this equation is restricted to amorphous polymers. In order to apply the above mentioned equations to semicrystalline (sc) polymers, where the crystallites impose constraints on the amorphous chains, and gathering of chain defects (ie, chain ends and chain branches) confers different properties to the amorphous phase of sc polymers (Salame, 1986), thus, an additional modification is required leading to:

$$\Pi_{sc} = \pi_a - 18 \ln \alpha = \pi_a - 41.5 \log(1 - X_c) \tag{14.8}$$

Where α is the amorphous volume fraction, and X_c is the crystallinity.

When Eqs. 14.5–14.7 were applied to starch, the values obtained for dry amorphous samples were 6.99×10^{-20} and 2.4×10^{-19} cm^2 s^{-1} Pa^{-1} for amylose and amylopectin, respectively (Arvanitoyannis et al., 1996a).

The water content notoriously influences the films containing hydrogen groups and the gas diffusivity increases with water content. Arvanitoyannis et al. (1996b) have shown how the glass transition temperature (T_g), can be determined from permeability measurements, which is the inflection of the line from permeability versus the reciprocal of temperature. A probable reason for such a change in the slope could be that a strong localization of the initially sorbed water takes place over a limited number of sites, whereas at higher water contents, the film matrix may swell, thus resulting in an even higher mobility of the sorbed water molecules. At high water content, it is believed that the polymer chains move farther apart and the number of intra- and interchain hydrogen bonds, decrease considerably (Trommsdorf and Tomka, 1995). The widening of the composite matrix structure due to incorporation of plasticizer molecules would probably enhance gas permeation.

2.1.3 Effect of Temperature on Diffusivity

The dependence of diffusivity on temperature satisfactorily follows the Arrhenius equation.

$$D = D_0 D = D_0 e^{(-E_D/RT)} \tag{14.9}$$

Where D_0 is the diffusion coefficient, E_D is the activation energy of diffusion and R is the universal gas constant. The logarithmic expression for Eq. 14.9 that describes the diffusivity is:

$$\ln D = \ln D_0 - E_D/RT \tag{14.10}$$

Thus, the temperature dependence of permeability can be expressed as:

$$P = P_0 e^{(-E_P/RT)} \tag{14.11}$$

Where E_P is apparent activation energy permeation, and by analogy:

$$\ln P = \ln P_0 - E_p/RT \tag{14.12}$$

The activation energy of diffusion (E_D) is considered to be the most important parameter in the diffusion process because it is the energy needed to enable the dissolved molecule to go through (Arvanitoyannis et al., 1996b).

The gas diffusion through the film matrix could be visualized as following two different modes depending on temperature. That is to say, at low temperatures it can be assumed that diffusion is affected by small oscillation of gas molecules, which are trapped in small local voids (Gelin, 1994). At higher temperatures, formation of new cavities and creation of interconnecting channels occurs, finally leading to complete destruction of the original cavities.

2.1.4 Barrier Properties Applications in Biodegradable Films

Biodegradable films have proved to be effective barrier to gas transport (O_2, CO_2), although they present high WVP (Forssell et al., 2002). As it is known, the main functional properties of these hydrophilic materials closely depend on their water content and, as a consequence, on the surrounding humidity.

WVP results may be useful to understand possible mass transfer mechanisms and solute and polymer interactions in biodegradable films. According to the thermodynamics of irreversible process, water chemical potential difference is the driving force of the water transfer through a film. When the process occurs at constant temperature and pressure, the water chemical potential difference, results proportional to water vapor concentration difference between the two faces (Morillon et al., 2002).

Materials used to produce edible films can be divided into four categories: (1) biopolymer hydrocolloids, (2) lipids, (3) resins, and (4) composites. Biopolymer hydrocolloids include proteins such as gelatin, keratin, collagen, casein, whey protein, wheat gluten, and corn zein as well as polysaccharides such as starch and starch derivatives, cellulose derivatives, and plant gums. In the categories of lipids, waxes, acylglycerols, and fatty acids are included. Resins include shellac and wood resin. Composites usually contain both lipid and hydrocolloid components in the form of a bilayer or an emulsion (Perez-Gago and Krochta, 2005). Two main techniques are used to produce edible films: (1) the "wet process" involves biopolymer dispersion or solubilization in a film-forming solution, followed by evaporation of the solvent; (2) the "dry process" is based on the thermoplastic behavior that some proteins and polysaccharides display at low moisture levels in compression molding and extrusion (Liu et al., 2006).

Compared to edible waxes and most synthetic polymers, protein films have high WVP values. Among several proteins, studies have shown that the lowest WVP corresponds to corn zein, wheat gluten, and fish miofibrillar protein films (Hernández-Izquierdo and Krochta, 2008). Sothornvit et al. (2003), studied the effect of plasticizers content and processing conditions on WVP of compression-molded whey protein films, plasticized with either glycerol or water. The WVP was not affected by plasticizer content, molding temperature, or molding pressure. However, films plasticized with glycerol had higher WVP values than those plasticized with water. This effect was attributed to the loss of moisture occurring during film-formation by compression molding.

2.1.5 Barrier Properties of Food Plastic Materials (FPM)

The testing of the barrier properties of food plastic materials (FPM), is also an important point. From the chemical viewpoint, all FPM and food-contact surfaces can be permeable to a variety of substances. This should be complied by the food packaging

(FP) and other parts of the food chain, while the normal consumer seems convinced of the virtual impenetrability of plastic items and similar packages (Parisi, 2013). Two different mechanisms can explain the permeability of FPM (Lorusso et al., 1991): (1) absorption of gaseous molecules on external surfaces, and (2) diffusion of gaseous compounds by means of microscopic pores and transfer of gaseous compounds from the one side of the FPM to the other side. Unlike coated metals and glass materials, gaseous species can easily penetrate polymeric structures by means of: (1) thermally enhanced vibrational motions, and (2) the rearrangement of molecular empty spaces. These two situations can be synergistic and concomitant. On this basis, the impermeability to gaseous substances can be discussed and demonstrated with reference to particular polymers and composite structures.

In reference to packaging materials, barrier properties can be estimated in comparison to one material and one gaseous substance by means of the already described mathematical equations. As shown earlier, five parameters can be inserted in this expression: (1) the amount of diffused gas, (2) the difference between the partial pressures on two sides of plastic FPM, (3) estimated surface of the plastic FPM, (4) thickness of the plastic FPM, (5) time of exposure to diffusive phenomena.

The explanation of the already mentioned mathematical expression may be omitted on condition that the following statement is shown: permeability will decrease proportionally with the simultaneous increase of thickness. In other words, the larger the thickness of the plastic containers, the higher the barrier property to a determined gaseous and diffusive substance. Naturally, the polymeric composition must be identical. Barrier properties are always considered in reference to different conventional gases: for example, water vapor (moisture), oxygen, carbon dioxide (Parisi, 2013).

3 Classifications and Interrelationship of Material Barrier Properties

The period from production to consumption of fruit, vegetables and food products usually range of 1 week to 6 months (Liu et al., 2011; Dias et al., 2014; Rizzolo et al., 2015). The food systems can deteriorated by loss or gain of water; another factor is the oxygen, it play an import role on respiration of vegetables and fruit during the postharvest period, due physiological changes (Escalona et al., 2006), and additionally, rise the oxidation process (Amodio et al., 2015; Sharma and Rao, 2015).

A selective barrier can improve the preservation effect and prolong the shelf life of products. The term material barrier properties on food applications, is applied to reduction or elimination on water vapor, oxygen and dioxide carbon transmission rate (DeLassus, 2000). The main function of food packaging is to provide this barrier material against adverse conditions. Petersen et al. (1999) on a review of biobased materials for food packaging, exhibited optimal conditions of storage to several foods, the required barrier properties focused on oxygen, carbon dioxide and water, is high. This means a material with barrier properties, to reduce significantly the mass transfer of gases and water vapor. Conventionally, the films made from synthetic polymer, provided excellent material barrier against these gas and water vapor (DeLassus, 2000; Rizzolo et al., 2015).

Nevertheless, today exists a global environmental concern due to low biodegradation of petroleum-based packaging material. Jambeck et al. (2015) estimated that in 2010 from 4.8 to 12.7 metric tons of plastic waste entered the ocean; the ingested plastic transfers hazardous chemicals to fish and bioaccumulation of chemicals on tissues (Rochman et al., 2013). Researches and industries try to overcome the drawback of petrochemical-based plastics, by means of design and test biodegradable films from renewable sources (Petersen et al., 1999; Félix et al., 2014; Peelman et al., 2014). Starch, gelatin, cellulose, and chitin are biodegradable biopolymers, renewable and cheap (Williams and Phillips, 2009); therefore, efforts have been made to obtain new materials combining with other biopolymers, polymers, nanomaterials, fillers, and additives (Ke and Sun, 2000; El Miri et al., 2015; Wang et al., 2015).

The European Bioplastic Association forecast for the production until 2018 of bioplastic: biobased/nonbiodegradable and biodegradable will be 5605 and 1126 kTon and owing to the concerns about accumulations in the oceans of plastics fossil-based or bioplastics nonbiodegradable, the analysis of the barrier properties will be focused on biodegradable plastics (Fig. 14.1).

The films made from biodegradable polymers show poor mechanical and barrier properties, compared to synthetic polymers to be used on packaging applications (Fig. 14.2), the use of additives is necessary to improve the processability, mechanical and barrier properties. The typical additives used are: alcohols (glycerol, xylitol, polyvinyl) and nanomaterials (cellulose, montee, graphene, etc.) (López et al., 2015). The first improve processability and barrier properties, while the second increase the mechanical properties and barrier properties.

The term nanomaterial appeared in scientific literature in 1994 (Peng et al., 1994), the nanomaterial definition today is still in

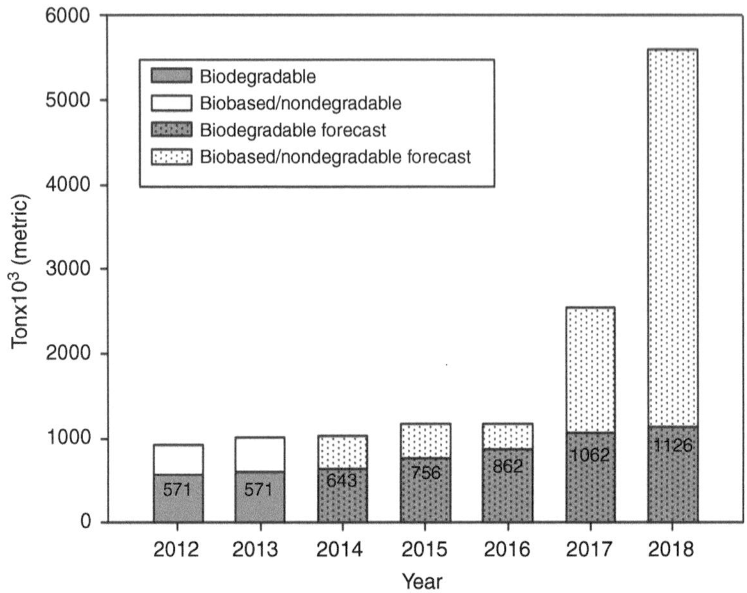

Figure 14.1. Global production of bioplastics. From European Bioplastics, Institute for Bioplastics and Biocomposites, nova-Institute, 2014. http://en.european-bioplastics.org/market/, http://ifbb.wp.hs-hannover.de/en/, and http://www.nova-institut.de/

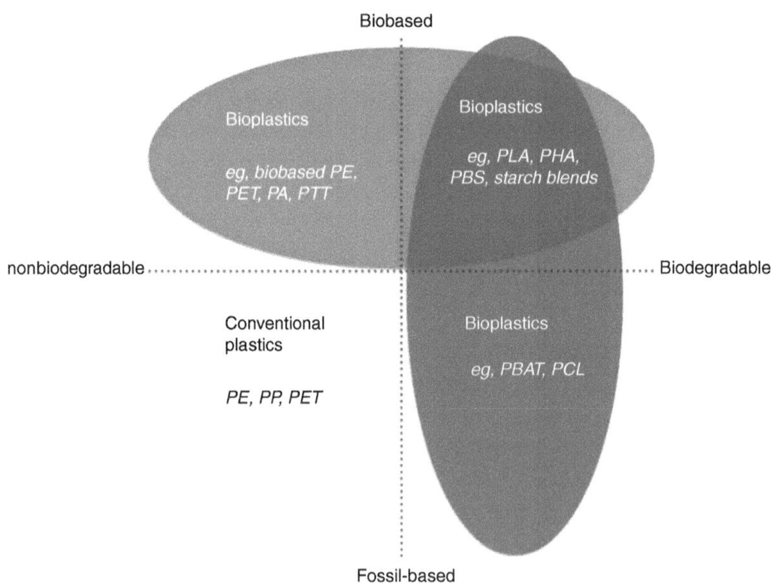

Figure 14.2. Classification of plastics based on raw material type of fabrication. From: European Bioplastics, Institute for Bioplastics and Biocomposites, nova-Institute, 2014. http://en.european-bioplastics.org/market/, http://ifbb.wp.hs-hannover.de/en/, and http://www.nova-institut.de/

debate, but various organizations have proposed the upper limit of 100 nm. Nanomaterials can be particles, agglomerated or aggregated. Additionally, the film term is used like the ASTM describes in brief films plastics, having a nominal thickness not greater than 250 μm. The next section describes the use of nanomaterials and additives to improve the material barrier properties of films, focused on the barrier property of permeability.

3.1 Water Vapor Permeability

The measurement of WVP is straightforward, and most of the researches related with biopolymers films include it. An analysis of papers in the database ScienceDirect show that the main biopolymers used to made films are starch, gelatin, and cellulose; the analysis of WVP primarily focuses on these biopolymers.

The mechanism of transport (permeability) of water across films is a complex process, due to the interactions of water molecules with the biopolymer molecular structure. Furthermore, as mentioned earlier, the permeability is affected by temperature, film thickness (30–150 μm), and plasticizer content. The addition of plasticizers to film-forming solutions is essential to avoid the film brittleness due intermolecular forces, reduce these forces, and add flexibility and extensibility of the films (Forssell et al., 2002).

The amylose, a linear biopolymer with repeated units of glucose, is a component of starch. On films of corn starch with high amylose content and glycerol as plasticizer, the WVP increased as temperature rose from 5 to 40°C, with values range between 2×10^{-10} and 1×10^{-9} $gm^{-1}s^{-1}$ Pa^{-1} (Bertuzzi et al., 2007). The WVP increased linearly with film thickness. McHugh et al. (1993) observed the same behavior, as film thickness increased, the film provided an increased resistance to mass transfer across it; and consequently, the equilibrium water vapor partial pressure at inner film surface increased. Other explanation about the thickness effect is due to film swelling as a result of attractive force between polymer, glycerol, and water (Park et al., 1993). The plasticizer effect of films is to reduce the intermolecular force among the biopolymer chains and to improve the extensibility and flexibility (Forssell et al., 2002). On films made of water soluble biopolymers, several plasticizers were tested, finding that the glycerol was the best (Jangchud and Chinnan, 1999), the glycerol effect over WVP of amylose films was to increase linearly as the concentration of glycerol increased (Bertuzzi et al., 2007).

The WVP of starch-gelatin films with two plasticizer (sorbitol and glycerol) were reported by Al-Hassan and Norziah (2012), sago starch with 32–34% amylose content and fish gelatin

(81.3% protein), using various ratios sago starch to gelatin fish (1:0, 2:1, 3:1, 4:1, and 5:1). The increase in gelatin content (from 10.9% to 21.6%) in film with glycerol addition, did not present differences on WVP, but higher values were obtained with sorbitol related to glycerol films, except for the starch-gelatin ratio 1:0, where WVP values were higher for glycerol plasticized films. Other research reported that the plasticizer modified the protein network structure and increased the WVP, when sorbitol or glycerol increased its concentration from 25% to 75% (Bourtoom et al., 2006). The values range of WVP were from 2.77×10^{-11} g m^{-1} s^{-1} Pa^{-1} to 9.72×10^{-11} g m^{-1} s^{-1} Pa^{-1} with thickness films from 50 to 70 μm, these values were lower than those obtained for amylose films reported previously (Bertuzzi et al., 2007), the starch films compared to amylose films seemed to reduce the WVP.

The gelatin is the second largest biopolymer sold and used on several areas (Gómez-Guillén et al., 2009; Haug and Draget, 2009). The gelatin is a protein obtained by a partial hydrolysis of collagen, it has been used as a source for films or edible films due to its abundance and biodegradability (Jongjareonrak et al., 2006). The films produced are transparent but the three-dimensional network formed produces brittle films (Vanin et al., 2005). Protein-based films report typically used polyols plasticizers to reduce the brittleness (Audic and Chaufer, 2005). Cao et al. (2009) described the effect of several plasticizer like organic acids, polyethylene glycol (PEG) onto the WVP of gelatin films, the lower molecular weight of PEG provided a better plasticizer effect and lower WVP values in the range 3.72×10^{-13} g m^{-1} s^{-1} Pa^{-1} to 6.36×10^{-13} g m^{-1} s^{-1} Pa^{-1} with thickness in the range 22–26 μm. The values WVP of gelatin films were lower than those reported for starch film (Bertuzzi et al., 2007). Other works reported the use the polyols (glycerol, propylene glycol, diethylene glycol, sorbitol) of gelatin films on WVP, showing values on the interval from 8.2×10^{-3}g m^{-1} s^{-1} Pa^{-1} to 1.6×10^{-2} g m^{-1} s^{-1} Pa^{-1} (Vanin et al., 2005; Jongjareonrak et al., 2006), WVP values were higher than those reported for starch films (Bertuzzi et al., 2007).

The natural or chemically modified cellulose confirm the third largest reports of films. The films produced are transparent and show oxygen and dioxide carbon barrier properties better than low and density polyethylene (PE) (Syverud and Stenius, 2009; Lavoine et al., 2012). The WVP cellulose films range from 1.8×10^{-13} to 3.6×10^{-10} g m^{-1} s^{-1} Pa^{-1}, the glycerol was used as plasticizer, the thickness was on the interval 20–45 μm (Minelli et al., 2010; Del Nobile et al., 2002).

Recently, it has been reported that for maize starch films with nanoparticles of maize starch, even the mechanical properties increase with maximal content of 1% of nanomaterial, the WVP minimal value was 3.04×10^{-8} g m^{-1} s^{-1} Pa^{-1} (Fan et al., 2016), the value reported was 1000 times greater than starch film without nanomaterial (Bertuzzi et al., 2007).

Although the main function of food packaging is to avoid or decrease the moisture transfer between the food and the adjacent atmosphere, WVP must be as low as possible, for example, 5.31×10^{-13} g m^{-1} $^{-1}$ Pa^{-1} (Gontard et al., 1992). The values of WVP reported for low density polyethylene (LDPE) and high density polyethylene (HDPE) films showed WVP of $1–1.5 \times 10^{-13}$ g m^{-1} s^{-1} Pa^{-1} and $3–4 \times 10^{-13}$ g m^{-1} s^{-1} Pa^{-1} at relative humidity of 90% and 38°C (Ashley, 1985), the starch and cellulose provided WVP in the range of those of HDPE and LDPE. The WVP of gelatin films had a broadly range from 3×10^{-13} to 8×10^{-3} g m^{-1} s^{-1} Pa^{-1}.

As general conclusion, WVP data of films must be provided with other characteristic parameters, such as plasticizer content, water activity gradient, film thickness and temperature, to be comparable (Bertuzzi et al., 2007).

4 Multilayer Film and Coating

In recent years, the use of synthetic materials for the production of films and coatings of multiple layers, known as multilayer packaging, is being promoted and investigated by the need to build coatings that combine their different properties and functionality, for better handling and applications, compared with those films and coatings that have a single layer or monolayer (Mitsoulis, 1988; Decker et al., 2015). These research topics have boomed in the European Union and countries such as Germany, France, Italy, and the United Kingdom, which are the main producing and consuming nations films and multilayer coatings. Among these countries, it is estimated a higher production at 160, 000 tons per year of this type of films addition, it was considered that the use of these products is steadily growing (Fang et al., 2005; Dixon, 2011; Decker et al., 2015).

4.1 Multilayer Films of Synthetic Polymers

Multilayer films and coatings made from synthetic polymers have a variety of applications, especially in the food industry. For example, they have been used as coatings to increase the lifetime of meats, fruits and vegetables minimally processed (Vidal et al., 2007; Dixon, 2011; Decker et al., 2015). On the other hand, the production of films and multilayer coatings for food applications, must meet all requirements and specifications established for the design and assembly of each of the components (monolayers) constituting the coating. For this, all foundations are based on the construction of a single structure that allows the assembly and combination of monolayer films of different materials, and with different properties, and that together fulfill the purpose and application requirements set (Dixon, 2011; Decker et al., 2015). Currently, the multilayer films sold consist of three to nine layers of different polymers (Mitsoulis, 1988; Vidal et al. 2007; Decker et al., 2015). Generally, multilayer coating films using synthetic polymers as raw material are produced by coextrusion processing (Mitsoulis, 1988). The internal design of this method has different channels, according to the number of layers that form the multilayer film. For each channel, a synthetic polymer melt flows at different temperatures of processing, which may involve different viscosity values of each polymer composing the coating, consisting of several layers (Mitsoulis, 1988; Wang et al., 2000).

Moreover, the monolayer and multilayer films made from synthetic materials have great advantages, particularly in resistance to microbial attack, degradation and biodegradation, and the transfer of moisture, odors, and flavors in products for human consumption (Fang et al., 2005; Vidal et al., 2007; Decker et al., 2015). However, resistance to degradation and biodegradation of synthetic polymers has led to the accumulation of solid waste in much of the land and sea territories, resulting in a serious environmental problem (Martin et al., 2001; Vidal et al., 2007). In addition to this problem, the habit in the reuse and recycling of products made from these materials products is almost nil. Due to environmental concerns associated with the management of plastic waste, many research groups from different countries have proposed technologies for the development and use of degradable and biodegradable polymers for the production of films and monolayer and multilayer coatings (Arvanitoyannis, 1999; Nayak 1999; Wang et al., 2000; Martin et al., 2001; Fang et al., 2005; Rhim et al., 2006; Vidal et al., 2007; Rivero et al., 2009; Martucci and Ruseckaite, 2010; Gu et al., 2013; Laufer et al., 2013).

4.2 Multilayer Films and Coatings of Biopolymers

Due to the properties of biopolymers exhibiting biodegradation of animal and vegetable origin (Vidal et al., 2007; Vartiainen et al., 2014), these materials have been used for the production of monolayer and multilayer films and coating (Rivero et al., 2009; Martucci and Ruseckaite, 2010). Most of the biopolymers used to prepare multilayer films include native starches and modified thermoplastics (Fang et al., 2005; Vidal et al., 2007), protein

(Ghanbarzadeh and Oromiehi, 2009), and other polysaccharides such as gelatin, chitosan, carrageenan (Laufer et al., 2013), combinations of biodegradable biopolymers with synthetic polymers to obtain multilayer coatings, have also been used (Gu et al., 2013; Laufer et al., 2013; Decker et al., 2015).

On the other hand, the most widely used methods for obtaining multilayer films use casting in plate biopolymers (Rhim et al., 2006; Rivero et al., 2009), extrusion, and coextrusion (Wang et al., 2000; Martin et al., 2001; Fang et al., 2005), compression, and thermoforming (Hulleman et al., 1999; Martucci and Ruseckaite, 2010) and layer-by-layer technology (Schneider et al., 2007; Jang et al., 2008; Dickinson, 2011; Gu et al., 2013; Laufer et al., 2013). The preparation of multilayer films with any of the aforementioned methods must satisfy the requirements for good design and assembly of each of the monolayers forming the multilayer film. These latter are generally formed of outer layers obtained from inexpensive materials, and with good barrier properties such as solubility and water permeability and good mechanical properties; and inner layers consisting of more expensive materials, which have good barrier properties to gases such as oxygen and carbon dioxide (Wang et al., 2000; Fang et al., 2005; Vidal et al., 2007). Moreover, in recent decades the production and consumption of films and multilayer coating has increased, but research related to the physicochemical properties of these products are few. The properties that have been more evaluated in the multilayer films are: breaking stress, elongation; which are related to material properties (Wang et al., 2000; Martin et al., 2001; Fang et al., 2005; Rhim et al., 2006; Rivero et al., 2009; Martucci and Ruseckaite, 2010). Also, the properties as a transmission barrier to water vapor, oxygen, carbon dioxide, and other gases have been studied (Martin et al., 2001; Rhim et al., 2006; Jang et al., 2008; Rivero et al., 2009; Martucci and Ruseckaite, 2010), and the preservation of the quality of food in the case of films with applications in consumer products. They have been evaluated, including the protection of odors and flavors, protection to visible light and UV, among others.

4.3 Barrier Properties of Multilayer Films and Coatings

One of the first works in the production of multilayer films was reported by Wang et al. (2000). These authors developed plates using the multilayer coextrusion method, thermoplastic corn starch was used as an inner plate and outer plates were used as polyester layers. Wang and coworkers characterized the plates obtained, by scanning electron microscopy, X-ray, and mechanical properties.

From studies of the multilayer boards, these researchers concluded that there was good compatibility between the plates of thermoplastic starch with polyester, and that this effect was given by hydrogen bonds between the two polymers. However, the multilayer plates obtained by these authors, using the method of coextrusion were thick (2–6 mm); this feature benefits the particular barrier properties to WVP, but applications of these plates may be limited.

Martin et al. (2001) developed multilayer films based on thermoplastic starch wheat and various biodegradable polyesters such as poly(lactic acid) (PLA), polyesteramide, poly(ε-caprolactone), poly(butylene succine adipate), and poly(hydroxybutyrate-*co*-valerate). They assessed the effect of the polyester in the adhesion and barrier properties. The inner layer of the multilayer films was made from thermoplastic starch and the outer layers were made of different polyesters. The multilayer films were prepared by two methods, coextrusion and compression molding. In respect to adhesion results, they found that the multilayer films produced by coextrusion, showed more adhesion between layers of multilayer films than that obtained by compression molding. This result was associated with the compatibility of different thermoplastic polyesters with starch. Also, they showed that when more plasticizers were added for processing starch, the adhesion with the polyesters decreased, because the plasticizers migrate to the surface of polymers. Barrier properties and water penetration to the multilayer films were measured via contact angle and water immersion tests, respectively. From such tests, Martin and coworkers obtained a very low water absorption in the multilayer films, indicating a strong hydrophobic character in the films, caused by the outer layer of polyester multilayer films. Regarding the immersion tests, they found that there were no drastic changes in the swelling of the multilayer films, which allowed them to conclude that these films have good resistance to moisture.

Fang et al. (2005) prepared multilayer films using potato starch modified with PLA by the method of coextrusion. Starch was used as inner film and the outer films as PLA. They found that the chemical modification carried starch that caused a water resistance compared with native starch and good extrusion processing. Furthermore, it was found good compatibility of potato starch modified with PLA for obtaining biodegradable multilayer films.

Rhim et al. (2006) using isolated soy protein and PLA-prepared multilayer films, the inner film being made of isolated soy protein and external PLA films. The multilayer films were prepared by the method of casting. The authors characterized separately films of soy protein isolate, PLA, and multilayer films by means of mechanical properties, permeability to water vapor and oxygen, the water contact angle, and the microstructure by scanning electron microscopy. During the production of the multilayer films, the authors found an easy separation of the multilayer films of the boxes, where there were discharged the solutions, that led them to conclude that films of isolated soy protein with PLA were compatible. From the results of WVP, it was found that multilayer films presented WVP values of films made with PLA ($4.66 \pm 0.25 \times 10^{-14}$ kg m m^{-2} s Pa), showing two orders of magnitude lower than those obtained with films of soy protein isolate ($268.37 \pm 011.02 \times 10^{-14}$ kg m/m^2 s Pa). Yet the WVP value of the multilayer film ($6.66 \pm 0.27 \times 10^{-14}$ kg m m^{-2} s Pa) showed an improvement compared to the WVP of the films from soy protein isolate being comparable with the value presented by PLA films. An improvement of nearly 40-fold the value of the multilayer films WVP with respect to films of soy protein isolate, was attributed to the hydrophobic character showed by films made from PLA. On the other hand, with respect to the evidence of water solubility (WS), these researchers noted that films made with soy protein isolate were 100% soluble for its hydrophilic character. However, the films made with PLA did not show water solubility for its hydrophobic property. The value of water solubility of the multilayer films was 45.4 ± 1.6 %, which was found between the values presented between PLA films and soy protein isolate. Rhim and coworkers attributed the value WS of the multilayer films to the effect of the solubility of the films from soy protein isolate, that is, among the films of PLA, as observed during the immersion of the multilayer films in water (24 h), they were separated into two films by the dissolution of the protein isolate film in water, as the latter possibly penetrated into the multilayer film around the edges.

Martucci and Ruseckaite (2010) obtained multilayer films of gelatin extracted from cattle, reinforced with nanoclay and gelatin, crosslinked by a chemical method. The multilayer films were obtained by the casting method and plating, followed by a compression molding process. The internal film was reinforced gelatin film with nanoclay, and the outer film was crosslinked gelatin. These authors characterized their multilayer films by scanning electron microscopy, water solubility, and WVP. From the results of the microstructure, the researchers observed a good compatibility between the three layers that constitute the multilayer film. With regard to the values of water solubility of the individual films and multilayer films, these authors obtained solubility values of 100% of the films made from gelatin reinforced with nanoclay and 31.2 ± 1.4% of the films obtained with the crosslinked gelatin, and in the case of the multilayer films, it was 31.0 ± 2.4%. This result was attributed to the solubility of external films, because due to the process of obtaining the multilayer films used by these authors, the edges of these were tightly sealed, preventing water penetration into the multilayer films, that is, the inner layer did not contact with water. Regarding the results of WVP, it was found that multilayer films showed the lowest value of about 0.8×10^{-13} kg m m^{-2} Pa s, compared with the values displayed by the individual films, gelatin reinforced with nanoclay and crosslinked gelatin, which were 10.9×10^{-13} and 12.1×10^{-13} kg m m^{-2} Pa s, respectively.

Furthermore, films made from polymers associated through hydrogen bonds or ionic, generally show good properties of oxygen permeability but are susceptible to water vapor (McHugh and Krochta, 1994). For example, the films obtained with isolated proteins having hydrophilic nature are good oxygen barriers at intermediate relative humidity conditions, but show poor water permeability. On the contrary, films with hydrophobic properties, that is, films of PE and polypropylene (PP), show high values of oxygen permeability and low water vapor values (Rhim et al., 2006).

5 Applications of Nanotechnology in Barrier Properties

In recent years, the field of nanotechnology is one of the most popular areas for current research and development. It is already proving to be a useful tool that is expected to have wide-ranging implications in all fields of science and technology, such as material science, mechanics, electronics, optics, medicine, energy and aerospace, plastics and textiles. Nanostructured materials may be defined as those materials whose structural elements, clusters, crystallites, or molecules, have at least 1 dimension in the order of the nanometers, that is, from 1 to 100 nm, more typically less than 50 nm. Nanostructured materials include atomic clusters, layered or lamellar films, filamentary structures, and bulk nanostructured materials. This obviously includes polymer science and technology and even in this field the investigations cover a broad range of topics (Paul and Roberson, 2008). This would include polymer-based biomaterials, nanoparticle drug delivery, miniemulsion particles, mixtures of polymers and composites in nanoscale. Indeed in nanoscale composites, there are numerous subjects that have been of interest, such as the study of barrier, electro-optical and bactericidal properties, and various applications such as cosmetic and resistance to catching fire.

On the other hand, nanotechnology is a technology that has the potential to revolutionize the food industry, being food packaging one of the earliest commercial applications of nanotechnology in the food sector. The significant purpose of nanopackaging is to set longer shelf life by improving the barrier function of food packaging, to reduce gas and moisture exchange and UV light exposure (Sorrentino et al., 2007). For example, after food was produced and will not be consumed in a short time, it is suggested to be stored using a packaging that serves and helps in several functions. Also, the care of food from impurity or dust, gas transfer (ie, oxygen), germs, humidity as well as other damaging substances, the bundle has to be safe under its intended conditions of use. A height argument in food packaging and still remaining, is the migration and permeability since no material may contain feature of being completely impermeable to gas migration or natural substances incorporated into a food being packaged, including the packaging material itself (Arvanitoyannis and Bosnea, 2004). The use of packaging material that promotes high barriers to migration or gas diffusion are not appropriate for fresh fruits and vegetables because this is detrimental to maintain the quality of food given that shelf life relies on a constant provision of oxygen for be a viable cellular respiration.

Due to these diversities, the food industry needs a coordinated system whereby products produced or harvested are accommodated within a set packaging for transportation from the production site to the place of consumption without suffering any damage. On the other hand, the demands on the packaging industry will only increase when food is protected from physical and environmental hazards during storage and transport, and thus, motivate the consumer to acquire assurance that it is the same quality at the time of collection or production.

Traditional materials for food packaging include metal, ceramic (glass), and paper (cardboard). While these materials are still used, the lightweight, low cost, ease of processing and formability, and remarkable diversity in physical properties of organic polymeric materials makes plastics attractive alternatives for the packaging of foods. Polymers have revolutionized the food industry for food packaging, and possess numerous advantages over conventional materials; but their major drawback is an inherent permeability to gases and other small molecules. Because no known pure polymer exhibits all the desired mechanical and barrier properties required for every conceivable food packaging application, complex multilayer films or polymer blends are often utilized.

The development of materials, devices, and innovative systems based on nanocomposites have allowed to generate an increase of attention particularly in a present-day scientific literature.

The application of nanocomposites promises to expand the use of edible and biodegradable films that come from renewable and normally biodegradable natural resources (Lagaron et al., 2005; Rivera-Corona et al., 2014). It will help to reduce the packaging waste (Rivera-Corona et al., 2014) associated with processed food and will support the preservation of fresh foods extending their shelf life (Labuza and Breene, 1988; Vermeiren et al., 1999). Polymer nanocomposites are created by dispersing an inert, nanoscale filler throughout a polymeric matrix. Filler materials can include clay and silicate, silica (SiO_2) nanoparticles (Wu et al., 2002; Vladimirov et al., 2006; Jia et al., 2007; Tang et al., 2008), carbon nanotubes (CNTs) (Zhou et al., 2004; Kim et al., 2009), starch nanocrystals (Le Corre and Bras, 2010; Lin et al., 2011), cellulose-based nanofibers (Radovanovic et al., 2008; Bhatnagar and Sain, 2005), and chitin or chitosan nanoparticles (Zhao et al., 2011).

The most important of the improved properties obtained from nanomaterials that differ from traditional materials is that nanomaterials are multifunctional, that is, simultaneously improve two or more properties without sacrificing any of the original properties, thus, have the potential to simplify the science of materials, since it has a material that makes the work diverse.

6 Preferred Barrier Material for Various Packaging Applications

Today, the key determinants of demand for packaging are: shifting consumer trends, lightweight, varied package sizes, barrier properties, shelf life of the packaged product, easy of recycling and transport costs. On the other hand, biobased barrier materials are slowly replacing the oil-based polymers that have long been used to package goods (Ting et al., 2013). A significant range of types and levels of barrier properties is now available, including oxygen barriers and grease resistance. On the other hand, it is aware of an extensive category of both polymers as film structured these polymers with marked physical properties in films, for example, barrier properties (encompass gas-barrier), wear and weathering resistances, and light-transmittance.

Regularly is referred the use and application of polymeric films either single or multilayer for specific purpose in packaging so that perform diverse and appropriate physical and chemical properties.

Multilayer films from versatile synthetic polymers, for example, PE, PP, and PET are used currently in packaging industry to storage and transportation diverse objects.

Some analogous synthetic polymers settled without having a recycling or incineration process, normally are taken to landfills where they hardly decompose and remain as waste. With this view, polymers having biodegradability and the articles constructed by these are experiencing a boom. Materials such as corn, starch, or sugarcane (naturally produced, modified, or synthetically produced) have been the basis for producing biodegradable polymers.

6.1 A Glance at Some Literature Available on the Use of Material for Packaging

Improving film properties such as barrier, mechanical, and chemical are becoming very popular in applications that are given both packaging and nonpackaging industries. One of the synthetic materials remains mainly used due to specialty and high-performance, is polyester and among other examples to elaborate films and sheets are nylon, polycarbonate, fluropolymers, CNTs, and others.

CNTs have been used by Sifuentes-Nieves et al. (2015) to prepare glycerol/gelatin based films, by doping the film with CNTs and sodium dodecyl sulfate (SDS).

Gelatin/glycerol films were made as follows: gelatin/glycerol (GG) incubated with equal concentrations of CNT and SDS; GG with 0.001% w/w CNT/SDS; GG with 0.002% CNT/SDS and GG with 0.004% CNT/SDS. They found in this study that mechanical tests showed a 30% increase in elongation at break of GG with 0.004% CNT/SDS, with respect to gelatin/glycerol/SDS control. Samples with CNT had increased WVP. The film fractal dimension indicated that, with the addition of the highest concentration of CNT, films with a homogeneous surface were obtained with probable nanotube inclusion in the protein matrix. According to the results, the easy method used to prepare gelatin composite materials gave place to films with better physical, mechanical, and thermal properties.

Specialty films such as EVOH (ethylene vinyl alcohol), PLA, PVDC (polyvinylidene chloride), and polyamide are increasingly used in a number of applications (Ginossatis, 2007; Ting et al., 2013). In a study by Mokwena et al. (2011), it was shown that the oxygen transmission rates (OTRs) of two multilayer EVOH films were found to increase with increasing water absorption at retort temperatures. They developed and analyzed two multilayer films (A and B) produced with EVOH, PET, PP, and nylon 6. OTRs were determined to the films once water absorption were measured at 121°C.

They concluded that OTRs of selected packaging films showed an increase when the amount of water absorbed in the film increases and, this being very marked, to obtain slope values of 6 and 4% for film A and film B, respectively.

Laminated high oxygen barrier film for packaging applications and especially for shrinkable lidding applications, has been developed by Ginossatis (2007). These authors pointed out that when a thermoplastic film is made and according to its orientation, it can be imparted a characteristic shrink feature, allowing the film to stretch or shrink when exposed to heat; stating these authors, for example, in a hot water bath or by exposure to hot air. Likewise, they also considered that the strong seal force between the film and the bottom web or tray that forms the package, is an important feature for thermoplastic film. In this work, a specific experimental example, where a multilayer heat shrinkable film composed of at least two films (named 1 and 2) was reported. In this example both films had a glycerol shrinkage; film 1 showed a shrinkage greater than 20%, for at least one direction (MD or TD) at 110°C for 5 s. The opposite case was for the film 2 because this presented a 20% less shrinkage for the same conditions of temperature (110°C) and time (5 s).

In this approach, the polyolefin used for the film manufacture, notably inhibited the shrinkability of the laminate and, therefore, it allowed a good combination of the desired properties, such as their degree of shrinkage, puncture, and abuse resistance, optical, and antifog properties.

As already mentioned, synthetic polymers are widely used as materials for manufacturing multilayer packaging materials for containing various objects. In view of these issues, suitable biodegradable polymeric materials and products formed there have been the basis for producing biodegradable polymeric films either single or multilayer from biodegradable polymers (naturally produced, modified or synthetically produced) formed from naturally renewable resources.

A recent study published by Argüello-González et al. (2014) investigated the influence of glycerol concentration and zein on the mechanical and structural properties of edible films from native and two oxidized banana starch (2.5% and 3.5% active chlorine based on starch), which were named as FNS, FOSA, and FOSB. They found that the oxidation reaction caused a decrease in the lipids, proteins, and amylose content. The mechanical properties of these films indicated that the film FOSB displayed high mechanical strength (5.05 MPa) and elongation values of about 27.1%, considered acceptable. Furthermore, this study showed that zein had a significant effect, showing a decrease on the results of the films WVP. Nevertheless, edible films from oxidized banana starch exhibited greater permeability than FNS and the inclusion of glycerol and the level of oxidation resulted in an increase in the films moisture.

García-Tejeda et al. (2013) evaluated the effect of the storage time (0–45 days) on mechanical and physicochemical properties of laminates produced by extrusion lamination and elaborates with native (NBS) and oxidized banana starch (OBS) chemically modified with 3 g/100 g of active chlorine.

This study concluded that little differences were observed in WVP between NBS and OBS laminates and laminates solubility, and decreased with storage time. The value range obtained in the measurement of modulus elasticity, tensile strength (TS), and increment of elongation were upper than the values attained on NBS films. In general, the experimental data shows the OBS is appropriate to produce laminates from extrusion process with functional properties similar to other laminates.

An invention of Ting et al. (2013) describes the incorporation of a biodegradable polymer layer on the high barrier multilayer films of an insulating material to enlarge the moisture barrier properties of the film composite. The polymeric layer constitutes for a PLA layer of 48 gauge (ga), 15.5 g/m^2 (gsm), showed a weight surrounding of 15.0–60 g/m^2. On the other hand, Ting et al. (2013) particularize that the polymeric layer may be created using traditional methods to elaborate multilayer films, which include coating (gravure and extrusion), lamination, and extrusion/coextrusion techniques.

However, for its excellent properties, the polymer nanocomposites and layered nanoclays have become a promising and attractive alternative to traditional composites. In the food and beverage packaging, they are used for their excellent barrier properties, which makes them very suitable in food packaging. In addition, the films of nanocomposite polymers are transparent, which does not happen with the respective conventional polymer composites. Incorporating nanoparticles into polymer matrices improves the barrier properties of the material, making it possible to reduce to 80% the permeability of the material incorporating only 4% of modified nanoclays. Nevertheless, when polymer nanocomposites prepared with phyllosilicates keep in mind that the nanoclays are not easily dispersed in most polymers, partly due to their tendency to agglomerate and the other to the intrinsic incompatibility of the phyllosilicates and plastics.

Tensile, water vapor barrier, and antimicrobial properties were tested by Rhim et al. (2009) in PLA-based composite films with different types of nanoclays, such as Cloisite Na^+, Cloisite 30B, and Cloisite 20A. They reported that Cloisite 20A showed better intercalation into and interaction with the PLA matrix than Cloisite 30B or Cloisite Na^+. Tensile, water vapor barrier, and antimicrobial properties of PLA-based composite films varied depending on the type and concentration of nanoclays. Packaging film properties can be improved by choosing the proper type of nanoclay and its optimum concentration. The authors note that the increased barrier property against WVT suggests a great potential of the PLA/clay composite films in the application of food and beverage packaging.

Recently, Müller et al. (2011) investigated the influence of nanoclay incorporation procedure on the mechanical and water vapor barrier properties of starch/nanoclay composite films that were prepared via two different methods: in D samples, the nanoclay was dispersed in glycerol and subsequently incorporated into the starch; and in ND samples, all ingredients were added in a single step before the extrusion. They used cassava starch to obtain films with (nanocomposite) and without nanoclay (control) using 0.25 g of glycerol/g of starch and 0.03 g of nanoclay per gram of starch.

This study concluded that the nanoclay improved the water vapor barrier properties of the starch film and this effect was more pronounced in D samples, where the WVP was 60% lower than that of the control samples. With respect to mechanical properties, the incorporation of nanoclay, increased the TS and the rigidity of the films and this effect was more significant when the nanoclay was dispersed in glycerol. Thus, the incorporation of nanoclay into starch-based films is a promising way to manufacture films with better mechanical and water vapor barrier properties.

Agar-based nanocomposite films have been used for Rhim (2011) to examine the mechanical, water vapor barrier, and water resistance properties when a nanoclay (Cloisite Na^+) was added to films at different levels (0, 2.5, 5, 10, 15, and 20 g clay/100 g agar), testing a method of solution intercalation. In general, an increase in TS property was observed with up to 10% of the clay content, while an incorporation of more than 10% of clay, explained the observed decrease in this property. Besides, decreases in WVP were observed with increasing amount of clay, which was attributed to the tortuous path and stable structure formed between the intercalated silicate layers that reduced the water vapor diffusion and the polymer matrix. This study concluded that the agar/clay nanocomposite films, with improved water vapor barrier and mechanical properties as well as controlled water resistance properties, could be potentially used as environmentally friendly food packaging materials, for extending the shelf life, or as hydrogels using high water holding capacity with enhanced gel strength of the nanocomposite film.

Although these studies provide encouraging results about nanopackaging, more research needs to be done with a combination of synthetic and biodegradable polymers to help to improve mechanical and barrier properties for commercial applications of nanotechnology in the food sector.

7 Conclusions

Films produced from biopolymers (obtained of proteins, polysaccharides, and lipids) have research opportunities due to availability, are low cost and renewable. The films elaborate of biopolymers are appropriate to be used on plastic wraps in food products for its protection from external factors, or as food coatings. Although the use of compatible plasticizers is necessary to achieve flexible films, such additives affect film barrier properties, which can be described by means of well-established mathematical equations. The difference on concentration of water across a film produce a chemical potential difference, it causes a driving force to water transfer through the film known like WVP. Its records are very useful to understand mass-transfer phenomena and solute-macromolecules interactions in degradable films. The films made of biopolymers have barriers properties appropriate to oxygen and aroma. The biopolymers have the functional groups OH along the main chain, these structural properties produce affinity to water molecules and then the biopolymer films tend to have hydrophilic nature, on elevated high relative humidity (RH) the biopolymer films have a tendency to absorb water, the water absorbs on film acts like plasticizer and produce films with poor barrier properties.

Variations in the WVP are of high importance when developing food packaging materials, where efficient barrier properties are desired to minimize moisture transfer between food and the outside packaging environment. The WVP of biodegradable films have a broadly range from 1.8×10^{-13} to 8×10^{-3} g m^{-1} s^{-1} Pa^{-1} and some these films have shown similar values of WVP of $1–4 \times 10^{-13}$ g m^{-1} s^{-1} Pa^{-1} when made with low- or high-density polyethylene.

A combination of synthetic and biodegradable polymers for the production of multilayer films and coatings are being promoted and investigated in recent years, because of the need to build coatings that do not contribute to environmental pollution, and also because they are expected to have better properties than the monolayer film, made from such materials. Multilayer coatings are composed of three or more layers of different natural polymers (degradable), where the outer layers are made of cheap materials and with adequate mechanical and barrier to moisture (water) properties; and the inner layers are made of more expensive materials and have good barrier properties to gases. These properties depend on the compatibility and adhesion existing between the materials constituting the layers, the transformation method used for the preparation of coatings and the characteristics of the hydrophobic polymer. Besides, further research needs to be done to improve film barrier properties, using novel techniques, such as those involving nanotechnology.

References

Al-Hassan, A.A., Norziah, M.H., 2012. Starch–gelatin edible films: water vapor permeability and mechanical properties as affected by plasticizers. Food Hydrocoll. 26, 108–117.

Amodio, M.L., Derossi, A., Mastrandrea, L., Colelli, G., 2015. A study of the estimated shelf life of fresh rocket using a non-linear model. J. Food Eng. 150, 19–28.

Argüello-González, E., Solorza-Feria, J., Rendón-Villalobos, R., Rodríguez-González, F., Jiménez-Pérez, A., Flores-Huicochea, E., 2014. Properties of edible films based on oxidized starch and zein. Int. J. Polym. Sci. 2014, 9.

Arvanitoyannis, I., Psomiadou, E., Nakayama, A., Aiba, S., Yamamoto, N., 1996a. Edible films from gelatin, soluble starch and polyols. In: Oliveira, C.J. (Ed.), Process Optimisation and Minimal Processing of Foods: European Commission. Copernicus Programme, pp. 13–32.

Arvanitoyannis, I., Psomiadou, E., Yamamoto, N., 1996b. Edible films made from sodium caseinate, starches, sugars or glycerol. Part I. Carbohydr. Polym. 31, 179–192.

Arvanitoyannis, I.S., 1999. Totally and partially biodegradable polymer blends based on natural and synthetic macromolecules: preparation, physical properties, and potential as food packaging materials. J. Macromol. Sci. RMC C39, 205–271.

Arvanitoyannis, I.S., Bosnea, L., 2004. Migration of substances from food packaging materials to foods. Crit. Rev. Food Sci. Nutr. 44, 63–76.

ASTM, 1989. Standard test methods for water vapour transmission of materials in sheet form, method E-96-80. American Society for Testing and Materials. Philadelphia, USA.

Ashley, R.J., 1985. Permeability and plastics packaging. In: Comyn, J. (Ed.), Polymer Permeability. Springer, Amsterdam, The Netherlands.

Audic, J.L., Chaufer, B., 2005. Influence of plasticizers and crosslinking on the properties of biodegradable films made from sodium caseinate. Eur. Polym. J. 41, 1934–1942.

Bertuzzi, M.A., Castro-Vidaurre, E.F., Armanda, M., Gottifredi, J.C., 2007. Water vapor permeability of edible starch based films. J. Food Eng. 80, 972–978.

Bhatnagar, A., Sain, M., 2005. Processing of cellulose nanofiber-reinforced composites. J. Reinf. Plast. Compos. 24, 1259–1268.

Bourtoom, T., Chinnan, M.S., Jantawat, P., Sanguandeekul, R., 2006. Effect of plasticizer type and concentration on the properties of edible film from water-soluble fish proteins in surimi wash-water. Food Sci. Technol. Int. 12, 119–126.

Cao, N., Yang, X., Fu, Y., 2009. Effects of various plasticizers on mechanical and water vapor barrier properties of gelatin films. Food Hydrocoll. 23, 729–735.

Debeaufort, F., Quezada-Gallo, J.A., Villey, A., 1998. Edible films and coatings: tomorrows packaging: a review. Crit. Rev. Food Sci. Nutr. 38, 299–313.

Decker, J.J., Meyers, K.P., Paul, D.R., Schiraldi, D.A., Hiltner, A., Nazarenko, S., 2015. Polyethylene-based nanocomposites containing organoclay: a new approach to enhance gas barrier via multilayer coextrusion and interdiffusion. Polymer 61, 42–54.

DeLassus, P., 2000. Barrier Polymers: Kirk-Othmer Encyclopedia of Chemical Technology. John Wiley & Sons, Inc., Hoboken, NJ.

Del Nobile, M.A., Fava, P., Piergiovanni, L., 2002. Water transport properties of cellophane flexible films intended for food packaging applications. J. Food Eng. 53, 295–300.

Dias, M.G., Camões, M.F.G.F.C., Oliveira, L., 2014. Carotenoid stability in fruits, vegetables and working standards:—effect of storage temperature and time. Food Chem. 156, 37–41.

Dickinson, E., 2011. Mixed biopolymers at interfaces: competitive adsorption and multilayer structures. Food Hydrocoll. 25, 1966–1983.

Dixon, J., 2011. Packaging Materials 9. Multilayer Packaging for Food and Beverages. ILSI Europe Report Series. Brussels Belgium, pp. 1-43.

El Miri, N., Abdelouahdi, K., Barakat, A., Zahouily, M., Fihri, A., Solhy, A., El Achaby, M., 2015. Bio-nanocomposite films reinforced with cellulose nanocrystals: rheology of film-forming solutions, transparency, water vapor barrier and tensile properties of films. Carbohydr. Polym. 129, 156–167.

Escalona, V.H., Verlinden, B.E., Geysen, S., Nicolaï, B.M., 2006. Changes in respiration of fresh-cut butterhead lettuce under controlled atmospheres using low and superatmospheric oxygen conditions with different carbon dioxide levels. Postharv. Biol. Technol. 39, 48–55.

Fan, H., Ji, N., Zhao, M., Xiong, L., Sun, Q., 2016. Characterization of starch films impregnated with starch nanoparticles prepared by 2,2,6,6-tetramethylpiperidine-1-oxyl (TEMPO)-mediated oxidation. Food Chem. 192, 865–872.

Fang, J.M., Fowler, P.A., Escrig, C., Gonzalez, R., Costa, J.A., Chamudis, L., 2005. Development of biodegradable laminate films derived from naturally occurring carbohydrate polymers. Carbohydr. Polym. 60, 39–42.

Félix, M., Martín-Alfonso, J.E., Romero, A., Guerrero, A., 2014. Development of albumen/soy biobased plastic materials processed by injection molding. J. Food Eng. 125, 7–16.

Forssell, P., Lahtinen, R., Lahelin, M., Myllärinen, P., 2002. Oxygen permeability of amylose and amylopectin films. Carbohydr. Polym. 47, 125–129.

García-Tejeda, Y.V., López-González, C., Pérez-Orozco, J.P., Rendón-Villalobos, R., Jiménez-Pérez, A., Flores-Huicochea, E., Bastida, A., Solorza-Feria, J., 2013. Effect of storage on the mechanical and physicochemical properties of extruded laminates from native and oxidized banana (*Musa paradisiaca* L.) starch. LWT 54, 1–9.

Gelin, B.R., 1994. Molecular Modeling of Polymer Structures and Properties. Hanser Publishers, Munich, Germany.

Gennadios, A., Park, H.J., Weller, C.L., 1993. Relative humidity and temperature effects on tensile strength of edible protein and cellulose ether films. Trans. ASAE 36, 1867–1872.

Gennadios, A., Weller, C.L., Gooding, C.H., 1994. Measurement errors in water vapor permeability of highly permeable, hydrophilic edible films. J. Food Eng. 21, 395–409.

Ghanbarzadeh, B., Oromiehi, A.R., 2009. Thermal and mechanical behavior of laminated protein films. J. Food Eng. 90, 517–524.

Ginossatis, D., 2007. United States Patent No. 7,160,604 B2. Laminated high barrier shrinkable film. Alexandria, VA.

Gómez-Guillén, M.C., Pérez-Mateos, M., Gómez-Estaca, J., López-Caballero, E., Giménez, B., Montero, P., 2009. Fish gelatin: a renewable material for developing active biodegradable films. Trends Food Sci. Tech. 20, 3–16.

Gontard, N., Guilbert, S., Cuq, J.L., 1992. Edible wheat gluten films: influence of the main process variables on film properties using response surface methodology. J. Food Sci. 57, 190–195.

Gu, C.H., Wang, J.J., Yu, Y., Sun, H., Shuai, N., Wei, B., 2013. Biodegradable multilayer barrier films based on alginate/polyethyleneimine and biaxially oriented poly(lactic acid). Carbohydr. Polym. 92, 1579–1585.

Haug, I.J., Draget, K.I., 2009. Gelatin. In: Phillips, G.O., Williams, P.A. (Eds.), Handbook of Hydrocolloids. Woodhead Publishing, Cambridge UK, pp. 142–163.

Hedenqvist, M., Gedd, U.W., 1996. Diffusion of small molecule penetrants in semicrystalline polymers. Prog. Polym. Sci. 21, 299–333.

Hernández-Izquierdo, V.M., Krochta, J.M., 2008. Termoplastic processing of proteins for film formation. Rev. J. Food Sci. 73, 30–39.

Hulleman, S.H.D., Kalisvaart, M.G., Janssen, F.H.P., Feil, H., Vliegenthart, J.F.G., 1999. Origins of B-type crystallinity in glycerol-plasticised compression-moulded potato starches. Carbohydr. Polym. 39, 351–360.

Jambeck, J.R., Geyer, R., Wilcox, C., Siegler, T.R., Perryman, M., Andrady, A., Narayan, R., Law, K.L., 2015. Plastic waste inputs from land into the ocean. Science 347, 768–771.

Jang, W.S., Rawson, I., Grunlan, J.C., 2008. Layer-by-layer assembly of thin film oxygen barrier. Thin Sol. Films 516, 4819–4825.

Jangchud, A., Chinnan, M.S., 1999. Properties of peanut protein film: sorption isothermal and plasticizer effect. LWT 32, 89–94.

Jia, X., Li, Y., Cheng, Q., Zhang, S., Zhang, B., 2007. Preparation and properties of poly(vinyl alcohol)/silica nanocomposites derived from copolymerization of vinyl silica nanoparticles and vinyl acetate. Eur. Polym. J. 43, 1123–1131.

Jongjareonrak, A., Benjakul, S., Visessanguan, W., Prodpran, T., Tanaka, M., 2006. Characterization of edible films from skin gelatin of brownstripe red snapper and bigeye snapper. Food Hydrocoll. 20, 492–501.

Ke, T., Sun, X., 2000. Physical properties of poly(lactic acid) and starch composites with various blending ratios. Cereal Chem. 77, 761–768.

Kim, J.Y., Han, S.I., Kim, D.K., Kim, S.H., 2009. Mechanical reinforcement and crystallization behavior of poly(ethylene 2,6-naphthalate) nanocomposites induced by modified carbon nanotube. Compos. Part A: Appl. Sci. Manuf. 40, 45–53.

Krochta, J.M., de Mulder-Johnston, C., 1997. Edible and biodegradable polymer films: challenges and opportunities. Food Technol. 51, 61–64.

Labuza, T.P., Breene, W.M., 1988. Applications of active packaging for improvement of shelf-life and nutritional quality of fresh and extended shelf-life foods. J. Food Process. Pres. 13, 1–69.

Lagaron, J.M., Cabedo, L., Cava, D., Feijoo, J.L., Gavara, R., Gimenez, E., 2005. Improving packaged food quality and safety. Part 2: Nanocomp. Food Addit. Contam. 22, 994–998.

Laufer, G., Kirkland, C., Cain, A.A., Grunlan, J.C., 2013. Oxygen barrier of multilayer thin films comprised of polysaccharides and clay. Carbohydr. Polym. 95, 299–302.
Lavoine, N., Desloges, I., Dufresne, A., Bras, J., 2012. Microfibrillated cellulose—its barrier properties and applications in cellulosic materials: a review. Carbohydr. Polym. 90, 735–764.
Le Corre, D., Bras, J., 2010. Dufresne starch nanoparticles: a review. Biomacromolecules 11, 1139–1153.
Lin, N., Huang, J., Chang, P.R., Anderson, D.P., Yu, J., 2011. Preparation, modification and application of starch nanocrystals in nanomaterials: a review. J. Nanomater. 20, 1–13.
Liu, H., Song, L., You, Y., Li, Y., Duan, X., Jiang, Y., Joyce, D.C., Ashraf, M., Lu, W., 2011. Cold storage duration affects litchi fruit quality, membrane permeability, enzyme activities and energy charge during shelf time at ambient temperature. Postharvest Biol. Technol. 60, 24–30.
Liu, L., Kerry, J.F., Kerry, J.P., 2006. Effect of food ingredients and selected lipids on the physical properties of extruded edible films/casings. Int. J. Food Sci. Technol. 41, 295–302.
López, O.V., Ninago, M.D., Lencina, M.M.S., García, M.A., Andreucetti, N.A., Ciolino, A.E., Villar, M.A., 2015. Thermoplastic starch plasticized with alginate–glycerol mixtures: melt-processing evaluation and film properties. Carbohydr. Polym. 126, 83–90.
Lorusso, S., Gramiccioni, L., Riganti, V., 1991. La Qualità nel Sistema Packaging. Franco Angeli S.r.l., Milan, Italy.
Martin, O., Schwach, E., Avérous, L., Couturier, Y., 2001. Properties of biodegradable multilayer films based on plasticized wheat starch. Starch/Stärke 53, 372–380.
Martucci, J.F., Ruseckaite, R.A., 2010. Biodegradable three-layer film derived from bovine gelatin. J. Food Eng. 99, 377–383.
McHugh, T.H., Avena-Bustillos, R., Krochta, J.M., 1993. Hydrophilic edible films: modified procedure for water vapor permeability and explanation of thickness effects. J. Food Sci. 58, 899–903.
McHugh, T.H., Krochta, J.M., 1994. Sorbitol vs. glycerol–plasticized whey protein edible films: integrated oxygen permeability and tensile properties evaluation. J. Agr. Food Chem. 42, 841–845.
Minelli, M., Baschetti, M.G., Doghieri, F., Ankerfors, M., Lindström, T., Siró, I., Plackett, D., 2010. Investigation of mass transport properties of microfibrillated cellulose (MFC) films. J. Memb. Sci. 358, 67–75.
Mitsoulis, E., 1988. Multilayer sheet coextrusion: analysis and design. Adv. Polym. Technol. 8, 225–242.
Mokwena, K.K., Tang, J., Laborie, M.P., 2011. Water absorption and oxygen barrier characteristics of ethylene vinyl alcohol films. J. Food Eng. 105, 436–443.
Morillon, V., Debeaufort, F., Blond, G., Capelle, M., Voilley, A., 2002. Factors affecting the moisture permeability of lipid-based edible films: a review. Crit. Rev. Food Sci. Nutr. 42, 67–89.
Müller, C.M., Laurindo, J.B., Yamashita, F., 2011. Effect of nanoclay incorporation method on mechanical and water vapor barrier properties of starch-based films. Ind. Crop. Prod. 33, 605–610.
Nayak, P.L., 1999. Biodegradable polymers: opportunities and challenges. J. Macromol. Sci. RMC 39, 481–505.
Parisi, S., 2013. Food Industry and Packaging Materials: Performance-Oriented Guidelines for Users. Smithers Rapra Technology Ltd., United Kingdom.
Park, H.J., Weller, C.L., Vergano, P.J., Testin, R.F., 1993. Permeability and mechanical properties of cellulose-based edible films. J. Food Sci. 58, 1361–1364.

Paul, D.R., Roberson, L.M., 2008. Polymer nanotechnology: nanocomposites. Polymer 49, 3187–3204.
Peelman, N., Ragaert, P., Vandemoortele, A., Verguldt, E., De Meulenaer, B., Devlieghere, F., 2014. Use of biobased materials for modified atmosphere packaging of short and medium shelf-life food products. Innov. Food Sci. Emerg. Technol. 26, 319–329.
Peng, Y., Wang, T., Zhou, S., Bu, W., Zhang, L., 1994. Investigation on the microstructure of nanomaterial Fe_2O_3 by positron lifetime spectroscopy. Acta Phys. Sin-Ch Ed. 43, 1208–1216.
Perez-Gago, M.B., Krochta, J.M., 2005. Emulsion and bi-layer edible films. In: Han, J.H. (Ed.), Innovations in Food Packaging. Elsevier Academic Press, San Diego, CA, pp. 384–402.
Petersen, K., Væggemose, N.P., Bertelsen, G., Lawther, M., Olsen, M.B., Nilsson, N.H., Mortensen, G., 1999. Potential of biobased materials for food packaging. Trends Food Sci. Technol. 10, 52–68.
Radovanovic, B., Markovic, G., Radovanovic, A., 2008. Wood flour as a secondary filler in carbon black filled of styrene butadiene/chlorosulphonated polyethylene rubber blend. Polym. Compos. 29, 692–697.
Rhim, J.W., 2011. Effect of clay contents on mechanical and water vapor barrier properties of agar-based nanocomposite films. Carbohydr. Polym. 86, 691–699.
Rhim, J.W., Hong, S.I., Ha, C.S., 2009. Tensile, water vapor barrier and antimicrobial properties of PLA/nanoclay composite films. LWT 42, 612–617.
Rhim, J.W., Mohanty, K.A., Singh, S.P., Ng, P.K.W., 2006. Preparation and properties of biodegradable multilayer films based on soy protein isolate and poly(lactide). Ind. Eng. Chem. Res. 45, 3059–3066.
Rivera-Corona, J.L., Rodríguez-González, F., Rendón-Villalobos, R., García-Hernández, E., Solorza-Feria, J., 2014. Thermal, structural and rheological properties of sorghum starch with cactus mucilage addition. LWT 59, 806–812.
Rivero, S., García, M.A., Pinotti, A., 2009. Composite and bi-layer films based on gelatin and chitosan. J. Food Eng. 90, 531–539.
Rizzolo, A., Grassi, M., Vanoli, M., 2015. Influence of storage (time, temperature, atmosphere) on ripening, ethylene production and texture of 1-MCP treated 'Abbé Fétel' pears. Postharv. Biol. Technol. 109, 20–29.
Rochman, C.M., Hoh, E., Kurobe, T., Teh, S.J., 2013. Ingested plastic transfers hazardous chemicals to fish and induces hepatic stress. Sci. Rep. 3, 3263.
Salame, M., 1986. Prediction of gas barrier properties of high polymers. Polym. Eng. Sci. 26, 1543–1546.
Schneider, A., Vodouhe, C., Richert, L., Francius, G., Le Guen, E., Schaaf, P., Voegel, J.C., Frisch, B., Picart, C., 2007. Multifunctional polyelectrolyte multilayer films: combining mechanical resistance, biodegradability, and bioactivity. Biomacromolecules 8, 139–145.
Schult, K.A., Paul, D.R., 1996. Techniques for measurement of water vapor sorption and permeation in polymer films. J. Appl. Polym. Sci. 61, 1865–1876.
Sharma, S., Rao, T.V.R., 2015. Xanthan gum based edible coating enriched with cinnamic acid prevents browning and extends the shelf-life of fresh-cut pears. LWT 62, 791–800.
Sifuentes-Nieves, I., Rendón-Villalobos, R., Jiménez-Aparicio, A., Camacho-Díaz, B., Gutiérrez-López, F., Solorza-Feria, J., 2015. Physical, physic-chemical, mechanical and structural characterization of films based on gelatin/glycerol and carbon nanotubes. Int. J. Polym. Sci. 2015, 8.
Sorrentino, A., Gorrasi, G., Vittoria, V., 2007. Potential perspectives of bio nanocomposites for food packaging applications. Trends Food Sci. Technol. 18, 84–95.

Sothornvit, R., Olsen, C.W., McHugh, T.H., Krochta, J.M., 2003. Formation conditions, water-vapor permeability, and solubility of compression-molded whey protein films. J. Food Sci. 68, 1985–1999.

Syverud, K., Stenius, P., 2009. Strength and barrier properties of MFC films. Cellulose 16, 75–85.

Takeuchi, H., 1990. A jump motion of small molecules in glassy polymers: a molecular dynamics simulation. J. Chem. Phys. 93, 2062–2067.

Takeuchi, H., Okazaki, K., 1990. Molecular dynamics simulation of diffusion of simple gas molecules in a short chain polymer. J. Chem. Phys. 93, 5643–5652.

Takeuchi, H., Roe, R.J., Mark, J.E., 1990. Molecular dynamics simulation of diffusion of small molecules in polymers. II. Effects of free volume distribution. J. Chem. Phys. 93, 9042–9048.

Tang, S., Zou, P., Xiong, H., Tang, H., 2008. Effect of nano-SiO_2 on the performance of starch/polyninyl alcohol blend films. Carbohydr. Polym. 72, 521–526.

Ting, Y.-P., R., Portex, S.J., Del Priore, A., 2013. United States Patent No. 8,513,144 B2. Property films from renewable polymers. Alexandria, VA.

Trommsdorf, U., Tomka, I., 1995. Structure of amorphous starch. 2 molecular interactions with water. Macromolecules 28, 6138–6150.

Vanin, F.M., Sobral, P.J.A., Menegalli, F.C., Carvalho, R.A., Habitante, A.M.Q.B., 2005. Effects of plasticizers and their concentrations on thermal and functional properties of gelatin-based films. Food Hydrocoll. 19, 899–907.

Van Krevelen, D.W., 1990. Properties of Polymers, third ed. Elsevier, Amsterdam, Netherlands.

Vartiainen, J., Vähä-Nissi, M., Harlin, A., 2014. Biopolymer films and coatings in packaging applications: a review of recent developments. Mater. Sci. Appl. 5, 708–718.

Vermeiren, L., Devlieghere, F., Van Beest, M., De Kruijf, N., Debevere, J., 1999. Developments in the active packaging of foods. Trends Food Sci. Technol. 10, 77–86.

Vidal, R., Martínez, P., Mulet, E., González, R., López-Mesa, B., Fowler, P., Fang, J.M., 2007. Environmental assessment of biodegradable multilayer film derived from carbohydrate polymers. J. Polym. Environ. 15, 159–168.

Vladimirov, V., Betchev, C., Vassiliou, A., Papageorgiou, G., Bikiaris, D., 2006. Dynamic mechanical and morphological studies of isotactic polypropylene/fumed silica nanocomposites with enhanced gas barrier properties. Compos. Sci. Technol. 66, 2935–2944.

Wang, L., Shogren, R.L., Carriere, C., 2000. Preparation and properties of thermoplastic starch-polyester laminate sheets by coextrusion. Polym. Eng. Sci. 40, 499–506.

Wang, Y., Guo, X., Pan, R., Han, D., Chen, T., Geng, Z., Xiong, Y., Chen, Y., 2015. Electrodeposition of chitosan/gelatin/nanosilver: a new method for constructing biopolymer/nanoparticle composite films with conductivity and antibacterial activity. Mater. Sci. Eng. 53, 222–228.

Williams, P.A., Phillips, G.O., 2009. Introduction to food hydrocolloids. In: Phillips, G.O., Williams, P.A. (Eds.), Handbook of Hydrocolloids. Woodhead Publishing Limited, Cambridge, U.K, pp. 1–22.

Wu, C.L., Zhang, M.Q., Rong, M.Z., Friedrich, K., 2002. Tensile performance improvement of low nanoparticles filled-polypropylene composites. Compos. Sci. Technol. 62, 1327–1340.

Zhao, L.M., Shi, L.E., Zhang, Z.L., Chen, J.M., Shi, D.D., Yang, J., Tang, Z.X., 2011. Preparation and application of chitosan nanoparticles and nanofibers. Braz. J. Chem. Eng. 28, 353–362.

Zhou, X., Shin, E., Wang, K.W., Bakis, C.E., 2004. Interfacial damping characteristics of carbon nanotube-based composites. Compos. Sci. Technol. 64, 2425–2437.

15

METAL NANOPARTICLES AND CARBON NANOTUBES—PERFECT ANTIMICROBIAL NANO-FILLERS IN POLYMER-BASED FOOD PACKAGING MATERIALS[a]

Iva Rezić*, Tatjana Haramina, Tonči Rezić†**

**University of Zagreb, Department of Applied Chemistry, Faculty of Textile Technology, Zagreb, Croatia; **University of Zagreb, Faculty of Mechanical Engineering and Naval Architecture, FAMENA, Department of Materials, Laboratory for Polymers and Composites, Zagreb, Croatia; †University of Zagreb, Faculty of Biotechnology and Food Technology, Department of Biochemical Engineering, Laboratory for Biochemical Engineering, Industrial Microbiology and Malting and Brewing Technology, Zagreb, Croatia*

1 Introduction

Tremendous development in nanotechnology has revolutionized the impact of nanoparticles in the scientific community and, more recently, in our society. Nanoparticles are found in many types of different technological and scientific applications and innovations, from delicate electronics to state-of-the-art medical treatments (Rezić, 2012). The food and textile industries have recognized the importance of polymer materials coated with nanoparticles, and utilize them widely due to their excellent antibacterial, antimicrobial, water resistant, and protective properties. Such materials are found in a wide variety of different products such as protective materials, food packaging, textiles, geotextiles, and automobile interior fabrics.

[a] IR wrote the first part of the chapter on nanoparticles in food packaging and their toxicity, TH wrote the part on polymers, and TR wrote the part on biodegradable polymers.

Food Packaging. http://dx.doi.org/10.1016/B978-0-12-804302-8.00015-7

Packaging materials can be divided into four categories:

1. *Passive packaging*—traditional packaging usage of covering materials.
2. *Active packaging*—packaging materials that actively react to internal and external environmental changes (eg, material can absorb oxygen and increase the shelf life of the packaging material).
3. *Intelligent packaging*—materials with innovative design convenient to users.
4. *Smart packaging*—usage of chemicals or technology (electrical, electronic, mechanical, or other) and their combination in creating products that are functional and useful, and that add novel functions to the whole product.

Many of the new sophisticated packaging materials contain nanoparticles. The list of most often used nanoparticles that are applied on polymer materials is presented in Table 15.1 (Rezić, 2011).

Food packaging materials containing nanoparticles are one of the most important current applications in the food industry. The emphasis is placed on: (1) contact materials such as nanomaterials that improve packaging properties (eg, flexibility, barrier,

Table 15.1 Nanoparticles Used on Polymer Materials (Rezić, 2011, Rezić, 2012)

Nanoparticle	Diameter	Nanoparticle	Diameter	Nanoparticle	Diameter
Ag	1.5–350 μm	Eu_2O_3	30–58 nm	Pr_6O_{11}	15–30 nm
Al	18 or 80 nm	Fe	25–250 nm	Si	30–70 nm
Au	50–150 nm	Gd_2O_3	15–80 nm	SiO_2	15–80 nm
B_2O_3	40–80 nm	In_2O_3	30–50 nm	Sm_2O_3	15–55 nm
$BaSO_4$	1–5 μm	$In(OH)_3$	20–70 nm	SnO_2	45–60 nm
C	3–400 nm	La_2O_3	15–30 nm	$SrTiO_3$	100 nm
CeO_2	15–105 nm	$Li_4Ti_5O_{12}$	20–60 nm	Ti	30–50 nm
Co	28 nm	MgO	20–100 nm	W	50 nm
Cr	50 nm	$Mg(OH)_2$	15 nm	Y_2O_3	20–40 nm
Cu	25–500 nm	Mn_2O_3	30–60 nm	YbF_3	40–80 nm
Dy_2O_3	30 or 55 nm	Mo	70 nm	Zn	80–130 nm
Er_2O_3	20–53 nm	Ni	20–50 nm	ZrC	30–60 m

temperature, and moisture stability) (Kotsilkova et al., 2001; Priolo et al., 2011); (2) active food contact materials (containing nanoparticles with antimicrobial properties) (Jin and He, 2011); (3) intelligent nanosensors that monitor and report the conditions of the food (Neethirajan et al., 2010); and (4) biodegradable polymer–nanomaterials composites (Yang et al., 2007; Echegoyen and Nerin, 2013). It is estimated that among all different engineered nanoparticles, silver nanoparticles have the largest degree of commercialization (Rezić, 2011, Rezić et al., 2011). Moreover, the most common nanocomposites used as antimicrobial films for food packaging containing silver nanoparticles due to the properties of ionic silver (strong toxicity and antimicrobial activity) or metallic silver with high temperature stability and low volatility (Echegoyen and Nerin, 2013; Castellano et al., 2007; de Azeredo, 2009; Lubick, 2008).

2 Nanoparticles in Food Packaging

Nanoparticles are synthesized or machined, and they are 2–100 nm in diameter. They are widely used in different industrial areas due to their physical and chemical characteristics (size, shape, surface charge, surface area, surface porosity, composition, structure, agglomeration, and particle size distribution), impact, and importance. A great diversity of synthetic bottom-up and top-down methods is used for developing and producing nanoparticles of varying chemical composition, size, and shape. Currently the Woodrow Wilson Center for Scholars presents the project: Project on Emerging Nanotechnologies (PEN) with its database containing 1015 commercially available nanotechnology products. The database contains products intended to be used in different markets (environmental, safety and security, industrial) and in different industrial segments (water, food, energy, medical, textiles, cosmetics, personal care products, defense, aerospace, automotive, recreation, manufacturing, sports equipment, and others).

Nanoparticles are not only present in food packaging materials, but also incorporated into our food products. For example, nanoparticles of titanium dioxide can be found in a wide variety of commercial food products (some of them being the bonbons and gums, puddings, tarts, and coffee creamers). Some investigators conclude that children under 10 years most likely consume the greatest amounts of nanoparticles, with the quantities reaching around 1 mg of titanium dioxide per kilogram body weight per day. The reason for such a high intake is children's consumption of different sweets (candy, gum, etc.).

It can be predicted that the food packaging market will reach $20 billion in the next 5 years (by the year 2020). Therefore, not only the food, but also the food packaging is one of the main routes for intake of nanoparticles.

It is known that nanotechnology in food packaging materials enables longer product life [with improved barrier properties, absorbing unwanted compounds and ultraviolet (UV) irradiation], flexibility and functionality of products (with antimicrobial, self-repair, sensors, and other properties).

Therefore, it is predicted that nanotechnology will change 25% of the food packaging business in the next decade, which corresponds to a market of $30 billion per year. This increase is the result of three of the most important advantages that nanoparticles and nanotechnology offer to food packaging materials: barrier resistance, functional performance provided by incorporated active compounds, and sensors of relevant information.

Barrier resistance materials need to fulfill some basic requirements. Such materials should protect the food against dirt, dust, oxygen, light, pathogenic organisms, moisture, and many other hazardous substances (Duncan, 2011). Moreover, all those materials need to be safe under the intended conditions of use, being inert, cheap to produce, lightweight, reusable, capable of withstanding the conditions during food producing and filling, inert during transport and storage, and resistant to physical and mechanical forces (Duncan, 2011).

Critical parameters for development and usage of food packaging materials are *migration* and *permeability* (Arvanitoyannis and Bosnea, 2004; Duncan, 2011; Finnigan, 2009). Permeability depends on many different parameters that are correlated, such as, polarity, structure of polymeric side chains, presence of hydrogen bonding, molecular weight, poly-dispersity, branching or cross-linking, processing methodology, method of synthesis, and degree of crystallinity (Duncan, 2011). Permeability cannot be fully achieved by any existing material, but this is not always a disadvantage; for example, in fruit and vegetable packaging some permeability is an advantage. In contrast, plastic materials for carbonated beverages should have strong oxygen and carbon dioxide barriers in order to preserve the beverage (Robertson, 2006; Duncan, 2011). The presence of microorganisms or other migrants can influence the permeability of food packaging materials. The desired degree of permeability can be achieved by using multilayer films, which can provide acceptable gas barrier properties. However, such materials have high production and material costs and use additional additives and adhesives, which complicates matters afterward due to the regulations about their safety and recycling. Therefore it is to be expected that the monolayer industry will have a more significant impact on development of future food packaging materials (Duncan, 2011).

Polymer nanocomposites are the newest materials that have been developed to resolve the aforementioned problems regarding permeability. Such materials are strong and flame resistant and have convenient thermal properties that control polymers that contain nanoscale filters. Materials used are clay and silica nano-plates, silica, nanoparticles (Tang et al., 2008; Vladimirov et al., 2006; Wu et al., 2002), carbon nanotubes (Kim et al., 2009; Prashantha et al., 2009), graphene, nanocrystals, cellulose-based nanofibers and nano-whiskers, chitin and chitosan nanoparticles, and other organic materials (Duncan, 2011).

Antimicrobial packaging is one of the most important types of packaging. It can be achieved by (1) addition of sachets containing antimicrobial agents, (2) incorporation of antimicrobial agents directly into the polymer materials, (3) coating onto polymer surfaces, (4) immobilization of antimicrobial agents by ion or covalent bonding, and (5) using polymer materials that have antimicrobial properties by themselves. For example, Aglon Technologies LLC has produced the silver-substituted zeolite Aglon, which is widely used for food storage containers (Appendini and Hotchkiss, 2002).

The antimicrobial activity of silver has been known for centuries. Some of the first antimicrobial investigations regarding silver were performed by Spadaro and his coworkers, who compared electrodes made of silver, platinum, gold, stainless steel, and copper. Their results have shown that silver had extremely high bacteriostatic effect, even at the lowest current, when used as the anode (Spadaro et al., 1974). Klabunde and coworkers showed in 1996 that metal nanoparticles have strong biocidal properties against bacteria, both Gram-negative bacteria and Gram-positive ones (Klabunde et al., 1996; Sondi and Salopek-Sondi, 2004). Silver nanoparticles have therefore been highly investigated as antimicrobial agents. The toxicity of silver was compared to the toxicity of germanium toward microorganisms by Slawson and his coworkers (Slawson et al., 1992). They reported that the toxicity of germanium was much lower than that found for silver nanoparticles. In addition, Zhao and Stevens presented an evaluative model that can serve as a reference method in the analysis of the impact of silver on bacteria. They have investigated the susceptibility of *Escherichia coli B* to the antibacterial activity of silver ions by measuring the following parameters: initial and complete inhibitory concentration, post agent effect for bacteriostatic susceptibility, minimum bactericidal concentration, and maximum tolerant concentration, as well as the time required for the bactericidal activity (Zhao and Stevens, 1998).

Recent investigations performed by Aymonier and his coworkers were focused on the investigations of antimicrobial activity of silver nanoparticles used as antimicrobial surface coatings on highly branched amphiphilically modified polyethyleneimines. Those researchers have applied the cells of the *E. coli* on polyethyleneimines and cultivated their growth agar overnight. The results were impressive—the number of grown bacterial colonies on the silver-nanoparticle-coated substrates were 98% less than in the surrounding uncoated parts of the substrate (Aymonier et al., 2002). Such properties are significant for making new environmentally friendly antimicrobial coatings (Aymonier et al., 2002). The mechanism of this reaction was partially explained by Sondi and Salopek-Sondi, who reported that silver nanoparticles accumulate in the bacterial membrane, which results with the death of the cell (Sondi and Salopek-Sondi, 2004).

Due to these stunning new properties of polymers coated with nanoparticles, it is to be expected that the global demand for new products will rise significantly. For example, the US National Science Foundation predicted that the total global market for nanotechnological products would reach around \$1 trillion in the year 2015 (Fig. 15.1) (Sargent, 2014). Estimations have shown that the European Union (EU) and Japan are investing approximately \$1.5 billion and \$1.8 billion, respectively, in nanotechnology. Among those, the active materials occupy the largest share of the existing market (\$4.35 billion in 2014), followed by the intelligent segment (\$2.47 billion). The total estimation of the market predicts that the packaging nanotechnology market will grow at an annual rate of 11.64%.

Traditionally, the most frequently used materials for food packaging are metal, ceramic, paper, and plastic. Today we have a broad range of different polymers: polyolefins such as polypropylene

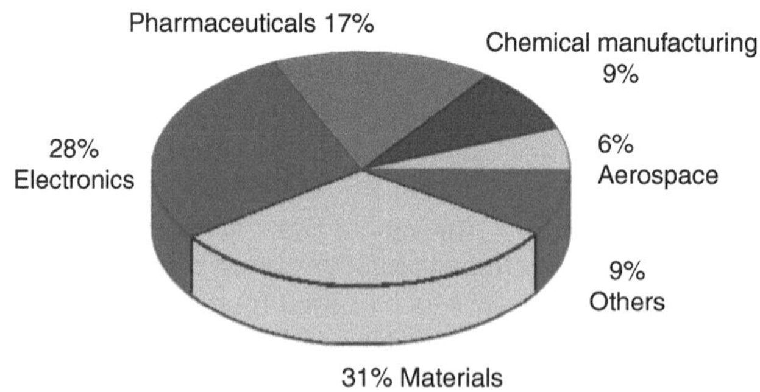

Figure 15.1. Global nanotechnology market reaching \$1 trillion by 2015 (Teixiera, 2010).

(PP) and various grades of polyethylene (HDPE, LDPE, etc.), polyethylene terephthalate (PET), polystyrene (PS), and polyvinyl chloride (PVC). It has to be emphasized; however, that the properties of those food packaging materials containing nanoparticles directly influence food taste, color, flavor, texture, consistency, absorption, and nutrients; provide mechanical and antimicrobial barriers; and monitor the food condition during transport and storage.

Since those new technologies are very often used in the industrial areas of barrier bottles and dual-ovenable packaging, it is estimated that their application will continue to grow. Current plastic rigid food packaging increases with an estimated average annual growth rate (AAGR) of 5.8%. Within this group of products, the dominant resin with 5.8 billion pounds is polyethylene terephthalate. Moreover, it can be predicted that the usage of polyethylene terephthalate will continue to rise faster than the total market, at an average annual growth rate of 6.4%. Polyethylene terephthalate is followed by polycarbonate (average annual growth rate 7.4%), high-density polyethylene (HDPE), polystyrene, and polypropylene, with an average annual growth rate of 7.3%.

It is predicted that nanotechnology will change 25% of the food packaging business in the next decade, which means a yearly income of over $30 billion for the industries involved with this new technology.

Table 15.2 and Fig. 15.2 show how the packaging market has risen in the past several years, the data being divided by the type of polymer materials being used.

Table 15.2 North American Rigid Food Packaging Market by Resin for the Period 2003–08, Expressed in Million kg (Teixiera, 2010)

Resin	2003	2004	2008
PET	2359	2676	3221
HDPE	925	953	1134
Polystyrene	612	680	894
Polypropylene	363	408	476
Others[a]	182	177	278

[a] Including LDPE, polycarbonate, SBCs, PVC, and others.

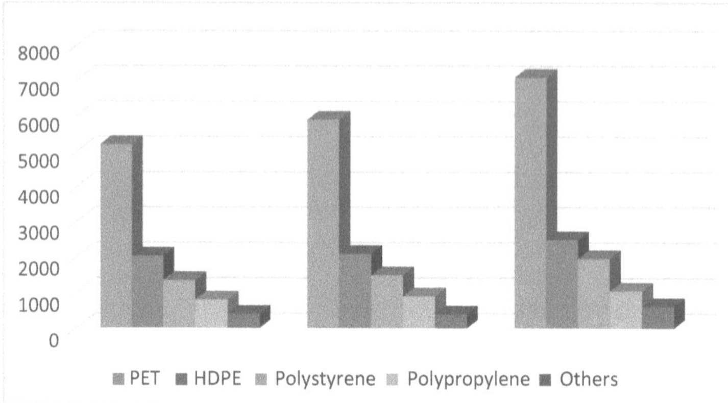

Figure 15.2. Graphical presentation of the North American rigid food packaging market by resin for the period 2003–08, expressed in million lbs (Teixiera, 2010).

Currently there are not many commercial products in the market because those should comply with very stringent conditions by being safe, effective, and not too expensive. Due to their acceptable prices, products that are more present in the market are gas barriers that can be produced by incorporation of nanoclay into polymers. Nevertheless, one of the most interesting groups of products that is currently present in the market consists of antimicrobial packaging materials with nanoparticles of zinc oxide, magnesium oxide, or silver. Such antimicrobial activity is effective against growth of pathogenic organisms on food surfaces.

As can be seen from Table 15.2 and Fig. 15.2, the global demand for different polymer packaging materials continues to grow significantly. However, it is to be expected that (according to the recent studies), the food production and consumption will be dramatically changed in this and the next decades. We are facing problems not only with insufficient natural resources such as water and energy, but also with the exponential growth of the total world population, expected to reach 9 billion people within the next 20 years.

Many polymer industries are aware of this development, so they are currently working on concepts of how to present their products in the scientific and technological fields. Therefore, nanotechnology already has a very important impact on the food and food packaging industries. Each step of the process is therefore important—from food production to preservation, packaging, transport, and consumption. By this, the nanotechnological products can enhance not only the taste of the food, but also its safety and health benefits.

It can be concluded that very big changes in the entire food and food packaging industries due to the developments of nanoproducts are expected. Nanotechnologically modified food packaging materials involve manipulation of the organic and inorganic particles at the nanoscale, which results in new materials, or materials with completely changed physical and chemical properties, with the goal to provide improvements in food quality and safety. We can see that nanotechnological development results in not only the creation of new material properties, but also in self-assembling, self-healing, and maintaining properties.

There are many examples of nanotechnological improvements of food packaging materials. Some of those are included in the following applications:

- Barriers against environmental conditions that prevent gas, UV irradiation, mechanical forces, and heat from penetrating into the product, antimicrobial materials, and multifunctional packaging.
- Sensors, including magnetic nanocomposites.
- Polymers filled with nanoparticles used in geotextiles.
- Packaging with enhanced durability and usage.
- Additives at the nanoscale for enhancing the functionality (with more effective mechanical properties such as strength, resistance, and absorption).
- Intelligent packaging.
- Materials with controlled release of nanoparticles, taste, or odor.
- Nanocoding of plastics and paper materials for identification purposes.
- Nanoparticles in beverage packaging.
- Food packaging for extreme conditions (extreme temperatures, aerospace).
- Self-cleaning materials packaging.

However, due to the great increase in the production of nanoparticles and their application (not only in the field of food packaging and production, but also in many other fields all over the world), their impact on the environment and human health needs to be considered. Moreover, the huge application of nanoparticles in the food industry and food packaging materials leads to a big concern that ingestion of nanoparticles and additives can present hazards to consumers' health. Such concerns are emphasized due to the fact that nanoparticles, being so small, can cross cellular barriers in the human body, after which follows increased production of oxyradicals and oxidative damage of the cells (Echegoyen and Nerin, 2013; Donaldson et al., 2004; Li et al., 2003).

Although we are not fully aware of nanoparticles in products that we use every day, as well as of those in the food that we eat, there are no regulatory forces strong enough to change this fact;

therefore we should be responsible for raising the awareness of government agencies and the whole population on this extremely important topic.

2.1 Toxicity of Nanoparticles

The implications for consumers' safety from the applications of nanoparticles in food packaging materials can be directly linked to the nature of applied nanoparticles. Moreover, the emphasis needs to be placed on the time of exposure of consumers during their consumption (Cockburn et al., 2012; Anthonissen et al., 2010).

Currently the toxicity and the potential harmfulness of currently applied nanoparticles in polymer food packaging materials are not thoroughly investigated and/or eliminated. Consequently, there is an urgent need for defining the most risky issues and to propose the methods for monitoring such particles and answering the question regarding their toxicity.

It can be assumed that we are currently exposed to different nanoparticles in our environment (Sánchez et al., 2011; da Silva et al., 2011). Many authors have been attempting to resolve the question about concentrations of different metal nanoparticles that are present in the human environment. For example, Pérez et al. specified concentrations of silver nanomaterials in the environment. They found 0.01–0.03 μg/L of silver nanoparticles in the water. The authors identified much lower concentrations in the air, with the range of 0.0014–0.0017 μg/m^2. In contrast, they reported higher concentrations of silver nanoparticles in the soil, ranging from 0.02 to 0.43 μg/kg (Pérez et al., 2009).

If we compare the concentrations reported for the silver nanoparticles with the concentrations predicted for the gold nanoparticles, we can see that gold nanoparticles can be found in much lower concentrations. The ranges reported by Pérez et al. were as follows: 0.14 μg/L in the water and 5.99 μg/kg in the soil samples. Slightly higher concentrations of nanoparticles were reported for zinc oxide (with 76 μg/L in the water and 3194 μg/kg in the soil), and slightly lower concentrations were determined for titanium oxide nanoparticles (0.7–24.5 μg/L in the water, 0.4–1030 μg/kg in the soil). Those concentrations are significant and need to be considered while thinking about introducing new products that contain similar nanoparticles into the food market (Rezić, 2011).

In contrast, much lower concentrations of other nanoparticles are reported for aluminum oxide nanoparticles, silica oxide nanoparticles, and carbon nanotubes. The ranges reported are the following: aluminum oxide, 0.0002 μg/L in the water and

0.01 μg/kg in the soil; silica oxide, 0.0007 μg/L in the water and 0.03 μg/kg in the soil; and carbon nanotubes, 0.0005 μg/L in the water and 0.001 μg/kg in the soil (Rezić, 2011; Pérez et al., 2009).

A big concern for the protection of the environment results from the concern that once the nanoparticles are washed and released into the wastewater they can destroy the microorganisms that are used in the wastewater treatment plants to treat sewage. The end result can be problems in drinking water safety (Choi et al., 2010).

The concern about human health arises not only from the aforementioned numbers, but also from the fact that very high concentrations of nanoparticles are added to our everyday products (cosmetics, food, textiles), so the real exposure and uptake are very hard or even impossible to predict. Moreover, we could estimate that the concentration of engineered nanoparticles in our environment will increase dramatically in the near future. For example, it is to be expected that the production of the nanoparticles will grow to 58,000 tons by the end of 2020 (Rezić, 2011). Some scientific studies that are monitoring the fate of nanoparticles after they enter the environment assume that all nanoparticle materials can accumulate in living organisms.

Human exposure to nanoparticles that are present in polymer food packaging originates from different products, and therefore we have different pathways of penetration into the human organism: the ingestion pathway (which is especially important for toddlers, who tend to taste everything from their environment), inhalation, and skin absorption (Hagens et al., 2007).

Since currently there are not enough data to precisely state all routes of exposure to nanoparticles, we can only assume that the most exposed are the workers in the nanotechnology industry, who are continuously in contact to extremely high quantities of nanoparticles.

Some studies relate the inhalation exposure to engineered nanoparticles (Born and Kreyling, 2004; Madl and Pinkerton, 2009). Asharani et al. report that the toxicity of silver nanoparticles is related to different kinds of organisms such as the liver and zebra fish (Asharani et al., 2011). The research study performed by Stampoulis et al. connects hazards of nanoparticles of silver to different edible plants such is the zucchini. Kumari et al. find that cell wall disintegration occurs in the onion plant (Stampoulis et al., 2009; Kumari et al., 2009). Moreover, reduced germination and shoot length can be observed in flax and ryegrass (El-Temsah and Joner, 2010).

Simon et al. performed a theoretical study in which they studied the migration of nanoparticles from different food packaging into the food itself (Simon et al., 2008). Their results show that the migration (which is present in a form of diffusion) can easily occur in the case if very small nanoparticles are present in the food packaging, particularly with the diameter of 1 nm and from polymer matrices that contain very low dynamic viscosity [polyolefins like low-density polyethylene (LDPE), high-density polyethylene (HDPE), and polypropylene (PP)]. Unfortunately, this work was focused only on diffusion processes, and not on the release of nanoparticles present as the coating on the surface of the polymers.

Other evidence of migration of nanoparticles from polylactide nanocomposites is reported by Schimdt et al. (2011). In contrast, Song et al. monitored the release and migration of silver ions and silver nanoparticles from polyethylene films (Song et al., 2011). Unfortunately, Song et al. could not monitor silver in a form of ions and nanoparticles individually, and does not differentiate between those species of different toxicities. The EU Regulation specifies a migration test in which different behaviors of dissolved silver and silver nanoparticles are detected by inductively coupled plasma–mass spectrometry (Laborda et al., 2011). Moreover, recent studies on the health effects of nanoparticles of silver that can penetrate into the human body through the gastrointestinal tract show that the colloidal silver can stay intact in the form of particles after digestion (Walczak et al., 2013). For this reason we can conclude that it is extremely important to assess the migration of nanoparticles from food containers and food packaging materials. Different diseases associated with human exposure to nanoparticles are summarized in Fig. 15.3 (Buzea et al., 2007).

Health consequences are still not known, but current research shows that different nanoparticles can possibly be accumulated in different organs. Nanoparticles, once ingested, can be accumulated in the human body in the following manner: They are taken up by the intestinal tract and pass into the lymph nodes, affecting the immune system, or into the capillaries, where they can settle in various organs. Afterward, nanoparticles can disrupt and impair biological, structural, and metabolic processes and weaken the immune system in humans. For example, inhaled silver nanoparticles cause lung toxicity or inflammation (when tested on a mice population). Similar consequences were found in a mice population after treating them with inhaled copper nanoparticles—the result was a high rate of pulmonary infections. Moreover, carbon nanotubes can be linked to possible damage to lungs like asbestos or cancer have been. It is estimated that different nanoparticles can possibly be accumulated in different organs, where they can disrupt and impair biological, structural, and metabolic processes and weaken the immune system.

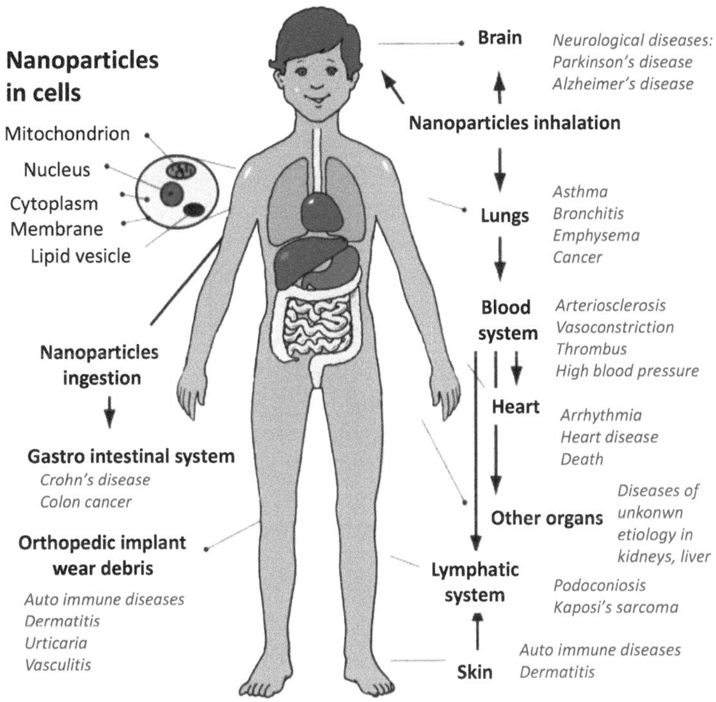

Figure 15.3. Diseases associated with human exposure to nanoparticles (Buzea et al., 2007).

As can be seen from Fig. 15.3, the most important routes for the uptake of nanoparticles into the human body are ingestion, skin, and respiratory tract. After penetrating into organs, nanoparticles have the potential to harmfully impact many of those organs and organ systems, resulting in different diseases.

For example, after the step of penetrating into the human body, the nanoparticles can accumulate directly in organs (liver, kidney, bone marrow, and spleen) (Rezić, 2011). Therefore many researchers and technologists are monitoring such processes and investigating the toxicokinetic properties of nanomaterials (Eom and Choi, 2009).

Since all nanoparticles are very small, 2–100 nm, they have the potential to penetrate easily into the cells in the body (Fig. 15.3) (Holgate, 2010). The toxicity of nanoparticles to the human body, after their penetration, will vary depending on their physical and chemical properties, like the size of nanoparticle, its shape and charge, the surface energy, chemical composition, and many others. More important, the toxicity of nanoparticles can also largely depend on the living organisms' properties and their different DNA covering ratios (Liu and Hurt, 2010; Marquis et al., 2009; Robbens et al., 2010; Tinkle et al., 2003).

Robbens showed that there is a growing concern from different researcher groups investigating nanoparticles regarding their potential safety (Robbens et al., 2010). Unfortunately, although there is a will from many different authorities (for example, the EPA in the European Union) to identify the risk of nanoparticles that are present in different products used by consumers in our everyday lives, there has been no regulation or law that prescribes the monitoring and analysis regarding their safety (EU Communication, 2004). However, recently the EPA has launched a recommendation to treat all nanoparticle-containing products like pesticides. Moreover, all cosmetic products in the EU need to have proper labeling if they contain nanoparticles. This is the first industry to have required labeling.

Up to now, there has been no requirement for any other industry anywhere in the world. Therefore it is to be expected that the same or even more rigorous rules will be valid for products related to food packaging. In the meantime one can access the guidelines issued by the European Food Safety Authority (EFSA) under the title *Guidance on risk assessment concerning potential risks arising from applications of nanoscience and nanotechnologies to food and feed*, as well as the Organization for Economic Cooperation and Development (OECD), under the title *compilation of nanomaterials exposure mitigation guidelines relating to laboratories.* However, there are no labeling regulations, nor is there legislation implemented that would require manufacturers to list nanoparticles used in their products. Moreover, there is a lack of knowledge and specific legislation for nanocomponents and nanomaterials that takes into account the differences in toxicity with respect to the bulk materials (Echegoyen and Nerin, 2013).

The question of whether we can have risk assessment that is comprehensive and integral enough, including the dosage, time of exposure, cell damage, and toxicity, or we still have to develop a proper one was raised by Casals and his coworkers (Casals et al., 2008). Moreover, there are other researchers like Blasco and Picó who have emphasized that currently we have only very limited studies concerning the risk assessment and the toxicology of nanoparticles, so much is yet to be done (Blasco and Picó, 2011).

Therefore it can be concluded that there is an urgent need to develop analytical methods and procedures for monitoring nanoparticles in food packaging materials and in other everyday products, since we need to learn many new things and collect data on physical and chemical behavior, occurrence levels, fate, and transport of nanoparticles that are present in food packaging materials and their possible migration into human organisms or the environment (Weinberg et al., 2011).

3 Polymers in Food Packaging

Packaging forms include bulk food storage containers, paperboard cartons, plastic or paper food wraps, jars, and bottles. Over the past few decades, the application of polymers as food packaging materials has increased tremendously due to their advantages over materials that are traditionally used such as metal, glass, or ceramic (Chawengkijwanich and Hayata, 2008; Lopez-Rubio et al., 2004). Polymers have a good combination of low density and good mechanical properties. They possess an extraordinary chemical stability, reducing the risk of chemical interaction of polymeric macromolecules with different food types. New technologies improve many properties, including strength and stiffness, barrier to oxygen and moisture, resistance to food component attack, and flexibility (Silvestre et al., 2011). The world's largest food packaging companies are active in their research of the potential of polymer nanotechnology. Besides mechanical and thermal properties, their goals include the improvement of barrier performance pertaining basically to gases such as oxygen and carbon dioxide and to UV light, as well as polymers' antimicrobial properties. Some packaging is even able to trace and monitor the food condition during transport and storage. Once the safety of a packaging material is proven and the material is produced at a competitive ratio of cost to performance, the new nanocomposite will be very attractive for broad applications (Silvestre et al., 2011; Plackett et al., 2006).

3.1 Polymer Nanotechnology

Over the past decades polymer nanotechnology has occupied many scientists, mainly from physics, chemistry, engineering, and biology. Research and development of polymer nanocomposites that are filled with particles or even devices that have one or more dimensions of the order of 100 nm or less requires an interdisciplinary approach. The extraordinary potential of these novel technologies has induced industrial activity that involves design, manufacture, processing, and application of the high-performance nanostructured materials.

By means of polymer nanotechnology packaging functions like containment, preservation, and protection, communication and marketing can be successfully extended and implemented (Silvestre et al., 2011; Lagaron et al., 2007; de Azeredo, 2009). In general, there are three fields where progress in polymer nanotechnology has been made (Silvestre et al., 2011):

1. Improved packaging where the presence of NPs in the polymer matrix materials improves the flexibility of the polymer matrix,

gas barrier properties, as well as temperature and moisture stability.

2. Active packaging where the presence of NPs allows the interaction of packages with food and the environment, having a dynamic role in food preservation.
3. Intelligent packaging where nanodevices embedded in the polymer matrix can monitor the food condition or the environment surrounding the packaged food and can also protect against fraudulent imitation.

Sometimes one additive has a double role (eg, oxygen or moisture absorbers). Both of them will improve barrier properties, but the first will actively inhibit the growth of aerobes, particularly molds, and the latter can reduce water activity, a_w, also indirectly affecting microbial growth (Appendini and Hotchkiss, 2002).

Polymers that are most frequently used for food packaging include, but are not limited to, polyolefins such as polypropylene (PP) and various types of polyethylene (HDPE, LDPE, etc.), polyethylene terephthalate (PET), polystyrene (PS), and polyvinyl chloride (PVC). However, polymers are rarely used as neat materials due to their limited performance. In order to improve their properties, additives in different forms are used.

Polymer composites are mixtures of polymers (that are used for matrix materials) with inorganic or organic fibers and fillers of certain shapes and geometries (flakes, spheres, particulates). Polymer nanocomposites are a new class of composites that are particle-filled polymers where at least one dimension of the dispersed particles is in the nanometer range. One can distinguish three types of nanocomposites, depending on how many dimensions of the dispersed particles are in the nanometer range: (1) isodimensional nanoparticles are those fillers where the three dimensions are in the order of nanometers; (2) when two dimensions are in the nanometer scale and the third is larger, forming an elongated structure, we speak about nanotubes or whiskers; and (3) the third type of nanocomposites is characterized by only one dimension in the nanometer range. In this case the filler is present in the form of sheets of one to a few nanometers thick and hundreds to thousands of nanometers long (Alexandre and Dubois, 2000).

New products that contain hybrid and nanocomposite organic and inorganic materials combine advantageous qualities of oxide and polymer components that are not simply physically blended but are very closely mixed together. The scale of mixing affects linearly the degree of homogeneity, which would be influenced when the mixture is reached at the nanometer range (Kubacka et al., 2009; Suppakul et al., 2011).

3.2 Polymers in the Antimicrobial Nanocomposites

At the moment, polymer-based nanocomposites are being mainly developed for antimicrobial packaging applications (Silvestre et al., 2011). Antimicrobial packaging is the packaging system that is able to kill or inhibit spoilage and pathogenic microorganisms that are contaminating foods. When the packaging system acquires antimicrobial activity, the packaging system (or material) limits or prevents microbial growth by extending the lag period and reducing the growth rate or decreases live counts of microorganisms (Han, 2003). Antimicrobials are impregnated into food packaging either to protect the package or to extend the shelf life of the food. Extensive research is under way on in-package enzymatic processing and nonmigratory antimicrobial packaging, including the grafting of active agents onto the polymer surface (Danielli et al., 2008; Goddard et al., 2007). The shelf-life extension and food safety improvement are regulated by reducing the rate of growth of specific microorganisms by direct contact of the package with the surface of solid food such as meats, cheese, and the like or in the bulk of liquids like milk and other beverages (Appendini and Hotchkiss, 2002). Surfaces of food processing equipment (eg, conveyers, filler gaskets, garments, gloves) may be coated by antimicrobial polymers so that they self-sanitize during use. For antimicrobial materials, close physical contact with the surface of the food is essential. Qualitative food contact or packaging materials extend the product's lifetime and enhance its quality and safety, which at the end leads to having much less food waste (Cushen et al., 2012; Vermeiren et al., 2002; Appendini and Hotchkiss, 2002).

It is very important to define target microorganisms and the food composition when packaging is considered. Antimicrobials that are incorporated in polymers have to be selected based on their spectrum of activity, mode of action, and chemical composition, and the rate of growth and physiological state of the targeted microorganisms (Appendini and Hotchkiss, 2002; Han, 2003). The activity of antimicrobials that diffuse from packaging to the food is at least in part determined by the diffusion kinetics, whereas the antimicrobials that are attached to macromolecules are immobilized and need to be active while attached to the polymer.

Concentration of antimicrobials in a polymer matrix may have a large impact on their effectiveness. The active substance may be immobilized on the material's surface or incorporated in the bulk. If the compound is entrapped in the matrix, the thickness of the material influences the diffusion and release dynamics.

The antimicrobial must diffuse toward the surface; therefore surface characteristics and diffusion kinetics of the antimicrobial agent become crucial. The diffusion has to be maintained at a minimum rate so that the surface concentration is above critical inhibitory concentration (Han, 2000). One possibility to achieve controlled release is by means of multilayer films containing a control layer, a matrix layer, and a barrier layer (Floros et al., 2000). The matrix layer contains the active substance, the inner layer controls the rate of the inward diffusion, and outward migration is prevented by the barrier layer.

Microbial growth in polymer-based packaging is rarely simultaneous with that in food components. Additionally, food components may inhibit the diffusion of antimicrobials to the food and thus limit their activity. For example, silver-substituted zeolites are not active in nutrient-rich media, since lysine, sulfates, sulfides, and other sulfur-containing amino acids weaken the antimicrobial activity. Hence, they are convenient for nutrient-poor beverages such as tea or mineral water (Appendini and Hotchkiss, 2002; Ishitani, 1995). Fillers, antifog and antistatic agents, lubricants, stabilizers, and plasticizers are polymer additives that can negatively affect activity of antimicrobial polymers. These additives may interact directly with the antimicrobial or may change polymer conformation, altering diffusion (Appendini and Hotchkiss, 2002).

Adding particles that carry antimicrobials in the polymer matrix may influence polymer properties such as mechanical, barrier (eg, oxygen and water vapor), optical, heat sealing, adhesion, and printing properties. The influence will be specific for each polymer–antimicrobial pair.

There are two main approaches to producing polymer nanomaterials. One is to manufacture polymer materials at a nanoscale, and the other is to incorporate particulate nanomaterials into a polymer matrix in order to make a nanocomposite. At the moment the latter is the most popular and most relevant approach, especially when food contact materials for food packaging are considered (Yang et al., 2007; Cushen et al., 2012).

All matter, independent of whether it is of biological or man-made origin, has the level of its organization at nanoscale. By different nanotechnological techniques, molecules can be assembled, and, in contrast, the objects can be disassembled into molecules, as is visible in nature itself (Roco, 2003; de Azeredo, 2009). There are currently two building strategies in nanotechnology: a "top-down" approach and the "bottom-up" approach. The first approach is currently a more common in the production of nanomaterials. It is performed by using milling, nanolithography, or precision engineering where small structures are obtained by size reduction of bulk components. This allows small (nano) structures to be created from atoms or even molecules that are capable of self-assembling (Föster and Konrad, 2003; Moraru et al., 2003).

Nanoparticles are usually incorporated into polymer matrix at 0.1–5 wt.% of packaging materials, particularly films. The particles can be incorporated into polymer matrix in melt, by in situ polymerization, or by solvent compounding. Melting and polymer processing are conducted at higher temperatures; hence, only those particles that can withstand higher temperatures will be mixed in a melt. The melting points of thermoplastic polymers that are used for packaging are between 400 and 600 K, and the temperatures for thermal processing methods such as extrusion and injection molding are usually about 20 K higher than the respective melting points. Silver-substituted zeolites, for example, can withstand temperatures up to 1073 K and therefore can be incorporated as a thin coextruded layer with other polymers (Ishitani, 1995; Appendini and Hotchkiss, 2002). In comparison with their microscale counterparts, nanoparticles will show their specific properties only if the surface-to-volume ratio remains on the nanoscale. However, clustering of nanoparticles is observed in many systems, especially those based on polyolefins and related nonpolar polymers. Silver nanoparticles tend to agglomerate at 773 K (Sun et al., 2006). Antimicrobials that cannot tolerate the temperatures used in polymer processing are often coated onto the material after forming or are added to cast films. The sensitivity of antimicrobial agents to elevated temperatures makes this kind of packaging inappropriate for microwave treatment.

In situ synthesis approach of nanocomposite preparation was reported by Toyota researchers for the synthesis of polyamide nanocomposites. Monomer or monomer solution is used to swell the filler interlayers. After swelling, polymerization of the monomer is initiated. As the monomer is present in and out of the filler interlayers, the generated structure is exfoliated or significantly intercalated. Nowadays, many polymers can be synthesized in the presence of filler, but there are still some limitations of the method; for example, when PET was synthesized in situ, only short molecular weight was achieved (Mittal, 2011).

In solvent compounding, both the polymer and the antimicrobial have to be soluble in the same solvent. The polymer matrix should not be soluble in the packed food; otherwise swelling will occur and mechanical, barrier, and optical properties will change. If the nanoparticles are nonvolatile, the packaging with antimicrobial effect needs to be in contact with the food.

3.3 Polymer-Based Nanocomposites

Polymeric materials that are dominant in polymer nanotechnology production, where they are used as matrices for nanocomposites, include polyolefins, PA, ethylene-vinyl alcohol copolymer

(EVOH), PS, epoxy resins, polyurethane, PET, and different biopolymers. There is a large variety of nanofillers that are nowadays added into polymer matrices, but not all of them are allowed in food technology. For food packaging based on polymer matrix, metal nanoparticles (silver, gold, and zinc), metal oxide nanomaterials [titanium dioxide (TiO_2), zinc oxide (ZnO), silicon oxide (SiO_2), and magnesium oxide (MgO)], and carbon nanotubes (CNT) are the most frequently used nanoparticles with antimicrobial effect. These particles function on direct contact, but they can also migrate slowly and react preferentially with organics present in food (Silvestre et al., 2011).

3.3.1 Silver Nanoparticles in Polymer Matrix

One of the biggest advantages of inorganic nanoparticles over molecular antimicrobials is the ease with which the former can be incorporated into polymers to form functional antimicrobial materials (Althues et al., 2007). This is especially true due to the controlled release properties of silver nanoparticles (AgNPs), which can be engineered to remain potent antimicrobial agents for long periods of time. Due to their capability to extend the shelf life, the systems AgNP/polymer are attractive for use not only as food packaging materials, but also for medical devices (Duncan, 2011).

The antimicrobial effectivity of polymer-based nanocomposite with AgNPs has been studied by several researchers. The results of comparing the efficiency of silver/polyamide 6 systems where particles were on nanometer scale showed that nanocomposites with lower Ag content were more efficient on *E. coli* than others that contained much more silver (Damm et al., 2008). Moreover, after those materials were immersed for 100 days in the mixture of water and polyamide 6 that was filled with 2 wt.% AgNPs was effective against *E. coli* (Damm et al., 2007). Moreover, the AgNPs absorb and decompose the plant-ripening hormone ethylene (Hu and Fu, 2003). Indeed the shelf life of the Chinese fruit jujube was extended when it was stored in food storage bags composed of AgNP/nanoparticulate TiO_2/polyethylene films. The fruits were firmer and heavier and had less decay, less browning, and slower ripening over a period of 12 days than fruit stored in control materials (Li et al., 2009). In a similar study, fresh asparagus spears coated with AgNP/polyvinylpyrrolidone (PVP) nanocomposite films had their shelf lives extended to 25 days when stored at 2°C. Asparagus spears suffered less weight loss; they retained their green color and tender texture. Coated asparagus also had reduced microorganism growth during this time period (An et al., 2008).

Silver nanoparticles improve mechanical properties of poly(vinyl alcohol) (PVOH) matrix, increasing its strength and stiffness and its thermal properties, enhancing its stability, and increasing its glass transition temperature T_g (Mbhele et al., 2003). A multilayer film (PE/tie/PA-6/tie/PE; tie-maleic anhydride grafted PE) was coated with an AgNP/PE nanocomposite layer, resulting in a material with antimicrobial activity against the fungus *Aspergillus. niger*, a common food contaminant (Sanchez-Valdes et al., 2009). Moreover, it was found that the activity was dependent on the surface roughness. Rough surfaces provide a larger surface area for silver ion release, leading to a higher activity than in the case of a smoother surface. The antimicrobial activity was also found to be dependent on factors that affect the silver ion release rate, such as degree of polymer crystallinity (Kumar and Münstedt, 2005), filler type (ie, silver particles, zeolites, etc.) (Kumar et al., 2005); hydrophobicity of the matrix (Radheshkumar and Münstedt, 2006); and particle size (ie, nanocomposite vs. microcomposite) (Damm et al., 2008).

While silver zeolites ensure antimicrobial effect of polymer-based composites for a certain time period, silver ion release rates into stored foods out of polymer-based nanocomposites with AgNPs will be slower, which is important for sustained antimicrobial activity (Duncan, 2011). For example, when the antimicrobial activity of an AgNP/SiO_2 nanocomposite material was compared to that of an Ag zeolite and an $AgNO_3$/SiO_2 composite, the latter two materials had more (about 10 times) potent acute antimicrobial responses, yet activity of the nanocomposite lasted longer (Egger et al., 2009). Thus when longtime protection is required such as for transportation and storage, AgNP/polymer systems are superior, whereas a zeolite-based material might offer a superior immediate effect.

A polymer composite based on AgNPs with SiO_2 will be effective against a broad spectrum of bacteria and fungi, yet it will be more effective against Gram-negative than Gram-positive bacteria. For food contact applications, these particles may be incorporated within a PP polymer matrix (Duncan, 2011).

Other AgNP/polymer nanocomposites exhibiting antimicrobial activity include versions in polyacrylamide (Tankhiwale and Bajpai, 2009), PVOH (Hong et al., 2006), poly(methyl methacrylate) (PMMA) (Kong and Jang, 2008), polyurethane (PU) (Sheikh et al., 2009), alginate (Mohammed Fayaz et al., 2009), cellulose (Fernández et al., 2010), and chitosan (Tankhiwale and Bajpai, 2009). Chitosan-based materials are particularly interesting, as chitosan is a polysaccharide with natural antimicrobial properties, which suggests that a cumulative antimicrobial effect is attainable. Moreover, chitosan films loaded with AgNPs have also been found to have better tensile and gas barrier properties than neat films composed of the same material (Yoksan and Chirachanchai, 2010). Multilayers have also been prepared layer-by-layer in order to create antimicrobial Ag-based polymer nanocomposites (Podsiadlo et al., 2005) with enhanced mechanical properties.

Despite all of these advances in the use of silver nanostructures for food packaging applications, comprehensive studies in various polymer systems are still lacking, and much work needs to be done to elucidate key relationships that influence the antimicrobial strength of various AgNP-based PNC materials (Duncan, 2011).

3.3.2 Titanium Dioxide Nanoparticles in Polymer Matrix

Titanium dioxide TiO_2 is widely used as photocatalytic disinfecting material for surface coatings (Fujishima et al., 2000). Nanoparticle TiO_2 powder (anatase-phase crystal structure) may be mixed with organic solvents (ethyl methyl ketone, toluene, isopropyl alcohol), and this TiO_2 suspension may be manually coated onto one side of an oriented-polypropylene (OPP) film used at room temperature and dried in air for 5–10 min (Chawengkijwanich and Hayata, 2008). The film has an ability to reduce *E. coli* contamination on food surfaces upon illumination with UV light, suggesting that it could be used for fresh-cut products. Doping TiO_2 with silver greatly improves photocatalytic bacterial inactivation (Page et al., 2007; Reddy et al., 2007). The good antimicrobial properties of TiO_2/Ag^+ were obtained in a nanocomposite based on PVC (Cheng et al., 2006). Orange juice stored at 4°C in LDPE films incorporating TiO_2 and 10 nm nanosilver mixture power at loading exhibited statistically significant reduction in *Lactobacillus plantarum* growth over a time period of 112 days (Emamifar et al., 2011).

3.3.3 Zinc Oxide Nanoparticles in Polymer Matrix

Antimicrobial properties of nano-ZnO have been discovered recently. Zinc oxide nanoparticles are more affordable and safer for food packaging solutions than silver (Silvestre et al., 2011). Unlike nanocomposites loaded with TiO_2 that require the presence of UV light, nanomaterials containing nano-ZnO-based light catalyst sterilize in interior lighting conditions (Jones et al., 2008). Antibacterial activity of ZnO increases with decreasing particle size (Yamamoto, 2001). ZnO nanoparticles have been incorporated into a number of different polymers, including polypropylene (Chandramouleeswaran et al., 2007).

3.3.4 Carbon Nanotubes in Polymer Matrix

Carbon nanotubes are manufactured as single-walled carbon nanotubes (SWCNT) or multiwalled carbon nanotubes (MWCNT) in variety of ways: using arc discharge, laser ablation, and chemical vapor deposition. Carbon nanotubes are the strongest and stiffest materials discovered, being more than 5 times stronger than the Kevlar fibers. Their specific strength is 300 times greater that that of stainless steel, although under compression CNTs can be a lot weaker.

Carbon nanotubes not only are able to improve properties of polymer matrix, but also possess antibacterial properties. Aggregates of CNTs in direct contact are destructive for *E. coli*, but their application at the moment is stopped due to the cytotoxic effect on human cells, at least when in contact to skin (Warheit et al., 2004).

4 Biopolymers

Biopolymers are important for the food industry, but have faced significant problems due to their high costs and poor overall. However, we have recently witnessed biopolymer-based materials being increasingly developed. For example, nanocomposite materials are expanded in their usage on edible as well as biodegradable films. Also, the addition of reinforcement materials being nanoparticles, improvements in the mechanical, thermal, and barrier properties of biopolymers were observed. Therefore, nanoparticles today have an important role in many applications, including food packaging (de Azeredo, 2009).

4.1 Inherently Antimicrobial Polymers

Some polymers are inherently antimicrobial and have been used in films and coatings (Appendini and Hotchkiss, 2002). Polymers that are inherently antimicrobial may have a synergic effect when combined with antimicrobial nanoparticles. Chitosan and poly-L-lysine are cationic polymers that act as cell adhesins since charged amines interact with negative charges on the cell membrane, causing leakage of intracellular constituents (Goldberg et al., 1990). Chitosan has been used as a coating and appears to protect fresh vegetables and fruits from fungal degradation (Appendini and Hotchkiss, 2002). In addition, chitosan-based antimicrobial films have been used to carry organic acids and spices (Ouattara et al., 2000). Calcium alginate films reduced the growth of natural flora and coliform inocula on beef (Cuq et al., 1995). Bactericidal acrylic polymers made by copolymerizing acrylic protonated amine comonomer were proposed as packaging materials for increased fruit and vegetable shelf life. Polymers containing biguanide substituents also yield antimicrobial activity (Appendini and Hotchkiss, 2002).

4.2 Physical Surface Modifications of Polymers

There are different types of physical surface modifications of polymers (flame, corona discharge, UV, gamma-ray, electron beam, ion beam, plasma, and laser treatments), some of them being effective in generating antimicrobial properties. For example, the antimicrobial potential of polyamide films treated with UV irradiation has been reported. Antimicrobial activity was presumably the result of an increase in amine concentration on the film's surface (Hagelstein et al., 1995; Appendini and Hotchkiss, 2002).

4.3 Antimicrobial Nanocomposites and Bioactive Food Packaging

Bioactive packaging has a direct impact on the health of the consumer by generating healthier packaged foods according to the potential to improve food quality and safety benefits. Bioactive (antimicrobial) packaging is a subcategory of active packaging and represents a promising form of active packaging. Active packaging is able to modify the condition of the packaged food while maintaining or increasing quality and safety of packaged foods. Active packaging elongates the shelf life of food, and avoids the contamination with microbial pathogens like *Listeria monocytogenes*, *Salmonella typhimurium*, pathogenic *E. coli*, or molds. From the definition of EC Regulation No. 450/2009: active materials are materials intended to extend the shelf life or to maintain or improve the condition of packaged food; they are designed to deliberately incorporate active components that would release or absorb substances into or from the packaged food or the environment surrounding the food. Most food packaging systems consist of the packaging materials, the food, and the headspace in the package. Basically, the antimicrobial system can be either a direct contact one using a nonmigratory antimicrobial agent or an indirect contact one using a volatile antimicrobial releasing system (Jin and Niemira, 2011). The majority of food packaging systems can be represented as a package–food or a package–headspace–food system (Fig. 15.4).

The new antimicrobial function can be achieved by adding antimicrobial agents in the packaging system and/or using antimicrobial polymers that satisfy conventional packaging requirements. The reason for incorporating antimicrobial agents into the packaging system is to prevent surface growth of pathogenic microorganisms in food by use of antimicrobial components and to allow a controlled release of antimicrobial components into the food surface during storage and distribution (Mastromatteo et al., 2010). Antimicrobial agents may be incorporated into the

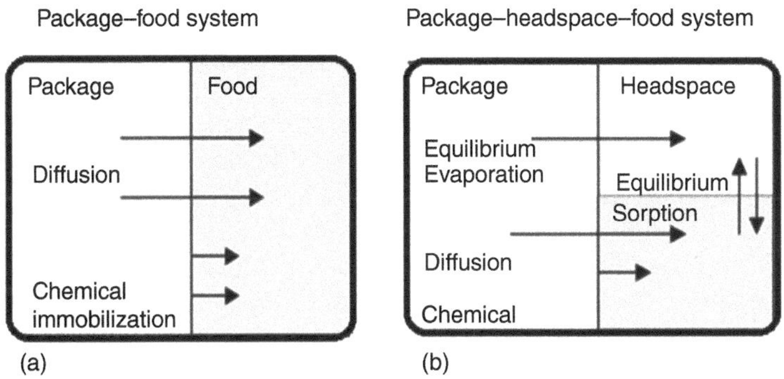

Figure 15.4. Food packaging system and transport phenomena in bioactive food packaging (Han, 2000). (a) Direct contact and (b) direct and indirect contact.

packaging materials and migrate into the food through diffusion and partitioning. The structure modification could increase the capacity of edible films to control the release of immobilized active compounds. Patterns of migration and action in antimicrobial food packaging systems are illustrated in Fig. 15.4. Diffusion between the packaging materials and products at the interface is the main migration phenomenon in the food packaging system. In these systems, mathematical diffusion modeling can be used to predict the release profile of the antimicrobial agents and optimization of the significant system parameters. The initial and boundary conditions can be used to determine the distribution coefficient. Moreover, a number of factors need to be considered in order to use the most appropriate diffusion model, including film thickness, the ways in which the antimicrobial agent diffuses, and the volume of food simulant into which the migration occurs. The various mass transfer models that can be used to analyze the migration of antimicrobial agents can also account for the influence of the molecular weight, ionic charge, and solubility of the agents (Han, 2003). According to extensive research findings, it has been postulated that the key parameters of mathematical modeling play a significant role in the controlled release of an antimicrobial agent from the polymeric network (Mastromatteo et al., 2010).

In the latter processes, antimicrobial activity may be rapidly lost due to inactivation of the antimicrobials by food components or dilution below active concentration due to migration into the bulk food matrix (Janjarasskul and Krochta, 2010). The concept of controlled release packaging is still being explored for food packaging. This is possible if the rate of dissipation of the antimicrobial agent or the controlled release during the shelf life of the system as a packaging material is fully understood, systematically performed, and accurately controlled. Therefore, the study of antimicrobial agent migration in the system will be very important in the future to ensure efficient food protection and safety (Tawakkal et al., 2014).

In general, methods of incorporating antimicrobial agents into the packaging system are their addition of into packages, incorporation directly to polymers, coating or adsorbing onto surfaces, and immobilization by ion or covalent linkages or at the end the usage of polymers that have antimicrobial properties by themselves. Antimicrobial coatings and films have a variety of advantages and constitute an innovation within the biodegradable active packaging concept. They have been developed in order to reduce and/or inhibit the growth of microorganisms on the surface foods. The use of appropriate coatings can impart antimicrobial effectiveness.

Numerous researchers have demonstrated that bioactive natural polymers (eg, alginate, chitosan, gelatin, etc.) and antimicrobial compounds such as organic acids (acetic, propionic, benzoic, sorbic, lactic, lauric), potassium sorbate, bacteriocins (nisin, lacticin), grape seed extracts, spice extracts (thymol, *p*-cymene, cinnamaldehyde), thiosulfinates (allicin), enzymes (peroxidase, lysozyme), proteins (conalbumin), isothiocyanates (allylisothiocyanate), antibiotics (imazalil), fungicides (benomyl), chelating agents (ethylenediaminetetraacetic acid-EDTA), metals (silver), or parabens (heptylparaben) could be added to reduce and inhibit microorganisms' growth on the food surface (Cutter, 2006). The choice of an antimicrobial agent is often restricted by the incompatibility of that agent with the packaging material or by its heat instability during extrusion. Another important factor is the concentration of an antimicrobial agent. If it is at or above the minimum inhibitory concentration (MIC) on the food surface, the system will maintain effective antimicrobial activity (Suppakul, 2004).

Unfortunately, the use of natural polymer (biopolymer)-based biodegradable films for food packaging has been very limited due to the limitations of poor barrier properties and weak mechanical and biochemical properties. With the antimicrobial delivery limitations properties, biodegradable active packaging materials have serious deficiencies for commercial packaging utilization. Recently, a new class of materials represented by nanocomposites (polymer matrices reinforced with nanoparticles) has proven to be promising option in the active food packaging systems. Nanocomposites as additives in active food packaging biopolymers system formulations improve mechanical and barrier properties of the polymers and could exhibit modifications in the properties of the materials or create novel properties and phenomena to the materials (Vermeiren et al., 1999). The primary goal of incorporating nanoparticles into packaging films is to enhance antimicrobial activity and improve food safety. There are some additional benefits that can be gained from this additive. For example, poly(hydroxyalkanoate) (PHA) biopolymers incorporated with a combination of silver nanoparticles and cellulose nanocrystals show improved mechanical strength compared with the unfilled PHA (Yu et al., 2014). Other systems involving PLA incorporated with microcrystalline cellulose (MCC) and silver nanoparticles show that a positive reinforcement is obtained, although this is attributed to the MCC with the silver nanoparticles imparting antibacterial activity (Fortunati et al., 2012a; Fortunati et al., 2012b).

Nanocomposite antimicrobial systems are particularly effective because of the high surface-to-volume ratio and enhanced surface reactivity of the nano-sized antimicrobial agents, making them able to inactivate microorganisms more effectively than their micro- or macro-scale counterparts. Commonly used or tested antimicrobial nano and nanocomposite materials include metal ions (silver, copper, gold, platinum), metal oxide (titanium dioxide, zinc oxide, magnesium oxide), and organically modified nanoclay [quaternary ammonium-modified montmorillonite (MMT)]. As mention before, natural biopolymers (chitosan), natural antimicrobial agents (nisin, thymol, carvacrol, isothiocyanate, antibiotics), enzymes (peroxidase, lysozyme), and synthetic antimicrobial agents and organic acids (quaternary ammonium salts, EDTA, propionic, benzoic, sorbic acids) are also utilized to provide antimicrobial function. In addition, various combinations of such antimicrobials are exploited by incorporating them into packaging materials, expecting their synergistic effect. The example of the synergistic effect is investigation of nanocomposite poly(lactic acid) films with particular nanocrystals (PLA-CNC) made of cellulose and nisin, which are used as an antimicrobial agent (Salmieri et al., 2014). Food packaging with antimicrobial films was tested in packages of sliced cooked ham as a food model. Those were afterward stored for 14 days at low temperatures (around 4°C) in order to be able to determine their capacity in the inhibition of bacteria *L. monocytogenes* as well as their physical, chemical, and structural properties. This study is also important since it was focused on the release of the antimicrobial films. It was achieved with agar diffusion bioassay. The results of the study showed that the values of mechanical properties in bioactive antimicrobial films (eg, tensile strength and modulus and elongation at break as well as vapor permeability) were stable during the investigated period of time (after 14 days of storage). The investigated antimicrobial bioactive films showed a significant reduction of bacteria, so these results clearly demonstrate the potential application of antimicrobial films on food packaging intended to be used for meat products (Salmieri et al., 2014).

Nisin is a 34-amino acid lanthionine-containing polypeptide produced by the lactic acid bacterium *Lactococcus lactis*. Used in antimicrobial packaging, nisin is one of the most effective agents against a number of Gram-positive bacteria such as *S. aureus*, *Streptococcus pneumonia*, *Clostridium difficile*, *L. monocytogenes*, *Bacillus cereus*, and *Clostridium botulinum* (Ahire and Dicks, 2014). The novel nanocomposite packaging materials with antimicrobial functions such as nisin have high potential for active food packaging (Lagaron et al., 2005; Jia et al., 2014). On the other hand, PLA is the most prominent biodegradable plastic, derived from biorenewable feedstock (Barlow and Morgan, 2013). Poly(lactic acid) (PLA), a poly(2-hydroxypropionic acid), is a biodegradable and semipolar biocompatible polymer. Lactic acid (2-hydroxy propionic acid), the single monomer of PLA, is produced via fermentation from 100% renewable resources such as corn, maize, wheat, wood residues, or other biomass. PLA has unique properties like good appearance, high mechanical strength (when compared to polyethylene, polypropylene, and polystyrene), and low toxicity; and good barrier properties have broadened its applications (Krishnamurthy et al., 2004). Approximately 25–55% less energy is required to produce PLA than petroleum-based polymers (Raquez et al., 2013). However, some mechanical and physical properties of PLA are not on the satisfactory level (Gorrasi et al., 2008). To improve these properties and fit the requirements, manufacture of packaging materials with different nanostructures has been developed in the recent years. The final goal was to develop various types of gas and water vapor/moisture permeability to flavor, color, pathogen preservation, and shelf life of the food products by means of active package materials using natural polymer-based formulations containing nanoparticles (Vermeiren et al., 1999). Therefore, the capabilities of PLA were studied for being a strong active packaging material in different areas requiring nanostructure implementation refined with functional antimicrobial characteristics. Nanocomposites tend to reduce the packaging waste associated with processed food products and serve to accomplish preservation requirements of fresh foods by extending their shelf lives. Additionally, nanotechnology enables the researchers to alter the structure of packaging materials at the molecular level. In the investigation of Oliveira Mori and his coworkers, the effect of the adding candeia (*Eremanthus erythropappus*) essential oil at different concentrations (5, 10, and 15% by mass) on the properties of nanostructured PLA mats was obtained by electrospinning (Oliveira Mori et al., 2015). Hydrophilic nisin-loaded polylactic acid of the known size was very successfully manufactured in a single-step emulsification procedure. The first test of shear stress as well as the temperature tests performed have shown no loss in nisin inhibition activity (Shaowen et al., 2014).

As mention in previous chapter for food contact applications, nanocomposite plastics offer many advantages over traditional polymers. However, while the unique properties of engineered nanomaterials (ENMs) may be harnessed for many positive ends, there are concerns about whether ENMs pose risks to human health. The primary areas of interest for assessing the safety of nanocomposite food contact materials (FCMs) are the potential for migration of ENMs into food and the potential toxicity of such released ENMs, discussed in further chapter.

5 Conclusions

The application of nanoparticles has the great potential to improve food quality and safety. Excellent new materials with fascinating properties can revolutionize today's food industry and food markets. However, the exposure of people to engineered nanoparticles from food packaging materials is increasing, and unfortunately this is not accompanied by the appropriate awareness and/or limits prescribed by safety regulation regarding their toxicological data. The results of many investigations show that the commercial application of nanoparticles has to be more carefully addressed and thoroughly assessed.

Acknowledgments

This work and collaboration have been fully supported by Croatian Science Foundation project UIP-2014-09-1534. Any opinions, findings, and conclusions or recommendations expressed in this material are those of the authors and do not necessarily reflect the views of the Croatian Science Foundation.

References

Ahire, J.J., Dicks, L.M.T., 2014. Nisin incorporated with 2,3-dihydroxybenzoic acid in nanofibers inhibits biofilm formation by a methicillin-resistant strain of *Staphylococcus aureus*. Probiot. Antimicrob. Proteins 7, 52–59.

Alexandre, M., Dubois, P., 2000. Polymer-layered silicate nanocomposites: preparation, properties and uses of a new class of materials. Mater. Sci. Eng. 28, 1–63.

Althues, H., Henle, J., Kaskel, S., 2007. Functional inorganic nanofillers for transparent polymers. Chem. Soc. Rev. 36 (9), 1454–1465.

An, J., Zhang, M., Wang, S., Tang, J., 2008. Physical, chemical and microbiological changes in stored green asparagus spears as affected by coating of silver nanoparticles-PVP. LWT Food Sci. Technol. 41 (6), 1100–1107.

Anthonissen, R., Toussaint, O., Vankoningsloo, S., Saout, C., Alfaro-Moreno, E., Hoet, P., Gonzalez, L., Dubruel, P., 2010. Eco-, geno- and human toxicology of bio-active nanoparticles for biomedical applications. Toxicology 269, 170–181.
Appendini, P., Hotchkiss, J.H., 2002. Review of antimicrobial food packaging. Innov. Food Sci. Emerg. Technol. 3, 113–126.
Arvanitoyannis, I.S., Bosnea, L., 2004. Migration of substances from food packaging materials to foods. Crit. Rev. Food Sci. Nutr. 44, 63–76.
Asharani, P.V., Lianwu, Y., Gong, Z., Valiyaveettil, S., 2011. Toxicity of silver nanoparticles on zebrafish models. Nanotoxicology 5, 43–54.
Aymonier, C., Schlotterbeck, U., Antonietti, L., Zacharias, P., Thomann, R., Tiller, J.C., Mecking, S., 2002. Chem. Commun. 24, 3018–3019.
Barlow, C.Y., Morgan, D.C., 2013. Polymer film packaging for food, an environmental assessment. Resour. Conser. Recy. 78, 74–80.
Blasco, C., Picó, Y., 2011. Determining nanomaterials in food. Trends Anal. Chem. 30, 84–99.
Bom, P.J., Kreyling, W., 2004. Toxicological hazards of inhalated nanoparticles, potential implications for drug delivery. J. Nanosci. Nanotech. 4, 521–531.
Buzea, C., Pacheco, I., Robbie, K., 2007. Nanomaterials and naoparticles: sources and toxicity. Biointerphases 2, MR17–MR71.
Casals, E., Vazquez-Campos, S., Bastus, N.G., Puntes, V., 2008. Distribution and potential toxicity of engineered inorganic nanoparticles and carbon nanostructures in biological systems. Trends Anal. Chem. 27, 672–683.
Castellano, J.J., Shafi, S.M., Donate, G., Wright, T.E., Mannari, R.J., Payne, W.G., Smith, D.J., Robson, M.C., 2007. Comparative evaluation of silver-containing antimicrobial dressings. Int. Wound J. 4, 114–122.
Chandramouleeswaran, S., et al., 2007. Functional behaviour of polypropylene/ZnO-soluble starch nanocomposites. Nanotechnology 18 (38), 385702.
Chawengkijwanich, C., Hayata, Y., 2008. Development of TiO_2 powder-coated food packaging film and its ability to inactivate *Escherichia coli* in vitro and in actual tests. Int. J. Food Microbiol. 123 (3), 288–292.
Cheng, Q., et al., 2006. Surface-modified antibacterial TiO_2/Ag^+ nanoparticles: preparation and properties. Appl. Surf. Sci. 252, 4154–4160.
Choi, O., Yu, C.P., Fernández, G.E., Hu, Z., 2010. Interactions of nanosilver with *Escherichia coli* cells in planktonic and biofilm cultures. Water Res. 44, 6095–6103.
Cockburn, A., Bradford, R., Buck, N., Constable, A., Edwards, G., Haber, B., Hepburn, P., Howlett, J., Kampers, E., Klem, C., Radomski, M., Stamm, H., Wijnhoven, S., Wildemann, T., 2012. Approaches tot he safety assessment of engineered nanomaterials in food. Food Chem. Toxicol. 50, 2224–2242.
Cuq, B., Gontard, N., Guilbert, S., 1995. Edible films and coatings as active layers. In: Rooney, M. (Ed.), Active Food Packaging. Blackie Academic and Professional, Glasgow, pp. 11–142.
Cushen, M., et al., 2012. Nanotechnologies in the food industry—recent developments, risks and regulation. Trends Food Sci. Technol. 24, 30–46.
Cutter, C.N., 2006. Opportunities for bio-based packaging technologies to improve the quality and safety of fresh and further processed muscle foods. Meat Sci. 74, 131–142.
Damm, C., Münstedt, H., Rösch, A., 2007. Long-term antimicrobial polyamide 6/silver nanocomposites. J. Mater. Sci. 42 (15), 6067–6073.
Damm, C., Münstedt, H., Rösch, A., 2008. The antimicrobial efficacy of polyamide 6/silver-nano- and microcomposites. Mater. Chem. Phys. 108, 61–66.
Danielli, D., et al., 2008. Active and intelligent food packaging: legal aspects and safety concerns. Trends Food Sci. Technol. 19 (1), 103–112.

da Silva, B.F., Pérez, S., Gardinalli, P., Singhal, R.K., Mozeto, A.A., Barceló, D., 2011. Analytical chemistry of metallic nanoparticles in natural environments. Trends Anal. Chem. 30 (3), 528–540.
de Azeredo, H.M.C., 2009. Nanocomposites for food packaging applications. Food Res. Int. 42, 1240–1253.
Donaldson, K., Stone, V., Tran, C.L., Kreyling, W., Borm, P., 2004. Deposition and effects of fine and ultrafine particles in the respiratory tract. J. Occup. Environ. Med. 61, 727–728.
Duncan, T.V., 2011. Applications of nanotechnology in food packaging and food safety: barrier. J. Colloid. Interf. Sci. 363, 1–24.
Echegoyen, Y., Nerin, C., 2013. Nanoparticle release from nano-silver antimicrobial food containers. Food Chem. Toxicol. 62, 16–22.
Egger, S., et al., 2009. Antimicrobial properties of a novel silver-silica nanocomposite material. Appl. Environ. Microbiol. 75 (9), 2973–2976.
El-Temsah, Y.S., Joner, E., 2010. Impact of Fe and Ag nanoparticles on seed germination and differences in bioavailability during exposure in aqueous suspension and soil. J. Environ. Toxicol. 27, 42–49.
Emamifar, A., Kadivar, M., Shahedi, M., Soleimanian-Zad, S., 2011. Effect of nanocomposite packaging containing Ag and ZnO on inactivation of Lactobacillus plantarum in orange juice. Food Cont. 22 (3), 408–413.
Eom, H.J., Choi, J., 2009. Oxidative stress of silica nanoparticles in human bronchial epithelial cell. Beas-2B Toxicol. Vitro. 23, 1326–1332.
EU Communication, 2004. Communication from the commission—towards a European strategy for nanotechnology. COM 338, 1.
Fernández, A., Picouet, P., Lloret, E., 2010. Cellulose-silver nanoparticle hybrid materials to control spoilage-related microflora in absorbent pads located in trays of fresh-cut melon. Int. J. Food Microbiol. 142 (1–2), 222–228.
Finnigan, B., 2009. Barrier polymers. In: Yam, K.L. (Ed.), The Wiley Encyclopedia of Packaging Technology. John Wiley & Sons, Hoboken, NJ, pp. 103–109.
Floros, J.D., Nielsen, P.V., Farkas, J.K., 2000. Advances in modified atmosphere and active packaging with applications in the dairy industry. Bull. Int. Dairy Fed. 246, 22–28.
Fortunati, E., Armentano, I., Iannoni, A., Barbale, M., Zaccheo, S., Scavone, M., Visai, L., Kenny, J.M., 2012a. New multifunctional poly(lactide acid) composites: mechanical, antibacterial, and degradation properties. J. Appl. Polym. Sci. 124, 87–98.
Fortunati, E., Armentano, I., Zhou, Q., Iannoni, A., Saino, E., Visai, L., Berglund, L.A., Kenny, J.M., 2012b. Multifunctional bionanocomposite films of poly(lactic acid), cellulose nanocrystals and silver nanoparticles. Carb. Polym. 87, 1596–1605.
Föster, S., Konrad, M., 2003. From self-organising polymers to nano- and biomaterials. J. Mater. Chem. 13, 2671–2688.
Fujishima, A., Rao, T.N., Tryk, D.A., 2000. Titanium dioxide photocatalysis. J. Photochem. Photobiol. C 1 (1), 1–21.
Goddard, J., Talbert, J., Hotchkiss, J., 2007. Covalent attachment of lactase to low-density polyethylene films. J. Food Sci. 72 (1), 36–41.
Goldberg, S., Doyle, R., Rosenberg, M., 1990. Mechanism of enhancement of microbial cell hydrophobicity by cationic polymers. J. Bacteriol. 172 (10), 5650–5654.
Gorrasi, G., Vittoria, V., Murariu, M., Ferreira Ada, S., Alexandre, M., Dubois, P., 2008. Effect of filler content and size on transport properties of water vapor in PLA/calcium sulfate composites. Biomacromolecules 9, 984–990.
Hagelstein, A., Hoover, D., Paik, J., Kelley, M., 1995. Potential of antimicrobial nylon as a food package. Proceedings of the IFT A.

Hagens, W.I., Oomen, A.G., de Jong, W.H., Cassee, F.R., Sips, A.J.A.M., 2007. What do we (need to) know about the kinetic properties of nanoparticles in the body? Regul. Toxicol. Pharmacol. 49, 217–229.
Han, J.H., 2000. Antimicrobial food packaging. Food Technol. 54 (3), 56–65.
Han, J.H., 2003. Antimicrobial Food Packaging. In: Ahvenainen, R. (Ed.), Novel Food Packaging Techniques. Elsevier, pp. 50–70.
Holgate, S.T., 2010. Exposure, uptake, distribution and toxicity of nanomaterials in humans. J. Biomed. Nanotechnol. 6, 1–19.
Hong, K., et al., 2006. Preparation of antimicrobial poly(vinyl alcohol) nanofibers containing silver nanoparticles. J. Polym. Sci. B 44 (17), 2468–2474.
Hu, A.W., Fu, Z.H., 2003. Nanotechnology and its application in packaging and. Packag. Eng. 24, 22–24.
Ishitani, T., 1995. Active packaging for food quality preservation in Japan. In: Ackerman, P.J.M.O.T. (Ed.), Food and Food Packaging Materials—Chemical Interactions. Royal Society of Chemistry, Cambridge, UK, pp. 177–188.
Janjarasskul, T., Krochta, J.M., 2010. Edible packaging materials. Annual Rev. Food Sci. Technol. 1, 415–448.
Jia, S., Lub, J., Liuc, Z., Srivastavaa, D., Songa, A., Liuc, Y., Leea, I., 2014. Dynamic encapsulation of hydrophilic nisin in hydrophobic poly(lactic acid) particles with controlled morphology by a single emulsion process. J. Colloid. Interf. Sci. 423, 85–93.
Jin, T., He, Y., 2011. Antibacterial activity of magnesium oxide (MgO) nanoparticles against foodborne pathogens. J. Nanosci. Nanotech. 13, 6877–6885.
Jin, T., Niemira, B.A., 2011. Application of polylactic acid coating with antimicrobials in reduction of *Escherichia coli* O157:H7 and *Salmonella Stanley* on apples. J. Food Sci. 76, 184–188.
Jones, N., Ray, B., Ranjit, K., Manna, A., 2008. Antibacterial activity of ZnO nanoparticle suspensions on a broad spectrum of microorganisms. FEMS Microbiol. Lett. 279 (1), 71–76.
Kim, J.Y., Han, S.I., Kim, D.K., Kim, S.H., 2009. Mechanical reinforcement and crystallization behavior of poly(ethylene 2,6-naphthalate) nanocomposites induced by modified carbon nanotube. Compos. Part A: Appl. Sci. Manuf. 40, 45–53.
Klabunde, K.J., Stark, J., Koper, O., Mohs, C., Park, D., Decker, S., Jiang, Y., Lagadic, I., Zhang, D., 1996. Nanocrystals as stoichiometric reagents with unique surface chemistry. J. Phys. Chem. 100, 12142–12153.
Kong, H., Jang, J., 2008. Antibacterial properties of novel poly(methyl methacrylate) nano fiber containing silver nanoparticles. Langmuir 24 (5), 2051–2056.
Kotsilkova, R., Petkova, V., Petovska, Y., 2001. Thermal analysis of polymer-silicate nanocpomsites. J. Therm. Anal. Calorim. 64, 591–598.
Krishnamurthy, K., Demirci, A., Puri, V.M., Cutter, C., 2004. Effect of Packaging Materials on Inactivation of Pathogenic Microorganisms on Meat during Irradiation. Am. Soc. Agr. Eng. 47 (4), 1141–1149.
Kubacka, A., et al., 2009. Plasmonic nanoparticle/polymer nanocomposites with enhanced photocatalytic antimicrobial properties. J. Phys. Chem. 113 (21), 9182–9190.
Kumar, R., Howdle, S., Münstedt, H., 2005. Polyamide/silver antimicrobials: effect of filler types on the silver ion release. J. Biomed. Mater. Res. 75B (2), 311–319.
Kumar, R., Münstedt, H., 2005. Polyamide/silver antimicrobials: effect of crystallinity on the silver ion release. Polym. Int. 54 (8), 1180–1186.
Kumari, M., Mukerjee, A., Chadreasekaran, N., 2009. Genotoxicity of silver nanoparticles in *Allium cepa*. Sci. Total Environ. 407, 5243–5246.
Laborda, F., Jiménez-Lamana, J., Bolea, E., Castillo, J.R., 2011. Selective identification, characterization and determination of dissolved silver (I) and

silver nanoparticles based on single particle detection by inductively coupled plasma mass spectrometry. J. Anal. At. Spectrom. 26, 1362–1371.
Lagaron, J.M., Cabedo, L., Cava, D., Feijoo, J.L., Gavara, R., Gimenez, E., 2005. Improving packaged food quality and safety. Part 2: nanocomposites. Food Addit. Contam. 22, 994–998.
Lagaron, J., et al., 2007. Novel Nanocomposites to Enhance Quality and Safety of Packaged Foods. Smithers Rapra Ltd., Shawbury, UK.
Li, N., Sioutas, C., Cho, A., Schmitz, D., Misra, C., Sempf, J., Wang, M., Oberley, T., Froines, J., Nel, A., 2003. Ultrafine particulate pollutants induce oxidative stress and mitochondrial damage. Environ. Health Perspect. 111, 455–460.
Li, H., et al., 2009. Effect of nano-packing on preservation quality of Chinese jujube (Ziziphus jujuba Mill. var. inermis (Bunge) Rehd). Food Chem. 114 (2), 547–552.
Liu, J., Hurt, R.H., 2010. Ion release kinetics and particle persistence in aqueous nano-silver colloids. Environ. Sci. Technol. 44 (6), 2169–2175.
Lopez-Rubio, A., et al., 2004. Overview of active polymer-based packaging technologies for food applications. Food Rev. Int. 20 (4), 357–387.
Lubick, N., 2008. Nanosilver toxicity: ions, nanoparticles—or both? Environ. Sci. Technol. 42, 8617.
Madl, A.K., Pinkerton, K.E., 2009. Health effects of inhalated engineered and incidental nanoparticles. Crit. Rev. Toxicol. 39, 629–658.
Marquis, B.J., Love, S.A., Braun, K.L., Haynes, C.L., 2009. Analytical methods to assess nanoparticle toxicity. Analyst 134, 425–439.
Mastromatteo, M., Conte, A., Del Nobile, M.A., 2010. Advances in controlled release devices for food packaging applications. Trends Food Sci. Technol. 21, 591–598.
Mbhele, Z.H., et al., 2003. Fabrication and characterization of silver–polyvinyl alcohol nanocomposites. Chem. Mater. 15 (26), 5019–5024.
Mittal., V., 2011. In-situ synthesis of polymer nanocomposites. In: Mittal, V. (Ed.), In-Situ Synthesis of Polymer Nanocomposites. John Wiley & Sons, Hoboken, NJ, pp. 1–25.
Mohammed Fayaz, A., et al., 2009. Mycobased synthesis of silver nanoparticles and their incorporation into sodium alginate films for vegetable and fruit preservation. J. Agric. Food Chem. 57 (14), 6246–6252.
Moraru, C.I., et al., 2003. Nanotechnology: a new frontier in food science. Food Technol. 57, 24–29.
Neethirajan, S., Freund, M.S., Shafai, C., Jayas, D.S., Thomson, D.S., White, N.D.G., 2010. Development of carbon dioxide (CO_2) sensor for grain quality monitoring. Biosyst. Eng. 106, 395–404.
Oliveira Mori, C.L.S., Passos, N.A., Oliveira, J.E., Altoé, T.F., Mori, F.A., Mattoso, L.H.C., Scolforo, J.R., Tonoli, G.H.D., 2015. Nanostructured polylactic acid/ candeia essential oil mats obtained by electrospinning. J. Nanomater., 1–9.
Ouattara, B., et al., 2000. Diffusion of acetic and propionic acids from chitosan-based antimicrobial packaging films. J. Food Sci. 65 (5), 768–773.
Page, K., et al., 2007. Titania and silver–titania composite films on glass-potent antimicrobial coatings. J. Mater. Chem. 17 (1), 95–104.
Pérez, S., la Farré, M., Barceló, D., 2009. Analysis, behavior and ecotoxicity of carbon-based nanomaterials in the aquatic environment. Trends Anal. Chem. 28, 820–832.
Plackett, D.V., Holm, V.K., Johansen, P., Ndoni, S., Nielsen, P.V., Sipilainen-Malm, T., Sodergard, A., Verstichel, S., 2006. Characterization of L-polylactide and L-polylactidepolycaprolactone copolymer films for use in cheese-packaging applications. Packag. Technol. Sci. 19, 1–24.
Podsiadlo, P., et al., 2005. Layer-by-layer assembly of nacre-like nanostructured composites with antimicrobial properties. Langmuir 21 (25), 11915–11921.

Prashantha, K., Soulestin, J., Lacrampe, M.F., Krawczak, P., Dupin, G., Claes, M., 2009. Masterbatch-based multi-walled carbon nanotube filled polypropylene nanocomposites: assessment of rheological and mechanical properties. Compos. Sci. Technol. 69, 1756–1763.
Priolo, M.A., Gamboa, D., Holder, K.M., Grunlan, J.C., 2011. Influence of clay concentration on the gas barrier of clay-polymer nanobrick wall thin film assemblies. Nano Lett. 10, 4970–4974.
Radheshkumar, C., Münstedt, H., 2006. Antimicrobial polymers from polypropylene/silver composites—Ag^+ release measured by anode stripping voltammetry. React. Funct. Polym. 66 (7), 780–788.
Raquez, J.M., Habibi, Y., Murariu, M., Dubois, P., 2013. Polylactide (PLA)-based nanocomposites. Prog. Polym. Sci. 38, 1504–1542.
Reddy, M.P., Venugopal, A., Subrahmanyam, M., 2007. Hydroxyapatite-supported Ag–TiO_2 as *Escherichia coli* disinfection photocatalyst. Water Res. 41, 379–386.
Rezić, I., 2011. Determination of engineered nanoparticles on textiles and in textile wastewaters. Trends Anal. Chem. 30, 1159–1167.
Rezić, I., Zeiner, M., Steffan, I., 2011. Determination of 28 selected elements in textiles by axially viewed inductively coupled plasma optical emission spectrometry. Talanta 83, 865–871.
Rezić, I., 2012. Engineered nanoparticles in textiles and textile wastewaters. In: Farre, M. Barcelo, D. (Eds.), Analysis and Risk of Nanomaterials in Environmental and Food Samples, 59, COAC. Elsevier, Chennai, India, pp. 235–264.
Robbens, J., Vanparys, C., Nobels, I., Blust, R., Van Hoecke, K., Janssen, C., De Schamphelaere, K., Roland, K., Blanchard, G., Silvestre, F., Gillardin, V., Kestemont, P., Anthonissen, R., Toussaint, O., Vankoningsloo, S., Saout, C., Alfaro-Moreno, E., Hoet, P., Gonzalez, L., Dubruel, P., Troisfontaines, P., 2010. Eco-, geno- and human toxicology of bio-active nanoparticles for biomedical applications. Arch. Toxicol. 269, 170–181.
Robertson, G.L., 2006. Food Packaging: Principles and Practice, second ed. CRC Press, Taylor & Francis Group, New York.
Roco, M.C., 2003. Nanotechnology: convergence with modern biology and medicine. Curr. Opin. Biotechnol. 14 (3), 337–346.
Salmieri, S., Islam, F., Khan, R.A., Hossain, F.M., Ibrahim, H.M.M., Miao, C., Hamad, W.Y., Lacroix, M., 2014. Antimicrobial nanocomposite films made of poly(lactic acid)-cellulose nanocrystals (PLA-CNC) in food applications: part A—effect of nisin release on the inactivation of Listeria monocytogenes in ham. Cellulose 21, 1837–1850.
Sánchez, A., Recillas, S., Font, X., Casals, E., González, E., Puntes, V., 2011. Ecotoxicity of, and remediation with, engineered inorganic nanoparticles in the environment. Trends Anal. Chem. 30 (3), 507–516.
Sanchez-Valdes, S., et al., 2009. Mechanical and antimicrobial properties of multilayer films with a polyethylene/silver nanocomposite layer. J. Appl. Polym. Sci. 111 (2), 953–962.
Sargent, J.F., Jr., 2014. The National Nanotechnology Initiative: Overview, Reauthorization, and Appropriations Issues. National Nanotechnology Initiative, Congressional Research Service report.
Schmidt, B., Katiyar, V., Plackett, D., Larsen, E.H., Gerds, N., Bender Koch, C., Petersen, J.H., 2011. Migration of nanosized layered double hydroxide platelets from polylactide nanocomposite films. Food Addit. Contam. 28, 956–966.
Shaowen, J., Lu, J., Liu, Z., Lee, I., 2014. Dynamic encapsulation of hydrophilic nisin in hydrophobic poly (lactic acid) particles with controlled morphology by a single emulsion process. J. Coll. Int. Sci. 423, 85–93.

Sheikh, F., et al., 2009. Electrospun antimicrobial polyurethane nanofibers containing silver nanoparticles for biotechnological applications. Macromol. Res. 17 (9), 688–696.

Silvestre, C., Duraccio, D., Cimmino, S., 2011. Food packaging based on polymer nanomaterials. Prog. Polym. Sci. 36, 1766–1782.

Simon, P., Chaudhry, Q., Bakos, D.J., 2008. Migration of nanosized layered double hydroxide platelets from polylactide nanocomposite films. Food Nutr. Res. 47, 105–113.

Slawson, R.M., Van Dyke, M.I., Lee, H., Trevors, J.T., 1992. Germanium and silver resistance, accumulation, and toxicity in microorganisms. Plasmid 27, 72–79.

Sondi, I., Salopek-Sondi, B., 2004. Silver nanoparticles as antimicrobial agent: a case study on *E. coli* as a model for Gram-negative bacteria. J. Coll. Interf. Sci. 275, 177–182.

Song, H., Li, B., Lin, Q.-B., Wu, H.-J., Chen, Y., 2011. Migration of silver from nanosilver-polyethylene composite packaging into food simulants. Food Addit. Contam. 28, 1758–1762.

Spadaro, J.A., Berger, T.J., Barranco, S.D., Chapin, S.E., Becker, R.O., 1974. Antibacterial effects of silver electrodes with weak direct current. Microb. Agents Chemother. 6, 637–642.

Stampoulis, D., Sinha, S.K., White, J.C., 2009. Assay-dependent phytotoxicity of nanoparticles to plants. Environ. Sci. Technol. 43, 9473–9479.

Sun, J., et al., 2006. Toward monodispersed silver nanoparticles with unusual thermal stability. J. Am. Chem. Soc. 128 (49), 15756–15764.

Suppakul, P., 2004; Study of antimicrobial polymeric packaging films containing basil extracts, PhD thesis, Victoria University of Technology. http://vuir.vu.edu.au/15423/1/Suppakul_2004compressed.pdf.

Suppakul, P., Sonneveld, K., Bigger, S.W., Miltz, J., 2011. Loss of AM additives from antimicrobial films during storage. J. Food Eng. 105, 270–276.

Tang, S., Zou, P., Xiong, H., Tang, H., 2008. Effect of nano-SiO_2 on the performance of starch/ polyvinyl alcohol blend films. Carbohydr. Polym. 72, 521–526.

Tankhiwale, R., Bajpai, S., 2009. Graft copolymerization onto cellulose-based filter paper and its further development as silver nanoparticles loaded antibacterial food-packaging material. Colloids Surf. B69 (2), 164–168.

Tawakkal, I.S.M.A., Cran, M.J., Miltz, J., Bigge, S.W., 2014. A review of poly(lactic acid)-based materials for antimicrobial packaging. J. Food Sci. 79, 1477–1490.

Teixiera, V., 2010. Opportunities and Challenges in Nanotechnology-Based Food Packaging Industry. NANOAGRI-2010. International Conference on Food and Agricultural Applications of Nanotechnologies, Brazil. Available from: <http://www.slideshare.net/teixeiravasco/opportunities-and-challenges-in-nanotechnologybased-food-packaging-industry>

Tinkle, S.S., Antonini, J.M., Rich, B.A., Roberts, J.R., Salmen, R., DePree, K., Adkins, E.J., 2003. Skin as a route of exposure and sensitization in chronic beryllium disease. Environ. Health Perspect. 111, 1202–1208.

Vermeiren, L., Devlieghere, F., Van Beest, M., De Kruijf, N., Debevere, J., 1999. Developments in the active packaging of foods. Trends Food Sci. Technol. 10, 77–86.

Vermeiren, L., Devlieghere, F., Debevere, J., 2002. Effectiveness of some recent antimicrobial packaging concepts. Food Addit. Contam. 19, 163–171.

Vladimirov, V., Betchev, C., Vassiliou, A., Papageorgiou, G., Bikiaris, D., 2006. Dynamic mechanical and morphological studies of isotactic polypropylene/fumed silica nanocomposites with enhanced gas barrier properties. Compos. Sci. Technol. 66, 2935–2944.

Walczak, A.P., Fokkink, R., Peters, R., Tromp, P.J., Herrera Rivera, Z.E., Rietjens, I.M.C.M.m, Hendriksen, P.J.M., Bouwmeester, H., 2013. Behaviour of silver

nanoparticles and silver ions in and in vitro human gastrointestinal digestion model. Nanotoxicology 7, 1198–1210.
Warheit, D., et al., 2004. Comparative pulmonary toxicity assessment of single-wall carbon nanotubes in rats. Toxicol. Sci. 77 (1), 117–125.
Weinberg, H., Galyean, A., Leopold, M., 2011. Evaluating engineered nanoparticles in waters. Trends Anal. Chem. 30, 72–83.
Wu, C.L., Zhang, M.Q., Rong, M.Z., Friedrich, K., 2002. Tensile performance improvement of low nanoparticles filled-polypropylene composites. Compos. Sci. Technol. 62, 1327–1340.
Yamamoto, O., 2001. Influence of particle size on the antibacterial activity of zinc oxide. Int. J. Inorg. Mater. 3 (7), 643–646.
Yang, K., Wang, X., Wang, Y., 2007. Progress in nanocomposite of biodegradable polymer. J. Ind. Eng. Chem. 13, 485–500.
Yoksan, R., Chirachanchai, S., 2010. Silver nanoparticle-loaded chitosan-starch based films: fabrication and evaluation of tensile, barrier and antimicrobial properties. Mater. Sci. Eng. C30 (6), 891–897.
Yu, H., Sun, B., Zhang, D., Chen, G., Yang, X., Yao, J., 2014. Reinforcement of biodegradable poly(3-hydroxybutyrate-*co*-3-hydroxyvalerate) with cellulose nanocrystal/silver nanohybrids as bifunctional nanofillers. J. Mater. Chem. B 2, 8479–8489.
Zhao, G.J., Stevens, S.E., 1998. Multiple parameters for the comprehensive evaluation of the susceptibility of *Escherichia coli* to the silver ion. Biometals 11, 27–32.

16

NEW APPROACHES AND FUTURE ASPECTS OF ANTIBACTERIAL FOOD PACKAGING: FROM NANOPARTICLES COATING TO NANOFIBERS AND NANOCOMPOSITES, WITH FORESIGHT TO ADDRESS THE REGULATORY UNCERTAINTY

Majid Montazer*, Tina Harifi**

**Department of Textile Engineering, Functional Fibrous Structures & Environmental Enhancement (FFSEE), Amirkabir Nanotechnology Research Institute (ANTRI), Amirkabir University of Technology, Tehran, Iran;*
***Amirkabir University of Technology, Department of Textile Engineering, Functional Fibrous Structures & Environmental Enhancement (FFSEE), Tehran, Iran*

1 Introduction

Nanotechnology has a strong influence on the food industry in almost all sections (Fig. 16.1), from agriculture (pesticide, fertilizer, or vaccine delivery, animal and plant pathogen detection) to food processing (encapsulation of flavor or odor enhancers, food textural or quality improvement, new gelation or viscosifying agents), food packaging (pathogen, gas, or abuse sensors, antibacterial, UV-protection), and nutrient supplements (nutraceuticals with higher stability and bioavailability). The majority of food companies are actively searching for the new potential of nanotechnology to be used in food applications (Duncan, 2011; Unalan et al., 2014). However, undeniably, the most active area of nano food research is packaging, forming the largest share of the

Food Packaging. http://dx.doi.org/10.1016/B978-0-12-804302-8.00016-9

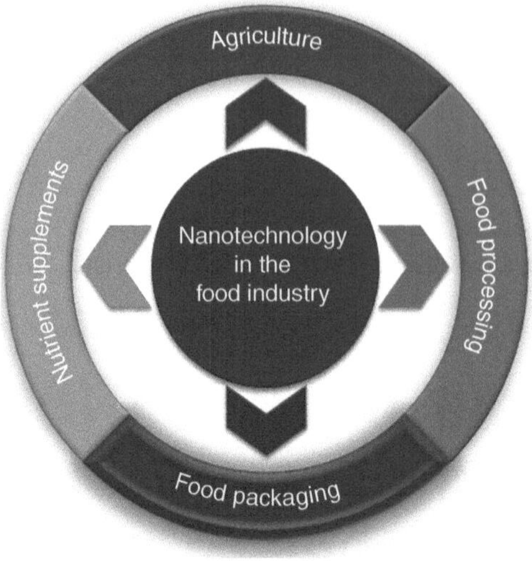

Figure 16.1. Applications of nanotechnology in the food industry.

food market with 4.13 in 2008 to 7.3 billion US dollars in 2014 with an estimated annual growth rate of 12% (Duncan, 2011; Unalan et al., 2014).

The first major step in modern food packaging was made in the 19th century with the advent of canning by Appert, followed by Prescott and Underwood, who worked on the application of bacteriology to the canning process (Brody et al., 2008). The development of polypropylene, polyester, and ethylene vinyl alcohol copolymers represented significant progress in replacing metal, glass, and paperboard packaging by plastic and flexible packaging (Brody et al., 2008).

Traditional passive food packages as barriers only delayed the adverse effects of the environment on the food product; however, more protective packaging was pioneered in the 20th century (Brody et al., 2008). The continued quest for innovation in food packaging has been answered by nanotechnology innovations in the 21st century, which is undoubtedly brightening the future of food packaging industry by introducing multifunctional active packaging with simultaneous enhanced barriers; structural, mechanical, and antibacterial properties; and sensing technologies (Brody et al., 2008).

Active packaging involves the interaction between the packaging and the food to provide desirable effects such as prolonged shelf-life, quality maintenance, and safety assurance (Bastarrachea et al., 2011). The food package interaction is achieved by adding active ingredients to the packaging system and/or using actively functional polymers, enhancing the performance of the system (Bastarrachea et al., 2011; Kim et al., 2008). Several applications of active packaging technologies have been commercialized and are currently in use by the food industry, including oxygen scavenging, carbon dioxide absorbing or emitting, moisture scavenging, and antimicrobial systems (Han, 2014).

Oxygen causes rotting of oxygen-sensitive foodstuffs and leads to a deterioration of quality characteristics such as color and freshness. Direct oxidation reactions are responsible for rancidity of vegetable oils and browning of fruits. Food deterioration can also be produced by indirect action of O_2 due to food spoilage by aerobic microorganisms (Echegoyen, 2015). The quality deterioration is more prominent in transparent packaged chilled products such as sausages, due to photo-induced oxidation process after exposure to light. It is necessary to reduce the oxygen content in the headspace of the packages by the addition of oxygen scavengers (Gibisa and Rieblingera, 2011). Sachets with the ability to remove oxygen were the first commercialized oxygen-scavenging systems, preventing food oxidation, rotting, and bacterial and mold growth. Oxygen-scavenging commercial technologies are mainly based on enzymatic and chemical systems, for instance catechol, ferrous salts, active zero-valent iron, ascorbic acid, and titania (Busolo and Lagaron, 2012). Xiao-e et al. (2004) introduced nano TiO_2/polymer films for the packaging of many oxygen-sensitive products due to the photocatalytic activity resulting in the oxygen removal in a closed environment.

Moreover, different polymers including polyethylene, polypropylene, fluorinated ethylene propylene, linear low-density polyethylene, and nylon have been infused with metal and metal oxide nanoparticles. It has been reported that films containing less than 1 wt.% Pd and Pt nanoparticles were active as oxygen scavengers (Echegoyen, 2015).

Carbon-dioxide-absorbing packaging system prevents packages from inflating, due to carbon dioxide formation after the packaging process. Carbon dioxide may be produced through nonenzymatic browning reactions in some packaged products such as coffee beans, pickles, fermented vegetables, and dairy products (Han, 2014). Carbon dioxide–absorbing systems are generally useful for products such as natural cheese or coffee that require fermentation or aging and produce gas after packaging (Han, 2014). CO_2 scavengers can be composed of a physical absorbent such as zeolite or an active carbon powder or a chemical absorbent such as $Ca(OH)_2$, Na_2CO_3 and $Mg(OH)_2$ (Charles et al., 2006).

A carbon dioxide–generating system can be classified as a technique complimentary to oxygen scavenging. The function of carbon dioxide with a packaging environment is to suppress microbial growth. A high carbon dioxide level (10–80%) is desirable for foods such as meat and poultry in order to inhibit surface microbial growth and extend shelf-life. Removal of oxygen from the package creates a partial vacuum resulting in the collapse of flexible packaging. Also, when a package is flushed with a mixture of gases including carbon dioxide, the carbon dioxide dissolves in the product, creating a partial vacuum. In such cases, the simultaneous release of carbon dioxide from inserted sachets, which consume oxygen, is desirable. Carbon dioxide–emitting systems are based on ferrous carbonate or a mixture of ascorbic acid and sodium bicarbonate (Kerry et al., 2006).

Excess water developed inside the packaging of high water activity products such as muscle-based foods (meat and fish) promotes bacterial and mold growth, resulting in quality loss. Moisture-scavenging systems are an effective way for controlling excess water accumulation in a food package and the most common moisture-absorbing systems have a superabsorbent polymer located between two layers of a nanoporous or nonwoven material (Realini and Marcos, 2014). Moisture-scavenging systems maintain a specific relative humidity inside the package by absorbing or releasing the moisture (Han, 2014).

It is generally accepted that microbial alternations are responsible for enormous losses in food supplies, and foodborne diseases are a global public health issue affecting public health, trade, and the economy. Incidents of foodborne diseases are increasing in developing as well as in the developed countries. Each year in the United States, there are an estimated 25–81 million foodborne illness cases, resulting in approximately 9,000 deaths caused by contaminated food consumption (Akbar and Anal, 2014). In Southeast Asia, approximately 22.8 million Salmonellosis cases with a death toll of 37,600 have been estimated each year (Akbar and Anal, 2014). *Salmonella, Listeria,* and *Toxoplasma* account for 75% of the cases of food-related diseases. *Campylobacter jejuni, Staphylococcus. aureus, Clostridium perfringens,* and *botulinum* and pathogenic strains of *Escherichia coli* as well as some parasites also have the potential to cause very large or lethal outbreaks (Coma, 2012). There are some other minor microbial hazards such as *Yersinia enterolitica* bacteria-forming biogenic amines, and also molds producing mycotoxins (Coma, 2012). Therefore, appropriate food-packaging products are a crucial factor in conservation and marketing (Dutta et al., 2009). In this regard, antibacterial active food packaging that inhibits spoilage and reduces pathogenic microorganisms has been developed as a novel packaging material with the potential to enhance food quality, safety, and shelf-life by prolonging the lag period of microorganisms, diminishing their growth (Dutta et al., 2009; Cruz-Romero et al., 2013; Bastarrachea et al., 2011). The antimicrobial function can be introduced by adding antimicrobial agents in the packaging system and/or using antimicrobial polymers (Han, 2003). Antimicrobial agents can be incorporated into a packaging system through simple blending with packaging materials, immobilizing or coating, depending on the characteristics of the packaging system, antimicrobial agent, and food (Han, 2003; Mangalassary, 2012). Antimicrobials may be incorporated into polymers in the melt or solvent. In the case of thermally stable antimicrobials such as inorganic compounds, thermal polymer processing methods such as extrusion and injection molding are mainly used (Emamifar, 2011). On the other hand, antimicrobials that cannot withstand the processing temperature are often coated on the polymers. Thus, surface immobilization requires the presence of functional groups on both the antimicrobial and the polymer and possibly spacer molecules linking the polymer surface to the antibacterial agent (Emamifar, 2011).

Organic compounds such as essential oils, organic acids, enzymes, and bacteriocins have been widely studied for their antimicrobial properties in polymeric matrices as antimicrobial packaging. However, organic compounds suffer from a drawback, namely, sensitivity to intense processing conditions such as high temperatures and pressure. On the other hand, inorganic nano compounds with strong antibacterial activity at low concentrations due to the high surface area to volume ratio, unique chemical and physical properties, and strong stability in extreme conditions outweighed the organic materials (Espitia et al., 2012). Metallic nanoparticles such as silver, copper, gold, platinum, and metal oxide nanoparticles including titanium oxide, zinc oxide, silica, alumina, iron oxide, and magnesium oxide are among the widely studied antibacterial inorganic materials (Lin et al., 2011; Espitia et al., 2012; Rhim et al., 2013). Moreover, polyamides, polyolefins, polystyrene, ethylene vinyl acetate copolymer, epoxy resins, polyurethane, polyimides, and polyethylene terephthalate have been used for food packaging (Espitia et al., 2012).

Combining the traditional food packaging material with nanomaterials known as nanocomposites expanded from less than 40 products in 2002 to more than 400 products in 2006 and is now a major active research field in the food packaging market, benefiting from excellent mechanical performance together with good antibacterial and strong resistant properties (Emamifar, 2011; Wei et al., 2011).

In spite of a wide range of antibacterial agents and polymers, new antimicrobial packaging materials can only be developed with authorization approval, following guidelines and regulations for the food industry (Han, 2003).

In this chapter, we offer an overview of all the different antibacterial nanostructures that have been used in food packaging, from the earliest application of single-layer nanoclay/polymer to the current use of nanofibers, polymer nanocomposites, and bio-nanocomposites that have been put together to create effective antibacterial food packaging systems. In each case, relative examples of the packaging materials for real food have been provided. The chapter finally concludes with specific health concerns related to the various nanomaterials applications in antibacterial food packaging.

2 Antibacterial Food Packaging Techniques

The first steps toward improving food safety and delaying spoilage were made by using antibacterial sprays or dips due to the primary contamination of foods at the surface (López-Rubio et al., 2004). However, direct surface application of antimicrobial materials ends in failure as the active substances are neutralized or diffuse rapidly from the surface into the food mass. Therefore, antimicrobial packaging has been used to control microbial growth in food products using packaging materials containing antimicrobial agents or using techniques for modifying the environmental conditions within the package (Dutta et al., 2009).

A thin layer of edible material can also be directly coated on food or formed into a film as a food wrap without changing the original ingredients or the processing method. Edible films and coatings have been used to improve the gas and moisture permeability, mechanical properties, sensory perceptions, microbial protection, and prolong the shelf life of various food products (Galus and Kadzińska, 2015). They are useful materials produced mainly from edible biopolymers and food-grade additives, usually classified according to the chemical structure including proteins, polysaccharides, lipids, or composites (Han, 2014; Galus and Kadzińska, 2015).

Use of packaging films containing antimicrobial agents is effective due to the possibility of controlled migration of the antimicrobial into the food, not only allowing for initial inhibition of undesirable microorganisms, but also residual activity over time, during the transport and storage of food during distribution (Emamifar, 2011).

Some packages were designed to modify the atmospheric composition to inhibit microbial growth. For instance, oxygen scavengers, CO_2 emitters, and active packages that reduce water content alter and reduce the kinetic growth of aerobic microorganisms (López-Rubio et al., 2004). Modified atmosphere packaging is one of the most important preservation techniques used to maintain the quality and extend the shelf-life of foods through replacing the surrounding air in the package with a specific gas mixture, mostly CO_2, O_2, and N_2 (Severino et al., 2014).

Packages containing polymers with inherent antimicrobial properties such as chitosan or polymers with immobilized antimicrobial agents are among the other antibacterial food packaging techniques. Most widely used polymer films for food packaging are polyethylene, polyvinyl chloride, polyvinyl alcohol, polylactic acid, and nylon (López-Rubio et al., 2004). Binding of antimicrobials to polymeric surfaces has been used by different methods, ranging from simply spreading antimicrobial solutions onto the polymer surface or combining the antimicrobials with binders of cellulosic or acrylic copolymer nature (López-Rubio et al., 2004). The antimicrobials could also bond to polymers by ionic or covalent linkages, which require the presence of functional groups on both the antimicrobial and the polymer. Some of the applied antibacterial agents with functional groups are peptides, enzymes, polyamines, and organic acids. In addition, some spacer molecules include dextrans, polyethylene glycol, ethylenediamine, and polyethyleneimine have been used to link the polymer surface with the bioactive agent (López-Rubio et al., 2004).

With the successful introduction of nanotechnology to the food packaging industry, polymer nanocomposites containing nano fillers have also gained attention.

Solution method, in situ, or inter-lamellar polymerization technique, and melt processing are three common methods used to produce polymer nanocomposites. Solution method contains nano filler swelling in a solvent, followed by addition to polymer solution and a final evaporation of solvent. In this method polymer molecules are allowed to extend between the layers of filler (Emamifar, 2011). The in situ or interlamellar method swells the fillers by absorption of a liquid monomer. After the monomer has penetrated in between the layers of filler, polymerization is initiated by heat, radiation, or incorporation of an initiator (Emamifar, 2011). In melt processing, the nanocomposite filler is incorporated into a molten polymer and then formed into the final material (Emamifar, 2011). Different techniques of antibacterial food packaging are summarized in Fig. 16.2.

3 Nanofibers

Nanofibers have desirable mechanical properties suitable for application in food packaging industry. They are incorporated into films to form a composite film with improved barrier properties. Moreover, electrospinning is a suitable technique for the

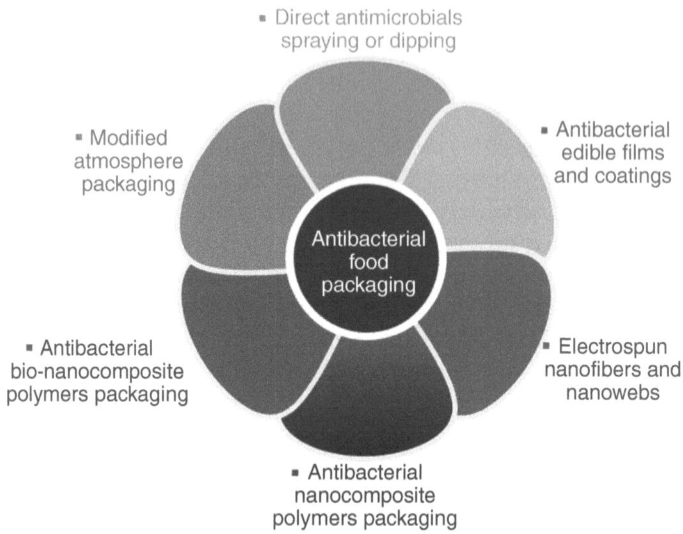

Figure 16.2. Different techniques for producing antibacterial food packaging.

fabrication of active packaging materials by forming nanofibers embedded with antimicrobial components (Rezaei et al., 2015). Electrospun nanofibers and nanowebs have been recognized for their potential in active food packaging, as active additives including antibacterials, antioxidants, essential oils, and probiotics. They can be effectively encapsulated into an electrospun nanofibrous matrix (Kayaci et al., 2013). ZnO incorporated into the cellulose acetate nanofiber exhibited strong antibacterial activity against *S. aureus*, *E. coli*, and *Citrobacter* and silver nanoparticles incorporated into the cellulose acetate nanofiber produced excellent antibacterial action against *S. aureus* and *E. coli*, *Klebsiella pneumonia*, and *Pseudomonas aeruginosa*, as some examples of antimicrobial immobilization, into cellulose nanofibers suitable for antibacterial food packaging systems (Rezaei et al., 2015). Antibacterial electrospun polylactic acid nanofibrous webs incorporating triclosan/cyclodextrin inclusion complexes were introduced as electrospun nanowebs with very high surface area and nanoporous structure as well as efficient antibacterial property applicable in active food packaging (Kayaci et al., 2013). High-density polyethylene/Cu-nanofiber nanocomposites with enhanced antibacterial and oxygen barrier properties were also prepared as suitable food packaging (Bikiaris and Triantafyllidis, 2013).

Recently, cellulose nanofibers with high aspect ratio were obtained and used as reinforcements in starch/polyvinyl alcohol cross-linked films. Starch films with 7 wt.% cellulose nanofibers showed high strength and stiffness comparable to that of polyolefins and was introduced as a low cost green substitute for application in food packaging and conservation (Panaitescu et al., 2015).

4 Nanocomposite Polymer

Polymer nanocomposites that are made by dispersing a nanoscale filler throughout a polymeric matrix offer new opportunities to the food packaging industry including cost saving and waste reduction due to the lower amount of polymer needed to produce packaging materials with identical or even better mechanical properties. Filler materials include clay and silicate nanoplatelets, silica nanoparticles, carbon nanotubes, graphene, starch nanocrystals, cellulose-based nanofibers or nanowhiskers, chitin or chitosan nanoparticles, and other inorganics (Duncan, 2011). In addition to the apparent enhanced polymer barrier properties, polymer nanocomposites possess antibacterial properties depending on the applied filler (Duncan, 2011). Nanocomposites used as food packaging materials are able to withstand the thermal stress of food processing as well as the mechanical stress produced during transportation and storage (Espitia et al., 2012). The nanosized fillers can be either organic or inorganic, such as clay; natural biopolymers such as chitosan; natural antimicrobial agents such as nisin; metals such as silver, and metal oxides such as TiO_2. The nanofillers have been studied for food packaging applications are shown in Fig. 16.3.

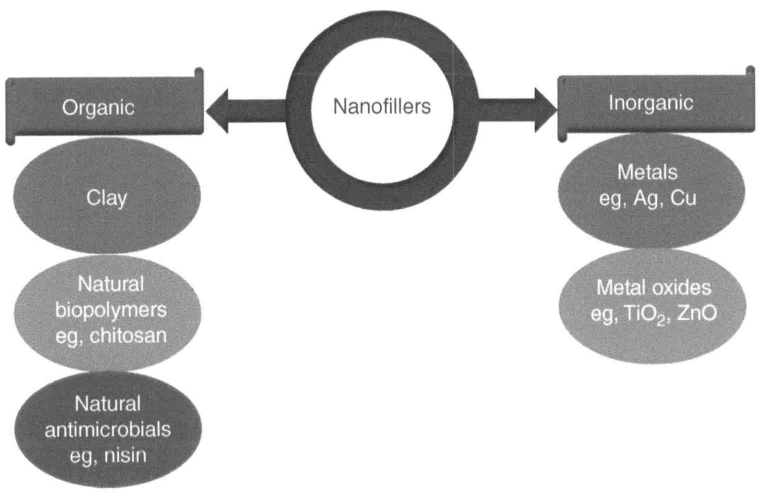

Figure 16.3. Classification of nanofillers used in food packaging.

4.1 Nanoclay

Use of polymer nanocomposites for food packaging dated back to 1990s with most of the researches focusing on the use of montmorillonite clay as the nanocomponent in a wide range of polymers such as polyethylene, nylon, polyvinyl chloride, and starch (Emamifar, 2011). Such vast number of studies led to the development of nanoclay polymer composites for potential use in a variety of food packaging applications including processed meats, cheese, confectionery, cereals, fruit juices, and dairy products (Chaudhry et al., 2008). Nanoclays are the widest commercial application of nanoparticles and constitute almost 70% of the market volume presumably due to the availability, low cost, simple process ability, and high stability (Huang et al., 2015). Nanoclays embedded in food packaging films decrease gas transmission rate to keep the freshness of oxygen-sensitive foods and prolong shelf life (Huang et al., 2015). Unmodified nanoclays have low antimicrobial properties, however; the functionalized nanoclays provide higher antimicrobial properties (Sadeghian Maryan and Montazer, 2014). For instance, organically modified nanoclays such as quaternary ammonium modified montmorillonite and Ag-zeolite nanoclay possess antibacterial property (Rhim et al., 2013). Researchers replaced the sodium ions of montmorillonite nanoclays with silver ions and showed antimicrobial activity of the silver nanoclays when dispersed in poly-ε-caprolactone and polylactic acid (Duncan, 2011; Busolo et al., 2010).

CASE STUDY: FRUIT SALAD

Silver/montmorillonite clay was used as active antimicrobial compound to improve the shelf life of fresh fruit salad (kiwifruit and pineapple) and the microbiological quality was determined by monitoring the principal spoilage microorganisms (mesophilic and psychrotrophic bacteria, coliforms, lactic acid bacteria, yeasts, and molds). A significant prolonged shelf life was indicated for fresh fruit salad using the new proposed packaging system (Costa et al., 2011).

4.2 Chitosan

Due to the good biocompatibility and biodegradability of chitosan and chitin, they have several biomedical applications (Tripathi et al., 2009). Chitosan has wide antimicrobial activity against a variety of pathogenic and spoilage microorganisms, including fungi, and Gram-positive and Gram-negative bacteria, mainly attributed to interactions of the cationic chitosan with the anionic cell membranes, increasing membrane permeability and subsequently rupture and leakage of the intracellular material (Azeredo, 2012; Dutta et al., 2009). Chitosan is a good candidate for food packaging systems due to its excellent film-forming capacity and barrier properties against gases in dry conditions (Tripathi et al., 2009). Chitosan nanostructures have been demonstrated to disperse well in biopolymers, having the potential to be used as antimicrobial agents in biodegradable food-packaging systems (Azeredo, 2012). Also, there have been studies focusing on antimicrobial effects of chitosan nanostructures combined with other antimicrobial agents (Azeredo, 2012). Chitosan-based antimicrobial films consisting of chitosan and polyvinyl alcohol were prepared by solution casting method and found effective in shelf-life extension of tomato (Tripathi et al., 2009). The microbiological screening has demonstrated the antimicrobial activity of the film against food pathogenic bacteria *E. coli*, *S. aureus*, and *Bacillus subtilis* (Tripathi et al., 2009). Based on a recent study chitosan/polyethylene was suggested as a suitable packaging material for chicken drumsticks to increase safety and quality. The active antibacterial bags' efficiency in inhibiting total aerobic mesophilic bacteria, coliforms, molds, and yeasts was confirmed while pH, color, and hardness of the samples were not affected significantly by the packaging film (Soysal et al., 2015). Chitosan/silver and chitosan/gold nanocomposite films were synthesized by a simple chemical method with significant antibacterial activity against *S. aureus* and *P. aeruginosa*, fungi (*Aspergillus niger*), and yeast (*Candida albicans*) (Youssef et al., 2014).

4.3 Nisin

Nisin is an amphiphilic cationic peptide belonging to bacteriocins, produced by certain strains of *Lactococcus lactis* (Imran et al., 2014). Nisin has a broad spectrum of antibacterial activity against Gram-positive foodborne pathogens and spoilage organisms. It is the only bacteriocin recognized as safe for the food industry by the World Health Organization (WHO). Moreover, the Food and Drug Administration (FDA) has recognized nisin as an approved food additive due to its nontoxicity (Imran et al., 2014). As an antibacterial agent, the release and antibacterial behavior of nisin have been investigated in films such as methyl cellulose-hydroxy propyl methyl cellulose, starch-based, cellulose, and heat-pressed and cast films (Wanga et al., 2015). For instance, the functional effectiveness of nisin-loaded chitosan/polyvinyl alcohol as an antibacterial packaging film has been reported against *S. aureus* recently (Wanga et al., 2015).

CASE STUDY: CHEESE

The potential application of sodium caseinate films containing nisin was introduced as a promising method to overcome problems associated with postprocess contamination, extending the shelf-life and improving the microbial safety of cheese. For this purpose, a sodium caseinate film containing nisin was produced and used to control *Listeria innocua* in an artificially contaminated mini red Babybel cheese as a model semisoft cheese. *L. innocua* was both surface- and in-depth inoculated to investigate the effectiveness of the antimicrobial film as a function of the distance from the surface in contact with the film. The presence of the active film resulted in a 1.1 log CFU/g reduction in *L. innocua* counts in surface-inoculated cheese samples after 1 week of storage at 4°C as compared to control (Cao-Hoang et al., 2010).

4.4 Metal Nanoparticles

4.4.1 Ag

The most common polymer nanocomposites used as antimicrobial films for food packaging are based on silver due to the broad spectrum toxicity to various bacteria, fungi, algae, and possibly some viruses (Nassar and Youssef, 2012; Duncan, 2011). Stable silver nanoparticles can be ex situ synthesized by the reduction of Ag^+ ions followed by dispersing into a polymerizable formulation, which is difficult due to the limitation of controlling the nanoparticle dispersity. On the other hand, nanoparticles can be in situ produced in a polymerizable medium from precursors with better dispersity (Huang et al., 2015).

Antibacterial efficiency of silver nanoparticles has been attributed to different mechanisms such as cell adhesion, lipopolysaccharide degradation, and pits formation in the membranes, penetration inside the bacterial cell, damage to bacteria DNA, and Ag^+ ions release (Tankhiwale and Bajpai, 2009; Silvestre et al., 2011; Harifi and Montazer, 2014a; Poortavasoly et al., 2014). The antimicrobial activity was also found to be dependent on factors affecting silver ions release rate, such as degree of polymer crystalline, hydrophobicity of the matrix, and particle size (Duncan, 2011). In case of antibacterial food packaging containing silver nanoparticles immobilized in a polymer film, the antibacterial effect can be performed without silver nanoparticle migration, with polymer acting as carriers of nano silver reservoirs (Huang et al., 2015).

Silver/polymer nanocomposites have been tested with real food systems to determine the effect of Ag nanoparticles antibacterial properties on food shelf life (Duncan, 2011). For instance, Chinese fruit stored in food storage bags composed of $Ag/TiO_2/$

polyethylene films were firmer, heavier, and had less decay, less browning, and slower ripening over a period of 12 days than fruit stored in control materials (Duncan, 2011). Fresh cut melon stored in silver-containing cellulose pads had lower microbial counts and longer microbial growth lag times (Duncan, 2011). The novel nano silver-loaded filter paper has been prepared with strong antimicrobial properties against *E. coli* suitable for possible use as antibacterial food packaging material (Tankhiwale and Bajpai, 2009). The effectiveness of silver nanoparticles on foodborne spoilage microorganisms was confirmed, demonstrating a significant prolonged shelf life for the fresh mozzarella cheese stored in the active packaging (Incoronato et al., 2011).

In addition to the antimicrobial activity, nano silver particles could be effective in retarding fruit and vegetable ripening through catalyzing the absorption and decomposition of the ethylene gas emitted from fruit metabolism (Huang et al., 2015).

CASE STUDY: CHICKEN BREAST FILLETS

In comparison to other types of meat, chicken has more pathogenic bacteria reducing the shelf-life to less than 4 days under aerobic chilled conditions 7 days under chilled and modified atmosphere packaging. In a recent study carried out in 2015, low-density polyethylene nanocomposite films containing different concentrations of silver nanoparticles were manufactured through extrusion and the antibacterial activity of chicken breast fillets wrapped with the prepared films were assessed. Shelf life and oxidative stability of the nanocomposite films were prominently improved and the increment was independent of silver nanoparticles concentration (Azlin-Hasim et al., 2015).

4.4.2 Cu

The efficiency of copper as an antibacterial agent traced back to ancient civilizations sterilizing drinking water (Longano et al., 2012). Today, copper is used as a water purifier, algaecide, fungicide, nematocide, molluscicide, antibacterial and antifouling agent (Delgado et al., 2011; Sedighi et al., 2014a). Moreover, copper is one of the metallic elements that are essential to human health and is considered as a safe material to humans (Delgado et al., 2011). It is an essential element presented in most foods such as meat, fish, pecans, and green vegetables in the form of ions or salts (Llorens et al., 2012). Copper has been used in various steps of food processing as a growth inhibitor of bacterial pathogens causing foodborne diseases (Delgado et al., 2011). Protein oxidation routes leading to oxygen reactive radical formation causing cleavage of DNA and RNA molecules, and membrane damage due to lipid peroxidation are the main mechanism of copper killing effect (Longano et al., 2012).

Copper has been embedded in high-pressure polyethylene and various biobased polymers, and also colloidal copper nanoparticles distributed in chitosan films for food packaging applications (Llorens et al., 2012). In 2012 the antibacterial food packaging for food preservation was prepared by nanocomposite polymer composed of copper nanoparticles embedded in polylactic acid, combining the antibacterial properties of copper nanoparticles with the biodegradability of the polymer matrix (Longano et al., 2012).

Novel montmorillonite modified with copper/low density polyethylene nanocomposites with strong antibacterial property against *E. coli* was prepared using the melt-blending technique in an extruder. In addition to antibacterial property the proposed nanocomposite satisfy the mechanical property requirements for food packaging (Bruna et al., 2012).

4.4.3 Fe

The iron-zero valent nanoparticles have been considered as oxygen scavenging agents and have also been applied for inactivation of bacteria and fungus (Busolo and Lagaron, 2012). Novel composites containing an iron-based Kaolinite incorporated into polyolefins were developed to produce oxygen-fighting nanocomposites to increase the shelf life of oxygen-sensitive food products. In addition to the scavenging activity, barrier properties and migration of the active agent into food simulants was confirmed, indicating the potential of the prepared nanocomposite in polyolefin-based active food packaging applications (Busolo and Lagaron, 2012).

4.5 Metal Oxide Nanoparticles

4.5.1 ZnO

ZnO has been generally recognized as safe material by US Food and Drug Administration (FDA) (Espitia et al., 2012). ZnO nanoparticles have been incorporated in polymeric matrices in order to provide antimicrobial activity to the packaging material. ZnO nanoparticles have advantages over silver nanoparticles incorporated in polymeric matrices, providing safer and more affordable antimicrobial food packaging systems (Espitia et al., 2012). In addition to the antibacterial efficiency, ZnO nanoparticles improve the packaging properties such as mechanical strength, barrier properties, and stability (Espitia et al., 2012). ZnO nanoparticles have been incorporated in different materials including glass, low-density polyethylene, polypropylene, polyurethane, paper, and chitosan using different methods (Espitia et al., 2012). The preparation of ZnO nanoparticles loaded starch-coated polyethylene film was developed as food packaging material to protect foodstuffs from bacterial contamination (Tankhiwale and Bajpai, 2012). The exact mechanism of action of ZnO nanoparticles is still unknown. However, the antimicrobial activity of these nanoparticles is attributed to several mechanisms including the release of antimicrobial ions, damaging the integrity of bacterial cell due to interaction of nanoparticles with microorganisms, and the formation of radicals by the effect of light irradiation (Espitia et al., 2012; Khosravian et al., 2015).

CASE STUDY: ORANGE JUICE

Orange juice as a popular juice is easily contaminated by acidophilic microorganisms, especially lactic acid bacteria and yeasts. *Lactobacillus plantarum* is one of the most important bacteria producing a "buttery" off-flavor and swelling of fresh fruit juice packages. In a recent research (Emamifar et al., 2011) Ag and ZnO nanoparticles were effective in increasing the shelf life of orange juice up to 28 days and decreasing the bacterial growth rate. For this purpose, packages were prepared from low-density polyethylene films incorporated with Ag and ZnO nanoparticles and were filled with orange juice preinoculated by 8.5 log CFU/mL *L. plantarum*. Good dispersion of nanomaterials in the polymer matrix is very important in the antimicrobial effects of the prepared packaging materials. The higher antimicrobial activity of nano silver against *L. plantarum* compared with ZnO nanoparticles, especially for longer storage. The prepared packages could be introduced to food industry for combining pasteurization with antimicrobial nanocomposite packages, resulting in cost-effective pasteurization method to control *L. plantarum* in orange juice and to preserve the desirable quality (Emamifar et al., 2011).

4.5.2 TiO_2

There has been considerable attention to the self-disinfecting property of TiO_2 for meeting the requirements in food processing and packaging surfaces (Bodaghi et al., 2013). TiO_2 nanoparticles have been found to be effective against common foodborne pathogens including *Salmonella choleraesuis* subsp., *Vibrio parahaemolyticus*, and *Listeria monocytogenes* under UV illumination, but not in the dark (Duncan, 2011). In food processing, the biocide effect of TiO_2 nanocomposites with polymer materials, such as ethylene vinyl alcohol copolymer and chitosan, has been reported. Moreover, TiO_2 nanoparticles can be used to produce oxygen scavenger films (Huang et al., 2015).

A food packaging study showed that polypropylene films coated with TiO_2 nanoparticles inhibited *E. coli* growth on fresh cut lettuce (Duncan, 2011). Food packaging films incorporated with

TiO_2 nanoparticles can be beneficially used for protecting food content from the oxidizing effects of UV irradiation in addition to antimicrobial efficiency (Duncan, 2011). A combination of anatase and rutile titanium dioxide incorporation into a low-density polyethylene polymeric matrix by a melt-blending method was found effective to be used in fruit packaging applications (Bodaghi et al., 2013). A high-density polyethylene/calcium carbonate film containing TiO_2 was prepared by blown film extrusion process and the positive effect of packaging photo-catalytic activity on the structural and microbiological stability of a short-ripened cheese was represented (Gumiero et al., 2013). Films with enhanced antimicrobial activity were also prepared by several researchers combining the antimicrobial properties of nano TiO_2 with silver nanoparticles (Duncan, 2011).

CASE STUDY: KIWIFRUIT

Kiwifruit is a fruit that is highly sensitive to exogenous ethylene promoting enzymatic activities, softening and ripening. Moreover, kiwifruit is decayed due to the soft texture, high ethylene concentration in atmosphere, and the susceptibility to infections by *Botrytis cinerea*, *Penicillium expansum*, and *Botryosphaeria dothidea*, limiting long-term cold storage. Introducing a preferable nanocomposite-based packaging to preserve kiwifruit has been searched. The effects of a novel nanocomposite-based packaging based on polyethylene blended with nano silver, nano $TiO_{2,}$ and montmorillonite on the quality parameters of kiwifruit was investigated during 42 days of storage at 4°C. The polyethylene-added nanoparticles significantly delayed the ripening of kiwifruit, reduced fruit decay, and maintained quality in kiwifruits during postharvest storage (Hu et al., 2011).

4.5.3 Other Metal Oxide Nanoparticles

Copper oxide is a semiconducting compound with a monoclinic structure that has attracted particular attention as the simplest member of the family of copper compounds and exhibits a range of potentially (Sedighi et al., 2014b). CuO is cheaper than silver and can be easily mixed with polymers. It is relatively stable in terms of both chemical and physical properties (Rena et al., 2009). CuO nanoparticles have been studied as potential antimicrobial agents against infectious organisms such as *E. coli*, *B. subtilis*, *Vibria cholera*, *P. aeruginosa*, *Syphillis typhus*, and *S. aureus* (Thekkae and Cernik, 2013). Copper oxide nanoparticles were embedded in a polypropylene matrix and the prepared nanocomposites presented strong antimicrobial activity against *E. coli*, dependent on the contact time between the sample and the bacteria. It has been reported that CuO nanoparticles were more effective fillers eliminating bacteria growth with a higher release rate than Cu nanoparticles (Delgado et al., 2011).

Magnesium oxide is a white hygroscopic solid mineral that occurs naturally as periclase with many advantages such as good reproducibility, suitability for large-scale production, simplicity, and low cost. It is used as nanofiller to improve the properties of materials and also presents good antibacterial properties (Sanuja et al., 2014). Magnesium oxide alone or in combination with other antimicrobials such as nisin could potentially be used as an effective antibacterial agent in food formulations directly or through the slow release from packaging materials to kill or inhibit the growth of foodborne pathogens (Sanuja et al., 2014). Chitosan-magnesium oxide-based nanocomposite film containing clove essential oil was prepared by solution cast method and introduced as a promising material for an active food packaging applications with good antibacterial activity against *S. aureus* (Sanuja et al., 2014).

In addition to the oxygen-scavenging effect of zero valent Fe nanoparticles and their potential for inactivation of bacteria and fungus, the bactericidal effect of iron oxide nanoparticles on *S. aureus* has been confirmed by researchers (Tran et al., 2010; Harifi and Montazer, 2014b) due to the binding to bacterial cell walls causing membrane disruption, making them suitable for food packaging applications.

4.6 Carbon Nanotubes

Carbon nanotubes are cylinders with nanoscale diameters that may consist of one-atom thick single-wall nanotubes, or a number of concentric tubes, which is called multiwall nanotube. They could be used in food packaging to enhance the mechanical properties of polymeric matrices as well as imparting antimicrobial activity, possibly due to the irreversible damage made in the microbial cells (Huang et al., 2015). However, carbon nanotubes showed cytotoxic effects and caused DNA damage in the lowest dose tested. Their usage in food packaging is controversial and is advisable only if used in very low concentrations (Ortiz-Zarama et al., 2014).

5 Bio-Nanocomposite Polymers

Eco-friendly packaging materials with the help of biopolymers and nanomaterials are the recent focus of researchers, replacing synthetic packaging materials suffering from drawbacks including being costly and nonbiodegradable, and releasing toxic gases.

During recorded history, there has been a vast number of studies on the use of nonbiodegradable materials or plastics as food packaging materials mainly derived from petroleum products, causing waste disposal problem (Othman, 2014; Avella et al., 2005). Concerning the increased demand for sustainability and environmental safety products, a growing number of investigations have been directed toward development of food packaging materials that could rapidly degrade and completely mineralize in the environment, such as food packaging materials produced from biopolymers capable of decomposing into organic by-products such as carbon dioxide and water, which are safe for the environment (Othman, 2014; Avella et al., 2005).

The drawbacks involved in the use of biopolymers as food packaging materials—such as poor mechanical, thermal, and barrier properties—have been solved with the addition of nano fillers producing bio-nanocomposite polymers. The incorporation of nano fillers such as silicate, clay, silver, and titanium dioxide to biopolymers has beneficial effects including enhanced mechanical and barrier properties and antibacterial activity (Siracusa et al., 2008; Sanchez-Garcia et al., 2010).

The most common type of biopolymers for food packaging applications includes natural biopolymers deriving from carbohydrates and proteins such as starch, cellulose, chitosan, agar, gelatin, gluten, alginate, whey protein, and synthetic biopoloymers produced from microbial fermentation, biomass, and petroleum such as poly hydroxy-alkanoates, poly lactic acid, polycaprolactone, polyglycolic acid and polyvinyl alcohol (Othman, 2014; Avella et al., 2005; Sorrentino and Gorrasi, 2007; Siracusa et al., 2008; Sanchez-Garcia et al., 2010; Mensitieri et al., 2011).

Bio-nanocomposites were developed by casting/evaporation of wheat gluten, cellulose nanocrystals, and TiO_2 nanoparticles with significant improvement in mechanical and antibacterial efficiency suggesting the potential use in active food packaging (El-Wakil et al., 2015). Nano zinc oxide at different concentrations (0.1, 0.3, and 0.5%) and neem essential oil were incorporated into the chitosan by solution cast method, indicating enhanced mechanical, physical, barrier, optical, and antibacterial properties, suggesting the potential of the prepared bionanocomposite film in food packaging for carrot (Sanuja et al., 2015). Biodegradable polymers including hydroxylpropyl methylcellulose, sodium caseinate, and polylactic acid incorporated with nisin were evaluated for antibacterial activity against *L. monocytogenes* and *S. aureus*, and the prepared packaging films could progressively release nisin to sustain antibacterial effect for prolonging shelf life of packed food (Imran et al., 2014).

CASE STUDY: DAIRY PRODUCT

Storage of fresh white cheese under aerobic conditions results in rapid spoilage, frequently in less than 7 days, while active packaging prolongs storage life to 20–30 days. Thus, controlling bacterial contamination of fresh white cheese during storage is an important safety issue and use of active packaging materials for both preservation and safety of white cheese is necessary.

For this purpose, soft white cheese was packaged by novel bio-nanocomposites of chitosan/polyvinyl alcohol/TiO_2 nanoparticles. The prepared bio-nanocomposite exhibited good mechanical properties and superior antibacterial activity against *S. aureus*, *P. aeruginosa*, *E. coli* bacteria, and *C. albicans* fungus. The soft white cheese was manufactured and packaged within the bio-nanocomposite films and stored at 7°C for 30 days. The total bacterial counts, mold, yeast, and coliform decreased with the increasing storage period and disappeared at the end of storage period compared with control (Youssef et al., 2015). The proposed novel bio-nanocomposite films could be used as an economic, environmentally friendly, and sustainable material for food packaging with enhanced antimicrobial activity (Youssef et al., 2015). Some of the different nanocomposite polymers recently applied in antimicrobial food packaging systems are summarized in Table 16.1.

Table 16.1 Different Nanocomposite Polymers Applied as Antimicrobial Food Packaging System

Antibacterial Agent	Packaging Material	Food	References
Clay	• Starch and cashew tree gum • Linear low density polyethylene • Carboxymethyl cellulose	• Nuts • Fresh and processed meats	Pinto et al. (2014) Tornuk et al. (2015) Shin et al. (2014)
Chitosan	• Low-density polyethylene	• Carp fish fillet • Chilled fish	Ramezani et al. (2015) Reesha et al. (2015)
Ag	• Polyethylene • Polyamide	• Iranian saffron (*Crocus Sativus* L.) • Sausages	Eslami et al. (2015) Patiño et al. (2014)
ZnO	• Low-density polyethylene • Low density polyethylene • Polylactic acid	• Chicken breast meat • Cheese	Panea et al. (2014) Beigmohammadi et al. (2015) De Silva et al. (2015)
TiO_2	• Low-density polyethylene	• Milk	Johnson et al. (2015)

6 Antibacterial Nano Packaging Applications in Different Food Products

6.1 Fruits and Vegetables

Nowadays, exotic fruit and vegetable products are in demand in local markets as well as for export purposes. However, long period of storage and transport conditions have negative impact on the quality of the products. Although fresh-cut fruits and vegetables are more desirable, they suffer from short shelf life due to microbial growth such as *L. monocytogenes*, *Salmonella* sp., and *E. coli* (Severino et al., 2014).

In addition, fruits produce moisture and gaseous metabolites continuously during storage due to respiration depending on storage time and temperature (Lin et al., 2015).

Modified atmosphere packaging has been used to control bacterial populations in several food products, especially ready-to-use vegetables, as the change in package atmosphere composition leads to the reduction of respiration of vegetables/fruits, increasing the shelf life (Severino et al., 2014).

Some of the applications of antibacterial food packaging materials in fruits and vegetables are as follows:

Cherry tomatoes were highly protected against *Alternaria alternata* using paper coatings with cinnamon essential oil. Chitosan and pectin incorporating β-cyclodextrin/cinnamaldehyde complex were used as layer-by-layer edible coating to increase the shelf life of fresh-cut papaya. Polypropylene/polyethylene film packages coated by soy protein containing cinnamaldehyde were highly effective in reducing microbial counts of radish, broccoli, and alfalfa sprouts (Rodriguez-Lafuente et al., 2010).

Ready-to-eat salads were packed by new food packaging systems based on polylactic acid/ *Allium* spp. extract. The best antibacterial activity tested in vivo in lettuce, prolonging the shelf life up to 5 storage days against aerobic bacteria and 7 storage days against molds has been reported for the package film containing 6.5% *Allium* spp. extract (Llana-Ruiz-Cabello et al., 2015).

The growth rate of mesophilic microorganisms in fresh cabbage was decreased through disc volatilization technique by using different essential oils (Hyun et al., 2015). The antibacterial activity of modified chitosan-based coatings containing nanoemulsion of essential oils, gamma irradiation, modified atmosphere packaging, alone or in combinations, against *E. coli* and *Salmonella Typhimurium* was evaluated on inoculated green bean samples. The combined treatment of antimicrobial coating, gamma irradiation, and modified atmosphere packaging reduced microbial population to undetectable level during the whole storage period for *E. coli* and from 7th day to the end of storage for *Salmonella Typhimurium* (Severino et al., 2014).

A novel nanocomposite of silver/titanium dioxide/chitosan was developed through photochemical reduction using a chitosan template. The prepared antibacterial packaging had potential for storage of fruits such as cantaloupes and lychees (Lin et al., 2015). The nanocomposite reduced the *E. coli* by 6 logs after 24 h of incubation and the minimum inhibitory concentration of Ag/TiO_2/chitosan was significantly lower than those of chitosan, $AgNO_3$, or nano silver particles (Lin et al., 2015).

Fresh broccoli was packed by bioactive films based on methyl cellulose and a blend of polycaprolactone/alginate with addition of antimicrobial formulations including organic acids and a mixture of rosemary extract and essential oil. The applied biodegradable active packaging was highly effective in reduction of *E. coli, Salmonella typhimurium,* and *L. monocytogenes* growth for 4 days (Takala et al., 2013).

6.2 Meat

Meat is characterized as a food with tissue structure and often treated by different types of technological processes that favor the development of microorganisms. Since the time that the United States Department of Agriculture imposed a zero tolerance policy against *L. monocytogenes* on processed and ready-to-eat food, large amounts of processed and ready-to-eat meats have been recalled due to *Listeria* contamination (Chen and Brody, 2013; Galus and Kadzińska, 2015). Out of 519 food product recalls between 2003 and 2011, 21% associated with *L. monocytogenes* according to United States Department of Agriculture Food Safety and Inspection Service (Chen and Brody, 2013). Bacteria of the species *L. monocytogenes* are especially inconvenient for the food industry. Apart from meat safety considerations, contamination of the products by spoilage microorganisms also leads to significant financial losses (Coma, 2012). The effectiveness of antimicrobial packaging structures in improving the microbial quality of ready-to-eat meat products was evaluated against *Listeria monocytogenes*. Cooked meat inoculated with a five-strain mixture of *Listeria monocytogenes* were packaged into three different antimicrobial packaging structures including a nylon/ethylene vinyl alcohol/polyethylene oxygen barrier bag, an antimicrobial film with the capacity of generating carbon dioxide and allyl isothiocyanate or scavenging molecular oxygen. The samples were stored at 4, 10, or 22°C, and populations of total aerobic bacteria were determined. Packaging structures with O_2 scavenger or CO_2 generator were more effective than the structure with an allyl isothiocyanate generator. This suggested the potential usage of the proposed packaging structures in controlling bacterial populations, particularly *Listeria monocytogenes* populations in ready-to-eat meat products (Chen and Brody, 2013).

Poultry belong to a class of highly perishable foods, and the main concern in poultry industries is extension in shelf life of products. In addition to microbiological safety, the color and lipid stability of the meat are important quality characteristics that influence product acceptance by consumers (Dias et al., 2013). Cellulose-based films incorporated with allyl isothiocyanate and carbon nanotube were developed and used for shredded cooked chicken meat inoculated with *S. choleraesuis*. Microbial counts, color characteristics, and changes in the oxidation of the meat as well as changes in the mechanical properties of the film, the structure of the antimicrobial film, and the diffusion of the antimicrobial agent into the food were all analyzed during 40 days of storage. The diffusion of allyl isothiocyanate from the film into the chicken reduced the microbial contamination, controlled oxidation, and reduced the color changes. The researchers recommend the use of allyl isothiocyanate at concentrations higher than 28% combined with carbon nanotube at concentrations higher than 0.02% to achieve microbial reduction, control oxidation, and reduce color changes. These packages were effective for 40 days of storage, which represents their great potential to increase shelf life and food safety (Dias et al., 2013). Zinc oxide nanoparticles loaded active film of calcium alginate was prepared for active packaging against the foodborne pathogens namely *S. typhimurium* and *S. aureus*. The film was also an active packaging in ready-to-eat poultry meat against the pathogens, and reduced the number of inoculated target bacteria from log 7 to 0 within 10 days of incubation at 8°C (Akbar and Anal, 2014).

6.3 Cheese

Fiordilatte cheese, a fresh pulled-curd cheese, typical of the center and south of Italy, is usually packaged in flexible films or trays made of polyethylene with brine solution. Due to the high moisture content and microbiological proliferation, this cheese is very perishable even when held in the refrigerator (Nardiello et al., 2014). Generally, the works available in the scientific literature, dealing with packaged fiordilatte, have focused on the effects of the packaging system on cheese quality. Determination of the storage time within microbiological and sensory properties remained acceptable is important (Nardiello et al., 2014). Researchers have reported that different dairy products were well protected from undesirable microorganisms attack and gas formation using modified atmosphere packaging with high concentrations of carbon dioxide. However, the use of carbon dioxide suffers from some drawbacks, mainly flavor-related problems. Due to the problems associated with modified atmosphere packaging, the idea of using silver nanoparticles incorporated into a biobased coating in combination with modified atmosphere packaging (50% CO_2, 50% N_2) was proposed recently to prolong the shelf life of Fiordilatte cheese to 10 days (Mastromatteo et al., 2015).

Extending the shelf life of Mozzarella cheese is an important issue for the dairy industry, due to the high interest in extending the distribution of traditional products beyond market borders. Eliot et al. (1998) reported that shredded Mozzarella cheese packaged in modified atmosphere packaging containing 75% CO_2 was well protected from undesirable organisms and gas formation. Alves et al. (1996) also found that microbial growth in sliced Mozzarella cheese packaged in modified atmosphere packaging was delayed with a high concentration of CO_2. Combined active coating based on sodium alginate and potassium sorbate and modified atmosphere packaging was also successfully achieved on prolonging shelf life of low-moisture Mozzarella cheese (Mastromatteo et al., 2015).

The occurrence of *E. coli* strains in sheep cheeses and cheeses made from raw milk has been reported by a number of authors all around the world. The antimicrobial activity of two packaging films of polypropylene and polyethylene terephthalate coated with different concentrations of essential oil and ethyl lauroyl arginate HCl was tested against *E. coli* strains using in vitro systems and a raw milk sheep cheese model (Otero et al., 2014).

6.4 Bread

Bread is one of the main foods around the world due to superior nutritional, sensorial and textural characteristics. However, bread is a food with rapidly degrading quality characteristics and the shelf life is only a few days at room temperature. Microbial spoilage and staling are the main factors limiting shelf life of bakery products. This reveals the importance of formulation, processing, packaging, and storage conditions of the products (Khoshakhlagh et al., 2014).

A most common method used to increase the shelf life of part-baked bread is modified atmosphere packaging. In spite of the positive effects of modified atmosphere packaging on minimizing microbial spoilage of bread, there have been some controversial reports on the effect of atmosphere inside a package on staling and physicochemical quality changes occurring during bread storage. While reducing crumb firmness of white and whole meal bread during storage in CO_2 was reported by Knorr and Tomlins (1985) and Cencic et al. (1996), Black et al. (1993) and Rasmussen and Hansen (2001) showed that bread firming is not affected by atmosphere. Contrary to these findings, Hematian Sourki et al. (2010) found a negative correlation coefficient between CO_2 concentration and hardness of flat bread.

The effect of part-baking and modified atmosphere packaging on quality and microbial stability of part-baked flat bread (Sangak) was investigated recently and the result demonstrated that it is possible to prolong the shelf life of Sangak bread from a few days to about 21 days by using bake-off technology and modified atmosphere packaging under high CO_2 concentration (Khoshakhlagh et al., 2014).

A paper coating based on cinnamon essential oil/wax paraffin was effectively used to inhibit the growth of *Rhizopus stolonifer* in bread (Balaguer et al., 2013).

The in vitro and real food effectiveness of bio-plastics films made from wheat gliadins incorporating cinnamaldehyde against food-contaminating fungi *Penicillium expansum* and *A. niger* was evaluated and the applicability in the design of active food packaging systems for sliced bread was achieved (Balaguer et al., 2013).

7 Health, Ethical and Regulatory Issues

With the increased number of researches on the potential impacts of nanotechnology in food packaging materials, great concerns has raised over consumer safety, human and environmental health, ethical, policy and regulation issues (Lin et al., 2011). Due to the theoretical and experimental possibility of nanoparticles migration from the packaging material into the food product, researchers are especially concerned whether the migration has a negative effect on the safety or quality of the food product (Dudo et al., 2011; Wei et al., 2011; Espitia et al., 2012). There is a possibility of nanomaterials accumulation in food, human body or environment from nano packaging (Coles and Frewer, 2013). The scientific reports conducted by United States, European Union, Japan, South Korea, and China indicated that nanomaterials have health and environment toxicities, which are dependent on the physicochemical characteristics such as size, morphology, and surface properties (Wei et al., 2011). Unfortunately nanofood materials are commercially used in some countries without any exact information about the possibility of nanoparticles migration. Until the year 2012, we could only find two papers assessing migration of nanoparticles into real food matrices. For instance, one published study on the migration of minerals (Fe, Mg, Si) from biodegradable starch/nanoclay nanocomposite films to the real food product, indicating an insignificant trend in the level of Fe and Mg in packaged vegetables, but a consistent increase in the amount of Si (Chaudhry et al., 2008). The level of silver in meat using silver nanoparticles incorporated into cellulose pad for use in modified atmosphere packaged was detected (Metak et al., 2015). Most researches on nanoparticles migration have been applied to food simulants without testing real food matrices, or commercial packaging (Metak et al., 2015).

According to the theoretical study carried out on the potential migration of nanoparticles from packaging to food, migration is more possible in the case of very small nanoparticles (around 1 nm) from polymer matrices with a low dynamic viscosity such as polyolefins (low-density polyethylene, high-density polyethylene and polypropylene) (Echegoyen and Nerin, 2013). However, there is no data available on the diffusion risk of nanoparticles coated on the surface of polymers.

Based on EU safety regulations the presence of silver ions in food matrices should be limited to 0.05 mg Ag/kg, which shows no biocide effect in food (Unalan et al., 2014). The understanding of the nanoparticles migration from packaging into food product is vital in order to evaluate the potential health impact of the products in contact with food substances (Metak et al., 2015).

A very recent study was conducted by Metak et al. (2015) testing commercial nanocomposite food packaging in two forms of nanosilver containers and coated films in real food to determine the degree of migration of nanoparticles using inductively coupled plasma-mass spectrometry and atomic absorption spectrometry (Metak et al., 2015). Based on the experimental result insignificant levels of nano silver were released and migrated from impregnated containers into real orange juice. On the other hand, a higher level of migration has been observed for samples exposed to nanosilver-coated films (Metak et al., 2015).

Another important issue is the adequate accuracy of analytical methods used to detect the migration of nanoparticles to the food product (Echegoyen and Nerin, 2013). The establishing analytical methods for detecting and characterizing antibacterial agents in polymer nanocomposites food packaging is crucial (Lin et al., 2011). Although performing the migration test with real food matrices is the best approach, the typical common route to assess the mass transport process is evaluating the specific and overall migration of targeted substances using food simulants that are selected model systems that produce similar interactions to those foods. Water is considered as a food simulant for a wide range of food products, such as bread, fresh fruits and vegetables, meat, and fish. The use of slightly acidified water solutions containing 3% acetic acid is also used as a more aggressive food simulant for acidic aqueous foods such as cola and carbonated drink with a pH lower than 4.6 (Huang et al., 2015).

Although vast number of studies has been published on the modeling of the transfer of substances from packaging materials into food product, there has been relatively little work reported on the prediction of migration in nanocomposite packaging. For instance, a mathematical modeling of the migration of carbon black nanoparticles from low density polyethylene and polystyrene into different food simulants based on the analytical solution of Fick's second law was proposed finding good agreement with the experimental data (Huang et al., 2015). A recent study used the theoretical modeling based on Fick's diffusion that permits the prediction of the nanoclay migration from a multilayer nanocomposite packaging film to aqueous and fatty food simulants (Huang et al., 2015).

While discussing the regulatory aspects related to active and intelligent packaging materials in food contact applications, the guiding legislation in Europe is represented by Regulation (EC) No. 1935/2004, 450/2009, 1935/2004, and 450/2009. Unlike Europe, the United States does not subject active and intelligent packaging materials to specific statutory or regulatory provisions. In this regard, any substance added to food by an active packaging material that is already approved as a food additive, does not need further clearance as long as all relevant food law requirements are fulfilled (Dainelli, 2015). The successful development of antimicrobial food packaging is still a challenging issue with relatively few commercialized products available in the market, presumably due to the strictness, and slow speed of legislative action on regulations (Azlin-Hasim et al., 2015).

The uncertainty about potential risks and regulatory issues has put food nanotechnology in a controversial position and the successful future of food nanotechnology seems dependent on to the balance of benefits, risks, and uncertainties against each other in order to gain public acceptance (Dudo et al., 2011). In order to broaden the applications of nanomaterials in food packaging, it is mandatory to maximize the safety regarding the leaching of nanomaterials into food products.

8 Conclusions

During the past 60 years different trends have been appeared in the evolution of food packaging from convenience in 1960s; lightweight, source reduction, and energy saving in the 1970s; safety and tampering in 1980s; environmental impact in the 1990s; safety and security in the 2000s; to carbon footprint reduction in the 2010s. Year after year, food packaging technology becomes more advanced.

A continuing trend in food packaging technology is the result of research and development of new nanomaterials possessing multifunctional properties providing antibacterial efficiencies along with other barrier properties, such as, UV protection.

Another important trend is natural and environmentally friendly food packaging materials. Renewed interest in food packages composed of natural biopolymers has started to emerge in recent years due to concerns about the environment and the requirement to reduce the consumption of disposable, nondegradable, petroleum-based conventional packaging materials. Replacing artificial chemical ingredients in food packaging materials with natural ingredients is attractive today and for consumers of the future.

New food packaging materials should be examined scientifically to ensure the safety and lack of toxicity and should be approved by regulatory agencies. In some cases, countries require new regulations and legislation for usage of new food packaging technologies. Three requirements are important to ensure the acceptability of commercial applications of new active food packaging technologies, including recognition of the benefits of active packaging technology by the food industry, development of economically viable packaging systems, and increased consumer acceptance.

References

Akbar, A., Anal, A.K., 2014. Zinc oxide nanoparticles loaded active packaging, a challenge study against *Salmonella typhimurium* and *Staphylococcus aureus* in ready to-eat poultry meat. Food Control 38, 88–95.

Alves, R.M.V., De Luca Sarantopoulos, C.I.G., Fernandes Van Dender, A.G., De Assis Fonseca Faria, J., 1996. Stability of sliced Mozzarella cheese in modified atmosphere packaging. J. Food Protect. 59, 838–844.

Avella, M., De Vlieger, J.J., Errico, M.E., Fischer, S., Vacca, P., Volpe, M.G., 2005. Biodegradable starch/clay nanocomposite films for food packaging applications. Food Chem. 93, 467–474.

Azeredo, H.M.C., 2012. Antimicrobial nanostructures in food packaging. Trends Food Sci. Technol. 30, 56–69.

Azlin-Hasim, S., Cruz-Romero, M.C., Morris, M.A., Cummins, E., Kerry, J.P., 2015. Effects of a combination of antimicrobial silver low density polyethylene nanocomposite films and modified atmosphere packaging on the shelf life of chicken breast fillets. Food Pack Shelf Life. 4, 26–35.

Balaguer, M.P., Lopez-Carballo, G., Catala, R., Gavara, R., Hernandez-Munoz, P., 2013. Antifungal properties of gliadin films incorporating cinnamaldehyde and application in active food packaging of bread and cheese spread foodstuffs. Int. J. Food Microbiol. 166, 369–377.

Bastarrachea, L., Dhawan, S., Sablani, S.S., 2011. Engineering properties of polymeric-based antimicrobial films for food packaging. Food Eng. Rev. 3, 79–93.

Beigmohammadi, F., Peighambardoust, S.H., Hesari, J., Azadmard-Damirchi, S., Peighambardoust, S.J., Karimian Khosrowshahi, N., 2015. Antibacterial properties of LDPE nanocomposite films in packaging of UF cheese. LWT Food Sci. Technol. 65, 106–111.

Bikiaris, D.N., Triantafyllidis, K.S., 2013. HDPE/Cu-nanofiber nanocomposites with enhanced antibacterial and oxygen barrier properties appropriate for food packaging applications. Mater. Lett. 93, 1–4.

Black, R.G., Quail, K.J., Reyes, V., Kuzyk, M., Ruddick, L., 1993. Shelf life extension of pita bread by modified atmosphere packaging. Food Aust. 45, 387–439.

Bodaghi, Mostofi, Y., Oromiehie, A., Zamani, Z., Ghanbarzadeh, B., Cristina, C., Conte, A., Del Nobile, M.A., 2013. Evaluation of the photocatalytic antimicrobial effects of a TiO_2 nanocomposite food packaging film by in vitro and in vivo tests. LWT Food Sci. Technol. 50, 702–706.

Brody, A., Bugusu, L.B., Han, J.H., Sand, C.K., Mchugh, T.H., 2008. Innovative food packaging solutions. J. Food Sci. 73, R107–R116.

Bruna, J.E., Pealoza, A., Guarda, A., Rodriguez, F., Galotto, M.J., 2012. Development of $MtCu^{2+}$/LDPE nanocomposites with antimicrobial activity for potential use in food packaging. Appl. Clay Sci. 58, 79–87.

Busolo, M.A., Lagaron, J.M., 2012. Oxygen-scavenging polyolefin nanocomposite films containing an iron modified kaolinite of interest in active food packaging applications. Innov. Food Sci. Emerg. Technol. 16, 211–217.

Busolo, M.A., Fernandez, P., Ocio, M.J., Lagaron, J.M., 2010. Novel silver-based nanoclay as an antimicrobial in polylactic acid food packaging coatings. Food Addit. Contam. A 27, 1617–1626.

Cao-Hoang, L., Chaine, A., Grégoire, L., Waché, Y., 2010. Potential of nisin-incorporated sodium caseinate films to control *Listeria* in artificially contaminated cheese. Food Microbiol. 27, 940–944.

Cencic, L., Bressa, F., DallaRosa, M., 1996. Influence of modified atmosphere on bread staling during storage. In: Industrie Alimentari, pp. 20–24.

Charles, F., Sanchez, J., Gontard, N., 2006. Absorption kinetics of oxygen and carbon dioxide scavengers as part of active modified atmosphere packaging. J. Food Eng. 72, 1–7.

Chaudhry, Q., Scotter, M., Blackburn, J., Ross, B., Boxall, A., Castle, L., Aitken, R., Watkins, R., 2008. Applications and implications of nanotechnologies for the food sector. Food Addit. Contam. A 25, 241–258.

Chen, J., Brody, A.L., 2013. Use of active packaging structures to control the microbial quality of a ready-to-eat meat product. Food Control. 30, 306–310.

Coles, D., Frewer, L.J., 2013. Nanotechnology applied to European food production: a review of ethical and regulatory issues. Trends Food Sci. Technol. 34, 32–43.

Coma, V., 2012. Antimicrobial and antioxidant active packaging for meat and poultry. In: Kerry, J.P. (Ed.), Advances in Meat Poultry and Seafood Packaging. Woodhead Publishing Limited, UK, pp. 477–503.

Costa, C., Conte, A., Buonocore, G.G., Del Nobile, M.A., 2011. Antimicrobial silver-montmorillonite nanoparticles to prolong the shelf life of fresh fruit salad. Int. J. Food Microbiol. 148, 164–167.

Cruz-Romero, M.C., Murphy, T., Morris, M., Cummins, E., Kerry, J.P., 2013. Antimicrobial activity of chitosan, organic acids, and nano-sized solubilisates for potential use in smart antimicrobially active packaging for potential food applications. Food Control. 34, 393–397.

Dainelli, D., 2015. Global legislation for active and intelligent packaging materials. In: Baughan, J.S. (Ed.), Global Legislation for Food Contact Materials. Woodhead Publishing, UK, pp. 183–199.
De Silva, R.T., Pasbakhsha, P., Lee, S. M., Kit, A. Y., 2015. ZnO deposited/encapsulated halloysite–poly (lactic acid) (PLA) nanocomposites for high performance packaging films with improved mechanical and antimicrobial properties. Appl. Clay Sci. 111, 10–20.
Delgado, K., Quijada, R., Palma, R., Palza, H., 2011. Polypropylene with embedded copper metal or copper oxide nanoparticles as a novel plastic antimicrobial agent. Lett. Appl. Microbiol. 53, 50–54.
Dias, M.V., Soares, N.F.F., Borges, S.V., Sousa, M.M., Nunes, C.A., Oliveira, I.R.N., Medeiros, E.A.A., 2013. Use of allyl isothiocyanate and carbon nanotubes in an antimicrobial film to package shredded, cooked chicken meat. Food Chem. 141, 3160–3166.
Dudo, A., Choi, D.H., Scheufele, D.A., 2011. Food nanotechnology in the news: coverage patterns and thematic emphases during the last decade. Appetite. 56, 78–89.
Duncan, T.V., 2011. Applications of nanotechnology in food packaging and food safety: barrier materials, antimicrobials, and sensors. J. Colloid Interface Sci. 363, 1–24.
Dutta, P.K., Tripathi, S., Mehrotra, G.K., Dutta, J., 2009. Perspectives for chitosan-based antimicrobial films in food applications. Food Chem. 114, 1173–1182.
Echegoyen, Y., 2015. Nano-developments for food packaging and labeling applications. In: Rai, M., Ribeiro, C., Mattoso, L., Duran, N. (Eds.), Nanotechnologies in Food and Agriculture. Springer International Publishing, Switzerland, pp. 141–168.
Echegoyen, Y., Nerin, C., 2013. Nanoparticle release from nano-silver antimicrobial food containers. Food Chem. Toxicol. 62, 16–22.
Eliot, S.C., Vuillemard, J.C., Emond, J.P., 1998. Stability of shredded Mozzarella cheese under modified atmospheres. J. Food Sci. 63, 1075–1080.
El-Wakil, N.A., Hassan, E.A., Abou-Zeid, R.E., Dufresne, A., 2015. Development of wheat gluten/nanocellulose/titanium dioxide nanocomposites for active food packaging. Carbohydr. Polym. 124, 337–346.
Emamifar, A., 2011. Applications of antimicrobial polymer nanocomposites in food packaging. In: Hashim, A. (Ed.), Advances in Nanocomposite Technology, InTech, Croatia, pp. 1–20.
Emamifar, A., Kadivar, M., Shahedi, M., Soleimanian-Zad, S., 2011. Effect of nanocomposite packaging containing Ag and ZnO on inactivation of *Lactobacillus plantarum* in orange juice. Food Control. 22, 408–413.
Eslami, M., Bayat, M., Mozaffari Nejad, A.S., Sabokbar, A., Anvar, A.A., 2015. Effect of polymer/nanosilver composite packaging on long-term microbiological status of Iranian saffron (*Crocus sativus* L.). Saudi J. Biol. Sci. 23, 341–347.
Espitia, P.J.P., Soares, N.F.F., Coimbra, J.S.R., Andrade, N.J., Souza Cruz, R., Medeiros, E.A.A., 2012. Zinc oxide nanoparticles: synthesis, antimicrobial activity and food packaging applications. Food Bioproc. Technol. 5, 1447–1464.
Galus, S., Kadzińska, J., 2015. Food applications of emulsion-based edible films and coatings. Trends Food Sci. Technol. 45, 273–283.
Gibisa, D., Rieblingera, K., 2011. Oxygen-scavenging films for food application, 11th International Congress on Engineering and Food (ICEF11). Procedia Food Sci. 1, 229–234.
Gumiero, M., Peressini, D., Pizzariello, A., Sensidoni, A., Iacumin, L., Comi, G., Toniolo, R., 2013. Effect of TiO_2 photocatalytic activity in a HDPE-based food packaging on the structural and microbiological stability of a short-ripened cheese. Food Chem. 138, 1633–1640.
Han, J.H., 2003. Antimicrobial food packaging. In: Ahvenainen, R. (Ed.), Novel Food Packaging Techniques. Woodhead Publishing Limited, UK, pp. 50–70.

Han, J.H., 2014. A review of food packaging technologies and innovations. In: Han, J.H. (Ed.), Innovations in Food Packaging. Elsevier Ltd., UK, pp. 3–11.
Harifi, T., Montazer, M., 2014a. Photo-, bio-, and magneto-active colored polyester fabric with hydrophobic/hydrophilic and enhanced mechanical properties through synthesis of TiO_2:Fe_3O_4/Ag nanocomposite. Ind. Eng. Chem. Res. 53, 1119–1129.
Harifi, T., Montazer, M., 2014b. In situ synthesis of iron oxide nanoparticles on polyester fabric utilizing color, magnetic, antibacterial, and sono-Fenton catalytic properties. J. Mater. Chem. B 2, 272–282.
Hematian Sourki, A., Ghiafeh Davoodi, M., Tabatabaei Yazdi, F., Mortazavi, S.A., Karimi, M., Razavizadegan Jahromi, S.H., Pourfarzad, A., 2010. Staling and quality of Iranian flat bread stored at modified atmosphere in different packaging. World Acad. Sci. Eng. Technol. 69, 390–395.
Hu, Q., Yong, F., Yang, Y., Ma, N., Zhao, L., 2011. Effect of nanocomposite-based packaging on postharvest quality of ethylene-treated kiwifruit (*Actinidia deliciosa*) during cold storage. Food Res. Int. 44, 1589–1596.
Huang, J.Y., Li, X., Zhou, W., 2015. Safety assessment of nanocomposite for food packaging application. Trends Food Sci. Technol. 45, 187–199.
Hyun, J.E., Bae, Y.M., Yoon, J.H., Lee, S.Y., 2015. Preservative effectiveness of essential oils in vapor phase combined with modified atmosphere packaging against spoilage bacteria on fresh cabbage. Food Control. 51, 307–313.
Imran, M., Klouj, A., Revol-Junelles, A.M., Desobry, S., 2014. Controlled release of nisin from HPMC, sodium caseinate, polylactic acid and chitosan for active packaging applications. J. Food Eng. 143, 178–185.
Incoronato, A.L., Conte, A., Buonocore, G.G., Del Nobile, M.A., 2011. Agar hydrogel with silver nanoparticles to prolong the shelf life of Fior di Latte cheese. J. Dairy Sci. 94, 1697–1704.
Johnson, D.S., Duncan, S.E., Bianchi, L.M., Chang, H.H., Eigel, W.N., O'Keefe, S.F., 2015. Packaging modifications for protecting flavor of extended-shelf life milk from light. J. Dairy Sci. 98, 2205–2214.
Kayaci, F., Umu, O.C.O., Tekinay, T., Uyar, T., 2013. Antibacterial electrospun poly (lactic acid) (PLA) nanofibrous webs incorporating triclosan/cyclodextrin inclusion complexes. J. Agr. Food Chem. 61, 3901–3908.
Kerry, J.P., O'Grady, M.N., Hogan, S.A., 2006. Past, current and potential utilization of active and intelligent packaging systems for meat and muscle-based products: a review. Meat Sci. 74, 113–130.
Khoshakhlagh, K., Hamdami, N., Shahedi, M., Le-Bail, A., 2014. Quality and microbial characteristics of part-baked Sangak bread packaged in modified atmosphere during storage. J. Cereal Sci. 60, 42–47.
Khosravian, S., Montazer, M., Malek, R.M.A., Harifi, T., 2015. In situ synthesis of nano ZnO on starch sized cotton introducing nano photo active fabric optimized with response surface methodology. Carbohydr. Polym. 132, 126–133.
Kim, Y.T., Kim, K., Han, J.H., Kimmel, R.M., 2008. Antimicrobial active packaging for food. In: Kerry, J., Butler, P. (Eds.), Smart Packaging Technologies for Fast-Moving Consumer Goods. John Wiley & Sons, US, pp. 99–110.
Knorr, D., Tomlins, R.I., 1985. Effect of carbon dioxide modified atmosphere on the compressibility of stored baked goods. J. Food Sci. 50, 1172–1173.
Lin, Q.B., Li, B., Song, H., Wu, H.J., Lin, Q.B., Li, B., Song, H., Wu, H.J., 2011. Determination of silver in nano-plastic food packaging by microwave digestion coupled with inductively coupled plasma atomic emission spectrometry or inductively coupled plasma mass spectrometry. Food Addit. Contam. A. 28, 1123–1128.
Lin, B., Luo, Y., Teng, Z., Zhang, B., Zhou, B., Wang, Q., 2015. Development of silver/titanium dioxide/chitosan adipate nanocomposite as an antibacterial coating for fruit storage. LWT Food Sci. Technol. 63, 1206–1213.

Llana-Ruiz-Cabello, M., Pichardo, S., Baños, A., Núñez, C., Bermúdez, J.M., Guillamón, E., Aucejo, S., Cameána, A.M., 2015. Characterization and evaluation of PLA films containing an extract of *Allium* spp. to be used in the packaging of ready-to-eat salads under controlled atmospheres. LWT Food Sci. Technol. 64, 1354–1361.
Llorens, A., Lloret, E., Picouet, P.A., Trbojevich, R., Fernandez, A., 2012. Metallic-based micro and nanocomposites in food contact materials and active food packaging. Trends Food Sci. Technol. 24, 19–29.
Longano, D., Ditaranto, N., Cioffi, N., Di Niso, F., Sibillano, T., Ancona, A., Conte, A., Del Nobile, M.A., Sabbatini, L., Torsi, L., 2012. Analytical characterization of laser-generated copper nanoparticles for antibacterial composite food packaging. Anal. Bioanal. Chem. 403, 1179–1186.
López-Rubio, A., Almenar, E., Hernandez-Muñoz, P., Lagarón, J.M., Catalá, R., Gavara, R., 2004. Overview of active polymer-based packaging technologies for food applications. Food Rev. Int. 20, 357–387.
Mangalassary, S., 2012. Antimicrobial food packaging to enhance food safety: current developments and future challenges. J. Food Proc. Technol. 3, 103–105.
Mastromatteo, M., Conte, A., Lucera, A., Saccotelli, M.A., Buonocore, G.G., Zambrini, A.V., Del Nobile, M.A., 2015. Packaging solutions to prolong the shelf life of Fiordilatte cheese: bio-based nanocomposite coating and modified atmosphere packaging. LWT Food Sci. Technol. 60, 230–237.
Mensitieri, G., Di Maio, E., Buonocorea, G.G., Nedi, I., Oliviero, M., Sansone, L., Iannace, S., 2011. Processing and shelf life issues of selected food packaging materials and structures from renewable resources. Trends Food Sci. Technol. 22, 72–80.
Metak, A.M., Nabhani, F., Connolly, S.N., 2015. Migration of engineered nanoparticles from packaging into food products. LWT Food Sci. Technol. 64, 781–787.
Nardiello, D., Conte, A., Natale, A., Lucera, A., Palermo, C., Centonze, D., Del Nobile, M.A., 2014. Effects of different packaging systems on microbiological, sensory, and peptide profile in Fiordilatte cheese. Food Res. Int. 62, 628–636.
Nassar, M.A., Youssef, A.M., 2012. Mechanical and antibacterial properties of recycled carton paper coated by PS/Ag nanocomposites for packaging. Carbohydr. Polym. 89, 269–274.
Ortiz-Zarama, M.A., Jiménez-Aparicio, A., Perea-Flores, M.J., Solorza-Feria, J., 2014. Barrier, mechanical, and morpho-structural properties of gelatin films with carbon nanotubes addition. J. Food Eng. 120, 223–232.
Otero, V., Becerril, R., Santos, J.A., Rodríguez-Calleja, J.M., Nerín, C., García-Lópe, M.L., 2014. Evaluation of two antimicrobial packaging films against *Escherichia coli* O157:H7 strains in vitro and during storage of a Spanish ripened sheep cheese (Zamorano). Food Control. 42, 296–302.
Othman, S.H., 2014. Bio-nanocomposite materials for food packaging applications: types of biopolymer and nano-sized filler. Agr. Agr. Sci. Procedia. 2, 296–303.
Panaitescu, D.M., Frone, A.N., Ghiurea, M., Chiulan, I., 2015. Influence of storage conditions on starch/PVA films containing cellulose nanofibers. Polym. Ind. Crops Prod. 70, 170–177.
Panea, B., Ripoll, G., González, J., Fernández-Cuello, Á., Albertí, P., 2014. Effect of nanocomposite packaging containing different proportions of ZnO and Ag on chicken breast meat quality. J. Food Eng. 123, 104–112.
Patiño, J.H., Henríquez, L.E., Restrepo, D., Mendoza, M.P., Lantero, M.I., García, M.A., 2014. Evaluation of polyamide composite casings with silver–zinc crystals for sausages packaging. Food Pack. Shelf-life 1, 3–9.
Pinto, A.M.B., Santos, T.M., Caceres, C.A., Lima, J.R., Henriette, E.N., Azeredo, M.C., 2014. Starch-cashew tree gum nanocomposite films and their application for coating cashew nuts. LWT Food Sci. Technol. 62, 549–554.

Poortavasoly, H., Montazer, M., Harifi, T., 2014. Simultaneous synthesis of nano silver and activation of polyester producing higher tensile strength aminohydroxylated fiber with antibacterial and hydrophilic properties. RSC Adv. 4, 46250–46256.

Ramezani, Z., Zarei, M., Raminnejad, N., 2015. Comparing the effectiveness of chitosan and nanochitosan coatings on the quality of refrigerated silver carp fillets. Food Control. 51, 43–48.

Rasmussen, P.H., Hansen, A., 2001. Staling of wheat bread stored in modified atmosphere. LWT Food Sci. Technol. 34, 487–491.

Realini, C.E., Marcos, B., 2014. Active and intelligent packaging systems for a modern society. Meat Sci. 98, 404–419.

Reesha, K.V., Panda, S.K., Bindu, J., Varghese, T.O., 2015. Development and characterization of an LDPE/chitosan composite antimicrobial film for chilled fish storage. Int. J. Biol. Macromol. 79, 934–942.

Rena, G., Hu, D., Cheng, E.W.C., Vargas-Reusc, M.A., Reipd, P., Allaker, R.P., 2009. Characterization of copper oxide nanoparticles for antimicrobial applications. Int. J. Antimicrob. Agents 33, 587–590.

Rezaei, A., Nasirpour, A., Fathi, M., 2015. Application of cellulosic nanofibers in food science using electrospinning and its potential risk. Com. Rev. Food Sci. Food Saf. 14, 269–284.

Rhim, J.W., Park, H.M., Ha, C.S., 2013. Bio-nanocomposites for food packaging applications. Prog. Polym. Sci. 38, 1629–1652.

Rodriguez-Lafuente, A., Nerin, C., Batlle, R., 2010. Active paraffin-based paper packaging for extending the shelf-life of cherry tomatoes. J. Agric. Food Chem. 58, 6780–6786.

Sadeghian Maryan, A., Montazer, M., 2014. Natural and organo-montmorillonite as antibacterial nanoclays for cotton garment. J. Ind. Eng. Chem. 22, 164–170.

Sanchez-Garcia, M.D., Lopez-Rubio, A., Lagaron, J.M., 2010. Natural micro- and nanobiocomposites with enhanced barrier properties and novel functionalities for food biopackaging applications. Trends Food Sci. Technol. 21, 528–536.

Sanuja, S., Agalya, A., Umapathy, M.J., 2014. Studies on magnesium oxide reinforced chitosan bionanocomposite incorporated with clove oil for active food packaging application. Int. J. Polym. Mater. Polym. Biomat. 63, 733–740.

Sanuja, S., Agalya, A., Umapathy, M.J., 2015. Synthesis and characterization of zinc oxide–neem oil–chitosan bionanocomposite for food packaging application. Int. J. Biol. Macromol. 74, 76–84.

Sedighi, A., Montazer, M., Hemmati nejad, N., 2014a. Copper nanoparticles on bleached cotton fabric: in situ synthesis and characterization. Cellulose. 21, 2119–2132.

Sedighi, A., Montazer, M., Samadi, N., 2014b. Synthesis of nano Cu_2O on cotton: morphological, physical, biological, and optical sensing characterizations. Carbohydr. Polym. 110, 489–498.

Severino, R., Ferrari, G., Vu, K.D., Donsì, F., Salmieri, S., Lacroix, M., 2014. Antimicrobial effects of modified chitosan based coating containing nanoemulsion of essential oils, modified atmosphere packaging, and gamma irradiation against *Escherichia coli* O157:H7 and *Salmonella typhimurium* on green beans. Food Control. 50, 215–222.

Shin, S.H., Kim, S.J., Lee, S.H., Park, K.M., Jaejoon, H., 2014. Apple peel and carboxymethylcellulose-based nanocomposite films containing different nanoclays. J. Food Sci. 79, 342–353.

Silvestre, C., Duraccio, D., Cimmino, S., 2011. Food packaging based on polymer nanomaterials. Prog. Polym. Sci. 36, 1766–1782.

Siracusa, V., Rocculib, P., Romani, S., Dalla Rosa, M., 2008. Biodegradable polymers for food packaging: a review. Trends Food Sci. Technol. 19, 634–643.

Sorrentino, A., Gorrasi, G., 2007. Potential perspectives of bio-nanocomposites for food packaging applications. Trends Food Sci. Technol. 18, 84–95.

Soysal, C., Bozkurt, H., Dirican, E., Güçlü, M., Bozhüyük, E.D., Uslu, A.E., Kaya, S., 2015. Effect of antimicrobial packaging on physicochemical and microbial quality of chicken drumsticks. Food Control. 54, 294–299.

Takala, P.N., Vu, K.D., Salmieri, S., Khan, R.A., Lacroix, M., 2013. Antibacterial effect of biodegradable active packaging on the growth of *Escherichia coli, Salmonella typhimurium, and Listeria monocytogenes* in fresh broccoli stored at 4°C. LWT Food Sci. Technol. 53, 499–506.

Tankhiwale, R., Bajpai, S.K., 2009. Graft copolymerization onto cellulose-based filter paper and its further development as silver nanoparticles loaded antibacterial food-packaging material. Colloids Surf. B 69, 164–168.

Tankhiwale, R., Bajpai, S.K., 2012. Preparation, characterization, and antibacterial applications of ZnO-nanoparticles coated polyethylene films for food packaging. Colloids Surf. B 90, 16–20.

Thekkae, V.V.P., Cernik, M., 2013. Green synthesis of copper oxide nanoparticles using gum karaya as a biotemplate and their antibacterial application. Int. J. Nanomed. 8, 889–898.

Tornuk, F., Hancer, M., Sagdic, O., Yetim, H., 2015. LLDPE-based food packaging incorporated with nanoclays grafted with bioactive compounds to extend shelf-life of some meat products. LWT Food Sci. Technol. 64, 540–546.

Tran, N., Mir, A., Mallik, D., Sinha, A., Nayar, S., Webster, T.J., 2010. Bactericidal effect of iron oxide nanoparticles on *Staphylococcus aureus*. Int. J. Nanomed. 5, 277–283.

Tripathi, S., Mehrotra, G.K., Dutta, P.K., 2009. Physicochemical and bioactivity of cross-linked chitosan–PVA film for food packaging applications. Int. J. Biol. Macromol. 45, 372–376.

Unalan, I.U., Cerri, G., Marcuzzo, E., Cozzolinoa, C.A., Farris, S., 2014. Nanocomposite films and coatings using inorganic nanobuilding blocks (NBB): current applications and future opportunities in the food packaging sector. RSC Adv. 4, 29393–29428.

Wanga, H., Zhang, R., Zhang, H., Jiang, S., Liu, H., Sun, M., Jiang, S., 2015. Kinetics and functional effectiveness of nisin-loaded antimicrobial packaging film based on chitosan/poly(vinyl alcohol). Carbohydr. Polym. 127, 64–71.

Wei, H., YanJun, Y., NingTao, L., LiBing, W., 2011. Application and safety assessment for nano-composite materials in food packaging. Chinese Sci. Bull. 56, 1216–1225.

Xiao-e, L., Green, A.N.M., Haque, S.A., Mills, A., Durrant, J.R., 2004. Light-driven oxygen scavenging by titania/polymer nanocomposite films. J. Photochem. Photobiol. A 162, 253–259.

Youssef, A.M., Abdel-Aziz, M.S., El-Sayed, S.M., 2014. Chitosan nanocomposite films based on Ag-NP and Au-NP biosynthesis by *Bacillus subtilis* as packaging materials. Int. J. Biol. Macromol. 69, 185–191.

Youssef, A.M., El-Sayed, M.S., Salama, H.H., El-Sayed, H.S., 2015. Evaluation of bionanocomposites as packaging material on properties of soft white cheese during storage period. Carbohydr. Polym. 132, 274–285.

17

NANOENCAPSULATION OF FLAVOR AND AROMAS IN FOOD PACKAGING

Nuria Burgos, Ana Cristina Mellinas, Esther García-Serna, Alfonso Jiménez

University of Alicante, Department of Analytical Chemistry, Nutrition & Food Sciences, Alicante, Spain

1 Introduction

The traditional concept of food packaging is based on the use of materials to protect food products from environmental agents, contamination and undesirable manipulation. No interaction between food and packaging material is considered, reducing the material's role as a passive defense system. But the rising trends in consumers' demands for high quality and more natural and safe products, joined to the extension of global markets and the high use of vending machines and ready-to-eat products, have promoted research in advanced packaging materials and technologies to extend the shelf life of foodstuffs.

In past decades new concepts in packaging technologies have been introduced, with *active packaging* being one of the most studied and fashionable. This term was first used by Labuza (1987), and has become one of the most updated trends in food packaging in the 21st century. Active packaging materials are designed to interact actively and continuously with the packaged food, with the purpose of extending shelf life while maintaining or improving food quality (EC Regulation 1935/2004; EC Regulation 450/2009). Active packaging includes systems that can eliminate undesirable compounds that can reduce the commercial life of food, for example, oxygen scavengers (Joven et al., 2015), and systems that can release directly to food active substances, such as antioxidants or antimicrobial compounds. In general terms, the use of antioxidant packaging systems helps to maintain nutritional quality, while antimicrobial packaging protects food from spoilage caused by pathogens.

Food Packaging. http://dx.doi.org/10.1016/B978-0-12-804302-8.00017-0

In addition, consumers' concerns about potential toxicological risks associated with synthetic additives have promoted the use of natural compounds for food preservation. In this context, the study of bioactive compounds, such as flavors and aromatic agents extracted from natural resources like plants or agricultural byproducts for their use in food protection against oxidation and/or spoilage, is a trending topic in the past few years (Vergis et al., 2015; Calo et al., 2015). It is well known that aromatic plants have been used for centuries because of their benefits to human health, but in recent years they have been extensively investigated as a source of aromatic agents and other flavoring compounds like essential oils. All these chemicals represent a natural alternative to synthetic preservatives (Aumeeruddy-Elalfi et al., 2015). However, most of these substances are very volatile and chemically unstable while suffering easy degradation from heat and light, resulting in a loss of activity in relatively short times. In addition, their direct incorporation to polymer matrices can change their structure, usually affecting negatively their mechanical properties (Joven et al., 2015; Júnior et al., 2014).

Encapsulation technologies could be useful alternatives to protect flavor agents and aromas from external agents while providing controlled release from the material to the packaged food. Three types of capsules have been proposed, depending on their particle size: macrocapsules, microcapsules, and nanocapsules, with this last group being the main subject of this chapter. Nanocapsule particles range from 0.1 nm to 1 μm in size (Khare and Vasisht, 2014). Micro and nanocapsules of aromatic compounds are widely used in many different applications, including medical, pharmaceutical, and food sectors. Encapsulation technologies permit the reduction in volatility of aromatic compounds, increasing their stability by protecting them from the aggressive actions of oxygen, humidity, and light. These technologies would also result in reductions of their degradation rates at high temperatures during material and food processing. The improvement in the efficiency of nanoencapsulated bioactive compounds has been reported (Woranuch and Yoksan, 2013a; Sobel and Su, 2014). The interaction mechanisms between capsules and food to deliver an active agent at a controlled rate are essentially three: (1) degradation of the capsule material, thus liberating the active agent; (2) diffusion of the active agent through the material structure; and (3) swelling followed by diffusion (Tuladhar and Anal, 2014).

Microencapsulation technologies are widely used in many industrial sectors, including textile, clinical, pharmaceutical, and food, to protect bioactive compounds (Papajani et al., 2015).

Nanoencapsulation represents one step forward by reducing particle sizes to the nanoscale and it shows great potential to develop new products for a wide range of applications, including food and packaging sectors. The main aim of using entities at the nanoscale, and particularly nanocapsules to contain active compounds, is the improvement of their availability and extension of their beneficial effect (FAO, 2010). Among the benefits obtained by using nanocapsules, the most significant lies on the increase of the control in the additives release and the possibilities for dispersion of water-insoluble compounds, as well as the improvement in efficiency of preservation of bioactive agents, requiring low amounts of these often-expensive compounds to achieve similar protective effects in food (Peng et al., 2014; Woranuch and Yoksan, 2013a,b). For example, it has been reported that the activity of cinnamaldehyde against carcinogenic agents increases by its nanoencapsulation (Wani et al., 2014). One important consequence of these advantages is the possibility to create lightweight but efficient active packaging materials (Carocho et al., 2014).

The incorporation of nanocapsules into food packaging materials could result in several benefits to the system, which can be summarized as follows:

- To improve the material intrinsic properties, such as resistance to high temperatures and higher barrier to oxygen and/or water vapor (Rhim et al., 2013).
- To develop a protective layer for active compounds: Nanocapsules provide good protection against antioxidant, antimicrobial, or antifungal chemicals, most of them obtained from natural sources. They could also help to modulate their interaction with food or package headspace (Rhim et al., 2013). This effect is particularly relevant in nanobiocomposites, where their biopolymer structures act as supporting materials of active compounds to extend the shelf life of perishable food (Severino et al., 2015; Mohammadi et al., 2015).
- To develop intelligent/smart packaging systems: Nanocapsules can interact with food to provide important information about the quality of the product (Neethirajan and Jayas, 2011; Carocho et al., 2014).

The aim of this chapter is to review the principal features in nanoencapsulation of flavors and aromatic compounds and application in food packaging, including the most common natural active agents, the different materials to obtain nanocapsules, and the main mechanisms to achieve safe systems with commercial potential. The last reported research on the effect of nanoencapsulation of flavors and aromas in active food packaging is revised.

2 Flavor and Aromatic Compounds

The general term *flavor* includes flavoring substances, preparations, processes, and mixtures. Flavoring substances and their mixtures are the main flavors used in food technologies and more specifically in food packaging. Flavoring substances are defined as chemicals with specific properties that are commonly extracted from plants or animals by physical, enzymatic, or microbiological methods (natural flavors); or obtained by chemical synthesis (synthetic flavors) (EC Directive, 1988). They can be used in a raw state or after some processing for permitting human consumption in culinary preparations. Flavoring substances should be registered before their use in the food industry (EC Decision, 2002/113/EC). This list includes compounds with different origins, such as limonene, champhor, menthol, allyl propionate, rosalva acetate, oleic acid, cinnamyl isobutyrate, amyl palmitate, 2,7-dimethylocta-5(trans),7-dieno-1,4-lactone (marmarolactone), and many others.

In general terms, flavoring systems are complex mixtures of volatile organic compounds, which may include natural and chemical extracts. For example, pineapple flavor is composed of nine chemicals and seven essential oils. In addition, essential oils are mixtures of different substances, resulting at the end in very complex systems. A complete knowledge of the flavors' intrinsic nature and composition is critical to determine the most adequate technique to perform their encapsulation (Sobel and Su, 2014).

Natural aromatic substances are synthesized by living organisms as a defense mechanism or as secondary metabolites in several organs, such as flowers (jasmine), herbs (rosemary), buds (clove), leaves (eucalyptus), fruits (citrus), barks (cinnamon), seeds (cardamom), or roots (ginger). The aromatic oily extracts obtained from specific parts of many plants are known as essential oils (EOs), which are defined as odorant products with complex composition. Due to their oily nature, essential oils have physical properties similar to other lipophilic compounds, being characterized by their low miscibility with water, but in contrast they are highly volatile. Their extraction yields are usually quite low and variable depending on species, organs, and methods, but they are highly valuable in their use in specific applications, particularly in the food sector (Asbahani et al., 2015).

2.1 Properties of Flavor and Aromatic Compounds

Aromatic plants and their components have been studied as potential antioxidants and inhibitors of bacterial growth. These properties are very valuable in food industries and packaging applications, since some important beneficial effects in active formulations have been linked to EOs and other secondary plant metabolites (Calo et al., 2015).

EOs have been extensively studied for their wide range of applications, one of the most important being an effective food preservative. This importance in food preservation is attributable to their antimicrobial and antioxidant performance supplied by the aromatic compounds in their composition (Aumeeruddy-Elalfi et al., 2015). Many EOs have shown in vitro activity against foodborne pathogens and spoilage organisms, and to a smaller degree in food. This effect is particularly relevant against Gram-positive bacteria, but they are also effective against Gram-negative, and it is attributable to the mixture of aromatic compounds in EOs composition and their synergies (Calo et al., 2015; Prakash et al., 2015). For example, the *lavandula* EO is composed of several aromatic substances, being fenchone (33–37%), camphor (16–24%), and eucalyptol (17–18%) the major components with lower concentrations of limonene, camphene, linalool, thymol, and E-β-caryophyllene. It has been reported that *lavandula* EOs have antiinflammatory, antioxidant, and antimicrobial properties, making them suitable in pharmaceutical or cosmetic industries for skin care preparations or in the food industry as natural preservatives (Carrasco et al., 2015). These EOs have been applied in food packaging by incorporation to gelatin films with the result of a functional material with high antimicrobial effect against *Staphylococcus aureus* and *Escherichia coli*, and good intrinsic properties even at low concentrations, with a significant improvement in water vapor permeability at high concentrations (Martucci et al., 2015). High antifungal activity of these EOs against *Candida albicans* has been also reported, such as other flavoring compounds (Liakos et al., 2014).

It is usual that plants belonging to the same family have similar aromatic compounds and EOs. For example, the major compounds identified in different EOs extracted from forest foliage are α-pinene, β-pinene, and limonene, all of them monoterpene hydrocarbons, and bornyl acetate. But these EOs have around 50 different compounds in their composition, among them camphene, myricene, δ-3-carene, terpinolene, and β-phellandrene as monoterpene hydrocarbons; borneol, terpin 1-en-4-ol, and α-terpineol as oxygenated monoterpenes; and β-carophyllene and δ-cardinene as sesquiterpenes. The EOs obtained from forest trees have shown antioxidant performance, bactericide effect against *E. coli*, *S. aureus*, *Staphylococcus enteriditis*, and *Staphylococcus typhymurium*, and antifungal effect against *Asperigus pullulans* and *Asperigus niger* (Poaty et al., 2015).

The huge majority of flavors shows bactericide effects against different foodstuff pathogens. Some examples are carvacrol and cinnamaldehyde against *Salmonella*; lactic acid, acetic acid, carvacrol, and cinnamaldehyde against *Lysteria monocytogenes* or linoleic acid, lemongrass and oregano essential oils against mesophylic microorganisms (Sanchez-Ortega et al., 2014).

2.2 Flavors Encapsulated for Food Packaging

Flavors are excellent compounds to be incorporated into active packaging systems, but their high volatility and unstable character represent a difficulty for use at the industrial scale. Nanoencapsulation is a promising technique to overcome these limitations by protecting them and to permit their application to polymer matrices. The main nanoencapsulated flavors and their application in food packaging are compiled in Table 17.1.

Table 17.1 shows that limonene and cinnalmaldehyde are the most studied flavors, while carvacrol, thymol, eugenol, and curcumin are other important natural additives used in active food packaging. Limonene is a highly volatile aromatic terpene, with important applications as fragrance in cosmetics and food industries. This compound, besides linalool, is the main aromatic substance in orange, lemon, and lemongrass EOs, which may be composed by more than 200 different components (Chang and McClements, 2014). It has found remarkable uses in the food industry due to its antimicrobial, antifungal, and antioxidant properties, which make it suitable to extend the shelf life of foodstuff. Limonene has shown great performance against *E. coli* and *C. albicans*, being also effective against molds (Liakos et al., 2014). The nanoencapsulation of limonene has shown several advantages over the direct application of this additive into a polymer matrix for active food packaging. It was reported that the release of limonene encapsulated in pullulan and β-cyclodextrin was high for the first 3 days of contact with food, but it was stabilized after 7 days, keeping the release for 45 days, the usual contact time for packaged food (Fuenmayor et al., 2013). The thermal stability of these formulations was maintained from room temperature to 170°C. In addition, it was observed that limonene remained stable and active at relative low humidity conditions. Camerlo et al. (2013) reported that limonene encapsulated in poly(vinyl alcohol) (PVA) showed an encapsulation efficiency of 74% at 1:8 ratio (limonene:PVA). The release rate increased at high humidity and temperature, with a complete release for 24 h at 37°C and 65% relative humidity. But release of limonene slowed down at room temperature and low humidity, making possible a controlled

Table 17.1 Recent Trends in Nanoencapsulated Flavors

Nanocarrier System	Flavor Agent	Material	Active Properties	Reference
Nanogels	*M. piperita* EO	Chitosan/cinnamic acid	Antifungal activity (*A. flavus*)	Beyki et al. (2014)
	Thyme EO	Chitosan/benzoic acid	Antifungal activity (*A. flavus*)	Khalili et al. (2015)
	Cuminum EO	Chitosan/caffeic acid	Antifungal activity (*A. flavus*)	Zhaveh et al. (2015)
	Lippia sodoides EO	Chitosan/cashew gum	Activity against larvae (*Aedes aegypti*)	Abreu et al. (2012)
	Lippia sodoides EO	Alginate/cashew gum	Activity against larvae (*A. aegypti*)	de Oliveira et al. (2014)
	Oregano EO; eugenol	Chitosan/sodium tripolyphosphate	Antioxidant activity	Hosseini et al. (2013); Woranuch and Yoksan (2013a,b)
	Z. multiflora EO	Chitosan/ sodium tripolyphosphate	Antifungal activity (*Botritys cinerea*)	Mohammadi et al. (2015)
	Cinnamon EO	Chitosan	Antioxidant and antimicrobial activity	Hu et al. (2015)
Nanoparticles	Eugenol and trans-cinnamaldehyde	PLGA	Antimicrobial activity (*Salmonella spp.* and *Listeria spp.*)	Gomes et al. (2011)
	Curcumin EO	PLGA	—	Grama et al. (2013)
	Cinnamon bark extract	PLGA	Antibacterial activity (*L. monocytogenes* and *S. typhimurium*)	Hill et al. (2013)
	Thymol	NaCas	Antimicrobial activity (*L. monocytogenes*)	Pan et al. (2014)
	Peppermint EO	Zein/GA	—	Chen and Zhong (2015)
	Thymol and carvacrol	Zein	Antioxidant and antimicrobial activity (*E. coli, Salmonella e., S. aureus* and *L. monocytogenes*)	da Rosa et al. (2015)
	Rosemary EO	Poly(ε-caprolactone) (PCL)	—	Ephrem et al. (2014)
	Jasmine EO	Gelatin/ GA	—	Lv et al. (2014)

(*Continued*)

Table 17.1 Recent Trends in Nanoencapsulated Flavors (*cont.*)

Nanocarrier System	Flavor Agent	Material	Active Properties	Reference
Nanofibers	R-(+)-limonene	PVA	—	Camerlo et al. (2013)
	Linalool, curcumin	PLA	—	Souza et al. (2015) Silva-Buzanello et al. (2014)
	Geraniol, eugenol	γ-Cyclodextrins/ PVA nanofibers	—	Kayaci et al., 2014, 2013
	Allyl isothiocyanate	β-Cyclodextrins/ PVA nanofibers	Antibacterial activity (*E. coli, S. aureus*)	Aytac et al. (2014)
Nanoemulsion	Clove bud, oregano EO; cinnamaldehyde	Water/Tween 80	Antifungal activity (*A. niger, Penicillium sp.*) Antibacterial activity (*E. coli, Salmonella, Listeria m., S. aureus*)	Otoni et al. (2014b) Otoni et al. (2014a)
	Carvacrol	Chain triglyceride/ water/Tween	Antifungal activity (*Z. bailii, S. cerevisiae, B. bruxellensis, B. naardenensis*)	Chang et al. (2013)
	Carvacrol, bergamot, mandarin and lemon EOs	Sunflower oil/glycerol monooleate/water/ Tween 20	Antibacterial activity (*E. coli, S. typhimurium*)	Severino et al. (2015)
	Lemongrass, clove, tea tree, thyme, geranium, marjoram, palmarosa, rosewood, sage, and mint EOs	Sodium alginate/ water/Tween 80	Antibacterial activity (*E. coli*)	Salvia-Trujillo et al. (2014)
Nanoliposomes	Cinanmon EO	Soybean lecithin/ cholesterol/Tween 80	Antimicrobial activity (*E. coli, S. aureus, A. niger*)	Wu et al. (2015)
	D-Limonene and orange EO	Rapeseed and soy lecithins/water	Antimicrobial activity	Jiménez et al. (2014)
	Cinnamaldehyde	Polydiacetylene/ N-hydroxysuccinimide	Antimicrobial activity (*E. coli, B. cereus*)	Makwana et al. (2014)
Solid Lipid Nanoparticles (SLNP)	Maltol	Stearic acid/ ethylcellulose	—	Eltayeb et al. (2013)

release (only 30%) during 15 days (Camerlo et al., 2013). These results showed the possibility to control the release of bioactive compounds as a function of the environmental conditions.

Cinnamaldehyde is one of the most widely studied aromatic active compounds, constituting the main component of the cinnamon EO and showing antimicrobial and antioxidant properties by reducing bacterial grown and being highly sensitive against Gram-positive bacteria. It was reported that 240 $\mu g/mL^{-1}$ cinnamaldehyde solutions were enough to reduce 50% of the *E. coli* concentration and *Bacillus cereus* to undetectable levels, since Gram-negative bacteria have a hydrophilic outer membrane with degradative enzymes able to react with cinnamaldehyde (Makwana et al., 2014). They also studied the effect of nanoliposomes in reduction of the bacteria activity and results showed a large increase in antimicrobial efficiency at very low concentrations (864 ng/mL^{-1}). These solutions were enough to reduce the activity of both bacteria to undetectable levels in just 8 h. This increase in the effect of cinnamaldehyde may be due to the interaction of this active agent with liposomes in the cell membrane, which promotes the effect of thc activc compound into thc ccll. PLA-based films with liposomes incorporated during casting showed positive antimicrobial effects, being more susceptible to *B. cereus* than to *E. coli*. The antimicrobial effect of cinnamaldehyde against other important food pathogens, such as *L. monocytogenes* and *Salmonella typhimurium*, has been also reported (Hill et al., 2013). In addition, active packaging materials with enhanced antimicrobial activity have been developed with cinnamon EO nanocapsulated into polymeric materials by spray drying to preserve high perishable foods, like fresh meat (Chen and Zhong, 2015). Other studies also suggested the potential use of nanoencapsulated cinnamaldehyde or cinnamon EO for preservation of minimally processed food (Ghosh et al., 2013).

Curcumin is the main bioactive compound extracted from turmeric plants, which have a distinctly earthy, slightly bitter, and hot peppery flavor and a mustardy smell. Curcumin shows anticancer, antiinflamatory, antimicrobial, and antioxidant properties (Sari et al., 2015) and it has been nanoencapsulated to enhance its antioxidant potential to be incorporated in food packaging materials.

Eugenol is a natural phenolic compound with flavoring properties and is the most abundant compound of clove EO, while it is also present in cinnamon EO. It is used in pharmaceuticals, cosmetics, food, and active food packaging applications by its antimicrobial and antioxidant properties. It has been reported that nanoparticles of chitosan loaded with eugenol show stronger radical scavenging activity than eugenol directly incorporated into chitosan films (Ghosh et al., 2013; Woranuch and Yoksan, 2013b).

Thymol and carvacrol are phenolic monoterpens very common in many active compounds, such as oregano and rosemary EO, but the oregano EO is much more effective against fungal infection than other polyphenols (Papajani et al., 2015). Both flavors show high antioxidant activity and significant antimicrobial effect against *E. coli, S. enteritidis, S. aureus,* and *L. monocytogenes* presenting high activity against Gram-positive bacteria. Both compounds were stabilized for 90 days by nanoencapsulation (da Rosa et al., 2015). Antioxidant films with the addition of thymol and carvacrol as bioactive compounds were developed and characterized, showing that thymol has higher antioxidant efficiency than carvacrol (Ramos et al., 2014). Other authors obtained nanocapsules with thymol to introduce this additive directly into food as an antibacterial agent against *L. monocytogenes* (Pan et al., 2014) and with carvacrol to stabilize food products against spoilage (Donsì et al., 2014). Antifungal nanoemulsions with carvacrol against *Zygosaccharomyces bailii, Saccharomyces cerevisiae, Brettanomyces bruxellensis,* and *Brettanomyces naardenensis* were also developed (Chang and McClements, 2014).

In summary, the vast majority of aromatic compounds and EO show antioxidant, antibacterial and antifungal properties and they are useful to reduce the most common spoilage pathogens. Nanoencapsulation in the adequate supporting matrices is essential to obtain good performance with controlled release rates in active packaging applications. The main materials and incorporation strategies to obtain adequate nanocapsules formulations will be considered in the following sections.

3 Wall Materials for Flavor Encapsulation

The selection of the most appropriate material for flavor or aroma nanoencapsulation mainly depends on the properties of the active compound(s), the desired release rate and the organoleptic characteristics of the packaged food. Other factors, such as encapsulation efficiency, defined as the percentage of the active agent effectively contained into nanocapsules, as well as raw materials and processing costs, should be taken into account when choosing the nanoencapsulation technologies (Wang et al., 2015; Zhu et al., 2012). In recent years, and considering the increase in social demand of environmentally-friendly materials, the interest in the use of biopolymers and biocomposites in the preparation of nanocapsules has increased. These biomaterials can be very useful in the nanoencapsulation of flavors due to their biocompatibility, low toxicity, and biodegradable characteristics (Abreu et al., 2012; Wang et al., 2015). There are some requirements that the ideal

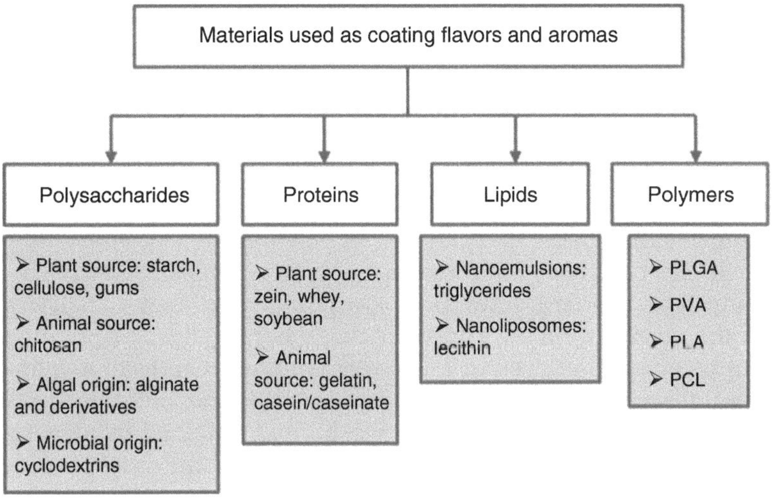

Figure 17.1. Materials used for flavors and aromas nanoencapsulation.

material for nanoencapsulation should fulfill, such as low viscosity and hygroscopic character at high solid contents, which could offer protection for the flavors against external factors. Other important requirements are their low reactivity with the active compound and solvents, low cost, no influence on the organoleptic characteristics of food, and good film-forming and emulsification properties (Blanco-Padilla et al., 2014; Zhu et al., 2012).

In the case of encapsulation of hydrophobic active agents, hydrophilic materials, such as polysaccharides and proteins, are required, while lipids or other polymers with hydrophobic character should be used for the encapsulation of hydrophilic compounds. It is important to remark that the encapsulation materials should be included in the positive list marked by the current legislation for substances which may be used in the manufacture of plastic materials intended to come into contact with food (EC Regulation 10/2011).

A wide variety of materials have been proposed to encapsulate flavors, including polysaccharides, proteins, lipids, and other polymers. Fig. 17.1 summarizes the most significant materials used in the nanoencapsulation of flavors and aromas.

3.1 Polysaccharides

Polysaccharides are natural carbohydrates obtained from the polymerization of monosaccharide units, which could vary depending on their biological origin, type, number, distribution, and bonding. Properties such as their molecular structure and ability to encapsulate active compounds, joined to their low cost and ubiquitous presence in nature, have attracted considerable attention as wall materials for active food-packaging systems (Fathi et al., 2014). Their functional properties could be tailored depending on their origin, isolation methods and physical, chemical, or biochemical modifications. Some examples in the use of polysaccharides in nanoencapsulated flavors have been summarized in Table 17.1.

In the past few years, natural polysaccharides, such as chitosan and alginate, have been proposed as polymer matrices in the formulation of nanoparticles where a flavor or aroma can be entrapped even at low concentrations. Chitosan is a linear polysaccharide formed by two monomeric building units (*N*-acetyl-glucosamine and D-glucosamine) obtained by alkaline deacetylation of chitin, which is found in the outer skeletons of crustaceans and insects. Chitosan is receiving important attention in the encapsulation of lipophilic bioactive compounds (Rinaudo, 2006; Nilsen-Nygaard et al., 2015). Recently, the effect of chitosan nanoparticles encapsulating cinnamon EO in contact with chilled pork was evaluated (Hu et al., 2015). Authors observed that the loading capacity of the active compound was improved by increasing the size of the nanoparticles (up to 527 nm), reaching 20.7% of encapsulation efficiency. Then, the application of these nanoparticles as coating in low-density polyethylene (LDPE) films led to excellent antimicrobial and antioxidant properties with a significant decrease of the organoleptic effect to pork.

The presence of amino and hydroxyl groups in chitosan structure makes possible its chemical modification by formation/rupture of covalent linkages with the active agents resulting in improvement of the functional properties and control of the release rate (Nilsen-Nygaard et al., 2015). Some of the hydrophobic compounds recently proposed to modify the chitosan structure by producing nanogels are cinnamic acid, benzoic acid, caffeic acid (Beyki et al., 2014; Khalili et al., 2015; Zhaveh et al., 2015) and sodium tripolyphosphate (TPP) to obtain nanoparticles for flavors encapsulation (Hosseini et al., 2013; Woranuch and Yoksan, 2013b; Mohammadi et al., 2015).

Nanogels are micelle-like nanoparticles that form aggregates in contact with water while their central region retains the hydrophobic active compounds. Therefore, nanogels are very promising carrier materials for flavors and aromas due to their abilities in the controlled release of these active compounds, keeping their high stability, and loading capacity (Khalili et al., 2015). In this sense, other authors synthesized nanosized polymer gels with narrow size distribution through the formation of amid linkages between amino groups present in chitosan and the carboxyl group of cinnamic acid (Beyki et al., 2014). This nanogel was used for the encapsulation of the hydrophobic peppermint EO enhancing its antimicrobial activity and stability against *Aspergillus flavus* and revealing the possibilities of the chitosan/cinnamic acid nanogel as a carrier for an EO. Similar results were obtained for the encapsulation of the thyme EO using a nanogel formed by the combination of chitosan and benzoic acid (Khalili et al., 2015) and *Cuminum cyminum* EO in chitosan/caffeic acid nanogels (Zhaveh et al., 2015). In other study, oregano essential oils (OEO) were successfully encapsulated in chitosan/TPP nanoparticles prepared by a two-step method, oil-in-water emulsion, and further ionic gelation. The success of this combination was based on the electrostatic interaction between the positively charged amino groups of chitosan and the negatively charged groups of TPP. The obtained nanoparticles showed spherical shape with size range between 40 and 80 nm, with high influence of the active agent content on the release rate. In addition, by increasing the OEO/chitosan mass ratio, the loading capacity was significantly improved (from 3 to 8%) while the encapsulation efficiency decreased (from 47 to 21%) (Hosseini et al., 2013). Other authors reported the capacity of the chitosan/TPP nanoparticles for the encapsulation and delivery of eugenol (Woranuch and Yoksan, 2013b) and *Zataria multiflora* EO (Mohammadi et al., 2015), concurring in the improvement of their release rate, in comparison to micro-size carriers, revealing the promising role of these nanoparticles to contain flavors to enhance the antioxidant/antimicrobial resistance of the packaged food in contact with them.

In the past few years, polysaccharides obtained from vegetal sources, such as cellulose, starch-based materials (cyclodextrins, maltodextrins, octenyl succinic anhydride, or porous starch), and gums have been widely used to encapsulate volatile compounds in food industry (Wang et al., 2015). Among them, cyclodextrins and gums have shown high potential as wall materials for the protection of flavors in food packaging, as can be observed in Table 17.1.

Cyclodextrins (CDs) are cyclic oligosaccharides formed by the enzymatic modification of starch molecules and they are composed of 6, 7, or 8 glucopyranose units (referred as α-, β-, or γ- CDs, respectively) connected through α-1,4-glycosidic linkages. They present a hydrophobic inner cavity (~6–10 Å) and a hydrophilic external surface, allowing the formation of inclusion complexes with compounds with low solubility in water, such as EOs and volatile flavors. The encapsulated molecules should have suitable dimensions to fit properly in the central cavity of CDs and to allow the interactions between both components by hydrophobic/hydrophilic forces joined to van de Waals and hydrogen bonding, resulting in inclusion complexes (Zhu et al., 2012, Wang et al., 2015, Fathi et al., 2014). Some beneficial effects on the physico-chemical properties of flavors were provided by their encapsulation into CDs cavities, enhancing their solubility and bioavailability and protecting them from degradation, oxidation, and evaporation. In addition, high-controlled release and enhanced thermal stability of flavors were achieved by means of CDs inclusion complexation, improving their retention during processes and storage, allowing the use of low amounts with significant effects in the packaged food (Marques, 2010). CDs are mainly applied for encapsulation of flavors that are not easy to stabilize in other supporting materials.

Recently, advanced processing techniques, such as electrospinning, were also applied for the encapsulation of flavors into CDs inclusion complexes. In this context, Kayaci et al. reported that electrospun PVA nanofibers incorporating eugenol/CDs or geraniol/CDs inclusion complexes are appropriate for active food packaging due to their very high surface area and nanoporous structure, providing enhanced durability and high thermal stability of flavors (Kayaci et al., 2013; Kayaci et al., 2014). A similar system based on the use of β-CDs allows the encapsulation of allyl isothiocyanate in PVA electrospun nanofibers, showing high antibacterial activity against *E. coli* and *S. aureus* due to the higher amount of the selected flavor available to interact with the packaged food (Aytac et al., 2014).

However, there are some factors to limit the industrial application of CDs in flavor encapsulation, such as their high cost ($\gamma > \alpha > \beta$) compared to conventional carrier materials, and the different legislation applicable to the use of CDs in direct contact with food being "Generally Recognized as Safe" (GRAS) in the United States, and partial approval in the European Union (α- and β-cyclodextrins are approved in Europe as food additives).

In general terms, the definition of gums corresponds to complex polysaccharides extracted from the resins of different trees that have the ability to form gels, to obtain viscous solutions or to stabilize emulsions. In the field of materials for nanoencapsulation, gum arabic (GA) (extracted from *Acacia* trees) and cashew gum (extracted from *Anacardium occidentale* trees) have been used as surface stabilizing agents in hydrophobic nanoparticles (Abreu et al., 2012; Chen et al., 2015). Both gums show similar properties, since their structures are composed by a main chain of galactose units with branches of arabinose, glucose, and rhamnose, which contribute to enhance the negative charge density of the polysaccharide.

The use of two different wall materials for flavors nanoencapsulation has been proposed to take profit of the main properties of each individual biopolymer. In this sense, the development of biopolymer blends formed by chitosan and cashew gum as encapsulating material for *Lippia sidoides* EO has been reported (Abreu et al., 2012). High loading (11.8%) and encapsulation efficiency (70%) were obtained when using gum:chitosan 1:1, matrix:oil 10:2 ratios, and 5% gum concentration. They found that nanoparticles with a higher size (from 335 to 558 nm) were obtained by increasing the chitosan proportion in the matrix, probably due to the electrostatic repulsion between chains caused by the presence of a high density of positive amino groups in nanoparticles.

L. sidoides EO was encapsulated in alginate/cashew gum nanoparticles by spray drying, with satisfactory loadings, good encapsulation efficiency (up to 55%), and average sizes in solution (223–399 nm) (de Oliveira et al., 2014). The release profiles were dependent on the relative ratio of both biopolymers and the addition of cashew gum to alginate in a 3:1 ratio allowed a faster release with satisfactory oil loading due to the increase of the hydrophilic character of the polymer matrix. Other authors reported that the nanoencapsulation of peppermint EO in and alginate matrix is a suitable way to protect flavors during processing and storage, retarding their antimicrobial and flavoring effects, and permitting a longer useful life for these systems (Ghayempour and Mortazavi, 2014).

3.2 Proteins

The use of proteins obtained from different sources, either animal (casein, gelatin), or vegetable (corn, whey, or soybean) as well as some of their derivatives, such as sodium caseinate (NaCas), has been recently proposed in the encapsulation of unstable flavor compounds and lipophilic EO by their high content in hydrophobic aminoacids, such as proline, tyrosine, leucine, or alanine (Pan et al., 2014; Patel and Velikov, 2014).

Nanocapsules containing thymol (1% w/v) were prepared by using NaCas as supporting material and following a high-shear homogenization method (Pan et al., 2014). These nanocapsules were stable for 30 days at room temperature and neutral pH showing higher solubility and antimicrobial activity for milk compared to thymol crystals directly incorporated to the polymer matrix. The addition of soluble polysaccharides extracted from soybean resulted in a clear increase in steric repulsion in these nanoparticles with the consequence of the aggregation and prevention of their potential precipitation at acidic conditions.

Prolamins are a group of plant storage proteins having high proline content and found in the seeds of cereal grains. The prolamin obtained from corn, also called zein, has been used as a supporting carrier of flavors and EOs (Patel and Velikov, 2014). Due to its inherent hydrophobic character, zein could form colloidal nanoparticles by the antisolvent precipitation method, where a suspension in a selected liquid phase produces the change in the solubility capacity of the primary solvent.

The combination between proteins and polysaccharides has been recently proposed to increase the interface stabilization with the aim to encapsulate flavors, protecting them from environmental conditions during processing or storage (Chen et al., 2015). In this context, authors reported that the addition of GA permits some increase in the stabilization of zein nanoparticles after antisolvent precipitation at a pH range between 3 and 8, while the protein solubility also increased. The excellent emulsifying properties of GA promoted its adsorption onto the hydrophobic zein surfaces, providing steric and electrostatic repulsions to increase stability. Under these conditions, the encapsulation of peppermint EO in zein/GA nanoparticles was successful, with good control of the flavor release.

Other authors reported the preparation of heat-resistant flavor nanocapsules based on gelatin:GA complexes (Lv et al., 2014). Spherical nanoparticles were obtained at pH 4.8 and 1:1 ratio. Stable nanocapsules based on these complexes with jasmine EO were prepared by cross-linking with transglutaminase at alkaline conditions, showing good resistance to humid heat (80°C) for 5 h. These results suggested that these electrostatic complexes may have potential application in the preparation of delivery vehicles for flavors and aromas in food industry.

3.3 Lipids

Lipid-based nanocapsules show some advantages over those obtained from polysaccharides and proteins, such as their high encapsulation efficiency, low toxicity, and the possibility of easier scale-up to industrial production (Fathi et al., 2012). Two main lipid-based nanoencapsulation systems have been recently proposed as carrier materials for flavors: nanoemulsions and nanoliposomes, as indicated in Table 17.1.

Nanoemulsions formed by dispersing nonpolar substances, such as EO, in water-soluble polymer solutions with suitable emulsifiers are stable colloidal systems with droplet sizes lower than 200 nm. They can be prepared by using high- or low-energy methods, depending on the final structure and desired functionalities

(Blanco-Padilla et al., 2014; Fathi et al., 2012). Komaiko and McClements showed that stable nanoemulsions with flavors can be produced by spontaneous emulsification processes (low-energy method), where an organic phase (food-grade oil and surfactant) was added to a stirred aqueous phase (Komaiko and McClements, 2015). They found that the size of the droplets decreased when increasing the surfactant-oil ratio, and was smaller when the surfactant was initially dispersed into the oil phase rather than into the aqueous one. The surfactant and oil types also influenced the droplets' size. Chang and McClements observed that the thermal stability of orange EO nanoemulsions was affected by the type and the ratio between the medium-chain triglyceride (carrier oil) and the nonionic surfactant (Chang and McClements, 2014). Other authors reported that small nanoemulsion droplets prepared by the direct oil-in-water method improved the antimicrobial properties of clove bud EO and oregano EO lipophilic flavors (Otoni et al., 2014b), cinnamaldehyde (Otoni et al., 2014a), and carvacrol (Severino et al., 2015), increasing their bioavailability and reducing the flavor content required to obtain high delivery efficiency.

Nanoliposomes are nanometric spherical core-shell structures where the hydrophobic parts of phospholipids are associated, forming a bilayer, and the polar head groups are in the inner and outer aqueous phases (Blanco-Padilla et al., 2014). Nanoliposomes show possibilities for encapsulating flavors with different solubility at the industrial scale. In fact, natural materials, such as rapeseed and soybean lecithins have been successfully applied as coring materials for selected flavors (Jiménez et al., 2014; Wu et al., 2015).

Recently, two new lipid-based encapsulation systems, called solid-lipid nanoparticles (SLNP) and nanostructured-lipid carriers (NLC) have been reported as showing high potential as nanocarriers for flavors by their high stability, longer times, and sustained release profiles, in comparison with the conventional lipid-based encapsulation systems (Eltayeb et al., 2013; Fathi et al., 2012).

3.4 Food Grade Biobased Polymers

Synthetic biobased polymers and copolymers, such as poly(lactide-*co*-glycolide) (PLGA) have shown some advantages over natural polymers because of their high purity and reproducibility in the processing conditions showing, in addition, biocompatibility and biodegradability. The nanoencapsulation of flavors in synthetic biopolymers is a promising way to improve their stability and solubility in water. In this sense, some authors showed that PLGA could efficiently encapsulate eugenol and trans-cinnamaldehyde with high efficiency (approximately 98 and 92%, respectively), forming spherical nanoparticles in the presence of poly (vinyl alcohol) (PVA) as surfactant by using the emulsion evaporation method (Gomes et al., 2011). PLGA nanoparticles containing cinnamon bark extracts were prepared with two different lactide to glycolide ratios (65:35 and 50:50) (Hill et al., 2013). Authors concluded that increasing the lactide concentration the nanoparticle size decreased while increasing the encapsulation efficiency and antimicrobial effect due to the slower release of the active compound. These authors performed release experiments and also tested the antimicrobial efficiency against *Salmonella* and *Listeria* bacteria. Both studies showed the potential application of PLGA as a nanocarrier in the food industry. Other authors reported the possibilities for scale-up of formulations based on curcumin EO encapsulated into PLGA nanoparticles and they showed their stability at chilling and room temperatures for six months (Grama et al., 2013).

The successful encapsulation of limonene in a PVA fibrous matrix by emulsion electrospinning with no other additives was also reported (Camerlo et al., 2013). These authors observed some influence of the emulsion formulation and the fibers morphology on the encapsulation efficiencies, but concluded with the potential application of these formulations in food packaging.

Other nanofibers or nanocapsules based in poly (lactic acid) (PLA) (Souza et al., 2015; Silva-Buzanello et al., 2014) and poly(ε-caprolactone) (PCL) (Ephrem et al., 2014) were successfully applied as coring materials for the encapsulation of flavors (linalool, curcumin and rosemary EO). In general terms, authors reported that these nanocapsules permitted a reduction in flavors' volatility, sensitivity to external conditions, and enhanced their water solubility.

4 Nanoencapsulation Methods

The selection of the most appropriate encapsulation technique is essential in the development of systems adequate for effective release of the active compounds. This selection should be based on the required particle size, physico-chemical properties, nature of the active agent and the nanocapsules' wall material (Ezhilarasi et al., 2013). The optimal encapsulation material to contain flavors should possess excellent emulsifying properties and good film-forming functionalities (Wang et al., 2015). As in the rest of nanomaterials, techniques for producing nanocapsules can be divided into two main categories following top-down and bottom-up

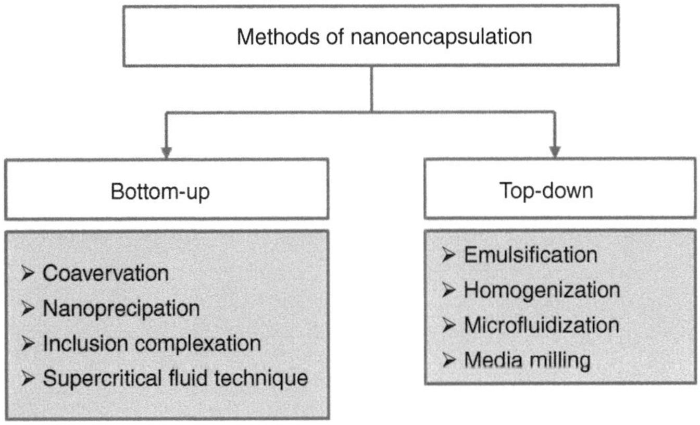

Figure 17.2. Summary of the most important methods for nanoencapsulation of active agents.

approaches, as shown in Fig. 17.2. Top-down processes consist of the size reduction of large particles into smaller ones by using different wet-milling techniques, such as high-pressure homogenization, microfluidization, and media milling, with no need of aggressive solvents. However, all top-down processes involve high-energy requirements, and they are expensive and tedious in some cases. In the bottom-up approach, the active agent is dissolved in an organic solvent and is further precipitated by the addition of an antisolvent in the presence of stabilizers. Some adaptations of thc bottom-up approach have become popular in the past few years, including supercritical fluid processes, emulsion-solvent evaporation, and spray drying (Venkataraman et al., 2011).

In general terms, techniques used to obtain nanoencapsulated systems are more complex than those used for microencapsulation. This situation is mainly due to difficulties in attaining a complex morphology for capsules at the nanoscale and the demand of highly controlled releasing rates for nanoencapsulated active compounds. Emulsification, coacervation, inclusion complexation, emulsification-solvent evaporation, nanoprecipitation, electrospining, and supercritical fluid techniques are considered as nanoencapsulation techniques since they can produce capsules in the nanometer range (10–1000 nm). All these methods are summarized in Table 17.2 and will be discussed in the following sections.

4.1 Coacervation

Coacervation is one of the most popular techniques for production of polysaccharides-based nanocapsules and delivery

Table 17.2 Summary of Methods Used for Flavor Nanoencapsulation

Method	Active Compound	Wall Material	Average Size (nm)	Encapsulation Efficiency (%)	Reference
Spray drying	*L. Sidoides* EO	Alginate/ cashew gum	223–399	55	de Oliveira et al. (2014)
	Eugenol & thymol EO	Zein/casein	107–182	21–26	Chen and Zhong (2015)
Oil-in-water emulsion and ionic gelation	Oregano EO	Chitosan	40–80	21–47	Hosseini et al. (2013)
Emulsification	Peppermint oil	Zein/GA	<150	54–89	Chen and Zhong (2015)
	Olive leaf extract	Pectin	6–675	—	Mohammadi et al. (2016)
	Fish oil	Chitosan	332	74	Esquerdo et al. (2015)
Complex coacervation	Jasmine EO	Gum arabic/ gelatin	74–384	—	Lv et al. (2014)
	L. Sidoides EO	Cashew gum/ chitosan	335–558	70	Abreu et al. (2012)
Ionic gelation	*M. piperita* EO	Chitosan/ cinamic acid	<100	—	Beyki et al. (2014)
	Oregano EO	Chitosan	40–80	—	Hosseini et al. (2013)
	Cuminum cyminun EO	Chitosan/ caffeic acid	100	85	Zhaveh et al. (2015)
Precipitation	*Rosemary EO*	PCL	230	99	Ephrem et al. (2014)

systems but is also used in systems based on proteins and cellulose. The coacervation process is simple and does not require high temperatures or toxic solvents. It involves the phase separation from a solution of a single polyelectrolyte or a mixture of them and the subsequent deposition of the newly-formed coacervation phase around the active agent (Ezhilarasi et al., 2013). It can be divided into two types: simple or complex coacervation.

The simple coacervation process involves the addition of water-miscible solvents and variations in pH or type of electrolyte used in the process, while the complex coacervation comprises a simultaneous desolvation induced by changes in pH caused by oppositely charged colloids. Complex coacervation is an excellent method for nanoencapsulation since it consists of only one process, shows high loading capacity and obtains highly stable nanocapsules due to the electrostatic attraction of oppositely charged molecules, eventually leading to a solvent-rich and a biopolymer-rich phase (Kaushik et al., 2015).

One example of the use of coacervation as a processing method to obtain nanocapsules for active agents delivery was focused on the preparation of nanogels based on chitosan and cashew gum with *L. sidoides* EO (Abreu et al., 2012). The obtained nanogels were characterized in their morphologies, particle size distribution, zeta potential, structure by Fourier transform infrared spectroscopy, and essential oil contents. Results showed that the optimal formulation was based on a 10:2 matrix:oil ratio with gum:chitosan 1:1 and 5% gum concentration resulting in high loading (11.8%) and encapsulation efficiency (70%). Nanogels exhibited average sizes in the 335–558 nm range. Polymer complexes based on gum:chitosan mixtures were prepared by adding the cashew gum to a chitosan solution by using a peristaltic pump, with a flow rate of 1.0 mL/min. The emulsion was separately prepared by using oil and Tween and was slowly added to the polymers complex. The turbid solution was then spray-dried operating at 170°C inlet temperature, 65°C outlet temperature, 5 mL/min pump feed flow, 35 m^3/h air volume flow, and 84 L/h aspirator flow meter.

4.2 Spray Drying

Spray drying is one of the most common technologies for encapsulation. It is a fast, relatively cheap and highly reproducible technique. The principle of spray drying is based on dissolving or dispersing the active ingredient in a solution of the biopolymer matrix. The dispersion is then atomized in a heated air chamber, which rapidly removes the solvent and produces dry particles consisting of the active agent embedded in a porous biopolymer structure. Nevertheless, this method has some limitations for highly volatile or thermo-sensitive active compounds (Fathi et al., 2014). de Oliveira et al. prepared alginate/cashew gum nanoparticles by spray drying, to develop a biopolymer blend for encapsulation of *L. sidoides* EO (de Oliveira et al., 2014). Nanoparticles in solution were found to have average sizes in the 223–399 nm range, and zeta potential values ranging from −30 to −36 mV. Encapsulated EO levels varied from 1.9 to 4.4% with encapsulation efficiency of up to 55%. Chen and Zhong studied the nanoencapsulation of eugenol and thymol in zein/casein blends at pH 6.0–8.0 (Chen and Zhong, 2015). 50 mL of 50% (v/v) ethanol aqueous solution was adjusted to the selected pH with 5 mM sodium phosphate and this solution was further mixed with zein, NaCas, thymol, and eugenol. Finally, pH was readjusted if needed and the mixture was homogenized at 10,000 rpm for 2 min. The dispersion was then spray-dried at 105°C inlet temperature, 60°C outlet temperature, 15% feed rate and 100% aspirator settings. The dry powder was collected and stored at –20°C. Authors concluded that the nanocapsules dimensions were larger at pH 6.0. The optimal conditions proposed by authors resulted in nanoparticles with a size lower than 200 nm. Preencapsulation in nanoparticles resulted in better retention of volatile EO in spray-dried powder than in direct incorporation (Chen et al., 2015).

4.3 Nanoprecipitation

The nanoprecipitation method is also called solvent displacement or interfacial deposition and it is a facile, mild, and low-energy input process for the preparation of polymeric nanoparticles. It is based on the mixture of active agents with adequate solvents through rapid mixing, stabilizing an insoluble low molecular weight compound in a nanosized, polymer-stabilized delivery vehicle (Pustulka et al., 2013). The solvent phase consists of a mixture of film-forming substances, such as polymers, proteins, or lipids, and active substances. Otherwise, the second phase consists of a nonsolvent supplemented with one or more natural or synthetic surfactants. In most cases, the solvent and nonsolvent phases are called organic and aqueous phases, respectively. However, it is possible to use either two organic phases or two aqueous phases as long as solubility and miscibility conditions are satisfied (Mora-Huertas et al., 2010). This procedure is based on the mixture of the organic and the aqueous phases to form a colloidal suspension. If the mixing conditions (flow and agitation of the aqueous phase) are optimized, it could be possible to obtain small-size particles, most of them smaller than 100 nm diameter (Tapia-Hernández et al., 2015).

Some authors have studied nanoprecipitation for nanoencapsulation of EO. Ephrem et al. applied the nanoprecipitation method to prepare PCL-based nanocapsules loaded with rosemary EO at laboratory scale and then the optimal formulation was scaled-up (×6) using a membrane contactor technique (Ephrem et al., 2014). Results showed that nanocapsules prepared at both scales were spherical, with an average size around 230 nm, negative zeta-potential around 20 mV, good stability over time, and encapsulation efficiency higher than 99% for most of the rosemary EO components. In general terms, nanoprecipitation can be a suitable method for the preparation of EO-loaded nanocapsules (Ephrem et al., 2014).

4.4 Inclusion Complexation

This processing technique is mainly used in the encapsulation of highly volatile organic compounds in order to mask undesirable odors while preserving aromas. Inclusion complexation generally refers to formation of a supra-molecular association of a ligand (active agent) in a highly porous polymer structure through van der Waals and hydrogen bonding as well as entropy-driven hydrophobic effects. In the food industry these kinds of polymers for inclusion complexation are unusual, but the use of this technique has increased since the introduction of cyclodextrins as encapsulant agents (Ezhilarasi et al., 2013). The inclusion complexes could increase the solubility and limit the undesired odors of guest molecules, while protecting them from oxidation, light-induced reactions, volatilization, sublimation, and heat-promoted decomposition (Zhang et al., 2015). The development of CD complexes is an interesting way to increase the solubility of EO in water solutions.

The ability of CD to form inclusion complexes with active molecules depends mainly on two factors, the size relationships between CD and the guest molecule and thermodynamic interactions occurring between both components (Wang et al., 2015). If there are important differences in size between CD and guest, it does not fit correctly into the CD cavity and no stable systems are obtained. The smaller the size of the active agent the more stable complex, but the agent's release is faster than in the case of other nanocapsules. As an example, the nanoencapsulation by inclusion complexation of *Citrus sinensis* EO (CSEO) with β-CD was studied (Galvão et al., 2015). The inclusion complexes were obtained with 1:1 β-CD:CSEO, molar ratio either in distilled water or 1:1 ethanol aqueous solution. β-CD and CSEO were treated with the selected solvent and further dried in a desiccator at room temperature until a glass film was formed. Authors reported that CSEO was effectively complexed in β-CD and complexes were fully characterized in their structural, thermal, and water absorption properties.

4.5 Emulsification Methods

Emulsification involves the formation of an oil–water emulsion by dispersion of a lipid in a polymer solution in the presence of surfactants. These methods were used, for example, to prepare chitosan nanocapsules containing concentrates of unsaturated

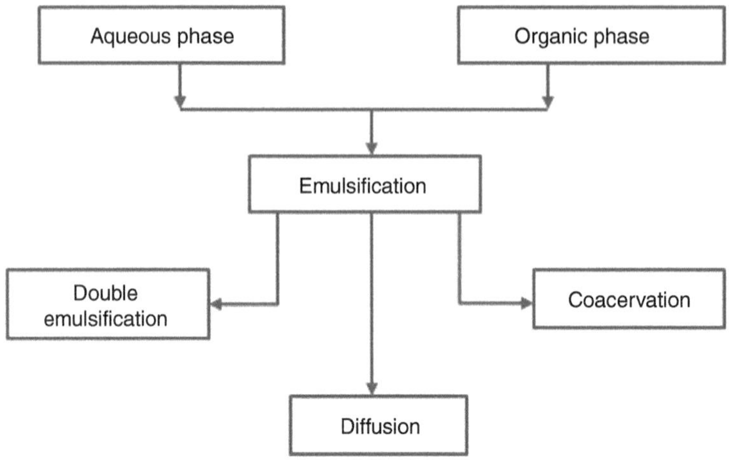

Figure 17.3. Nanoencapsulation of active agents by the emulsification method.

fatty acids (Esquerdo et al., 2015). Emulsification can be combined with different processes, such as the diffusion or coacervation methods (Fig. 17.3) where the first step is common, consisting of the emulsification of the wall material solution containing the substance to be encapsulated.

Multiple emulsions can be used to improve the encapsulation efficiency. In particular, double emulsions are used, where the dispersed phase is an emulsion itself formed by small droplets. Two types of double emulsions can be considered, oil-in-water-in-oil (O/W/O) and water-in-oil-in-water (W/O/W). In O/W/O, oil is first dispersed in an aqueous phase and then this emulsion is dispersed in a second oil phase. In the same way, in W/O/W water is first dispersed in oil and then this emulsion is dispersed in water. Two different strategies to prepare double emulsions have been proposed. One of them consists of heating one of these emulsions with a nonionic emulsifier or a mixture of them, resulting in a phase inversion and leading to the formation of a double emulsion. Another process can be carried out in two stages where the initial step consists in the formation of a primary emulsion, involving the use of a surfactant with low hydrophilic-lipophilic balance (HLB) in the case of W/O/W emulsions, and high HLB in the case of O/W/O emulsions. In the second step, water/oil emulsions are dispersed in an aqueous continuous phase containing a surfactant with high HLB to form W/O/W double emulsions (Lamba et al., 2015). A recent report by Mohammadi et al. indicated that nanoencapsulation by W/O nanoemulsions and W/O/W double emulsions could be a suitable novel technique to improve the antioxidant efficiency of natural ingredients as the olive leaf extract (Mohammadi et al., 2016).

In the case of the emulsion-diffusion method, the emulsification process involves the addition of an organic solvent partially soluble in water to promote the initial thermodynamic equilibrium between the two phases. The wall material and the active compound are dissolved in the saturated organic solvent phase, while the aqueous phase is previously saturated with the solvent and a stabilizer. The addition of water to the system causes the solvent to diffuse into the external phase, resulting in the deposition of the interfacial material to form the nanocapsules. The organic solvent can be further evaporated under reduced pressure (Kakran and Antipina, 2014). The advantages of this method include high reproducibility and encapsulation efficiencies, better control of the particle size with low dispersion, and possibilities for scaling-up. One example was proposed for the nanoencapsulation of eugenol, using PCL as wall material (Choi et al., 2009).

Finally, the emulsion-coacervation process is mainly used as the most adequate strategy for the preparation of nanocapsules in biopolymers formulations. This procedure involves the formation of an O/W emulsion formed by an organic phase (oil, active substances, and solvents if necessary) with an aqueous phase (water, biopolymers, and stabilizing agents) by mechanical or ultrasound stirring. Then, a simple coacervation process is performed by using electrolytes (Mora-Huertas et al., 2010). The advantages of this processing strategy include high encapsulation efficiencies, high reproducibility, better control of the particle size and ease for scaling-up (Choi et al., 2009).

4.6 Supercritical Fluids

Nanoencapsulation of active compounds by using processing methods based on the use of supercritical fluids have attracted increasing attention in the past few years. Carbon dioxide is widely used in this application because the supercritical region can be achieved at moderate temperatures and pressures making it suitable for many processes at a reasonable cost. Supercritical fluids are normally used for the encapsulation of thermally sensitive active compounds, where they are dissolved together with the polymer matrix in the supercritical fluid and the resulting solution is further expanded through a nozzle. Then, the supercritical fluid evaporates by a spraying process, similar to spray drying, and the obtained nanoparticles precipitate. This technique has been widely used in encapsulation strategies due to its low critical temperature and the minimum use of organic solvents, making it an environmentally friendly processing technique (Ezhilarasi et al., 2013). On the other hand, the application of this method facilitates the separation of the supercritical fluids from the product by depressurization, while the high solubility of organic solvents in the most usual supercritical fluids allows obtaining solvent-free nanocapsules and producing particles with controlled morphology and narrow size distribution (Fathi et al., 2014). Several recent studies focused on the supercritical fluid-assisted synthesis of functional nanostructured materials, especially in the field of inorganic and hybrid materials, such as nanocoatings, nanofibers, nanowires, nanotubes, dendrimers, mesoporous nanoparticles, and nanostructures (Gutiérrez et al., 2013). These techniques have been also used to encapsulate β-carotene and lycopene in modified starches obtaining particle sizes around 100 nm without significant degradation of the active agent. Furthermore, the carotenoid was present in the amorphous part of the polymer structure, rather than crystalline form, which could improve its bioavailability (Santos et al., 2012).

4.7 Electrospray and Electrospinning

The principle of the electrospray is based on the capacity of an electric field to deform the interface of one particular material to get drops in the micrometer or nanometer range, depending on the experimental parameters. Although the electrospray technique is simple to perform and easy to implement in encapsulation strategies, different experimental variables must be controlled to obtain the desirable particle dimensions and size distributions. These variables involve the main system parameters, such as viscosity of solutions, collector distance, flow rates, applied electric potential, the desirable molecular weight for the capsules, conductivity, surface tension, and polymer concentration (Tapia-Hernández et al., 2015).

The principle of electrospinning is based on the use of electrostatic forces to generate polymer fibers. The equipment for electrospinning consists of a capillary through which the liquid polymer solution is pumped; a high-voltage source with positive or negative polarity to generate a charge in the polymer solution, and a grounded collector, which is brought into contact with the counter electrode. Different authors have applied this technique for the encapsulation of active agents in recent years, for example by the incorporation of ethyl vanillin to a PVA matrix (Fathi et al., 2014). In this case, the working solution (20% m/v) of ethyl vanillin and PVA solution in water was fixed horizontally on the syringe pump and the solutions were electrospun by high-voltage power supply to obtain homogeneous films stable at room temperature. Authors concluded that the morphology and fibers diameters were strongly influenced by the immobilized flavor.

4.8 Ionic Gelation

This method is based in the addition of an anionic cross-linking agent to polymer aqueous solutions. This technique has been successfully applied to the preparation of nanocapsules in chitosan matrices, where positively charged nanoparticles are formed through the inter- and intra-cross-linking of the chitosan amino groups and form links with the negatively charged anionic cross-linking agent (Hassani et al., 2015). This method has been extensively studied, including the effect of several parameters on the particle size distributions and the stability of the chitosan nanoparticles. Their size is dependent on the ionic strength and pH of the solution.

Ionic gelation has been applied to the incorporation of *Mentha piperita* cinnamic acid to chitosan (Beyki et al., 2014). Nanogels were prepared for the preservation of the active agent and to increase control over the release of *M. piperita* EO. Nanogels are formed spontaneously by formation of covalent bonds between the primary amino groups in chitosan chains and the carboxyl groups of the fatty acids in the EO. The nanosized polymeric gel with narrow size distributions was obtained by self-aggregation in aqueous medium and was further characterized by electronic microscopy and infrared spectroscopy. The superior performance of *M. piperita* EO when encapsulated in chitosan nanogels under sealed or nonsealed conditions in comparison with the direct addition of free oils to food was demonstrated.

5 Nanoencapsulation in Food Packaging Applications

Many examples of the use of nanocapsules to contain and release active agents in food packaging applications have been introduced all through this chapter. In this section, a summary of the most relevant reported work in this area is discussed. It is known that flavors and aromas have been historically incorporated by direct addition during processing into natural or synthetic polymers to form active packaging films. However, it is usual that high temperatures during processing and environmental external factors may result in the partial or complete degradation, oxidation, or evaporation of these volatile substances from the polymer matrices. It has been stated all through this chapter that nanoencapsulation technologies provide beneficial effects on the flavors' properties, enhancing their solubility in water-soluble polymers, ensuring good stabilization during processing, storage, and distribution, reducing their impact on food organoleptic characteristics and also offering high loading capacity with effective release properties. However, the mechanical, morphological, thermal, and gas barrier properties of the active films could be altered by the presence of nanocapsules and should be studied.

In the past few years, a limited number of studies about the effects of the incorporation of nanoencapsulated flavors into the physical and functional properties of polymer films for food packaging applications have been reported. In this context, nanoemulsions of clove bud EO and oregano EO (diameters 180–250 nm) have been successfully incorporated into methylcellulose matrices, improving their antimicrobial efficiency against yeast and molds in sliced bread during 15 days (Otoni et al., 2014b). Their efficiency in protecting bread was much higher than results obtained with coarse emulsions (diameters 1.3–1.9 μm) and other commercial agents currently used in the bakery industry. However, nanoemulsions reduced the rigidity of films due to their plasticizing effect, resulting in a loss of mechanical properties and resistance to environmental factors. In the same way, the incorporation of cinnamon EOs nanoliposomes to fish gelatin resulted in more flexible films accompanied with a decrease of the tensile strength and increase in water vapor permeability and flavor release rate that promoted some improvement in the antimicrobial stability (Wu et al., 2015). An opposite effect on the mechanical properties of polymer films was observed with the incorporation of orange EO and limonene nanoliposomes to starch-NaCas (50:50) films and cinnamaldehyde nanoemulsions in edible films based on pectin and papaya puree (Jiménez et al., 2014). In the last case, nanoemulsions reduced the plasticizer effect of papaya pure and increased the barrier effect to water vapor in pectin films. Antimicrobial properties of cinnamaldehyde against *E. coli, Salmonella enterica, L. monocytogenes,* and *S. aureus* were also improved by the droplet size reduction achieved by increasing stirring energies. The increased surface area produced the higher delivery of flavors (Otoni et al., 2014a).

The incorporation of cinnamon EO/β-CD to PVA produced electrospun nanofibrous films to improve their hydrophilic character and the thermal stability of the EO, masking their special flavor. In addition, these films resulted in some improvement in the antimicrobial activity against both Gram-positive and Gram-negative bacteria, compared with results obtained by casting. These nanoencapsulation strategies resulted in the effective extension of shelf life of strawberries, indicating its potential application in active food packaging (Wen et al., 2016).

A recent trend to improve the positive effects of natural active agents in food is the use of active coatings on packaging materials. In this sense, the effect of coating the inner surface of LDPE with cinnamon EO encapsulated in chitosan nanoparticles (CE-NPs) at three different particle sizes (112, 215, and 527 nm) was evaluated on the preservation of chilled pork (Hu et al., 2015). Authors reported that CE-NPs with the largest size exhibited the highest antimicrobial and antioxidant performance after 15 days of refrigerated storage, showing high potential as a natural preservative of meat products. In another study, cinnamaldehyde nanoliposomes showed significant antimicrobial effect against *E. coli* and *B. cereus* when coated on glass surfaces, showing a potential application in preserving liquid foods (Makwana et al., 2014). However, no antimicrobial effect was observed when these nanoliposomes were coated onto PLA surfaces.

Nanoparticles of eugenol encapsulated in chitosan were incorporated into thermoplastic flour (TPF) by using extrusion at temperatures above 150°C (Woranuch and Yoksan, 2013b). It was observed that 3% (w/w) of these nanoparticles provided antioxidant activity and high water vapor permeability to TPF films, while decreased their flexibility, oxygen gas barrier properties, and degradation temperatures.

In another study, carvacrol nanoemulsions incorporated into modified edible chitosan and coated directly onto green beans provided antimicrobial activity against two pathogens during storage (Severino et al., 2015). It was reported that the combination of a treatment with gamma irradiation, coating, and modified atmosphere packaging (60% O_2, 30% CO_2, 10% N_2) resulted in high antimicrobial efficiency.

6 Concluding Remarks

All results presented in this chapter indicate that the incorporation of nanoencapsulated flavors into polymer and biopolymer films is a highly promising field to increase food shelf life and quality. This potential is highlighted by the reduction of the undesirable effects of oxidation and microbiological degradation that deplete the potential of packaged food to be preserved safely for longer times than those currently used in food distribution from sources to the consumer. The encapsulation strategies have shown their possibilities in the protection of volatile flavors and aromas and make it easier to control their sustained release effect from the packaging material to food in active packaging formulations. Nevertheless, much work is still necessary to overcome the inherent problems of these strategies, such as an increase in costs and loss of some properties essential for the packaging materials, for an effective scale-up to food industry.

Acknowledgments

Authors thank the Spanish Ministry of Economy and Competitiveness for financial support (Ref. MAT2014-59242-C2-2-R).

References

Abreu, F.O.M.S., Oliveira, E.F., Paula, H.C.B., de Paula, R.C.M., 2012. Chitosan/cashew gum nanogels for essential oil encapsulation. Carbohyd. Polym. 89, 1277–1282.

Asbahani, A.E., Miladi, K., Badri, W., Sala, M., Addi, E.H.A., Casabianca, H., Mousadik, A.E., Hartmann, D., Jilale, A., Renaud, F.N.R., Elaissari, A., 2015. Essential oils: from extraction to encapsulation. Int. J. Pharm. 483, 220–243.

Aumeeruddy-Elalfi, Z., Gurib-Fakim, A., Mahomoodally, F., 2015. Antimicrobial, antibiotic potentiating activity and phytochemical profile of essential oils from exotic and endemic medicinal plants of Mauritius. Ind. Crop. Prod. 71, 197–204.

Aytac, Z., Dogan, S.Y., Tekinay, T., Uyar, T., 2014. Release and antibacterial activity of allyl isothiocyanate/β-cyclodextrin complex encapsulated in electrospun nanofibers. Coll. Surf. B 120, 125–131.

Beyki, M., Zhaveh, S., Khalili, S.T., Rahmani-Cherati, T., Abollahi, A., Bayat, M., Tabatabaei, M., Mohsenifar, A., 2014. Encapsulation of *Mentha piperita* essential oils in chitosan-cinnamic acid nanogel with enhanced antimicrobial activity against *Aspergillus flavus*. Ind. Crop. Prod. 54, 310–319.

Blanco-Padilla, A., Soto, K.M., Hernández-Iturriaga, M., Mendoza, S., 2014. Food antimicrobials nanocarriers. Sci. World J. 2014, 1–11.

Calo, J.R., Crandall, P.G., O'Bryan, C.A., Ricke, S.C., 2015. Essential oils as antimicrobials in food systems: a review. Food Control 54, 111–119.

Camerlo, A., Vebert-Nardin, C., Rossi, R.M., Popa, A.M., 2013. Fragrance encapsulation in polymeric matrices by emulsion electrospinning. Eur. Polym. J. 49, 3806–3813.

Carocho, M., Barreiro, M.F., Morales, P., Ferreira, I.C.F.R., 2014. Adding molecules to food, pros and cons: a review on synthetic and natural food additives. Compr. Rev. Food Sci. F. 13, 377–399.

Carrasco, A., Ortiz-Ruiz, V., Martinez-Gutierrez, R., Tomas, V., Tudela, J., 2015. *Lavandula stoechas* essential oil from Spain: Aromatic profile determined by gas chromatography-mass spectrometry, antioxidant, and lipoxygenase inhibitory bioactivities. Ind. Crop. Prod. 73, 16–27.

Chang, Y., McClements, D.J., 2014. Optimization of orange oil nanoemulsion formation by isothermal low-energy methods: influence of the oil phase, surfactant, and temperature. J. Agric. Food. Chem. 62, 2306–2312.

Chang, Y., McLandsborough, L., McClements, D.J., 2013. Physicochemical properties and antimicrobial efficacy of carvacrol nanoemulsions formed by spontaneous emulsification. J. Agric. Food. Chem. 61, 8906–8913.

Chen, H., Zhang, Y., Zhong, Q., 2015. Physical and antimicrobial properties of spray-dried zein-casein nanocapsules with coencapsulated eugenol and thymol. J. Food Eng. 144, 93–102.
Chen, H., Zhong, Q., 2015. A novel method of preparing stable zein nanoparticle dispersions for encapsulation of peppermint oil. Food Hydrocoll. 43, 593–602.
Choi, M.J., Soottitantawat, A., Nuchuchua, O., Min, S.G., Ruktanonchai, U., 2009. Physical and light oxidative properties of eugenol encapsulated by molecular inclusion and emulsion-diffusion method. Food Res. Int. 42, 148–156.
da Rosa, C.G., de Oliveira Brisola Maciel, M.V., de Carvalho, S.M., de Melo, A.P.Z., Jummes, B., da Silva, T., Martelli, S.M., Villetti, M.A., Bertoldi, F.C., Barreto, P.L.M., 2015. Characterization and evaluation of physicochemical and antimicrobial properties of zein nanoparticles loaded with phenolics monoterpenes. Coll. Surf. A 481, 337–344.
de Oliveira, E.F., Paula, H.C.B., Paula, R.C.M., 2014. Alginate/cashew gum nanoparticles for essential oil encapsulation. Coll. Surf. B 113, 146–151.
Donsì, F., Cuomo, A., Marchese, E., Ferrari, G., 2014. Infusion of essential oils for food stabilization: unraveling the role of nanoemulsion-based delivery systems on mass transfer and antimicrobial activity. Innov. Food Sci. Emerg. 22, 212–220.
EC Decision 2002/113/. Commission Decision of 23 January 2002 amending Commission Decision 1999/217/EC as regards the register of flavoring substances used in or on foodstuffs. Official Journal of the European Communities, February 20, 2002.
EC Directive 88/388/. Council Directive of 22 June 1988 on the approximation of the laws of the Member States relating to flavorings for use in foodstuffs and to source materials for their production. Official Journal of the European Communities, July 15, 1988.
EC Regulation 10/2011 on plastic materials and articles intended to come into contact with food. Official Journal of the European Communities, January 15, 2011, 1.
EC Regulation 450/2009 on active and intelligent materials and articles intended to come into contact with food. Official Journal of the European Union, June 29, 2009.
EC Regulation 1935/2004 on materials and articles intended to come into contact with food and repealing Directives 80/590/EEC and 89/109/EEC. Official Journal of the European Union, October 27, 2004.
Eltayeb, M., Bakhshi, P.K., Stride, E., Edirisinghe, M., 2013. Preparation of solid lipid nanoparticles containing active compound by electrohydrodynamic spraying. Food Res. Int. 53, 88–95.
Ephrem, E., Greige-Gerges, H., Fessi, H., Charcosset, C., 2014. Optimization of rosemary oil encapsulation in polycaprolactone and scale-up of the process. J. Microencap. 31, 746–753.
Esquerdo, V.M., Dotto, G.L., Pinto, L.A.A., 2015. Preparation of nanoemulsions containing unsaturated fatty acid concentrate-chitosan capsules. J. Colloid Interf. Sci. 445, 137–142.
Ezhilarasi, P.N., Karthik, P., Chhanwal, N., Anandharamakrishnan, C., 2013. Nanoencapsulation techniques for food bioactive components: a review. Food Bioproc. Tech. 6, 628–647.
FAO., 2010. Expert meeting on the application of nanotechnologies in the food and agriculture sectors: potential food safety implications. Meeting report.
Fathi, M., Martín, Á., McClements, D.J., 2014. Nanoencapsulation of food ingredients using carbohydrate based delivery systems. Trends Food Sci. Tech. 39, 18–39.
Fathi, M., Mozafari, M.R., Mohebbi, M., 2012. Nanoencapsulation of food ingredients using lipid-based delivery systems. Trends Food Sci. Tech. 23, 13–27.

Fuenmayor, C.A., Mascheroni, E., Cosio, M.S., Piergiovanni, L., Benedetti, S., Ortenzi, M., Schiraldi, A., Mannino, S., 2013. Encapsulation of R-(+)-limonene in edible electrospun nanofibers. Chem. Eng. Trans. 32, 1771–1776.

Galvão, J.G., Silva, V.F., Ferreira, S.G., França, F.R.M., Santos, D.A., Freitas, L.S., Alves, P.B., Araújo, A.A.S., Cavalcanti, S.C.H., Nunes, R.S., 2015. β-cyclodextrin inclusion complexes containing *Citrus sinensis* (L) Osbeck essential oil: an alternative to control *Aedes aegypti* larvae. Thermochim. Acta 608, 14–19.

Ghayempour, S., Mortazavi, S.M., 2014. Antibacterial activity of peppermint fragrance micro-nanocapsules prepared with a new electrospraying method. J. Essent. Oil Res. 26, 492–498.

Ghosh, V., Saranya, S., Mukherjee, A., Chandrasekaran, N., 2013. Cinnamon oil nanoemulsion formulation by ultrasonic emulsification: investigation of its bactericidal activity. J. Nanosci. Nanotechnol. 13, 114–122.

Gomes, C., Moreira, R.G., Castell-Perez, E., 2011. Poly (DL-lactide-*co*-glycolide) (PLGA) nanoparticles with entrapped trans-cinnamaldehyde and eugenol for antimicrobial delivery applications. J. Food Sci. 76, N16–N24.

Grama, C., Venkatpurwar, V., Lamprou, D., Ravi Kumar, M.N.V., 2013. Towards scale-up and regulatory shelf-stability testing of curcumin-encapsulated polyester nanoparticles. Drug Deliv. Transl. Res. 3, 286–293.

Gutiérrez, F.J., Albillos, S.M., Casas-Sanz, E., Cruz, Z., García-Estrada, C., García-Guerra, A., García-Reverter, J., García-Suárez, M., Gatón, P., González-Ferrero, C., Olabarrieta, I., Olasagasti, M., Rainieri, S., Rivera-Patiño, D., Rojo, R., Romo-Hualde, A., Sáiz-Abajo, M.J., Mussons, M.L., 2013. Methods for the nanoencapsulation of β-carotene in the food sector. Trends Food Sci. Technol. 32, 73–83.

Hassani, S., Laouini, A., Fessi, H., Charcosset, C., 2015. Preparation of chitosan-TPP nanoparticles using microengineered membranes: effect of parameters and encapsulation of tacrine. Colloid Surf. A 482, 34–43.

Hill, L.E., Taylor, T.M., Gomes, C., 2013. Antimicrobial efficacy of poly (DL-lactide-*co*-glycolide) (PLGA) nanoparticles with entrapped cinnamon bark extract against *Listeria monocytogenes* and *Salmonella typhimurium*. J. Food Sci. 78, N626–N632.

Hosseini, S.F., Zandi, M., Rezaei, M., Farahmandghavi, F., 2013. Two-step method for encapsulation of oregano essential oil in chitosan nanoparticles: preparation, characterization, and in vitro release study. Carbohyd. Polym. 95, 50–56.

Hu, J., Wang, X., Xiao, Z., Bi, W., 2015. Effect of chitosan nanoparticles loaded with cinnamon essential oil on the quality of chilled pork. LWT – Food Sci. Technol. 63, 519–526.

Jiménez, A., Sánchez-González, L., Desobry, S., Chiralt, A., Tehrany, E.A., 2014. Influence of nanoliposomes incorporation on properties of film forming dispersions and films based on corn starch and sodium caseinate. Food Hydrocoll. 35, 159–169.

Joven, R., Garcia, A., Arias, A., Medina, J., 2015. Development of an active thermoplastic film with oxygen scavengers made of activated carbon and sodium erythorbate. Packag. Technol. Sci. 28, 113–121.

Júnior, A.V., Fronza, N., Foralosso, F.B., Dezen, D., Huber, E., dos Santos, J.H.Z., Machado, R.A.F., Quadri, M.G.N., 2014. Biodegradable duo-functional active film: antioxidant and antimicrobial actions for the conservation of beef. Food Bioproc. Tech. 8, 75–87.

Kakran, M., Antipina, M.N., 2014. Emulsion-based techniques for encapsulation in biomedicine, food, and personal care. Curr. Opin. Pharmacol. 18, 47–55.

Kaushik, P., Dowling, K., Barrow, C.J., Adhikari, B., 2015. Complex coacervation between flaxseed protein isolate and flaxseed gum. Food Res. Int. 72, 91–97.

Kayaci, F., Ertas, Y., Uyar, T., 2013. Enhanced thermal stability of eugenol by cyclodextrin inclusion complex encapsulated in electrospun polymeric nanofibers. J. Agric. Food. Chem. 61, 8156–8165.

Kayaci, F., Sen, H.S., Durgun, E., Uyar, T., 2014. Functional electrospun polymeric nanofibers incorporating geraniol-cyclodextrin inclusion complexes: high thermal stability and enhanced durability of geraniol. Food Res. Int. 62, 424–431.

Khalili, S.T., Mohsenifar, A., Beyki, M., Zhaveh, S., Rahmani-Cherati, T., Abdollahi, A., Bayat, M., Tabatabaei, M., 2015. Encapsulation of thyme essential oils in chitosan-benzoic acid nanogel with enhanced antimicrobial activity against *Aspergillus flavus*. LWT—Food Sci. Technol. 60, 502–508.

Khare, A.R., Vasisht, N., 2014. Nanoencapsulation in the food industry: technology of the future. In: Gaonkar, A., Vasisht, N., Khare, A., Sobel, R. (Eds.), Microencapsulation in the Food Industry: A Practical Implementation Guide. Academic Press, San Diego, pp. 151–156.

Komaiko, J., McClements, D.J., 2015. Low-energy formation of edible nanoemulsions by spontaneous emulsification: factors influencing particle size. J. Food Eng. 146, 122–128.

Labuza, T.P., 1987. Food research. Cereal Food World 32, 338–339.

Lamba, H., Sathish, K., Sabikhi, L., 2015. Double emulsions: emerging delivery system for plant bioactives. Food Bioproc. Technol. 8, 709–728.

Liakos, I., Rizzello, L., Scurr, D.J., Pompa, P.P., Bayer, I.S., Athanassiou, A., 2014. All-natural composite wound dressing films of essential oils encapsulated in sodium alginate with antimicrobial properties. Int. J. Pharma. 463, 137–145.

Lv, Y., Yang, F., Li, X., Zhang, X., Abbas, S., 2014. Formation of heat-resistant nanocapsules of jasmine essential oil via gelatin/gum arabic-based complex coacervation. Food Hydrocoll. 35, 305–314.

Makwana, S., Choudhary, R., Dogra, N., Kohli, P., Haddock, J., 2014. Nanoencapsulation and immobilization of cinnamaldehyde for developing antimicrobial food packaging material. LWT—Food Sci. Technol. 57, 470–476.

Marques, H.M.C., 2010. A review on cyclodextrin encapsulation of essential oils and volatiles. Flavour Frag. J. 25, 313–326.

Martucci, J.F., Gende, L.B., Neira, L.M., Ruseckaite, R.A., 2015. Oregano and lavender essential oils as antioxidant and antimicrobial additives of biogenic gelatin films. Ind. Crop. Prod. 71, 205–213.

Mohammadi, A., Hashemi, M., Hosseini, S.M., 2015. Nanoencapsulation of *Zataria multiflora* essential oil preparation and characterization with enhanced antifungal activity for controlling *Botrytis cinerea*, the causal agent of gray mould disease. Innov. Food Sci. Emerg. 28, 73–80.

Mohammadi, A., Jafari, S.M., Esfanjani, A.F., Akhavan, S., 2016. Application of nano-encapsulated olive leaf extract in controlling the oxidative stability of soybean oil. Food Chem. 190, 513–519.

Mora-Huertas, C.E., Fessi, H., Elaissari, A., 2010. Polymer-based nanocapsules for drug delivery. Int. J. Pharm. 385, 113–142.

Neethirajan, S., Jayas, D.S., 2011. Nanotechnology for the food and bioprocessing industries. Food Bioproc. Tech. 4, 39–47.

Nilsen-Nygaard, J., Strand, S.P., Vårum, K.M., Draget, K.I., Nordgård, C.T., 2015. Chitosan: gels and interfacial properties. Polymers 7, 552–579.

Otoni, C.G., Moura, M.R., Aouada, F.A., Camilloto, G.P., Cruz, R.S., Lorevice, M.V., Soares, N.F.F., Mattoso, L.H.C., 2014a. Antimicrobial and physical-mechanical properties of pectin/papaya puree/cinnamaldehyde nanoemulsion edible composite films. Food Hydrocoll. 41, 188–194.

Otoni, C.G., Pontes, S.F.O., Medeiros, E.A.A., Soares, N.D.F.F., 2014b. Edible films from methylcellulose and nanoemulsions of clove bud (*Syzygium aromaticum*) and oregano (*Origanum vulgare*) essential oils as shelf life extenders for sliced bread. J. Agric. Food. Chem. 62, 5214–5219.

Pan, K., Chen, H., Davidson, P.M., Zhong, Q., 2014. Thymol nanoencapsulated by sodium caseinate: physical and antilisterial properties. J. Agric. Food. Chem. 62, 1649–1657.

Papajani, V., Haloci, E., Goci, E., Shkreli, R., Manfredini, S., 2015. Evaluation of antifungal activity of *Origanum vulgare* and *Rosmarinus officinalis* essential oil before and after inclusion in β-cyclodextrine. Int. J. Pharm. Pharm. Sci. 7, 270–273.

Patel, A.R., Velikov, K.P., 2014. Zein as a source of functional colloidal nano- and microstructures. Curr. Opin. Coll. In. 19, 450–458.

Peng, X.Q., Wei, M.J., Wang, L., Gu, L.P., 2014. Study on microcrystalline chitin cinnamon nanoemulsion. Appl. Mech. Mater. 525, 53–57.

Poaty, B., Lahlah, J., Porqueres, F., Bouafif, H., 2015. Composition, antimicrobial, and antioxidant activities of seven essential oils from the North American boreal forest. World J. Microbiol. Biotechnol. 31, 907–919.

Prakash, B., Kedia, A., Mishra, P.K., Dubey, N.K., 2015. Plant essential oils as food preservatives to control moulds, mycotoxin contamination, and oxidative deterioration of agri-food commodities: potentials and challenges. Food Control 47, 381–391.

Pustulka, K.M., Wohl, A.R., Lee, H.S., Michel, A.R., Han, J., Hoye, T.R., McCormick, A.V., Panyam, J., Macosko, C.W., 2013. Flash nanoprecipitation: particle structure and stability. Mol. Pharmacol. 10, 4367–4377.

Ramos, M., Beltrán, A., Peltzer, M., Valente, A.J.M., Garrigós, M.C., 2014. Release and antioxidant activity of carvacrol and thymol from polypropylene active packaging films. LWT—Food Sci. Technol. 58, 470–477.

Rhim, J.W., Park, H.M., Ha, C.S., 2013. Bio-nanocomposites for food packaging applications. Prog. Polym. Sci. 38, 1629–1652.

Rinaudo, M., 2006. Chitin and chitosan: properties and applications. Prog. Polym. Sci. 31, 603–632.

Salvia-Trujillo, L., Rojas-Grau, M.A., Soliva-Fortuny, R., Martín-Belloso, O., 2014. Impact of microfluidization or ultrasound processing on the antimicrobial activity against *Escherichia coli* of lemongrass oil–loaded nanoemulsions. Food Control 37, 292–297.

Sanchez-Ortega, I., Garcia-Almendarez, B.E., Santos-Lopez, E.M., Amaro-Reyes, A., Barboza-Corona, J.E., Regalado, C., 2014. Antimicrobial edible films and coatings for meat and meat products preservation. Sci. World J. 2014, 1–18.

Santos, D.T., Martín, Á., Meireles, M.A.A., Cocero, M.J., 2012. Production of stabilized sub-micrometric particles of carotenoids using supercritical fluid extraction of emulsions. J. Supercrit. Fluids 61, 167–174.

Sari, T.P., Mann, B., Kumar, R., Singh, R.R.B., Sharma, R., Bhardwaj, M., Athira, S., 2015. Preparation and characterization of nanoemulsion encapsulating curcumin. Food Hydrocoll. 43, 540–546.

Severino, R., Ferrari, G., Vu, K.D., Donsì, F., Salmieri, S., Lacroix, M., 2015. Antimicrobial effects of modified chitosan based coating containing nanoemulsion of essential oils, modified atmosphere packaging and gamma irradiation against *Escherichia coli* O157:H7 and *Salmonella typhimurium* on green beans. Food Control 50, 215–222.

Silva-Buzanello, R.A.D., Ferro, A.C., Bona, E., Cardozo-Filho, L., Araújo, P.H.H.D., Leimann, F.V., Gonçalves, O.H., 2014. Validation of an ultraviolet-visible (UV-Vis) technique for the quantitative determination of curcumin in poly(l-lactic acid) nanoparticles. Food Chem. 172, 99–104.

Sobel, R., Su, C.P., 2014. Novel concepts and challenges of flavor microencapsulation and taste modification. In: Gaonkar, A., Vasisht, N., Khare, A., Sobel, R. (Eds.), Microencapsulation in the Food Industry: A Practical Implementation Guide. Academic Press, San Diego, pp. 1–15.

Souza, M.A., Oliveira, J.E., Medeiros, E.S., Glenn, G.M., Mattoso, L.H.C., 2015. Controlled release of linalool using nanofibrous membranes of poly(lactic acid) obtained by electrospinning and solution blow-spinning: a comparative study. J. Nanosci. Nanotechnol. 15, 5628–5636.

Tapia-Hernández, J.A., Torres-Chávez, P.I., Ramírez-Wong, B., Rascón-Chu, A., Plascencia-Jatomea, M., Barreras-Urbina, C.G., Rangel-Vázquez, N.A., Rodríguez-Félix, F., 2015. Micro- and nanoparticles by electrospray: advances and applications in foods. J. Agric. Food. Chem. 63, 4699–4707.

Tuladhar, A., Anal, A.K., 2014. Recent Advances in Applications of Encapsulation Technology for the Bioprotection of Phytonutrients in Complex Food Systems. Functional Foods and Dietary Supplements. John Wiley & Sons, New York.

Venkataraman, S., Hedrick, J.L., Ong, Z.Y., Yang, C., Ee, P.L.R., Hammond, P.T., Yang, Y.Y., 2011. The effects of polymeric nanostructure shape on drug delivery. Adv. Drug Deliv. Rev. 63, 1228–1246.

Vergis, J., Gokulakrishnan, P., Agarwal, R.K., Kumar, A., 2015. Essential oils as natural food antimicrobial agents: a review. Crit. Rev. Food Sci. Nutr. 55, 1320–1323.

Wang, X., Yuan, Y., Yue, T., 2015. The application of starch-based ingredients in flavor encapsulation. Starch/Staerke 67, 225–236.

Wani, K.D., Kadu, B.S., Mansara, P., Gupta, P., Deore, A.V., Chikate, R.C., Poddar, P., Dhole, S.D., Kaul-Ghanekar, R., 2014. Synthesis, characterization and in vitro study of biocompatible cinnamaldehyde functionalized magnetite nanoparticles (CPGF Nps) for hyperthermia and drug delivery applications in breast cancer. PLoS ONE 9, e107315.

Wen, P., Zhu, D.H., Wu, H., Zong, M.H., Jing, Y.R., Han, S.Y., 2016. Encapsulation of cinnamon essential oil in electrospun nanofibrous film for active food packaging. Food Control 59, 366–376.

Woranuch, S., Yoksan, R., 2013a. Eugenol-loaded chitosan nanoparticles: I. Thermal stability improvement of eugenol through encapsulation. Carbohyd. Polym. 96, 578–585.

Woranuch, S., Yoksan, R., 2013b. Eugenol-loaded chitosan nanoparticles: II. Application in bio-based plastics for active packaging. Carbohyd. Polym. 96, 586–592.

Wu, J., Liu, H., Ge, S., Wang, S., Qin, Z., Chen, L., Zheng, Q., Liu, Q., Zhang, Q., 2015. The preparation, characterization, antimicrobial stability, and in vitro release evaluation of fish gelatin films incorporated with cinnamon essential oil nanoliposomes. Food Hydrocoll. 43, 427–435.

Zhang, W., Li, X., Yu, T., Yuan, L., Rao, G., Li, D., Mu, C., 2015. Preparation, physicochemical characterization and release behavior of the inclusion complex of trans-anethole and β-cyclodextrin. Food Res. Int. 74, 55–62.

Zhaveh, S., Mohsenifar, A., Beiki, M., Khalili, S.T., Abdollahi, A., Rahmani-Cherati, T., Tabatabaei, M., 2015. Encapsulation of *Cuminum cyminum* essential oils in chitosan-caffeic acid nanogel with enhanced antimicrobial activity against *Aspergillus flavus*. Ind. Crop. Prod. 69, 251–256.

Zhu, G., Xiao, Z., Zhou, R., Yi, F., 2012. Fragrance and flavor microencapsulation technology. Adv. Mat. Res. 2, 440–445.

18

POLYPROPYLENE/TALC NANOCOMPOSITES FILMS AS LOW-COST BARRIER MATERIALS FOR FOOD PACKAGING

Karolth Espinosa, Luciana A. Castillo, Silvia E. Barbosa
Planta Piloto de Ingeniería Química, PLAPIQUI (UNS-CONICET), Bahía Blanca, Argentina

1 Role of Food Packaging

The development of packaging has been particularly propelled in the second half of the twentieth century in industrialized and developed countries, even though it has been in existence for centuries. Packaging has evolved from a relatively limited range of heavy and rigid containers using wood, glass, and steel to a broad array of rigid, semi-rigid and flexible packaging options made from lightweight materials. Particularly, the use of flexible packaging grown continuously in the last years. The flexible packaging industry had $31.1 billion sales in the United States in 2014, covering the second-largest packaging segment. The largest market for flexible packaging is food (retail and institutional), accounting for about 60% of shipments. The aforementioned data evidence the important role of packaging in the food market, mainly as help for food preservation. The use of adequate packaging helps to conserve quality, minimize waste, and reduce the amount of preservatives. In this sense, besides serving to contain food, packaging protects it against chemical and physical damage, and also provide information essential to consumers and marketers (Flexible Packaging Association, 2015). On the other hand, packaging is a key component of the modern lifestyle. Examples of the functionalities that it offers are the possibility to package individually small servings as well as microwavable meals. However, rising population demands concerning food quality make the packaging market

Food Packaging. http://dx.doi.org/10.1016/B978-0-12-804302-8.00018-2

more competitive and innovative, without losing its original functionality and characteristics.

Several materials such as paper, aluminum, or plastic are used for flexible packaging, but plastics are most often used, constituting around 70% of the market (Page, 2011). The main reason for this tendency proceeds from their unique combination of properties, their low cost, and sustainability. Plastic packaging in general is lightweight and strong—different plastics can be molded, extruded, cast, and blown into seemingly limitless shapes and films or foams. This resourcefulness often delivers results while using minimal resources, creating less waste, consuming fewer resources, and creating fewer CO_2 emissions than alternative materials. Plastics make packaging more efficient, which ultimately conserves resources. Also it meet expectations regarding aesthetics, merchandising, cost, ease of use, ease of opening and resealing, weight, fuel savings, greenhouse gas emissions, and so on (Prnewswire, 2015).

However, packaging in general and food packaging in particular became more and more specific, performing roles other than providing an inert barrier between the product and the outside environment. This kind of package, namely active packaging, differs from conventional passive ones in one or more forms of interactions with the product. In this sense, the main goal of active packaging is to extend shelf life, to be produced with low-cost materials in order to be less expensive, to be easy to process and handle, to use sustainable materials, and to have good presentation (easy paintability and printability), among other goals (Paine and Paine, 2012).

In order to cover most of the desired characteristics, a lot of materials or a combination of them, mainly polymer based, have been investigated and used in past decades. Promising materials with a capability of adjusting their properties as required, produced with low-cost components and completely recyclable are the polyolefin-based nanocomposites. In this chapter a detailed analysis of flexible packaging for fresh products obtained from polypropylene and nanotalc is provided in next sections.

2 Polymer Nanocomposites as Material for Food Packaging Development

Food packaging is a multifunctional system because it provides numerous functions, as it was aforementioned. In this sense, the package serves as a medium of contention of food, as well as, it must protect it from dirt or dust, oxygen, light exposition, pathogenic microorganisms, moisture, and a variety of other destructive or harmful substances. The design of a proper packaging should guarantee food safety along with fulfilling the intended conditions of use. In addition, the package must be inert, cheap, and easy to produce; lightweight; capable of being disposed, reused, or recycled; as well as able to withstand extreme conditions during processing or filling, storage, and transport.

One of the disadvantages associated with food packaging is its permeability to gasses, water vapor, and natural substances contained within the food being packaged or even in the packaging material itself (Arvanitoyannis and Bosnea, 2004; Mercea, 2008; Robertson, 2006; Finnigan, 2009). Depending on the application, high barrier packages to gas diffusion are not suitable, especially in the case of fresh fruits and vegetables, which require a continuous supply of oxygen in order to maintain cellular respiration. However, carbonated beverage containers must be made of plastics with high carbon dioxide and oxygen barriers to avoid oxidation and decarbonation of the products (Finnigan, 2009).

Considering all these scenarios, it is necessary to design packaging with specific and appropriated functions according to the food requirements, considering that the products are transported over long distances between producers and consumers.

In past decades, the development of polymer nanocomposites offers a solution of these aforementioned problems related to food packaging. These materials are the result of the incorporation and dispersion of an inert, nanoscale filler within a polymeric matrix. In this sense, silicate nanoplatelets, silica nanoparticles (Tang et al., 2008; Jia et al., 2007), carbon nanotubes (Prashantha et al., 2009; Kim et al., 2009), graphene (Wakabayashi et al., 2008; Borriello et al., 2009), starch nanocrystals (Chen et al., 2008; Kristo and Biliaderis, 2007), cellulose-based nanofibers or nanowhiskers (Azeredo et al., 2010; Bilbao-Sáinz et al., 2010), chitin or chitosan nanoparticles (De Moura et al., 2011), and other inorganics (Yang et al., 2008; Zhang et al., 2008), are different alternatives of fillers used in nanocomposites development.

The incorporation of particles with at least one nanometric dimension throughout polymer matrix induces better final properties than control polymers which contain no filler. In this sense, nanocomposite materials are generally stronger (Li et al., 2009; Gu et al., 2007), more flame resistant (Kandola et al., 2008; Smart et al., 2008), and possess better thermal properties (Yoo et al., 2004; Bertini et al., 2006). In addition, changes in surface wettability and hydrophobicity have also been reported as a consequence of nanoparticles presence. But the most important change that promotes nanocomposite materials is their barrier properties, being the main reason for their application in the food industry. Consequently, polymer nanocomposites are interesting and promising materials to be used in food applications in order to develop adequate packaging, with the ability to tailor their final properties according to the specific requirements of food products. In this sense, nanocomposites offer to the packaging industry the possibility of reducing material thickness, in addition to cost savings and waste reductions, due to the smaller amounts of polymer that need to be used to develop packaging materials with identical or even better mechanical attributes than control polymers.

3 Potential Perspectives of Polypropylene/Talc Nanocomposites Films for Food Packaging Applications

In order to develop materials with tailored properties focused on food packaging, several polymer nanocomposites have been studied (Duncan, 2011). Generally, these materials are based on polyolefins and mineral particles due to their appropriated relationship between cost and final properties. However, nanocomposites based on polypropylene (PP) and talc particles is an alternative that is less investigated within the packaging field. These materials have a wide range of applications, comprising household appliances, automotive, and construction industries. Their final performance is a consequence of the synergetic combination of properties between PP and talc nanoparticles. In this sense, it is important to note that talc is a good nucleating agent that favors the increment of PP crystallinity, improving final properties (Ferrage et al., 2002; Varga et al., 1994; Castillo et al., 2012a, 2013a,b,c). In addition, the fact that talc is a naturally occurring mineral, as well as its harmlessness and chemical inertness, makes talc suitable for food packaging applications. In this sense, talc is approved by the US Food and Drug Administration for use in polymeric compounds in contact with food (Clark and Steen, 2003). Moreover, the presence of these particles does not interfere with the recyclability of the polymer. As a consequence, the development of flexible food packaging from PP/talc nanocomposites is an interesting alternative which could provide a very cost-efficient solution to overcome barrier deficiencies of traditional food packaging (Castillo and Barbosa, 2011).

In order to reach an optimum performance of PP/talc nanocomposites, a high degree of dispersion and distribution as well as particle-matrix interfacial adhesion must be guaranteed. These requirements are relevant to obtain a proper design of final nanocomposites and to prevent their premature failure (Jana and Jain, 2001). Taking into account that talc nanoparticles have a strong tendency to agglomerate, an homogeneous particle dispersion and distribution is a very difficult issue when talc is compounded with PP, even at the laboratory scale (Ericson, 2000; Jana and Jain, 2001). In order to promote an efficient and close mixing between nanoparticles and polymer matrix, the academic sector has devoted a lot of interest to process PP/talc nanocomposites at laboratory scale. However, if these materials are to be inserted in the packaging market, it must be proved they can be processed at a higher scale. In this sense, the prospective future of packaging based on these nanocomposite is huge, but one of the biggest challenges is the development of processing-manufacturing technologies at a high scale according to the requirements in terms of quantity, quality, and cost of the commercial products.

Focusing on food packaging applications, the blow extrusion is one of the most used methods for the manufacture of plastic films at industrial scale. In addition, it provides biaxial stretching which could induce additional characteristics in final film performance (Abu-Zurayk et al., 2009; Soon et al., 2009). However, considering the aforementioned obstacles related to PP/talc nanocomposites development at laboratory scale, the processing of these nanocomposites at higher level production is scarcely investigated, even less the report of good nanoparticle dispersion and distribution.

4 Blown Nanocomposite Films From Polypropylene and Talc

In order to promote a good dispersion and distribution of talc nanoparticles within polymer matrix, nanocomposite film processing was designed in a two-step methodology. Taking into account that PP/talc compounding is a key factor to obtain optimal and homogeneous properties, the first step of nanocomposite processing was performed in a twin-screw extruder. In order to achieve intimate close and effective contact between particles and matrix, nanocomposites formulations were based in PP powder and talc with a high purity degree (A10). In this sense, a premixture containing both components was obtained after feeding a powder rotary mixing machine. This physical mixture was then introduced in a twin-screw counter-rotating extruder to achieve an effective mixing. An optimal dispersive and distributive mixing in nanocomposite was achieved after varying the barrel temperature profile and the screw rotation speed.

Once optimal mixing was reached, the second stage of the proposed processing was the blow extrusion in order to obtain PP/talc nanocomposite films. In this sense, a downward blow film machine at a pilot scale was used. Since polypropylene displays a tension thinning flow in the melt and also its crystallization rate during cooling is rather slow, it is difficult to use the conventional film-blowing process for it's processing (Morton-Jones, 1989). In this sense, polypropylene process is run vertically downwards, in which the extrudate is quenched from the melt with a water bath (Fig. 18.1).

Fig. 18.2 shows the downward blown nanocomposite film line at different sections, which were named using the nomenclature *PP(talc weight percentage)A10*. From the results obtained by the designed process, the possibility could be demonstrated of blowing nanocomposite films with good and consistent quality based on polypropylene containing up to 5% w/w talc nanoparticles.

Initially, the blowing conditions were set with pure PP, but the same values were used for all nanocomposites, obtaining films that were both thin and of good appearance. This result is a first confirmation of a good talc nanoparticle dispersion in PP taking

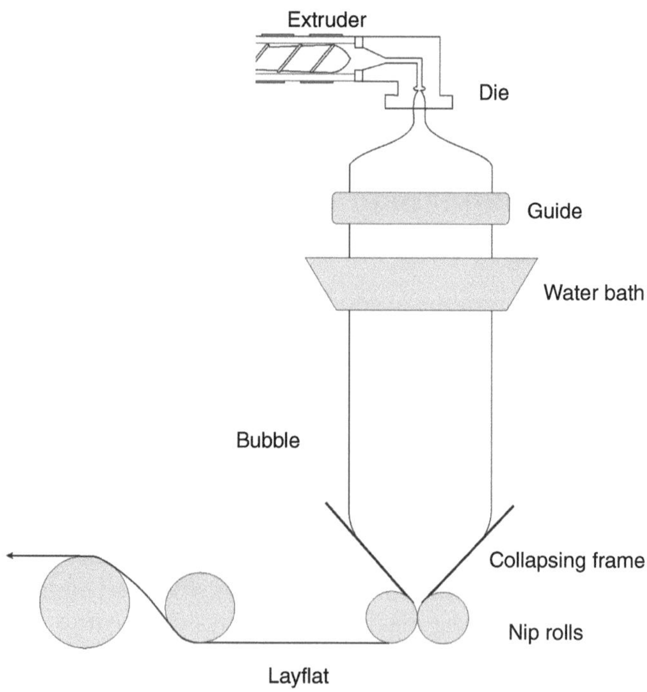

Figure 18.1. A schematic draw of a downward blow film machine for polypropylene.

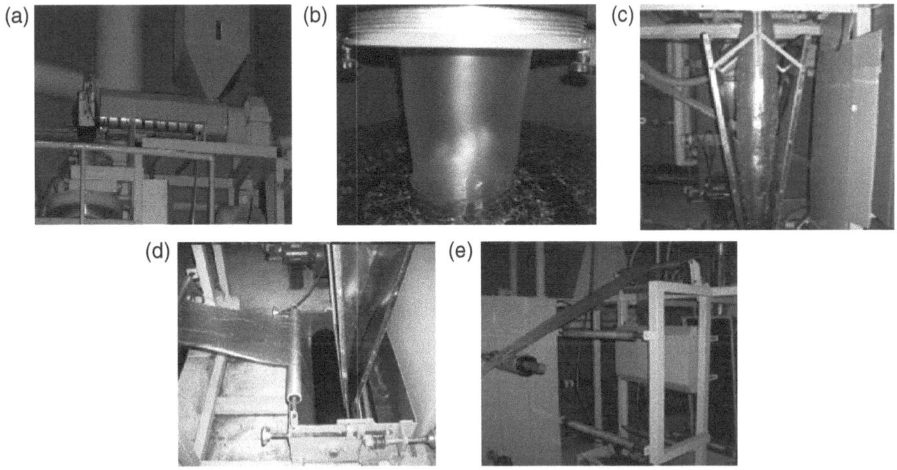

Figure 18.2. Blown nanocomposite film line: (a) extruder, (b) bubble entrance in water bath, (c) bubble, (d) collapsing frame, and (e) layflat.

into account that agglomerates or impurities perturb the bubble stability because they act as stress concentrators (Nielsen and Landel, 1994). Another confirmation of a good talc nanoparticle dispersion and distribution was the fact that nanocomposite films were translucent without inhomogeneities presence, being talc nanoparticles indistinguishable by sight. Films presented good appearance, being homogeneous and easy to handle even when 5% w/w of particles was incorporated. These films had a mean layflat width of 30 ± 2 cm, which was very close to the nominal value for the machine (30 cm). Moreover, film thickness of developed nanocomposites was 68 ± 5 μm, showing a dispersion less than 10% which demonstrates the homogeneity in film thickness values.

Particularly, during the blow extrusion process, molten material is subjected to a biaxial stretching for the occurrence of longitudinal and transverse stresses, which induce a macromolecular orientation (Yuksekkalayci et al., 1999). In this sense, besides the macroscopic evaluation, a systematic microstructural analysis was performed on nanocomposite films in the machine (MD) and transverse (TD) direction. According to this nomenclature, MD is the direction that the films moves through the blow film machine, and TD is the direction perpendicular to the machine direction. In this sense, the morphological analysis of blown films was performed using a scanning electronic microscope (F), exploring the fracture surfaces in both directions. PP films presented a continuous and smooth fracture surface (Fig. 18.3a). All blown nanocomposite films showed a good particle distribution in both studied

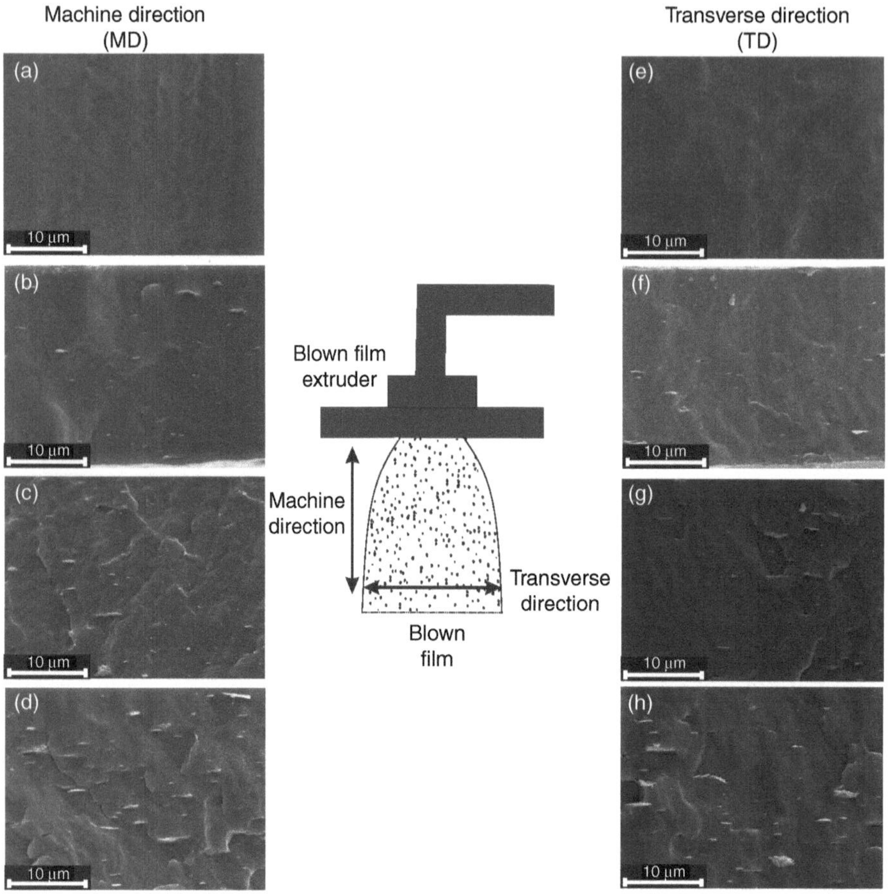

Figure 18.3. SEM micrographs (5000×) of blown films. (a and b) PP, (c and d) PP1A10, (e and f) PP3A10 and (g and h) PP5A10; in MD and TD directions, respectively, according to the schematic representation of blown films.

directions. These results at microscopic scale corroborated the good mixing of nanocomposites which was previously suspected from the macroscopic observation of blown films (Fig. 18.3b–h).

Particle orientation was observed in blown films, independently of the analyzed direction (MD and TD) and talc concentration. This nanocomposite morphology was a consequence of the processing method and particle shape, particularly, during film processing, high-shear stresses in the die and large elongational flow conditions of particle movement within the viscous molten polymer. In this sense, taking into account that the flow strength depends on the area exposed to it, particles are positioned offering the least possible surface to the actual stresses. This particle orientation is accomplished by talc morphology, which consists of platelets having micrometric basal surfaces and nanometric thicknesses. In this sense, particles were oriented with their basal surfaces parallel to the film surface in order to offer the least resistance to the flux during blow extrusion.

A good matrix–particle adhesion was observed in all nanocomposite films, even when talc particles were not surface modified (Fig. 18.3). Talc particles seemed to be embedded in the fracture surface of nanocomposites, being an indication of good wetting behavior between filler and matrix. An ulterior confirmation of the good compatibility is the continuous interphase between talc and PP matrix caused by the voids' absence around particles.

Regarding particle dispersion, a small number of small aggregates were detected, only in nanocomposite films with 5% w/w talc. In this sense, the state of particles dispersion within blown films was also examined by transmission electron microscopy (Fig. 18.4). These micrographs revealed the talc morphology as platelike planes which were uniformly distributed and predominantly aligned along the flow direction of the films. In addition, the micrometric length as well as the nanometric thickness of talc particles could be corroborated from Fig. 18.4. Even though the presence of particle agglomerates was detected, they are small enough and compact enough to bear the biaxial deformation during blow extrusion, preventing the bubble break. The increment of talc concentration in blown nanocomposite films is in accordance with the higher presence of particle embedded within the polymeric matrix.

The presence of talc particles and PP in blown films was corroborated through the functional groups detected by fourier transform Infrared Spectroscopy (FTIR). Fig. 18.5 shows the spectra of all developed films, where signals corresponding to PP as well as those related to talc particles are clearly identified. The presence of talc in nanocomposites is detected at 3676 cm^{-1} (Mg_3–OH), 1019 cm^{-1} (Si–O–Si stretching), 670 cm^{-1} (–OH deformation), 536 and 466 cm^{-1} (Mg–O), 452 cm^{-1} (Si–O–Mg), and 426 cm^{-1} (Si–O) (Castillo et al., 2011). When talc concentration increases, an

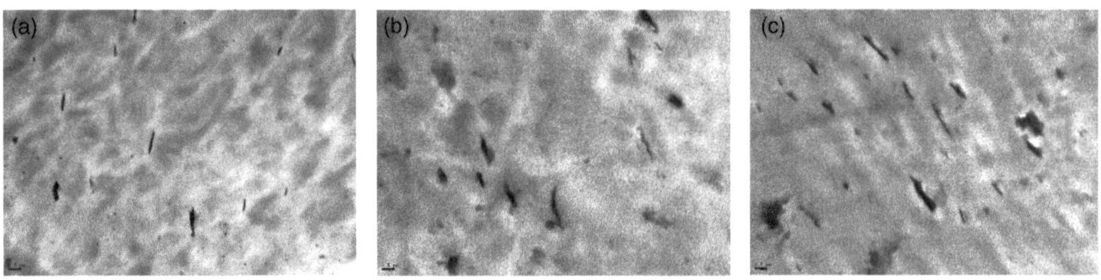

Figure 18.4. TEM micrographs (10,000x) of blown nanocomposites films. (a) PP1A10; (b) PP3A10; and (c) PP5A10.

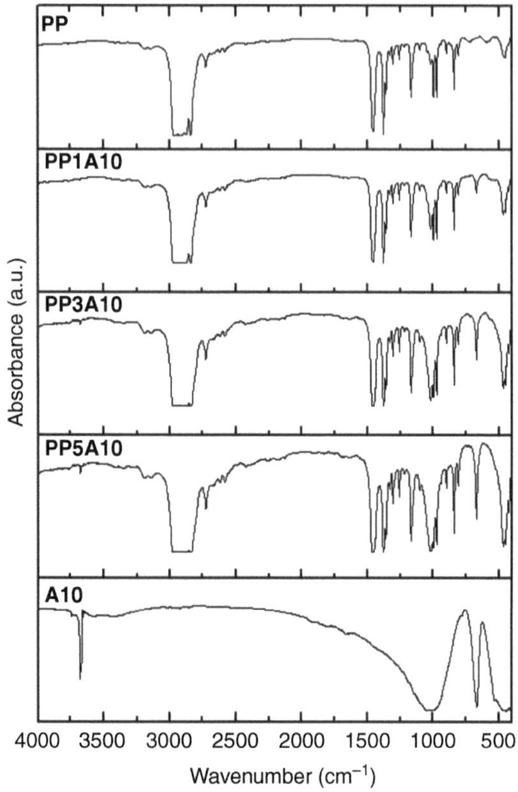

Figure 18.5. FTIR spectra of PP and PP/talc nanocomposites blown films.

increment on intensity of bands corresponding to this mineral is detected in the nanocomposite spectra, as expected. In addition, FTIR provides indirect information about particle distribution and the possibility to estimate the macroscopic concentration in nanocomposite films. In this sense, actual talc concentration could be assessed through a ratio between talc and PP band areas. Prior to this estimation, a calibration curve should be built from different values of bands areas ratio using films with known talc concentrations. In order to define this ratio, the signals located at 670 and 2720 cm^{-1}, corresponding to talc and PP, respectively, were chosen as reference bands. The fact that these band ratio values were similar at several points on blown films was indicative of a homogeneous talc concentration, corroborating indirectly the good particle distribution, previously observed for each film formulation by visual evaluation, SEM, and TEM. Moreover, the actual particle concentration in nanocomposite films determined from band ratio values of FTIR spectra were very close to nominal ones.

In addition, FTIR offers a useful tool to detect and estimate thermal-oxidative degradation in polymers, one of the reasons for the detriment in mechanical properties as a consequence of the decrease of molecular weight induced by this phenomenon. Particularly, the developed PP/talc films were exposed twice to thermal processes, such as melt and blow extrusion, where high temperatures and shear fields as well as oxygen presence could induce thermooxidative degradation on polymeric matrix. Regarding to nanocomposites processing, this phenomenon is also favored by the presence of talc that could induce a viscous heating more intense than in pure PP as a consequence of particle–particle interactions. As a consequence, PP degradation in oxygen is manifested by the occurrence of carbonyl groups in its structure. In this sense, FTIR not only provides a qualitative information concerning to degradation state through the presence or absence of bands associated to carbonyl groups, but also supplies a quantitative result to estimate the extension of the thermooxidative degradation after material processing. For this purpose, a carbonyl index is defined as the ratio of the areas under the band assigned to carbonyl group (1700–1800 cm^{-1}) and a reference band corresponding to PP, which is not affected by the degradation (2720 cm^{-1}). Carbonyl index determined for blown films was lower than 0.54, which is an indicative of insignificant degradation, being in accordance with values reported for same materials and processing conditions (Rabello and White, 1996). These results confirmed the absence of yellowish in films, which could be detected by visual examination.

Films optical properties are important concerns related to the applications in the packaging sector that can affect consumer acceptability. Table 18.1 presents the color indexes as well as haze values determined for blown films. Talc nanoparticles presence increased significantly the achromatic component, *L*, related to a higher lightness of blown films. Concerning to chromaticity

Table 18.1 Color Parameters and Haze of Blown Films

Film Formulations	Color Parameters				
	L	*a*	*b*	ΔE	Haze (%)
PP	94.87 ± 0.17	0.02 ± 0.00	0.71 ± 0.05	3.18 ± 0.01	2.1 ± 0.2
PP1A10	96.58 ± 0.26	0.12 ± 0.01	0.15 ± 0.01	2.25 ± 0.04	2.2 ± 0.2
PP3A10	97.13 ± 0.16	0.22 ± 0.02	0.14 ± 0.01	2.06 ± 0.05	3.7 ± 0.4
PP5A10	97.95 ± 0.43	0.35 ± 0.03	0.13 ± 0.01	2.09 ± 0.08	5.6 ± 0.5

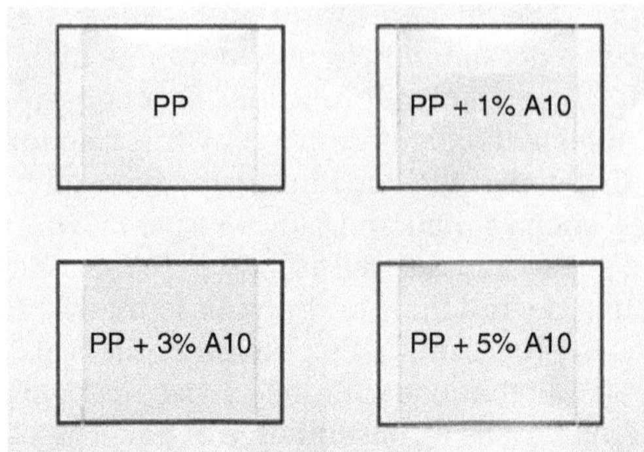

Figure 18.6. Photographs of PP and PP/talc nanocomposites films.

parameters, an increment in *a* values was observed when particle concentration raised. Even though the *b* value was decreased by talc incorporation at 1% w/w, this parameter was unchanged by further increment on talc concentration. This result is an indication that there is no positive tendency toward yellowness with increasing particle concentration, confirming the absence of thermooxidative degradation induced by viscous heating due to particle–particle interactions. Regarding the total color difference, lower ΔE values were detected with an increased talc concentration. Despite talc presence causing twice the haze values of blown films, they were optically clear in the visible range, taking into account that talc nanoparticles are well dispersed in the polymeric matrix, as shown in Fig. 18.6.

Another relevant issue in food packaging applications is the UV absorption capability of developed films in order to protect products from the effect of this radiation (Kanatt et al., 2012). In this sense, this property is useful to develop packages for food which must be exposed to UV light in order to diminish their microbial charge. For this purpose, films having this capability prevent the oxidative rancidity in food that are vulnerable to this phenomenon catalyzed by UV light. The UV-vis absorption spectra for blown films are presented in Fig. 18.7a. It is important to note that all curves presented an absorption peak located between 270 and 300 nm, revealing the UV absorption capacity of blown films. The absorption intensity reached higher values according to talc concentration increase due to the characteristic absorbing properties of the filler. Concerning the blocking effect of talc contained in nanocomposites films against UV and visible radiation, there is observed an increased tendency for particle concentration in all studied wavelengths (Fig. 18.7b), as expected. This increment is more notorious for the visible range (700 and 800 nm), although there is no significant difference between the values measured for both these wavelengths. Despite the

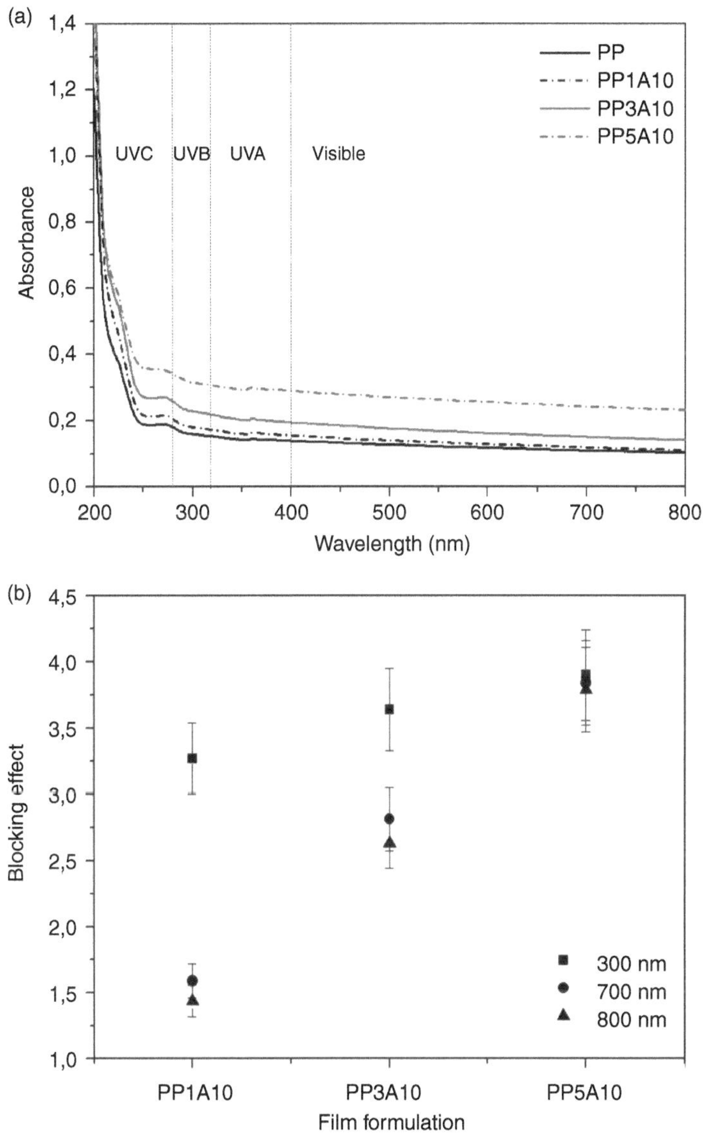

Figure 18.7. (a) UV-vis spectra of blown films; (b) blocking effect of talc nanoparticles against UV and visible radiation in nanocomposite blown films.

higher blocking effect being detected for 300 nm that corresponds to the UV range, the incorporation of 5% w/w talc allows obtaining the same optical barrier in the visible as in the UV wavelengths.

Crystalline morphology constitutes an important structural characteristic that influences the final properties of packaging materials. Particularly, in blow extrusion there are at least two main contributions on final crystalline morphology: particles presence and crystallinity induced by processing. In this sense, the analysis of X-ray diffraction spectra of blown films on MD and TD directions could provide useful information concerning to the influence of biaxial deformation produced by the blow extrusion and the presence of talc particles as nucleating agents on crystalline structure. All diffractograms show the occurrence of stable monoclinic α-phase of PP, identified mainly by four reflections corresponding to (110), (040), (130), and (060) planes in Fig. 18.8.

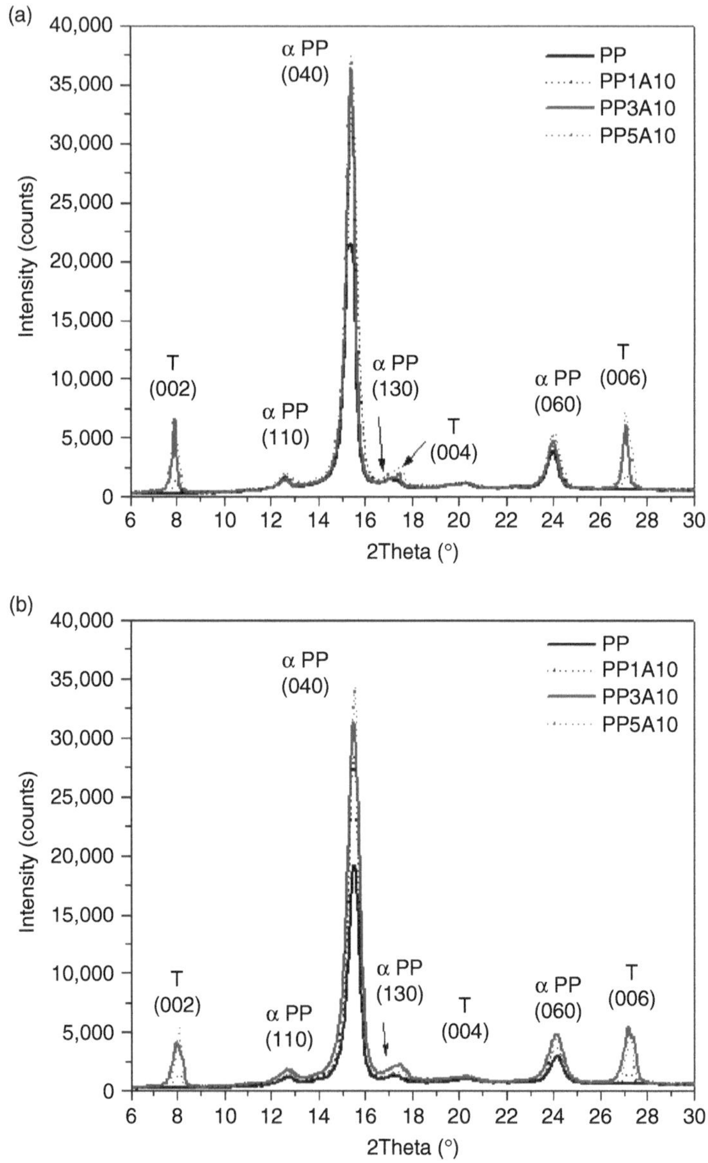

Figure 18.8. XRD spectra of PP and PP/talc nanocomposites films. (a) MD and (b) TD.

Table 18.2 Crystallinity Degree (X_c) Obtained by XRD of PP and PP/Talc Nanocomposites Blown Films Determined for Machine (MD) and Transverse (TD)

	X_c (%)	
Film Formulation	**MD**	**TD**
PP	45.9	41.0
PP1A10	51.8	45.6
PP3A10	53.8	46.1
PP5A10	54.2	49.4

It was also detected that there was a lack of β- and γ-phases corresponding to PP in blown films. Particularly, the absence of these crystalline phases in nanocomposites films corroborated the α-nucleating capability, as reported by other authors (Varga et al., 1994). The reflections of (002), (004), and (006) planes revealed talc presence in nanocomposite films. Increasing of talc concentration was corresponded to higher intensities of these reflections in MD and TD, as expected.

Despite the same crystalline structure that was observed in MD and TD, there was a notable difference related to the intensities of reflections among spectra performed in these directions for blown films. This result evidenced a higher crystalline development in the blow direction. In order to quantify the PP crystals in the structure of developed films, crystallinity degree (X_c) was obtained by XRD data (Table 18.2). These values showed an increasing tendency according to talc concentration evolution in MD and TD. The fact that talc acts as a nucleating agent partially justified the crystalline behavior of blown nanocomposites films. In this context, few PP chains are aligned on talc basal surfaces that allow the formation of a nucleus during crystallization, from which PP crystallites grow into a particular direction (Ferrage et al., 2002). Because talc particles are mainly oriented with their major dimension in MD, as previously mentioned, the epitaxial crystals grow preferentially in TD, revealed by (040) reflection in XRD spectra (Radhakrishnan et al., 1995).

On the other hand, the occurrence of high elongational flows during blow extrusion promotes a fibrillar-like structure which is oriented in the stress direction (MD). This structure serves as nucleation site for radial growth of chain-folded lamellae perpendicular to stress direction (TD), also mainly detected by the intensity of (040) plane (Meijer, 1997). Particularly, the orientation in isotactic polypropylene tubular blown films has been deeply studied by Shimomura et al. (1982) and by Choi and White (2001). They stressed that the (040) plane, or the b-axis, is preferentially oriented perpendicular to the film surface in high blow-up ratio films. In this context, the crystal induction by talc nucleating action as well as by blow extrusion process is observed in a higher intensity of (040) reflection in MD than in TD, as corroborated in Figure 18.8. According to this, a higher crystallinity degree of blown films is detected for MD than TD, accounting mainly this crystalline contribution.

Thermal properties of PP and nanocomposites-blown films obtained by differential scanning calorimetry (DSC) are presented in Table 18.3. Melting temperature remained constant when talc concentration in film formulation rises (Castillo et al., 2012b). However, the influence of talc presence is observed on the crystallization temperature, due to the increment of particle concentration led to higher values. This behavior is intimately related with the nucleating character of talc that allows increasing the nucleation sites for PP crystallization (Ferrage et al., 2002). In this sense, the incorporation of 5% w/w talc increased the crystallization temperature in 6°C. Concerning to the melting and crystallization enthalpies, these thermal properties showed an increasing tendency according to talc concentration. Consequently, the presence of talc induced an increment in crystallinity degree of blown films, being in accordance with the results obtained by XRD data (Table 18.2). However, the comparison between the values obtained by DSC and XRD shows a numerical difference. In this sense, XRD spectra provide information only at a few Angstroms of depth penetration on blown films, meanwhile DSC corresponds to the overall crystallinity of the material. By

Table 18.3 Thermal Properties and Crystallinity Degree (X_c) of Blown Films

Film Formulation	Melting Temperature (°C)	Melting Enthalpy (J/g)	Crystallization Temperature (°C)	Crystallization Enthalpy (J/g)	X_c (%)
PP	165.4	84.3	118.0	94.1	40.7
PP1A10	165.0	88.1	122.7	98.1	42.5
PP3A10	164.5	88.7	123.3	99.0	42.8
PP5A10	163.7	89.1	124.2	99.3	43.0

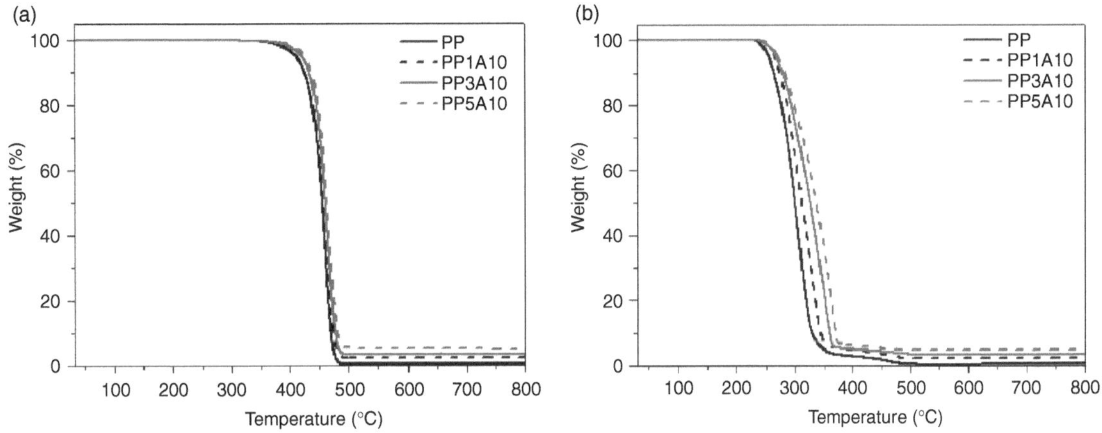

Film formulation	T_d (°C)
PP	461
PP1A10	461
PP3A10	464
PP5A10	458

Film formulation	T_d (°C)
PP	307
PP1A10	320
PP3A10	350
PP5A10	363

Figure 18.9. TGA curves of blown films. (a) Nitrogen and (b) air atmosphere.

the analysis of crystallinity degree determined by these techniques, nanocomposites-blown films developed a higher crystallinity on their surface than in the bulk. Therefore, the information provided by DSC and XRD allows detecting the presence of a gradient in crystallinity degree along film thickness.

Thermal stability of pure PP and its nanocomposites-blown films was evaluated by thermogravimetrical analysis (TGA) in terms of the decomposition temperature (T_d), both in nitrogen and air atmosphere (Fig. 18.9). All blown films showed only one temperature value, corresponding to PP decomposition, taking into account that in the studied range (30–800°C), talc is thermally stable (Schone et al., 2012). Talc presence offered to nanocomposites films a higher thermal stability than PP ones, under both atmospheres. This effect was also increased according to talc concentration, as was expected due to the increment of crystallinity degree leading to a decrease of polymer entropy, giving a more stable film.

The comparison between T_d values obtained in both atmospheres revealed that thermal decomposition of films was accelerated in air presence due to its more favorable formation of free radicals, which further induces chain fragmentation (Allen and Edge, 1992). Particularly, talc concentration at 5% w/w raised decomposition temperature in 6°C respect to the corresponding value to PP under nitrogen, meanwhile T_d is strongly increased in 56°C if nanocomposites films were exposed to thermal degradation in air. Additionally, talc presence allowed a slower thermal degradation than PP film, being mainly evident under air atmosphere. On the other hand, TGA could provide the actual talc concentration in blown films, analyzing the weight of residue at 800°C in nitrogen, when PP was completely degraded and only talc was the remnant material. In this way, particle concentration was corroborated, being these values very close to the nominal ones.

Food packaging should withstand wear, pressure, and damage along the shelf-life. In this sense, it requires resistance to high stress with deformation according to the intended application. For this reason, packaging materials must present an adequate mechanical performance in order to provide to food products the protection against external forces.

During blow extrusion of nanocomposites films, an orientation of nanoparticles within the polymeric matrix occurs, as was demonstrated previously, which mostly induces changes in mechanical properties. Fig. 18.10 show the stress–strain curves for PP and PP/talc nanocomposites films along the MD and TD directions. Mechanical pattern of PP films revealed a clear yield point, which is detected at the local maximum located at low strain values. In addition, it was observed a strain hardening mechanism developed at the end of the curve as a consequence of an oriented crystalline structure, both in MD and TD. In this sense, from mechanical behavior of PP films in MD and TD directions, the appropriate biaxial orientation of the downward blowing machine used was corroborated.

Nanocomposites films tested in MD show a similar behavior than PP. Stress–strain curves has a linear initial section and a local maximum when strain is increased, which is assumed to be the yield strength of nanocomposites films. After yield point, stress remains nearly constant up to high elongations, where strain hardening is detected before film fracture occurs. In this sense, nanocomposites films demonstrate the elastic, capable to form neck, flexible, tough, and ductile in MD.

Young modulus of nanocomposites films, determined from the zero-strain extrapolation of stress–strain curves and associated to the material stiffness, increases as talc content increases in MD (Table 18.4). This behavior could be attributed to different contributions. First, mainly at relative low talc concentrations (less than 1% w/w), the nucleating effect of mineral particles dominates on Young modulus increment compared to talc stiffness contribution by rule of mixtures (Utracki, 1991). In this sense, talc induces the formation of smaller-sized PP crystals, which have an increased

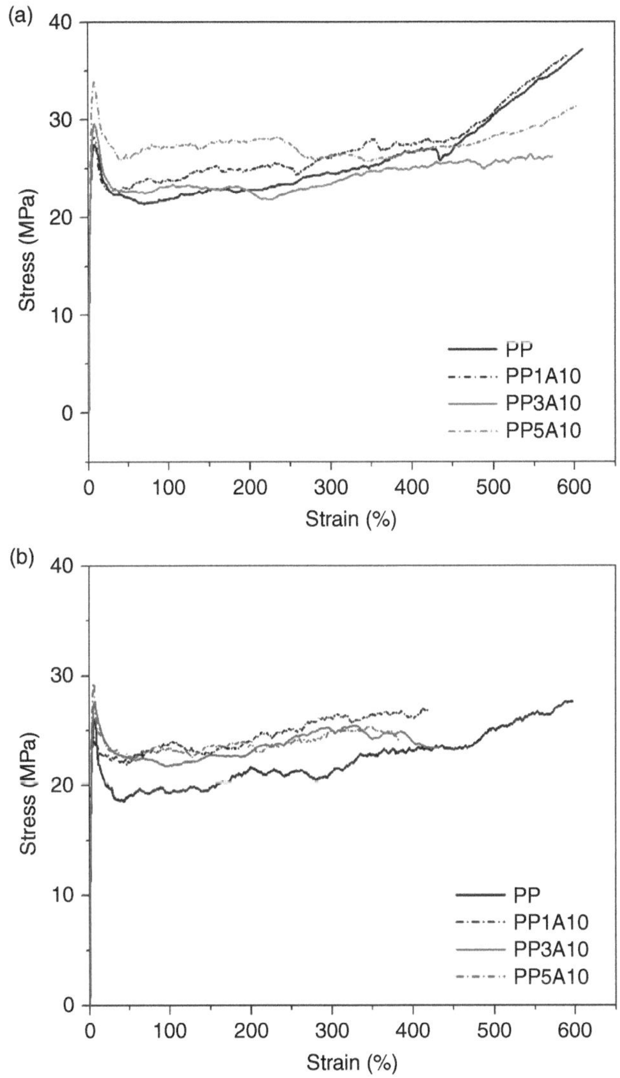

Figure 18.10. Stress–strain curves for PP and nanocomposite blown films. Performed in (a) MD and (b) TD.

stiffness (Menczel and Varga, 1983; Utracki, 1991). On the other hand, at higher particle concentrations, the greater stiffness of talc (170 GPa) compared to PP one (1.2 GPa) governs over the modulus increment (Martinatti and Ricco, 1994; Denac et al., 1998). Moreover, the contribution of particle orientation in PP matrix, observed by SEM (Figure 18.3), could also increase nanocomposite stiffness.

Yield stress (σ_y) increases with talc concentration when this property is evaluated in MD. Regarding ductility, it should be highlighted that even incorporating 5% w/w talc, particles presence has a negligible effect on elongation at break of nanocomposites films on MD.

Mechanical behavior in TD is similar to those in MD taking into account that all films present necking and cold-drawing mechanisms but nanocomposites films break at significantly lower strain than PP ones. In this sense, particle affects the ductility of the matrix, shifting the mechanical pattern to a more brittle behavior, but behaving as ductile films. The incorporation of 5% w/w talc reduces 1.5 times in respect to elongation at break value for PP (Table 18.4). The decrease in breaking elongation by talc addition is associated to mechanical restraint imposed by mineral particles to the mobility and deformability of polymer chains, but also is constrained by the presence of crystals induced by particles and by the processing. Regarding elastic modulus and yield strength of nanocomposites films, these properties show the same behavior with talc concentration as was observed for MD. In this sense, Young modulus and yield strength presents an increasing tendency with talc concentration.

Nowadays, polymer films are widely used in the food packaging industry primarily because of their advantageous mechanical properties, such as the combination of stiffness, strength, and ductility. Particularly, polymer films are exposed to different loading conditions during transportation, handling, and use of packages along with the shelf-life of the product. In this sense, among the functionalities that must be satisfied, tear resistance is a relevant requirement in packaging film. This mechanical property

Table 18.4 Mechanical Properties of PP and PP/Talc Nanocomposites Films Assayed in Machine (MD) and Transverse (TD) Direction

Film Formulation	MD			TD		
	E (MPa)	σ_y (MPa)	ε_b (%)	*E* (MPa)	σ_y (MPa)	ε_b (%)
PP	1315 ± 113	27.4 ± 2.1	610 ± 80	1312 ± 112	26.0 ± 2.2	597 ± 55
PP1A10	1373 ± 122	28.3 ± 2.4	590 ± 64	1390 ± 110	25.9 ± 2.8	420 ± 36
PP3A10	1502 ± 141	29.5 ± 2.0	573 ± 55	1420 ± 146	27.5 ± 2.3	432 ± 48
PP5A10	1563 ± 137	33.9 ± 3.0	603 ± 61	1479 ± 125	29.4 ± 2.5	384 ± 42

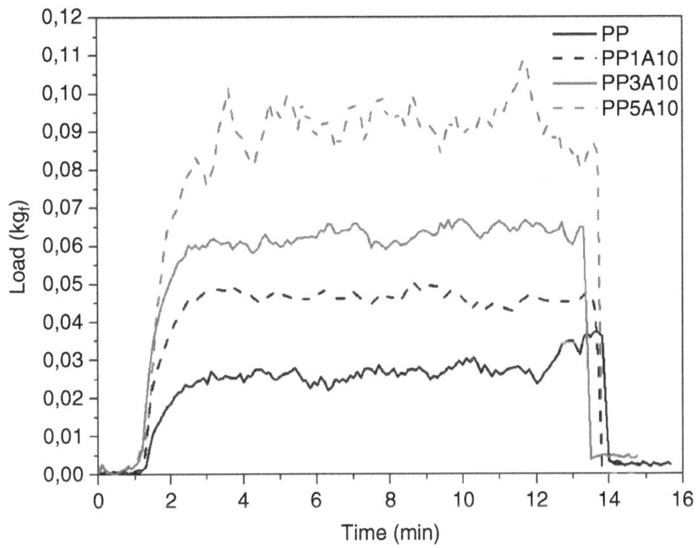

Figure 18.11. Tear test for PP and PP/talc blown films.

is strongly influenced by the processing conditions, which affects crystalline and amorphous orientation, as well as material characteristics. In order to evaluate the tear resistance of blown films, the trouser test was chosen. The curves obtained from tear test for PP and nanocomposites-blown films is presented in Fig. 18.11. The pattern observed for all films revealed that they are low extensible, according to Andreasson et al. (2013). When talc concentration increases, it is observed that the blown films showed a higher tear force. This behavior could be explained in terms of how particles generate a zig-zag path that improves the tear resistance in nanocomposites films, and this phenomenon is favored with talc concentration increment. It is important to note that according to the particle increment, a higher noise in the tear propagation force has been detected as a consequence of talc presence. Additionally, the rate of start propagation, which is determined by the initial part of load–time curves, was influenced by talc particles, inducing higher rate values.

Generally, flexible packages made from polymer films use a heat-sealing process to ensure closure in order to provide protection to the food product. This process involves the join of two film surfaces using heat and pressure during a certain time. In this sense, polymeric chains are able to flow across the sealing interface and mix with the adjacent surface. Meanwhile the conditions of the heat-sealing process remain constant, and mixing will continue until the interface is indistinguishable.

Table 18.5 Seal Strength of PP and PP/Talc Nanocomposites Films

Film Formulation	Maximum Force (N)
PP	14.96 ± 1.74
PP1A10	19.42 ± 1.62
PP3A10	22.16 ± 1.48
PP5A10	26.72 ± 1.74

Typically, a problem associated to flexible packaging is the inability to heat seal properly. An open or weak seal can be the consequence of an improperly heat seal. In this sense, tensile heat-seal strength is a primary method to evaluate seal performance of flexible films. The values of the maximum force for blown films are presented in Table 18.5. It is important to mention that PP and PP/talc nanocomposites films were able to be thermosealed, obtaining a good quality seal.

The maximum force to pull apart the two films showed an increased tendency with talc concentration in nanocomposite films. Particularly, the incorporation of 5% w/w talc nanoparticles induced an increase of the maximum force of around 79% in respect to neat PP. This result implies that talc presence leads to a more resistant heat-seal of nanocomposite films.

The study and comprehension of barrier properties of polymer films has acquired more relevance in recent years, especially for those used in the design of food packaging. Basically, depending on the characteristics and requirements of the food products, it is defined as a specific packaging material. In this sense, the selection of this material is a crucial point for food market because, besides its suitable gas barrier properties, it needs to be versatile enough to withstand handling and to maintain its physical and chemical integrity. Particularly, barrier properties to oxygen and carbon dioxide are important issues in order to extend the shelf life of food considering that they are associated to the respiration of fresh products. Meanwhile, water vapor permeability of the packaging materials has great influence on the quality of the packed food since it is directly related to its dehydration.

Water vapor permeation of blown films decreased with an increasing talc concentration (Fig. 18.12). The moisture barrier property of nanocomposite films could be attributed to different contributions of talc particles. In this sense, the particle

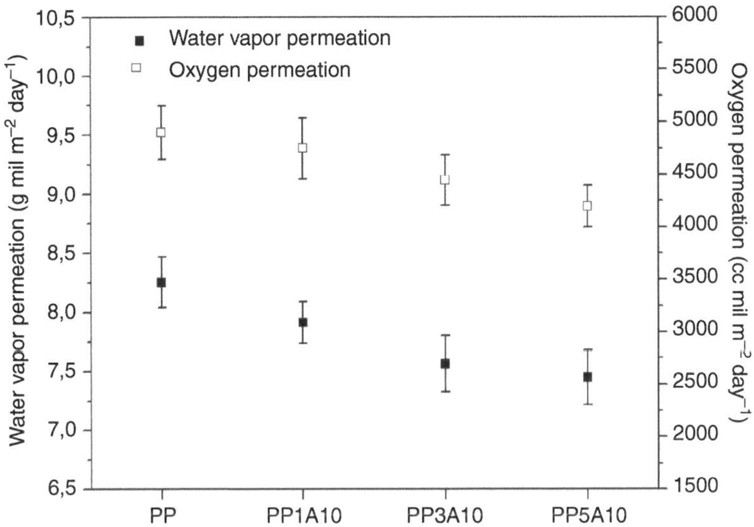

Figure 18.12. Water vapor and oxygen permeation of blown films.

distribution induced a tortuous path within the polymer matrix which forces water molecules to travel through the film thickness, avoiding talc particles because they pose impermeable obstacles in their diffusion travel. As a consequence, this tortuosity increases the time necessary for gas molecules to diffuse through the film. In this context, the efficiency of a nanocomposite functional barrier relies on hindering or prolonging the diffusion of gases molecules along the film. The platelet structure and the high aspect ratio (length/thickness) of talc nanoparticles add more length to the diffusion path of the molecules (Hancock and Rothon, 2003; Dearmitt and Hancock, 2003). Additionally, the nature of talc surface also plays an important role in increasing water vapor barrier in nanocomposite films. In this sense, the presence of siloxanes (Si–O–Si) bonds in basal surfaces imparts to particles an hydrophobic character, which favors the repulsion of water molecules. Another factor that contributes to a low water vapor permeability is the crystalline structure of PP. Polymer crystals play the same role as talc particles, being impermeable to the flow of water molecules. It is important to note that talc presence increased the degree of crystallinity of nanocomposite films, as it was demonstrated earlier, favoring once again the reduction of water vapor permeation.

Taking into account that talc nanoparticles were well distributed within the polymer film for all studied particle concentrations, the degree of tortuosity increased, according to talc content. In addition, the good wettability of particles by the matrix implies that the adhesion between talc and PP does not favor the formation of air bubbles or microvoids. Therefore, increasing talc concentration allows that talc particles act as a real barrier to water molecules' diffusion in nanocomposite films.

Oxygen permeation showed a decreasing tendency with talc concentration of blown films, as it was observed for water vapor permeation. The tortuous path generated by talc particles, which was previously mentioned, hinders the flow of oxygen molecules through the packaging. In this sense, talc provides the opportunity to reduce transmission rates of water vapor and oxygen in nanocomposite films, which allows control over the shelf life of products in food packaging.

5 Development of Flexible Food Packaging From Blown PP/Talc Films

The use of flexible packaging has acquired great importance, reaching a noticeable growth over the past decade throughout the world. The future trend is toward flexible packaging that will continue contributing to the expansion of the market. The development of this kind of packaging is impelled by consumer demand (convenience packaging, portion packaging, longer shelf life), the marketing demand (improved product appearance, enhanced graphics, security features), and the production demand (energy efficient packaging, higher output, cost effective). Additionally, flexible packaging has become popular from the viewpoint of environmental concerns because it favors a reduction of waste generation and offers a recyclable character. Among the major uses of flexible packaging, food packaging comprises about 50% of the worldwide applications (Rolando, 2000). The need of the development of flexible packaging is often derived from either a new use for packaging, such as for fresh-cut fruits and vegetables, or as a replacement for more conventional rigid packaging, for example, for snack food and dry products. In this context, the development of new and improved materials as well as methods of producing innovative flexible packaging is the challenge that the scientific and industrial sector must face.

In order to provide a very cost efficient alterative to overcome barrier deficiencies of traditional food packaging, flexible packaging were developed from blown PP/talc nanocomposites. Bags were obtained by thermosealing different film formulations. The selected fruits to be packaged were cherry tomatoes which were sorted for uniform size, similar conditions of maturity, and without visual physical damage. It is important to note that cherry tomatoes were not exposed to any physicochemical treatment before being packaged. The flexible packaging containing cherry tomatoes was stored under controlled conditions of temperature and relative humidity (25°C and 50% HR). Tomato control products were exposed to the same conditions as the packaged fruits. It is important to highlight that the packaging and the control tomatoes were not refrigerated during the storage time. The flexible packaging was developed from blown films, as fruit control is presented in Fig. 18.13 for initial time and after 150 days of storage.

The appearance of packaged food is one of the first quality factors that the consumer evaluates at the moment of choosing a product. A qualitative analysis concerning the evolution of tomato appearance during storage time can be appreciated from Fig. 18.13. At the initial time, control tomato presented the firmness and color typical of fresh fruit; meanwhile, after 150 days of storage it suffered a complete dehydration (Fig. 18.13a and d). Concerning to the tomato contained within packaging based on PP film, the fruit showed a wrinkled skin as a signal of an incipient dehydration after the storage time (Fig. 18.13e). However, packages made from nanocomposites with 5% w/w talc allowed that tomato appearance remain unaltered (Fig. 18.13c and f) after 150 days of storage. Color of packaged tomatoes did not suffer any perceptible change regarding the initial condition at the end of the storage time, regardless of film packaging. In this sense,

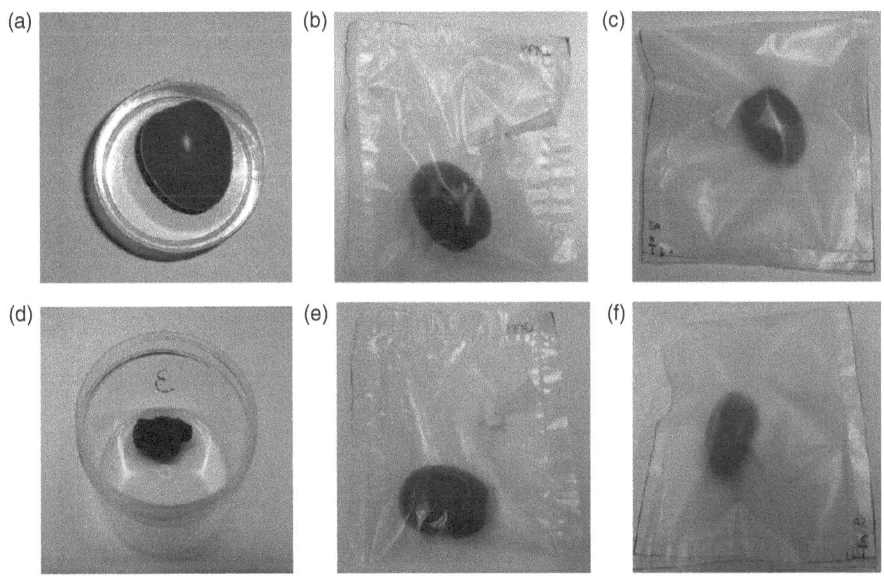

Figure 18.13. Photographs: (a and d) control tomato and packaged tomatoes in packaging from (b and e) PP; (c and f) PP5A10, for the initial time and after 150 days of storage, respectively.

considering the classification of maturity grades for tomatoes according to the color, the packaged fruits correspond to the red category for which more than 90% of skin is not green (USDA, 1991; FAO, 2011). In this sense, the color conservation as well as the absence of wrinkles on external skin of packaged tomatoes were two favorable aspects that the packages made from nanocomposite films offer concerning the conservation of fresh fruits during prolonged times, without refrigeration.

Normalized weight difference as a function of the storage time for control and packaged tomatoes is presented in Fig. 18.14. Control tomato showed an abrupt increment in the normalized weight difference compared to the curves corresponding to packaged tomatoes. Control fruits completely lost their aqueous content after 110 days. This behavior was evidenced by the plateau in the curve corresponding to the normalized weight difference, which is extended up to 150 days (Fig. 18.14). However, packaged tomatoes presented a lineal behavior in the studied storage time, showing a lower slope, which is associated with a significant lower rate of the loss weight than those of the control tomatoes. The maximum value of the normalized weight difference for the fruits contained in nanocomposite packaging, at the end of the storage, was only 0.37. Taking into account this observation and the fact that neither of the packaged tomatoes reached a constant value of weight

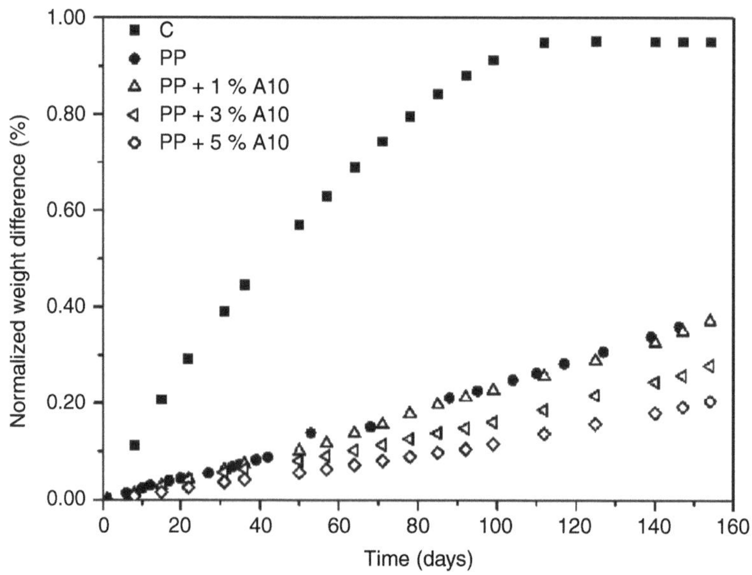

Figure 18.14. Normalized weight difference as a function of storage time of control tomato and packaged fruits in flexible packaging made from PP and nanocomposite films.

difference, these results are evidence that the packages based on nanocomposite films allowed tomatoes to maintain their weight during a considerably longer time, without refrigeration.

Tomatoes contained in packaging developed from PP and nanocomposite with 1% w/w talc showed similar values for the normalized weight difference. Particularly, the behavior of these nanocomposites was mainly dominated by the matrix, taking into account that particle contribution was poor due to low talc concentration. However, when the effect of talc content in blown films was evaluated, tomatoes lost weight with a rate increasingly lower according to the increment on talc concentration. It is important to note that the loss in weight that tomatoes have experienced during storage is mainly related with the loss of water. Consequently, the analysis of this behavior is intimately related with the results of water vapor permeability of nanocomposite films. In this sense, as previously mentioned, several factors associated with the packaging material contributed to a lower loss weight for tomatoes. First, the dispersion and distribution of talc nanoparticles within the polymer matrix could allow the establishment of a physical barrier against water vapor flow through the film. Second, talc acts as a nucleating agent, leading to an increase in PP crystals, which pose obstacles to gas diffusion. Finally, the predominant hydrophobic character of talc particles offers an additional obstacle caused by the repulsion of water molecules.

It is important to highlight that the results obtained from the quantitative analysis through the evaluation of loss weight corresponded to the qualitative analysis performed on the tomatoes' appearance. In this sense, the normalized weight difference is in agreement with the firmness loss on tomatoes.

6 Conclusions

The purpose of using polymer nanocomposites in food packaging applications was in order to overcome mainly the barrier properties deficiencies presented by conventional plastics, considering that packaging is crucial in the shelf life of the products. According to the specific requirements of different kinds of food, it is necessary to tailor final properties of packaging materials in order to enhance and extend the shelf life of products. In this scenario, a polymer nanocomposite plays an important role because it allows adaptation of properties with relative ease, depending on the product specifications. Although several nanocomposite materials have been extensively studied over past decades for different application fields, polypropylene/talc nanocomposites are scarcely investigated, particularly as packaging materials. Despite the lack of reports related to the use of these nanocomposites in the packaging sector, they meet with several compulsory requirements to be used in contact with food. The fact that talc is a naturally occurring mineral as well as its harmlessness and chemical inertness makes polypropylene/talc nanocomposites suitable for these kinds of applications. Concerning polypropylene, this polymer is also suitable for food packaging, and is used for biscuits, crisps, sweets, bread, pasta, dairy products, convenience foods, dried fruits and snacks, among other products. Moreover, the presence of these mineral particles does not interfere with the recyclability of the polymer. However, the most important characteristic offered by the nanocomposites based on polypropylene and talc particles is their combination of good optical, thermal, barrier, and mechanical properties. Taking into account all these advantages, the development of flexible food packaging from PP/talc nanocomposites is an interesting alternative that could provide a very cost-efficient solution to enhance and extend the shelf life of food products. However, in order to insert these materials within the packaging market, they must prove they can be processed at a higher scale. In this sense, the blow extrusion is one of the most-used methods for the manufacture of plastic films at industrial scale. Nevertheless, considering the obstacles related to PP/talc nanocomposites processing at laboratory scale due to particle agglomeration, the processing of these materials at a higher level of production is a big challenge.

From the results presented in this chapter, it was demonstrated that:

- Nanocomposite films based on polypropylene containing up to 5% w/w talc nanoparticles were obtained by using a downward blow film machine at a pilot scale.
- Blown films presented good appearance, homogeneity, and ease in handling even when 5% w/w of particles was incorporated. They were translucent without inhomogeneities present, being talc nanoparticles indistinguishable by sight.
- Good particle distribution was observed at a microscopic scale, corroborating the macroscopic observation of blown films. In addition, a particle orientation and a good matrix-particle adhesion were evidenced.
- Actual talc concentration in nanocomposites was corroborated by using FTIR and TGA. The analysis of results obtained by these both techniques leads to values very close to the nominal ones.
- Despite the exposure of developed films to different thermal processes, such as melt and blow extrusion, negligible thermooxidative was determined by carbonyl index from FTIR

spectra. This result agreed with the absence of yellowish in films by visual examination.

- Regarding to color determination of blown films, talc nanoparticles presence increased significantly the achromatic component, *L*, related to a higher lightness of blown films. In addition, an increment in *a* values was observed when particle concentration was raised. Even though the *b* value was decreased by talc incorporation at 1% w/w, this parameter was unchanged by incremental increases in talc concentration.
- PP and PP/talc blown films nanocomposites were optically clear in the visible range, revealed by the haze values, taking into account that talc nanoparticles were well dispersed in the polymeric matrix.
- The UV absorption capacity of blown films was demonstrated due to talc nanoparticles acting as blocking agents to this radiation.
- Blown films showed a higher crystalline development in the blow direction due to the crystal induction by talc nucleating action as well as by the blow extrusion process.
- Concerning thermal properties of nanocomposites blown films, the influence of talc presence was observed mainly on the crystallization temperature. This behavior is intimately related with the nucleating character of talc that allows for increase in the nucleation sites for PP crystallization. In addition, a higher crystallinity degree was detected by DSC, in accordance with XRD data.
- Talc presence offered to nanocomposites films a higher thermal stability than PP ones, under nitrogen and oxygen, considering that the increment of crystallinity degree induced by talc leads to a decrease of polymer entropy, giving a more thermally stable film. Talc concentration at 5% w/w raised decomposition temperature 6°C in respect to the corresponding value to PP under nitrogen; meanwhile, T_d is strongly increased in 56°C if nanocomposites films were exposed to thermal degradation in air.
- Nanocomposites films tested in the MD showed a behavior similar to PP and they were demonstrated to be elastic, capable to form neck, flexible, tough, and ductile in MD.
- Young modulus of nanocomposites films showed an increased behavior with talc content in MD. The presence of smaller-sized PP crystals (induced by talc), the greater stiffness of talc particles compared to the PP one and the contribution of particle orientation in PP matrix could increase nanocomposite stiffness.
- Despite the incorporation of 5% w/w talc, particles presence has a negligible effect on elongation at break of nanocomposites films in MD.

- Mechanical behavior of blown films tested in TD revealed that talc particles affected the ductility of the matrix, shifting the mechanical pattern to a more brittle behavior, but remaining ductile films. This result is associated to the mechanical restraint imposed by mineral particles to the mobility and deformability of polymer chains, but by the presence of crystals induced by particles.
- Nanocomposite-blown films showed a higher tear force due to particles generating a zig-zag path that improves the resistance.
- PP and PP/talc nanocomposites films were able to be thermosealed, obtaining a good quality seal. However, talc presence led to a more resistant heat-seal of nanocomposite films.
- The moisture barrier property was enhanced by the talc nanoparticles presence in blown films attributed to: (1) the tortuous path induced by particle orientation within the polymer matrix which forces to water molecules to travel through the film thickness, (2), the hydrophobic character of talc surface that repels water molecules, and (3) the increased crystallinity degree, which act as obstacles for diffusion.
- Nanocomposite films provided the opportunity to reduce transmission rates of water vapor and oxygen, which allows control of the shelf life of products in food packaging.
- Flexible packages were developed from blown PP/talc nanocomposites by thermosealing with relatively low cost. They are easy to handle and convenient for the manufacturer, retailer, and consumer.
- Cherry tomatoes were packaged and stored under controlled conditions of temperature and relative humidity (25°C and 50% HR), without refrigeration.
- Packages made from nanocomposites with 5% w/w talc allowed that tomato appearance remain unaltered after 150 days of storage, conserving its weight during a considerably longer time, without refrigeration.

References

Abu-Zurayk, R., Harking-Jones, E., Mcnally, T., Menary, G., Martin, P., Armstrong, C., 2009. Biaxial deformation behavior and mechanical properties of a polypropylene/clay nanocomposite. Compo. Sci. Technol. 69, 1644–1652.

Allen, N.S., Edge, M., 1992. Fundamentals of Polymer Degradation and Stabilization. Springer Science & Business Media, Cambridge, UK.

Andreasson, E., Mehmood, N., Kao-Walter, S., 2013. Trouser tear tests of two thin polymer films. In: 13th International Conference on Fracture. June 2013, Beijing, pp. 16–21.

Arvanitoyannis, I.S., Bosnea, L., 2004. Migration of substances from food packaging materials to foods. Crit. Rev. Food Sci. Nutr. 44, 63–76.
Azeredo, H.M.C., Mattoso, L.H.C., Avena-Bustillos, R.J., Filho, G.C., Munford, M.L., Wood, D., Mchugh, T.H., 2010. Nanocellulose-reinforced chitosan composite films as affected by nanofiller loading and plasticizer content. J. Food Sci. 75, N1–N7.
Bertini, F., Canetti, M., Audisio, G., Costa, G., Falqui, L., 2006. Characterization and thermal degradation of polypropylene montmorillonite nanocomposites. Polym. Degrad. Stabil. 91, 600–605.
Bilbao-Sáinz, C., Avena-Bustillos, R.J., Wood, D.F., Williams, T.G., Mchugh, T.H., 2010. Composite edible films based on hydroxypropyl methylcellulose reinforced with microcrystalline cellulose nanoparticles. J. Agric. Food Chem. 58, 3753–3760.
Borriello, C., De Maria, A., Jovic, N., Montone, A., Schwarz, M., Antisari, M.V., 2009. Mechanochemical exfoliation of graphite and its polyvinyl alcohol nanocomposites with enhanced barrier properties. Mater. Manuf. Process. 24, 1053–1057.
Castillo, L., Barbosa, S., 2011. Plastic-based nanocomposite films for active packaging with tailor diffusion properties. In: Wythers, M.C. (Ed.), Advances in Materials Science Research, vol. 12. Nova Publishers, New York.
Castillo, L.A., Barbosa, S.E., Maiza, P., Capiati, N.J., 2011. Surface modifications of talcs: effects of inorganic and organic acid treatments. J. Mater. Sci. 46, 2578–2586.
Castillo, L., Barbosa, S., Capiati, N., 2012a. Improved performance of polypropylene/talc composites. Plastic Research Online. Available from: http://www.4spepro.org/view.php?article=004405-2012-08-30&category=
Castillo, L., Barbosa, S., Capiati, N., 2012b. Influence of talc genesis and particle surface on the crystallization kinetics of polypropylene/talc composites. J. Appl. Polym. Sci. 126, 1763–1772.
Castillo, L., Barbosa, S., Capiati, N., 2013a. Influence of talc morphology on the mechanical properties of talc-filled polypropylene. J. Polym. Res. 126, 151–160.
Castillo, L., Barbosa, S., Capiati, N., 2013b. Mechanical properties of polypropylene/talc composites: role of the mineral morphology and particle surface modification. In: Silva, L.P., Fernandes Barbosa, E. (Eds.), Polypropylene: Synthesis, Applications and Environmental Concerns. Nova Publishers, New York.
Castillo, L., Barbosa, S., Capiati, N., 2013c. Surface modified talc particles by acetoxy groups grafting: effects on mechanical properties of polypropylene/talc composites. Polym. Eng. Sci. 53, 89–95.
Chen, Y., Cao, X., Chang, P.R., Huneault, M.A., 2008. Comparative study on the films of poly(vinyl alcohol)/pea starch nanocrystals and poly(vinyl alcohol)/native pea starch. Carbohydr. Polym. 73, 8–17.
Choi, D., White, J.L., 2001. Crystal structures and orientation development in tubular film extrusion of syndiotactic polypropylene and isotactic polypropylene. Polym. Eng. Sci. 41, 1743–1751.
Clark, R., Steen, W., 2003. Talc in polypropylene. In: Karian, H. (Ed.), Handbook of Polypropylene and Polypropylene Composites. Marcel Dekker, New York.
De Moura, M.R., Lorevice, M.V., Mattoso, L.H.C., Zucolotto, V., 2011. Highly stable, edible cellulose films incorporating chitosan nanoparticles. J. Food Sci. 76, N2–N29.
Dearmitt, C., Hancock, M., 2003. Filler thermoplastics. In: Rothon, R.N. (Ed.), Particulate-Filled Polymer Composites. second ed. Rapra Tech. Lim., United Kingdom.

Denac, M., Musil, V., Makarovie, M., 1998. Study of mechanical properties and morphology of polypropylene and talc composites. Metal. Alloy. Technol. 32 (1–2), 69–72.

Duncan, T.V., 2011. Applications of nanotechnology in food packaging and food safety: barrier materials, antimicrobials and sensors. J. Colloid Interf. Sci. 363, 1–24.

Ericson, L., 2000. Strength characterization of suspended single-wall carbon nanotube ropes. Master Thesis, Rice University.

FAO., 2011. Manual Técnico para las Buenas Prácticas Agrícolas (BPA) en la Producción de Tomate bajo Condiciones Protegidas. Corporación Colombiana de Investigación Agropecuaria. Mejoramiento Alimentario y Nutricional de Antioquia. Gobernación de Antioquía. FAO. http://ftp.fao.org/docrep/fao/010/a1374s/a1374s07.pdf

Ferrage, E., Martin, F., Boudet, A., Petit, S., Fourty, G., Jouffret, F., Micoud, P., De Parseval, P., Salvi, S., Bourgerette, C., Ferret, J., Saint-Gerard, Y., Buratto, S., Fortune, J.P., 2002. Talc as nucleating agent of polypropylene: morphology induced by lamellar particles addition and interface mineral-matrix modelization. J. Mater. Sci. 37, 1561–1573.

Finnigan, B., 2009. Barrier polymers. In: Yam, K.L. (Ed.), The Wiley Encyclopedia of Packaging Technology. John Wiley and Sons, New York.

Flexible Packaging Association, U.S., 2015 Available from: http://www.flexpack.org

Gu, S.-Y., Ren, J., Dong, B., 2007. Melt rheology of polylactide/montmorillonite nanocomposites. J. Polym. Sci. B 45, 3189–3196.

Hancock, M., Rothon, R.N., 2003. Principal types of particulate fillers. In: Rothon, R.N. (Ed.), Particulate-Filled Polymer Composites. second ed. Rapra Tech. Lim., United Kingdom.

Jana, S.C., Jain, S., 2001. Dispersion of nanofillers in high performance polymers using reactive solvents as processing aids. Polymer 42, 6897–6905.

Jia, X., Li, Y., Cheng, Q., Zhang, S., Zhang, B., 2007. Preparation and properties of poly(vinyl alcohol)/silica nanocomposites derived from copolymerization of vinyl silica nanoparticles and vinyl acetate. Eur. Polym. J. 43, 1123–1131.

Kanatt, S.R., Rao, M.S., Chawla, S.P., Sharma, A., 2012. Active chitosan-polyvinyl alcohol films with natural extracts. Food Hydrocoll. 29, 290–297.

Kandola, B.K., Smart, G., Horrocks, A.R., Joseph, P., Zhang, S., Hull, T.R., Ebdon, J., Hunt, B., Cook, A., 2008. Effect of different compatibilisers on nanoclay dispersion, thermal stability, and burning behavior of polypropylene–nanoclay blends. J. Appl. Polym. Sci. 108, 816–824.

Kim, J.Y., Han, S.I., Kim, D.K., Kim, S.H., 2009. Mechanical reinforcement and crystallization behavior of poly(ethylene 2, 6-naphthalate) nanocomposites induced by modified carbon nanotube. Compos. A 40, 45–53.

Kristo, E., Biliaderis, C.G., 2007. Physical properties of starch nanocrystal-reinforced pullulan films. Carbohydr. Polym. 68, 146–158.

Li, Y.-C., Schulz, J., Grunlan, J.C., 2009. Polyelectrolyte/nanosilicate thin-film assemblies: influence of pH on growth, mechanical behavior, and flammability. ACS Appl. Mater. Interf. 1, 2338–2347.

Martinatti, F., Ricco, T., 1994. High-rate fracture toughness of polypropylene-based, hybrid, particulate composites. J. Mater. Sci. 29, 442–448.

Meijer, H.E.H., 1997. Materials Science and Technology: Processing of Polymer. VCH, New York.

Menczel, J., Varga, J., 1983. Influence of nucleating agents on crystallization of polypropylene: I. Talc as a nucleating agent. J. Therm. Anal. Calorim. 28, 161–174.

Mercea, P., 2008. Models for diffusion in polymers. In: Piringer, O.G., Baner, A.L. (Eds.), Plastic Packaging. second ed. Wiley-VCH GmbH & Co KGa, Weinheim.

Morton-Jones, D.H., 1989. Polymer Processing. Chapman and Hall, New York.

Nielsen, L.E., Landel, R.F., 1994. Mechanical Properties of Polymers and Composites. Marcel Dekker., New York.

Page, A., 2011. Pira Packaging Summit, Nice, France.

Paine, F., Paine, H., 2012. A Handbook of Food Packaging, second ed. Springer Science & Business Media, London.

Prashantha, K., Soulestin, J., Lacrampe, M.F., Krawczak, P., Dupin, G., Claes, M., 2009. Masterbatch-based multiwalled carbon nanotube-filled polypropylene nanocomposites: assessment of rheological and mechanical properties. Compos. Sci. Technol. 69, 1756–1763.

Prnewswire, 2015. Global Flexible Packaging Market (Plastic, Paper, Aluminum Foil and Others) for Food & Beverages, Retail Non-Food, Pharmaceutical and Other Applications—Global Industry Analysis, Size, Share, Growth, Trends and Forecast, 2013–2019. http://www.prnewswire.com/

Rabello, M.S., White, J.R., 1996. Photodegradation of talc-filled polypropylene. Polym. Compos. 17, 691–704.

Radhakrishnan, S., Kane, K., Kadu, A., Natu, H.P., 1995. Structure development in processing of polypropylene films with additives. J. Appl. Polym. Sci. 58, 571–577.

Robertson, G.L., 2006. Food Packaging: Principles and Practice, second ed. CRC Press, Taylor & Francis Group, New York.

Rolando, T.E., 2000. Flexible Packaging: Adhesives, Coatings and Processes. RAPRA Technology Limited, United Kingdom.

Schone, J., Kotter, I., Grellman, W., 2012. Properties of polypropylene talc compounds with different talc particle size. J. Plastic. Technol. 8, 231–251.

Shimomura, Y., Spruiell, J.E., White, J.L., 1982. Orientation development in the tubular film extrusion of polypropylene. J. Appl. Polym. Sci. 27, 2663–2674.

Smart, G., Kandola, B.K., Horrocks, A.R., Nazaré, S., Marney, D., 2008. Polypropylene fibers containing dispersed clays having improved fire performance. Part II: characterization of fibers and fabrics from PP–nanoclay blends. Polym. Adv. Technol. 19., 658–670.

Soon, K.H., Harkin-Jones, E., Rajeev, R.S., Menary, G., Mcnally, T., Martin, P.J., Armstrong, C., 2009. Characterization of melt processed poly(ethylene terephthalate) (PET)/synthetic clay nanocomposite sheet and its biaxial deformation behavior. Polym. Int. 58, 1134–1141.

Tang, S., Zou, P., Xiong, H., Tang, H., 2008. Effect of nano-SiO_2 on the performance of starch/polyvinyl alcohol blend films. Carbohydr. Polym. 72, 521–526.

USDA., 1991. U.S. Standards for Grades of Fresh Tomatoes. USDA, Agr. Mktg. Serv., Washington, DC, EE.UU.

Utracki, L.A., 1991. Two-Phase Polymer Systems. Hanser Publishers, Munich.

Varga, J., Schulektoth, F., Mudra, I., 1994. Blends of the β-modification of polypropylene. Macromol. Symp. 78, 229–241.

Wakabayashi, K., Pierre, C., Dikin, D.A., Ruoff, R.S., Ramanathan, T., Brinson, L.C., Torkelson, J.M., 2008. Polymer-graphite nanocomposites: effective dispersion and major property enhancement via solid-state shear pulverization. Macromolecules 41, 1905–1908.

Yang, D., Hu, Y., Song, L., Nie, S., He, S., Cai, Y., 2008. Catalyzing carbonization function of α-ZrP based intumescent fire retardant polypropylene nanocomposites. Polym. Degrad. Stabil. 93, 2014–2018.

Yoo, Y., Kim, S.-S., Won, J.C., Choi, K.-Y., Lee, J.H., 2004. Enhancement of the thermal stability, mechanical properties, and morphologies of recycled PVC/ clay nanocomposites. Polym. Bull. 52, 373–380.

Yuksekkalayci, C., Yilmazer, U., Orbey, N., 1999. Effects of nucleating agent and processing conditions on the mechanical, thermal, and optical properties of biaxially oriented polypropylene films. Polym. Eng. Sci. 39 (7), 1216–1222.

Zhang, F., Zhang, H., Su, Z., 2008. In situ preparation of $(Mg(OH)_2$/polystyrene nanocomposite via soapless emulsion polimerization. Polym. Bull. 60, 251–257.

19

THE INFLUENCE OF NANOFILLERS ON PHYSICAL–CHEMICAL PROPERTIES OF POLYSACCHARIDE-BASED FILM INTENDED FOR FOOD PACKAGING

Aleksandra R. Nesic*, Sanja I. Seslija**

**University of Belgrade, Vinca Institute for Nuclear Sciences, Belgrade, Serbia; **University of Belgrade, Institute of Chemistry, Technology and Metallurgy, Belgrade, Serbia*

1 Introduction

With an ever-increasing world population, the question is not whether the global environment will be impacted by our presence but how and to what degree. The implementation of sustainable practices will help to minimize our impact on the environment and conserve resources for future generations. To that end, there is a need to perpetuate the culture of environmental stewardship and sustainability that has grown stronger in recent years. Consumers and municipalities must continue to demand more sustainable packaging materials and practices. Retailers must continue to provide suppliers incentives for greater sustainability in their packaging choices. Although some of the biobased materials may not currently be cost-competitive with petroleum plastics, this could change as petroleum prices increase.

Due to of the desired features such as softness, lightness, and transparency, polymers represent the most common feedstock for production of packaging materials. However, increased use of polymers obtained from petrochemical resources has led to the serious ecological problems due to their poor biodegradability. Although their complete replacement with ecofriendly packaging

Food Packaging. http://dx.doi.org/10.1016/B978-0-12-804302-8.00019-4

films is hard to achieve, at least for specific applications like food packaging the use of natural polymers, such as polysaccharides, should be the future. This approach will provide the quality improvement of many products and it is satisfying from the standpoint of the consumers demands. In order to obtain the required properties, the polysaccharides are usually reinforced by the addition of various nanofillers and the main goal of this chapter is to provide a complete view of the state of the art on polysaccharide-based nanocomposite materials intended for food packaging applications.

2 Relevant Properties for Food Packaging Applications

Traditional food packaging provides mechanical supporting and protection of food from external influences. Striving to meet the requirements of the consumer society, food packaging technologies must be directed toward keeping the food safe and reducing the environmental impact. Furthermore, food packaging must be designed to ensure product deterioration, retain the beneficial effects of processing, extend shelf life, and maintain or increase the quality and safety of food (Marsh and Bugusu, 2007). Regarding the different external influences, the protection provided by food packaging materials is classified in three categories: chemical, biological, and physical. Chemical protection minimizes compositional changes caused by environmental influences such as exposure to gases (typically oxygen), moisture (gain or loss), or light (visible, infrared, or ultraviolet). Biological protection provides a barrier to microorganisms (pathogens and spoiling agents), insects, rodents, and other animals while the physical protection shields food from mechanical damage and various hazards (Restuccia et al., 2010). The properties of food packaging are closely related to their structure and the widened application depends on various functional parameters such as mechanical strength, barrier to gases, moisture and water, transparency, temperature resistance, and biodegradability. The main precondition for selection of the suitable material is a minimum of interaction between food and packaging (Lagaron et al., 2005).

2.1 Barrier Properties

One of the main functions of food packaging is to act as a barrier that separates and protects the product from exposure to the environment. Barrier functions include moisture, oxygen and other gases, water vapor, fats and oils. The barrier functions of the package depend on the chemical properties of the used material. However, environmental conditions, such as temperature, relative humidity, and handling of the product by consumers also affect the performance of the package. Good gas barrier properties prolong the shelf life of the products. Material with high oxygen barriers helps to reduce lipid oxidation and nutrition loss of fresh products (fruits, vegetables, and salads) and high fat (eg, nuts, donuts). In general, materials with higher crystallinity have better oxygen barrier properties. The water vapor permeability of a packaging material indicates the rate of moisture transfer between the storage environment and the internal environment of the package. The moisture-absorbing rate of the film is also important in comparison with the intermediate moisture of the food and on how this influences the drying of the product during storage (Siracusa et al., 2008).

2.2 Optical Properties

The optical properties of packaging materials are very important in the prevention of photodegradation of food components during the storage. The food exposed to light causes the shortness of the shelf life of food products due to degradation reaction of sensitive components in foods such as lipids, flavors, vitamins, and pigments. Therefore, the light exposure and light transmittance of packaging materials can significantly affect the food quality. The transmission rate of visible light (400–700 nm) and ultraviolet radiation (100–400 nm) play important roles in designing the right packaging in terms of preservation and protection of food products during the handling, storage, transport and until they reach the consumer (Turhan and Sahbaz, 2001).

2.3 Mechanical Properties

Protection of packaging materials against external forces depends on its mechanical properties. In packaging technology, mechanical properties should be considered and evaluated at the level of the packaging material, the formed empty package, the product package assembly, and the outer packages. The main mechanical characteristics of packaging materials are related to tensile strength (TS), elongation at break (E), elastic modulus (EM), and tearing resistance (TR). TS is related to the measure of the material ability to resist breaking under tension. E is a quantitative representation of the material's ability to stretch, while EM presents the fundamental measure of material stiffness. TR corresponds to the average force applied during the tearing operation. Therefore, the stress/strain curves obtained from mechanical testing provide information about the flexibility, toughness, and elongation that can be used to predict the material performance during handling.

2.4 Chemical Reactivity

Almost all packaging materials may react in a certain way with food and/or environment without causing serious consequences towards safety and food quality. Possible chemical compound-migrants that are incorporated within packaging materials to improve functionality include plasticizers, antioxidants, thermal stabilizers, slip compounds, and monomers (Ferrara et al., 2001). Chemical migration from food packaging is affected by various parameters such as concentration of compound-migrants in packaging materials, nature of the foods, storage condition, temperature of the system, and the time period over which duration of contact occurs. Some chemical components contained within packaging may transfer to food contact surfaces via the "set-off" process (Muncke, 2014). This is a direct transfer from the external surface of the packaging to the food contact surface during stacking and storage of packaging. Transfer can also occur via evaporation and then leaching into food via the gaseous phase. The type of packaging material largely determines the potential and migration level of chemicals into food (Muncke, 2014).

2.5 Materials for Food Packaging

The selection of suitable packaging materials for foods is primarily determined by the properties of food, which should be packed. Also, the chemical composition and physical properties of packaging materials are important factors to fulfill the various functions expected from the package in order to preserve food and extend its shelf-life. The most common food package materials on market belong to the following classes: metal, glass, paper, and polymers. Nowadays, food packages often combine several materials to exploit each material's functional or aesthetic properties.

2.5.1 Metal

Metal can provide good protection against physical damage during transportation, distribution, handling, and storage of food products due to its superior mechanical strength. Its high barrier properties to gases, water vapor, and light and good thermal conductivity makes this material suitable for thermal in-packaging processes (such as pasteurization). Tinplate is one of the most commonly used packaging materials for production of metal cans and canisters that preserves liquid foods and extends their shelf life. Tinplate consists of a thin sheet of steel which is coated with tin in order to reduce risk of corrosion. The traditional method for coating the steel plates with tin is the "hot dip" method (Robertson, 2005). However, in order to produce lighter and less expensive cans with improved performance, the hot dip method has now been replaced by a process of electrolytic deposition (Robertson, 2005).

Among all metal packaging materials, the most important material next to tinplate is aluminum. The main advantage of aluminum compared to tinplate is that aluminum does not require protective anticorrosion layer due to formation of the thin aluminum oxide film on the surface, which protects the metal against oxygen and mild acids corrosion. Aluminum is much lighter and more ductile than tinplate, but it is more expensive. Aluminum is mostly used for the production of cans to preserve the liquids such as beer and soft beverages, or in form of aluminum foil. Aluminum foils are used to pack takeaway meals, ready snacks, and long-life pet foods.

2.5.2 Glass

The glass is commonly used for making bottles and jars for food and beverages. The main advantages of glass as a packaging material are transparence, inertness, high barrier properties against gases and chemicals. Also, its thermal resistance provides insulation effect, which keeps foods fresh during storage. Glass containers can be recycled (remelted), which is technically and economically feasible. Despite the good properties, it does have some downsides. Glass containers are very fragile and the strong impact can fracture a container, making it useless. Glass is also vulnerable to sharp changes in temperature. Heating and then exposing the headed glass to cool water or air could be enough to shatter it. The second main disadvantage of glass is its weight. Glass is heavier compared to paper, plastic, or even some metal containers, which directly affects the price of shipping costs (Robertson, 2005).

2.5.3 Paper

Paper and paperboard have become major packaging materials around the world due to their low cost. Paperboard presents the most popular rigid material used for food packaging. It is mainly produced with primary fiber and secondary fiber derived from old corrugated containers. The main advantages of paper as a packaging material are wide availability, low weight, printability, mechanical strength, reuse, and recycling after use. However, the most serious shortcoming is its sensitivity to moisture. In order to improve the barrier properties against water and gases, as well as mechanical strength of paper and paperboard, they are combined with other materials such as oil, wax, polymers (plastics), and metals through coating or lamination. Such multilayer packaging materials have been used in a variety of food products (Miltz, 1992).

2.5.4 Polymers

Among the four basic packaging materials, polymer-based materials have been widely used since the middle of the 20th century, mainly due to less expensive technology and excellent physical–chemical properties (thermosealability, optical properties, and unlimited sizes and shapes) over traditional materials such as glass and metal. However, unlike metal or glass, polymers are permeable to small molecules to a greater or lesser extent. There have been some health concerns regarding migration of monomers, stabilizers, and plasticizers from package to food. Until now synthetic polymers such as polyethylene terephthalate (PET), polyvinylchloride (PVC), polyethylene (PE), polypropylene (PP), polystyrene (PS), and polyamide (PA) have been increasingly used as packaging materials (Teck Kim et al., 2014). Due to their large availability at relatively low cost, good mechanical performance such as tensile and tear strength, good barrier to oxygen, carbon dioxide, anhydride, and aroma compound as well as heat sealability, these materials are widely abundant in food packaging application. High-density PE is used as a material for milk bottles and bags, while low-density PE is used in production of general-purpose containers. PP obtains the lowest density of the plastics used for packaging application, high melting point and an excellent chemical resistance. PET is transparent, tough, and has good gas and moisture barrier properties. It exhibits good resistance to heat, mineral oils, solvents, and acids. Hence, it is becoming the packaging material of choice for many food products, especially beverages and mineral waters (Willige et al., 2002). The main functional characteristics of synthetic polymer-based food packaging materials are presented in Table 19.1.

Despite their enormous versatility, polymeric materials used for food packaging application exhibit numerous limitations related to their inherent permeability to gases and vapors, including oxygen, carbon dioxide, and organic vapors. Furthermore, the sustainability of these products is low, and renewability and degradability have become key words in the debate over sustainable production and utilization of plastic.

Table 19.1 The Main Functional Characteristics of Synthetic Polymer-Based Food Packaging Materials

Packaging Material	Water Vapor Permeability ($\times 10^{14}$) (g/m s Pa)	O_2 Permeability ($\times 10^{7}$) (mL/m day Pa)	Tensile Strength (MPa)	Elastic Modulus (GPa)	Impact Strength (J/m)	Light Transmission (%)
PVC	18.279	0.198–0.790	42–55	2.8	180–290	90
PVDC	1.161	0.001–0.030	49–98	0.2–0.6	—	90
PP	2.321–4.642	4.936–9.869	27–98	1.18	43	80
HDPE	1.741–3.482	7.127	19–31	0.98	373	—
LDPE	6.673–8.704	44.756	7–25	0.5–0.34	375	65
EVA	6.673–17.118	21.220	6–19	—	45	55–75
PA	5.803–114.314	0.010–0.098	49–69	0.7–0.98	50–60	88
PET	5.803–22.921	0.098–0.494	157–177	3.5	100	88
PS	11.315–45.552	9.869–14.805	31–49	2.7–3.4	59	92

Reproduced with permission from Sung, S.Y., et al., 2013. Trends Food Sci. Technol. 32, 110–123. Copyright © 2013, Elsevier B.V.

The current global consumption of plastics exceeds 200 million tons, with an annual grow of approximately 5%, which represents the largest field of application for crude oil. Plastic packaging materials usually are contaminated by food stuff and/or biological substance; therefore, recycling is not practical nor economically feasible. Due to the lack of a comprehensive solution related to disposal, several thousands of tons of goods, made on plastic materials, are landfilled. These materials, being resistive to degradation (depending on the polymer, additives, and conditions, etc.) tend to accumulate as plastic waste, creating a serious problem of plastic waste management (Sorrentino et al., 2007). The utilization of renewable resources in material applications is receiving increasing attentions due to serious concerns regarding environmental sustainability. The current streams related to the packaging materials are directed on development and promotion of biodegradable and biobased materials. Therefore, strong efforts are made to identify suitable materials derived from renewable sources, which could be used for food packaging application (Souza et al., 2010).

Biodegradable polymers are defined as materials, which are able to decompose into carbon dioxide, methane, inorganic compounds or biomass as a result of being exposed to the enzymatic action of microorganisms. Intensive researches paved the way to biopolymers as materials to be used as edible coatings in food packaging (Azeredo et al., 2012). This approach provides the quality improvement of many products and it is satisfying from the standpoint of the consumers demands for more environmentally friendly packaging. Biopolymers can be broadly divided into different categories based on the origin of the raw materials and their manufacturing processes. The first category, based on the origin of the raw materials, contains biopolymers derived from plants such as starch, cellulose, chitosan, alginate, agar, carrageenan, and biopolymers derived from animals or plant origin proteins such as soy protein, corn zein, wheat gluten, gelatin, collagen, whey protein, casein. The polymers polymerized from renewable monomers, such as poly(lactic) acid (PLA), poly(glycolic acid) (PGA), poly(ε-caprolactone) (PCL), poly(butylene succinate) (PBS) are also biodegradable and belong to the second category. The third category includes polymers produced by microbial fermentation like microbial polyesters, such as poly(hydroxyalkanoates) (PHAs) including poly(β-hydroxybutyrate) (PHB), poly(3-hydroxybutyrate-*co*-3-hydroxyvalerate) (PHBV), and microbial polysaccharides, such as pullulan and curdlan (Bordes et al., 2009; Clarinval and Halleux, 2005). The main characteristics of biobased food packaging materials obtained from literature are presented in Table 19.2.

The biopolymers used in food packaging have numerous advantages that are in accordance with environmental issues: availability from renewable resources, biocompatibility, and biodegradability (Prashanth and Tharanathan, 2007). The other features of importance such as: thermal stability, flexibility, good barrier to gases and chemicals, are directly affected by the structure of polymer used in production of package. However, biopolymers have poor mechanical or barrier properties toward moisture and water vapor compared to the synthetic polymers, which is a major limitation concerning its widened applicability in food packaging. It has been suggested that inherent shortcomings of biopolymer-based packaging materials may be overcome by implementation of nanocomposite technology. Nanocomposites exhibit improved barrier properties, mechanical strength, as well as the heat resistance compared to their neat polymers and conventional composites (Sinha Ray et al., 2006).

2.6 New Trends in Food Packaging

Most developments in the field of food technology have been focused on improvement of safety, shelf life, and convenience of the food product. Three broad innovation groups of packaging

Table 19.2 The Main Functional Properties of Biobased Films Intended for Food Packaging Application

			Mechanical and Permeability Properties		
Film	Concentration in the Film-Forming Solution	Plasticizer	Tensile Strength (MPa)	Strain at Break (%)	Water Vapor Permeability (g mm/m^2 day kPa)
Cellulose ethers	Hydroxypropylcellulose 3% w/w	—	16–18	60–110	0.043–0.056
	Hydroxypropyl methylcellulose 6% w/v	Glycerol	20–43	26–41	3.6–89.9
	Hydroxypropyl methylcellulose 1–6% w/w	Poly(ethylene glycol)	3–86	1.9–60.9	4.1–13.7
	Methylcellulose 1.5–6% w/w	Poly(ethylene glycol)	17–44	14–97	2–10
Starches	Tapioca 5% w/w	Glycerol	0.16–2.3	70	54–139
	Cassava 5% w/w	Sugars	1.0–4.7	39–164	1.5–7.2
	Sweet potato 4% w/w	Glycerol	8–43	0.6–3.2	14–86
Seaweed extracts	κ-Carrageenan 0.1–1% w/w	—	57	7	—
	Alginate 1% w/w	Glycerol	39–66	2.7–4.8	18.7–30.9
	Alginate 1.5% w/w	Glycerol	23–160	2.2–32.2	7–14
Gums	Locus bean 0.9% w/v	Poly(ethylene glycol)	4–40	0.5–10	—
	Gellan 0.5% w/v	Glycerol	—	—	18–23
Pectin	Pectin citric fruit 2.5–5.8 mg/cm	—	13–25	0.8–1.2	1.0–4.7
Chitosan	2% w/v	Glycerol	5.5–21.3	7–43	4.8–7.3
	1% w/v	—	7–12	11–17	60–138
	1.5% w/v	—	18–106	5–20	14–80
Proteins	Sodium caseinate 4% w/w	Sorbitol	2–77	2–130	85–564
	Sodium caseinate 5% w/w	Glycerol	3.4–4.1	78–125	—
	Soy protein isolate 10% w/w	Glycerol	4.7–10.7	—	—
	Whey protein isolate 5% w/w	Sorbitol	4–16	1–10	204–264
Blends	Whey protein/hydroxypropyl methylcellulose	Glycerol	4–61	16–112	—
	Starch/chitosan	Glycerol	0.36–19.7	61–152	0.15–3.3
	Pectin/fish skin gelatin or soybean flour protein	Glycerol	17–99	1.7–6.4	98–226
	Sago starch/sodium alginate	Glycerol	13–16	3.7-13.2	20.7–34.6

materials include active packaging, modified atmosphere packaging (MAP), and edible films and coatings. Recently, food packaging technologies have been shifted toward the use of more sustainable systems and materials in order to reduce energy consumption and effect on environment, as well as to promote greater use of recyclable, renewable, or biodegradable packaging materials.

2.6.1 Active Packaging

Active packaging employs the packaging materials that interact with the environment and with the food, playing an active role in prolonging the shelf-life of products due to reduced oxidation and microbial spoilage. Most important active packaging technologies include oxygen and ethylene scavenging, carbon dioxide scavengers, moisture regulators, antimicrobial packaging, and antioxidant release technologies. Oxygen scavengers are in the form of small sachets inserted into the package to reduce the level of oxygen within the package. An oxygen-free environment prevents food oxidation and the growth of aerobic bacteria and mold. Ethylene scavenging systems include packaging material with incorporated ethylene-absorbing material intended to slow the process of ripening and senescence of fruits and vegetables which are triggered by ethylene. Carbon dioxide scavengers have been used for products which are subjected to the fermentation or aging and produce gas after packaging. An active system that scavenges moisture has been used for packaging of dried and moisture-sensitive foods and usually can be found in the form of a sachet, which includes desiccant material (Brody, 2009).

Advanced active packaging systems include the development of systems designed to release agents referring to undesirable activity within the package. Antimicrobial agents incorporated in packaging prevent the growth of spoilage and pathogenic microorganisms that directly influence the shelf life of products. Antimicrobial packaging has more advantages compared to direct addition of antimicrobial agents onto foods by sprays or drips. The direct addition of antimicrobial agents on the surface of food is not effective enough to inhibit microorganisms due to their rapid diffusion into foods and denaturation of the active substances by food constituents that reduce the reactivity of antimicrobial agent. On the contrary, antimicrobial packaging offers slow and continuous release of antimicrobial agent from packaging material to food surfaces, preventing them to maintain high concentration over a long period (Quintavalla and Vicini, 2002). The commercialized antimicrobial packaging is packaging which contains antimicrobial agents in volatile gas-form: chlorine dioxide, ethanol, or sulfur dioxide. The most commonly investigated antimicrobial agents are metal and metal oxide nanomaterials such as silver, zinc oxide (ZnO), silica, titanium dioxide (TiO_2). Besides them, various enzymes, bacteriocins, essential oils, and anhydrides, as well as weak organic acids are also studied in terms of antimicrobial activity of food packaging systems (Corrales et al., 2014).

In order to protect the film and the product from degradation, similarly to antimicrobial agents, antioxidants can be incorporated into packaging films. Antioxidant agents inhibit the lipid oxidation, which is of the great importance in protecting the foodstuffs with high amounts of unsaturated fatty acids from possible quality deterioration and for providing the required shelf life. The synthetic antioxidants such as butylated hydroxyanisole (BHA), butylated hydroxytoluene (BHT), and Irganox have been widely used in the packaging industry. However, these antioxidants migrate into foods and induce off-flavors in the packaged food, so there is growing interest for finding new natural antioxidants. Antioxidants derived from natural sources, which have been mostly studied in terms of the packaging development are tocopherol, ascorbic acid, curcumin, tyrosine, essential oils, and plant extracts of barley husks, borage, cinnamon, citronella, clove, ginger, green tea, marigold, murta leaves, rosemary, oregano, and thyme (Sun Lee, 2014).

2.6.2 Modified Atmosphere Packaging

Modified atmosphere packaging (MAP) is a package with adjusted and carefully balanced mix of normal atmospheric gases: pure oxygen, carbon dioxide, and nitrogen. The main role of MAP is to preserve the freshness of fresh-cut processed fruits and vegetables, nonfrozen chilled meats/fish/poultry, ready-to-eat meals, and semiprocessed bulk foods by controlling their biochemical metabolism (respiration or fermentation). In contrast to the other types of food packaging, MAP does not require the addition of chemical preservatives or stabilizers to be efficient in prolongation of shelf life of foods. However, MAP has some limitations for wide commercial use due to the high cost of gas packaging machinery and the potential growth of food-borne pathogens due to temperature abuse by retailers and consumers. Also, the benefits of MAP are lost once the pack is opened or leaks. The most investigated packs for MAP are based on PVC, PET, PE, and PP, depending on the features desired for the intended use (McMillin, 2008).

2.6.3 Edible Films and Coatings

Edible films and coatings are thin layers that are coated on the surface of food products. Beyond those coming from the conventional ones, the main functions of this type of packaging are edibility and biodegradability. Edible films may potentially improve barrier and mechanical properties, convenience, and prolong the shelf life of various products. Edible films and coatings are produced mainly from biopolymers, including proteins, polysaccharides (carbohydrates and gums), lipids, and from food-grade additives. Produced by casting methods, edible films are very easy to handle since they require less heat and no toxic solvents compared to traditional petroleum-based food packaging. Polysaccharide-based edible films possess good oxygen barriers and have the potential to be crosslinked with various micronutrients or flavors in order to enhance functionalities of packaging material. However, these films, as well as protein-based edible films, have poor moisture barriers because of their hydrophilic nature. On the contrary, lipid-based edible films have good moisture barrier, but low mechanical properties due to their hydrophobicity. The addition of nanofillers is one of the approaches that offer improved performances of the edible films and coatings along with environmental acceptability (Krochta, 1997).

3 Nanoscale Reinforcements for Food Packaging

Despite all the advantages, the usage of biopolymers has been limited due to the poor mechanical and barrier properties. Biopolymers are generally brittle, low in heat distortion temperature, and low resistance to prolonged process operations. For instance, the use of the most favorable synthetic biopolymers, PLA, as food packaging material exhibits low performances due to low heat distortion temperature, low resistance to extreme heat and humidity, and low flexibility (Sorrentino et al., 2007). In order to tailor the properties of materials obtained from renewable resources to meet specific processing requirements, functional and structural demands, the certain drawback could be improved by addition of nanosized compounds or fillers. According to the available literature data, nanocomposite was established as a promising route to enhance mechanical and barrier properties of biopolymers. Namely, nanocomposites are defined as a multiphase solid material system that is formed by a combination of two or more components that include polymers as matrices (continuous phase) and nanostructured materials (discontinuous phase) with at least 1 dimension in nanosize (<100 nm). The role of nanosized fillers is mainly structural, in which they act as reinforcement to improve the mechanical and barrier properties of the matrix. The material properties are improved by transferring of the matrix tension to the nanoparticles through the boundary between them. Remarkable improvement of the properties obtained in certain nanocomposite formulations have expanded the application of nanocomposite films in various industrial applications such as biomedical and food packaging sectors (Duncan, 2011). Additionally, inorganic particles may be used to introduce multiple functionalities like color and odors but also to act as reservoirs for the controlled release functions of drugs or fungicides (Rhim et al., 2013).

At nanoscale level, the size of the nanophase or filler is significantly reduced, leading to the dramatic increment in the surface area of the fillers. The high surface area of the nanosized fillers combined with the corresponding matrix results in large interfacial or boundary area between the components of the composite material. The molecular mobility is strongly affected by these interactions so the relaxation behavior, as well as mechanical, thermal, and barrier properties of the bionanocomposites are changed (Azeredo, 2009). Especially for food packaging applications, bionanocomposite materials are designed to support mechanical and thermal stress applied during food processing, transportation, and storage (Rhim et al., 2013).

Many types of nanosized fillers have been used to enhance the performance of the biopolymers intended for food packaging applications. They can be classified depending on dimension in nanometer range into few types, which include nanoparticles (isodimensional-silica, metals, metal oxide), nanorods, nanowires, and nanotubes (2 dimensions in nanometer range—carbon nanotubes, cellulose nanoparticles) and nanofibrils (3 dimensions in nanometer range) (Azeredo, 2009). The comparison between biocomposite materials based on literature turns to be difficult due to different combinations of filler and biopolymer used for the material preparation.

The nanosized fillers can be generally classified as organic and inorganic. Table 19.3 represents some of the nanofillers that have been studied for food packaging applications.

The most common nanofiller type that has been used for preparation of bionanocomposite materials intended for food packaging was clay (Abdollahi et al., 2013a,b; Kim and Cha, 2014; Mondal et al., 2013; Pinto et al., 2015; Shin et al., 2014). The main reason for the heightened interest related to the clay nanoparticles are: presence in abundant, low cost, relative process ability, and significant reinforcement of bionanocomposite materials in the same time. Due to the high surface area and large aspect ratio (50–100) the most extensively studied clay is montmorillonite (MMT), which is defined as a hydrated alumina-silicate layered clay consisting of

Table 19.3 Brief Overview of the Nanofillers That Have Been Studied for Food Packaging Applications

Classification	Type of Nanofiller	Example
Organic	Clay	MMT, cloisite–Na^+, Cloisite 30B, kaolinite, brucite, hectorite, sepiolite, palygorskite, laponite
	Natural biopolymers	Chitosan, starch, cellulose, chitin
Inorganic	Metal	Silver, copper, gold
	Metal oxide	ZnO, TiO_2, Ag_2O, MgO, $nSiO_2$

an edge-shared octahedral sheet of aluminum hydroxide between two silica tetrahedral layers (Park et al., 2013; Pinto et al., 2015; Weiss et al., 2006).

The unique antimicrobial properties of silver nanoparticles put them in the line of the most commercialized nanomaterials, especially in food packaging application (Benn and Westerhoff, 2008; Loutfi et al., 2015). On the other hand, this antibacterial activity is undesirable when nanosilver is disposed of and ends up dissolving and leaching ions (Laban et al., 2010), which act against aquatic microorganisms (Erickson, 2009).

The investigation directed toward detailed understanding of its toxicity has been promoted according to the requirements of regulatory agencies (Navarro et al., 2008; Nel et al., 2009). In this way, correct risk assessments can be made and its safe use can be established for minimal, if any, environmental impact. Furthermore, there is an ongoing debate regarding the role of released Ag^+ ions from nanosilver and its impact toward microorganisms (Miao et al., 2009; Xiong et al., 2008). Opposed to the streamy trend related to use of nanosilver, the application of nanogold particles is not so widespread, due to the initially high costs of implementation. Besides nanosilver, the most investigated antimicrobial nanofillers are zinc and copper nanoparticles and metal oxide nanomaterials such as TiO_2, ZnO, and magnesium oxide (MgO) nanoparticles (Premanathan et al., 2011). The results obtained in research conducted by Conte et al. indicated a good antibacterial activity of copper nanoparticles incorporated in active polylactic acid film, and represent a step forward in their possible application in the food packaging industry (Conte et al., 2013).

TiO_2 has been widely applied as surface coating agent with photocatalytic disinfecting properties. Jing et al. (2011) demonstrated the antibacterial efficiency of both TiO_2 nanoparticles and nanotubes against *Escherichia coli*. Beside the antimicrobial activity, metal oxides were reported to be effective like ultraviolet (UV)-blocking agents due to the increased surface area and intense absorption in the UV region (Vigneshwaran et al., 2006). Kanmani and Rhim showed that the addition of ZnO nanoparticles increased color, UV barrier, moisture content, hydrophobicity, elongation, and thermal stability of the various polysaccharides films, while decreased water vapor permeability, TS, and EM (Kanmani and Rhim, 2014). The observations made by Tripathi et al. (2011) suggested that CS based nanocomposite film containing silver oxide obtain an excellent antibacterial ability for food packaging applications, while the thermal properties of the film remained the unchanged after the addition of nanoparticles.

Silica nanoparticles ($nSiO_2$) have been reported to improve the mechanical and/or barrier properties of several polymer matrices, both natural and synthetic (Zhang and Rong, 2003). For instance, it was noted that the addition of $nSiO_2$ decreased water absorption and enhanced thermal and mechanical properties of PVOHand starch-based nanocomposite films (Navarro et al., 2008).

Polysaccharide-based nanoparticles have appeared as one of the most promising nanofillers due to biocompatibility, biodegradability, and nontoxicity. Namely, chitosan, starch, and CNs have been investigated for food packaging application. Martelli et al. (2013) have studied the effect of the chitosan nanoparticles on the mechanical properties, water vapor transmission rate, thermal stability, and antimicrobial activity of the edible films based on over-ripe banana puree, with small addition of pectin and glycerol as plasticizers. It was found that the additions of pectin (0.5%) and glycerol could promote a significant improvement in elongation, implying better plasticity and handability. Small amounts of chitosan nanoparticles (0.2 wt.%) were found to be sufficient as a reinforcement material, enhancing the overall mechanical properties. Additionally, the presence of nanoparticles in the film matrix reduced significantly the water vapor permeation. However, for the amount of chitosan nanoparticles used (0.2 wt.%) in this study, any detectable antimicrobial activity was not found.

The chitosan nanoparticles crosslinked with tripolyphosphate were incorporated into starch films plasticized with glycerol (Chang et al., 2010). It was found that chitosan nanoparticles up to 6 wt.% were uniformly dispersed in starch matrix, which resulted in improvement of mechanical, thermal and water vapor barrier stability, while the higher amounts (8 wt.%) of chitosan nanoparticles into starch films plasticized with glycerol led to the aggregation of nanoparticles.

Teodoro et al. (2015) reported that inclusion of acetylated starch nanoparticles as nanoreinforcement into starch matrix enhanced mechanical, thermal and water vapor properties. The best water vapor barrier properties were observed by addition of 1.5% (w/w) of acetylated starch nanoparticles, while the optimum amount of acetylated starch nanoparticles for the moisture barrier properties showed to be 10% (w/w). Water vapor permeability and moisture adsorption were reduced by addition of nanoparticles for 41 and 33%, respectively. The addition of 0.5% (w/w) of nanoparticles enhanced the Young's modulus and thermal stability by 162 and 15%, respectively.

Azeredo et al. (2010) have conducted the study in order to evaluate the effect of different concentrations of nanofiller (cellulose nanofibers) and plasticizer (glycerol) on tensile properties (TS, E, and Young's modulus), water vapor permeability, glass transition temperature of chitosan edible films, and to establish a formulation to optimize their properties. In addition to the advantages related to environmental benefits, the chitosan-based film prepared using 15% cellulose nanofibers and plasticized with 18% glycerol was comparable to the synthetic polymers in terms of strength and stiffness. Despite the fact that mechanical and water vapor barrier properties of chitosan films were improved by the addition of cellulose nanofibers, their elongation and water vapor barrier were poorer, indicating that the film may be used only for applications that do not require a great flexibility and/or water vapor barrier.

A similar study was conducted by El-Wakil et al. (2015) wherein TiO_2 nanoparticles together with cellulose nanocrystals (CNCs) were used for reinforcement of wheat gluten films. With regard to the TS, and water resistance it was found that the optimal content of CNCs and TiO_2 nanoparticles was 7.5 and 0.6% by weight, respectively. Under certain conditions, the coated craft paper with wheat gluten/CNCs 7.5%/0.6% TiO_2 exhibited excellent antimicrobial activities, that is, 100, 100, and 98.5% against *Saccharomyces cervisiae*, *E. coli*, and *S. aureus*, respectively.

Although promising results were obtained, the road to the consummate bionanocomposites is still long. Nanotechnologies promise many stimulating changes to enhance health, wealth and quality of life, while reducing the environmental impact. The use of nanofillers in materials' production requires proper considerations in terms of safety for end use because there are limited investigations done on the toxicological effects and migration of nanofillers into food from the packaging films. It is mandatory that consumers have adequate information about food and its packaging (that includes type and size of nanoparticle, physicochemical properties of nanoparticle/nanocomposite, nanocomposite processing method, and safety concerns of nanomaterial/nanocomposite). The absence of a clear definition, public debate, food safety assessments, and proper food regulations may endanger the benefits of nanotechnologies implemented in food industry. Two specific hurdles in performing risk assessments on nanomaterials have been identified by Scientific Committee under European Food Safety Authority (EFSA) difficulty in characterizing, detecting, and measuring nanomaterials and insufficient information on toxicology data (Committee, 2009). The classification, labeling, and packaging of substances and mixtures are regulated by the Regulation (EC) No. 1272/2008. Users of hazardous substances and mixtures must be informed about the dangers by means of a new labeling system along with the safety data sheets, which must be clearly subjected. The information about the properties of the substance produced at the nanoscale must be reported to the European Chemical Agency (ECHA). This regulation is not strictly related to nanomaterials but definitely represents a new hurdle with which chemical producers must contend in order to produce new substances. There are only a few international regulations related to nanotechnology and nanoproducts. The standards and regulation for defining the application of nanotechnology have been established only by few government agencies or organizations from different countries. The existing regulations of nanotechnology in some cases are discussed in Table 19.4.

Studies concerning the safety of nanotechnology and the development of nanoregulation among most countries are still mainly focusing on nonfood nanomaterials and aspects. In accordance with the fact that application of nanotechnology in the food industry is growing rapidly, it is advisable for researches to be directed into safety, development of regulatory standards, public awareness and acceptance of the use related to nanomaterials (Othman, 2014).

4 Polysaccharide

Polysaccharides are biopolymers comprised of multiple saccharide units joined together by means of glycosidic linkages, offering a number of unique features, which makes them divers compared to the other families of biopolymers. Most polysaccharides are easily and inexpensively recovered and can be produced by microbes (dextran, xanthan gum, pullan), plants (cellulose,

Table 19.4 Some Regulations Stated on Food Nanotechnology

Year	Organization/Directive	Regulation
2004	European Commission/Regulation on Materials and Articles Intended for Food Contact (EC, No. 1935/2004)	The Regulation on Materials and Articles Intended for Food Contact, (EC) No. 1935/2004, regulates food packaging including new types of materials which actively maintain or improve the condition of the food (active FCMs), which is widely viewed as a promising use of nanomaterials
2008	Food Additives Directive (89/109/EEC)	It was the first piece of legislation to mention nanotechnologies explicitly. Article 12 of this document states that if there is a change in the starting material used or in the production method of an additive (for example, a change of the particle size), it must undergo a new authorization process and safety evaluation
2009	Active and Intelligent Materials and Articles Intended to Come in Contact With Food (EC) No. 450/2009	It is stated that if legislation limits the quantity of a substance in a food, the total quantity should not exceed this limit regardless of the source, that is, originally included in the foodstuff or following release of that substance from the FCM. If a substance is released into the food in this way, it is required to be included in the ingredients list. For active food substances that are not designed to be released from the packaging and have no function in the food, there is a risk that these substances may migrate into the food. The approach with these types of materials is the same as for plastic materials, which is that these substances should undergo a safety assessment by European Food Safety Authority (EFSA) and a Community authorization. Once authorized, the substance can be placed on a positive list and used within specific constraints
2010	Novel Food Regulation/(EC) No. 258/97	In the Novel Food Regulation ((EC) No. 258/97), novel foods are defined as "foods and food ingredients that have not been used for human consumption to a significant degree within the Community" (European Commission). Under this Regulation, foods deemed to be novel must undergo a safety assessment prior to being placed on the market. It is thought that the soon-to-be-revised Novel Food Regulation ((EC) No. 258/97) will be more specific in terms of nanotechnologies than previous documents
2011	National Industrial Chemicals Notification and Assessment Scheme (NICNAS)	NICNAS, which regulates chemicals for the protection of human health and the environment, has recently introduced new administrative processes to address nanotechnology. NICNAS determines volumes, types, and data holdings of nanomaterials being used in Australia, and has the responsibility of determining whether legislation is sufficient to protect people from potential risks arising from nanotechnology

Reproduced with permission from Farhoodi, M., 2015. Food Eng. Rev. 2015, doi: 10.1007/s12393-015-9114-2. Copyright © 2015, Springer Science and Business Media, NY.

starch, pectin, guar), and animals (chitosan, hyaluronic acid). Polysaccharides may also be classified by their charge: cationic (chitosan), anionic (hyaluronic acid, heparin), and nonionic (dextran). They are highly stable, safe, nontoxic, hydrophilic, and biodegradable. There are numerous criteria concerning the structural diversity of polysaccharides: low, intermediate, and high molecular weights with different polydispersities; linear or branched structures; monofunctional, having only hydroxyl groups, or polyfunctional, having hydroxyl, carboxyl, and/or amino groups; a high level of chirality; water soluble or insoluble properties; low toxicity, environmentally safe, and nonimmunogenic. In order to adjust their properties to the real needs of the consumer society, polysaccharides can be easily modified chemically and biochemically, which gives tremendous possibilities related to further application.

A wide range of polysaccharides and their derivatives have already found applications in various technologies:

1. in food products as thickeners due to their ability to increase product viscosity when hydrated (alginate, pectin, starch),
2. medicine-wound dressing, drug carrier, tissue engineering (starch, cellulose, chitosan, dextran, hyaluronan, pectin, carrageenan, alginates, bacterial cellulose),
3. cosmetics (hyaluronan, chitosan),
4. textile industry (cellulose),
5. packaging industry (starch, cellulose, chitosan),
6. agriculture industry (starch, chitosan).

The main driving force for the development and implementation of polysaccharide materials is their ubiquitous presence and renewable character, making them the primary CO_2 neutral candidates for the global transformation to a more sustainable world. Utilization of polysaccharides has the potential of increasing biodiversity and sustainability, decreasing of CO_2 emissions and pollution, showing the huge benefit over the petrochemical products. Due to growing environmental waste problems caused by nonbiodegradable petrochemical products, the main target in recent years was to design the new materials with satisfied functional properties, in such a way, that after their service they can be reused, recycled, or disposed of without negative impact on the environment. Disposal in landfill implies that the materials can be fully degraded by influences of environment such as temperature, UV light, water, and/or microbial attack. Most polysaccharides are biodegradable and prone to bacterial and enzymatic degradation, and degrade also under high temperature conditions; therefore, they will not accumulate in the environment, which is one of the most important advantages related to their use (Persin et al., 2011).

Polysaccharide films become increasingly important in food packaging industry due to their edibility, good barrier properties toward gases and antimicrobial activity which contributes to the improving food quality and extending shelf life through minimization of the microbial growth in the product. However, there are some limitations regarding the commercial use of polysaccharide-based films due to their poor mechanical properties. Also, they possess high sensitivity to moisture and water vapor because of their hydrophilic nature. Despite the numerous efforts aimed to improve the physical, thermal, and mechanical properties of polysaccharide-based films, these films are still not meet all criteria for wider application in food packaging industry. A new class of materials, polysaccharide-based nanocomposite films, offers enhanced barrier, mechanical, and thermal properties and therefore, they are considered to be a promising alternative for petrochemical-based package (Rhim et al., 2013). The most commonly used method for the synthesis of polysaccharide-based nanocomposite films is solution intercalation method. In the first place, this method involves swelling of clay as well as the dissolution of polymer (intercalat), both in an appropriate solvent. Both solutions are then combined, and the polymer chains intercalate and displace the solvent within the interlayer of the clay. The most investigated polysaccharides in this field are starch, chitosan, cellulose, pectin, and alginate. The next sections present the main characteristics of these polysaccharides and give the overview of developments related to polysaccharide-based nanocomposite films intended for application in food packaging.

5 Starch

5.1 Molecular Structure

Starch granules are mainly found in seeds, roots, and tubers of different origins such as corn (maize), wheat, potato, and rice. The granules obtain relatively smooth surface and concentric 3-dimensional architecture from the hilum with a total crystallinity varying from 15% to 45% depending on the botanical source (Zobel, 1988).

Starch is a polysaccharide consisting of repeating 1,4-α-D glucopyranosyl units: amylose and amylopectin. The relative amounts of amylose and amylopectin depend on the plant source (Avella et al., 2005). Generally, the functional properties of a starch, as well as the degradation of those properties, are governed by the starch's amylose and amylopectin content. Amylose is predominantly linear polymer, composed of 500–20,000 α-(1–4) linked

D-glucopyranosyl units. Due to the arrangement of hydrogen groups exposed in the interior of the helix, amylose forms a helical structure with a hydrophobic core (Almeida et al., 2010). According to some literature data, amylose is localized in the amorphous regions within the starch granule (Shamekh et al., 1999). Contrary to the previous allegation, Atkin et al. have reported the difficulty related to the specific location of amylose and suggested that it may be randomly dispersed in the granule matrix (Atkin et al., 1998). Depending on the source, amylose content can be varied from <1% to 83% (Tan et al., 2007), and therefore, has a great impact on the thermal, rheological, and processing properties of starch (Li et al., 2011).

Amylopectin is one of the largest known natural polymers (10^7–10^9 Da), made up of α-(1–4) D-glucopyranose chains with α-(1–6) branches (Kirwan and Strawbridge, 2003). Amylopectin provides the bulk of the matrix for the starch granule and the double helices coming from its branches are arranged in crystalline domains (Oates, 1997; Sarko and Wu, 1978). The crystalline regions are believed to be hard and brittle, but may be interspersed with amorphous regions of amylopectin that are assumed to have greater mobility and behave more like a rubbery polymer (Morgan et al., 1995). Generally, native starch granules can be classified into three crystalline forms: the A type is associated with cereal starches, the B type is mostly characteristic for tuber and amylose-rich starches and C-type, which represents a mixture of A- and B-type, and is obtained mostly from legume starches (Zhao et al., 2008). The crystalline V-form conformation appears during complexation of amylose with fatty acids and monoglycerides. Depending on the botanical resource, starch granules also contain very small amounts of proteins, lipids, and phosphorus (Jane, 2009). These components can interact with the carbohydrate chains during processing (eg, Maillard reaction), and therefore, affect the behavior of the starchy materials. The amylose and amylopectin structure is shown in Fig. 19.1.

5.2 Sources and Extraction

Corn, rice, wheat, and potato are the main sources of starches, which differ significantly in composition, morphology, thermal, rheological, and retrogradation properties. Physical qualities of the starch are highly affected by environmental conditions such as growth temperature of the plant and water stress (Geigenberger et al., 1998). Generally, the extraction of starch is carried out by autoclaving in an inert atmosphere in water at neutral pH.

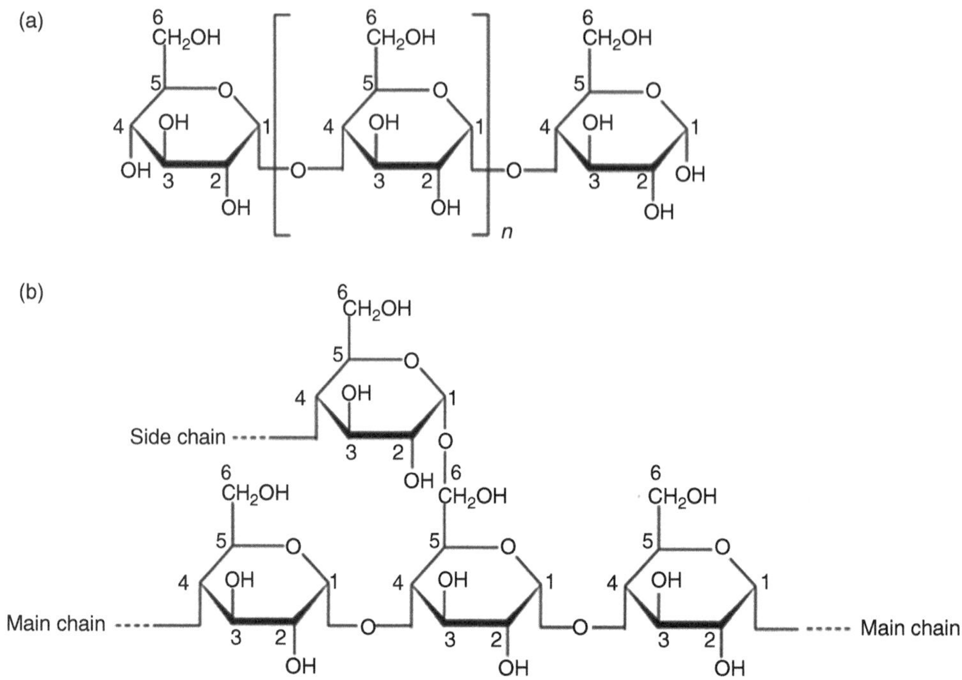

Figure 19.1. Chemical structure of amylose and amylopectin. Reproduced with permission from Herrero-Martinez, J.M., et al., 2004. J. Chromatogr. A. 1053, 227–234. Copyright © 2004, Elsevier B.V.

Separation of the starch component from protein and other substances is achieving by addition of salts or by employing mixtures of dimethyl sulfoxide and water (up to 15%) under milder conditions. The extraction of the granular starch is conducted by processes in which incrusting substances are removed by dissolving. Fractionation of the starch is facilitated due to starch properties such as high density and cold water insolubility (Stephen and Phillips, 2006).

The procedure of the starch isolation from wheat starts up with milling and sifting of the grains to remove husk, with the final result of obtaining flour particles composed of starch and gluten. By kneading the flour and water dough, starch granules can be separated. In order to inhibit fermentation and promote liberation of the polysaccharide, water used for this process contains low levels of sulfur dioxide. Before the air drying at about 450°C, the starch is exposed to the several washing and centrifuging steps (Mondal et al., 2013).

Nowadays, concerning genetic manipulation, the corn starch is in the middle of attention. The starch content of the kernel goes up to 75%, with an approximately ratio 1:3 between amylose and amylopectin. Corn starch is produced by traditional wet-milling processes (Takeda et al., 1984). Also, numerous effective methods for the extraction of good quality starch by applying mild alkaline treatment have been reported (Mistry et al., 1992).

Compared with wheat and corn starches, isolation of rice starch is difficult, and therefore costly. The starch nucleuses in rice endosperm are surrounded by protein bodies I and II (PB I and PB II). PB I presents 20% of rice protein, whereas PB II is a glutelin accounting for 60% of total protein (Adoracion et al., 1993). Beside problems related to the rice protein, the tiny granules of rice starch exhibit slow sedimentation in water with the final result of increasing losses during the separation and purification. Because of the small particle size and low starch damage, wet-milled rice flour is commonly used for the isolation of starch. The purification of the rice starch from endosperm is performed by means alkali, detergent, and protease digestions (Kim et al., 1978; Yamamoto et al., 1973).

Potato starch, derived from subterranean tuber, is very sensitive in terms of climate and agricultural conditions. The extraction method is consisted of following steps: the tubers are washed and cut into small pieces, milled, blended in water acidified with sulfur dioxide, and screened (200 μm) to allow the passage of small suspended particles. Starch granules settle and after slurring with water and centrifugation, a product with very low noncarbohydrate content is obtain in about 95% yield (Kim and Cha, 2014). The other most studied tuber starches are those isolated from sweet potato and cassava (Huang et al., 2015; Teodoro et al., 2015; Xijun et al., 2012). Although the quantities of deposited starch are not very high, plantain and chestnut are known and reliable sources (Bello-Flores et al., 2014; Lemos et al., 2015). Generally, laboratory procedures for the isolation of starch from plant tissue imply the minimization of enzymatic activity by working at low temperatures and with the addition of inhibitors. Previously homogenized mass is passed through a wide-mesh filter whereupon the granules are separated by centrifugation. Centrifugation is essential for the small-size granules. For example, granules derived from corn are 10 times smaller than those from potato whereas wheat and barley starches have a bimodal size distribution (Moon and Giddings, 1993). The most common procedure for the purification of granules is suspension treatment using 0.1 M sodium chloride, shaking in presence of toluene and sedimentation. The usage of detergents may cause structural changes.

5.3 Physical Properties

The starches derived from diverse sources are significantly different in terms of composition, morphology, thermal, rheological, and retrogradation properties. Regarding the structural properties, cereal starches are rich in phospholipids, while potato starch contains a significant quantity of esterified phosphorus (Singh et al., 2003). Numerous authors examined the dissolution process of starch and found that granules do not simply dissolve in an "outside to inside" (Glenn et al., 2008) way as might be expected. Instead, they appear to dissolve from the "inside out" (Glenn et al., 2008; Li et al., 2004). This pattern of starch dissolution may be due more to the insolubility of the granule surface and underlying material, commonly called starch granule remnant or envelope. The starch granule envelope is believed to be comprised mostly of amylopectin (Glenn et al., 2014).

Aqueous slurries of starch form a viscous paste when heated and the viscosity increases along with swelling and hydration of granules. The internal pressure caused by hydration and solubilization of the granule interior may result in rupture and release of its content into the continuous phase. Probably due to the rupture of the granule envelope, starch pastes typically reach the maximum followed by the drop in viscosity (Conte et al., 2013). The granule envelopes that remain after the granules have ruptured are called "ghosts" (Conte et al., 2013), and they turned out to be very difficult to dissolve. Potato starch exhibits higher swelling power, solubility, paste clarity, and viscosity than wheat, rice, or corn starches.

Morphological properties of the starch granules, such as shape and size, also depend on the plant source. Potato starch granules have relatively smooth surface, oval and irregular or cuboidal-shaped while corn, rice, and wheat starch granules are angular, pentagonal, and angular; and spherical- and lenticular-shaped, respectively. The surfaces of the corn, rice, and wheat starch granules are less smooth than potato starch granules. Potato starch granules are the largest (<110 mm) in size, followed by wheat (<30 mm), corn (<25 mm), and rice (<20 mm) granules (Tang et al., 2008).

The disruption of the hydrogen bonds between adjacent glucose units results in progressively destruction of the crystalline structure of the granule (Rodriguez-Gonzalez et al., 2003). The general term used to describe the breakdown and dissolution of the starch granule in aqueous media is gelatinization. Namely, starch is considered completely gelatinized when the starch granules are dissolved to the point in which only starch ghosts remain in the solution (Atkin et al., 1998).

Retrogradation is related to the cooling process of the hot gelatinous starch melt wherein dispersed amylose molecules form elastic gels (Blaszczak et al., 2005; Richardson et al., 2004). Gelatinization temperatures, as well as the enthalpies of gelatinization (ΔH_{gel}) of corn, rice, and potato starches also differ. Generally, corn and rice starches show higher transition temperatures than wheat and potato starches, while the ΔH_{gel} values are higher for potato and wheat starches. Potato starch shows a higher tendency toward retrogradation than the cereal starches. During heating of all the starches, the rheological properties, such as storage modulus (G') and loss modulus (G'') increase to a maximum and then drop. Potato starch exhibits the highest peak G', G'', and lower tan δ than corn, rice, and wheat starches during the heating cycle (Tang et al., 2008).

5.4 Application

The hydrophilic nature of starch and its tendency to embrittle with age do not make it very suitable as a replacement for commonly used materials. In recent years, starch has become viable alternative for petroleum-based plastics primarily because of its low cost and abundance and because it is a renewable material that degrades readily under different conditions (composting, landscapes, and aquatic environments) (Law et al., 2010). Native starch is not a true thermoplastic but it can be converted into a plastic-like material called "thermoplastic starch" (TPS). Several authors have shown the possibility to transform native starch into thermoplastic resin-like products under destructure and plasticization conditions (Ollett et al., 1991; Slade and Levine, 1993). Starch can be plasticized at high temperature (90–180°C) and under pressure, which lead to its melting and flowing and allow to be extruded, injection molded, or blow molded to give foams or biobased thermoplastic molded materials, similar to the procedure for obtaining the most conventional synthetic thermoplastic materials.

Mainly due to high hydrophilicity, there were numerous attempts of overcoming drawbacks related to the use of starch as a food packaging material. Some authors tried to modify the starch structure, for example, by acetylation, to reduce the hydrophilic character of the chains (Fringant et al., 1996). This chemical process resulted in inferior mechanical properties and greater product cost (Averous et al., 2000). Furthermore, some authors have described changes in the mechanical properties of TPS in relation with the crystallinity and the contents of plasticizer and water, during the aging (Van Soest and Knooren, 1997; Van Soest and Vliegenthart, 1997).

Despite certain advantages, TPS still has the same limitations as native starch, such as high water-sensitivity and poor barrier and mechanical properties at high moisture contents (Vigneshwaran et al., 2006).

The addition of nanoscale particles into native and TPS is one of the approaches that offer improved performances along with environmental acceptability. Furthermore, the incorporation of appropriately tuned nanoparticles into starch as a biopolymer with complex structures and special properties will provide a rich new class of polymer nanocomposites, able to be designed for a wide range of both conventional and emerging applications. The starch-based nanocomposites are very promising in terms of food packaging application due to low cost and complete degradation in soil and water. Therefore, extensive research work, preparation, and application regarding starch-based nanocomposite have been performed recently.

Avella et al. (2005) reported the synthesis and characterization of biodegradable starch/clay nanocomposite films intended for food packaging application. The films were obtained by homogeneously dispersing MMT nanoparticles in different starch-based materials via polymer melt processing techniques. The nanocomposite showed good exfoliation, but still, films were quite brittle without addition of polyester. Also, the conformity of starch/clay nanocomposite films with actual regulations and European directives on biodegradable materials was verified, demonstrating that these materials can be utilized in the food packaging sector.

Wilhelm et al. (2003) investigated the influence of different nanoclays (natural kaolonite and hectorite and synthetic layered double hydroxide, LDH, and brucite) on the physical–chemical properties of glycerol plasticized and oxidized starch films. The nanocomposite films were obtained by casting method. The effects of the filler type and the plasticizer were analyzed by X-ray diffraction, dynamic mechanical analysis, and thermogravimetry, and it was shown that the storage modulus was higher for kaolinite > brucite > hectorite than for LDH starch composites. A film obtained with 30% in w/w of clay showed an increase of more than 70% in the Young's modulus compared to nonreinforced plasticized starch. It was reported that glycerol could be intercalated into the clay galleries and that oxidized starch exhibited greater intercalation of clay layers. Clay exfoliation was detected in unplasticized starch/clay mixtures.

Park et al. (2002) reported the synthesis and characterization of the biodegradable TPS/clay hybrids, prepared by melt intercalation. Three organically modified MMT with different ammonium cations and one unmodified Na^+ MMT (Cloisite Na^+) were used. It was noted that Cloisite Na^+ MMT was the best dispersed in the nanostructure of the TPS hybrids. The hydrophilicity of the organoclays, as well as the polar interaction between the silicate layers and the TPS, highly affected the dispersion of the organically modified MMT in the TPS matrix. The strong interaction between the TPS and Cloisite Na^+ resulted in higher tensile properties and lower water vapor permeability compared to the pristine TPS. The improvements in TS and barrier properties of TPS were achieved by addition of only 5 wt.% of the clay to the TPS matrix. For the TPS/Cloisite Na^+ hybrid, the storage modulus was higher than any other TPS containing hybrid systems, as well as the pristine TPS in the temperature ranges above c. –20°C.

Due to the advances in mechanical and barrier properties, and highlights to importance of starch–plasticizer–clay interactions, the studies carried out by Park et al. (2002) were very important in terms of the deeper insights and development of starch based nanocomposites.

Pandey and Singh (2005) also have investigated the film-forming possibilities of the heated solutions composed of starch, plasticizer (glycerol), and clay (sodium MMT). Namely, applying the casting method, they noted that the mechanical properties of the obtained films depend on the addition sequence of the component. The most homogenous films and therefore the highest mechanical stability were obtained when prior to plastification procedure, clay was added to starch.

The method of melt extrusion, used for the preparation of glycerol–plasticized thermoplastic starch (GTPS) reinforced with MMT was reported by Huang et al. (2005). It was found that mechanical properties of material were greatly enhanced, even after 14 days of storage. Also, differential thermal analysis (DTA) and water absorption testing showed that the thermal stability and water-resistant properties of composite were better than those of GTPS.

A similar study was conducted by Huang and Yu (2006). They reported synthesis and characterization of composites based on thermoplastic cornstarch (TPCS) and activated-MMT, prepared by blend extrusion. The thermoplastic cornstarch was plasticized using urea and ethanolamine as plasticizers, while MMTs were activated using ethanolamine as a solvent. The addition of nanonfiller up to 8 wt.% led to enhanced TS and elasticity of films. The stability of bionanocomposite films in water was also improved.

Significant improvement of the starch-based composites reinforced using MMT was achieved in the study conducted by

Cyras et al. (2008). Namely, new corn starch-based films with addition of glycerol as plasticizer and three different loadings of MMT aqueous suspensions were obtained. It was found that the Young's modulus was increased up to 500% for the nanocomposite containing 5 wt.% of clay. The addition of MMT to the starch resulted in the decrease of water absorbed by the nanocomposites.

Chiou et al. (2006) investigated the dispersion of various nanoclays [Na^+–cloisite (MMT) and Cloisite 30B (a MMT modified with a quaternary ammonium salt of hydrogenated tallow)] on extruded wheat starch films varying different parameters during procedure: moisture content, screw speed, and temperature. Moisture content had the significant effect on Na^+–cloisite dispersion, where cloisite– Na^+ samples with high content of moisture during extrusion contained exfoliated nanoclays, whereas cloisite–Na^+ samples with low moisture content contained intercalated nanoclays. Furthermore, the results showed that starch was incompatible with more hydrophobic Cloisite 3B, which resulted in absence of intercalation. In addition, water absorbance tests indicated the cloisite– Na^+ sample with exfoliated nanoclay had the lowest water uptake. Obtained results were in agreement with previous work by the same authors (Chiou et al., 2005), wherein better exfoliation and higher shear moduli were observed using Na^+–cloisite mixtures in wheat, potato, corn, and waxy corn starch compared to the results obtained by use of organically treated clays incorporated into starch-based materials.

Chen and Evans (2005) investigated the influence of different clays (MMT, hectorite, organically treated hectorite, kaolinite) on characteristics of potato starch nanocomposites plasticized with glycerol. The nanocomposites showed increases in exfoliation and modulus for a given volume fraction of clay, although the results indicate that a plasticizer other than glycerol is preferable.

Bagdi et al. (2006) reported synthesis and characterization of the composites prepared from TPS and layered silicates organophilized with different surfactants. Sodium MMT and silicates were treated with aminododecanoic acid, stearyldihydroxyethyl ammonium chloride, and distearyl dimethyl ammonium chloride surfactants and added to plasticized starch in amounts of 0–7 vol.%. The results indicated that competitive dissolution and adsorption of the surfactant and the plasticizer took place in the composites. The reinforcing effect of the silicates was definitely observed: the amino acid modified clay improved strength almost 10 times as much as the silicate treated with the aliphatic amine surfactant.

6 Chitosan

6.1 Molecular Structure

Chitosan can be found in different polymeric forms that have unique structures, multifunctional properties, and a wide range of applications (Muzzarelli and Muzzarelli, 2005). Beside numerous benefits concerning environmental issues, chitosan is interesting because of its compatibility and efficiency related to different bioapplications (Kumar et al., 2004). Chitosan structure consists of 2-acetamido-D-glucose and 2-amino-D-glucose units linked with glycosidic linkages, as shown in Fig. 19.2. Chitosan can be produced by deacetylation reaction of chitin under alkaline conditions or through enzymatic hydrolysis in the presence of chitin deacetylase (Kumar, 2000).

The primary —NH_2 groups of chitosan are important structural elements in terms of their diverse application. Chitosan is insoluble at neutral and alkaline pH values, whereas under acidic conditions amine groups from chitosan become protonated resulting in soluble polysaccharide. In strong acidic medium hydrolysis could occur, which led to decreased viscosity and molecular weight of chitosan. In recent years, high attention has been given to correlation of physical–chemical properties of chitosan-based materials with respect to the chitosan origin, molecular weight and degree of deacetylation (Shukla et al., 2013).

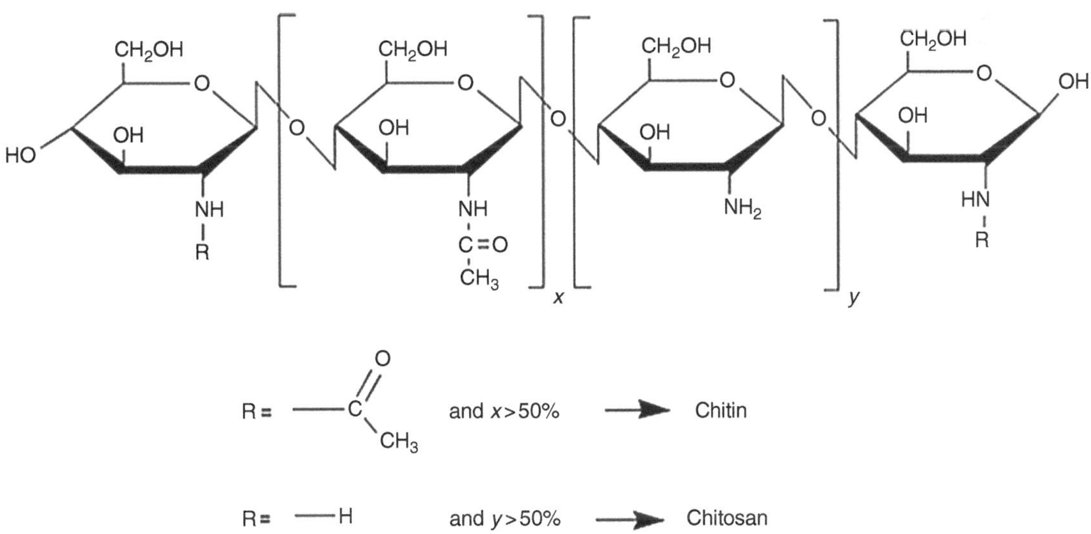

Figure 19.2. Chemical structure of chitin and chitosan. Reproduced with permission from Shaari, N., et al., 2015. J. Power Sources. 289, 71–80.Copyright © 2015, Elsevier B.V.

6.2 Sources and Extraction

Chitosan is natural biodegradable polymer obtained in deacetylation process of chitin. It is derived from naturally occurring sources, which is the exoskeleton of insects, crustaceans such as crabs, shrimps, prawns, lobsters, and cell walls of some fungi such as aspergillus and mucor (Hirano et al., 1990). The production of chitosan by amide and enzymatic hydrolysis of chitin is shown in Fig. 19.3.

The reaction is generally carried out heterogeneously, using concentrated (40–50%) NaOH or KOH solutions at temperatures above 100°C, preferably in an inert atmosphere or in the presence of reducing agents, such as $NaBH_4$ or thiophenol, in order to avoid depolymerization. The reaction conditions are coupled with several factors such as the starting material, the previous treatment,

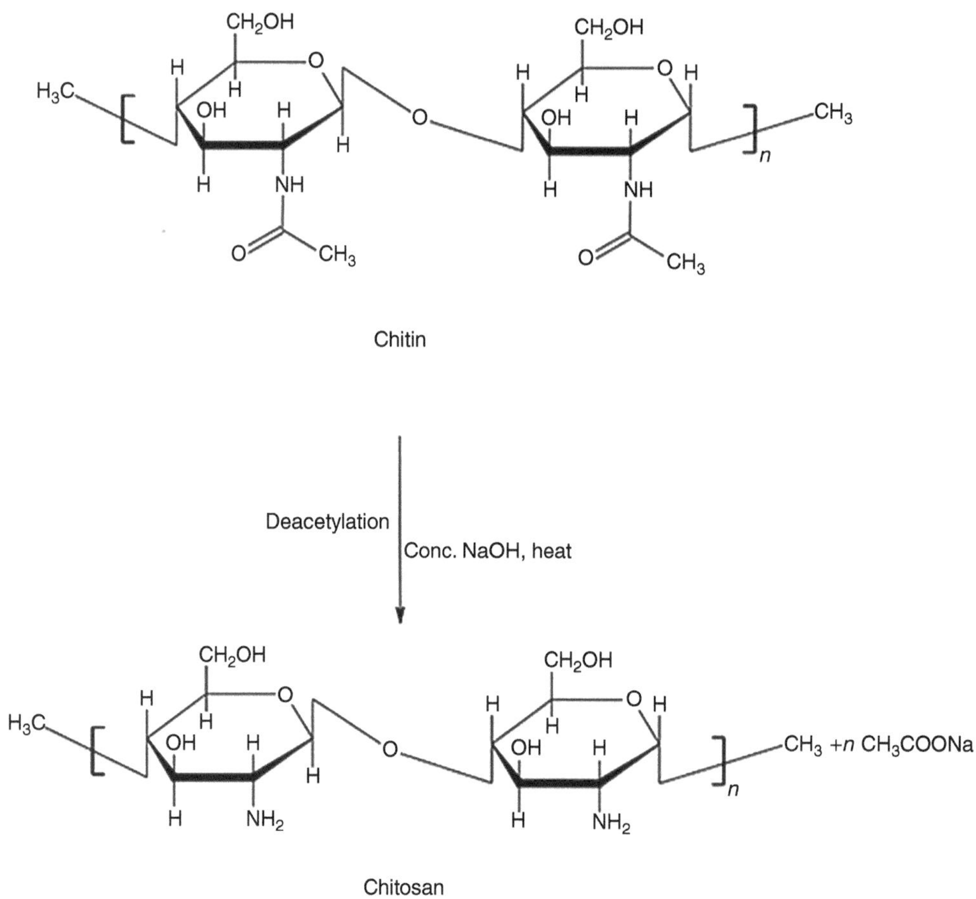

Figure 19.3. Deacetylation of chitin to chitosan. Reproduced with permission from Shukla, S.K., et al., 2013. Int. J. Biol. Macromol. 59, 46–58. Copyright © 2013, Elsevier B.V.

and the desired degree of acetylation. If the raw material is exposed only to one alkaline treatment, the maximum deacetylation degree attained will not surpass 75–85%, while prolonged treatments cause the degradation of the polymer without resulting in an appreciable increase in the deacetylation degree (Tsaih and Chen, 2003). To the best of our knowledge, there are several treatments developed to obtain fully deacetylated chitosan (Domard and Rinaudo, 1983; Mima et al., 1983).

In order to optimize the process, multistep heterogeneous deacetylation process for the production of chitosans with high deacetylation degrees and molecular weights was reported by Lamarque et al. The procedure was applied to α- and β-chitins in the presence of 50% (w/v) NaOH, at temperatures ranging from 80°C to 110°C under an argon atmosphere (Lamarque et al., 2004).

With regard to the semicrystalline structure of chitin, deacetylation reaction performed in heterogeneous conditions takes place mainly at the amorphous regions, whereas homogeneous conditions lead to more uniform modification of the polymer. Also, it is reported that chitosans obtained by the heterogeneous process are polydispersed in terms of the acetylation degree of their chains, while the chitosans obtained under homogeneous conditions do not exhibit chain compositional dispersion (Sannan et al., 1976).

The main drawback of the common methods employed to produce chitosan are related to the long processing times and large consumption of alkali. Numerous unconventional deacetylation methods are developed in order to overcome deficiencies: thermomechanical processes using a cascade reactor operating under alkali conditions (Pelletier et al., 1990); flash treatment undersaturated steam (Focher et al., 1990); microwave dielectric heating (Goycoolea et al., 1997); and intermittent water washings (Mima et al., 1983).

6.3 Physical Properties

Chitosan is a linear polycationic polymer composed of one —NH_2 group and two —OH groups on each glucosidic residue (Agrawal et al., 2010). The primary amino group is involved in the attachment of the side group under mild reaction conditions, which gives many possibilities related to chitosan application (Yi et al., 2005). The side group on chitosan provides flexibility to materials with specific functionality, and highly affects biological and physical properties (Suh and Matthew, 2000). Also, chitosan shows ability to form chelate complex by interacting with various metal ions. This phenomenon is mostly pronounced below dissociation constants (pK_a) value and in acidic medium, because of protonation of amine groups from chitosan. Basically, chitosan is a weak base with a pK_a value of the D-glucosamine residue of about 6.2–7.0, depending on degree of deacetylation, ionic strength, and charge neutralization of amine groups (Guibal, 2005). Along with other factors that have a great impact on metal, sorption properties are also affected by polymer porosity and material crystallinity. Crystallinity, and therefore hydration of chitosan are variable, and in the first place depends on the origin of the polysaccharide and its extraction procedure from the raw materials. The maintenance and establishment of chitosan crystallinity can be kept by freeze-drying, air and oven-drying of the chitosan solution. Croisier and Jerome (2013) discussed the structure–properties relationship of chitosan and remarkable futures of chitosan as novel biomaterial for widen application.

Recently, chitosan gelation has been investigated by classic experimental methods such as rheometric or light scattering studies (Montembault et al., 2005). Chitin is completely insoluble in aqueous media, while the chitosan can be dissolved under acidic conditions that provide sufficient protonation of its amino groups. The obtained aqueous solutions are usually stable at pH up to 6.2. Neutralization of chitosan solutions by common alkali solutions leads systematically to the formation of a hydrated gel-like precipitate when the pH exceeds approximately 6.2 (Chenite et al., 2006). Chitosan gelation occurs by exposing the polymeric solution to basic media, involving three main mechanisms: (1) mass transport of the basic molecule (ammonia or sodium hydroxide for instance) to the chitosan solution surface; this step is mainly controlled by local hydrodynamics in the surrounding bulk solution; (2) diffusion of the basic molecule through the polymeric system (from the top to the bottom); and (3) chemical reaction between the basic molecule and the chitosan polymer chains, which are initially under protonated form and are deprotonated by the chemical reaction. The nature of the interface between the basic media and the polymeric solution can be liquid/liquid or gas/liquid. In the first case, the process is the so called wet process (Venault et al., 2012). In addition, chitosan is considered to be in line with the most valuable natural polymers due to valuable inherent properties such as antibacterial, antifungal, antiviral, antacid, nontoxic, total biodegradable as well as film formation, fiber formation, and hydrogel formation properties. By virtue of these properties, chitosan has prospective applications in many fields such as medical, waste water treatment, cosmetics, dentifrices, food, agriculture, pulp and paper, and textile industries (Prashanth and Tharanathan, 2007).

6.4 Application

Biologically active biomolecules such as chitosan and its derivatives have a significant potential in the food industry in view of contaminations associated with food products and the increasing concerns in relation with the negative environmental impact of conventional packaging materials such as plastics. Great potential of chitosan related to food packaging application derives from its distinct physical–chemical properties, short time biodegradability, biocompatibility with human tissues, antimicrobial an antifungal activities, and nontoxicity. In addition, chitosan exhibits high antimicrobial activity against pathogenic and spoilage microorganisms, including fungi, and both Gram-positive and Gram-negative bacteria.

Thus, chitosan-based films have attracted serious attention in food preservation and packaging technology. Still, as well as for the other biopolymers, the same is true for the chitosan: the usage is limited because of unsatisfactory mechanical and barrier properties compared to those of synthetic polymers. The addition to reinforcing fillers, chitosan has proven to be effective in enhancing thermal, mechanical, and barrier properties.

Khan et al. (2012) reported chitosan films reinforced with 1–10% (dry wt basis) of CNC. It was found that mechanical properties of the nanocomposite were improved by the formation of a percolating network and strong filler–matrix interaction. A 27% reduction in water vapor permeability was obtained by incorporation of only 5% CNC. The authors believe that CNC reinforced nanocomposite films, because of their excellent mechanical and barrier properties should have a promising impact on food packaging in the coming years.

Nanocomposites of chitosan (CS) with varying levels of MMT and encapsulated tocopherol were developed by Dias et al. (2014). The tocopherol content up to 10% influenced the intercalation of MMT in the CS matrix, resulting in films with reduced water vapor permeability [3.48×10^{-11} (g/m s Pa)], increased hydrophobicity ($\Delta G_{Hydroph}$ |7.93–59.54| mJ/m^2) and lower equilibrium moisture content, thus showing potential for active food packaging materials.

Kasirga et al. (2012) have studied chitosan /MMT-K10 (MMTK-10) clay composite films with different amounts of the clay MMTK-10 (0.5, 1, 2.5, and 5%) prepared using a solution-casting method. It was found that water vapor and gas permeability values of the composite films decreased significantly with increasing filler concentration. TS of the composites increased significantly with the addition of clay, and E decreased with increasing clay concentration. Namely, the tensile modulus of nanocomposites was improved for 74.63%

with addition of only 5 wt.% of clay when compared to the neat chitosan. The resulting films had an opaque appearance, which depended on the amount of MMTK-10 added. The oxygen permeability decreased with an increase in MMTK-10. The minimum oxygen permeability (1.54 cm^3/m^2 day atm) was observed for film with 5% MMTK-10. The dispersed clay improved the thermal stability and enhanced the hardness and EM of the matrix systematically with the increased loading of clay.

Lavorgna et al. (2014) reported bionanocomposite films with incorporated silver–MMT nanoparticles into chitosan matrix, where silver-MMT nanoparticles were obtained by ion-exchange reaction. It has been demonstrated that both Ag metallic and AgO/Ag_2O nanoparticles are mainly located on the surface of MMT platelets, with a preferential location on the edges. The addition of nanoparticles of MMT modified with silver into chitosan matrix improved thermal, mechanical, and water uptake stability. Also, the modification of MMT with silver showed significant impact on antimicrobial activity toward the *Pseudomonas* spp. after 24 h.

Within the study reported by Sanuja et al. (2015) nano-ZnOat different concentrations (0.1, 0.3, and 0.5%) and neem essential oil were incorporated into the chitosan matrix by solution cast method. An increase in the ZnO concentration (0.1–0.5%) resulted in a decrease of water solubility from 60.3 to 43.2%. Incorporation of oil into the chitosan-based films shows further decrease in water solubility (30.11%). Furthermore, by increasing the concentration of ZnO in chitosan film as 0.1, 0.3, and 0.5% its transparency increases from 4.065, 4.576, and 5.400, respectively. Among the different concentration, 0.5% nano-ZnO incorporation was found to give the most satisfactory properties of the film.

Youssef et al. (2014) reported the synthsis of the novel chitosan-based nanocomposite films improved with silver and gold nanoparticles. Beside the completely green methodology for production of gold and silver nanoparticles, the final result of this study are biobased nanocomposites with good activity against Gram-positive (*S. aureus*) and Gram-negative bacteria (*Pseudomonas aeruginosa*), fungi (*Aspergillus niger*), and yeast (*Candida albicans*), which could be used in edible packaging applications.

Youssef et al. (2015) proposed the synthesis of novel bionanocomposites based on chitosan/poly(vinyl alcohol)/titanium nanoparticles (CS/PVA/TiO_2 nanocomposite) intended for the packaging of soft white cheese. The obtained CS/PVA/TiO_2 bionanocomposites showed good mechanical properties and hydrophobic character. Concerning the antibacterial activity, the CS/PVA/TiO_2 nanocomposites are proved to be very effective against Gram-positive (*S. aureus*), Gram-negative (*P. aeruginosa, E. coli*) bacteria and fungi (*C. albicans*). The results showed that the total bacterial counts (mold, yeast, and coliform) reduced with the rising storage period and completely disappeared at the end of the storage period compared with control. Accordingly, this novel CS/PVA/TiO_2 bionanocomposite films can be used as an economic, environmentally friendly, and sustainable material for food packaging with enhanced antimicrobial activity.

The excellent antimicrobial properties were also found for nanocomposite of silver/TiO_2/chitosan adipate (Ag/TiO_2/CS) proposed by Lin et al. (2015). It was concluded that chitosan adipate not only acted as a reducing agent for the metal ions but also stabilized the product possibly through the Ag—N coordination bonds. The nanocomposite demonstrated reduced microbial growth of the *E. coli* population by 6 logs after 24 h of incubation, with minimum inhibitory concentration of 0.38 mg Ag/mL. The results obtained in this article gave potential to Ag/TiO_2/CS materials as an antibacterial protective coating for fruit storage.

Pradhan et al. (2015) investigated barrier properties of chitosan/SiC nanocomposites synthesized by a low-cost green technique where water was used as the solvent. The obtained results indicated improved thermal stability and TS with increasing concentration of nano-SiC. The oxygen flow rate through all the nanocomposites was observed to be less compared to the virgin chitosan at constant pressures, while it decreases with increase in the percentage of SiC loading. Degradation tests showed that the percentage weight losses in composites were decreased in comparison with virgin chitosan. The nanocomposites having enhanced thermal stability and TS with substantial reduction in oxygen permeability and resistance to alkali may enable the materials suitable for packaging applications.

Interesting, advances in mechanical properties and water resistivity of chitosan film were also attained by incorporation of nanosized chitin whiskers (individual chitin whiskers range was from 20 to 40 nm). Rubentheren et al. (2015) have shown that the addition of chitin whiskers increased the rigidity and stiffness of the chitosan film and reduced the water uptake and solubility of the film.

Chitin nanocrystals (CHNC) and chitin nanofibers (CHNF) were used for the preparation of chitosan-based nanocomposite films within the study reported by Salaberria et al. (2015). The obtained results demonstrated that the final properties were dependent on the content and nanonature (size and shape) of chitin nanofillers incorporated in the chitosan matrix. The chitosan-based bionanocomposite films reinforced with CHNF showed superior mechanical properties than those prepared with CHNC.

Moreover, antifungal assessment demonstrated that CS/CHNF bionanocomposite films presented an inhibitory effect (FGI > 80%) against *A. niger*. Unquestionably, these bionanocomposite films showed a great potential to be used in packaging applications.

7 Cellulose

7.1 Molecular Structure

Cellulose is a linear polymer consisting of β-D-glucose residues linked together by 1.4-links to form a flattened ribbon structure (Fig. 19.4). The molecular size of polymer molecules can be conveniently described in terms of degree of polymerization (DP), which is an average value of the number of monomer units. The DP varies from 300 for bacterial cellulose, up to 5000 for raw cotton.

There are four different polymorphs of cellulose: cellulose I, II, III, and IV. Native cellulose is in a form of cellulose I, which has a unit cell containing two cellulose molecules in line with the b-axis, but arranged counter currently to each other. Cellulose II, or regenerated cellulose (RC), emerges after recrystallization or mercerization with aqueous sodium hydroxide. It is considered the most thermodynamically stable polymorph of cellulose. Cellulose III is obtained by ammonia treatment of cellulose I or II, and cellulose IV is finally produced from cellulose III by treatment with glycerol (Aulin et al., 2009).

7.2 Sources and Extraction

Cellulose is the most abundant polymer available worldwide and presents the main constituent of lignocellulosic plants cell wall. Generally, the cellulose content within lignocellulosic plant varies from 23% to 53% on a dry-weight basis, which is less to the content found in cotton plants, made of almost pure fibrous cellulose (Knauf and Moniruzzaman, 2004). Cellulose is mostly obtained from wood pulp, but can also be produced from many other plants, such as corn cobs or stalks, soybean hulls, bagasse (sugar

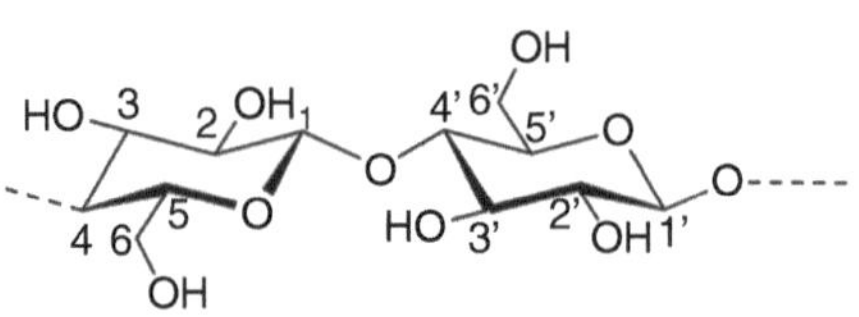

Figure 19.4. Chemical structure of cellulose. Reproduced with permission from Kang, H. et al., 2015. Polymer 70, A1–A16 (2015). Copyright © 2015, Elsevier B.V.

cane stalks), oat hulls, rice hulls, wheat straw, sugar beet pulp, bamboo, and fibers such as jute, flax, and ramie among others.

The isolation of highly pure cellulose has been subject of the extensive studies for many years because of the complexity of cell wall structure. The first step in isolation of cellulose is delignification of lignins, hemicelluloses and other noncellulosic substances by pulping and bleaching method. In order to isolate cellulose fibers from lignocellulosic materials, chemical pulping involves the treatment of wood with NaOH, Na_2S, H_2SO_4, Na_2SO_3, $NaHSO_3$, and/or SO_2. The cellulose produced from pulping usually contains the small amount of lignin, which affects the structure and properties of cellulose. Therefore, the pulping process is followed by bleaching process. The bleaching is carried out over a number of consecutive stages using chlorine dioxide, hydrogen peroxide, ozone, and/or per acetic acid as bleaching agents. The final step is alkali extraction, washing and drying of the cellulose fibers (Sun et al., 2004).

7.3 Physical Properties

Cellulose is a hygroscopic material, insoluble in water, as well as in most common organic solvents due to the stiff molecules and close chain packing. Solubility can be achieved in concentrated acids but at the expense of considerable degradation through acetal (glycosidic) hydrolysis, while alkali solutions lead to considerable swelling and dissolution of hemicelluloses present. Cellulose can be utilized in many different forms, including the usage of original unaltered fibers for textile and papers. Cellulose displays strong mechanical strength and high resistance to biological degradation. However, its low solubility in water and organic solvents imposes certain limitations related to the wide application. In order to enhance solubility of raw cellulose, a chemical modification has to be performed. The properties of the derivatives depend on the type, distribution, and uniformity of the substitution group. The average number of hydroxyl groups replaced by the substituent is known as the degree of substitution (DS) (maximum is three substituents per D-anhydroglucopyranose unit).

In the field of food packaging, the most investigated cellulose derivates are cellulose acetate (CA), methylcellulose (MC), and carboxymethyl cellulose (CMC). CA is formed by acetylation of cellulose with a mixture of acetic acid, acetic anhydride, and sulfuric acid (Balser et al., 1986). In contrast to cellulose, CA possesses a less crystalline structure and thus, exhibits better solubility in common organic solvents such as acetone. CMC and MC are ether derivatives of cellulose prepared by reaction of alkylation

(Brandt, 1986). DS significantly influences the solubility of cellulose. CMC and MC with a DS < 0.3 are only soluble in alkali. However, at higher DS, the interchain associations are more disrupted, yielding partially soluble material. CMC and MC with DS > 0.7, have a highly disrupted crystalline region, which makes them completely soluble in water. Average chain length and DS determine molecular weight of the CMC and MC.

7.4 Application

For a long time, cellulose and its derivatives have been used in additives' fabrication intended for food application due to their rheology, textural attributes, and insolubility in water. The applicability of cellulose and its derivatives for specific food applications can be determined according to the physical and chemical properties. The remarkable stability under physiological conditions, the diversity of chemical functionality of the polysaccharide and overall, the abundance of the raw material have resulted in expansion of the technological development of cellulose-based materials. One of the emerging and rapidly growing technologies belongs to the group of cellulose-based barrier films, which are especially formulated for frozen foods to prevent migration of water from areas of high relative humidity to the areas of low relative humidity, and to preserve the textural qualities associated with fresh products.

Ghaderi et al. (2014) synthesized cellulose nanocomposite films from sugarcane bagasse nanofibers using *N,N*-dimethylacetamide/lithium chloride as a solvent. The effect of dissolution time (ranging from 5 to 120 min at room temperature) on physical properties of cellulose nanocomposite films was investigated. Due to diminution in apparent crystallinity, the thermal stability of cellulose nanocomposite films decreased according to the partial dissolution process. Water vapor permeability as well as the TS of films increased with increasing dissolution time and it reached 140 MPa after 10 min. Over longer dissolution time, TS of films decreased because the ratio of the noncrystalline phase among undissolved parts increased.

Yang et al. (2014) reported the use of LiOH/urea as a solvent for preparation of cellulose/MMT nanocomposite film. The nanocomposite film with 15% of MMT showed the highest TS and Young's modulus and the values were greater than those of the pristine cellulose film for 161 and 180%. The coefficient of thermal expansion and oxygen permeability at 50–75% RH decreased to 60 and 42–33%, respectively. These effects were attributed to the high aspect ratio of the MMT platelets and their nanolayered structure in the composites.

Soheilmoghaddam et al. (2014) prepared RC/sepiolite nanocomposite films using an environmentally friendly ionic liquid, 1-butyl-3-methylimidazolium chloride. Young's modulus and TS of nanocomposite film with 8 wt.% of sepiolite increased by 166 and 74%, respectively, compared to that of the pure RC film. O_2 permeability decreased by about 56% by addition of 8 wt% of sepiolite. Also, the addition of sepiolite in RC film significantly influenced thermal stability of nanocomposite films, which was attributed to thermal insulation effect of sepiolite and entrapment of volatile RC products inside the porous sepiolite structure.

Mahmoudian et al. (2012) reported that addition of 6 wt.% of MMT into RC matrix, using the same solvent as in RC/sepiolite system, improved the TS and Young's modulus for 12 and 40%, respectively. By addition of 8 wt% of MMT, thermal stability and TS of RC/MMT nanocomposites decreased due to the aggregation of MMT platelets at higher content.

On the other side, Abdul Hanid et al. (2014) used halloysite as nanofiller for RC and 1-ethyl-3-methylimidazolium chloride ionic liquid as a solvent. When 5 wt.% of halloyisite was added, the TS and Young's modulus increased for 67.3 and 54.8%, respectively.

Rimdusit et al. (2008) the investigated the synthesis and characterization of methyl cellulose (MC)/MMT films. When compared with pure MC film, the E slightly decreased for about 16%, due to the high surface area of MMT during exfoliation. The TS was enhanced for 65% with addition of 0.1 wt.% of MMT in film, while moisture absorption property decreased 19%. The biodegradability of MC/MMT films was retarded compared to the pristine MC film, probably due the interaction and adhesion of MC and layered silicate surfaces of MMT, which restricted segmental motion at the interface.

Perez Espitia et al. (2013) reported that the addition of ZnO nanoparticles and pediocin into MC matrix did not significantly influence the TS and load at break as well as that the presence of ZnO nanoparticles made films more rigid. Developed MC/ZnO/pediocin nanocomposite films exhibited antimicrobial activity against *Listeria monocytogenes* and *S. aureus*. The optimal concentrations of tested antimicrobials in nanocomposite films with satisfied criteria were 20% (w/w) of ZnO nanoparticles and 15% (w/w) of pediocin.

El Miri et al. (2015) developed nanocomposite films of CMC/starch matrix reinforced with CNC, at different CNC loading levels (0.5–5 wt.%) (El Miri et al., 2015). With respect to the CMC/starch blend film, the Young's modulus and TS of the CMC/starch–CNC bionanocomposite films increased gradually with an increase of CNC content from 0.5 to 5 wt.%. With an addition of 5 wt.% CNC, the Young's modulus and TS were enhanced for 94.77 and 65.86%, respectively. The water vapor permeability of bionanocomposite films decreased to 4.22×10^{-7}, 4.01×10^{-7}, and 5.24×10^{-7} g m/m^2 h Pa when 0.5, 2.5, and 5 wt.% CNC were added, respectively.

Yadollahi et al. (2014) reported that incorporation of LDH into CMC matrix significantly decreased water vapor permeability (up to 37%) and transparency of the bionanocomposite films. CMC/LDH nanocomposite film with the content of 3 wt.% LDH showed the increase in the TS and tensile modulus for 148 and 143%, as well as a decrease in elongation for 62% in comparison with the pure CMC film. CMC films reinforced with laponite showed a decrease in water vapor permeability, achieving its minimum permeability at the content of laponite of 17 wt.%, while the highest TS and Young's modulus at 10 wt.% of laponite was obtained (Leite de Oliveira et al., 2015).

The influence of the molecular weight of CA (M_w = 30,000 and 50,000) and various amount of organoclay Cloisite30B on physical properties of nanocomposite films was investigated by Rodríguez et al. (2012a). The addition of 10 wt.% of organoclay into CA 30,000 and 50,000 matrices resulted in increase of tensile modulus for 84 and 60%, decrease of the TS for 40 and 14% and decrease of elongation for 89 and 62%, respectively. Also, a significant decrease in the oxygen permeability of the nanocomposites was evidenced with the increase of the organoclay content and the molecular weight of cellulose. These differences could be explained by means of different intercalation levels of CA 30,000 and 50,000 inside the organoclay structure. The same trend in oxygen permeability was observed in case when 5 wt.% of MMT/alkylammonium and MMT/chitosan was added into CA matrix (Rodríguez et al., 2012b).

8 Alginate

8.1 Molecular Structure

Alginate belongs to the family of unbranched polysaccharides consisting of 1, 4-linked β-D-mannuronic acid (M) and its C-5 epimer α-L-guluronic acid (G) (Pawar and Edgar, 2012). It contains homopolymeric sequences of M units (M–M blocks), homopolymeric sequences of G units (G–G blocks), and regions of alternating sequences of M and G units (M–G blocks) (Fig. 19.5). The relative amount of the G and M units, sequential arrangement along the polymer chain and molecular weight depend widely on alginate source and can strongly affect the physical properties of alginate. G units are linked by diaxial bonding and

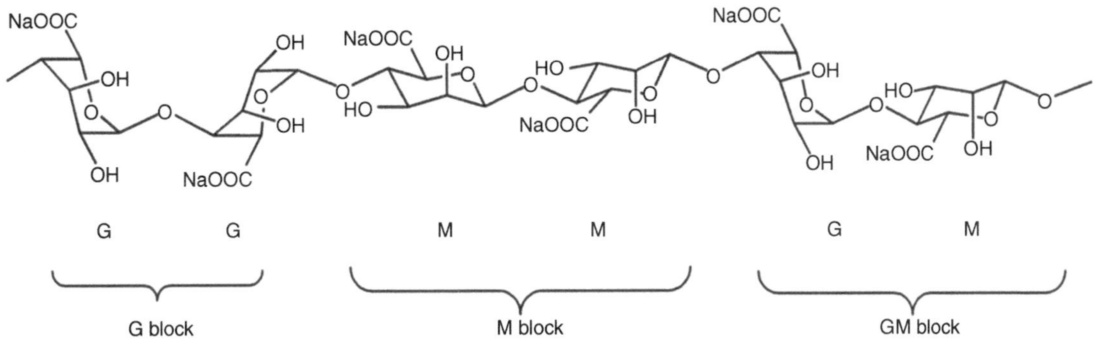

Figure 19.5. Chemical structure of alginate. Reproduced with permission from Daemi, H., Barikani, M., Barmar, M., 2013. Carbohydr.Polym. 92 (1), 490–496.Copyright © 2013, Elsevier B.V.

therefore G–G blocks have shorter chains than M–M blocks and M–G blocks. The diaxial linkages in G–G blocks prevent free rotation around the glycosidic linkage, which results in buckled alginate chains. M–M blocks have diequatorial linkages and longer chains forming an extended ribbon structure, while M–G blocks have axial–equatorial and equatorial–axial linkages, which enable the highest flexibility of the glycosidic linkages in alginates (Dentini et al., 2005; Draget et al., 2001). Alginates rich in G–G blocks generally possess a low M/G ratio and give transparent, stiffer and more brittle films, while alginates with higher content of M–M or M–G blocks form flexible films, with low elastic moduli (Nyvall et al., 2003).

8.2 Sources and Extraction

Alginates are naturally present in brown seaweed cell walls. Commercial alginates are mainly extracted from species of *Laminaria, Macrocystis, Ascophyllum, Eclonia, Lessonia, Durvillea,* and *Sargassum*. Although the main source of alginates is brown seaweed, alginate-like polysaccharides can be produced as an extracellular material by bacteria (Llanes et al., 1997). The most commonly used bacteria for production of this type of alginate are *Pseudomonas* and *Azotobacter*, which are abundant in vegetative growing cells (Rehm and Valla, 1997). Alginates obtained from algae may have variety of molecular weight, while alginates obtained from bacteria possess high DP and high molar masses. Also, bacterial alginate does not have oligomeric G sequences in their structure (Windhues and Borchard, 2003).

Extraction of alginate from brown algae is comprised of five main steps. As the alginate is present in cell wall of brown algae in the form of alginate salts, the first step of extraction is acidification in order to convert the alginate salts into water insoluble alginic acids. The second step is alkaline extraction using sodium hydroxide solution, to provide conversion of insoluble alginic acid into soluble sodium alginate. The third step is related to the separation of seaweed residues from the sodium alginate solution using floatation/flocculation and filtration. This step is followed by precipitation of sodium alginate in the acidification process of sodium alginate solution. The final step includes pressing and drying of sodium alginate (Vauchel et al., 2009).

8.3 Physical Properties

Alginate is negative charged polyelectrolite, with high dissociation degree of carboxylic groups at pH7.4. The acid pK_a of alginate polymers differ only slightly from those of the monomeric residues, which were determined to be 3.38 and 3.65 for mannuronic and guluronic acid, respectively (Haug et al., 1966). Electrostatic repulsive forces between the charged groups give high viscosity for alginate solution. The viscosity of alginate solution is mainly influenced by alginate concentration, molecular weight, and ionic strength in the solution. Due to a large number of carboxyl group, alginates can easily be manipulated changing the pH of solution. By lowering pH below the pK_a of the carboxyl group (about 3.5), alginic acid can be obtained. This protonated form is insoluble in water and, therefore, precipitates from solution.

Alginates have the ability to form gel in the presence of divalent cations, where divalent cations interact mainly with G–G blocks. Two adjacent G–G blocks make up a cavity wherein divalent cation binds forming tightly held junctions that has been called “egg box” arrangement (Fig. 19.6). M–G blocks may also participate in interaction with divalent cations, but it results in formation of weak junctions. Hence, alginates with high content of G–G blocks yield gels with high strength, stability, and low shrinkage.

8.4 Application

Nowadays, alginates have found commercial applications as thickeners, stabilizers, gel-forming and film-forming agents because of the ability to retain water, to form viscous solutions at low concentration and to gel under different conditions. Due to the large number of free hydroxyl and carboxyl groups distributed along the backbone, alginate can be easily modified into diverse structural forms such as hydrogels, microspheres, microcapsules, sponges, foams, fibers, and films. By modification of alginate, various properties such as solubility, hydrophobicity, physical,

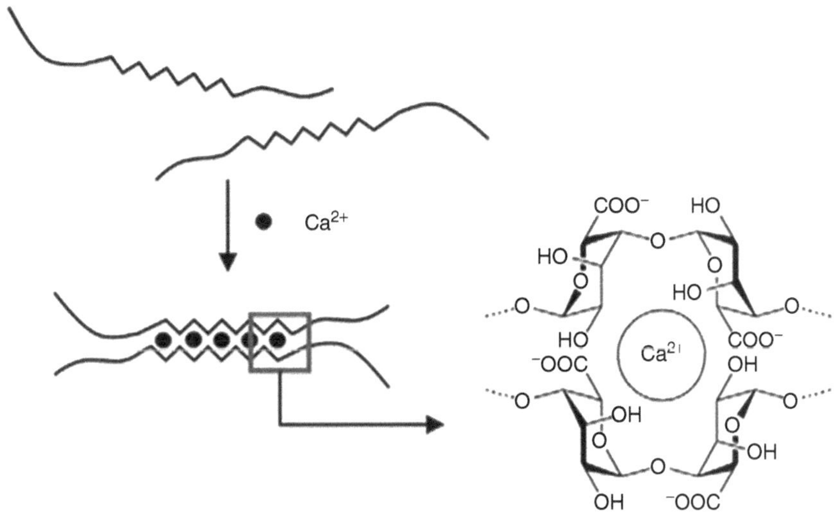

Figure 19.6. Formation of an alginate gel by calcium cations, resulting in "egg-box" calcium linked junctions. Reproduced with permission from Paques, J., et al., 2014. Adv. Colloid Interface 209, 163–171. Copyright © 2014, Elsevier B.V.

chemical, and biological characteristics may be enhanced. Modification potential coupled with wide application spectrum is significantly important from the standpoint of its usage in the fields of tissue engineering, drug delivery, wound healing, wastewater treatment, and food packaging.

CN have attracted significant interest as potential nanoreinforcements during the past decade. Abdollahi et al. (2013a) investigated the influence of CN loading content (1, 3, 5, and 10 wt.%) on microstructural, physical, mechanical, and optical properties of the alginate/glycerol/CN nanocomposite films. The results showed that water vapor permeability of the nanocomposites decreased for 17% with an increase of the CN content up to 10%. The TS increased with increasing NC content from 0 to 5%, but further increase of CN content resulted in decrease of TS. It was noticed that the addition of CN caused significant reduction in the film transparency, which was an indication of the partial agglomeration of the fillers at higher content.

The similar observation was found by Huq et al., wherein alginate/CN/$CaCl_2$ nanocomposite film reinforced with 5 wt.% CN exhibited the highest TS, which was increased by 37% compared to the pure alginate film. The further increases in CN content, up to 8%, resulted in decrease of TS (Huq et al., 2012). The highest water vapor barrier showed the alginate/CNC/$CaCl_2$ film with 5 wt.% of CN. Compared to the pure alginate film the decrease in water vapor permeability of 31% was achieved.

Azeredo et al. (2012) investigated the influence of two nanofillers: MMT and CN in combination with acerola/alginate edible film on shelf life of acerola fruit. Both nanofillers (cellulose whiskers and MMT) improved water vapor barrier of standalone alginate–acerola puree films. When applied as coatings to fresh acerolas, the alginate–acerola puree film was effective in extending fruit stability by decreasing weight loss, ascorbic acid loss, and decay incidence, and by delaying the ripening process. The nanocomposite coating containing MMT was the most effective in reducing weight loss of acerolas.

Abdollahi et al. (2013b) investigated also the influence of MMT and nanocellulose on various physical–chemical properties of alginate nanocomposite films. The addition of both nanoreinforcements resulted in enhanced water vapor barrier stability of the alginate-nanocomposite films. An increase in CN content up to 5 wt.% led to improved TS and Young's modulus of the alginate nanocomposite films. On the contrary, the maximum TS for alginate/MMT films was observed with 1 wt.% of MMT loading, while the further increase in nanofiller content caused decrease in TS. Also, with all loading contents Young's modulus of the alginate/CN nanocomposites was higher than that of alginate/MMT films, suggesting that CNwere better dispersed in alginate matrix than MMT.

Alboofetileh et al. (2014) found that optimum content of MMT in alginate/glycerol/MMT nanocomposite films at which the best mechanical properties are achieved was 3 wt.%, while the films containing 5 wt.% of MMT showed the highest reduction in water vapor permeability (19%), with respect to neat alginate film. The same authors were studied antimicrobial effects of pure essential oils: [clove (CO), coriander, caraway, marjoram (MO), cinnamon, and cumin] encapsulated into alginate/glycerol/MMT nanocomposite films against *E. coli*, *Staphylococcus aureus*, and *L. monocytogenes* (Alboofetileh et al., 2014). Considering all investigated nanocomposites, marjoram showed the highest antimicrobial activity. Film containing 1.5% of marjoram decreased the numbers of *L. monocytogenes*, *E. coli*, and *S. aureus* populations with respect to the control up to 6.33, 4.52, and 5.80 log, respectively.

Alcantara et al. (2015) reported the use of sepiolite and palygorskite fibrous clays as nanoreinforcing fillers in alginate matrix. The TS and Young's moduli of films incorporated with sepiolite were higher than those loaded with palygorskite. EM of single sepiolite crystals, as well as the high external surface area available in sepiolite in comparison to palygorskite could be the main reason for synergetic interactions toward the alginate chains. Fibrous clays modified with the hydrophobic protein, zein, were also tested as nanofillers. The presence of zein assembled to the clay fibers had a crucial role in the water vapor barrier, gas permeation, and mechanical properties of the alginate films. An improvement related to the plastic behavior of alginate matrices loaded with both zein-sepiloite and zein-palygorskite was obtained. Plasticizing effect that was attributed to zein resulted in an increase in the E values up to 20%. Zein-based organoclays afforded good transmittance in the visible region and inhibited the passage of UV light in the same time, which is very important from the standpoint of food protection.

9 Pectin

9.1 Molecular Structure

Pectin is composed of 1, 4-linked α-D-galacturonic acid (GalA) residues and at least 17 different monosaccharides containing more than 20 linkages. The main building blocks in pectin structure are homogalacturonan (HG), rhamnogalacturonan-I (RG-I), and rhamnogalacturonan-II (RG-II) (Pellerin et al., 1996). HG is linear and unbrunched homopolymer composed of a 1, 4-linked galacturonic acid and it is the most prevalent type in the pectin structure, accounting for approximately 60% of the total pectin amount. HG can be partially methyl esterified at the C6 atom of carboxyl group and/or acetylated at O-2 or O-3 atom (Fig. 19.7) (Mohnen, 2008). RG-I comprises 25–30% of pectin and contrary to HG, RG-I, it is characterized by highly brunched structure. RG-I backbone is consisting of the repeating rhamnose-galacturonic acid disaccharide $[\text{-}\alpha\text{-D-GalA-1,2-a-L-Rha-1-4-}]_n$. The rhamnosyl residues of RGI can be substituted at O-4 with neutral sugars side chains which are mainly composed out of galactosyl and/or arabinosyl residues (Fig. 19.8) (O'Neill et al., 1990). Rhamnogalacturonan II accounting only 10% of pectin, but it is the most structurally complex pectic polysaccharide. It is consisted of at least eight 1, 4-linked α-D-GalA residues with side branches that contain 12 different monosccharides, including the rare sugar species (Ridley et al., 2001).

The chemical constitution of pectin is given on the basis of esterification degree (DE). The degree of esterification presents the ratio of esterified GalA groups per total amount of GalA in pectin structure. Therefore, pectins can be divided in to majors groups based on their degree of esterification: high methylated (HM) pectin, where more than 50% of carboxylic groups exist as methyl ester and low methylated (LM) pectin with less than 50% of methyl ester group.

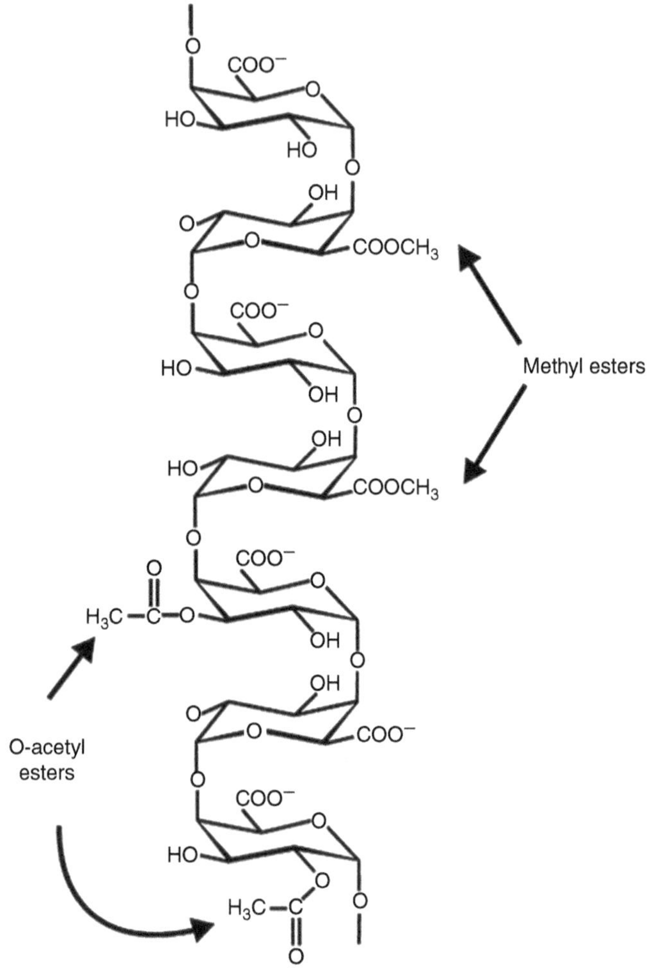

Figure 19.7. The primary structure of homogalacturonan. Reproduced with permission from Ridley, B.L., et al., 2001. Phytochemistry 57, 929–967. Copyright © 2001, Elsevier B.V.

9.2 Sources and Extraction

Pectins are naturally present in the middle lamella of plant cell walls. Commercial pectins are usually obtained by extraction from citrus peel or apple pomace. The content of dry pectin matter in apple pomace and citrus peel amounts 10–15% and 20–30%, respectively. There are numerous alternative sources of pectin such as sugar beet waste, sunflower heads, and mango waste. However, the citrus fruits contain the highest amount of pectin. Typical levels of pectin in plants are: apples 1–1.5%, apricot 1%, cherries

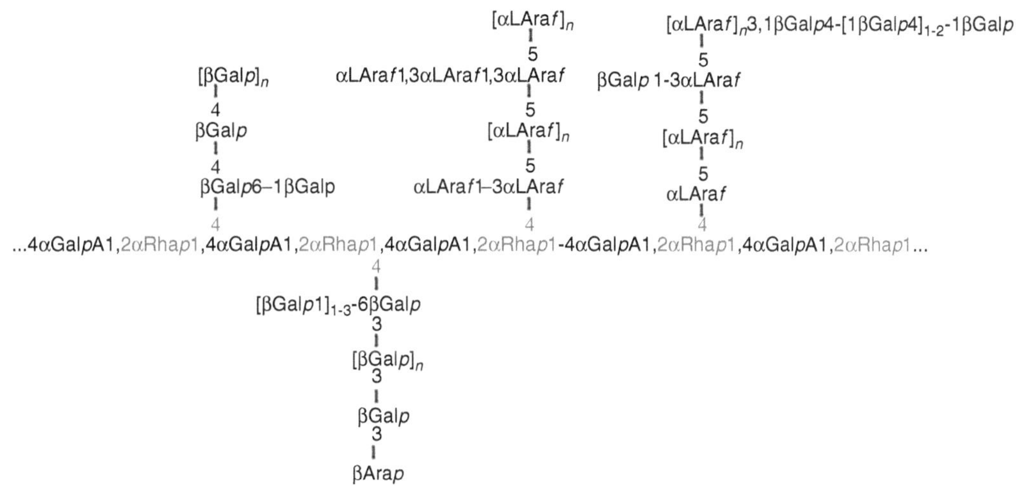

Figure 19.8. The structure of RG-I backbone and representative side chains. Reproduced with permission from Caffal, K.H., et al., 2009. Carbohydr. Res. 344, 1879–1900. Copyright © 2009, Elsevier B.V.

0.4%, oranges 0.5–3.5%, carrots 1.4%, and citrus peel 30% (Thakur et al., 1997).

The conventional method for pectin extraction involves four different steps. The first step is acidification of raw material using hot dilute mineral acids (sulfuric, hydrochloride, or phosphoric acid) at pH 2. This step is followed by separation of solid residue from pectin extract. The third step includes the precipitation of pectin with alcohol, or by addition of polyvalent cations, usually aluminum. The last step involves washing, drying, and milling of the pectin powder (May, 1990). The yield and degree of esterification of extracted pectin depends on operating temperature and time. The described process is applicable only for HM pectin, while the production of LM pectins includes addition of acid in order to hydrolyze some ester groups of pectin, either before or during a prolonged extraction.

The described traditional extraction method requires long extraction time, high solvent consumption and it is thermally unsafe. Recently, strong efforts have been made in order to develop new extraction techniques. Microwave heating extraction turned to be efficient method that provides reduced process time as well as the higher yields of pectin (US patent, 2007).

9.3 Physical Properties

The most important property of pectin is its ability to form gels in the presence of divalent and trivalent cations or in acidic solutions. The gelation mechanism of pectin depends mainly on its degree of esterification. LM pectin can form gel in the presence of divalent cations, and independently of pH value, making intermolecular ionic junction zones between homogalacturonic smooth regions of different chains. The structure of the junction zones is generally ascribed to the so called egg box arrangement, originally suggested for alginates (Morris et al., 1982).

Contrary to LM pectin, HM pectin can form gels only in the presence of sugar and at acidic condition. However, under certain conditions HM pectin are showing gelling ability in presence of aluminum and few divalent cations such as copper, nickel, zinc, cadmium (it cannot gel in presence of calcium ions). The junction zones are formed by means of hydrophobic interactions between methoxyl ester groups and hydrophilic interactions between undissociated carboxyl groups and/or hydroxyl groups via hydrogen bonds (Oakenfull and Scott, 1984).

9.4 Application

Pectins are mainly used as gelling agents, thickeners, water binders, and stabilizers in food industry. Also, pectin is an interesting constituent of pharmaceutical use as a carrier for controlled release of various drugs, because of its ability to degrade by colonic enzymes (Malviya et al., 2010). Although there is a growing interest toward investigation and development of pectin-based films intended for food packaging due to its nontoxicity, edibility, and high level of biodegradability, only few systems based on pectin nanocomposites have been elaborated, studied and reported in the literature.

Cavallaro et al. (2013a,b) prepared films based on low methoxyl pectin and low methoxyl pectin/PEG 2000 blend, filled with inorganic natural halloysite nanotubes. The effect of halloysite nanotubes on the mechanical and thermal properties of the blend pectin/PEG20000 films was investigated. Results showed that an increase in content of halloysite content in pectin/PEG2000 films led to the increase of Young's modulus and decrease in E, while TS was not influenced by content of nanoparticles. The good dispersion of the filler within the polymer matrix was proved by an increase in E values upon the addition of halloysite nanotube. Additionally, the uniform distribution of halloysite inside the pectin/PEG films enhanced the affinity and the adhesion of the polymers to the filler surface. Thermal analysis of pectin/halloysite and pectin/PEG2000/halloysite films confirmed good distribution of halloysite due to delayed pectin degradation after the nanoparticeles were incorporated. The homogeneous distribution of halloysite within the LM matrix improved barrier properties, as well as thermal stability of the films at high content of halloysite (30%). The high thermal stabilization of pectin/halloysite and pectin/PEG2000/halloysite films was ascribed to the entrapment of the volatile products (obtained from the polymer thermal decomposition) into the halloysite nanoparticle.

Gorrasi (2015) investigated films based on high methoxyl pectin and nanohybrids composed of halloysite nanotubes loaded with rosemary essential oil for application in active packaging, where controlled release of active species is required. A nanohybrid composed of halloysite and rosemary essential oil was prepared by mixing 27 wt.% of halloysite nanotube and 73 wt.% of essential rosemary oil. The nanohybrid suspension was added to the pectin/glycerol/water solution at different amounts, in order to obtain nanohybrid concentrations equal to 5, 10, and 20 wt.%. Thermal analysis of obtained films indicated that the essential oil molecules incorporated into the nanohybrid postponed the thermooxidation, probably due to successful entrapment of oil molecules into the nanotubes. The degradation of the pectin matrix was not greatly influenced by addition of nanohybrid filler. Mechanical properties showed an improvement in EM and stress at break with content of 5 and 10% of nanohybrid when compared to the neat pectin. However, the higher content of nanohybrid (up to 20%) resulted in a decrease of mechanical properties. An increase of nanohybrid content in films (ie, 20%) caused the formation of clusters and deterioration of the mechanical parameters. Presence of the nanohybrid filler increased the tortuosity of the system and led to a decrease in the water vapor permeability.

Mangiacapra et al. (2006) prepared two composites of pectin with 3% of both, a sodium MMT and an organically modified MMT with stearyl-dietoxamine, using high energy ball milling method that involves mixing of the organic and inorganic components by mechanical milling. Afterward, the mixed powders were cast into films and characterized. Incorporation of sodium MMT into pectin matrix resulted in improved EM, which increased from 1630 to 2962 MPa, while milling the pectin with modified MMT did not show significant influence on mechanical properties. The results indicated the formation of hydrogen bonds between pectin matrix and silicate layers of sodium MMT, whereas organically modified MMT hindered the formation of hydrogen bonds with pectin matrix due to steric effects. Thermal analysis showed that degradation of pectin chain was not significantly influenced by addition of clay into the structure, while barrier analysis demonstrated lower water vapor and oxygen permeability compared to the pure pectin.

Vartiainen et al. (2010) investigated the pectin films with MMT concentrations of 0, 10, 20, and 30 wt.%, where the nanoclay was dispersed in an aqueous pectin solution using high pressure fluidizer. The films were obtained by casting method. The addition of MMT at 30% reduced the water vapor permeability for 23%, while oxygen transmission rate was reduced over 80% when compared to the pure pectin film. All of the films turned to be totally impermeable to grease. Authors explained barrier improvements by tortuous path theory, which is related to the alignment of the nanoclay platelets. As a result of the delamination, the effective path length for molecular diffusion increased enough to reduce the effect of gas transmission through the film.

Yu et al. (2014) synthesized the LM pectin-CMC-based MMT nanocomposite films with nine ratios of LM pectin: CMC (from 10:0 to 0:10) and different MMT contents (1–8 wt.%) The mechanical and water vapor barrier properties were significantly improved by addition of MMT. The maximum of TS (39.85 MPa) was obtained at LMP: CMC ratio of 4:6 and 4 wt.% of MMT, which confirmed the formation of hydrogen bonds between clay and LMP–CMC. The reduction in water vapor permeability of the same nanocomposite film was 333% compared to the pure CMC film due to the tortuous path caused by MMT incorporation.

Moreira et al. (2013) prepared nanocomposite films of low methoxyl pectin and high methoxyl pectin with addition of [$Mg(OH)_2$] nanoplates. The obtained results showed that degree of pectin methylation had influence on mechanical properties. The mechanical and thermal properties of the nanocomposite films were significantly improved compared to the control. $Mg(OH)_2$ nanoplates were more effective in improving properties of low methoxyl pectin due to the greater number of polar groups that favor the cohesion between the chains through hydrogen bonding. Preliminary migration studies using arugula leaves proved that pectin–$Mg(OH)_2$ nanocomposites could release magnesium hydroxide by contact, demonstrating their potential for magnesium supplementation in bioactive packaging.

10 Conclusions

Nanocomposite concept has been considered as a stimulating route for creating innovative materials with improved performance properties. Renewable biopolymers such as polysacharides are some of the most widely used in production of nanocomposites intended for food packaging applications. From the stand point of environmental protection, the utilization of bionanocomposites will help to reduce the packaging waste associated with processed food. Antimicrobial activity, enhanced barrier properties against O_2, CO_2, and flavor compounds, as well as improved mechanical and thermal properties of the bionanocomposites, put them in an enviable position among the candidates for the replacement of petroleum-based plastics. Such property enhancements are generally attained at low nanoclay content (less than 5%) compared to that of conventional fillers (in the range of 10–50%). Thus, nanocomposites are far lighter in weight than conventional composite materials, which makes competitive compared to the other materials for specific applications such as food packaging.

However, despite the numerous benefits deriving from bionanocomposites, these materials still do not meet all required criteria. The main drawback is related to their poor water resistance, inadequate for utilization of the bionanocomposites as packaging materials, especially in wet environmental conditions. Also, there are only limited investigations done on the toxicological effects and migration of nanofillers into food from the packaging films. The challenges related to safe usage of bionanocomposites must be solved by means of a multidisciplinary approach that includes the efforts of scientists and technicians with expertise in different fields, as well as the implementation of suitable international laws recommended by the governments and international organizations.

Acknowledgment

The authors acknowledge funding from the Ministry of Science, Technology and Development of the Republic Serbia, Science Projects No. 43009 and No. 172062.

References

Abdollahi, M., Alboofetileh, M., Behrooz, R., Rezaei, M., Miraki, R., 2013a. Reducing water sensitivity of alginate bio-nanocomposite film using cellulose nanoparticles. Int. J. Biol. Macromol. 54, 166–173.

Abdollahi, M., Alboofetileh, M., Rezaei, M., Behrooz, R., 2013b. Comparing physico-mechanical and thermal properties of alginate nanocomposite films reinforced with organic and/or inorganic nanofillers. Food Hydrocoll. 32, 416–424.

Abdul Hanid, N., UzirWahit, M., Guo, Q., Mahmoodian, S., Soheilmoghaddam, M., 2014. Development of regenerated cellulose/halloysitesnanocomposites via ionic liquids. Carbohydr. Polym. 99, 91–97.

Adoracion, P.R., Li, X.L., Okita, T.W., Juliano, B.O., 1993. Characterization of poorly digested protein of cooked rice protein bodies. Cereal Chem. 70, 101–105.

Agrawal, P., Strijkers, G.J., Nicolay, K., 2010. Chitosan-based systems for molecular imaging. Adv. Drug Deliv. Rev. 62, 42–58.

Alboofetileh, M., Rezaei, M., Hosseini, H., Abdollahi, M., 2014. Antimicrobial activity of alginate/clay nanocomposite films enriched with essential oils against three common foodborne pathogens. Food Control. 36, 1–7.

Alcantara, A.C.S., Darder, M., Aranda, P., Ayral, A., Ruiz-Hitzky, E., 2015. Bionanocomposites based on polysaccharides and fibrous clays for packaging applications. J. Appl. Polym. Sci. 42362, 1–14.

Almeida, M., Alves, R., Nascimbem, L., Stephani, R., Poppi, R., De Oliveira, L., 2010. Determination of amylose content in starch using Raman spectroscopy and multivariate calibration analysis. Anal. Bioanal. Chem. 397, 2693–2701.

Atkin, N.J., Abeysekera, R.M., Cheng, S.L., Robards, A.W., 1998. An experimentally based predictive model for the separation of amylopectin subunits during starch gelatinization. Carbohydr. Polym. 36, 173–192.

Aulin, C., Ahola, S., Josefsson, P., Nishino, T., Hirose, Y., Österberg, M., Wagberg, L., 2009. Nanoscale cellulose films with different crystallinities and mesostructures: their surface properties and interaction with water. Langmuir 25 (13), 7675–7685.

Avella, M., De Vlieger, J.J., Errico, M.E., Fischer, S., Vacca, P., Volpe, M.G., 2005. Biodegradable starch/clay nanocomposite films for food packaging applications. Food Chem. 93, 467–474.

Averous, L., Moro, L., Dole, P., Fringant, C., 2000. Properties of thermoplastic blends: starch polycaprolactone. Polymer 41 (11), 4157–4167.

Azeredo, H.M.C., 2009. Nanocomposites for food packaging applications. Food Res. Int. 42, 1240–1253.

Azeredo, H., Mattoso, L.H.C., Avena-Bustillos, R.J., Munford, M.L., Wood, D., McHugh, T.H., 2010. Nanocellulose reinforced chitosan composite films as affected by nanofiller loading and plasticizer content. J. Food Sci. 75, N1–N7.

Azeredo, H.M.C., Miranda, K.W.E., Ribeiro, H.L., Rosa, M.F., Nascimento, D.M., 2012. Nanoreinforced alginate–acerola puree coatings on acerola fruits. J. Food Eng. 113, 505–510.

Bagdi, K., Müller, P., Pukánszky, B., 2006. Thermoplastic starch/layered silicate composites: structure, interaction, properties. Compos. Interf. 13, 1–17.

Balser, K., Hoppe, L., Eicher, T., Wandel, M., Astheimer, H.J., 1986. Cellulose esters. In: Gerhartz, W., Stephen, Y.Y., Thomas, C.F., Pfefferkorn, R., James, F. (Eds.), Ullmann's Encyclopedia of Industrial Chemistry. Wiley, New York, pp. 419–459.

Bello-Flores, C.A., Nunez-Santiago, M.C., San Martín-Gonzalez, M.F., BeMiller, J.N., Bello-Pérez, L.A., 2014. Preparation and characterization of octenylsuccinylated plant in starch. Int. J. Biol. Macromol. 70, 334–339.

Benn, T.M., Westerhoff, P., 2008. Nanoparticle silver released into water from commercially available sock fabrics. Environ. Sci. Technol. 42, 4133–4139.

Blaszczak, W., Fornal, J., Valverde, S., Garrido, L., 2005. Pressure-induced changes in the structure of corn starches with different amylose content. Carbohydr. Polym. 61, 132–140.

Bordes, P., Pollet, E., Avérous, L., 2009. Nano-biocomposites: biodegradable polyester/nanoclay systems. Prog. Polym. Sci. 20, 125–155.

Brandt, L., 1986. Cellulose ethers. In: Gerhartz, W., Stephen, Y.Y., Thomas, C.F., Pfefferkorn, R., James, F. (Eds.), Ullmann's Encyclopedia of Industrial Chemistry. Wiley, New York, pp. 461–487.

Brody, A.L., 2009. Innovations in fresh prepared meal delivery systems. Food Technol. 63, 84–86.

Cavallaro, G., Donato, G.D., Lazzara, G., Milioto, S., 2013a. Films of halloysite nanotubes sandwiched between two layers of biopolymer: from the morphology to the dielectric, thermal, transparency, and wettability properties. J. Phys. Chem. C 115, 20491–20498.

Cavallaro, G., Lazzara, G., Milioto, S., 2013b. Sustainable nanocomposites based on halloysite nanotubes and pectin/polyethylene glycol blend. Polym. Degrad. Stabil. 98, 2529–2536.

Chang, P.R., Jian, R., Yu, J., Ma, X., 2010. Fabrication and characterisation of chitosan nanoparticles/plasticised-starch composites. Food Chem. 120, 736–740.

Chen, B., Evans, J.R., 2005. Thermoplastic starch-clay nanocomposites and their characteristics. Carbohydr. Polym. 61, 455–463.

Chenite, A., Gori, S., Shive, M., Desrosiers, E., Buschmann, M.D., 2006. Monolithic gelation of chitosan solutions via enzymatic hydrolysis of urea. Carbohydr. Polym. 64, 419–424.

Chiou, B.S., Yee, E., Glenn, G.M., Orts, W.J., 2005. Rheology of starch-clay nanocomposites. Carbohydr. Polym. 59, 467–475.

Chiou, B.S., Yee, E., Wood, D., Shey, J., Glenn, G., Orts, W., 2006. Effects of processing conditions on nanoclay dispersion in starch-clay nanocomposites. Cereal Chem. 83, 300–305.

Clarinval, A.M., Halleux, J., 2005. Classification of biodegradable polymers. In: Smith, R. (Ed.), Biodegradable Polymers for Industrial Applications. Woodhead Publishing Ltd, Cambridge, UK, pp. 3–31.

Committee, E.S., 2009. The potential risks arising from nanoscience and nanotechnologies on food and feed safety. EFSA J. 958, 1–39.

Conte, A., Longano, D., Costa, C., Ditaranto, N., Ancona, A., Cioffi, N., Scrocco, C., Sabbatini, L., Contò, F., Del Nobile, M.A., 2013. A novel preservation technique applied to fiordilatte cheese. Innov. Food Sci. Emerg. 19, 158–165.

Corrales, M., Fernandez, A., Han, J.H., 2014. Antimicrobial packaging systems. In: Han, J. (Ed.), Innovations in Food Packaging. Elsevier, Amsterdam, pp. 133–169.

Croisier, F., Jerome, C., 2013. Chitosan-based biomaterials for tissue engineering. Eur. Polym. J. 49, 780–792.

Cyras, V.P., Manfredi, L.B., Ton-That, M.T., Vázquez, A., 2008. Physical and mechanical properties of thermoplastic starch/montmorillonite nanocomposite films. Carbohydr. Polym. 73, 55–63.

Dentini, M., Rinaldi, G., Risica, D., Barbetta, A., Skjak-Bræk, G., 2005. Comparative studies on solution characteristics of mannuronan epimerized by C-5 epimerases. Carbohydr. Polym. 59, 489–499.

Dias, M.V., Azevedo, V.M., Borges, S.V., Soares, N.D.F.F., de Barros Fernandes, R.V., Marques, J.J., Medeiros, É.A.A., 2014. Development of chitosan/montmorillonite nanocomposites with encapsulated α-tocopherol. Food Chem. 165, 323–329.

Domard, A., Rinaudo, M., 1983. Preparation and characterization of fully deacetylated chitosan. Int. J. Biol. Macromol. 5, 49–52.

Draget, K.I., Gåserod, O., Aune, I., Andersen, P.O., Storbakken, B., Stokke, B.T., Smidsrod, O., 2001. Effects of molecular weight and elastic segment flexibility on syneresis in Ca alginate gels. Food Hydrocoll. 15 (46), 485–490.

Duncan, T.V., 2011. Applications of nanotechnology in food packaging and food safety: barrier materials, antimicrobials and sensors. J. Colloid Interface Sci. 363, 1–24.

El Miri, N., Abdelouahdi, K., Barakat, A., Zahouily, M., Fihri, A., Solhy, A., El Achaby, M., 2015. Bio-nanocomposite films reinforced with cellulose nanocrystals: rheology of film-forming solutions, transparency, water vapor barrier and tensile properties of films. Carbohydr. Polym. 129, 156–167.

El-Wakil, N.A., Hassan, E.A., Abou-Zeid, R.E., Dufresne, A., 2015. Development of wheat gluten/nanocellulose/titanium dioxide nanocomposites for active food packaging. Carbohydr. Polym. 124, 337–346.

Erickson, B., 2009. Nanosilverpesticides. Chem. Eng. News 87, 25–26.
Ferrara, G., Bertolodo, M., Scoponi, M., Ciardelli, F., 2001. Diffusion coefficient and activation energy of irganox 1010 in poly(propylene-*co*-ethylene) copolymers. Polym. Degrad. Stabil. 73, 411–416.
Focher, B., Beltrame, P.L., Naggi, A., Torri, G., 1990. Alkaline N-deacetylation of chitin enhanced by flash treatments: reaction kinetics and structure modifications. Carbohydr. Polym. 12, 405–418.
Fringant, C., Desbrieres, J., Milas, M., Rinaudo, M., Joly, C., Escoubes, M., 1996. Characterisation of sorbed water molecules on neutral and ionic polysaccharides. Int. J. Biol. Macromol. 18, 281–286.
Geigenberger, P., Geiger, M., Stitt, M., 1998. High-temperature perturbation of starch synthesis is attributable to inhibition of ADP-glucose pyrophosphorylase by decreased levels of glycerate-3-phosphate in growing potato tubers. Plant Physiol. 117, 1307–1316.
Ghaderi, M., Mousavi, M., Yousefi, H., Labbafi, M., 2014. All-cellulose nanocomposite film made from bagasse cellulose nanofibers for food packaging application. Carbohydr. Polym. 104, 59–65.
Glenn, G.M., Klamczynski, A., Chiou, B.S., Orts, W.J., Imam, S.H., Wood, D.F., 2008. Temperature related structural changes in wheat and corn starch granules and their effects on gels and dry foam. Starch-Starke 60, 476–484.
Glenn, G.M., Orts, W., Imam, S., Chiou, B.S., Wood, D.F., 2014. Starch plastic packaging and agriculture applications. In: Halley, P., Averous, L. (Eds.), Starch Polymers: From Genetic Engineering to Green Applications. Elsevier, pp. 421–452.
Gorrasi, G., 2015. Dispersion of halloysite loaded with natural antimicrobials into pectins: characterization and controlled release analysis. Carbohydr. Polym. 127, 47–53.
Goycoolea, F.M., Higuera-Ciapara, I., Hernández, G., Lizardi, J., García, K.-D., 1997. Preparation of chitosan from squid (*Loligo* spp.) pen by a microwave-accelerated thermochemical process. Adv. Chitin Sci., 78–83.
Guibal, E., 2005. Heterogeneous catalysis on chitosan-based materials: a review. Prog. Polym. Sci. 30, 71–109.
Haug, A., Larsen, B., Smidsrod, O., 1966. A study of constitution of alginic acid by partial acid hydrolysis. Acta Chem. Scand. 20 (1), 183–190.
Hirano, S., Seino, H., Akiyama, Y., Nonaka, I., 1990. Chitosan: a biocompatible material for oral and intravenous administrations. In: Gebelein, C.H., Dunn, R.L. (Eds.), Progress in Biomedical Polymers. Plenum Press, New York, pp. 283–290.
Huang, M., Yu, J., 2006. Structure and properties of thermoplastic corn starch/montmorillonite biodegradable composites. J. Appl. Polym. Sci. 99, 170–176.
Huang, M.F., Yu, J.G., Ma, X.F., Jin, P., 2005. High performance biodegradable thermoplastic starch—EMMT nanoplastics. Polymer 46, 3157–3162.
Huang, T.T., Zhou, D.N., Jin, Z.Y., Xu, X.M., Chen, H.Q., 2015. Effect of debranching and heat-moisture treatments on structural characteristics and digestibility of sweet potato starch. Food Chem. 187, 218–224.
Huq, T., Salmieri, S., Khan, A., Khan, R.A., Tien, C.L., Riedl, B., Fraschini, C., Bouchard, J., Uribe-Calderon, J., Kamal, M.R., Lacroix, M., 2012. Nanocrystalline cellulose (NCC) reinforced alginate based biodegradable nanocomposite film. Carbohydr. Polym. 90, 1757–1763.
Jane, J.L., 2009. Structural features of starch granules II. In: BeMiller, J., Whistler, R. (Eds.), Food Science and Technology. Elsevier, pp. 193–236.
Jing, Z., Guo, D., Wang, W., Zhang, S., Qi, W., Ling, B., 2011. Comparative study of titania nanoparticles and nanotubes as antibacterial agents. Solid State Sci. 13, 1797–1803.

Kanmani, P., Rhim, J.W., 2014. Properties and characterization of bionanocomposite films prepared with various biopolymers and ZnO nanoparticles. Carbohydr. Polym. 106, 190–199.

Kasirga, Y., Oral, A., Caner, C., 2012. Preparation and characterization of chitosan/montmorillonite-K10 nanocomposites films for food packaging applications. Polym. Compos. 33, 1874–1882.

Khan, A., Khan, R.A., Salmieri, S., Le Tien, C., Riedl, B., Bouchard, J., Lacroix, M., 2012. Mechanical and barrier properties of nanocrystalline cellulose reinforced chitosan based nanocomposite films. Carbohydr. Polym. 90, 1601–1608.

Kim, S.W., Cha, S.H., 2014. Thermal, mechanical, and gas barrier properties of ethylene-vinyl alcohol copolymer-based nanocomposites for food packaging films: effects of nanoclay loading. J. Appl. Polym. Sci. 131, 40289.

Kim, S.K., Hahn, T.R., Lee, Y.H., D'Appolonia, B.L., 1978. Physicochemical properties of tongil (indica type) and paldal (japonica type) rice starch. Korean J. Food Sci. Technol. 10, 157–161.

Kirwan, M.J., Strawbridge, J.W., 2003. Plastics in food packaging. Food Packag. Technol., 174–240.

Knauf, M., Moniruzzaman, M., 2004. Lignocellulosic biomass processing: a perspective. Int. Sugar J. 106, 147–150.

Krochta, J.M., 1997. Edible protein films and coatings. In: Damodaran, S., Paraf, A. (Eds.), Food Proteins and Their Applications. Marcel Dekker, New York, pp. 529–549.

Kumar, M.N.R., 2000. A review of chitin and chitosan applications. React. Funct. Polym. 46, 1–27.

Kumar, M.R., Muzzarelli, R., Muzzarelli, C., Sashiwa, H., Domb, A.J., 2004. Chitosan chemistry and pharmaceutical perspectives. Chem. Rev. 104, 6017–6084.

Laban, G., Nies, L.F., Turco, R.F., Bickham, J.W., Sepúlveda, M.S., 2010. The effects of silver nanoparticles on fathead minnow (*Pimephales promelas*) embryos. Ecotoxicology 19, 185–195.

Lagaron, J.M., Cava, D., Cabedo, L., Gavara, R., Gimenez, E., 2005. Increasing packaged food quality and safety: (II) nanocomposites. Food Addit. Contam. 22, 994–998.

Lamarque, G., Viton, C., Domard, A., 2004. Comparative study of the second and third heterogeneous deacetylations of α- and β-chitins in a multistep process. Biomacromolecules 5, 1899–1907.

Lavorgna, M., Attianese, I., Buonocore, G.G., Conte, A., Del Nobile, M.A., Tescione, F., Amendola, E., 2014. MMT-supported Ag nanoparticles for chitosan nanocomposites: structural properties and antibacterial activity. Carbohydr. Polym. 102, 385–392.

Law, K.L., Moret-Ferguson, S., Maximenko, N.A., Proskurowski, G., Peacock, E.E., Hafner, J., Reddy, C.M., 2010. Plastic accumulation in the North Atlantic subtropical gyre. Science 329, 1185–1188.

Leite de Oliveira, R., da Silva Barud, H., Bonemer De Salvi, D.T., FrigiPerotti, G., Lima Ribeiro, S.J., LeopoldoConstantino, V.R., 2015. Transparent organic–inorganic nanocomposites membranes based on carboxymethylcellulose and synthetic clay. Ind. Crops Prod. 69, 415–423.

Lemos, A.M., Abraão, A.S., Cruz, B.R., Morgado, M.L., Rebelo, M., Nunes, F.M., 2015. Effect of granular characteristics on the viscoelastic and mechanical properties of native chestnut starch (*Castanea sativa* Mill). Food Hydrocoll. 51, 305–317.

Li, J.H., Vasanthan, T., Hoover, R., Rossnagel, B.G., 2004. Starch from hull-less barley: IV. Morphological and structural changes in waxy, normal and high-amylose starch granules during heating. Food Res. Int. 37, 417–428.

Li, M., Liu, P., Zou, W., Yu, L., Xie, F., Pu, H., Chen, L., 2011. Extrusion processing and characterization of edible starch films with different amylose contents. J. Food Eng. 106, 95–101.
Lin, B., Luo, Y., Teng, Z., Zhang, B., Zhou, B., Wang, Q., 2015. Development of silver/titanium dioxide/chitosan adipate nanocomposite as an antibacterial coating for fruit storage. LWT Food Sci. Technol. 63, 1206–1213.
Llanes, F., Sauriol, F., Morin, F.G., Perlin, A.S., 1997. An examination of sodium alginate from *Sargassum* by NMR spectroscopy. Can. J. Chem. 75, 585–590.
Loutfi, A., Coradeschi, S., Mani, G.K., Shankar, P., Rayappan, J.B.B., 2015. Electronic noses for food quality: a review. J. Food Eng. 144, 103–111.
Mahmoudian, S., Uzir Wahit, M., Ismailc, A.F., Yussuf, A.A., 2012. Preparation of regenerated cellulose/montmorillonite nanocomposite films via ionic liquids. Carbohydr. Polym. 88, 1251–1257.
Malviya, R., Srivastava, P., Bansal, M., Sharma, P.K., 2010. Mango peel pectin as super disintegrating agents. J. Sci. Ind. Res. 69, 688–690.
Mangiacapra, P., Gorrasi, G., Sorrentino, A., Vittoria, V., 2006. Biodegradable nanocomposites obtained by ball milling of pectin and montmorillonites. Carbohydr. Polym. 64, 516–523.
Marsh, K., Bugusu, B., 2007. Food packaging—roles, materials, and environmental issues. J. Food Sci. 72, 39–55.
Martelli, M.R., Barros, T.T., de Moura, M.R., Mattoso, L.H., Assis, O.B., 2013. Effect of chitosan nanoparticles and pectin content on mechanical properties and water vapor permeability of banana puree films. J. Food Sci. 78, N98–N104.
May, C.D., 1990. Industrial pectins: sources, production and applications. Carbohydr. Polym. 12 (1), 79–99.
McMillin, K.W., 2008. Where is MAP Going? A review and future potential of modified atmosphere packaging for meat. Meat Sci. 80, 43–65.
Miao, A.J., Schwehr, K.A., Xu, C., Zhang, S.J., Luo, Z., Quigg, A., Santschi, P.H., 2009. The algal toxicity of silver engineered nanoparticles and detoxification by exopolymeric substances. Environ. Pollut. 157, 3034–3041.
Miltz, J., 1992. Food packaging. In: Heldman, D.R., Lund, D.B. (Eds.), Handbook of Food Engineering. Marcel Dekker, New York, NY, pp. 677–740.
Mima, S., Miya, M., Iwamoto, R., Yoshikawa, S., 1983. Highly deacetylated chitosan and its properties. J. Appl. Polym. Sci. 28, 1909–1917.
Mistry, A.H., Schmidt, S.J., Eckhoff, S.R., Sutherland, J.W., 1992. Alkali extraction of starch from corn flour. Starch-Starke 44, 284–288.
Mohnen, D., 2008. Pectin structure and biosynthesis. Curr. Opin. Plant Biol. 11, 266–277.
Mondal, D., Bhowmick, B., Mollick, M.M.R., Maity, D., Mukhopadhyay, A., Rana, D., Chattopadhyay, D., 2013. Effect of clay concentration on morphology and properties of hydroxypropylmethylcellulose films. Carbohydr. Polym. 96, 57–63.
Montembault, A., Viton, C., Domard, A., 2005. Rheometric study of the gelation of chitosan in aqueous solution without cross-linking agent. Biomacromolecules 6, 653–662.
Moon, M.H., Giddings, J.C., 1993. Rapid Separation and measurement of particle size distribution of starch granules by sedimentation/steric field-flow fractionation. J. Food Sci. 58, 1166–1171.
Moreira, F.K.V., De Camargo, L.A., Marconcini, J.M., Mattoso, L.H.C., 2013. Nutraceutically inspired pectin–$Mg(OH)_2$ nanocomposites for bioactive packaging applications. J. Agric. Food Chem. 61, 7110–7119.
Morgan, K.R., Furneaux, R.H., Larsen, N.G., 1995. Solid-state NMR studies on the structure of starch granules. Carbohydr. Res. 276, 387–399.
Morris, E.R., Powell, D.A., Gidley, M.J., Rees, D.A., 1982. Conformations and interactions of pectins. I. Polymorphism between gel and solid states of calcium polygalacturonate. J. Mol. Biol. 155 (4), 507–516.

Muncke, J., 2014. Hazards in food contact material: food packaging contaminants. Montarjemi, Y., Moy, G.G., Todd, E.C.D. (Eds.), The Encyclopedia of Food Safety, vol. 2, Elsevier, Boston, pp. 430–437.
Muzzarelli, R.A.A., Muzzarelli, C., 2005. Chitosan chemistry: relevance to the biomedical sciences. In: Heinze, T. (Ed.), Polysaccharides I. Springer, Berlin, Heidelberg, pp. 151–209.
Navarro, E., Piccapietra, F., Wagner, B., Marconi, F., Kaegi, R., Odzak, N., Behra, R., 2008. Toxicity of silver nanoparticles to *Chlamydomonas reinhardtii*. Environ. Sci. Technol. 42, 8959–8964.
Nel, A.E., Mädler, L., Velegol, D., Xia, T., Hoek, E.M., Somasundaran, P., Thompson, M., 2009. Understanding biophysicochemical interactions at the nano–bio interface. Nat. Mater. 8, 543–557.
Nyvall, P., Corre, E., Boisset, C., Barbeyron, T., Rousvoal, S., Scornet, D., Kloareg, B., Boyen, C., 2003. Characterization of mannuronan C-5-epimerase genes from the brown alga *Laminaria digitata*. Plant Physiol. 133 (2), 726–735.
O'Neill, M., Albersheim, P., Darvill, A., 1990. The pectic polysaccharides of primary cell walls. In: Dey, P.M. (Ed.), Methods in Plant Biochemistry. Academic Press, London, pp. 415–441.
Oakenfull, D., Scott, A., 1984. Hydrophobic interaction in the gelation of high methoxyl pectins. J. Food Sci. 49, 1093–1098.
Oates, C.G., 1997. Towards an understanding of starch granule structure and hydrolysis. Trends Food Sci. Technol. 8, 375–382.
Ollett, A.L., Parker, R., Smith, A.C., 1991. Deformation and fracture behaviour of wheat starch plasticized with glucose and water. J. Mater. Sci. 26, 1351–1356.
Othman, S.H., 2014. Bio-nanocomposite materials for food packaging applications: types of biopolymer and nano-sized filler. Agric. Agric. Sci. Proc. 2, 296–303.
Pandey, J.K., Singh, R.P., 2005. Green nanocomposites from renewable resources: effect of plasticizer on the structure and material properties of clay-filled starch. Starch-Starke 57, 8–15.
Park, H.M., Li, X., Jin, C.Z., Park, C.Y., Cho, W.J., Ha, C.S., 2002. Preparation and properties of biodegradable thermoplastic starch/clay hybrids. Macromol. Mater. Eng. 287, 553–558.
Park, J.H., Park, S.M., Kim, Y.H., Oh, W., Lee, G.W., Karim, M.R., Park, J.H., Yeum, J.H., 2013. Effect of montmorillonite on wettability and microstructure properties of zein/montmorillonite nanocomposite nanofiber mats. J. Compos. Mater. 47, 251–257.
Pawar, S.N., Edgar, K.J., 2012. Alginate derivatization: a review of chemistry, properties and applications. Biomaterials 33, 3279–3305.
Pellerin, E., Doco, T., Vidal, S., Williams, P., Brillouet, J.M., 1996. Pectins and pectinases. In: Voragen, A.G.J., Visser, J. (Eds.), Progress in Biotechnology. Elsevier, Amsterdam, pp. 67–79.
Pelletier, A., Lemire, I., Sygusch, J., Chornet, E., Overend, R., 1990. Chitin/chitosan transformation by thermo-mechano-chemical treatment including characterization by enzymatic depolymerization. Biotechnol. Bioeng. 36, 310–315.
Perez Espitia, P.J., Ferreira Soares, N.F., Teófilo, R.F., dos Reis Coimbra, J.S., Vitor, D.M., Andrade Batista, R., Olavo Ferreira, S., de Andrade, N.H., Alves Medeiros, E.A., 2013. Physical–mechanical and antimicrobial properties of nanocomposite films with pediocin and ZnO nanoparticles. Carbohydr. Polym. 94, 199–208.
Persin, Z., Stana-Kleinschek, K., Foster, T.J., Van Dam, J., Boeriu, C.G., 2011. Challenges and opportunities in polysaccharides research and technology: the EPNOE views for the next decade in the areas of materials, food and health care. Carbohydr. Polym. 84 (1), 22–32.

Pinto, A.M.B., Santos, T.M., Caceres, C.A., Lima, J.R., Ito, E.N., Azeredo, H.M.C., 2015. Starch-cashew tree gum nanocomposite films and their application for coating cashew nuts. LWT- Food Sci. Technol. 62, 549–554.
Pradhan, G.C., Dash, S., Swain, S.K., 2015. Barrier properties of nano silicon carbide designed chitosan nanocomposites. Carbohydr. Polym. 134, 60–65.
Prashanth, K.V.H., Tharanathan, R.N., 2007. Chitin/chitosan: modifications and their unlimited application potential: an overview. Trends Food Sci. Technol. 18, 117–131.
Premanathan, M., Karthikeyan, K., Jeyasubramanian, K., Manivannan, G., 2011. Selective toxicity of ZnO nanoparticles toward Gram-positive bacteria and cancer cells by apoptosis through lipid peroxidation. Nanomed. Nanotechnol. 7, 184–192.
Quintavalla, S., Vicini, L., 2002. Antimicrobial food packaging in meat industry. Meat Sci. 62, 373–380.
Rehm, B.H.A., Valla, S., 1997. Bacterial alginates: biosynthesis and applications. Appl. Microbiol. Biotechnol. 48 (3), 281–288.
Restuccia, D., Spizzirri, U.G., Parisi, O.I., Cirillo, G., Curcio, M., Iemma, F., Picci, N., 2010. New EU regulation aspects and global market of active and intelligent packaging for food industry applications. Food Control 21, 1425–1435.
Rhim, J.W., Park, H.M., Ha, C.S., 2013. Bio-nanocomposites for food packaging applications. Prog. Polym. Sci. 38, 1629–1652.
Richardson, G., Kidman, S., Langton, M., Hermansson, A.M., 2004. Differences in amylase aggregation and starch gel formation with emulsifiers. Carbohydr. Polym. 58, 7–13.
Ridley, B.L., O'Neill, M.A., Mohnen, D.A., 2001. Pectins: structure, biosynthesis, and oligogalacturonide-related signaling. Phytochemistry 57 (6), 929–967.
Rimdusit, S., Jingjid, S., Damrongsakkul, S., Tiptipakorn, S., Takeichi, T., 2008. Biodegradability and property characterizations of methyl cellulose: effect of nanocompositing and chemical crosslinking. Carbohydr. Polym. 72, 444–455.
Robertson, G.L., 2005. Food Packaging, Principles and Practice, 2nd ed. CRC Press, New York, NY.
Rodríguez, F.J., Coloma, A., Galotto, M.J., Guarda, A., Bruna, J.E., 2012a. Effect of organoclay content and molecular weight on cellulose acetate nanocomposites properties. Polym. Degrad. Stabil. 97, 1996–2001.
Rodríguez, F.J., Galotto, M.J., Guarda, A., Bruna, J.E., 2012b. Modification of cellulose acetate films using nanofillers based on organoclays. J. Food Eng. 110, 262–268.
Rodriguez-Gonzalez, F.J., Ramsay, B.A., Favis, B.D., 2003. High performance LDPE/thermoplastic starch blends: a sustainable alternative to pure polyethylene. Polymer 44, 1517–1526.
Rubentheren, V., Ward, T.A., Chee, C.Y., Tang, C.K., 2015. Processing and analysis of chitosan nanocomposites reinforced with chitin whiskers and tannic acid as a crosslinker. Carbohydr. Polym. 115, 379–387.
Salaberria, A.M., Diaz, R.H., Labidi, J., Fernandes, S.C., 2015. Preparing valuable renewable nanocomposite films based exclusively on oceanic biomass: chitin nanofillers and chitosan. React. Funct. Polym. 89, 31–39.
Sannan, T., Kurita, K., Iwakura, Y., 1976. Studies on chitin, 2. Effect of deacetylation on solubility. Macromol. Chem. Phys. 177, 3589–3600.
Sanuja, S., Agalya, A., Umapathy, M.J., 2015. Synthesis and characterization of zinc oxide–neem oil–chitosan bionanocomposite for food packaging application. Int. J. Biol. Macromol. 74, 76–84.
Sarko, A., Wu, H.C.H., 1978. The crystal structures of A-, B- and C-polymorphs of amylase and starch. Starch-Starke 30, 73–77.

Shamekh, S., Forssell, P., Suortti, T., Autio, K., Poutanen, K., 1999. Fragmentation of oat and barley starch granules during heating. J. Cereal Sci. 30, 173–182.
Shin, S.H., Kim, S.J., Lee, S.H., Park, K.M., Han, J., 2014. Apple peel and carboxymethylcellulose-based nanocomposite films containing different nanoclays. J. Food Sci. 79, E342–E353.
Shukla, S.K., Mishra, A.K., Arotiba, O.A., Mamba, B.B., 2013. Chitosan-based nanomaterials: a state-of-the-art review. Int. J. Biol. Macromol. 59, 46–58.
Singh, N., Singh, J., Kaur, L., Sodhi, N.S., Gill, B.S., 2003. Morphological, thermal and rheological properties of starches from different botanical sources. Food Chem. 81, 219–231.
Sinha Ray, S., Easteal, A., Quek, S.Y., Chen, X.D., 2006. The potential use of polymer–clay nanocomposites in food packaging. Int. J. Food Eng. 2 (4), 1–11, 5.
Siracusa, V., Rocculi, P., Romani, S., Dalla Rosa, M., 2008. Biodegradable polymers for food packaging: a review. Trends Food Sci. Technol. 19, 634–643.
Slade, L., Levine, H., 1993. Water relationships in starch transitions. Carbohydr. Polym. 21, 105–131.
Soheilmoghaddam, M., UzirWahit, M., Ali Yussuf, A., Al-Saleh, M.A., Tuck Whye, W., 2014. Characterization of bio regenerated cellulose/sepiolite nanocomposite films prepared via ionic liquid. Polym. Test. 33, 121–130.
Sorrentino, A., Gorrasi, G., Vittoria, V., 2007. Potential perspectives of bio-nanocomposites for food packaging applications. Trend. Food Sci. Technol. 18, 84–95.
Souza, B.W.S., Cerqueira, M.A., Martins, J.T., Casariegoa, A., Teixeira, J.A., Vicente, A.A., 2010. Influence of electric fields on the structure of chitosan edible coatings. Food Hydrocoll. 24, 330–335.
Stephen, A.M., Phillips, G.O., 2006. Food polysaccharides and their applications, second ed. CRC Press, New York.
Suh, J.K.F., Matthew, H.W., 2000. Application of chitosan-based polysaccharide biomaterials in cartilage tissue engineering: a review. Biomaterials 21, 2589–2598.
Sun, R.C., Sun, X.F., Tomkinson, J., 2004. Hemicelluloses and their derivatives. In: Gatenholm, P., Tenhanen, M. (Eds.), Hemicelluloses: Science and Technology. American Chemical Society Publishing, Washington, DC, pp. 2–22.
Sun Lee, D., 2014. Antioxidative packaging system. In: Han, J. (Ed.), Innovations in Food Packaging. Elsevier, Amsterdam, pp. 111–131.
Takeda, Y., Shirasaka, K., Hizukuri, S., 1984. Examination of the purity and structure of amylose by gel-permeation chromatography. Carbohydr. Res. 132, 83–92.
Tan, I., Flanagan, B.M., Halley, P.J., Whittaker, A.K., Gidley, M.J., 2007. A method for estimating the nature and relative proportions of amorphous, single, and double-helical components in starch granules by 13C CP/MAS NMR. Biomacromolecules 8, 885–891.
Tang, S., Zou, P., Xiong, H., Tang, H., 2008. Effect of nano-SiO_2 on the performance of starch/polyvinyl alcohol blend films. Carbohydr. Polym. 72, 521–526.
Teck Kim, Y., Min, B., Won Kim, K., 2014. General characteristics of packaging materials for food system. In: Han, J. (Ed.), Innovations in Food Packaging. Elsevier, Amsterdam, pp. 13–35.
Teodoro, A.P., Mali, S., Romero, N., de Carvalho, G.M., 2015. Cassava starch films containing acetylated starch nanoparticles as reinforcement: physical and mechanical characterization. Carbohydr. Polym. 126, 9–16.
Thakur, B.R., Singh, R.K., Handa, A.K., 1997. Chemistry and uses of pectin–a review. Crit. Rev. Food Sci. Nutr. 37, 47–73.

Tripathi, S., Mehrotra, G.K., Dutta, P.K., 2011. Chitosan-silver oxide nanocomposite film: preparation and antimicrobial activity. B. Mater. Sci. 34, 29–35.
Tsaih, M.L., Chen, R.H., 2003. The effect of reaction time and temperature during heterogenous alkali deacetylation on degree of deacetylation and molecular weight of resulting chitosan. J. Appl. Polym. Sci. 88, 2917–2923.
Turhan, K.N., Sahbaz, F., 2001. A simple method for determining light transmittance of polymer films used for packaging foods. Polym. Int. 50, 1138–1142.
US patent 6,143,337, 2007. Extraction of pectin by microwave heating under pressure.
Van Soest, J.J.G., Knooren, N., 1997. Influence of glycerol and water content on the structure and properties of extruded starch plastic sheets during aging. J. Appl. Polym. Sci. 64, 1411–1422.
Van Soest, J.J., Vliegenthart, J.F., 1997. Crystallinity in starch plastics: consequences for material properties. Trends Biotechnol. 15, 208–213.
Vartiainen, J., Tammelin, T., Pere, J., Tapper, U., Harlin, A., 2010. Biohybrid barrier films from fluidized pectin and nanoclay. Carbohydr. Polym. 82, 989–996.
Vauchel, P., Leroux, K., Kaas, R., Arhaliass, A., Baron, R., Legrand, J., 2009. Kinetics modeling of alginate alkaline extraction from *Laminariadigitata*. Bioresour. Technol. 100, 1291–1296.
Venault, A., Bouyer, D., Pochat-Bohatier, C., Vachoud, L., Faur, C., 2012. Investigation of chitosan gelation mechanisms by a modeling approach coupled to local experimental measurement. AIChE J. 58, 2226–2240.
Vigneshwaran, N., Kumar, S., Kathe, A.A., Varadarajan, P.V., Prasad, V., 2006. Functional finishing of cotton fabrics using zinc oxide–soluble starch nanocomposites. Nanotechnology 17, 5087–5095.
Weiss, J., Takhistov, P., McClements, D.J., 2006. Functional materials in food nanotechnology. J. Food Sci. 71, R107–R116.
Wilhelm, H.M., Sierakowski, M.R., Souza, G.P., Wypych, F., 2003. Starch films reinforced with mineral clay. Carbohydr. Polym. 52, 101–110.
Willige, R.V., Linssen, J.P.H., Meinders, M.B.J., Stege, H.V.D., Voragen, A.G.J., 2002. Influence of flavour absorption on oxygen permeation through LDPE, PP, PC and PET plastics food packaging. Food Addit. Contam. 19, 303–313.
Windhues, T., Borchard, W., 2003. Effect of acetylation on physicochemical properties of bacterial and algal alginates in physiological sodium chloride solutions investigated with light scattering techniques. Carbohydr. Polym. 52 (1), 47–52.
Xijun, L., Lin, L., Kunsheng, Z., Ying, X., Jiaxin, L., 2012. A new proposed sweet potato starch granule structure—pomegranate concept. Int. J. Biol. Macromol. 50, 471–475.
Xiong, H., Tang, S., Tang, H., Zou, P., 2008. The structure and properties of a starch-based biodegradable film. Carbohydr. Polym. 71, 263–268.
Yadollahi, M., Namazia, H., Barkhordari, S., 2014. Preparation and properties of carboxymethyl cellulose/layered double hydroxide bionanocomposite films. Carbohydr. Polym. 108, 83–90.
Yamamoto, K., Sumie, S., Toshio, O., 1973. Properties of rice starch prepared by alkali method with various conditions. J. Jpn. Soc. Starch Sci. 20, 99–104.
Yang, Q., Wu, C.N., Saito, T., Isogai, A., 2014. Cellulose–clay layered nanocomposite films fabricated from aqueous cellulose/LiOH/urea solution. Carbohydr. Polym. 100, 179–184.
Yi, H., Wu, L.Q., Bentley, W.E., Ghodssi, R., Rubloff, G.W., Culver, J.N., Payne, G.F., 2005. Biofabrication with chitosan. Biomacromolecules 6, 2881–2894.

Youssef, A.M., Abdel-Aziz, M.S., El-Sayed, S.M., 2014. Chitosan nanocomposite films based on Ag-NP and Au-NP biosynthesis by *Bacillus subtilis* as packaging materials. Int. J. Biol. Macromol. 69, 185–191.
Youssef, A.M., El-Sayed, S.M., Salama, H.H., El-Sayed, H.S., Dufresne, A., 2015. Evaluation of bionanocomposites as packaging material on properties of soft white cheese during storage period. Carbohydr. Polym. 132, 274–285.
Yu, W.X., Wang, Z.W., Hu, C.Y., Wang, L., 2014. Properties of low methoxyl pectin-carboxymethyl cellulose based on montmorillonite nanocompositefilms. Int. J. Food Sci. Technol. 49, 2592–2601.
Zhang, M., Rong, M., 2003. Performance improvement of polymers by the addition of grafted nano-inorganic particles. Chin. J. Polym. Sci. 21, 587–602.
Zhao, R., Torley, P., Halley, P.J., 2008. Emerging biodegradable materials: starch- and protein-based bio-nanocomposites. J. Mater. Sci. 43, 3058–3071.
Zobel, H.F., 1988. Molecules to granules: a comprehensive starch review. Starch-Starke 40, 44–50.

20

HIGH BARRIER GRAPHENE/ POLYMER NANOCOMPOSITE FILMS

Ahmad Allahbakhsh
Department of Polymer Engineering, Faculty of Chemical Engineering, Tarbiat Modares University, Tehran, Iran; Young Researchers and Elite Club, Shiraz Branch, Islamic Azad University, Shiraz, Iran

1 Introduction

Based on the application conditions, the required properties of food packaging systems vary in a wide range of barrier performances. In some cases, such as packages for fresh fruits and vegetables, packages should be partially permeable to atmospheric gases, as the shelf life of these materials is dependent on access to oxygen for sustained cellular respiration (Duncan, 2011). On the other hand, high oxygen and carbon dioxide impermeability is needed for carbonated beverage containers (Duncan, 2011). Therefore, there is a growing demand for the fabrication of engineered packaging systems to fulfil requirements in any application as great as possible.

Polymers are the most-used materials in food packaging industry, due to their low cost, light weight, high performance, and availability (Duncan, 2011). Polypropylene (PP), polyethylene (PE), polyethylene terephthalate (PET), polystyrene, and polyvinyl chloride are some of the most interesting polymers in the food packaging industry. However, most of these polymers have some drawbacks for use in food packaging systems. Regarding barrier performance, some of them have better oxygen impermeability (such as polystyrene), while others present better barrier performances against water vapor and moisture (high-density polyethylene, HDPE, is an example).

Graphene nanosheets, with high aspect ratio and large platelet surface area, can provide a balance in barrier performance of polymer films by increasing permeation torture path and forming strong polymer-graphene interactions. Moreover, crystallinity

Food Packaging. http://dx.doi.org/10.1016/B978-0-12-804302-8.00020-0

of polymer chains is expected to increase in the presence of graphene nanosheets. Therefore, a great interest has been devoted to improving the fabrication methods and barrier properties of high-performance graphene/polymer nanocomposite films. These concepts are discussed throughout this chapter. First, structures and characteristics of graphene and graphene oxide (GO) nanosheets are covered through the next section. Polymer-graphene interactions are also discussed at the end of this section. Gas and moisture barrier properties of polymer/graphene nanocomposite films are investigated through the third section. Finally, a summary of presented discussions and future perspective of graphene polymer nanocomposites in food packaging industry is presented in the last section of this chapter. Discussions presented throughout this chapter can provide initiating information regarding fabrication mechanisms and barrier properties of graphene/polymer nanocomposite films for those interested in food packaging applications of graphene-based polymer nanocomposites.

2 Graphene and Graphene Oxide

Graphene, as a two-dimensional single sheet of carbon atoms packed densely in a honeycomb lattice, has a high potential for use in the fabrication of food packaging films to improve both air and moisture barrier performance of polymer films. Contradistinctive arrangement of carbon atoms in the highly ordered structure of graphene nanosheets results in a high degree of selectivity for gas and water atoms to permeate. Moreover, the extremely high specific surface area of graphene nanosheets leads to a high applicability of this nanomaterial to increase the barrier properties of polymer films, at even very low loading contents.

Graphene nanosheets, however, cannot be mixed directly with polymer macromolecules, as these nanosheets have a high tendency to agglomerate and even restack through van der Waal's interactions in their reduced form (Allahbakhsh and Mazinani, 2015). This is due to the hydrophobicity of graphene nanosheets, which increases through reduction process and hence, the propensity of these nanosheets to return to their stable form (graphite lattice) improves. Therefore, surface modification of graphene nanosheets is a requirement for the fabrication of high barrier graphene/polymer nanocomposite films.

Functionalization of graphene nanosheets has been reported as a routine for the fabrication of highly dispersed graphene suspensions (Allahbakhsh et al., 2013b). Large surface areas of graphene nanosheets can be decorated with a variety of functional groups, from oxygen-containing functional groups to even polymer macromolecules. GO, however, can be considered as the most interested graphene derivatives, which has a high compatibility with most of the polymeric systems (Allahbakhsh et al., 2013a).

2.1 Fabrication of Graphene Oxide

Generally, GO is the product of a two-step process: (1) chemical oxidation of graphite to graphite oxide, and (2) micromechanical exfoliation of graphite oxide to GO. There are three well-known methods for the chemical oxidation of graphite: Brodie, Staudenmaier, and Hummers (Allahbakhsh et al., 2013b). Almost all other methods reported in the literature for the oxidation of graphite (such as Hofmann and Tour methods) are complete (or partial) modifications to these three methods (Chua et al., 2012). Brodie technique (Brodie, 1859) was the first reported method for the oxidation of graphite flake into graphite oxide. Brodie method was based on oxidation via potassium chlorate and nitric acid. The first modification to this method was reported by Staudenmaier (1898), who used concentrated sulfuric acid (H_2SO_4) to increase the acidity of reaction mixture (Dreyer et al., 2010). However, the most interesting method for the large-scale production of graphite oxide was the method reported by Hummers and Offeman. Our focus through this part is on Hummer method, and we study this method in detail.

The original reported method by Hummers (Hummers and Offeman, 1958) was initiated with the preparation of a mixture of powdered flake graphite, sodium nitrate ($NaNO_3$), and H_2SO_4, which the graphite/$NaNO_3$ weight ratio was 1/0.5. Potassium permanganate ($KMnO_4$) was then added to the suspension with a graphite/$KMnO_4$ weight ratio of 1/3. The mixture was placed in an ice-bath through this step and the rate of $KMnO_4$ addition was controlled carefully to prevent rise of mixture temperature above 20°C. Then, the temperature was maintained at around 35°C. The color of the mixture was changed during this step to brownish gray. Then, water was added to the mixture (about twice the volume of the mixture), and finally the oxidation process was terminated by addition of hydrogen peroxide solution to the mixture. The color of the mixture changed from brown to bright yellow during this step.

This method is surely the most efficient method for the oxidation of graphite. There are plenty of modifications in the literature of this method, and most of them are concerned about the graphite/$KMnO_4$ ratio and the influence of this ratio in the oxidation process. The efficiency of the oxidation process can be increased by increasing the amounts of $KMnO_4$ in the reaction mixture. Moreover, excluding the $NaNO_3$ from the reaction mixture besides increasing the amount of $KMnO_4$ will lead to an increase in the efficiency of the process (Marcano et al., 2010). H_2SO_4, which plays the oxidation medium role through Hummers method, eases the intercalation of $KMnO_4$ into the three-dimensional gallery of graphite. Therefore, more H_2SO_4 treatment of graphite powder before the addition of $KMnO_4$ will lead to more efficiency of the oxidation process.

Suggested reaction mechanism for the formation of oxidizing agents in the $KMnO_4$–H_2SO_4 mixture is presented in Eq. 20.1. It has been suggested that diamanganese heptoxide (Mn_2O_7) is the predominant oxidizing agent in the $KMnO_4$–H_2SO_4 oxidation mixture (Koch, 1982). Mn_2O_7 detonates in temperatures around 55°C, or in contact with organic compounds (Simon et al., 1987). However, Koch believed that $HMnO_4$ is also an oxidizer in this mixture, which cannot be ignored (Koch, 1982).

$$KMnO_4 + 3H_2SO_4 \rightarrow K^+ + MnO_3^+ + H_3O^+ + 3HSO_4^- \qquad (20.1)$$

$$MnO_3^+ + MnO_4^- \rightarrow Mn_2O_7$$

Based on the suggested reaction mechanism for the oxidation process, a number of molecular models for the orientation of oxygen-containing functional groups on the basal space of GO nanosheets have been proposed. Some models only consider the orientation of atoms, and some other more recent models also consider the presence of voids and imperfections (also known as point defects) in the proposed molecular structures. Some of these well-known and most-recent models are presented in Fig. 20.1. Presence of defects in the structure of GO nanosheets is confirmed both in theory as well as atomic-scale transmission electron microscopy images (GóMez-Navarro et al., 2010). The content of these defects is highly related to the used oxidation method (discussed latter). These defects can influence the surface behavior of GO nanosheets, as well as their conductivity performance, as they act as a barrier against energy transfer through nanosheets.

2.2 Graphene Oxide Nanosheets Characteristics

GO is highly dispersible in common solvents and is compatible with the most polymers. Stability of dispersed GO nanosheets, however, varies from solvent to solvent. It has been reported that GO can be dispersed stably in some common solvents such as water, *N,N*-dimethylformamide (DMF), *N*-methyl-2-pyrrolidone (NMP), tetrahydrofuran (THF), and ethylene glycol (Paredes et al., 2008). However, the contents of functional groups on the

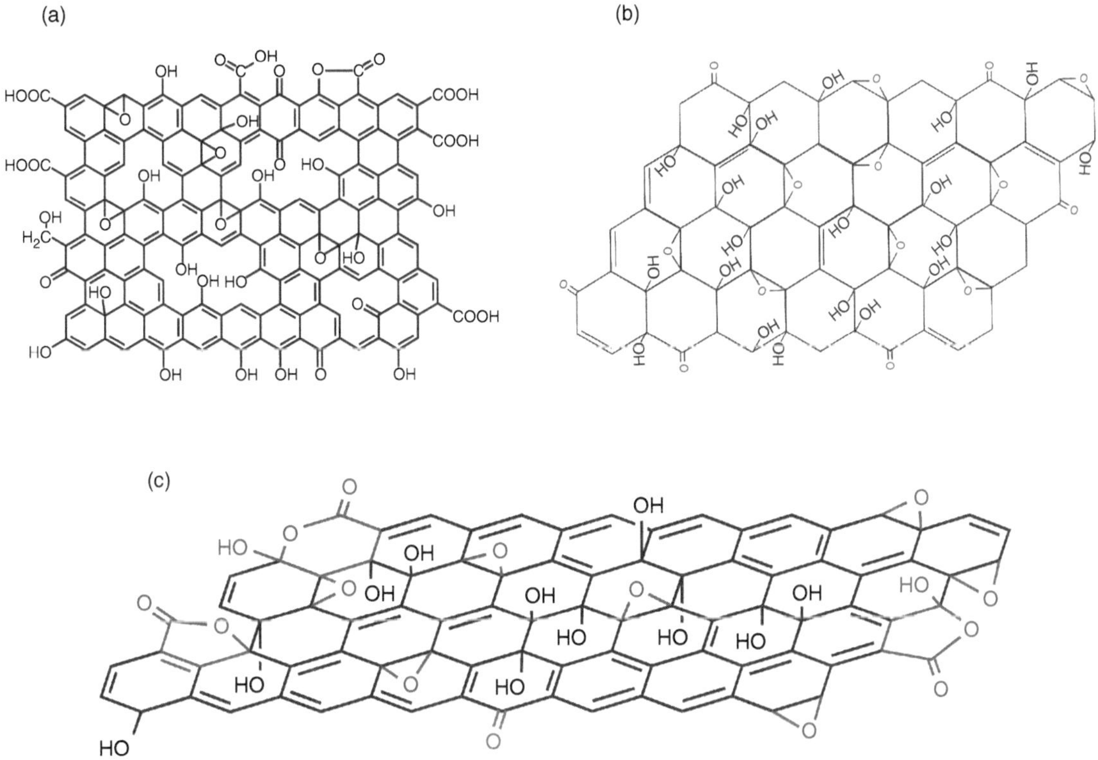

Figure 20.1. Suggested models for the molecular structure of GO nanosheets. (a) Li model: with missing atoms and functional groups like epoxide, hydroxyl, phenol, carbonyl, carboxyl, lactone, and quinone. (b) Lee model: with hydroxyl groups located randomly and far from each other, epoxy groups separated far from hydroxyl groups and ketone groups. (c) Gao model: with five- and six-membered lactol rings, and ester, hydroxyl, epoxy and ketone functional groups. Part (a) Reprinted from Li et al. (2009), with permission. Copyright (2009) American Chemical Society; (b) reprinted from Lee et al. (2010), with permission. Copyright (2010) American Chemical Society; (c) reproduced with permission from Gao et al. (2009). Copyright (2009) Macmillan Publishers Ltd., Nature Chemistry.

basal plane and edge area of GO nanosheets play a key role in the formation of stable colloidal GO suspensions.

Presence of epoxide (—C—O—C—), ketone (—C═O), hydroxyl (—OH), and carboxyl (—COOH) oxygen-containing functional groups on the basal plane and edge region of GO nanosheets have been reported in the literature (Haghighi et al., 2014). Therefore, dispersibility of GO is related to the negative charge of GO surface area, which increases with the amount of located functional groups (Allahbakhsh and Mazinani, 2015). Ionization of the carboxylic acid and phenolic hydroxyl groups on the basal plane and edge of GO nanosheets results in electrostatic repulsive forces, which stabilize colloidal GO nanosheets (Li et al., 2008). Moreover, it has been confirmed that the removal of carboxyl and hydroxyl edge-groups, through chemical reduction of GO, is the main reason for the agglomeration and restack of nanosheets (Li et al., 2008).

The content of oxygen-containing functional groups on the basal plane and edge region of GO nanosheets is closely related to the method used for the oxidation of graphite (Botas et al., 2013). As discussed, methods developed by Brodie, Staudenmaier, and Hummers are three well-known procedures for the oxidation of graphite. Fuming nitric acid and potassium chlorate are two oxidizers used by Brodie for the oxidation of graphite (You et al., 2013). The C:O ratio of graphite oxide prepared through Brodie method is about 2.9, and the interlayer spacing is about 0.61 nm (Botas et al., 2013). The most important disadvantage of Brodie method is the explosion danger, which can occur during the addition of potassium chlorate. Therefore, this method is mostly considered as a dangerous method for the fabrication of graphite oxide.

In Staudenmaier method, on the other hand, oxidizers are fuming nitric acid and concentrated H_2SO_4 (Poh et al., 2012). The C:O ratio of graphite oxide oxidized through this method is about 2.52 and an interlayer spacing of about 0.71 nm is achievable (Poh et al., 2012). The Hummers method is the third and the most widely used method for the fabrication of graphite oxide through a chemical oxidation procedure. This method has a high potential for the industrial large-scale production of GO. The graphite oxide powder, prepared through Hummers method, has the largest interlayer spacing between discussed three methods and lowest C:O ratio. The C:O ratio of graphite oxide prepared through hummers method is about 1.12, and the interlayer spacing is about 0.81 nm (Poh et al., 2012).

The amounts of oxygen containing functional groups on the basal space and edge region of GO nanosheets have a direct impact on nanoroughness of nanosheets, as well as covalent and noncovalent interactions with polymer macromolecules and surfactant molecules. Nanoroughness of GO nanosheets is in close relation with the wettability of nanosheets (Allahbakhsh et al., 2013b). Higher nanoroughness of GO nanosheets leads to more nanosheets wetting by polymer macromolecules. Nanoroughness of nanosheets increases with an increase in the concentration of oxygen containing functional groups on the surface of GO nanosheets. Therefore, higher amounts of oxygen functional groups result in more facial interaction between nanosheets and polymer chains and hence, more barrier performance of final GO/polymer nanocomposites. Structural characteristics of oxidized graphite fabricated using different oxidation methods are compared in Fig. 20.2.

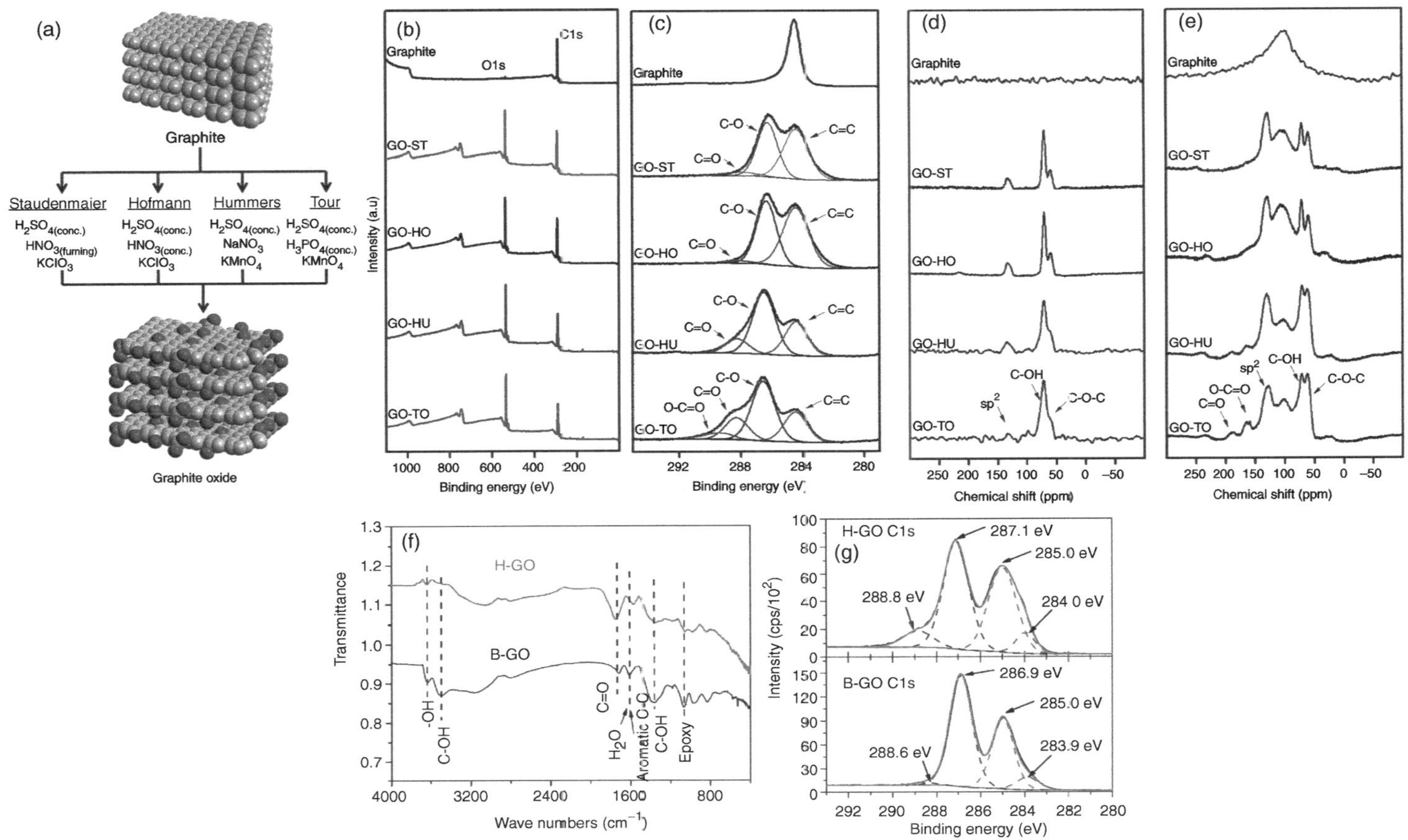

Figure 20.2. Methods for the oxidation of graphite. (a) Schematic presentation of the graphite oxidation process using the Staudenmaier (GO-ST), Hofmann (GO-HO), Hummers (GOHU), and Tour (GO-TO) methods. (b) X-ray photoelectron spectroscopy (XPS) spectra. (c) C1s core-level spectra. (d and e) Solid-state 1D^{13}C magic-angle spinning nuclear magnetic resonance spectra of fabricated graphite oxide. (f) FT-IR spectra. (g) XPS spectra for C1s of as-prepared Hummers' (H-GO) and Brodie's GO (E-GO). Parts (a–e) Reprinted from Chua et al. (2012), with permission. Copyright (2012) John Wiley & Sons. Parts (f and g) Reprinted from You et al. (2013), with permission. Copyright (2013) Elsevier.

As discussed, Hummers method results in formation of graphite oxide with more oxygen-containing functional groups. Therefore, it can be concluded that the Hummers method is the most efficient method for the preparation of graphite oxide through chemical oxidation of natural graphite, when high nanoroughness is needed. Regarding the density of point (especially vacancy) defects, however, Hummers method results in formation of more defects, compared to Staudenmaier method. Raman spectroscopic results demonstrate that the I_D/I_G ratio (the ratio of D-band and G-band integrals at 1335 and 1587 cm^{-1}, respectively) of graphite oxide prepared using Staudenmaier, Brodie, and Hummers oxidation procedures are about 0.77, 0.88, and 0.88, respectively (Botas et al., 2013; Poh et al., 2012). This ratio represents the contents of defects in the structure of nanosheets. Higher D-band intensities mean the presence of more defects in the structure of nanosheets. Therefore, it can be concluded that Staudenmaier method results in fewer defects in the structure of final product.

The content of defects in the structure of GO can directly affect the final barrier performance of nanosheets, as well as GO reinforcing capability. Transmission electron microscopy images of graphene nanosheets with atomic defects are presented in Fig. 20.3. Based on density functional theory investigations, graphene packed aromatic structure has high enough electron density to prevent atomic penetrations. The penetration barriers decrease exponentially with the size of the defects in the presence of defects. The penetration barriers, however, are still high for small defects and very large defects are needed to make the graphene sheet permeable for small atoms and molecules (Leenaerts et al., 2008).

Although gas molecules can penetrate through large grain defects on the graphene nanosheets, but gas permeability of graphene nanosheets can be reduced (and not completely avoided) by an increase in the number of graphene layers (Yoo et al., 2014). Therefore, multilayered graphene nanosheets are more applicable for gas barrier applications when large defects are expected in the structure of nanosheets. However, formation of gas flow channels and pores through high-temperature processing of GO can increase gas permeability of nanosheets, even in the multilayered form. CO and CO_2 gases, which release from oxygen-containing functional groups on basal planes or at edges of GO nanosheets at temperatures around 140°C, can form irreversible defects and gas flow channels (Kim et al., 2013). Therefore, an increase in the gas permeability of graphene/polymer nanocomposite films processed at temperatures above 140°C is unavoidable. As a result, higher gas barrier performance is expected in the case

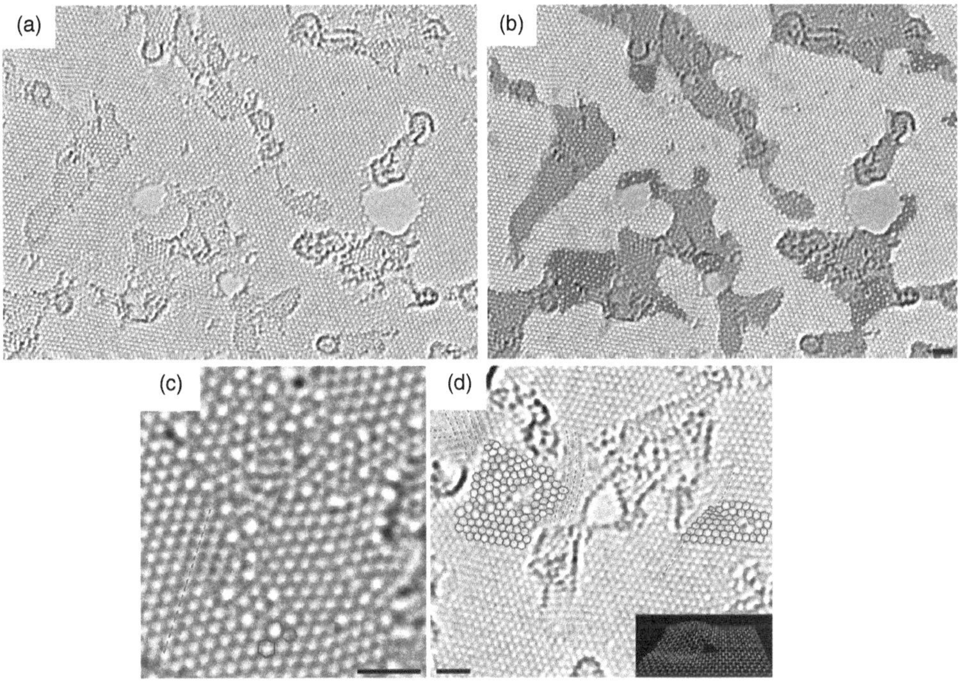

Figure 20.3. Atomic-resolution TEM images of graphene nanosheets. (a) Original image; (b) highlighted image of a single graphene nanosheet with defect-free, contaminated, disordered, substitutions-containing, and isolated topological defects-containing (single bond rotations or dislocation cores) regions (scale bar 1 nm); (c) image of a nonperiodic defect configuration; and (d) partial assignment of the configurations in defective areas. Reprinted from GóMez-Navarro et al. (2010) with permission. Copyright (2010) American Chemical Society.

of GO-loaded polymer nanocomposite films, rather than reduced GO loaded ones, as the reduction process of GO (thermally or chemically) is mostly accompanied by formation of volatile gases.

2.3 Graphene Oxide-Polymer Interactions

GO is a perfect candidate for improving functional performance of polymer materials. Broad surface area of GO provides a suitable ground for physical and chemical interactions between oxygen-containing functional groups on the basal plane of GO, and polymeric macromolecules. Formation of such interactions can enhance barrier performance of graphene/polymer nanocomposites, as these interactions increase the molecular integration between graphene nanosheets and polymer macromolecules. As discussed, voids and imperfections in the structure of graphene nanosheets lead to an increase in the permeability of nanosheets.

When composite structures are considered, however, these structural imperfections can be due to both basal plane defects and voids, which can form between polymer and nanosheets. Consequently, molecular compatibility between polymer macromolecules and nanosheets is important.

Generally, graphene-polymer interactions can be classified into two main categories: covalent and noncovalent (physical) interactions. There are two well-known routes for bonding polymer macromolecules and graphene nanosheets covalently. The first is based on growing a polymer on the graphene nanosheets (also known as "grafting-from" route). In this route, graphene is the initiating site for the polymerization of an initiator. The second, however, is based on coupling graphene and polymer macromolecules through a chemical reaction (also known as "grafting-to" route). Both routes result in formation of polymer-grafted graphene nanosheets, which can be used directly in the fabrication of polymer/graphene nanocomposites.

2.3.1 Covalent Interactions

Covalent interactions and bonding mostly take place either between free radicals and graphene C═C bonds, or between functional groups within the structure of polymer chains or initiator and oxygen-containing functional groups of GO nanosheets (Georgakilas et al., 2012). Considering growing polymer chains on graphene nanosheets, both GO and reduced GO can be used for the fabrication of covalently bonded polymer-grafted graphene nanosheets. In this route, typical initiators are first placed on the basal planes and the edges of graphene nanosheets. Then, polymer chains are grown from each initiator site through a polymerization process. Atom transfer radical polymerization (ATRP) is the most interesting method for the polymerization of initiators in this route (Salavagione et al., 2011). This polymerization method presents a fast initiation step and the development of a dynamic equilibrium. Moreover, different monomers with controlled chain length can be polymerized with this polymerization method and hence; a high control over the polymerization process is accessible with ATRP.

In ATRP method, the amount of initiator molecules, fixed on the GO nanosheets, controls the density of active sites and the distribution of polymer chains along the nanosheets. Through different initiators for the covalent bonding of polymer macromolecules to GO surface, 2-bromo-2-methylpropionyl bromide (BMPB) is the most widely used initiator (Salavagione et al., 2011), which has the ability to chemically interact with both the basal planes and the edges of GO nanosheets. BMPB, located in the basal planes of GO nanosheets, has been successfully employed for the polymerization of different monomers such as styrene, methyl methacrylate, and butyl acrylate (Salavagione et al., 2011). BMPB can be located on the basal planes of GO nanosheets through an esterification-like reaction of the acyl bromide in BMPB with the hydroxyl groups in GO (Salavagione et al., 2011).

In ATRP, the chain length of grafted polymer is controllable via controlling the ratio of monomer and initiator. Steric effect is not a controlling parameter in this route as polymer chains growth away from GO surface through ATRP and hence, no steric hindrance is expected. Moreover, distribution of grafted polymeric chains is directly related to the fixed active initiator sites on the basal planes and edges of GO nanosheets. The amount of grafted initiators can be controlled through controlling the concentration of diazonium salts. Highly activated free radicals are produced through heating of diazonium salts, which can attack sp^2 carbon atoms of reduced GO as well as carboxylic groups of GO. In addition, the reaction of other initiators such as benzoyl peroxide, benzyl chloride, and methacryloyl chloride with GO nanosheets can also lead to formation of required free radicals for polymerization process. Polycondensation, ring opening polymerization, miniemulsion polymerization, and Ziegler–Natta polymerization are other polymerization methods, which can be used for the covalent bonding of polymer chains to the GO surface through this route (Salavagione et al., 2011). A typical in situ polymerization of PP on the GO surface using Ziegler–Natta polymerization is presented in Fig. 20.4.

The second route for grafting polymer chains to the GO nanosheets surface is based on covalent interactions between functional groups of polymer chains on one side, and oxygen-containing functional groups on the basal plane and edges of GO nanosheets on the other side (Salavagione et al., 2011). There are two strategies for covalent bonding of polymer chains and GO nanosheets in this route: (1) functionalization of GO surface with required functional groups, which can react with specific polymer groups, and (2) functionalization of polymer chains with appropriate functional groups, which have the ability to react with GO surface. In both strategies, however, steric hindrance becomes an effective factor as the mobility of polymer chains in this route is extremely lower than the mobility of chains through grafting-from route. If the first strategy is used, the number of covalent interactions between polymer chains and GO nanosheets is limited by the efficiency of the functionalization process as well as mobility of polymer macromolecules. If the second strategy is used, however, the number of available sites for covalent interactions could

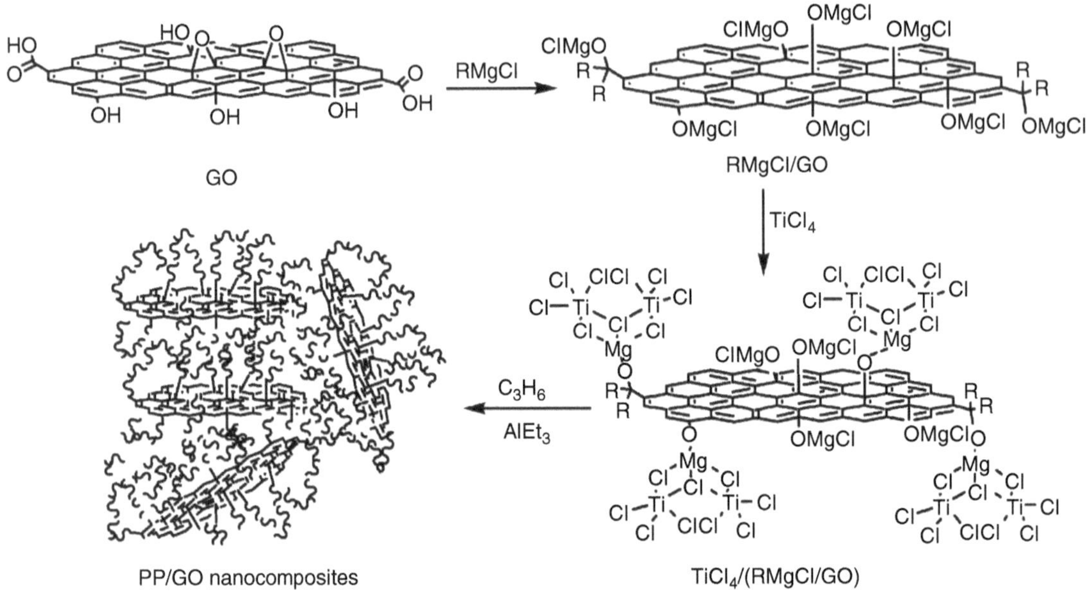

Figure 20.4. In situ Ziegler–Natta polymerization of PP/GO nanocomposites using $TiCl_4$/(BuMgCl/GO), GO-supported Ziegler–Natta catalyst. Reprinted from Huang et al. (2010) with permission. Copyright (2010) American Chemical Society.

be more, as the active sites are distributed throughout the bulk of polymer. In both cases, the perfect distribution of active sites and as a result, the perfect distribution of covalently interacted GO nanosheets strongly depends on the distribution of active sites.

Esterification and amidation of carboxylic groups at the edges of GO nanosheets with hydroxyl or amine groups of polymer macromolecules are most widely used reactions in this route (Salavagione et al., 2011). By attaching appropriate functional group to polymer macromolecules, different polymers such as poly(vinyl alcohol) (PVA), poly(ethyleneimine), and poly(vinyl chloride) can be covalently bonded to the GO surface through this route (Salavagione and Martínez, 2011). It should be noted here that a change in tacticity of polymer chains has been observed in this route (Salavagione et al., 2011). The observed change in tacticity is related to the immobilization of polymer macromolecules in the presence of graphene laminates. Conformations that contain more available functional groups in the structure of polymer macromolecules for the covalent bonding are more preferred in this route, as the degree of covalent interactions is closely related to the availability and concentration of reaction sites on both sides (graphene and polymer).

2.3.2 Noncovalent Interactions

In addition to discussed covalent interactions, there are physical interactions (noncovalent) between GO surface and polymer macromolecules, which should not be undertaken, as these interactions have a high impact on crystallinity and molecular dynamic of polymer chains. It has been reported that the influence of physical interactions on molecular mobility of polymer macromolecules is less than impact of covalent interactions when polymer macromolecules are covalently bonded to GO surface (Georgakilas et al., 2012). However, the clear impact of these interactions on thermal performance and crystallinity behavior of polymer chains in the absence of any covalent bonding reveals that these interactions should not be underestimated in scientific discussions.

When we talk about physical interactions between GO surface and polymer macromolecules, one can classify these interactions into three major categories (Georgakilas et al., 2012): (1) π–π interaction, (2) H–π interaction, and (3) anion/cation–π interaction. These interactions are actually the source of repulsive/attractive interactions between GO surface and polymer macromolecules, which in turn control the mobility of polymer chains near GO interface. Moreover, these interactions can be used to locate appropriate functional groups and surfactant agents on the basal plane and edges of GO nanosheets. Functional groups can be located on the GO surface to increase the compatibility between GO nanosheets and polymeric macromolecules.

Regarding π–π interactions, these interactions are the source of attractive interactions due to the negatively charged and diffuse electron clouds of π–π systems (Georgakilas et al., 2012). However, the type of interactions in this case is closely related to the density of electron between interaction parties. Attractive interactions are expected when both parties have similar or close electron densities, and induction interactions are expected when one side is electron-rich, and the other is electron-deficient. It has been reported that the interaction between two adjacent benzene rings is predominated by π–π attractive interactions (same electron densities). This can be considered as the main reason for the observed tendency of graphene nanosheets to agglomerate in their nonfunctionalized form, in combination with polymer macromolecules (similar electron densities between graphene nanosheets and dissimilar electron densities between graphene nanosheets and polymer macromolecules).

H–π interactions are one of the hydrogen bonds (Georgakilas et al., 2012). These interactions are predominated when an appropriate receptor group exists within the system. It has been suggested that the polarizability of the π-electron system is the factor that controls the nature of H–π interactions. In the case of C=C bonds, induction energies have been reported to be more important. However, electrostatic, induction, and dispersion energies become more important for the benzene complexes. On the other hand, it has been suggested that the dispersion energies become more important in the polymer structures. Therefore, the nature of H–π interactions is strongly depended on the nature of donor and receptor groups and as a result, the nature and polarizability of the system.

Regarding anion/cation–π interactions, cation–π interactions are more interested, as aromatic rings are considered as electron sources and anion–π interactions are supposed to result in repulsive interactions (Georgakilas et al., 2012). However, recent studies revealed that anion–π interactions could lead to a new type of anion recognition, host architecture, and supramolecular self-assembly. Recent studies demonstrated that interaction energies of anion–π interactions are even comparable to the energies of cation–π interactions. Dispersion and exchange-repulsive energies are predominant strategies in anion–π interactions. It has been reported that, based on the medium conditions, anions can form covalent or noncovalent interactions with aromatic rings. In the gas phase, anions can covalently react with π rings to form "Meisenheimer" complexes, which are the key intermediate in nucleophilic aromatic substitution. In the solution phase, however, anion–π type or displaced anion–π type complexes are more likely to form.

Interactions between a metal cation and a π-system result in formation of a cation–π interaction, with a combination of electrostatic and induction energies as dominant energies (Georgakilas et al., 2012). In the case of positively charged organic cation, however, polarizability of the π systems dictates the nature of cation–π interactions. It should be noted here that interaction energies are more susceptible to the dielectric of the environment when there are large electrostatic energies in the system. If the cation source is an aromatic cation, the dispersion energies increase and π_{cation}–π interactions take place. These interactions are weaker in terms of binding energies, compared to the normal metal cation–π interactions. However, these interactions are much stronger than π–π interactions.

Both dispersing agents and polymer macromolecules can form physical interactions with C=C bonds in the aromatic rings, as well as functional groups on the basal plane and edges of GO nanosheets (Haghighi et al., 2014). Formation of these interactions can increase the compatibility between polymer matrix and GO nanosheets (Allahbakhsh and Mazinani, 2015). Moreover, these interactions can affect the molecular mobility of polymer chains and as a result, change the crystallinity of the polymer matrix. It has been proved that polymer macromolecules are immobilized by GO surface, either through attractive interactions or repulsive ones (Allahbakhsh and Mazinani, 2015). GO nanosheets can provide an extremely high surface area when homogeneously are dispersed, distributed, and mixed in polymer melt. This high surface area presents a promising floor for covalent and/or noncovalent reactions with polymer macromolecules. During the molding process, these interactions immobilize polymer macromolecules and hence, affect the crystallinity of nanocomposite products.

GO nanosheets can interact physically with polymer backbone through π–π interactions, anion–π interactions, H–π interactions, or even hydrogen bonding. Noncovalent interactions between GO surface and two polymer systems, hydrophilic PVA and hydrophobic poly(methyl methacrylate), are depicted in Fig. 20.5. These interactions result in formation of attractive or repulsive energies

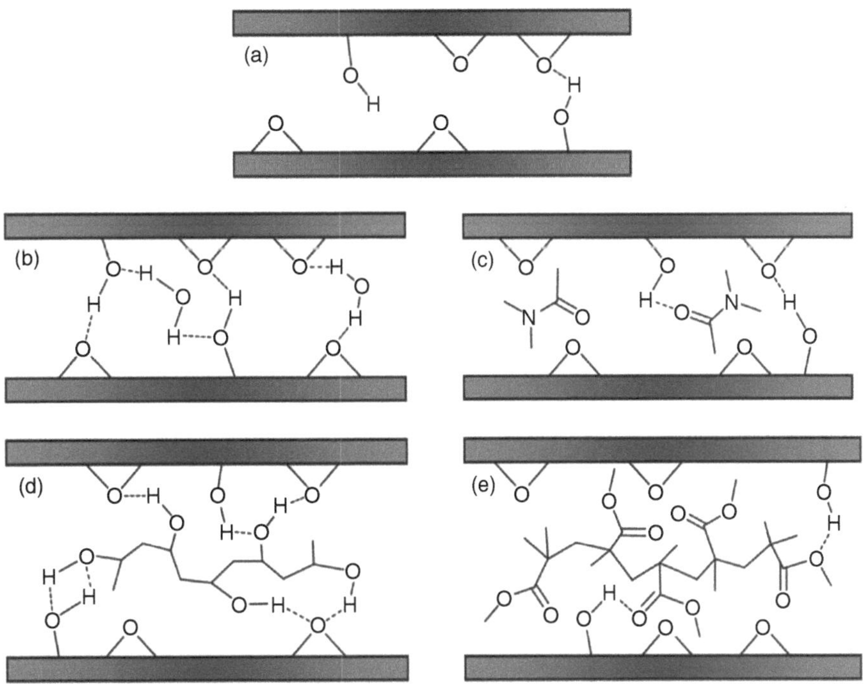

Figure 20.5. Proposed mechanism for noncovalent interactions between GO and polymer chains. (a) Anhydrous stacked GO sheets with hydrogen bonds, (b) hydrogen bonds between GO and water molecules, (c) hydrogen bonds in the presence of DMF molecules, (d) interactions between GO and PVA chains, and (e) interactions between GO and PMMA chains. Reprinted from Putz et al. (2010) with permission. Copyright (2010) John Wiley & Sons.

between polymer macromolecules and GO surface. For instance, sulfonated polyaniline can interact with GO surface through π–π stacking between polymer backbone and GO aromatic rings (Georgakilas et al., 2012). Moreover, electrostatic repulsion between polyaniline and negatively charged GO surface can also result in formation of immobilized polyaniline macromolecules on the surface region of GO nanosheets (Georgakilas et al., 2012).

Polystyrene macromolecules can also form strong π–π interactions with GO basal plane with a high graft ratio. GO nanosheets can form repulsive interactions with ethylene or propylene long segments (Allahbakhsh and Mazinani, 2015). These interactions affect mobility of polymer macromolecules significantly. Moreover, presence of sulfur species in the structure of polymer chains can lead to formation of immobilized macromolecules on the basal plane of GO nanosheets through noncovalent interactions between GO nanosheets and sulfur-containing segments (Haghighi et al., 2014).

Different kinds of covalent and noncovalent interactions, which can present between polymer chains and GO surface, were covered through this section. These interactions become highly important when one is studying polymer/graphene nanocomposite structures, as these interactions have a great influence on the final barrier performance of these nanostructures, which is the main subject of this chapter. Formation and type of these interactions are strongly dependent on the nature of polymer backbones, and differ from polymer to polymer.

3 Barrier Performance of Graphene/Polymer Nanocomposites

Presence of oxygen and the rate of oxygen diffusion are important in the food packaging industry. Oxygen presence results in direct and indirect deterioration of many foods. Direct oxidation leads to browning of fruits and rancidity of vegetable oils, and food spoilage by aerobic microorganisms is an indirect influence of oxygen presence in food packaging industry (Azeredo, 2009).

Generally, oxygen and moisture impermeabilities of polymers depend on many factors such as structure and nature of polymer macromolecules, conformation, and configuration of polymeric chains, polarity, molecular weight, and crystallinity of polymer. Interestingly, permeation of one migrate depends on the presence of other migrates in the packaging system (Duncan, 2011). Some polymers present high oxygen impermeability in dry conditions, but poor barrier performances against oxygen in humid situations, or vice versa. This is the main reason behind the introduction of complex packaging systems to provide packages with appropriate barrier characteristics, accompanied by desired mechanical, structural, and functional performances. Blends, multilayer films, composites, and nanocomposites are some of the known introduced complex systems to overcome this problem. Based on the growing publications in the field, polymer nanocomposites can be considered as the most interested solution, both academically and industrially (Duncan, 2011).

Polymer nanocomposites form through homogenous dispersion and distribution of nanomaterials in the polymer matrix (Allahbakhsh et al., 2013a). Different nanomaterials, such as clay nanoplatelets, silica nanoparticles, nature-based nanofibers, and nanowhiskers, as well as, carbon-based nanomaterials, with different aspect ratios and shapes can be used for the fabrication of polymer nanocomposites. Based on the selected polymer matrix and dispersed nanomaterial, polymer nanocomposites can perform different properties for a variety of applications. Regarding food packaging applications, however, nanomaterials with platelet shapes and high specific surface area are more favored. Moreover, polymers with high gas barrier performance and resistance against water vapor permeation are more desirable in this application field. It will be more interesting if the polymeric matrix is selected from biopolymers with sustainable resources.

Due to their poor mechanical and barrier properties, the use of biopolymer in food packaging industry is limited. These properties can be improved through incorporation of nanomaterials. Through formation of appropriate polymer-nanofiller interactions, nanomaterials can improve not only functional properties of the polymeric matrix, but antimicrobial activity, enzyme immobilization as well as moisture and gas barrier performance of polymers (Azeredo, 2009).

In situ polymerization, melt compounding, and solution mixing are the three most common methods for the fabrication of polymer nanocomposites. Among them, melt compounding method is the most industrially and environmentally efficient method. No solvents are required through this method, and compounds are mixed using an extruder in polymer melt phase. This process can be a continuous-process, and large-scale production is achievable using this method. High shear stress is applied to the mixing system through this technique and as a result, high dispersion of nanofillers can be obtained by this process.

Solution mixing process, on the other hand, is based on the mixing of compounds in a solvent medium, and therefore this process is extremely less environmentally efficient. The dispersion state of nanofillers is strongly dependent on the intercalation of polymer macromolecules into nanofiller agglomerates. Therefore, the dispersion state of nanofillers is less controllable in this method. This method is more academic than commercial, and with exceptions for some specific cases, is less economically effective than melt mixing process. High solubility of both nanofiller and polymer in solvent is the key to reach homogenous distribution and dispersion states. Moreover, this is a time-consuming process, and required time should be given for the intercalation process.

In situ polymerization is the third technique for the fabrication of polymer nanocomposites. Different polymerization processes can be used in this route, based on the nature of the polymer matrix (Allahbakhsh et al., 2013c). The polymerization process is almost always initiated in the presence of nanofiller in the polymerization reactor. This process is also time-consuming, as monomers should first intercalate into nanofiller agglomerates to achieve high nanofiller dispersions. However, intercalation of monomers into nanofiller gallery is easier than intercalation of long polymer chains. Therefore, nanofillers are mostly more dispersed uniformly through the polymer matrix, using this technique for the fabrication of polymer nanocomposites.

Regarding gas permeability improvement of polymers, layered nanostructures can be considered as the best nanomaterials for improving gas barrier performance of polymers, as these nanomaterials can increase the tortuosity of the diffusive path for a penetrant molecule and enhance barrier performance of polymers (see Fig. 20.6). Based on the dispersion extent,

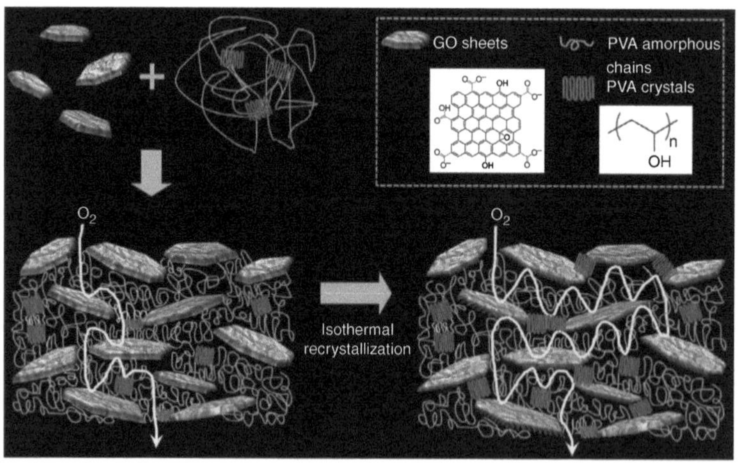

Figure 20.6. Formation of gas barrier PVA/GO nanocomposite film and influence of crystallinity on oxygen permeability tortuous path. Reprinted from Chen et al. (2014) with permission. Copyright (2014) Elsevier.

layered-nanomaterials-reinforced polymer nanocomposites can be classified under two categories: (1) intercalated and (2) exfoliated nanocomposites. Penetration of polymeric chains into the layered-nanomaterials interlayer region results in formation of the intercalated nanocomposites. In this case, the polymer macromolecules presence leads to an increase in the interlayer spacing between nanomaterial layers, but there are still some degrees of order in the structure of nanomaterial. In the case of exfoliated morphology, however, extensive penetration of polymer macromolecules results in delamination of nanomaterial layers and random dispersion of nanosheets. Although both morphologies can present enhanced performance, but exfoliated morphology can be considered as the optimum morphology for layered-nanomaterials-reinforced polymer nanocomposites, as nanocomposites with this morphology present extremely enhanced functional properties (Azeredo, 2009).

The absorption rate of gas molecules and the diffusion rate of the penetrating gas molecules through the polymeric matrix are two factors usually used for the assessing of polymer nanocomposites permeability. The absorption rate is depended on the mobility of polymer chains and as a result, the content of free volume holes in the structure of the polymer matrix. The diffusion rate is also related to the rate of gas molecules diffusion through adjacent free volume holes. Therefore, the permeability generally depends on the free volume of the polymer matrix as well as polymer–polymer and polymer–gas interactions.

As discussed, large polymer-filler interfacial region, which can be achieved through uniform dispersion of nanofillers in the polymer matrix, results in high immobility of polymer macromolecules near the interface region, and as a consequence, improved relaxation and crystallization behavior. In this case, nanomaterials with a high aspect ratio are more interested, as uniform dispersion and distribution of these materials in the polymeric matrix can provide high surface area and enhanced reinforcing effect (Azeredo, 2009).

As discussed, polymer-nanofiller interactions have a high impact on permeability of polymer nanocomposites. This impact is high enough that new models have been suggested for the prediction of polymer nanocomposites permeability by considering this phenomenon. Based on these models, three regions can be defined around polymer-nanofiller interfacial region: the surface modifier region, the constrained polymer region, and the unconstrained polymer region (the polymer bulk region) (Adame and Beall, 2009). The first region involves covalent and noncovalent bonding between polymer macromolecules and nanomaterial surface. This region is assumed to be small enough (1–2 nm) to have little influences on the overall permeability. Indirect (induced) interactions between polymer macromolecules and nanomaterial surface region take place through the constrained polymer region. This region is assumed to have low contents of free volumes and as a result, lower diffusion coefficient, compared to the bulk polymer. As this region is more responsible for the variations in crystallinity of polymer nanocomposite, this region has more impact in crystalline polymers.

As mentioned, dispersal of nanomaterials into the polymer matrix affects barrier performance of polymer nanocomposites in two ways: an increase in gas diffusion tortuous path and an impact on crystallinity of the polymer matrix. As discussed, layered-nanomaterials with high specific surface area and aspect ratio (such as clay and graphene nanosheets) are perfect candidates for enhancing gas impermeability of polymer materials (polymers and biopolymers) through fabrication of highly dispersed polymer nanocomposites. In the presence of graphene nanosheets, for instance, gas molecules must diffuse around nanosheets, as graphene nanosheets are highly impermeable. Consequently, gas molecules must diffuse through a longer mean diffusion path to permeate through polymer-graphene nanocomposites. Moreover, covalent and noncovalent interactions between polymer macromolecules and graphene surface results in formation of induced crystalline regions in the polymer matrix, which directly affect permeability of polymer–graphene nanocomposites.

Plenty of models have been presented for the prediction of permeability performance of polymeric systems. Most of the presented models are modifications to the well-known Nielsen model. Nielsen proposed a simple model for the permeability prediction of polymer nanocomposites reinforced with rectangular-shape fillers. Nielsen assumed that the tortuosity is the only factor influencing the diffusion rate and presented Eq. 20.2 for the ratio of composite and polymeric matrix permeability values:

$$\frac{K_{\text{composite}}}{K_{\text{matrix}}} = \frac{1-\phi}{1-(\alpha/2)\phi} \tag{20.2}$$

where $K_{\text{composite}}$ and K_{matrix} represent the predicted permeability values of the polymer nanocomposite and polymeric matrix, respectively. Moreover, ϕ and α values in Eq. 20.2 represent the volume fraction and the aspect ratio (length divided by width) of the reinforcing nanofiller, respectively. Based on this model, the impermeability of polymer nanocomposites increase when anisotropic and platelet-shape nanoparticles are used. This model is valid only for nanocomposites with low nanoparticle loading ($\phi < 0.1$). There are a large number of models in the literature, which mostly present modifications on the Nielsen model. These modifications are on randomness of nanoparticles dispersion and size uniformity (Lape et al., 2004), different nanoparticle shapes (Moggridge et al., 2003), agglomerations and high nanofiller loading (Dunkerley and Schmidt, 2010) modifications.

Besides the Nielsen model, the model proposed by (Cussler et al., 1988) is also a widely used model for the prediction of polymer nanocomposite permeability behavior. In this model, the impermeable dispersed nanoplatelets are proposed to be well-ordered and oriented through the polymer matrix. Moreover, the decrease of the diffusion available area and the increase of in the tortuous pathway for permeating molecules result in the increase in barrier properties of polymer nanocomposite in this model. The relative permeability of polymer nanocomposite, R_p, which presents the ratio of $K_{composite}$–K_{matrix}, is defined in the Cussler model in Eq. 20.3 (Huang et al., 2012):

$$R_p = \left(1 + \frac{\alpha^2\phi^2}{1-\phi}\right)^{-1} \tag{20.3}$$

Moreover, the permeation of gas molecules through a polymer nanocomposite film depends on two phenomena: solubility and diffusion. Based on Fick and Henry law, the permeability coefficient of a polymer nanocomposite film, P_g, can be expressed as in Eq. 20.4 (Huang et al., 2012):

$$P_g = D \times S \tag{20.4}$$

where D and S represent the diffusion and solubility coefficients of a gas in a polymer film, respectively. Therefore, the impermeability of a polymer nanocomposite film is always described by these two factors and the contribution of these factors in the overall gas permeation process.

3.1 Graphene/Polypropylene Nanocomposites

PP is one of the most commercially used thermoplastic polymers with easy processability, low-cost and interesting functional properties. PP has good thermal, mechanical, and physical properties at ambient conditions, and has been widely used in food packaging, automotive, and other industrial sectors. Regarding food packaging application, PP presents high structural and mechanical performance as well as acceptable oxygen and water vapor impermeability (Johansson and Leufvén, 1994). The gas barrier of PP, however, can be more enhanced via incorporating impermeable nanomaterials such as graphene and GO nanosheets.

Graphene/PP nanocomposites can be fabricated using different compounding methods such as melt mixing (El Achaby et al., 2012), solution mixing (Yang et al., 2013b), in situ polymerization (Huang et al., 2010), and layer-by-layer method (Lee et al., 2015). Kalaitzidou et al. (2007) prepared exfoliated graphite nanoplatelets/PP nanocomposites by melt mixing using a twin-screw extruder followed by injection molding. Based on their results, graphene nanosheets have a high potential as an alternative to nanoclays for improving gas barrier performance of PP nanocomposites (Kalaitzidou et al., 2007). They compared influences of different nanomaterials on the oxygen permeability of PP nanocomposites. They also fabricated PP nanocomposites using graphite nanosheets with different structural properties to investigate the influences of nanosheets aspect ratio on barrier performance of PP nanocomposites. Compared to nanoclays, exfoliated graphite nanosheets with a high aspect ratio can improve oxygen impermeability of PP more effectively. As used nanosheets were nonfunctionalized, these nanosheets presented a high tendency to form agglomerates through melt mixing of PP and nanosheets. Kalaitzidou and coworkers suggested that these nanosheets could slide against each other and align parallel to the mold plats, during the molding process. Moreover, they suggested that incorporation of graphene nanosheets with a high aspect ratio can result in formation of more crystals and increase the final oxygen impermeability of PP nanocomposite.

It has been revealed that the crystallinity of reduced GO/PP nanocomposites, fabricated using melt mixing method, change with the content of graphene nanosheets (El Achaby et al., 2012). Reduced GO nanosheets can act as nucleating agents for the crystallization of PP. Moreover, mechanical characteristics of reduced GO/PP nanocomposites increase with reduced GO loading content. In addition to the reduced form of GO, functionalized GO nanosheets can also be used for the fabrication of graphene-based PP nanocomposites. Silane-based coupling agents can be used for the functionalization of GO nanosheets, to improve compatibility between PP and GO (Yang et al., 2013b). Functionalized GO nanosheets can then be mixed with PP using solution mixing method.

PP/graphene nanocomposites can also be prepared through in situ polymerization of propylene monomer in the presence of graphene nanosheets (Huang et al., 2010; Milani et al., 2012). Huang et al. (2010) synthesized PP/graphene nanocomposites using in situ Ziegler–Natta polymerization method. They used a GO-supported $TiCl_4$ Ziegler–Natta catalyst for the synthesis process. This process has the advantage of bypassing the thermodynamic or kinetic limitations of direct polymer intercalation, as olefin monomers can penetrate easily into the interlayer spaces of GO nanosheets and then be synthesized into polymeric chains. In this route, more intercalations are expected and the chance of exfoliated morphology is very high.

Layer-by-layer assembly method can also be used to fabricate PP/GO multilayer films with enhanced barrier properties. Layer-by-layer assembly is a simple process can be used for the fabrication of multilayer thin polymer nanocomposite films. This process involves deposition of alternating layers with opposite charges. Dip-coating, spin-coating, and spray-coating (see Fig. 20.7) are three main strategies for the fabrication of multilayer nanocomposite

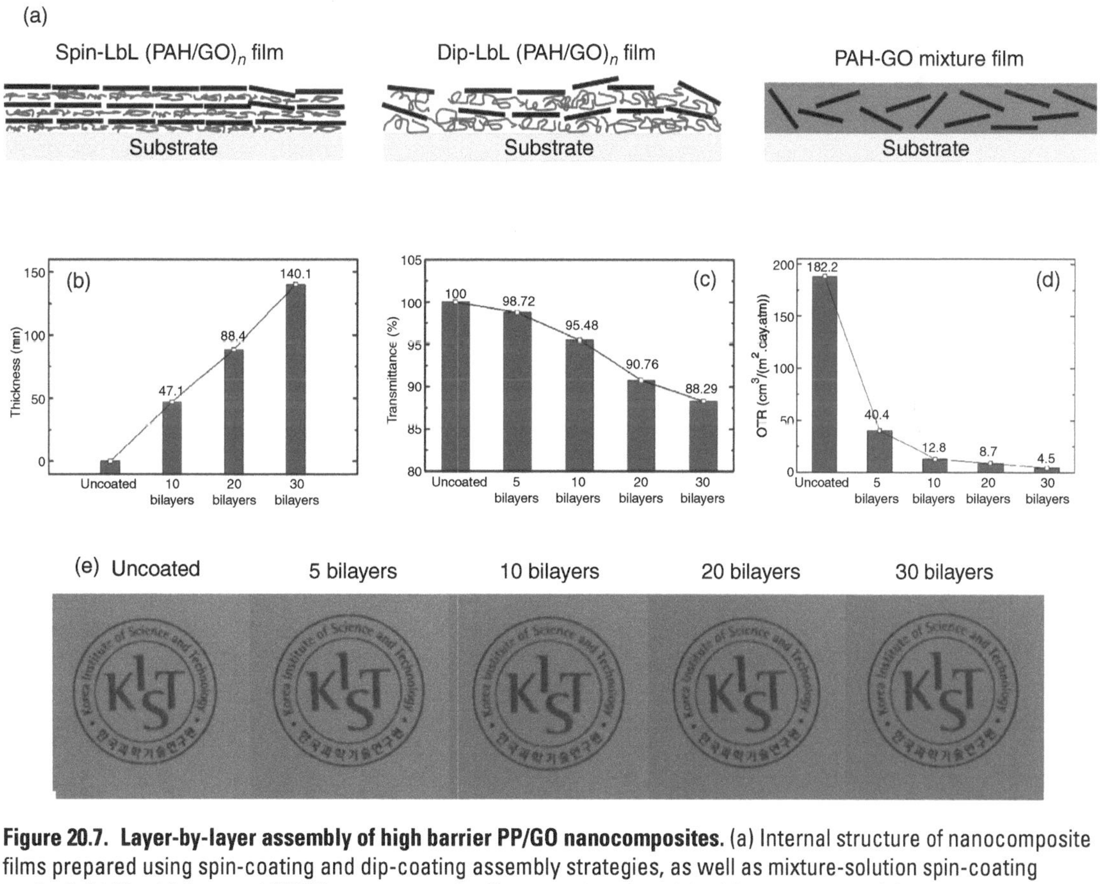

Figure 20.7. Layer-by-layer assembly of high barrier PP/GO nanocomposites. (a) Internal structure of nanocomposite films prepared using spin-coating and dip-coating assembly strategies, as well as mixture-solution spin-coating method; (b) film thickness of PP/GO nanocomposite films as a function of the bilayer numbers; (c) light transmittance of nanocomposite films as a function of the bilayer numbers; (d) oxygen-transmission rate as a function of the bilayer numbers; and (e) images of PP/GO films with different bilayer numbers and film transparency. Reprinted from Lee et al. (2015) with permission. Copyright (2015) Elsevier.

films through layer-by-layer assembly process (Lee et al., 2015). Due to the ease of process, dip-coating strategy is more used through the literature. However, spin-assisted strategy has the capability of producing highly ordered and dense multilayer films, as centrifugal and capillary forces are introduced through this strategy.

Lee et al. (2015) used layer-by-layer compounding method for the fabrication of a PP/GO nanocomposite by employing poly(allylamine hydrochloride) (PAH) as the positively charged layer. They used dip-coating and spin-coating strategies for the fabrication of nanocomposites and compared the gas-barrier performance of nanocomposites prepared through these strategies. Their results revealed that the oxygen-transmission rate of PP films can be reduced up to 98% by coating 30 PAH/GO bilayers. The total film thickness of this nanocomposite will increase only about 0.46%, compared to the thickness of noncoated PP film. They attributed the observed enhanced barrier performance of nanocomposite to the nanometer-scale thickness of binary layers, the high aspect-ratio of GO nanosheets and high order of nanosheets, which is achievable through spin-coating strategy. These factors all lead to an increase in the permeation path of oxygen molecules and as a result, an increase in impermeability of PP/GO nanocomposite. Through these factors, however, the high order of GO nanosheets is the key factor that leads to the observed interesting gas barrier performance. Moreover, they suggested the Cussler model as the most appropriate model for the permeability of fabricated nanocomposites gas permeability.

3.2 Graphene/Polyethylene Nanocomposites

PE is the second widely interested industrial polymer, which has been used widely in food and drug packaging industry (Ren et al., 2014). PE is a nonpolar thermoplastic polymer, with high flexibility, lightweight, low-cost, and easy processing feature. Although PE presents acceptable water vapor impermeability, the gas permeability of PE is not high enough to prevent food and drug oxidation damages (Ren et al., 2014). Moreover, the mechanical characteristics of PE are another disadvantage of this polymer for food packaging applications.

HDPE is a linear thermoplastic polymer with balanced mechanical properties, chemical resistance, and water vapor impermeability (Ferreira et al., 2014). High water vapor transmission barrier (or low water vapor transmission rate, WVTR) is one of the most important factors, which made HDPE films perfect candidates for food packaging applications. It is believed that transformation of randomly distributed spherulitic crystals into anisotropic fibrillar morphology through stretching of the HDPE chains is the main reason of the improved moisture barrier performance of this material (Chatterjee et al., 2014). Like other members of PE family, however, the gas permeability of HDPE can be considered as the main drawback of this material for packaging applications. Consequently, HDPE/graphene nanocomposites have been proposed as a promising solution for the observed low gas barrier performance of HDPE food packaging films.

HDPE/graphene nanocomposites can be fabricated through solution mixing (Ren et al., 2014), melt mixing (El Achaby and Qaiss, 2013) and in situ polymerization (Shehzad et al., 2014) techniques. Xylene, as a nonpolar solvent, has been used as the solution medium for the fabrication of HDPE/GO nanocomposites using solution mixing method. Ren et al. (2014) functionalized GO nanosheets with dodecyl amine to increase compatibility between HDPE macromolecules and GO nanosheets. They also used a combination of amidation and chemical reduction processes to prepare dodecyl amine functionalized reduced GO nanosheets, to fabricate HDPE/reduced GO nanocomposites.

Their results revealed that highly dispersed graphene reinforced HDPE nanocomposites, with uniform dispersion of graphene nanosheets and strong interfacial interactions between HDPE macromolecules and graphene nanosheets can be fabricated by incorporating either functionalized GO or functionalized-reduced-GO nanosheets (Ren et al., 2014). However, their results also demonstrated that mechanical performance, crystallinity, gas permeability, and thermal stability of fabricated nanocomposites do not change through the reduction process (Ren et al., 2014). Therefore, they suggested that the reduction process in not necessary for the fabrication of high-performance graphene-based HDPE nanocomposites.

Ren et al. demonstrated that the oxygen permeability of HDPE films strongly depends on the loaded graphene nanosheets content. About 62% decrease in oxygen permeability of HDPE/graphene nanocomposites is achievable via incorporating only 0.1 wt.% dodecyl amine functionalized GO nanosheets. Moreover, the observed decrease in oxygen permeability of nanocomposite films is mostly attributed to the decrease of the oxygen diffusion coefficient, rather than solubility. Increases in HDPE crystallinity and diffusion tortuous path, due to immobilization of HDPE macromolecules on the GO nanosheets interfacial region, are two reasons suggested for the observed decrease in the diffusion coefficient. In addition, 7% increase in storage modulus and 5°C increase in decomposition temperature is expected with incorporating 0.1 wt.% GO nanosheets into HDPE matrix.

Low-density polyethylene (LDPE) is another member of the PE thermoplastic resins family, mostly used for the manufacturing of film blowing products (Hong et al., 2000). Compared to HDPE (with about 90% crystallinity), LDPE is less crystalline (about 60% crystallinity), as LDPE is a branched polymer while HDPE is a more crystalline nonbranched polymer (Marcilla et al., 2005). Gas barrier performance of LDPE can also be improved by introducing graphene nanosheets into the polymer matrix. Checchetto et al. (2014) used film blowing technology to fabricate LDPE/graphene nanocomposites. Based on their results, graphene nanosheets can be iso-oriented through the LDPE matrix with nanosheets oriented parallel to the membrane surface. Moreover, they demonstrated that N_2, H_2, and CO_2 permeability of LDPE/graphene nanocomposites reduce by a factor of 2, compared to nonreinforced LDPE polymer (Checchetto et al., 2014).

In addition to blow extrusion technique (Pinto et al., 2013b), LDPE/graphene nanocomposites can also be fabricated through solution mixing method, by using toluene as the solvent. Wang et al. (2010) used the solution mixing method for the fabrication of vinyl triethoxysilane-functionalized-graphene/LDPE nanocomposites. Compared to the nonreinforced LDPE film, fabricated nanocomposites present up to 27.0 and 92.8% increase in tensile strength and Young's modulus values, respectively. LDPE/ethylene vinyl acetate (EVA)/graphene nanocomposites can also be fabricated using the same solution casting method (Tayebi et al., 2015). By loading only 1 wt.% of GO nanosheets in the LDPE/EVA copolymer matrix, oxygen permeability decreases up to 62%, compared to pristine copolymer film.

Graphene nanosheets can also be used to enhance gas barrier properties of linear low-density polyethylene (LLDPE) polymer films, as shown in Fig. 20.8. Compared to other members of PE family, industrial interest on LLDPE has grown most rapidly, as LLDPE presents a good balance between rigidity and strength and more flexible processing conditions in comparison with LDPE or HDPE (Kim et al., 2011a). Compared to LDPE, LLDPE has shorter branches and therefore, different structural characteristics. As this polymer is also a nonpolar polymer, miscibility of this material and graphene nanosheets can be improved by using one of the functionalization methods, discussed through prior sections. Kuila et al. (2012) used xylene as the medium for the solution mixing of dodecyl amine-modified graphene nanosheets and LLDPE polymer. They show that the crystallinity of LLDPE films decrease in the presence of functionalized GO nanosheets. Moreover, they reported an elevation in dynamic

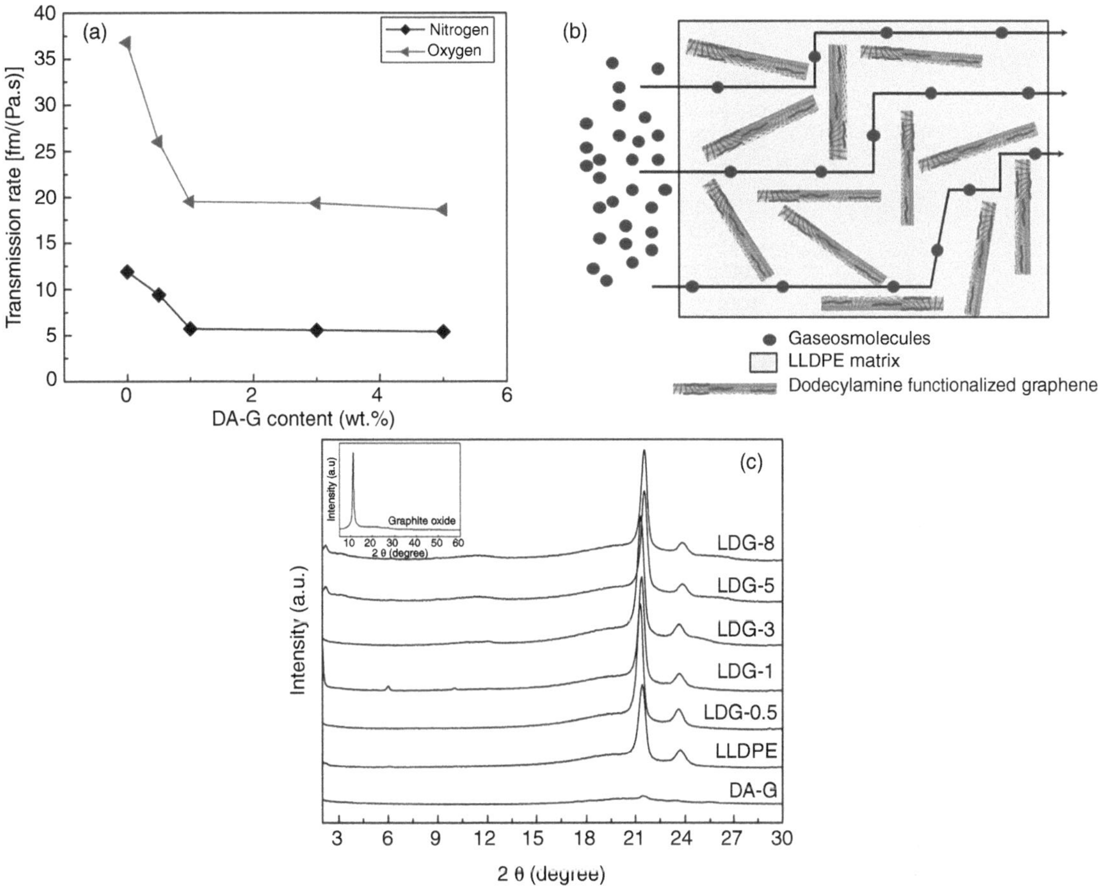

Figure 20.8. Gas permeability of LLDPE/GO nanocomposites. (a) Oxygen and nitrogen permeability of LLDPE/functionalized-GO nanocomposites at room temperature and 1 atm pressure (DA-G: dodecyl amine-functionalized-GO), (b) "Tortuous path" model for LLDPE/functionalized-GO nanocomposite films gas permeation, and (c) XRD patterns of fabricated films. Reprinted from Kuila et al. (2012) with permission. Copyright (2012) Elsevier.

mechanical properties of LLDPE films by loading small amounts of functionalized GO.

Interestingly, the oxygen transmission rate of LLDPE/graphene nanocomposite decrease up to 45%, by incorporating 1 wt.% functionalized GO nanosheets. In addition, the nitrogen permeability of LLDPE film decreased from 11.9 to 5.7 by loading 1 wt.% of dodecyl amine-modified graphene nanosheets into LLDPE matrix (Kuila et al., 2012). With an increase in the content of loaded nanosheets, however, no significant change in gas permeability was observed. Kim et al. (2015) used solution mixing technique for the fabrication of a graphene/LLDPE nanocomposite with superior water vapor impermeability performance. Their results revealed that the lowest reported WVTR for a polymer-graphene nanocomposite (1.2×10^{-1} g m^{-2} day^{-1}) can be achieved via incorporating only 1 wt.% nonoxidized graphene nanosheets into LLDPE matrix.

The fabricated nanocomposite film shows about 58% increase in the moisture barrier properties compared to the pure LLDPE film. Moreover, their results demonstrated that nonoxidized graphene nanosheets can enhance moisture barrier performance of LLDPE films more effectively than GO nanosheets (Kim et al., 2015). Furthermore, the barrier performance of fabricated nanocomposites was closer to the Cussler model predictions. In addition to solution mixing and film blowing methods, LLDPE/graphene nanocomposites can also be fabricated via melt mixing technique (Vasileiou et al., 2014). Additionally, LLDPE matrix can be functionalized with pyridine aromatic groups to increase noncovalent interactions between graphene nanosheets, and the aromatic moieties grafted onto the polymer matrix (Vasileiou et al., 2014).

3.3 Graphene/Polyethylene Terephthalate Nanocomposites

PET was first introduced in 1946 and shortly has become as one of the most widely used polymers in food packaging industry. Due to low cost, high transparency, structural stability, and high mechanical properties, PET has been widely used in many forms, including films, bottles, and fibers (Paszkiewicz et al., 2016). PET has good gas barrier performance against oxygen and carbon dioxide and consequently; it is known as the best polymer for beverage and water bottles. In the case of carbonated beverage containers, however, further enhancement of gas barrier performance is desirable, as high pressure bottles are needed for this application (Shim et al., 2012). This is the main reason for the growing interest on PET/graphene nanocomposites with high barrier performance, during recent years.

PET/graphene nanocomposites can be fabricated through all common polymer nanocomposite fabrication techniques. Shim et al. (2012) fabricated PET/GO nanocomposites through solution blending method, using chlorophenol as the medium. Their results revealed that the oxygen permeability coefficients of PET films can be decreased up to 85% by loading only 1 wt.% of functionalized GO nanosheets. Moreover, about 96% decrease in oxygen permeability can be expected by increasing the content of loaded nanosheets to 3 wt.%. They suggested that the observed superior enhancement in mechanical and gas barrier properties of functionalized GO/PET nanocomposite is related to the interactions between the alkyl groups in functionalized GO nanosheets and the PET matrix (Shim et al., 2012).

Paszkiewicz et al. (2016) used in situ polymerization method for the fabrication of PET/graphene nanocomposites with oxygen and water vapor impermeability. They also attributed the observed improvement in gas barrier performance of PET films in the presence of graphene nanosheets to the increase in the tortuosity of the diffusion path for a penetrant molecule. They also compared the influences of graphene and carbon nanotubes presence on the barrier performance of PET films and found that graphene nanosheets can enhance barrier performance of PET films more efficiently. However, they also proposed a graphene/carbon nanotubes hybrid nanofiller system as the most efficient gas barrier enhancer nanofiller system.

Layer-by-layer assembly method can also be used for the fabrication of high barrier PET/graphene transparence films (Yu et al., 2015, 2013c). Polyethylenimine (PEI), as a branched and positively charged polymer, can be used in the assembly process. Yu et al. (2015) used dip-coating strategy for the fabrication of PET/graphene nanocomposite bilayer films. Negatively charged GO nanosheets can be alternatively deposited with positively charged PEI. The light transmittance of fabricated nanocomposite multilayer film was 85.6%. Moreover, the oxygen transmittance rate of nanocomposite film was a function of the number of deposition layers and the transmittance rate of nanocomposites increases from 8.229 to less than 0.05 cc/m^2 day, by increasing the number of layers from 1 to 8 layers.

3.4 Graphene/Poly(Vinyl Alcohol) Nanocomposites

PVA, as a water-soluble biodegradable polymer with good transparency and high gas barrier properties, has attracted a great interest for packaging applications (Huang et al., 2012). Although PVA films perform great gas impermeability, high affinity to water is known as the most important shortcoming of PVA films in packaging applications. The poor water and water vapor barrier performance of highly polar PVA is attributed to the presence of hydroxyl groups in the structure of PVA (Kim et al., 2011b). Water acts as a plasticizer for PVA and its presence result in an increase in the elongation and tear strength, but also a reduction in the tensile strength (Wang et al., 2011). To overcome this problem, graphene/PVA nanocomposites have been introduced as a promising solution. Noncovalent interactions between PVA and graphene/GO nanosheets are expected to result in enormous enhancements in the functional properties of PVA films (Bao et al., 2011).

Wang et al. (2011) fabricated PVA/GO nanocomposites using solution mixing, by employing water as the solvent. Up to 212% enhancement in tensile strength and more than 34% improvement in elongation performance of nanocomposite was achieved with only 0.5 wt.% loading of GO. Interestingly, the water absorption ratio decreased from 105.2% for pure PVA film, to 48.8% for GO/PVA nanocomposite, with 1 wt.% GO loading. These results reveal that GO nanosheets can be used to reduce water permeability of PVA films, which is known as a shortcoming of PVA films for food packaging applications. PVA/GO noncovalent interactions, which form between functional groups of GO nanosheets and PVA hydroxyl groups, can be considered as the main reason for the discussed effects.

Huang et al. (2012) fabricated PVA/GO nanocomposites using solution blending. Their results revealed that oxygen and water vapor permeability of PVA films can be reduced up to 98 and 68%, respectively, by incorporating about 0.72 vol.% GO nanosheets into the PVA matrix. As expected, the improved barrier performance of fabricated nanocomposite was attributed to the high dispersion and orientation of graphene nanosheets through PVA matrix. Moreover, high interfacial interactions between polymer macromolecules and GO nanosheets were proposed as the main important reason for the observed enhancement in gas and moisture barrier of PVA/GO nanocomposites. These interactions have a direct impact on PVA crystallinity, and the crystalline structures play a critical role in enhanced barrier properties of semicrystalline PVA chains (Huang et al., 2012).

Chen et al. (2014) also investigated the role of PVA/GO interactions on the crystallinity and gas barrier performance of PVA/GO nanocomposites. They suggested the formation of new crystal regions through PVA matrix in the presence of GO nanosheets. They proposed that GO nanosheets could act as a nucleating agent and as a result, affect crystalline behavior of PVA macromolecules. Moreover, they reported that the oxygen permeability of PVA/GO nanocomposites presents a close fit to the Cussler model (Chen et al., 2014). Consequently, transparent and highly impermeable films with enhanced moisture impermeability PVA films can be fabricated by incorporating low contents of GO nanosheets (see Fig. 20.9). These films can have oxygen permeability values 17 times lower than the oxygen permeability of PET films and 92% light transmittance performance, by loading only 0.3 wt.% GO nanosheets (Kim et al., 2011b).

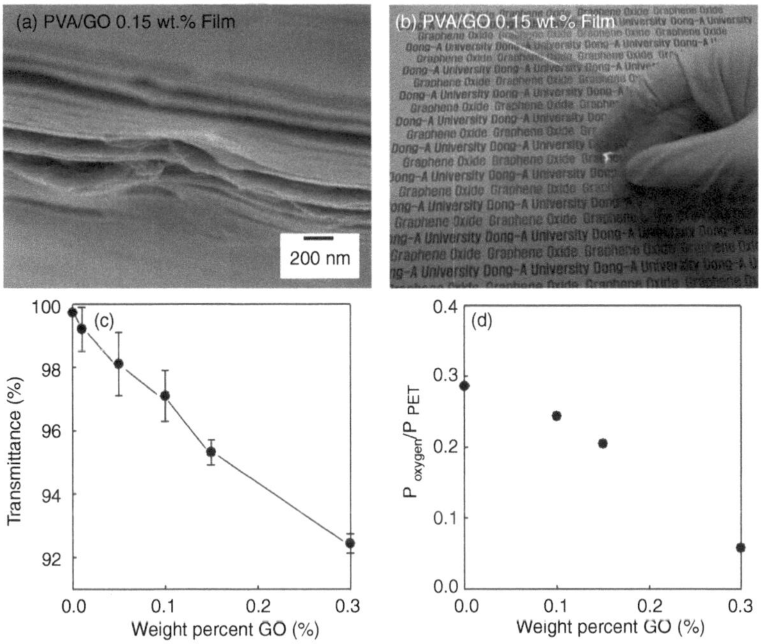

Figure 20.9. Transparent high barrier PVA/GO nanocomposite films. (a) SEM image of the PVA/GO film cross-section, (b) picture of transparent a PVA/GO film, (c) light transmittance of PVA/GO films as a function of GO contents, and (d) oxygen permeability of PVA/GO films. Reprinted from Kim et al. (2011b) with permission. Copyright (2011) Elsevier.

3.5 Other High Barrier Graphene/Polymer Nanocomposites

In addition to discussed high barrier nanocomposites, other packaging polymers such as poly(vinyl chloride) (Hasan and Lee, 2014), polystyrene (Compton et al., 2010), polyimide (Lim et al., 2015), polyethylene vinyl alcohol (EVOH) (Yang et al., 2013a), and poly(lactic acid) (PLA) (Huang et al., 2014), can be employed for the fabrication of graphene/polymer nanocomposites with improved gas impermeability performance. Compton et al. (2010) used dimethylformamide as solvent for solution mixing of phenyl isocyanate decorated GO nanosheets and polystyrene. Fabricated nanocomposite film is a perfect candidate for food packaging applications, especially where the shelf life of foods can be further extended by keeping them out of light. Moreover, fabricated nanocomposite performed about 80% reduction in oxygen permeability by loading only 0.02 vol.% functionalized GO nanosheets.

In situ polymerization of polyimide, in the presence of low contents of GO nanosheets, not only results in an enhancement in gas impermeability, but also leads to a clear improvement in moisture barrier performance of polyimide films. Tseng et al. (2012) demonstrated that the water vapor barrier of polyimide films can be reduced up to 17%, by loading only 0.001 wt.% of graphene nanosheets in the polyimide matrix. Lim et al. (2015) also fabricated high barrier graphene/polyimide nanocomposites by loading small amounts of aminophenyl functionalized graphene nanosheets in the polyimide matrix, using in situ polymerization method. They reported that incorporating 5 wt.% of functionalized graphene nanosheets into polyimide results in about 99% reduction in oxygen permeability performance of nanocomposite. Therefore, the barrier performance of graphene/polyimide nanocomposites can be enhanced noticeably using graphene nanosheets.

Solution mixing method can also be used for the fabrication of graphene reinforced EVOH nanocomposites with enhanced gas barrier performance. Yang et al. (2013a) reported that more than 280% increase in tensile strength and 400% increase in Young's modulus of EVOH can be achieved by incorporating 4 wt.% of graphene nanosheets (see Fig. 20.10). Moreover, about 1671-time reduction in the oxygen permeability coefficient of EVOH films can be achieved by loading only 0.5 wt.% graphene nanosheets in EVOH matrix. This is a great achievement for food packaging applications by loading very low contents of graphene nanosheets. This amazing impact of reduced graphene nanosheets on the oxygen permeability of EVOH films is probably due to high interactions between EVOH and reduced graphene carboxyl groups, as well as between EVOH and the basal plane of graphene nanosheets.

Gas and moisture barrier performance of biopolymers, such as PLA and starch, can also be enhanced by introducing GO nanosheets in the biopolymer matrix. Pinto et al. (2013a) used solvent-casting method for the fabrication of GO reinforced PLA nanocomposites, with improved mechanical and gas impermeability properties. Permeabilities of fabricated nanocomposites towards oxygen and nitrogen decreased 3 and 4 times with loading GO nanosheets, respectively. Moreover, more than 100% enhancement in mechanical characteristics was obtained in the presence of low contents (between 0.5 and 1 wt.%) of graphene nanosheets. Huang et al. (2014) also fabricated GO reinforced PLA nanocomposite films, using solution mixing technique. Their results also demonstrated significant improvement in oxygen and CO_2 barrier performances of PLA films (about 45 and 68%) by loading small amounts of GO nanosheets (about 1.37 vol.%).

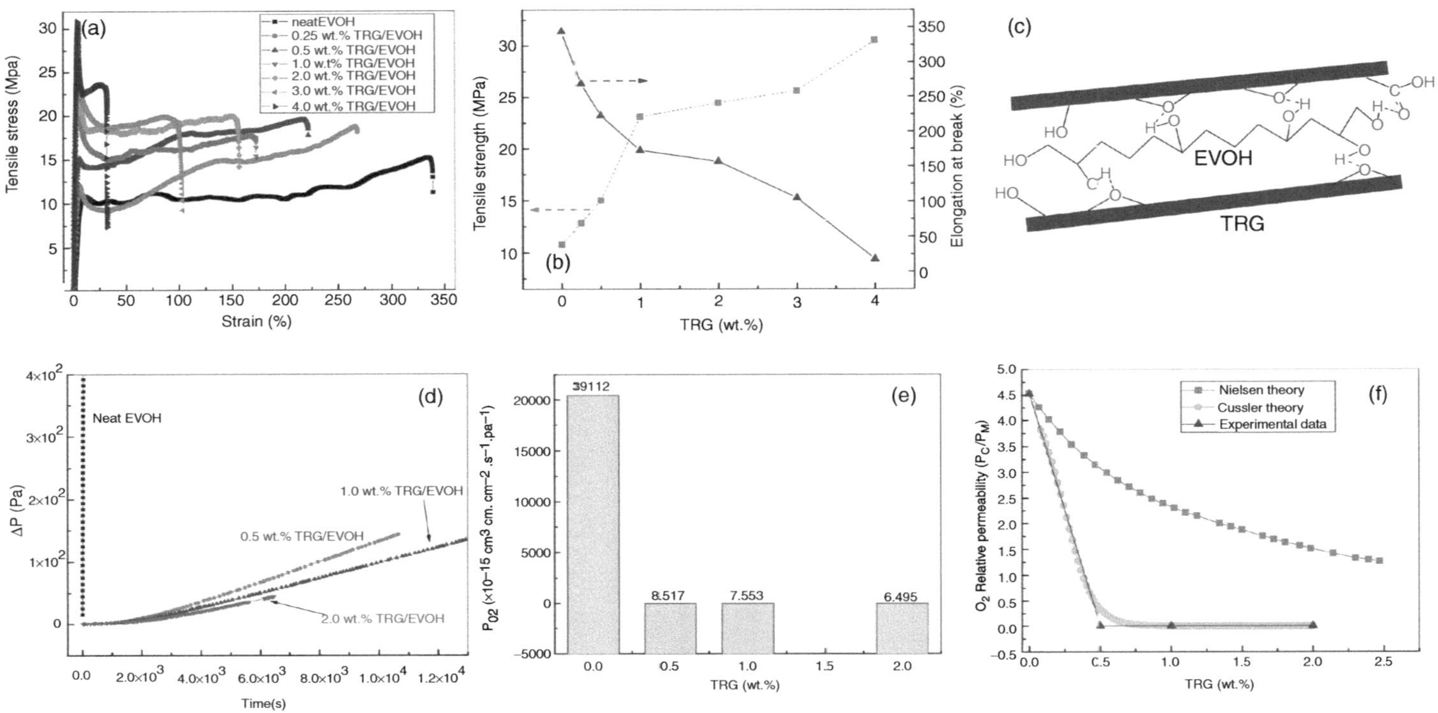

Figure 20.10. EVOH/GO nanocomposites with superior mechanical and gas barrier performance. (a) Stress–strain curves of EVOH/GO nanocomposites, (b) tensile strength and elongation at break versus GO loadings, (c) hydrogen bonds between GO and EVOH, (d) oxygen transmission rates of EVOH/GO nanocomposites, (e) oxygen permeability coefficients of EVOH/GO nanocomposites as a function of GO loading, and (f) relative permeability plots of experimental values and predicted Nielsen and Cussler values as a function of GO contents. Reprinted from Yang et al. (2013a) with permission. Copyright (2013) American Chemical Society.

Unalan et al. (2015) used solution casting method for the fabrication of pullulan/GO nanocomposites. Pullulan is a water soluble, nontoxic and biodegradable nonionic polysaccharide, with good adhesive and binding properties. Due to extensive hydroxyl groups on its backbone and high oxygen barrier properties under dry conditions, pullulan has attracted a great interest for the fabrication of high barrier graphene/biopolymer nanocomposites, with a potential for food packaging applications (Unalan et al., 2015). Mechanical properties and oxygen barrier performance of pullulan films can be enhanced noticeably by loading very small contents of GO nanosheets. With loading only 0.2 wt.% GO in pullulan matrix, about 40% increase in tensile strength of pullulan films is obtainable. Moreover, homogeneous dispersion of GO nanosheets (about 0.3 wt.%) in pullulan matrix results in significant improvement in oxygen barrier performance of final nanocomposite (about 78% decrease in oxygen permeability). Therefore, GO/pullulan nanocomposites can be considered as a candidate for the production of biopolymer-based food packaging films.

4 Conclusions and Perspectives

Replacing conventional polymer films with graphene/polymer nanocomposite films can be considered as a promising strategy for the enhancement of gas and moisture barrier performance of food packaging films, by conserving (and even improving) interesting mechanical and functional properties of polymer films. However, high dispersion of graphene nanosheets, as well as homogenous distribution of them through the polymer matrix strongly affects the gas barrier performance of final nanocomposite films. Moreover, in most cases crystallinity of the polymer matrix, which has a direct impact on the barrier performance of polymer films, has a direct relation with the polymer-graphene interface. Polymer-graphene interactions on this interface region can either be covalent, or noncovalent. Strong interactions (even covalent or noncovalent bonding) can improve mechanical, barrier, and functional properties of final nanocomposite films. Therefore, presence of required functional groups on graphene nanosheets is essential for the fabrication of high-performance graphene/polymer nanocomposites. High surface area of graphene nanosheets provides the perfect conditions for decorating basal plane and edges of graphene nanosheets with required functional groups. Based on the nature of the polymer matrix, required functional groups should be placed on the surface area of graphene nanosheets. However, in most cases GO can provide strong interactions (noncovalent interactions) with polymer chains, and there is no need for further functionalization steps. This is the area that demands a great effort in the future. Although there are many works on the barrier performance of graphene/polymer nanocomposites, but there are just few works with sufficient details regarding the influences of graphene-polymer interactions on the gas barrier performance of nanocomposite films.

Moreover, as an alternative to the conventional polymers used widely in food packaging industry, recently biopolymer-based food packaging systems have attracted a great interest. The concept of sustainability can be introduced in food packaging industry by using sustainable polymers as the matrix in graphene/polymer nanocomposite systems. These biopolymers mostly present high gas impermeability, but are not strong against water and water vapor. This is another area that needs more effort and research, especially regarding low moisture resistance issues. In addition, the migration of graphene nanosheets from graphene/polymer nanocomposite films into eatable foods and beverages is another concept that demands a great interest. Although it has been reported that no significant harm will be expected by incorporating graphene nanosheets into polymer food packaging films, but more works in this field are needed.

Based on the growing interest on graphene/polymer nanocomposites with high barrier performance, clearly, these materials will find their way through food packaging industry, sooner or later. Fascinating gas and moisture impermeability of biopolymer-based graphene/polymer nanocomposites, even with loading very small amounts of graphene nanosheets, will protect foods for longer periods of time and as a result, increase the durability of foods. In addition, perfect gas barrier performance of graphene/polymer nanocomposites, even at elevated pressures, can prolong the storage period of carbonated beverages. These performances cannot be achieved with any other nanomaterial, but graphene. These are all the reasons that we cannot underestimate the role of graphene-based polymer nanocomposites in the future of the food packaging industry.

References

Adame, D., Beall, G., 2009. Direct measurement of the constrained polymer region in polyamide/clay nanocomposites and the implications for gas diffusion. Appl. Clay Sci. 42, 545–552.

Allahbakhsh, A., Mazinani, S., 2015. Influences of sodium dodecyl sulfate on vulcanization kinetics and mechanical performance of EPDM/graphene oxide nanocomposites. RSC Adv. 5, 46694–46704.

Allahbakhsh, A., Mazinani, S., Kalaee, M.R., Sharif, F., 2013a. Cure kinetics and chemorheology of EPDM/graphene oxide nanocomposites. Thermochim. Acta 563, 22–32.

Allahbakhsh, A., Sharif, F., Mazinani, S., 2013b. The influence of oxygen-containing functional groups on the surface behavior and roughness characteristics of graphene oxide. Nano 08, 1350045.

Allahbakhsh, A., Sheydaei, M., Mazinani, S., Kalaee, M., 2013c. Enhanced thermal properties of poly(ethylene tetrasulfide) via expanded graphite incorporation by in situ polymerization method. High Perform. Polym. 25, 576–583.

Azeredo, H.M.C.D., 2009. Nanocomposites for food packaging applications. Food Res. Int. 42, 1240–1253.

Bao, C., Guo, Y., Song, L., Hu, Y., 2011. Poly(vinyl alcohol) nanocomposites based on graphene and graphite oxide: a comparative investigation of property and mechanism. J. Mater. Chem. 21, 13942.

Botas, C., Álvarez, P., Blanco, P., Granda, M., Blanco, C., Santamaría, R., Romasanta, L.J., Verdejo, R., López-Manchado, M.A., Menéndez, R., 2013. Graphene materials with different structures prepared from the same graphite by the Hummers and Brodie methods. Carbon 65, 156–164.

Brodie, B.C., 1859. On the atomic weight of graphite. Philos. Trans. R. Soc. London 149, 249–259.

Chatterjee, T., Patel, R., Garnett, J., Paradkar, R., Ge, S., Liu, L., Forziati, K.T., Shah, N., 2014. Machine direction orientation of high density polyethylene (HDPE): barrier and optical properties. Polymer 55, 4102–4115.

Checchetto, R., Miotello, A., Nicolais, L., Carotenuto, G., 2014. Gas transport through nanocomposite membrane composed by polyethylene with dispersed graphite nanoplatelets. J. Membr. Sci. 463, 196–204.

Chen, J.-T., Fu, Y.-J., An, Q.-F., Lo, S.-C., Zhong, Y.-Z., Hu, C.-C., Lee, K.-R., Lai, J.-Y., 2014. Enhancing polymer/graphene oxide gas barrier film properties by introducing new crystals. Carbon 75, 443–451.

Chua, C.K., Sofer, Z., Pumera, M., 2012. Graphite oxides: effects of permanganate and chlorate oxidants on the oxygen composition. Chem.: Eur. J. 18, 13453–13459.

Compton, O.C., Kim, S., Pierre, C., Torkelson, J.M., Nguyen, S.T., 2010. Crumpled graphene nanosheets as highly effective barrier property enhancers. Adv. Mater. 22, 4759–4763.

Cussler, E.L., Hughes, S.E., Ward, W.J., Aris, R., 1988. Barrier membranes. J. Membr. Sci. 38, 161–174.

Dreyer, D.R., Park, S., Bielawski, C.W., Ruoff, R.S., 2010. The chemistry of graphene oxide. Chem. Soc. Rev. 39, 228–240.

Duncan, T.V., 2011. Applications of nanotechnology in food packaging and food safety: barrier materials, antimicrobials and sensors. J. Coll. Interf. Sci. 363, 1–24.

Dunkerley, E., Schmidt, D., 2010. Effects of composition, orientation and temperature on the O_2 permeability of model polymer/clay nanocomposites. Macromolecules 43, 10536–10544.

El Achaby, M., Arrakhiz, F.-E., Vaudreuil, S., El Kacem Qaiss, A., Bousmina, M., Fassi-Fehri, O., 2012. Mechanical, thermal, and rheological properties of graphene-based polypropylene nanocomposites prepared by melt mixing. Polym. Compos. 33, 733–744.

El Achaby, M., Qaiss, A., 2013. Processing and properties of polyethylene reinforced by graphene nanosheets and carbon nanotubes. Mater. Des. 44, 81–89.

Ferreira, M.S., Sartori, M.N., Oliveira, R.R., Guven, O., Moura, E.a.B., 2014. Short vegetal-fiber reinforced HDPE: a study of electron-beam radiation treatment effects on mechanical and morphological properties. Appl. Surf. Sci. 310, 325–330.

Gao, W., Alemany, L.B., Ci, L., Ajayan, P.M., 2009. New insights into the structure and reduction of graphite oxide. Nat. Chem. 1, 403–408.

Georgakilas, V., Otyepka, M., Bourlinos, A.B., Chandra, V., Kim, N., Kemp, K.C., Hobza, P., Zboril, R., Kim, K.S., 2012. Functionalization of graphene: covalent and non-covalent approaches. Deriv. Appli. Chem. Rev. 112, 6156–6214.
GóMez-Navarro, C., Meyer, J.C., Sundaram, R.S., Chuvilin, A., Kurasch, S., Burghard, M., Kern, K., Kaiser, U., 2010. Atomic structure of reduced graphene oxide. Nano Lett. 10, 1144–1148.
Haghighi, A.H., Sheydaei, M., Allahbakhsh, A., Ghatarband, M., Hosseini, F.S., 2014. Thermal performance of poly(ethylene disulfide)/expanded graphite nanocomposites. J. Therm. Anal. Calorim. 117, 525–535.
Hasan, M., Lee, M., 2014. Enhancement of the thermo-mechanical properties and efficacy of mixing technique in the preparation of graphene/PVC nanocomposites compared to carbon nanotubes/PVC. Prog. Nat. Sci.: Mater. Int. 24, 579–587.
Hong, Y., Coombs, S.J., Cooper-White, J.J., Mackay, M.E., Hawker, C.J., Malmström, E., Rehnberg, N., 2000. Film blowing of linear low-density polyethylene blended with a novel hyperbranched polymer processing aid. Polymer 41, 7705–7713.
Huang, H.-D., Ren, P.-G., Chen, J., Zhang, W.-Q., Ji, X., Li, Z.-M., 2012. High barrier graphene oxide nanosheet/poly(vinyl alcohol) nanocomposite films. J. Membr. Sci. (409–410), 156–163.
Huang, H.-D., Ren, P.-G., Xu, J.-Z., Xu, L., Zhong, G.-J., Hsiao, B.S., Li, Z.-M., 2014. Improved barrier properties of poly(lactic acid) with randomly dispersed graphene oxide nanosheets. J. Membr. Sci. 464, 110–118.
Huang, Y., Qin, Y., Zhou, Y., Niu, H., Yu, Z.-Z., Dong, J.-Y., 2010. Polypropylene/graphene oxide nanocomposites prepared by in situ Ziegler–Natta polymerization. Chem. Mater. 22, 4096–4102.
Hummers, W.S., Offeman, R.E., 1958. Preparation of graphitic oxide. J. Am. Chem. Soc. 80, 1339–11339.
Johansson, F., Leufvén, A., 1994. Influence of sorbed vegetable oil and relative humidity on the oxygen transmission rate through various polymer packaging films. Pack. Technol. Sci. 7, 275–281.
Kalaitzidou, K., Fukushima, H., Drzal, L.T., 2007. Multifunctional polypropylene composites produced by incorporation of exfoliated graphite nanoplatelets. Carbon 45, 1446–1452.
Kim, H., Kobayashi, S., Abdurrahim, M.A., Zhang, M.J., Khusainova, A., Hillmyer, M.A., Abdala, A.A., Macosko, C.W., 2011a. Graphene/polyethylene nanocomposites: effect of polyethylene functionalization and blending methods. Polymer 52, 1837–1846.
Kim, H.M., Lee, J.K., Lee, H.S., 2011b. Transparent and high gas barrier films based on poly(vinyl alcohol)/graphene oxide composites. Thin Sol. Films 519, 7766–7771.
Kim, H.W., Yoon, H.W., Yoon, S.M., Yoo, B.M., Ahn, B.K., Cho, Y.H., Shin, H.J., Yang, H., Paik, U., Kwon, S., Choi, J.Y., Park, H.B., 2013. Selective gas transport through few-layered graphene and graphene oxide membranes. Science 342, 91–95.
Kim, J., Song, S.H., Im, H.-G., Yoon, G., Lee, D., Choi, C., Kim, J., Bae, B.-S., Kang, K., Jeon, S., 2015. Moisture barrier composites made of non-oxidized graphene flakes. Small 11, 3124–3129.
Koch, K.R., 1982. Oxidation by Mn207: an impressive demonstration of the powerful oxidizing property of dimanganeseheptoxide. J. Chem. Educ. 59, 973.
Kuila, T., Bose, S., Mishra, A.K., Khanra, P., Kim, N.H., Lee, J.H., 2012. Effect of functionalized graphene on the physical properties of linear low density polyethylene nanocomposites. Polym. Test. 31, 31–38.

Lape, N.K., Nuxoll, E.E., Cussler, E.L., 2004. Polydisperse flakes in barrier films. J. Membr. Sci. 236, 29–37.
Lee, D.W., De Los Santos, V.L., Seo, J.W., Felix, L.L., Bustamante, D.A., Cole, J.M., Barnes, C.H.W., 2010. The structure of graphite oxide: investigation of its surface chemical groups. J. Phys. Chem. B 114, 5723–5728.
Lee, K.-H., Hong, J., Kwak, S.J., Park, M., Son, J.G., 2015. Spin self-assembly of highly ordered multilayers of graphene-oxide sheets for improving oxygen barrier performance of polyolefin films. Carbon 83, 40–47.
Leenaerts, O., Partoens, B., Peeters, F.M., 2008. Graphene: a perfect nanoballoon. Appl. Phys. Lett. 93, 193107.
Li, D., Müller, M.B., Gilje, S., Kaner, R.B., Wallace, G.G., 2008. Processable aqueous dispersions of graphene nanosheets. Nat. Nanotechnol. 3, 101–105.
Li, X., Wang, H., Robinson, J.T., Sanchez, H., Diankov, G., Dai, H., 2009. Simultaneous nitrogen doping and reduction of graphene oxide. J. Am. Chem. Soc. 131, 15939–15944.
Lim, J., Yeo, H., Goh, M., Ku, B.-C., Kim, S.G., Lee, H.S., Park, B., You, N.-H., 2015. Grafting of polyimide onto chemically-functionalized graphene nanosheets for mechanically-strong barrier membranes. Chem. Mater. 27, 2040–2047.
Marcano, D.C., Kosynkin, D.V., Berlin, J.M., Sinitskii, A., Sun, Z., Slesarev, A., Alemany, L.B., Lu, W., Tour, J.M., 2010. Improved synthesis of graphene oxide. ACS Nano 4, 4806–4814.
Marcilla, A., Beltrán, M.I., Navarro, R., 2005. Effect of regeneration temperature and time on the activity of HUSY and HZSM5 zeolites during the catalytic pyrolysis of polyethylene. J. Anal. Appl. Pyroly. 74, 361–369.
Milani, M.A., Quijada, R., Basso, N.R.S., Graebin, A.P., Galland, G.B., 2012. Influence of the graphite type on the synthesis of polypropylene/graphene nanocomposites. J. Polym. Sci. A 50, 3598–3605.
Moggridge, G.D., Lape, N.K., Yang, C., Cussler, E.L., 2003. Barrier films using flakes and reactive additives. Prog. Org. Coat. 46, 231–240.
Paredes, J.I., Villar-Rodil, S., Martínez-Alonso, A., Tascón, J.M.D., 2008. Graphene oxide dispersions in organic solvents. Langmuir 24, 10560–10564.
Paszkiewicz, S., Kwiatkowska, M., Rosłaniec, Z., Szymczyk, A., Jotko, M., Lisiecki, S., 2016. The influence of different shaped nanofillers (1D, 2D) on barrier and mechanical properties of polymer hybrid nanocomposites based on PET prepared by in situ polymerization. Polym. Compos., [e-pub ahead of print].
Pinto, A.M., Cabral, J., Tanaka, D.a.P., Mendes, A.M., Magalhães, F.D., 2013a. Effect of incorporation of graphene oxide and graphene nanoplatelets on mechanical and gas permeability properties of poly(lactic acid) films. Polym. Int. 62, 33–40.
Pinto, F., Carotenuto, G., Meo, M., 2013b. Preparation and thermomechanical characterisation of graphene nanoplatelets/low-density polyethylene composites. J. Thermoplast. Compos. Mater. 28, 745–761.
Poh, H.L., Šaněk, F., Ambrosi, A., Zhao, G., Sofer, Z., Pumera, M., 2012. Graphenes prepared by Staudenmaier, Hofmann and Hummers methods with consequent thermal exfoliation exhibit very different electrochemical properties. Nanoscale 4, 3515.
Putz, K.W., Compton, O.C., Palmeri, M.J., Nguyen, S.T., Brinson, L.C., 2010. High-nanofiller-content graphene oxide-polymer nanocomposites via vacuum-assisted self-assembly. Adv. Funct. Mater 20, 3322–3329.
Ren, P.-G., Wang, H., Huang, H.-D., Yan, D.-X., Li, Z.-M., 2014. Characterization and performance of dodecyl amine functionalized graphene oxide and dodecyl amine functionalized graphene/high-density polyethylene nanocomposites: a comparative study. J. Appl. Polym. Sci. 131, 39803.

Salavagione, H.J., MartíNez, G., 2011. Importance of covalent linkages in the preparation of effective reduced graphene oxide–poly(vinyl chloride) nanocomposites. Macromolecules 44, 2685–2692.
Salavagione, H.J., Martínez, G., Ellis, G., 2011. Recent advances in the covalent modification of graphene with polymers. Macromol. Rap. Comm. 32, 1771–1789.
Shehzad, F., Thomas, S.P., Al-Harthi, M.A., 2014. Non-isothermal crystallization kinetics of high density polyethylene/graphene nanocomposites prepared by in-situ polymerization. Thermochim. Acta 589, 226–234.
Shim, S.H., Kim, K.T., Lee, J.U., Jo, W.H., 2012. Facile method to functionalize graphene oxide and its application to poly(ethylene terephthalate)/graphene composite. ACS Appl. Mater. Interf. 4, 4184–4191.
Simon, A., Dronskowski, R., Krebs, B., Hettich, B., 1987. The Crystal Structure of Mn2O7. Angew. Chem. Int. Ed. Eng. 26, 139–140.
Staudenmaier, L., 1898. Verfahren zur Darstellung der Graphitsäure. Ber. deutsch. chem. Gesell. 31, 1481–1487.
Tayebi, M., Ramazani, S.A.A., Hamed Mosavian, M.T., Tayyebi, A., 2015. LDPE/EVA/graphene nanocomposites with enhanced mechanical and gas permeability properties. Polym. Adv. Technol. 26 (9), 1083–1090.
Tseng, I.H., Liao, Y.-F., Chiang, J.-C., Tsai, M.-H., 2012. Transparent polyimide/graphene oxide nanocomposite with improved moisture barrier property. Mater. Chem. Phys. 136, 247–253.
Unalan, I.U., Wan, C., Figiel, Ł., Olsson, R.T., Trabattoni, S., Farris, S., 2015. Exceptional oxygen barrier performance of pullulan nanocomposites with ultra-low loading of graphene oxide. Nanotechnology 26, 275703.
Vasileiou, A.A., Kontopoulou, M., Docoslis, A., 2014. A noncovalent compatibilization approach to improve the filler dispersion and properties of polyethylene/graphene composites. ACS Appl. Mater. Interf. 6, 1916–1925.
Wang, J., Wang, X., Xu, C., Zhang, M., Shang, X., 2011. Preparation of graphene/poly(vinyl alcohol) nanocomposites with enhanced mechanical properties and water resistance. Polym. Int. 60, 816–822.
Wang, J., Xu, C., Hu, H., Wan, L., Chen, R., Zheng, H., Liu, F., Zhang, M., Shang, X., Wang, X., 2010. Synthesis, mechanical, and barrier properties of LDPE/graphene nanocomposites using vinyl triethoxysilane as a coupling agent. J. Nanopart. Res. 13, 869–878.
Yang, J., Bai, L., Feng, G., Yang, X., Lv, M., Zhang, C.A., Hu, H., Wang, X., 2013a. Thermal reduced graphene based poly(ethylene vinyl alcohol) nanocomposites: enhanced mechanical properties, gas barrier, water resistance, and thermal stability. Ind. Eng. Chem. Res. 52, 16745–16754.
Yang, X., Wang, X., Yang, J., Li, J., Wan, L., 2013b. Functionalization of graphene using trimethoxysilanes and its reinforcement on polypropylene nanocomposites. Chem. Phys. Lett. 570, 125–131.
Yoo, B.M., Shin, H.J., Yoon, H.W., Park, H.B., 2014. Graphene and graphene oxide and their uses in barrier polymers. J. Appl. Polym. Sci. 131, 39628.
You, S., Luzan, S.M., Szabó, T., Talyzin, A.V., 2013. Effect of synthesis method on solvation and exfoliation of graphite oxide. Carbon 52, 171–180.
Yu, L., Lim, Y.S., Han, J.H., Kim, K., Kim, Y.K., Choi, S.Y., Shin, K., 2015. A graphene oxide oxygen barrier film deposited via a self-assembly coating method. Synth. Met. 162, 710–714.

SUBJECT INDEX

FSC
www.fsc.org
MIX
Papier | Fördert
gute Waldnutzung
FSC® C083411

Kolibri 360 GmbH
Ferdinand-Jühlke-Straße 7
99095 Erfurt, Deutschland
produktsicherheit@kolibri360.de